PRINCIPLES OF TURBOMACHINERY IN AIR-BREATHING ENGINES

This book is intended for advanced undergraduate and graduate students in mechanical and aerospace engineering taking a course commonly called Principles of Turbomachinery and Aerospace Propulsion. The book begins with a review of basic thermodynamics and fluid mechanics principles to motivate their application to aerothermodynamics and real-life design issues. This approach is ideal for the reader who will face practical situations and design decisions in the gas turbine industry. Among the features of the book are:

- an emphasis on the role of entropy in the process of performance assessment;
- a timely review of different flow structures;
- revisiting the subsonic and supersonic De Laval nozzle as it applies to bladed turbomachinery components;
- an applied review of the boundary layer principles; and
- highlighting the importance of invariant properties across a turbomachinery component in carrying out real computational tasks.

The text is fully supported by 398 figures, numerous examples, and homework problems.

Erian Baskharone is a Professor Emeritus of Mechanical and Aerospace Engineering at Texas A & M University and a member of the Rotordynamics/Turbomachinery Laboratory Faculty. He is a member of the ASME Turbomachinery Executive Committee. After receiving his PhD from the University of Cincinnati, Dr. Baskharone became a senior engineer with Allied Signal Corporation, responsible for aerothermodynamics design of various turbofan and turboprop engines. His research covered a wide spectrum of turbomachinery topics including unsteady stator/rotor flow interaction, and the fluid-induced vibration Space Shuttle Main Engine Turbopumps. His perturbation approach to the problem of turbomachinery fluid-induced vibration was a significant breakthrough. He is the recipient of the General Dynamics Award of Excellence in Engineering teaching (1991) and the Amoco Foundation Award for Distinguished Teaching (1992).

Cambridge Aerospace Series

Editors
Wei Shyy
and
Michael J. Rycroft

1. J. M. Rolfe and K. J. Staples (eds.): *Flight Simulation*
2. P. Berlin: *The Geostationary Applications Satellite*
3. M. J. T. Smith: *Aircraft Noise*
4. N. X. Vinh: *Flight Mechanics of High-Performance Aircraft*
5. W. A. Mair and D. L. Birdsall: *Aircraft Performance*
6. M. J. Abzug and E. E. Larrabee: *Airplane Stability and Control*
7. M. J. Sidi: *Spacecraft Dynamics and Control*
8. J. D. Anderson: *A History of Aerodynamics*
9. A. M. Cruise, J. A. Bowles, C. V. Goodall, and T. J. Patrick: *Principles of Space Instrument Design*
10. G. A. Khoury and J. D. Gillett (eds.): *Airship Technology*
11. J. Fielding: *Introduction to Aircraft Design*
12. J. G. Leishman: *Principles of Helicopter Dynamics*
13. J. Katz and A. Plotkin: *Low Speed Aerodynamics*, 2nd Edition
14. M. J. Abzug and E. E. Larrabee: *Airplane Stability and Control: A History of the Technologies that made Aviation Possible*, 2nd Edition
15. D. H. Hodges and G. A. Pierce: *Introduction to Structural Dynamics and Aeroelasticity*
16. W. Fehse: *Automatic Rendezvous and Docking of Spacecraft*
17. R. D. Flack: *Fundamentals of Jet Propulsion with Applications*
18. J. G. Leishman: *Principles of Helicopter Dynamics*, 2nd Edition
19. E. A. Baskharone: *Principles of Turbomachinery in Air-Breathing Engines*

Principles of Turbomachinery in Air-Breathing Engines

ERIAN A. BASKHARONE

Texas A&M University

CAMBRIDGE UNIVERSITY PRESS
Cambridge, New York, Melbourne, Madrid, Cape Town, Singapore, São Paulo

Cambridge University Press
40 West 20th Street, New York, NY 10011-4211, USA

www.cambridge.org
Information on this title: www.cambridge.org/9780521858106

First published 2006

Printed in the United States of America

A catalog record for this publication is available from the British Library.

Library of Congress Cataloging in Publication Data

Baskharone, Erian A., 1947–
Principles of turbomachinery in air-breathing engines / Erian A.
Baskharone.
p. cm. – (Cambridge aerospace series; 19)
Includes bibliographical references and index.
ISBN-13: 978-0-521-85810-6 (hardback)
ISBN-10: 0-521-85810-0 (hardback)
1. Gas-turbines. 2. Turbomachines. I. Title. II. Series.
TJ778.B33 2006
621.43′3–dc22 2005031259

ISBN-13 978-0-521-85810-6 hardback
ISBN-10 0-521-85810-0 hardback

To

Magda

Daniel

Christian

Richard

and

Robert

Contents

Preface

Beginning with the class-notes version, this book is the outcome of teaching the courses of Principles of Turbomachinery and Aerospace Propulsion in the mechanical and aerospace engineering departments of Texas A&M University. Over a period of fourteen years, the contents were continually altered and upgraded in light of the students' feedback. This has always been insightful, enlightening, and highly constructive.

The book is intended for junior- and senior-level students in the mechanical and aerospace engineering disciplines, who are taking gas-turbine or propulsion courses. In its details, the text serves the students in two basic ways. First, it refamiliarizes them with specific fundamentals in the fluid mechanics and thermodynamics areas, which are directly relevant to the turbomachinery design and analysis aspects. In doing so, it purposely deviates from such inapplicable subtopics as external (unbound) flows around geometrically standard objects and airframe-wing analogies. Instead, turbomachinery subcomponents are utilized in such a way to impart the element of practicality and highlight the internal-flow nature of the subject at hand. The second book task is to prepare the student for practical design topics by placing him or her in appropriate real-life design settings. In proceeding from the first to the second task, I have made every effort to simplify the essential turbomachinery concepts, without compromising their analytical or design-related values.

Judging by my experience, two additional groups are served by the book. First, practicing engineers including, but not necessarily limited to, those at the entry level. As an example, the reader in this category will benefit from the practical means of estimating the stage aerodynamic losses. These stem from special well-explained flow behavioral features within the blade-to-blade, hub-to-casing flow passages and are hardly known a priori in a real-life design procedure. Another example of practical topics involves the different means of improving the overall efficiency through minor hardware adjustments, without having to undergo a tedious and expensive redesign procedure. To the same group of young engineers, the mere exposure to the different stacking patterns of stationary and rotating blades, as well as the drastic aerodynamic consequences of seemingly minor blade-surface irregularities, should both bring the student much closer to real problems in the turbomachinery industry.

Also served by the book contents are those who are considering a research career in the turbomachinery area. The book, in many sections, makes it clear that there are topics which are in need of research contributions, not only for improving the aerodynamic performance, but also for elongating the life of a turbomachine or avoiding a premature mechanical failure. Two such topics are the unsteady fatigue-causing stator/rotor flow interaction and the role of secondary (casing-to-shroud) flow stream in causing an unstable rotor operation. The latter has to do, in particular, with the swirling motion of this flow stream. Publicized by frequent mechanical failures in an early development phase of the Space Shuttle Main Engine turbopumps, extensive exploratory work identified this motion component as the major contributor in an unstable and potentially catastrophic shaft motion, termed *whirl*, where the shaft centerline undergoes an eccentric motion around the housing centerline at a finite "whirl" frequency. Other research-needing topics, also indicated in the text, involve the design and performance-related area. Examples of these include proper casing (or housing) treatment for tip-leakage control and yet unevaluated ways to improve the blading efficiency and flow guidedness. With the foundation laid in Chapter 3 and through a heavy exploitation of graphical means, another research-worthy subject I am identifying is that of variable-geometry turbomachines. These, in my opinion, remain today as poorly researched and insufficiently tested.

Solved and unsolved, the problems contained in the text are so constructed to underscore the different design and analysis topics within each individual chapter. Because of their critical role, the objective of each solved problem is clearly stated ahead of the problem or during the numerical solution. On more than just a few occasions, I chose to write specific problems in such a format to impart specific concepts from previous chapters, in an effort to contrast the performance of, and limitations on, a given component to those of another. Unique about many solved problems is that common mistakes are intentionally made, upon warning the student, with the wrong solution segment clearly marked. The segment is, at an appropriate point, terminated, and the lastly obtained unrealistic result examined, prior to taking the student back to the correct sequence of calculations.

Briefly outlined in Chapter 11, my experience in the research area of turbomachinery fluid-induced vibration led me to conclude that the efforts of competent vibration investigators often remain incomplete in the absence of contributions from modest-level fluid-dynamicists. Perhaps the most revealing example of this fact is the occasion when I undertook, as a side issue, the task of designing a swirl "brake" for use in the Space Shuttle Main Engine booster impeller. The stated objective, then, was to practically destroy the circumferential velocity component of the secondary-flow stream just upstream from its inlet station. Given that virtually any serious reader of Chapter 10 is capable of designing such a stationary device, I had the device T.L. (technical layout) ready in a rather short time. Upon fabrication and testing, it was clear that this simple component helped stabilize the impeller operation. This particular incident reminded me of a long-held principle; that thoughtfully presented fundamentals can pave the student's way to positively contribute to serious turbomachinery issues, for even the simplest of ideas may well escape the minds of specialists.

Some of the book chapters are notably different in size by comparison. Chapter 3, for instance, is one of the largest, as it sets the aerothermodynamic foundation for the subsequent chapters to utilize. The chapter begins with the most essential flow-governing equations, with emphasis on the need for a rotating frame of reference, and the corresponding set of relative thermophysical properties. For a rotor blade-to-blade passage, as indicated in the text, many physical phenomena, including the boundary-layer "blockage" effect and the onset of passage choking, lend themselves to such a reference frame. The chapter then proceeds to introduce the student to the flow total and total-relative properties in the stage stationary and rotating sub-domains, respectively. In doing so, the concept of critical velocity is presented as a turbomachinery-suited deviation from the traditional sonic speed in the Mach number definition. Furthermore, this same chapter exclusively covers the flow structure in an exhaust diffuser, a critical flow passage which, in a typical turboshaft engine, determines whether a much-feared flow reversal, back to as far upstream as the turbine section, could indeed take place. Despite its obvious classification as a non-turbomachinery component, an exhaust diffuser is potentially capable of causing an unmatched performance deterioration, under off-design operation modes depending, primarily, on the turbine-exit swirl angle. I came to better comprehend and appreciate the role of this component during my design of it for the TPE 331-14 turboprop engine, in the early 1980s. Prior denial of the FAA certification, in this case, had to do with unacceptable power-decline magnitudes under FAA-specified off-design modes, with the exhaust-diffuser pressure-recovery characteristics being at the heart of the problem.

Also longer than average are Chapters 8 and 10, covering the theory and design aspects of axial and radial-inflow turbines, compared to their compressor counterparts. One of the reasons here is the substantial arbitrariness of the turbine-blade geometrical features, as a result of the streamwise-favorable pressure gradient, compared to the traditionally standard compressor-blade configurations. Moreover, there is this severe environment surrounding the turbine operation, where a bladed component, spinning at a speed that is rarely below 40,000 rpm, is exposed to a gas-stream temperature which, for an early stage, may very well be in excess of 1300 K. Such environment would naturally give rise to truly punishing thermal and mechanical stress fields which, together with stress-concentration subregions, can be devastating from a mechanical standpoint. With virtually unavoidable early-stage blade and endwall-cooling, serious tip-leakage challenges in early (short-span) rotor blades, and various issues concerning the blade-stacking pattern, the larger emphasis placed on these two turbine chapters is perhaps justified.

The book contents, in many ways, were influenced by my turbomachinery industrial experience, at Garrett Turbine Engine Co. (later Allied-Signal Aerospace Co., and currently Honeywell Aerospace Co.), in the early and mid-1980s, with duties in the aerothermodynamic turbine design of different turboprop and turbofan engines, as well as such power systems as turbochargers and auxiliary power units. As I now revisit the introductory chapters, as well as design-oriented topics in later chapters, it becomes clear to me that these, as well as many other subjects, do share a common thread, namely the element of practicality, one that is gained from hands-on field experience. This same experience taught me to thoroughly examine the downside of a simplifying assumption before adopting it. Perhaps the clearest example of this topic is the introduction to the Radial Equilibrium Theory (Chapter 6). The objective there is to clearly underscore the fact that a casually stated assumption such as "no stator/rotor flow interaction" implies, in plain language, an *infinitely long* distance between the stator vanes and rotor blades. The introduction of Chapter 6 is but one of many commentary sections that address the all important issue of when an assumption can be "engineeringly" acceptable and when its unrealistic implications can destroy its very own credibility.

The bigger question, however, is whether a turbomachinery analyst can "live" with an unrealistic assumption, such as that of the infinitely long stator/rotor gap (above). The short answer to this is yes, depending on the intended use of the results. In this particular instance, a largely simplified form of the radial-momentum conservation principle, namely the radial equilibrium equation, is solved with the hardly demanding objective of having a first look at the flow variables away from the mean radius, in one of these unbladed subregions. During a typical preliminary design phase, the solution of this simple equation would be in the interest of achieving a crude estimate of the hub and tip degrees of reaction. These are then examined, only to see if they are between zero and 100%, a fittingly modest objective with a rather limited final-design impact.

As I look back at the very early educational settings which helped me voice out and gainfully discuss turbomachinery-related issues, it is only fair to acknowledge a group of professionals with whom I was fortunate to interact, almost on a daily basis. This is a group of young engineers I supervised during my tenure at Garrett Turbine Engine Company. In my mind, these individuals perhaps had as much of an educational impact on me as I had on them. For that, I owe them all a great deal of appreciation, one that is definitely overdue.

I am also grateful to Honeywell Aerospace Co. (Phoenix, Arizona) for granting me the permission to publish several hardware pictures, along with relevant design features. The effort of Mr. Robert Desmond, Chief of the Intellectual Property Department, in securing this permission, is greatly appreciated.

In my early years at Texas A&M University, I was indeed fortunate to have Professor Dara Childs, director of the Texas A&M Turbomachinery Laboratory, as a colleague and a mentor. His dedication to, and extreme enthusiasm about, the turbomachinery rotordynamics topic had an impact on me to the point of adopting it as one of my primary research activities for more than a decade.

Finally, I will always be indebted to my wife Magda for her patience and unlimited support. Her relentless effort in typing the major part of this text, including its mathematically involved segments, has been a major contributor to the mere existence of it.

Erian A. Baskharone

CHAPTER ONE

Introduction to Gas-Turbine Engines

Definition

A gas turbine engine is a device that is designed to convert the thermal energy of a fuel into some form of useful power, such as mechanical (or shaft) power or a high-speed thrust of a jet. The engine consists, basically, of a gas generator and a power-conversion section, as shown in Figures 1.1 and 1.2.

As is clear in these figures, the gas generator consists of the compressor, combustor, and turbine sections. In this assembly, the turbine extracts shaft power to at least drive the compressor, which is the case of a turbojet. Typically, in most other applications, the turbine will extract more shaft work by comparison. The excess amount in this case will be transmitted to a ducted fan (turbofan engine) or a propeller (turboprop engine), as seen in Figure 1.2. However, the shaft work may also be utilized in supplying direct shaft work, or producing electricity in the case of a power plant or an auxiliary power unit (Fig. 1.1). The fact, in light of Figures 1.1 through 1.5, is that different types of gas-turbine engines clearly result from adding various inlet and exit components, to the gas generator. An always interesting component in this context is the thrust-augmentation devices known as afterburners in a special class of advanced propulsion systems (Fig. 1.4).

Gas-turbine engines are exclusively used to power airplanes because of their high power-to-weight ratio. They have also been used for electric-power generation in pipeline-compressor drives, as well as to propel trucks and tanks. In fact, it would be unwise to say that all possible turbomachinery applications have already been explored.

Advantages of Gas-Turbine Engines

Of the various means of power production, the gas turbine is in many ways the most efficient because of specific exclusive features. As mentioned earlier, the high power-to-weight ratio makes gas turbines particularly suited for propulsion applications. The absence of reciprocating and rubbing members, in comparison with internal-combustion engines, means fewer balancing problems and less lubricating-oil

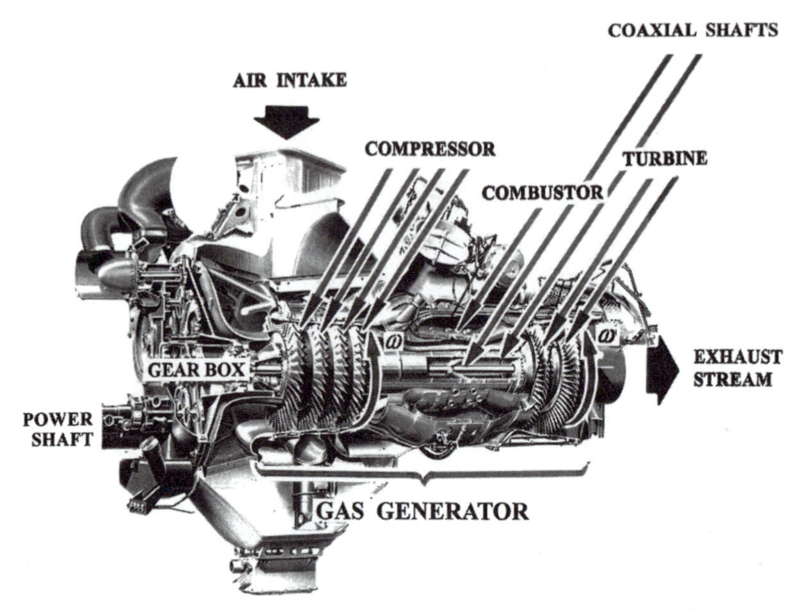

Figure 1.1. Example of an auxiliary power unit. GTCP 660-4 Auxiliary power unit (Garrett Turbine Engine Co.).

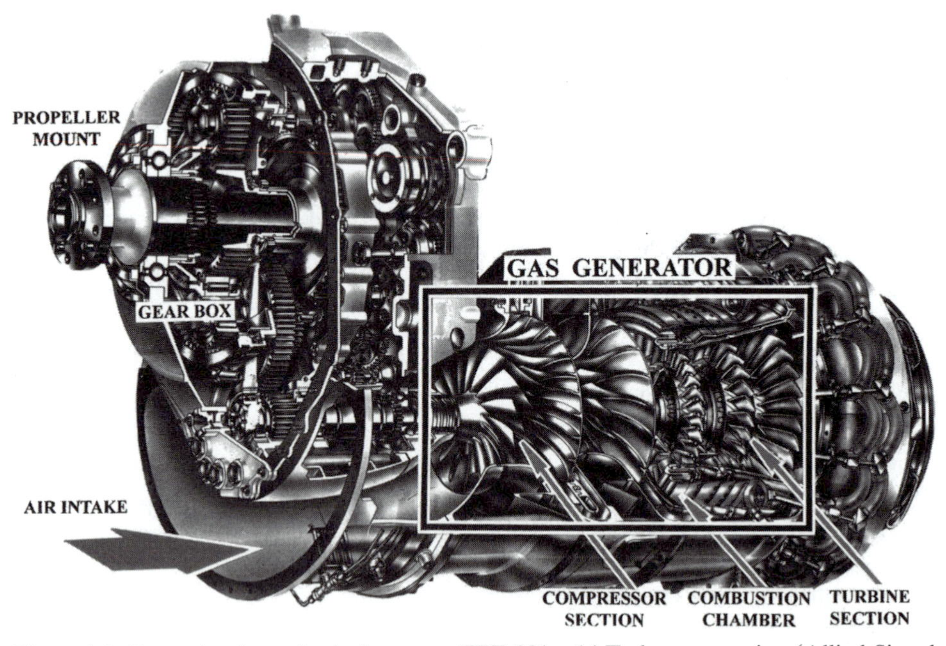

Figure 1.2. Example of a turboshaft system. TPE 331 – 14 Turboprop engine (Allied-Signal Technologies).

consumption. The reliability of gas-turbines, as a result, can be much higher by comparison.

Applications of Gas-Turbine Engines

1) Power-system applications: This category includes such items as auxiliary power units (Fig. 1.1), gas-turbine power plants, and turbochargers (Fig. 1.5). Note that heavy and "bulky" components, such as radial turbines and centrifugal compressors, are typically tolerated in this type of application because they are usually ground applications. Although it is not usually classified as such, a turboprop engine (Fig. 1.2) belongs to the turboshaft-engine category, a phrase that is synonymous with power-system applications.

2) Propulsion applications: Included in this category are the turbojet, turboprop, and turbofan engines, without and with afterburning (Fig. 1.4). These are illustrated in Figures 1.2 through 1.4. With the exception of turboprop engines, the thrust force in this engine category is generated by high-speed gases at the exhaust-nozzle outlet. In fact, turboprop engines are functionally closer to power-system turboshaft engines in the sense that they also provide a net output in the form of propeller-transmitted shaft power.

The Gas Generator

Referring to Figures 1.1 and 1.2, the engine's gas generator is composed of the compressor section, followed by the combustor, and leading to the turbine section. In jet engines, the responsibility of a gas generator is to produce a high-pressure, high-temperature stream of combustion products (predominantly air), which are allowed to expand down to (ideally) the local ambient pressure in an exhaust nozzle. This nozzle gives rise to a high-momentum gas stream generating, in turn, the thrust force that is necessary to propel the airframe. The final outcome in a power system, however, is different, as it is the gas generator's function to transmit shaft work to the power shaft, usually through a gear box, as shown in Figure 1.5. Because the gear box frequently needs maintenance, it can often be replaced by a "free turbine" type of arrangement with the same speed-reduction objective. The free-turbine stage, typically the last in the turbine section, is separately mounted on the inner part of a coaxial twin-shaft assembly. One of the desirable features in using a free turbine is the ease with which the engine can be started. The major drawback, however, is a significant deterioration in performance under off-design operating modes.

Air Intake and Inlet Flow Passage

Attached to the gas generator, on the upstream end, is an air-intake section, which substantially differs from one engine category to another. In auxiliary power units (Fig. 1.1) and turbochargers (Fig. 1.5), the air-intake section is normally covered by a fine-mesh screen. This has the function of protecting the engine, particularly the

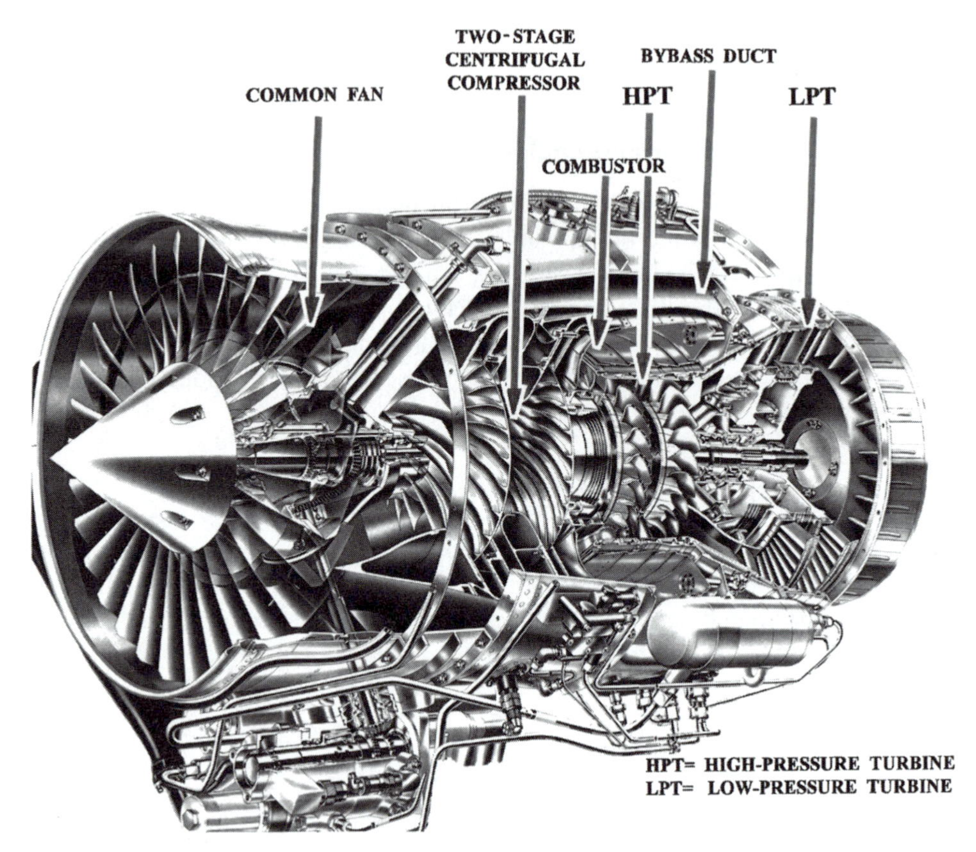

Figure 1.3. A small turbofan engine for a subsonic trainer. F-109 Turbofan engine (Allied-Signal Technologies).

turbine rotors, from the erosion effects that can be imparted by solid objects, such as sand particles, in the inlet air stream.

As for propulsion systems, the majority of air-intake sections are as simple as annular ducts (Fig. 1.3). An exception to this rule is shown in Figure 1.2 for a typical turboprop engine. In this case, the inlet section appears like a "smiling mouth," a clearly odd shape that is primarily caused by the existence of the propeller in this region. As shown in Figure 1.2, a cross-section conversion, ultimately leading to a perfect annular duct, quickly follows. Upstream from the compressor section in turbofan engines is a single or multistage ducted fan, which, by definition, is a smaller compressor in the sense of pressure ratio. The fan may partially exist in the "core" flow stream (Fig. 1.3) or the secondary bypass duct (Fig. 1.4).

Engine Exhaust Component

In a turboshaft engine such as the turboprop in Figure 1.2 and the turbocharger in Figure 1.5, the engine typically ends with an exhaust diffuser. This device converts the incoming kinetic energy into a pressure rise before releasing it at the local ambient pressure. The exhaust diffuser, typically an annular-cross-section duct, may in part

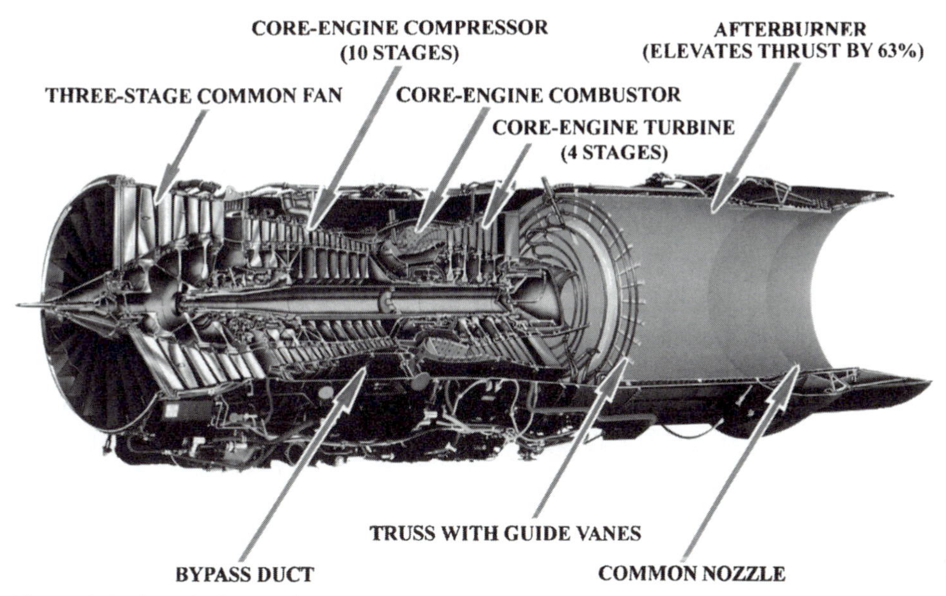

Figure 1.4. A turbofan engine with a bypass ratio of 42%. F-16 Turbofan engine (United Technologies).

rely on what is termed the "dump" effect, which is a result of abruptly ending the center body. Although this sudden area enlargement does cause a pressure rise, it is perhaps one of the most undesirable means of diffusion because of the aerodynamic degradation that it imparts in this sensitive engine segment.

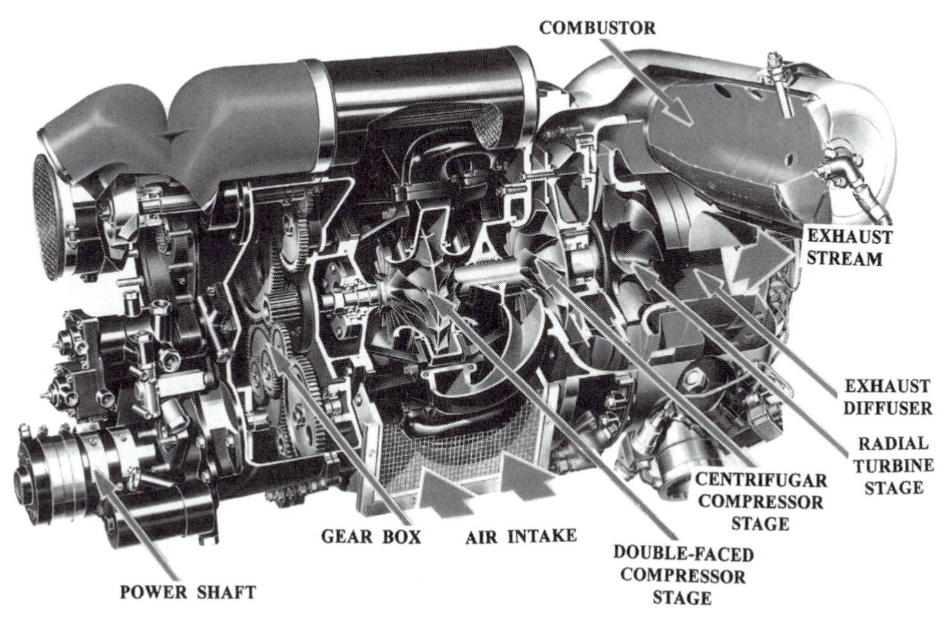

Figure 1.5. Example of the use of radial turbomachinery in a turbocharger. GTCP-85 Turbocharger (Garrett Turbine Engine Co.).

Aside from turboprop engines, propulsion systems will obviously have to end with a flow-accelerating device, namely one or more nozzles. As is well-known, it is the converging-diverging De Laval nozzle that is uniquely capable of producing the highest exit kinetic energy. Under normal operating conditions, the nozzle(s) will then produce a supersonic exit velocity. Referring to the turbofan engine in Figure 1.3 as an example, a viable exhaust system may very well be composed of two separate annular nozzles. Of these, the inner nozzle handles the core (or primary) flow stream. The outer nozzle, on the other hand, concerns the secondary (or bypass-duct) flow stream. Returning to Figure 1.3, note that both exhaust nozzles are of the converging (or subsonic) type. In many other turbofan configurations, a single "mixer" nozzle is utilized for both primary and secondary flow streams. The decision whether to leave these flow streams separate or to mix them is by no means obvious. Such a decision will typically be based on such variables as the secondary-to-primary total-to-total pressure ratio at the point where the two streams can join one another, as well as the bypass ratio. There is, however, an acoustic incentive in mixing these two flow streams because the engine noise in this case is notably less than would otherwise be the case.

Multispool Engine Arrangements

If the gas turbine is required to operate under fixed-speed, fixed-load conditions, then the single-shaft arrangement (whereby all rotating components spin at the same speed) may be suitable. In this case, flexibility of operation (i.e., the speed with which the machine can lend itself to changes of load and rotational speed) is not important.

However, when flexibility of operation is of primary importance, which is typically the case, the use of a mechanically independent (or free) power turbine becomes desirable. In the twin-shaft arrangement in Figure 1.1, the high-pressure turbine drives the upstream compressor. Such a coaxial shaft arrangement could theoretically alleviate the need for an expensive and rather bulky gear box in turboshaft engines where the net power output is to be delivered at a low speed. An example is a turboprop engine, where the load is naturally the propeller, a component that spins at a small fraction of the core engine rotational speed.

Twin-shaft (or twin-spool) configurations are particularly critical in large-scale electricity-generating units in the turboshaft category of gas turbines. In this case, the free turbine is designed to run at the alternator speed, again eliminating the need for an expensive speed-reduction gear box or considerably reducing its size. Note that an additional advantage here is that the starter unit need only be sized to run over the gas generator. Nevertheless, the free-turbine engine configuration is known to cause major performance degradation under off-design operating modes.

Thermodynamic Cycle in a Single-Combustor Engine

Composing a conceptually Brayton-type cycle, the sequence of thermodynamic events in a simple turbojet engine is sketched on the Mollier enthalpy-entropy (h - s) diagram in Figure 1.6. For reference purposes, the ideal cycle is shown first, where all of the compression and expansion processes are isentropic and no total

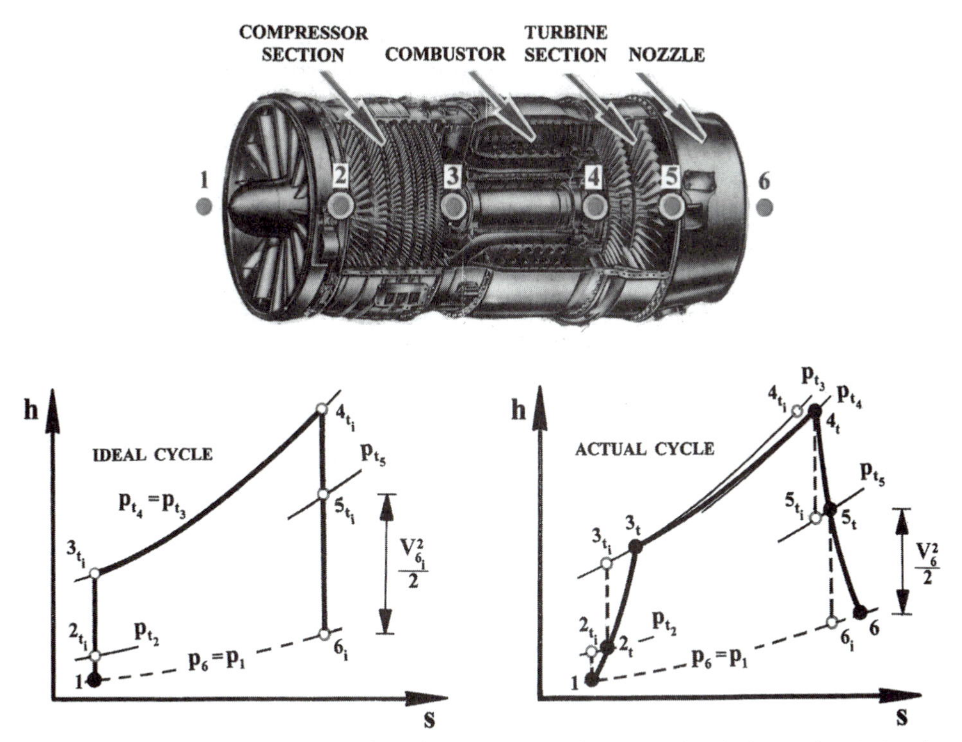

Figure 1.6. Brayton thermodynamic cycle for a typical single-spool turbojet engine. Subscript "i" refers to an ideal state, or the end state of an isentropic process. Subscript "t" refers to a total (stagnation) property, or a state that is defined by its total properties.

(or stagnation) pressure loss occurs over the combustor. The real-life sequence of processes is shown separately in the same figure. As seen, the principle of entropy production, a quantitative interpretation of the Second Law of Thermodynamics, is clearly represented in this figure. This is clearly applicable to the compressor and turbine flow processes, where the ideal (isentropic) flow processes are shown alongside the real-life process. Note that the net useful outcome of the engine is the production of high-velocity gases (products of combustion), which are responsible for propelling the airframe. In Figure 1.6, this velocity is identified as V_6. As would be expected, the ideal-cycle "theoretical" operation gives rise to a higher V_6 magnitude by comparison. Finally, note that the combustor-exit/turbine-inlet total temperature (T_{t4}) is maintained fixed in the entire figure. The reason is that this particular variable is a strong function of, among other items, the turbine metallurgical strength and shaft speed. In practice, this temperature is therefore treated as a design constraint.

Importance of Metallurgical Progress

Since their inception in the 1940s, gas-turbine engines have been under virtually uninterrupted development and upgrade. Design refinements have progressively been at a pace that is proportional to advancements in the strength-of-materials area. These metallurgical advancements have been focused on and continually implemented in the turbine area. The reason is that turbine stages, especially the earlier ones, are exposed to a critically high temperature (typically above 1300 K) while spinning at a high speed that exceeds, in some applications, 100,000 rpm. Combined together, these factors simply constitute a recipe for a potentially disastrous mechanical environment. Nowadays, however, ceramic turbine rotors are capable of spinning at more than 120,000 rpm and under temperatures as high as 1800 K. As a result, turbomachinists find it much easier to maximize the turbine's power output to levels that in the past were simply unthinkable.

CHAPTER TWO

Overview of Turbomachinery Nomenclature

A brief introduction to gas-turbine engines was presented in Chapter 1. A review of the different engines, included in this chapter, reveals that most of these engine components are composed of "lifting" bodies, termed airfoil *cascades*, some of which are rotating and others stationary. These are all, by necessity, bound by the hub surface and the engine casing (or housing), as shown in Figures. 2.1 through 2.5. As a result, the problem becomes one of the internal-aerodynamics type, as opposed to such traditional external-aerodynamics topics as "wing theory" and others. Referring, in particular, to the turbofan engines in Chapter 1 (e.g., Fig. 1.3), these components may come in the form of ducted fans. These, as well as compressors and turbines, can be categorically lumped under the term "turbomachines." Being unbound, however, the propeller of a turboprop engine (Fig. 1.2) does not belong to the turbomachinery category.

The turbomachines just mentioned, however, are no more than a subfamily of a more inclusive category. These only constitute the turbomachines that commonly utilize a compressible working medium, which is totally, or predominantly, air. In fact, a complete list of this compressible-flow subfamily should also include such devices as steam turbines, which may utilize either a dry (superheated) or wet (liquid/vapor) steam mixture with high quality (or dryness factor). However, there exists a separate incompressible-flow turbomachinery classification, where the working medium may be water or, for instance, liquid forms of oxygen or hydrogen, as is the case in the Space Shuttle Main Engine turbopumps. This subcategory also includes power-producing turbomachines, such as water turbines. Presented next is a summary of turbomachinery classifications in accordance with some specific criteria, beginning with the very definition of a turbomachine.

Definition of a Turbomachine

A turbomachine is a device where mechanical energy, in the form of shaft work, is transferred either to or from a continuously flowing fluid by the dynamic action of rotating blade rows.

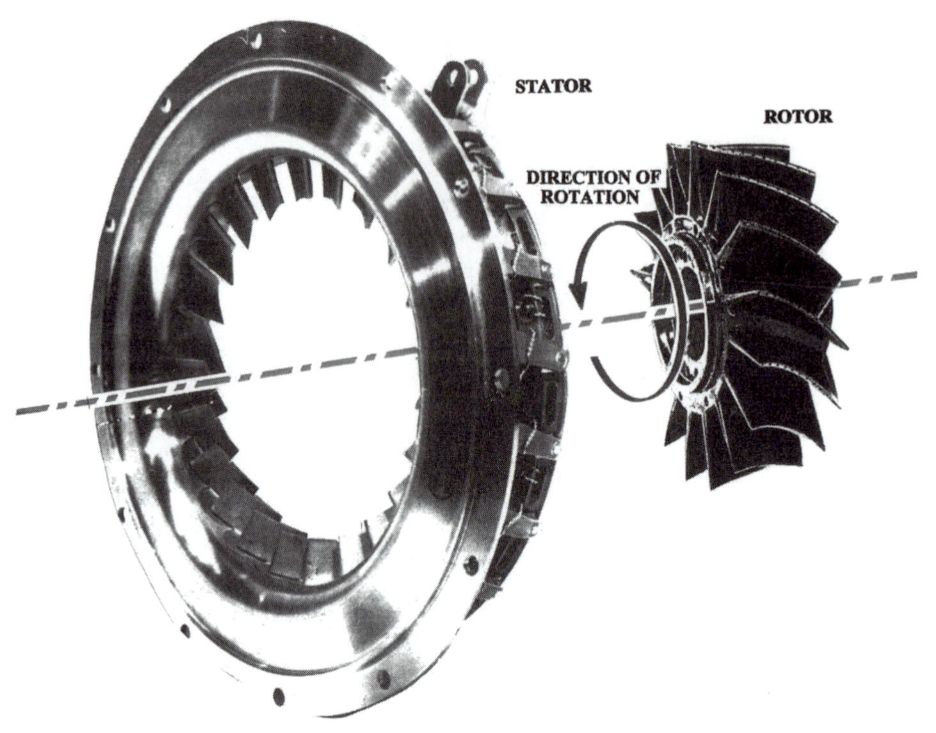

Figure 2.1. A typical axial-flow turbine stage.

General Classification of Turbomachines

1) By their functions:

   - Work-absorbing turbomachines: such as compressors and fans
   - Work-producing turbomachines: generally known as turbines

   Figures 2.2 and 2.3 show examples of these two turbomachinery categories.

2) By the nature of the working medium:

   a) Compressible-flow turbomachines: where the incoming fluid is totally air, as in fans and compressors, or the products of combustion, as in gas turbines. In the latter category, and in the absence of an afterburner, the working medium will still be treated as predominantly air.
   b) Incompressible-flow turbomachines: where the working medium may be water (hydraulic pumps) or any single-phase substance in the liquid form.

3) By the type of meridional flow path:

   Using a cylindrical frame of reference, the projection of a turbomachine onto the axial-radial (z-r) plane is called the meridional view (also termed the meridional flow path). Referring to the two compressor examples in Figure 2.4 and the corresponding turbine examples in Figure 2.5, it is only the meridional projection of the rotating blades that is relevant in this particular classification. Should the

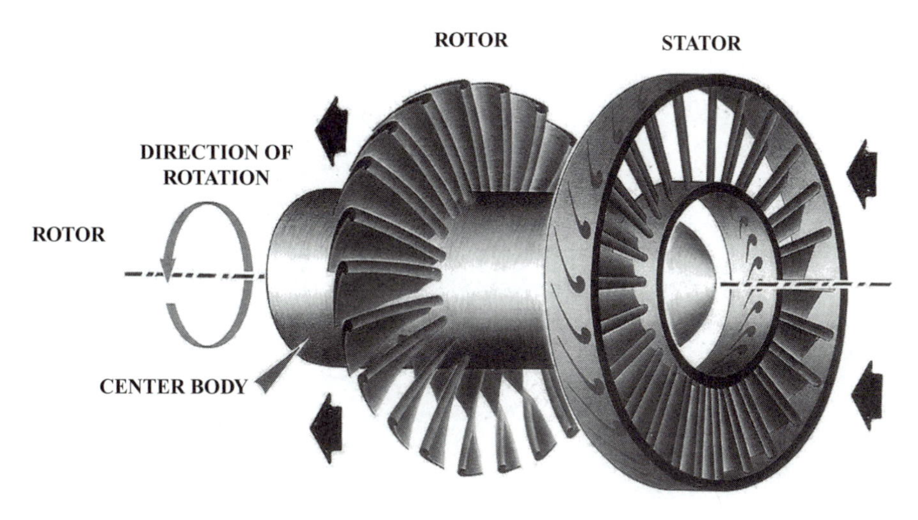

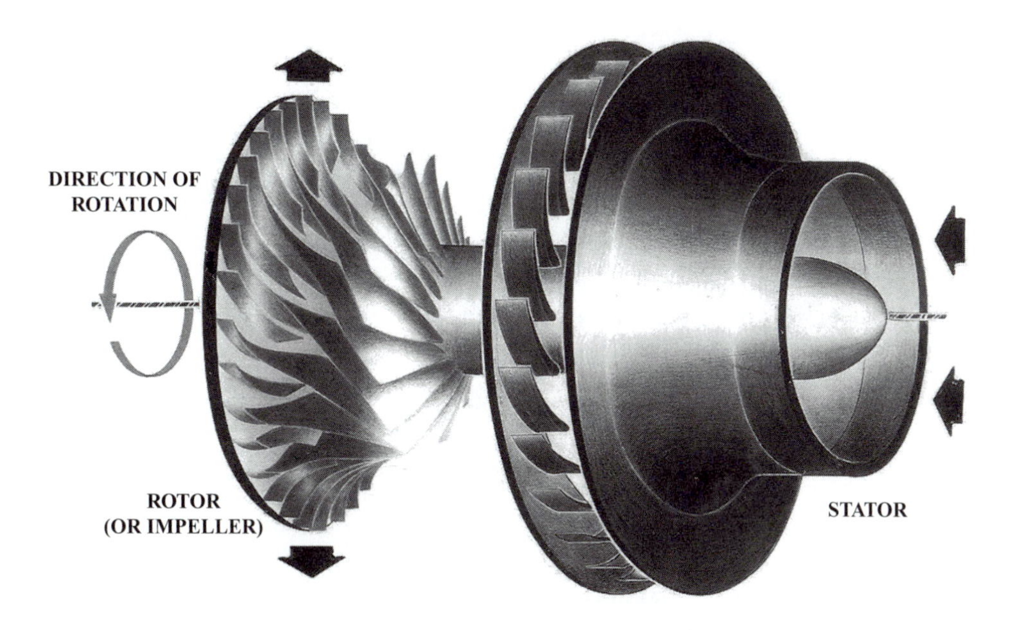

Figure 2.2. Comparison between axial and centrifugal compressor stages.

rotor's flow path remain (exactly or nearly) parallel to the axis of rotation, the entire stator-rotor assembly (termed a *stage*) is said to be of the axial-flow type. However, if the meridional flow path changes direction from axial to radial, or vice versa, the stage is referred to as a centrifugal compressor or a radial turbine, as shown in Figures 2.4 and 2.5, respectively.

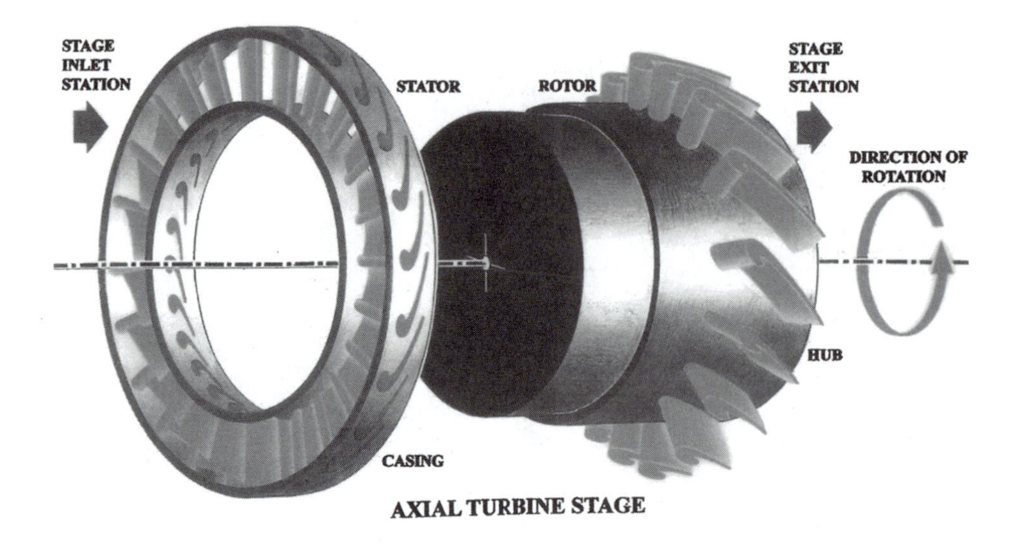

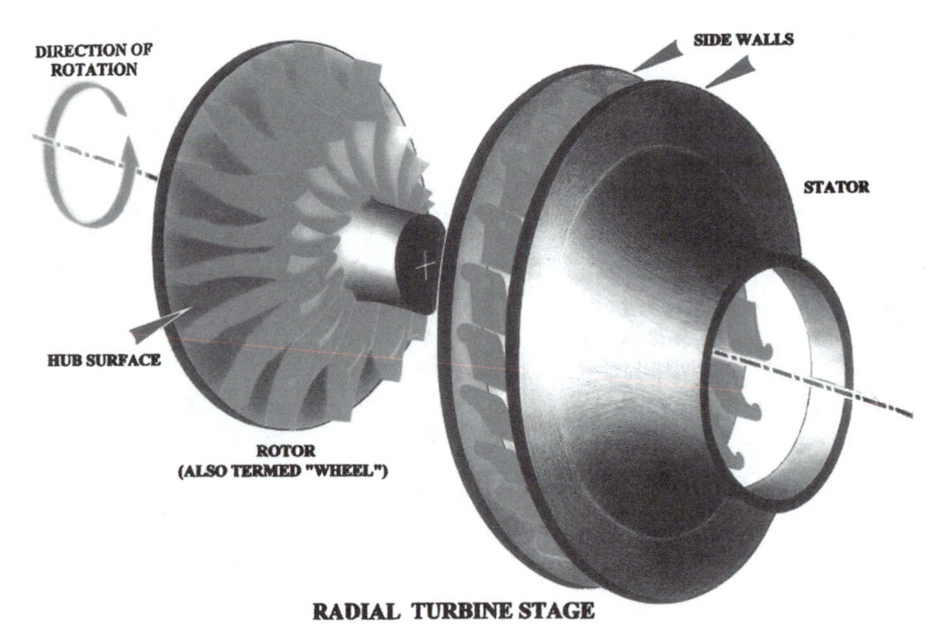

Figure 2.3. Comparison between axial- and radial-inflow turbine stages.

It is perhaps fitting to emphasize the fact that terms such as "purely" axial or "purely" radial meridional-view flow direction represent ideally perfect flow guidedness by the endwall. These terms imply the total lack of such real-life flow effects as viscosity and secondary cross-flow migration. In reality, the viscosity-related boundary-layer buildup will give rise to a normally slight cross-flow velocity component away from the endwalls. Such a phenomenon is part of what is known as the boundary-layer "displacement" effect. This flow behavior, by itself, brings weight to the fact that the term "flow" in the phrase "meridional flow path" is probably

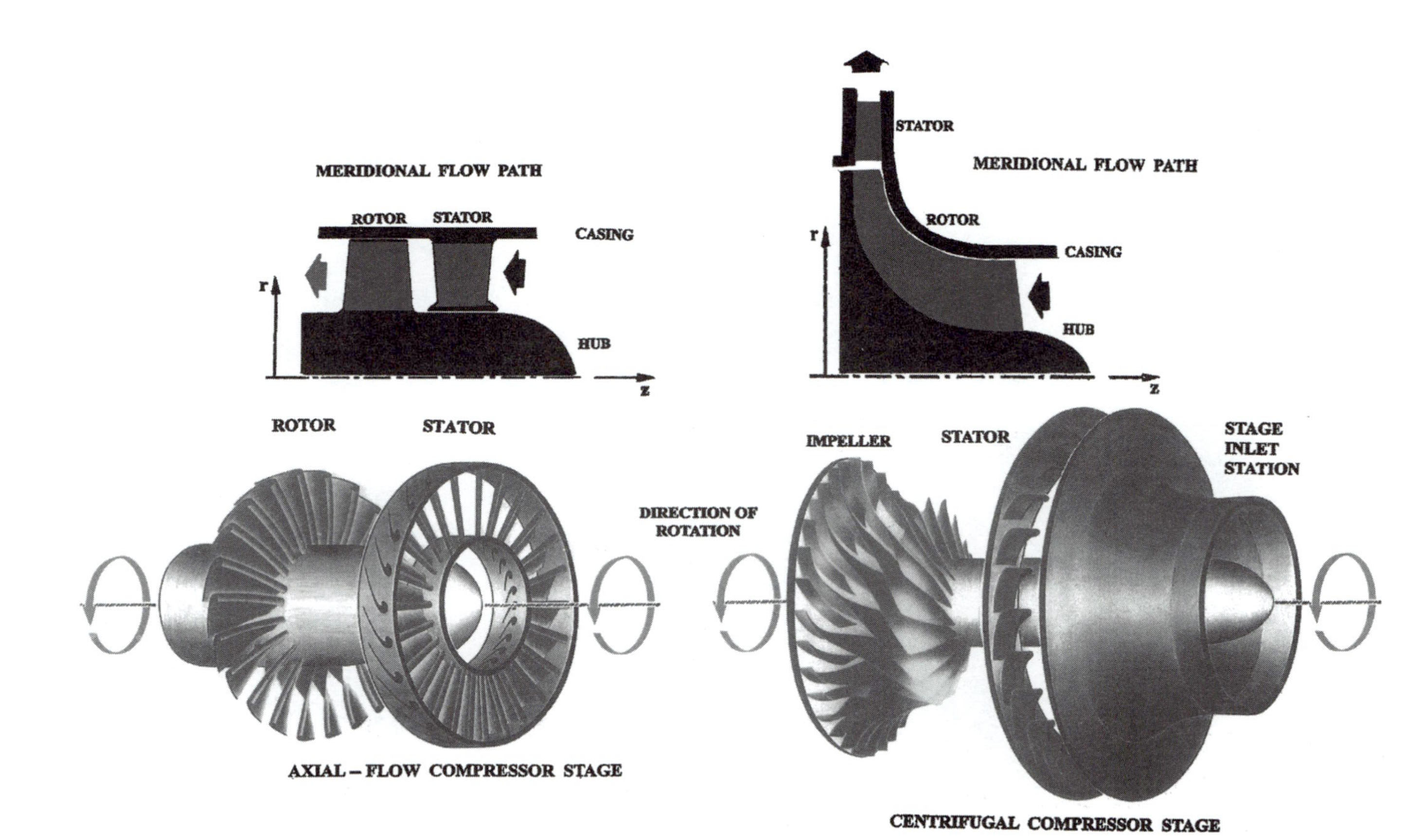

Figure 2.4. Meridional projections of axial and centrifugal compressor stages.

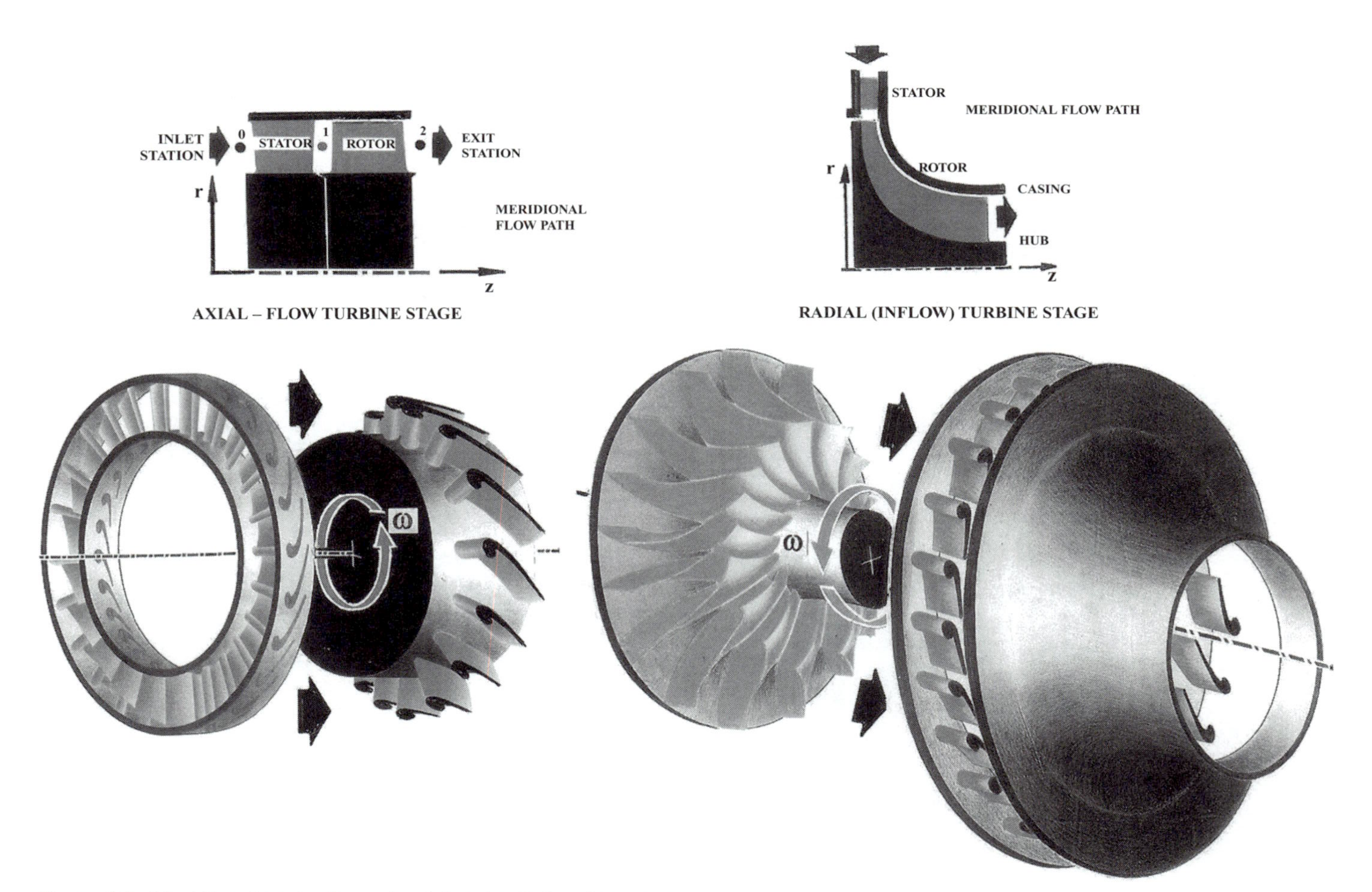

Figure 2.5. Meridional projections of axial and radial turbine stages.

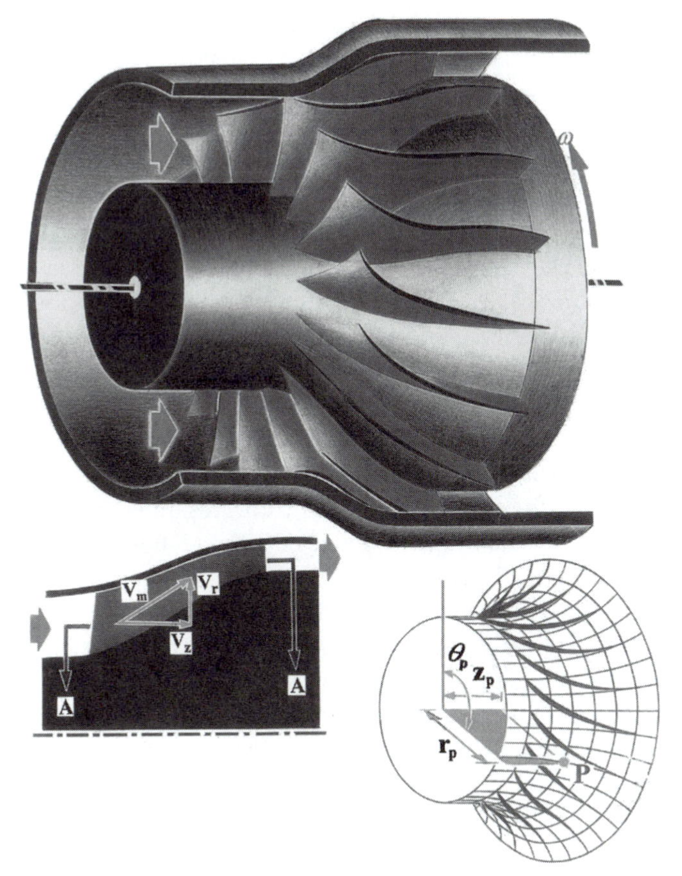

Figure 2.6. A mixed-flow compressor and cylindrical frame of reference.

a "loose" usage of the word. Nevertheless, we will always override such real flow effects, making use of (only) the bounding endwalls when it comes to this particular turbomachinery classification.

As one would naturally expect, there exists a separate subfamily of turbomachines that lies somewhere between the two (axial and radial) extremes. Referred to as mixed-flow turbomachines, a compressor rotor that is representative of this category is shown in Figure 2.6. As seen in the figure, the meridional rotor projection indicates the existence of neither axial nor radial rotor segments. In a rigorous way of viewing things, one may never come across a "purely" axial or radial meridional rotor projection. It therefore becomes a matter of "geometrical dominance" that sets the rule in this case.

Stage Definition

A turbomachinery stage is typically composed of a stationary cascade of "vanes," termed a stator, followed by a rotating "blade" row, the rotor, as shown in Figures 2.1

through 2.5. Across the stator, no shaft-work exchange occurs to or from the fluid, but rather a conversion from thermodynamic energy content (namely enthalpy) into kinetic energy, as in a turbine stator, or vice versa in the case of a compressor. This energy transfer is primarily effected by the vane-to-vane passage shape, which can be diffuser-like, as in a compressor stator, or nozzle-like, as in its turbine counterpart. An additional flow deceleration or acceleration component would normally arise from the divergence or convergence of the (hub and casing) endwalls. Applicable only to radial stators is the effect of the streamwise radius shift, as the cross-flow area is clearly radius-dependent. Except for relatively minor losses in total (or stagnation) temperature (caused by cooling, for example) and in total pressure (caused by friction and other degrading mechanisms), the total flow properties (enthalpy, temperature, pressure, density, etc.) can be assumed constant. As is further discussed in Chapter 3, the flow acceleration in this case creates what is categorically defined as a "favorable" pressure gradient, under which the total pressure loss is comparatively low.

Work is produced (or consumed) only across the stage rotor, resulting in a substantial total (or stagnation) enthalpy change. Knowledge of the mass-flow rate and the flow kinematics (so-called velocity diagrams) associated with the rotor makes it possible to calculate the shaft-exerted torque, as well as the fluid/rotor shaft-work interaction. The process quality, in terms of losses, is notably different between compressors and turbine rotors. The blade-to-blade passage convergence in the latter produces an accelerating "relative" flow stream, which leads to a smaller total "relative" pressure loss across turbine rotors by comparison. The quoted terms in this as well as previous statements will be thoroughly defined and examined in the remainder of this chapter and (more so) in Chapters 3 and 4.

Coordinate System

Because the stage flow is, by definition, confined between two surfaces of revolution, the most convenient frame of reference to describe both the turbomachine geometry and the flow kinematical variables is clearly cylindrical. Referring to the compressor rotor in Figure 2.6, a fluid particle at position P will be defined by its axial, tangential, and radial coordinates (z_P, θ_P, r_P). With no mathematical penalty, the arc-length, $r\theta$ axis, in some later references, will only carry the circumferential coordinate θ. Despite its mathematical legitimacy, however, this choice may lead to airfoil-shape distortion, particularly in radial (or centrifugal) turbomachines, because of the streamwise change of radius in this case.

For purely or predominantly axial-flow turbomachines, the worthiest, most revealing flow/hardware-interaction picture to examine is that on the axial-tangential (z-θ) surface (Fig. 2.7). Coordinate axes in this case are the longitudinal distance (z) and the tangential coordinate θ. Figure 2.7 provides a graphical means of visualizing and "flattening" a surface carrying the flow trajectories. Depending on the radial location of interest, a cylinder is allowed to cut through all of the rotor blades. Traces of these blades, as well as the local set of streamlines on the cylindrical surface, are preserved and the cylinder itself unwrapped, as illustrated in Figure 2.7.

Figure 2.7. Unwrapping the axial-tangential surface with the airfoil cascade traces.

For all three types of flow-viewing planes, another meaningful frame of reference is the meridional plane (Figs. 2.4 and 2.5). This is simply the projection of the endwalls and, if required, the streamlines, on the axial-radial (z-r) plane. As already stated, this viewing plane exclusively separates axial, radial, and mixed-flow turbomachines from one another.

The third of the turbomachinery-viewing projections is the radial-tangential (r-θ) plane of reference. This plane, shown in Figures 2.8 and 2.9, is relevant exclusively in centrifugal compressors and radial turbines. Depending on the blade segment where the flow path is purely radial, this plane will always show the true blade-to-blade passage shape in this segment and the fluid/structure interaction outcome.

Velocity Diagrams

A velocity diagram (or triangle) is simply a graphical vector representation of the well-known kinematical principle that the absolute velocity vector **V** of a fluid particle is composed of the velocity vector **W** relative to the rotating blade, plus the linear velocity vector **U** of the blade itself.

In an equation form, this vector relationship can be expressed (by reference to Fig. 2.10), as

$$\mathbf{V} = \mathbf{W} + \mathbf{U} \tag{2.1}$$

where

$$\mathbf{U} = \omega r \mathbf{e}_\theta \tag{2.2}$$

with ω and r being the rotor speed (in radians/s) and the local radius, respectively.

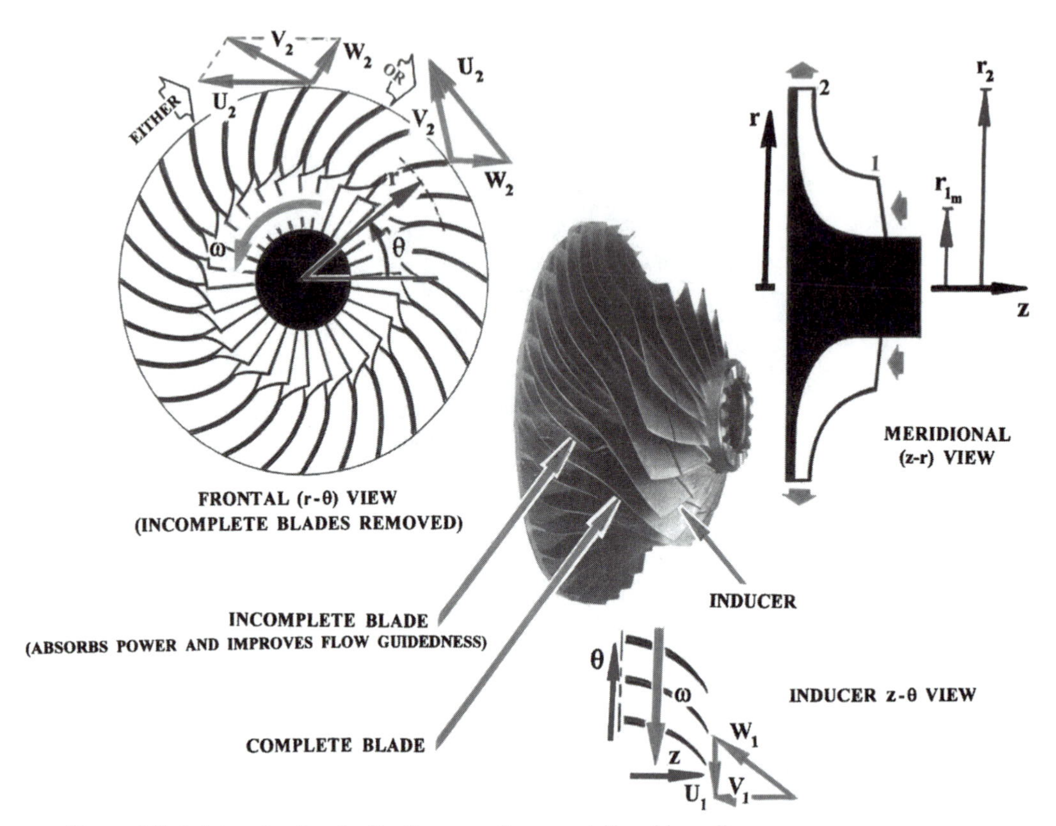

Figure 2.8. Inlet and exit velocity diagrams for a centrifugal impeller.

The velocity diagrams (or triangles) corresponding to equation (2.1) are often required at the flow inlet and exit stations 1 and 2 in Figure 2.3. Incorporation of these diagrams in Eulers theory (as will be discussed in Chapter 4) enables the engineer to calculate such parameters as the torque, shaft work (w_s), and power produced (or absorbed) by the flow stream. Note that the velocity diagrams at the inlet and exit stations of a stator are rather obvious. The reason is that this is a nonrotating component for which $\mathbf{U} = 0$. Consequently, the velocity relationship becomes

$$\mathbf{V} = \mathbf{W} \tag{2.3}$$

Let us now examine the velocity diagrams associated with typical turbomachinery stages, starting with the axial compressor and turbine stages in Figure 2.11. The sign conventions of flow angles and tangential-velocity components are shown on the same figure. In short, all angles and tangential-velocity components will be positive in the direction of rotation and negative otherwise. Figures 2.8 and 2.9 present the same information, but for a centrifugal compressor and a radial turbine, respectively. Referring to these two figures, the rotor inlet and exit segments in the compressor

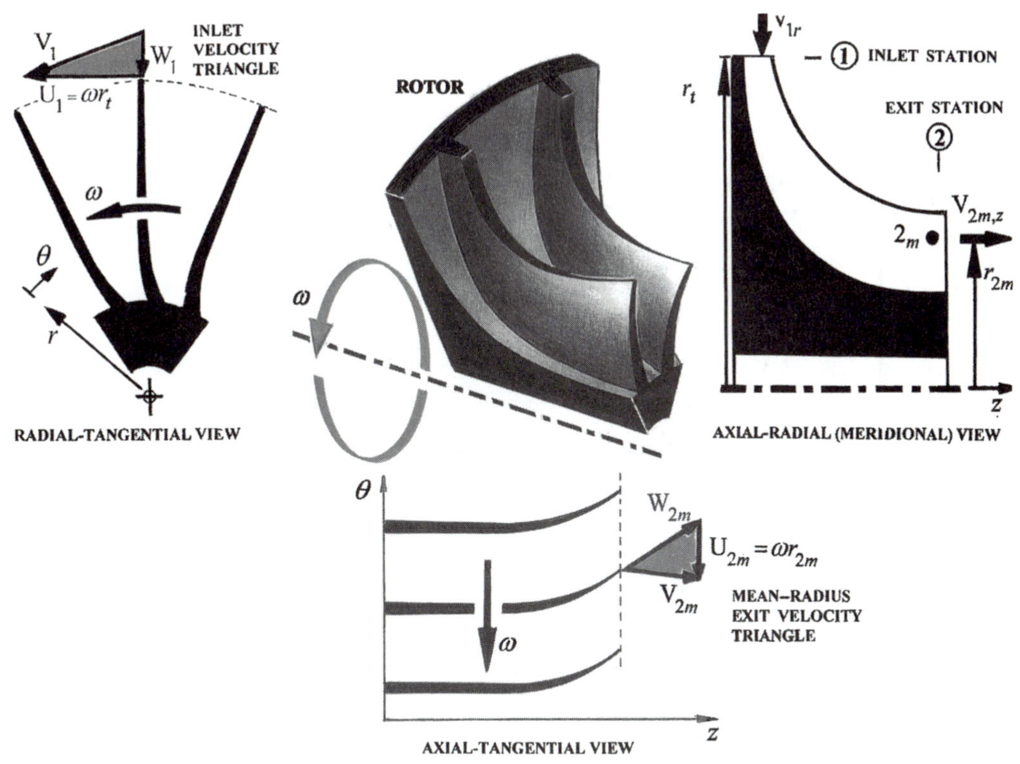

Figure 2.9. Inlet and exit velocity diagrams for a radial-turbine rotor.

and turbine cases, respectively, are both of the axial-flow type and are handled as such. The centrifugal compressor rotor, in Figure 2.8, is usually referred to as an "impeller," which is the proper terminology in this case.

Velocity components in the axial-radial (or meridional) projections reflect, for the most part, the flow-path effects, meaning (in particular) the hub and casing geometrical features. Convergence or divergence of these two surfaces relative to one another will only affect the acceleration or deceleration through the flow stream. Because this reference plane is perpendicular to the tangential direction, the velocity-triangle concept is inapplicable.

Although the emphasis in this text is placed on the inlet and exit velocity triangles, one may very well choose any location inside the rotor subdomain to create the local velocity triangle. Note that the direction of the relative velocity vector (**W**) is predominantly implied by the local shape of the blade-to-blade passage. Such velocity triangles would lead, in effect, to the variation of the accumulative shaft work up to a given streamwise location. This is instrumental in making such decisions as front versus rear blade loading (Chapter 8). In a turbine rotor, for instance, such information will guide the decision of whether to turn the flow and then accelerate it, or vice versa.

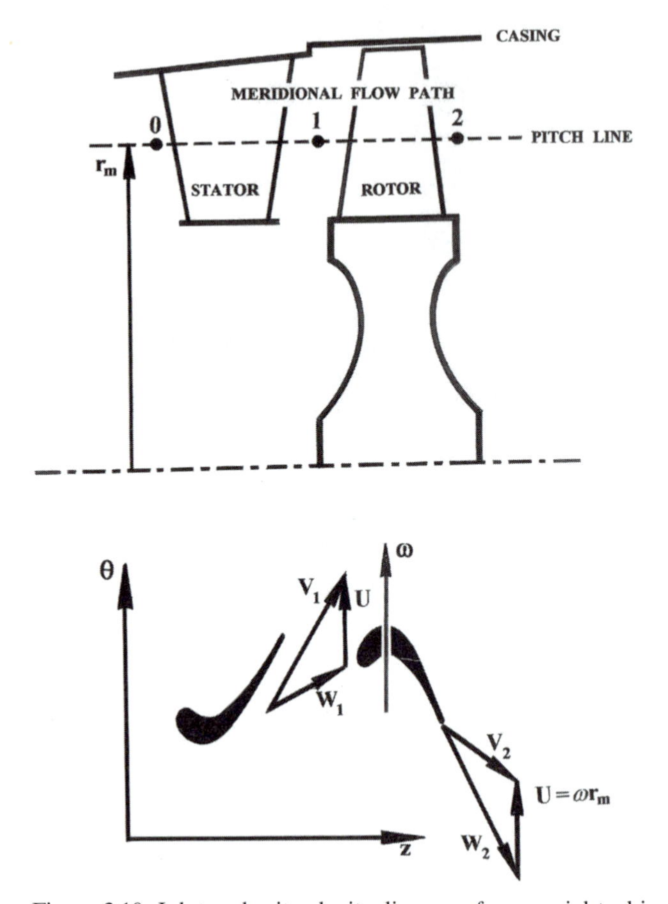

Figure 2.10. Inlet and exit velocity diagrams for an axial-turbine stage.

In reference to, say, the rotor-inlet velocity triangle in Figure 2.11, it is not exactly silly to wonder which of the two velocity vectors, **V** or **W**, is really the flow inlet velocity. The answer is *either one*, depending on the intention of the analyst. For instance, a full-scale flow analysis inside the blade-to-blade passage is better served by adopting the relative velocity vector (**W**) throughout the rotor subdomain. After all, many physical criteria, such as choking and fluid/structure viscous interaction (including the viscosity-dictated "no-slip" condition over solid boundaries), apply exclusively to the relative velocity vector (**W**). Despite its legitimacy as a flow velocity vector, the absolute velocity vector (**V**), however, would appear to penetrate the blade surface, which we know is impossible. The reason is that the rotor-blade surface itself is rotating.

Multiple Staging

In most instances, the amount of shaft work would force the designer to divide it over more than one stage. This is frequently the case in compressor sections in particular.

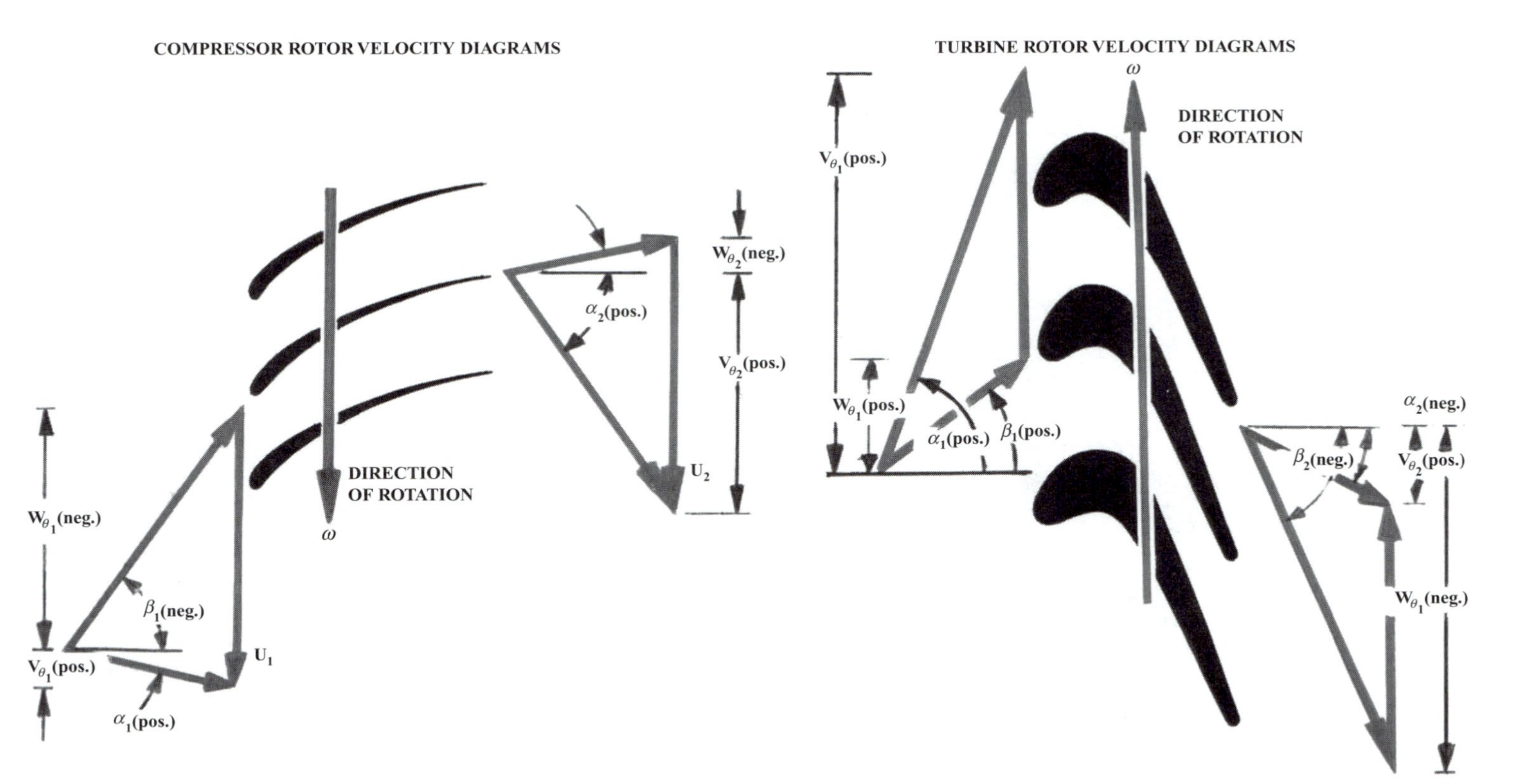

Figure 2.11. Comparison between axial compressor and turbine velocity diagrams. *Sign convention*: All angles and velocity components in the direction of rotation are positive.

Notes:

- $U = \omega$ r is the rotor "linear" velocity, where "r" is the local radial location,
- in a purely axial turbomachine, U_1 and U_2 are identical.

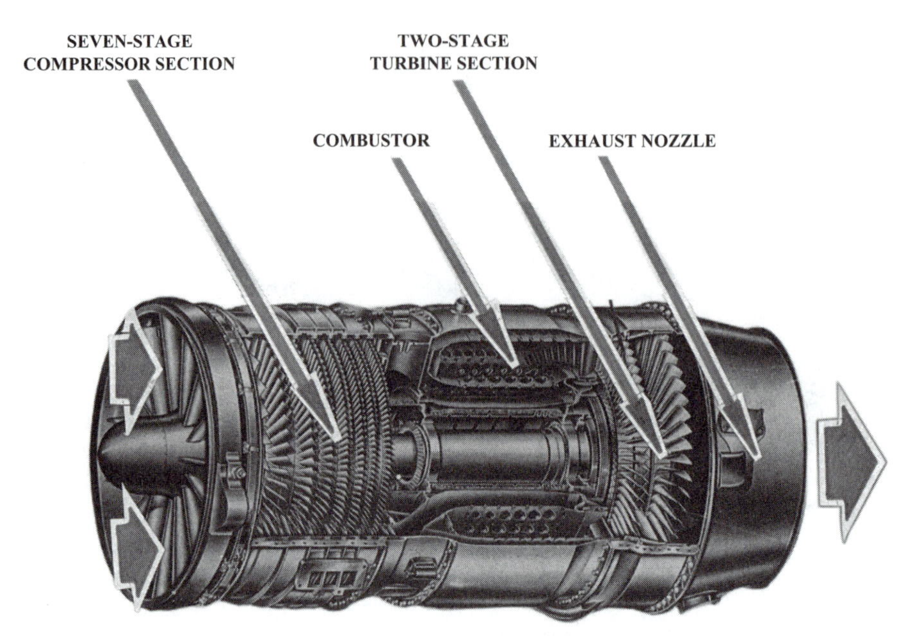

Figure 2.12. Multiple staging of axial-flow turbomachines in a General Electric turbojet engine.

The motive in this case is to avoid the adverse effects of what is known as an "unfavorable" pressure gradient. This is true in the sense that the rising pressure, to which the compressor subcomponents are exposed, will typically aggravate the boundary layer buildup on the blade and endwall surfaces. This buildup can be so severe as to cause highly degrading flow separation and recirculation. Figures 2.12 and 2.13, respectively, highlight a two-stage axial-flow turbine and a centrifugal compressor for the purpose of comparison. Examination of the former reveals the ease with which multiple staging of axial-flow turbomachines is typically achieved. Contrasting this is the centrifugal-compressor double staging in Figure 2.13. In this case, the so-called "return duct" is a flow-turning device that, by category, may cause a substantial magnitude of total (or stagnation) pressure loss. Worse yet are radial-turbine return ducts, for they not only turn the interstage flow stream but expose it to a larger magnitude of unfavorable pressure gradient as well. The reason is that the duct in this case would be of the diffuser-like type, for it causes a considerable streamwise radius growth, with the result being a monotonic rise in its cross-flow area.

Viscosity and Compressibility Factors

As noted earlier, there are definitely some real-life effects on the flow structure through an airfoil cascade that cause varied magnitudes of losses. These are mostly caused by the flow viscosity and, to a lesser extent, its compressibility.

The influence of flow viscosity is normally assessed on the basis of the all-familiar Reynolds number (Re), which is generally defined as

$$Re = \frac{VL}{\nu} \tag{2.4}$$

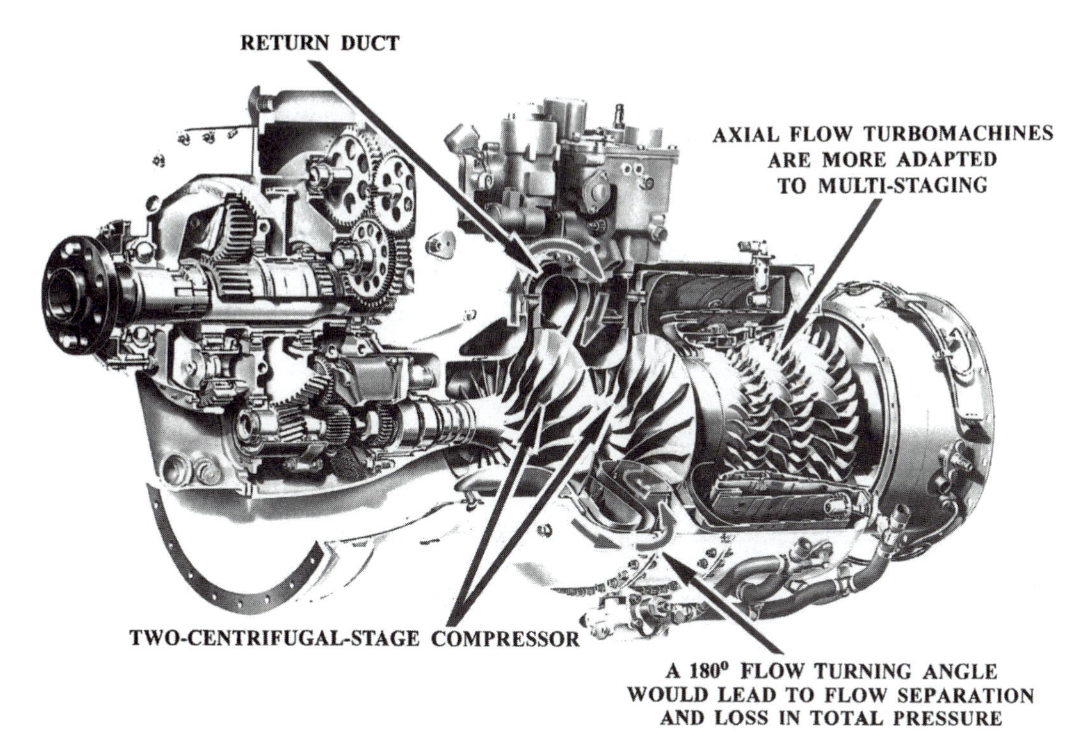

Figure 2.13. Return duct in multiple staging of centrifugal compressors (aerodynamic degradation source).

where

- V is the cascade-inlet absolute, or relative, velocity for a stator or a rotor, respectively.
- L is a characteristic length, normally chosen to be the airfoil true chord shown in Figures 2.14 (for a compressor airfoil) and 2.15 (for a turbine airfoil).
- ν is the kinematic viscosity.

The Reynolds number is viewed, at least computationally, as the ratio between the inertia and viscosity forces in the cascade flow field. As an example, a flow regime with a significantly low Reynolds number (lower than, say, 2×10^5, on the basis of the airfoil mean camber-line length) is referred to as being viscosity-dominated.

The high magnitude of total (or stagnation) pressure losses caused by the flow viscosity is not limited only to the solid walls. Similar undesirable effects exist in the so-called wake of the airfoil downstream from the trailing edge, where two streams, with two different histories, mix together. This open-ended subregion with significantly high lateral velocity gradients fits under the "free-shear" layer classification. Details of this viscosity-dominated subregion are shown in Figure 2.16.

The flow compressibility, in a Mach number sense, can also be a degrading factor should the flow become supersonic. This may occur over a blade-to-blade cross-flow plane or be in the form of supersonic pockets near one endwall or another.

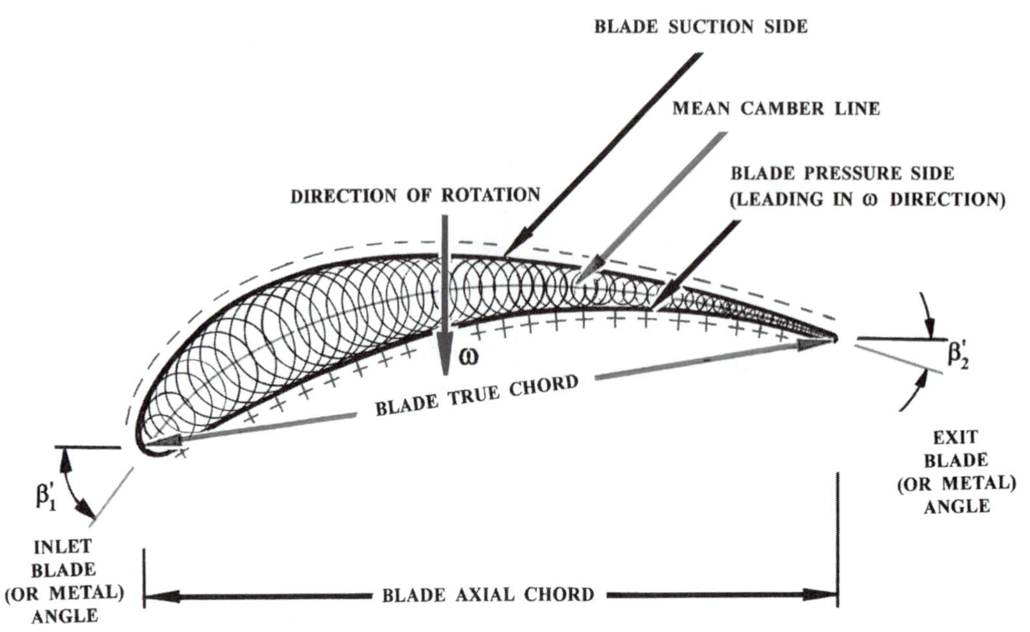

Figure 2.14. Geometrical configuration of a compressor rotor blade. Mean camber-line and airfoil angles in axial compressors. *Note*: In axial-flow turbomachines, all (flow and metal) angles are measured from the axial direction.

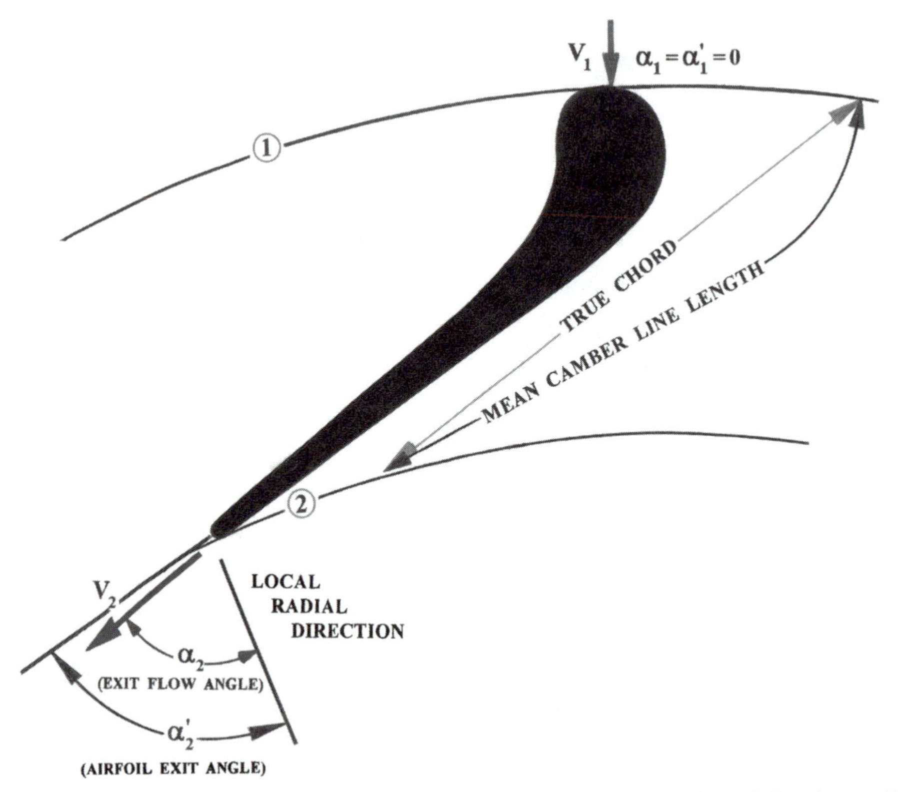

Figure 2.15. Metal and flow angles in a radial-turbine stator. *Note*: In radial turbomachines, all (flow and metal) angles are measured from the local radial direction.

Figure 2.16. Stator/rotor interaction in an axial-flow turbine: a velocity vector plot.

The potential for a usually oblique shock under such circumstances may very well materialize, where an abrupt loss in total pressure would occur.

Stator/Rotor Interaction

Figure 2.16 shows the stator-domain velocity distribution in an axial-flow turbine stage as well as the downstream rotor blades at the mean radius of an experimental turbine (MacGregor and Baskharone 1996). In particular, the figure illustrates the numerically obtained velocity-deficit regions (termed wakes) downstream from the stator vanes. The rotor blades, being the recipients of such a nonuniform velocity distribution, will be exposed to time-dependent pressure fluctuations as they spin in the presence of these wakes. The result will naturally be excessive fatigue stresses, not only on the rotor blades but on the stator vanes themselves. This stator/rotor flow-interaction phenomenon is a function of such variables as the stator-to-rotor axial gap and the difference, if any, between the vane and blade counts. These, as well as other variables, are usually subject to trade-off studies, where aerodynamic objectives can very well be at odds with material strength limitations.

CHAPTER THREE

Aerothermodynamics of Turbomachines and Design-Related Topics

In this chapter, the flow-governing equations (so-called conservation laws) are reviewed, with applications that are purposely turbomachinery-related. Particular emphasis is placed on the total (or stagnation) flow properties. A turbomachinery-adapted Mach number definition is also introduced as a compressibility measure of the flow field. A considerable part of the chapter is devoted to the so-called total relative properties, which, together with the relative velocity, define a legitimate thermophysical state. Different means of gauging the performance of a turbomachine, and the wisdom behind each of them, are discussed. Also explored is the entropy-production principle as a way of assessing the performance of turbomachinery components. The point is stressed that entropy production may indeed be desirable, for it is the only meaningful performance measure that is accumulative (or addable) by its mere definition.

The flow behavior and loss mechanisms in two unbladed components of gas turbines are also presented. The first is the stator/rotor and interstage gaps in multistage axial-flow turbomachines. The second component is necessarily part of a turboshaft engine. This is the exhaust diffuser downstream from the turbine section. The objective of this component is to convert some of the turbine-exit kinetic energy into a static pressure rise. Note that it is by no means unusual for the turbine-exit static pressure to be less than the ambient magnitude, which is where the exhaust-diffuser role presents itself.

In terms of the flow-governing equations, two nonvectorial equations will be covered in this chapter. These are the energy- and mass-conservation equations (better known as the First Law of Thermodynamics and the continuity equation, respectively). The third equation, namely the momentum equation, is a vector equation and will be presented and discussed in Chapter 4.

Assumptions and Limitations

In applying the conservation laws (e.g., those of mass and energy), a bulk-flow-analysis approach will be taken. This is where the average flow properties are assumed to prevail over the so-called pitch line, which is midway between the endwalls. In axial-flow turbomachines, the average properties will exist on an axial, constant-radius line, with a "mean" radius (r_m), two plausible definitions of which will be discussed.

In doing so, several simplifications will be made. Among them, the following two are particularly important:

(1) *An adiabatic flow process:* In turbomachinery applications, the flow domain is composed of stationary and rotating blade-to-blade channels. Even if the blades are cooled (normally the case for the first stage of a high-pressure turbine), the coolant/primary flow ratio is often ignorable (typically 1–3%). With this being the case and particularly during the preliminary design phase, it would be wise, from an engineering viewpoint, to simply ignore the heat-energy exchange. Note that an adiabatic flow process is very hard to attain, whereas an isentropic process is simply impossible.

(2) *An inviscid flow field:* Based on a mean-radius leading-to-trailing-edge distance of a typical blade (as a characteristic length), the Reynolds number is normally well above 5×10^5, which would classify the flow field as dominated by inertia (rather than viscosity). Therefore excluding what would be a far-off design operating mode, with massive flow separation and recirculation, and within the framework of a bulk-flow model, the inviscid-flow assumption would be fitting. Of course, invoking this or any other simplification reflects how accurate a flow-analysis outcome is desired. It is important, nevertheless, to identify those subdomains where the flow viscosity is pronounced. First is the near-wall thin subregion, known as the boundary layer, where the shear stresses caused by viscosity are predominant. Equally degrading in cascade aerodynamics is the airfoil wake (Fig. 3.1). As discussed in Chapter 2, this subregion, downstream from the trailing edge, is where two (suction and pressure) flow streams mix together in what is categorically termed a free-shear layer, which is a significant source of viscosity-related losses. This layer could also have a serious, even catastrophic, impact on the blade's life. Depending on the length of the stator-to-rotor gap, the rotor blades may be exposed to a superimposed cyclic loading as they sweep through the stator-airfoil wakes (Fig. 3.2), causing the likelihood of fatigue failure to be a real threat. As odd as it may appear, this same phenomenon poses an equally dangerous effect on the upstream stator, for it is basically a stator/rotor-interaction problem in an all-subsonic flow field. However, and with the exception of some possible "chipping" of the vane's trailing-edge segments, the stator's possible damage is less catastrophic by comparison with that of the rotor. The reason is that the stator vanes are comparatively safe, from a mechanical standpoint, because they are typically welded to the hub and casing endwalls. As for the "lossy" viscosity-related layers and from a pure steady-state aerodynamic standpoint, there will be different occasions in this text where the influence of these layers will be simulated, but not when expressing the basic conservation laws in their "bulk" control-volume forms.

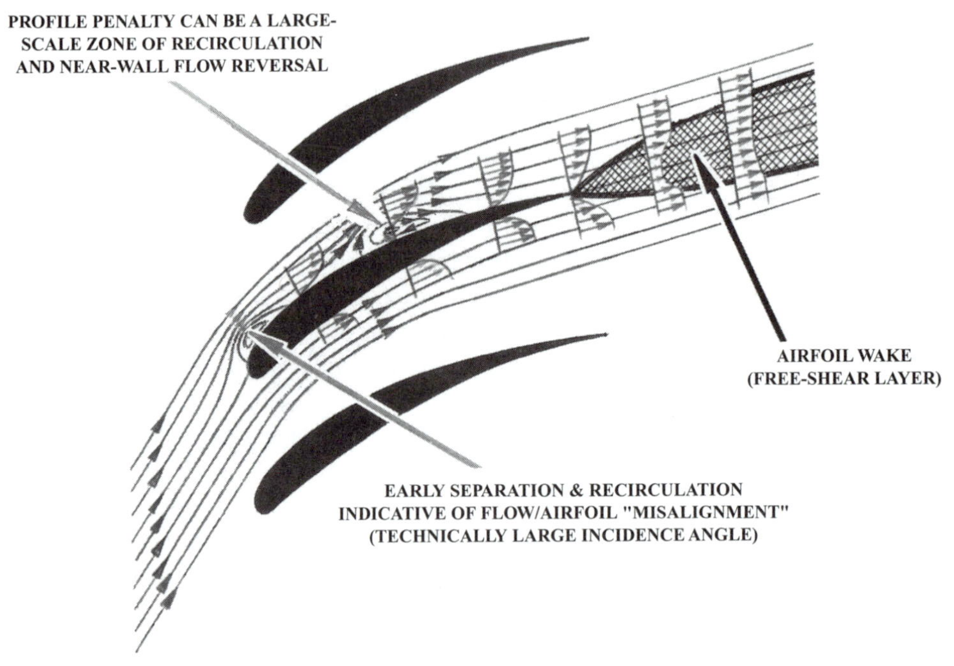

Figure 3.1. Boundary-layer mixing region in a compressor cascade.

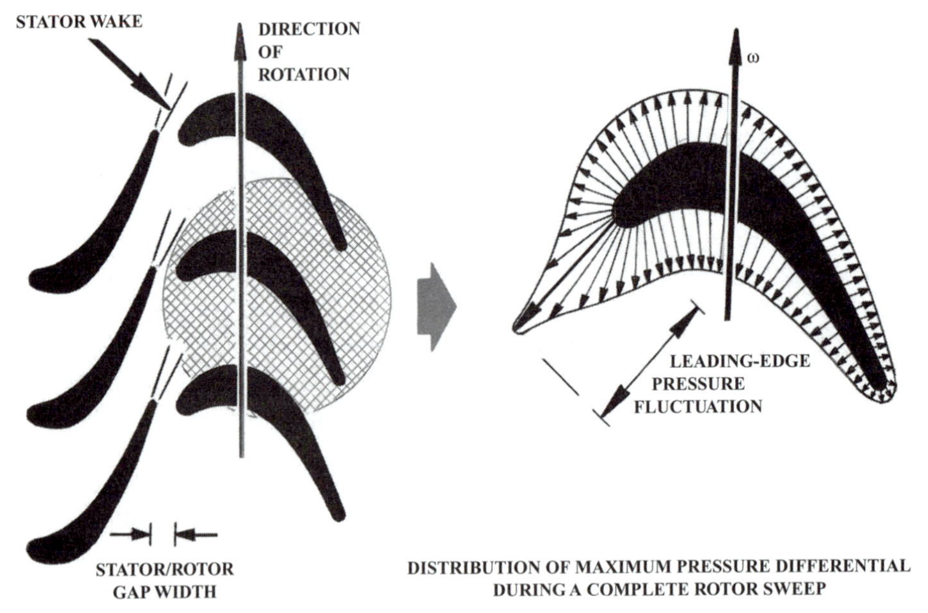

Figure 3.2. Mechanical danger of stator/rotor flow interaction in a turbine stage: rotor-blade pressure fluctuation.

Energy-Conservation Law

Across a turbomachine, or any of its components, the following statement of energy conservation applies:

$$q + w_s = (h_2 - h_1) + \frac{1}{2}({V_2}^2 - {V_1}^2) \tag{3.1}$$

where 1 and 2 refer to the flow inlet and exit stations, respectively. The variables in equation (3.1) are defined as follows:

q is the heat energy gained per unit mass of the working medium,
w_s is the shaft work exerted per unit mass,
h is the specific (per unit mass) enthalpy, and
V is the local velocity.

Ignored in equation (3.1) is a familiar right-hand-side term representing the change in potential energy (caused by, say, the streamwise change in elevation), as it is immaterial in dealing with gas-turbine-engine applications.

Introduction of Total Properties

The energy equation (3.1) can be compactly rewritten as

$$q + w_s = h_{t2} - h_{t1} \tag{3.2}$$

where the total enthalpy h_t is composed of a static part (h) and a dynamic part that is proportional to the local velocity as follows:

$$h_t = h + \frac{V^2}{2} \tag{3.3}$$

The specific dynamic enthalpy is recognized elsewhere as the kinetic energy per unit mass. Note that the adiabatic-flow assumption nullifies the specific heat-energy exchange q.

Ideal Gas as a Working Medium

Under a sufficiently low (theoretically zero) pressure, a gas will behave ideally, whereby the so-called equation of state applies:

$$\frac{p}{\rho} = RT \tag{3.4}$$

where R is referred to as the gas constant and is defined as

$$R = \frac{R_u}{M_g}$$

with R_u referring to the universal gas constant [approximately 8315 J/(kMole K)] and M_g being the gas molar mass (commonly known as the molecular weight).

It is appropriate at this point to stress a few important items that will influence the majority of numerical examples virtually everywhere in this text. These have to do with the nature of the working medium, its conformity to the ideal-gas simplification, and the general topic of numerical precision in this and later chapters.

1) Throughout a gas-turbine engine, the gas pressure will change so dramatically from an ambient magnitude at inlet to typically more than 10 atmospheres at the turbine's inlet station. Under a pressure limitation too theoretical for the gas to behave ideally, in this case, it would appear logical to use gas tables rather than the equation of state to obtain the appropriate magnitudes of thermodynamic properties. Despite how fitting this may be, it has been customary to employ the equation of state, even within high-pressure subdomains. Of course, this will always produce errors of varying magnitudes. The question, nevertheless, is whether a preliminary sizing designer, in particular, can "live" with these errors. Perhaps a numerical example (presented next in this section) can clarify this point from an engineering viewpoint.

2) In traversing a gas-turbine engine, the flowing medium will change its chemical composition in the so-called hot engine segment, beginning with and downstream from the combustor, from pure air to simply products of combustion. In the absence of an afterburner, however, it is traditionally acceptable to refer to the gas mixture in this case as predominantly air and treat it as such. Stated differently, the molecular weight of the flowing medium in turbine-section applications will be set at 29 (the molecular weight of air), giving rise to a gas constant of 286.7 J/kg K (approximated to 287 J/kg K throughout the remainder of this text).

3) Approximating the gas constant may very well be a subject of criticism in terms of the few significant digits retained. However, in an age where computer utilization is as common as it is, it takes virtually no effort to save a more accurate value of this, as well as other constants, to be used anywhere during the computational process.

4) In all of the remaining chapters, we will make componentwise approximations of many other constants. These overwhelmingly involve temperature-dependent variables (such as the specific heat ratio, to be defined later) whereby the constant will be based on an average temperature within that component.

In the general area of gas dynamics, the static part of enthalpy h is usually expressed in terms of the local static temperature as

$$h = c_p T \tag{3.5}$$

where c_p is termed the specific heat under constant pressure, but this in no way limits its validity to constant-pressure flow processes. The variable c_p is generally a function of temperature and pressure for a nonideal gas behavior. However, the ideal-gas version of c_p is dependent only on temperature and is normally tabulated as such. The c_p–temperature dependency is normally passed to another variable, namely the specific-heat ratio γ, where

$$c_p = \left(\frac{\gamma}{\gamma - 1}\right) R$$

The variable γ is normally tabulated in precisely the same manner as c_p itself, namely as a function of temperature. For air, γ will change between a compressor-inlet value of roughly 1.4 and a turbine-inlet value of roughly 1.33. For a separate turbomachinery stage, it is customary to work with a fixed average magnitude of c_p (or γ), whereby the per-stage change in static enthalpy is expressed in the following simple form:

$$\Delta h = c_p(T_2 - T_1) \tag{3.6}$$

Similar to the static-enthalpy definition, the total enthalpy h_t, being a legitimate thermophysical property, can be expressed as

$$h_t = c_p T_t \tag{3.7}$$

where T_t is consistently termed the total temperature. This can also be split into two (static and dynamic) components as follows:

$$T_t = T + \left(\frac{V^2}{2c_p}\right) \tag{3.8}$$

In this expression, the static temperature T may be viewed as the physically measurable temperature, whereas the dynamic part is simply calculable. Substituting T_t into the energy equation yields

$$q + w_s = c_p(T_{t2} - T_{t1}) \tag{3.9}$$

The total temperature is but one of several total (sometimes called stagnation) properties, all of which define a thermodynamically legitimate state that we will refer to as the total (or stagnation) state. Because the state of an ideal gas is defined by (only) two independent properties, it is appropriate to seek to define another total property, say the total pressure p_t at a state (or flow station) that is defined by its static properties (p and T) and the velocity V contributing its share to the total temperature (3.8). Contribution of the same to the total pressure p_t, however, is not as easy. In order to visualize a state where the dynamic pressure has already been added just as the dynamic temperature was, we have to consider an ideally isentropic (adiabatic and reversible) process that proceeds from our physically existing state and in the manner explained next.

Referring to Figure 3.3, let us focus on the stator exit station 1, which is defined by the static properties p_1 and T_1 as well as the velocity V_1. Also shown in the figure is a *fictitious* isentropic diffuser, which, by definition, is ideally frictionless with an ideally gradual increase in the cross-sectional area in such a way as to allow the flow stream to proceed from one equilibrium state to another. Aimed at slowing the flow stream down to a stagnation state at exit, it is this diffuser exit station that will carry the total (or stagnation) state label. In other words, the temperature at this exit station will be numerically identical to T_{t1} and the pressure identical to p_{t1}. Shown by a dotted vertical line in Figure 3.3 is such a fictitious diffuser's flow process, on the T-s diagram, with the end state appropriately labeled 1_t. In this diagram, we find the magnitude of the static temperature (i.e., T_1), as well as the magnitude of the corresponding total temperature (i.e., T_{t1}). The difference between these two

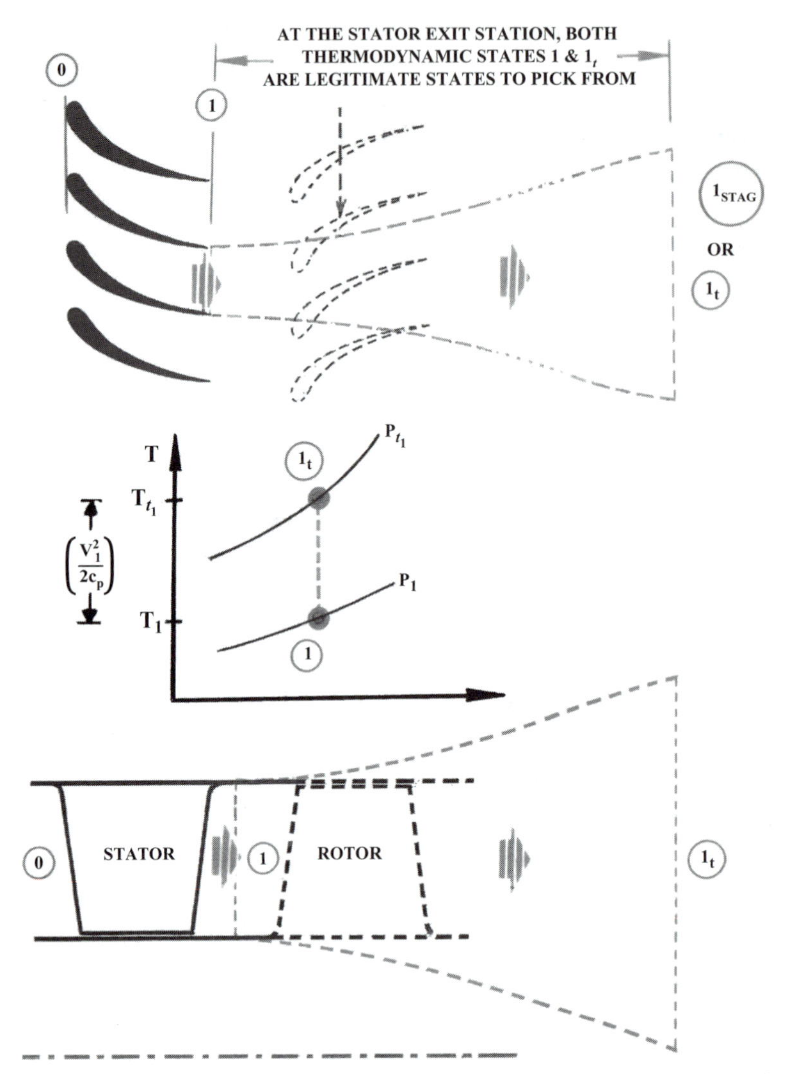

Figure 3.3. Simulation of the stagnating process to obtain the total properties.

magnitudes is the stator-exit dynamic temperature $(V_1{}^2/2c_p)$, which was essentially destroyed across the fictitious diffuser.

As shown in Figure 3.3, the isentropic process described earlier ends on a constant-pressure line that carries the magnitude of the sought-after total pressure, namely p_{t1}, at the stator's exit station. In order to calculate it, we simply refer to the same fictitious isentropic process in Figure 3.3 and apply one of the well-recognized isentropic relationships, namely

$$\frac{p_{t1}}{p_1} = \left(\frac{T_{t1}}{T_1}\right)^{\frac{\gamma}{\gamma-1}} \tag{3.10}$$

Having gone through this *fictional* exercise, we can now make the following *real* conclusions:

1) Corresponding to any component inlet or exit station, where the static properties and velocity are known, there exists a so-called total state on the T-s diagram that is vertically above the state at hand and is defined by the total properties. Of these, the total temperature can easily be computed by simple substitution in expression (3.8). The total pressure, however, can be calculated by substitution in expression (3.10). Other total properties (e.g., ρ_{t1}) can be computed either by applying the equation of state (3.3) or by employing an isentropic relationship similar to (3.10). Although reaching this total state is achieved (only) through a fictitiously isentropic diffusion process, the total properties themselves are real, usable, and important, for they represent the overall thermal and pressure energies of the flow stream by recognizing the contributions of the kinetic energy to the corresponding static properties.

2) The specific-heat ratio in equation (3.10) is itself a function of temperature, meaning the static (or directly measurable) temperature T. It is customary in this case to simply use the γ magnitude that corresponds to some average static temperature magnitude.

3) Throughout the major part of this text, any turbomachinery component will be assumed adiabatic, meaning that the component is perfectly insulated. Although such a condition is ideal, it is practically achievable, in contrast with an isentropic process, which is much tougher even to simulate and theoretically impossible to exist. In turbomachinery applications, the adiabatic-flow condition will fail to exist should the component be cooled, an action that usually applies to the first stage(s) of the high-pressure turbine. In a case such as this, the cooled component is normally treated as a heat exchanger, whereby the coolant-flow rate and its entry-point static temperature would pave the way to apply the appropriate form of energy equation. Referring to the (stationary) flow-decelerating and flow-accelerating passages in Figure 3.4, an adiabatic flow process is one where the total temperature T_t will remain constant. This is clear in reference to the energy equation (3.2) by substituting zeros for both q and w_s. Of course, the static temperature will change across any of the flow passages in Figure 3.4. However, it is the sum of this and the velocity-dependent dynamic temperature (i.e., the total temperature) that will remain constant. The question of what happens to the total pressure is answered next.

4) Figure 3.4 offers a detailed look at the thermophysical flow process through two stationary passages, namely those of adiabatic compressor and turbine stators. In either of these two real-life flow passages, we know that the entropy-production principle (a version of the Second Law of Thermodynamics) will apply. In an inequality form, this principle can be cast as

$$\Delta s = c_p \ln\left(\frac{T_{t1}}{T_{t0}}\right) - R \ln\left(\frac{p_{t1}}{p_{t0}}\right) > 0 \tag{3.11}$$

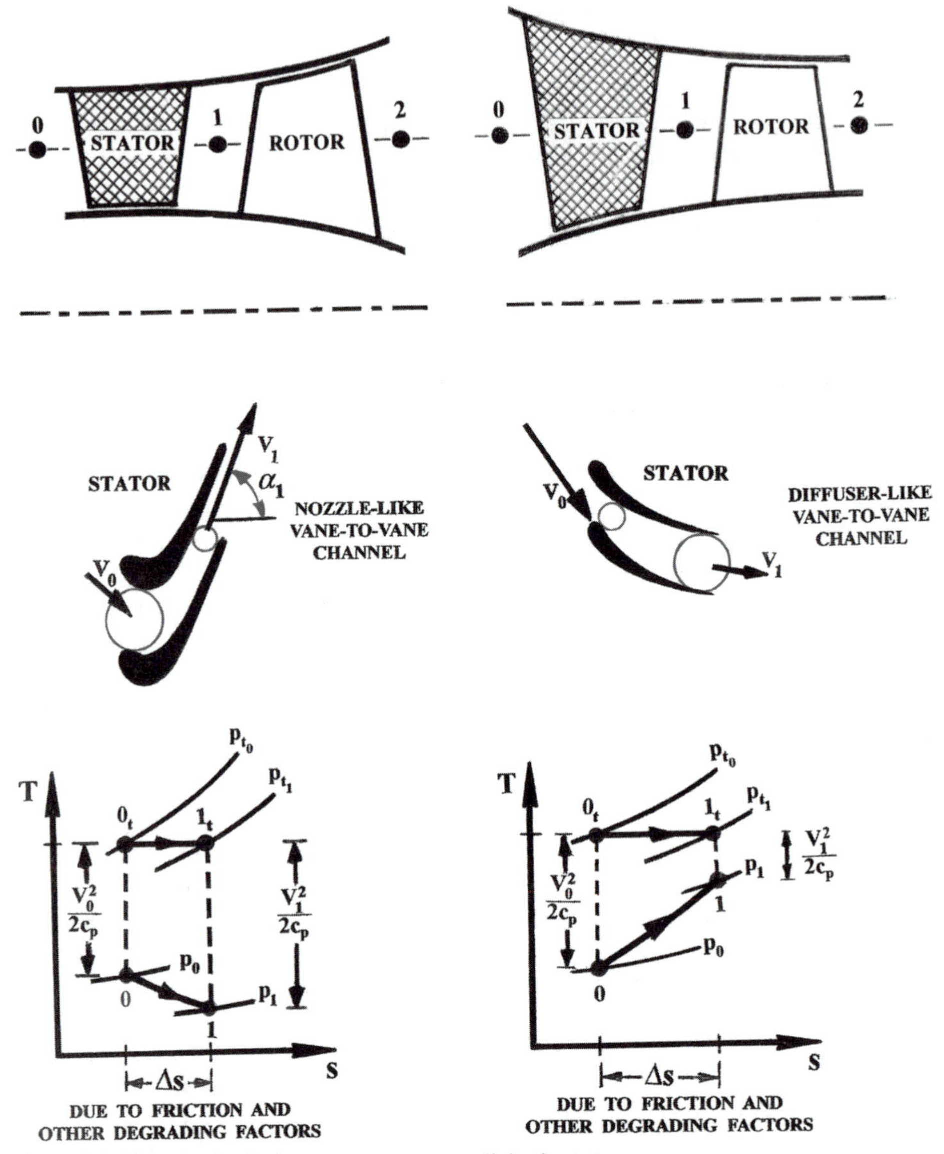

Figure 3.4. Thermophysical process across an adiabatic stator.

Of course, we could have used the static magnitudes of pressures and temperatures instead of the total properties in equation (3.11). Reference to the T-s process representation in Figure 3.4 quickly reveals the fact that either choice would produce the same entropy-production result. However, using the total properties here gives us an advantage, namely that of eliminating the first term in equation (3.11) for the reason that $T_{t1} = T_{t0}$ across either of the two adiabatic stators where no shaft work is produced or consumed. With this in mind, examination of the inequality (3.5) reveals

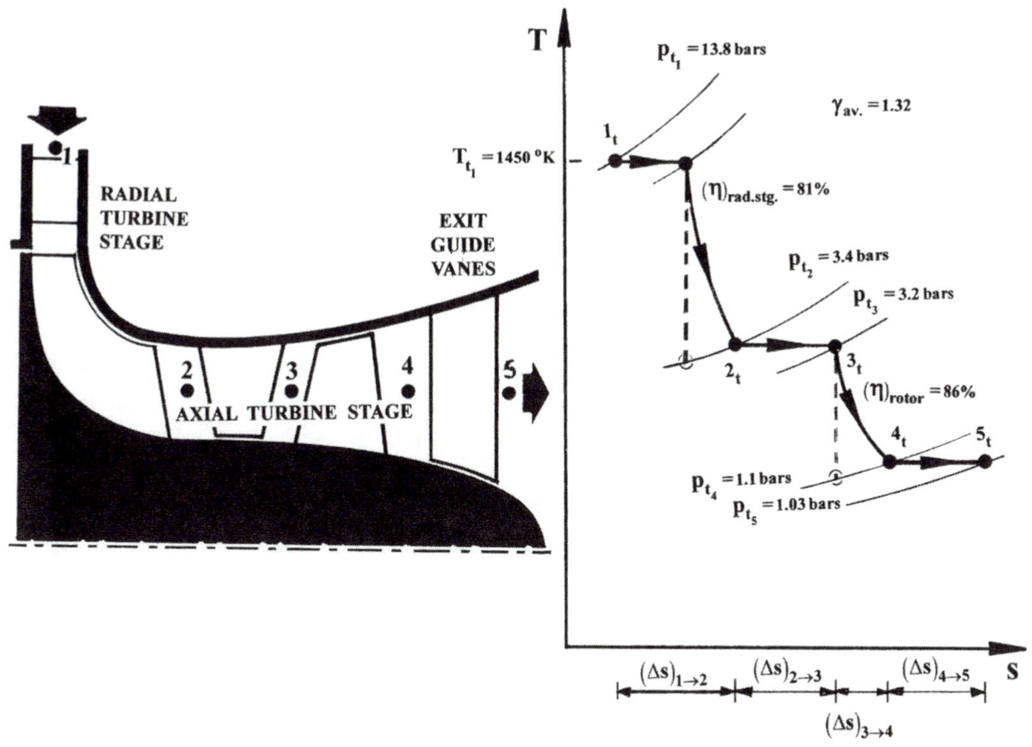

Figure 3.5. Thermodynamic representation of a multistage turbine.

that for Δs to be positive, the ratio $(\frac{p_{t1}}{p_{t0}})$ will have to be less than 1. Stated differently, in a real-life flow process with no shaft-work or heat-transfer interactions, the flow-process irreversibilities (or imperfections) will present themselves in the form of a loss in total pressure. This fact is clear by reference to Figure 3.5. This conclusion will hold true irrespective of whether the stationary flow passage is nozzle-like or diffuser-like.

5) Item 4 (above) defines one of the most meaningful loss-gauging parameters in turbomachinery stators, namely the total pressure loss. In a rotor subdomain, however, we know that the total temperature will have to change significantly (even for an adiabatic-flow field) if shaft work is to be produced or absorbed, a fact that is obvious by reference to the energy equation (3.9). It is also known that the total pressure in this case will either have to decrease (within a turbine rotor) or rise (within a compressor rotor). The total pressure in these cases would therefore fail to be a reflection of such effects as friction and other real-life effects. As will be explained later, the fictitious-diffuser exercise can still be valid, but this time only in a rotating frame of reference. The loss-representative property in this case will (appropriately) be termed the total relative pressure loss, Δp_{t_r}. Unfortunately, both the analysis and implications here will apply only to an axial-flow rotor. In radial (or centrifugal) rotors, however, the mere streamwise radius change will

cause p_{t_r} to decline (in radial-turbine rotors) or rise (in centrifugal-compressor rotors). In Chapters 10 and 11, we will have to identify and use other loss-gauging parameters as we get to examine real-life flow processes in this turbomachinery category.

Entropy-Based Loss Coefficient

Figure 3.5 shows one of the most general turbine sections. The flow stream in this figure progresses from a radial stage to an axial stage and finally through a stationary cascade of exit guide vanes. The *T-s* diagram in the same figure highlights a very important fact: Regardless of the component shape, and whether or not it is associated with shaft work, only one variable, namely Δs, is both loss-related and addable. It is therefore only sensible to devise an entropy-production-based loss coefficient that, by reference to Figure 3.5, is as capable of representing the overall multistage flow process as it is when applied to a single component in the flow path. This rather universal loss coefficient is referred to by the symbol $\bar{q}$, where

$$\bar{q} = 1 - e^{-\left(\frac{\Delta s}{R}\right)} \tag{3.12}$$

In an adiabatic flow process, across a *stationary* component, the loss coefficient $\bar{q}$ is identical to the traditional total pressure loss coefficient, $\bar{\omega}$. The latter is simply defined as the total pressure loss, Δp_t, divided by the inlet total pressure, p_{t1}. Noting that the total temperature in this case remains constant, we can prove the equality of $\bar{\omega}$ and $\bar{q}$ by recasting the entropy-production expression as

$$\Delta s = -R \ln\left(\frac{p_{t2}}{p_{t1}}\right)$$

which means that

$$e^{-\left(\frac{\Delta s}{R}\right)} = \left(\frac{p_{t2}}{p_{t1}}\right)$$

or

$$1 - e^{-\left(\frac{\Delta s}{R}\right)} = \bar{q} = 1 - \left(\frac{p_{t2}}{p_{t1}}\right) = \frac{\Delta p_t}{p_{t1}} = \bar{\omega}$$

Tracking the sequence of events on the *T-s* diagram in Figure 3.5, the path of the process is continually shifting to the right, as would be expected. The entropy rise in each component represents the deviation of the actual (adiabatic) process from being isentropic. In instances where nonconsistent sets of data are provided, the overall loss coefficient $\bar{q}$ can be easily computed by utilizing the given data for each component to compute the individual entropy productions, summing them all up, and substituting the final magnitude in expression (3.12). This procedure is numerically clarified in Example 1.

EXAMPLE 1

Referring to the sequence of turbomachinery components in Figure 3.5, the system's operating conditions are as follows:

Radial stage

- Inlet total pressure (p_{t1}) = 13.8 bars
- Inlet total temperature (T_{t1}) = 1450 K
- Stage-exit total pressure (p_{t2}) = 3.4 bars
- Stage total-to-total (isentropic) efficiency ($\eta_{stg\,1}$) = 0.81%

Axial stage

- Stator-exit total pressure (p_{t3}) = 3.2 bars
- Rotor total-to-total efficiency ($\eta_{rot.\,2}$) = 86%
- Rotor-exit total pressure (p_{t4}) = 1.1 bars

Exit guide vanes

- Exit total pressure (p_{t5}) = 1.1 bars

Assuming an adiabatic flow process throughout the entire system, and an average specific-heat ratio γ of 1.32:

(a) Compute the entropy-based overall loss coefficient ($\bar{q}$).
(b) By computing the system-exit total temperature (T_{t5}), calculate the overall total-to-total efficiency ($\eta_{sys.}$).

SOLUTION

(a) In order to compute the overall entropy production, Δs, we consider each component individually as follows:

$$\eta_{rad.\ stg.} = \frac{1 - \frac{T_{t2}}{T_{t1}}}{1 - \left(\frac{p_{t2}}{p_{t1}}\right)^{\frac{\gamma-1}{\gamma}}} = 0.81$$

which yields

$$T_{t2} = 1111.8\,\text{K}$$

Now

$$\Delta s_{rad.\ stg.} = c_p \ln\frac{T_{t2}}{T_{t1}} - R\ln\frac{p_{t2}}{p_{t1}}$$

where

$$c_p = \frac{\gamma}{\gamma - 1} R \approx 1184\ \text{J/(kg K)}$$

and

$$T_{t2} = T_{t1}\ \text{(adiabatic flow)}$$

Direct substitution in the Δs expression yields

$$(\Delta s)_{rad.\ stg.} = 87.6 \text{ J/(kg K)}$$

Next, we calculate the entropy production of the axial-flow stage, as

$$(\Delta s)_{stator} = -R \ln \frac{p_{t2}}{p_{t1}} = 17.4 \text{ J/(kgK)}$$

$$(\Delta s)_{rotor} = c_p \ln \frac{T_{t2}}{T_{t1}} - R \ln \frac{p_{t2}}{p_{t1}} = 47.9 \text{ J/(kg K)}$$

As for the exit guide vanes, we have

$$(\Delta s)_{G.V.} = -R \ln \frac{p_{t2}}{p_{t1}} = 18.9 \text{ J/(kg K)}$$

Adding these individual entropy-production magnitudes, the overall system's entropy production may now be obtained:

$$(\Delta s)_{sys.} = 171.8 \text{ J/(kg K)}$$

which yields the entropy-based loss coefficient ($\bar{q}$) as follows:

$$\bar{q}_{sys.} = 1 - e^{-\frac{(\Delta s)_{sys.}}{R}} = 0.45$$

(b) Having computed the system's exit total temperature, we may now compute the overall system's total-to-total (or isentropic) efficiency as follows:

$$\eta_{sys.} = \frac{1 - \frac{T_{t5}}{T_{t1}}}{1 - \left(\frac{p_{t5}}{p_{t1}}\right)^{\frac{\gamma-1}{\gamma}}} = 82.3\%$$

COMMENTS

1) The preceding example clearly identifies the (specific) entropy production as the only accumulative (or addable) loss-reflecting parameter in virtually any arrangement of turbomachines. Although $\bar{q}$ appears to be a rather abstract variable (just as the entropy itself might be viewed and, perhaps, just as dislikable), it could aid us tremendously in comparing one system with another to (qualitatively) weigh the effect of adding more components to the system or to single out the most process-degrading component(s).

2) Although it was relatively easy, in the preceding example, to compute the traditional total-to-total system efficiency (by computing the total exit temperature), such a luxury is by no means available in most industrial settings. A typical engineer may end up with the unfortunate task of assessing the performance of a system that consists of, say, a scaled-up (or scaled-down) version of an existing stage followed by a new stage about which a limited amount of information is known, and so on. In a case such as this, the total inlet and exit properties corresponding to a certain component may not be readily available, but ratios of these properties (across the component) could be. In most cases, the given data can still be sufficient for computing the component-wise entropy production. In the end, these increments can all be

summed up and the overall entropy production used to compute $\bar{q}$. This, in turn, can later be used in any of a number of comparative studies, which may very well be the engineer's task.

Compressibility of the Working Medium

The propagation speed of a sound wave in any fluid is referred to as the sonic speed "a" and is defined as

$$a = \sqrt{\left(\frac{\partial p}{\partial \rho}\right)_s} \tag{3.13}$$

The subscript s in equation (3.13) refers to the fact that the change in applied pressure and the resulting change in density will have to be within a fixed-entropy (i.e., isentropic) process. In practical terms, and in light of expression (3.13) one could achieve a rough estimate of the sonic speed by enclosing the fluid, say, in an apparatus such as a well-insulated piston-cylinder device in which the surfaces moving relative to one another are highly polished in order to nearly eliminate friction effects. Two very close states of the fluid can then be allowed to slowly and gradually take place, whereby the applied pressure is slightly increased (relative to the initial-state pressure) and the new volume measured. Knowing the mass of the fluid, the density can be computed in both states. Finally, the small changes in pressure and density can be substituted in an approximate equivalent to expression (3.13) as follows:

$$a = \sqrt{\left(\frac{\delta p}{\delta \rho}\right)_{id}} \tag{3.14}$$

In the preceding process description, you may have noticed that we were simply trying to come as close to an isentropic process as possible. As silly as it may appear, the foregoing exercise may help explain why the sonic speed in liquids can be exceedingly high. After all, no matter what the amount of pressure change might be, the density-change response of the liquid will be very small. Noting that the latter constitutes the denominator in expression (3.14), it is evident that the sonic speed through liquids is much higher than that in gases. It is therefore safe to say that the fluid compressibility is inversely proportional to the sonic speed through it. Finally, note that we have thus far spoken of a fluid at rest. In the case of a fluid in motion, however, there is another gauge, namely the Mach number, which is a compressibility measure of the flow field, as presented later in this chapter.

Sonic Speed in Ideal Gases

As represented earlier, an ideal gas is one that is under an extremely small (ideally zero) static pressure. The thermodynamic state of the gas in this case is defined by only two independent properties. This is a result of the existence of a thermodynamic "closure," namely the equation of state:

$$p = \rho R T \tag{3.15}$$

Assuming that the gas undergoes thermodynamic processes with "sufficiently" small-temperature processes, the specific-heat ratio γ can be treated as constant. Among such processes, let us consider in particular the isentropic process, which eventually leads to the sonic-speed definition. With the assumptions stated earlier, it is rather easy to prove that the "path" of the process is controlled by the following relationship:

$$\frac{p}{\rho^{\gamma}} = \text{constant}$$

Differentiating this relationship, and invoking the equation of state, the sonic-speed expression (3.13) becomes

$$a = \sqrt{\gamma RT} \tag{3.16}$$

Examination of expression (3.16) reveals that the sonic speed in an ideal gas is a function of its static temperature only. Again, the fact should be emphasized that the sonic speed represents the compressibility of a working medium that is at rest.

Mach Number and Compressibility of a Flow Field

For a fluid in motion, the term "compressibility" is defined as the tendency for the flow to undergo disproportionately high density changes in response to local pressure gradients of finite magnitude. The flow property that would profess such a flow behavior is referred to as the Mach number (M), which is defined anywhere in the flow domain as the ratio between the local velocity (V) and the speed of sound (a).

$$M = \frac{V}{a} \tag{3.17}$$

Magnitudes of M that are less than unity classify the flow field as "subsonic." Above unity, the Mach number is associated with a "supersonic" flow field. The importance of the sonic state, where $M = 1.0$, stems from the fact that it is the interface between two drastically different flow structures. For example, a subsonic flow field exclusively possesses the so-called downstream effect, whereby the fluid particle senses the existence of an obstacle that is further downstream and accordingly adjusts its direction before it physically reaches the obstacle. No similar or equivalent flow characteristic exists in a supersonic flow field.

Total Properties in Terms of the Mach Number

Let us begin with the total-temperature definition (3.4), reexpressing it in terms of the Mach number:

$$T_t = T + \frac{V^2}{2c_p T} = T\left(1 + \frac{V^2}{2c_p T}\right) \tag{3.18}$$

where

$$c_p = \left(\frac{\gamma}{\gamma - 1}\right) R \tag{3.19}$$

Rearranging terms, and recalling the definition of the Mach number (3.17), the following expression can easily be obtained:

$$T_t = T\left(1 + \frac{\gamma - 1}{2} M^2\right) \tag{3.20}$$

Following the same logic outlined in the earlier segment of this chapter, the following expressions for p_t and ρ_t can be achieved:

$$p_t = p\left(1 + \frac{\gamma - 1}{2} M^2\right)^{\frac{\gamma}{\gamma-1}} \tag{3.21}$$

$$\rho_t = \rho\left(1 + \frac{\gamma - 1}{2} M^2\right)^{\frac{1}{\gamma-1}} \tag{3.22}$$

Equations (3.20) through (3.22) are better viewed as simply relating the total properties to the static flow properties at the same thermophysical state, as opposed to visualizing a fictional isentropic process that stagnates the flow. Furthermore, equations (3.21) and (3.22) are often referred to as "isentropic" relationships, which is misleading, as the term *isentropic* may wrongly imply the existence of an isentropic process somewhere. Again, the phrase *total-to-static* property relationship is the phrase of choice in this text.

Definition of the Critical Mach Number

Perhaps the only undesirable characteristic of the Mach number is that it is not directly indicative of the local flow velocity. This is true in the sense that the sonic speed itself is a function of temperature. To clarify this point, consider two fluid particles that are proceeding within the first stage of a compressor and that of a turbine section with the *same* velocity. Because the compressor particle is exposed to a much lower temperature than the turbine particle, the sonic speeds at these locations will be drastically different, and so will the Mach numbers.

From a turbomachinery-analysis standpoint, it would therefore be advantageous to replace the Mach number with another nondimensional-velocity ratio, the magnitude of which would be directly indicative of the local velocity. In the following, a new nondimensional-velocity ratio, namely the critical-velocity ratio (or critical Mach number), will be derived not only to achieve this objective but also to be a tool in classifying the flow regime as subsonic or supersonic.

Instead of the local sonic speed, let us now define the so-called critical velocity (V_{cr}) as a velocity nondimensionalizer. To better comprehend this parameter physically, let us visualize a fictitious subsonic nozzle that admits the flow stream at the location where the V_{cr} magnitude is sought. The objective of this nozzle is to choke the flow isentropically. It is the flow velocity at the choking location (where $M = 1.0$) that is referred to as the "critical" velocity, V_{cr}. The name here is appropriate in the sense that most aerodynamics textbooks refer to the throat state (where $M = 1$) as the "critical" state.

Referring to expression (3.20), the way to calculate a sonic speed (which is a local property) is first to compute the static temperature. Appropriately termed the "critical" temperature (T_{cr}), we use expression (3.16) to compute this fictitious throat temperature by simply substituting 1.0 for the Mach number, as follows:

$$T_{cr} = \frac{T_t}{\left[1 + \frac{\gamma-1}{2}(1.0)^2\right]} = \left(\frac{2T_t}{\gamma+1}\right) \tag{3.23}$$

The sonic speed corresponding to T_{cr} in equation (3.23) is precisely the critical velocity V_{cr} that we are after. In order to compute this velocity, we simply substitute T_{cr} for the static temperature T in expression (3.16) as follows:

$$V_{cr} = \sqrt{\gamma R T_{cr}} = \sqrt{\gamma R\left(\frac{2T_t}{\gamma+1}\right)}$$

or

$$V_{cr} = \sqrt{\frac{2\gamma}{\gamma+1} R T_t} \tag{3.24}$$

As stated earlier, this particular velocity will replace the (local) sonic speed in nondimensionalizing the velocity. The result is the appropriately named "critical"-velocity ratio, or simply the "critical" Mach number M_{cr}:

$$M_{cr} = \frac{V}{V_{cr}} = \frac{V}{\sqrt{\frac{2\gamma}{\gamma+1} R T_t}} \tag{3.25}$$

The advantage of using V_{cr} and M_{cr} in most of our computations from now on is twofold:

a) In an adiabatic flow domain with no shaft work involved (e.g., an adiabatic stator), the combination of equations (3.2) and (3.9) leads to a total temperature T_t that is *constant* throughout the flow domain. It follows, by reference to expression (3.25), that the denominator, namely V_{cr}, will also remain constant over the stator. It is in this sense that the newly defined critical Mach number M_{cr} will be directly indicative of the local velocity magnitude anywhere in the flow field. An implied assumption here is that the static temperature variation, ΔT, within the flow field is "sufficiently" small (from an engineering standpoint) to justify a constant γ magnitude.

b) As the "traditional" Mach number (M) gets to be 1.0 (the sonic state), so does the critical Mach number (M_{cr}). In fact, the M_{cr} expression (3.25) becomes precisely the M expression as the sonic state is attained. It therefore becomes obvious why the use of M_{cr} does not compromise the important feature of classifying the flow field as subsonic or supersonic.

Total Properties in Terms of the Critical Mach Number

We now repeat the exercise producing expressions 3.20 through 3.22, except that the nondimensional-velocity ratio is no longer the Mach number but the critical Mach number. Again, we begin with the total temperature:

$$
\begin{aligned}
T_t &= T + \frac{V^2}{2c_p} \\
&= T + \frac{V^2}{2\left(\frac{\gamma}{\gamma-1}R\right)} \\
&= T + \frac{(\gamma-1)V^2}{2\gamma R} \times \frac{T_t}{T_t} \\
&= T + \frac{\frac{\gamma-1}{\gamma+1} \times V^2 \times T_t}{\frac{2\gamma}{\gamma+1} \times R \times T_t}
\end{aligned}
$$

This relationship can be manipulated further to yield

$$T = T_t - \frac{\gamma-1}{\gamma+1}\frac{V^2}{V_{cr}^2}T_t$$

or

$$T = T_t\left(1 - \frac{\gamma-1}{\gamma+1}{M_{cr}}^2\right) \tag{3.26}$$

Note that it is the *static* temperature that is on the left-hand side now, as opposed to the total temperature in expression (3.20). The interchange here should also help in justifying the negative sign inside the bracket of expression (3.26). That expression is virtually the cornerstone in deriving the rest of the property relationships. In the process, it also helps to recall that the "static" state (defined by T, p and ρ) is vertically underneath the "total" state (defined by T_t, p_t, and ρ_t) in the T-s diagram (Fig. 3.3), which should suggest an *isentropic* type of relationship between corresponding pairs of properties. With this in mind, we can express the static-to-total pressure relationship as

$$\frac{p}{p_t} = \left(\frac{T}{T_t}\right)^{\frac{\gamma}{\gamma-1}} = \left(1 - \frac{\gamma-1}{\gamma+1}{M_{cr}}^2\right)^{\frac{\gamma}{\gamma-1}}$$

or

$$p = p_t\left(1 - \frac{\gamma-1}{\gamma+1}{M_{cr}}^2\right)^{\frac{\gamma}{\gamma-1}} \tag{3.27}$$

Similarly, the static-to-total density relationship can be expressed as

$$\rho = \rho_t\left(1 - \frac{\gamma-1}{\gamma+1}{M_{cr}}^2\right)^{\frac{1}{\gamma-1}} \tag{3.28}$$

The relationships (3.23) through (3.28) are as applicable to flow stations within a rotor as they are within a stator. However, the simplicity offered by the critical

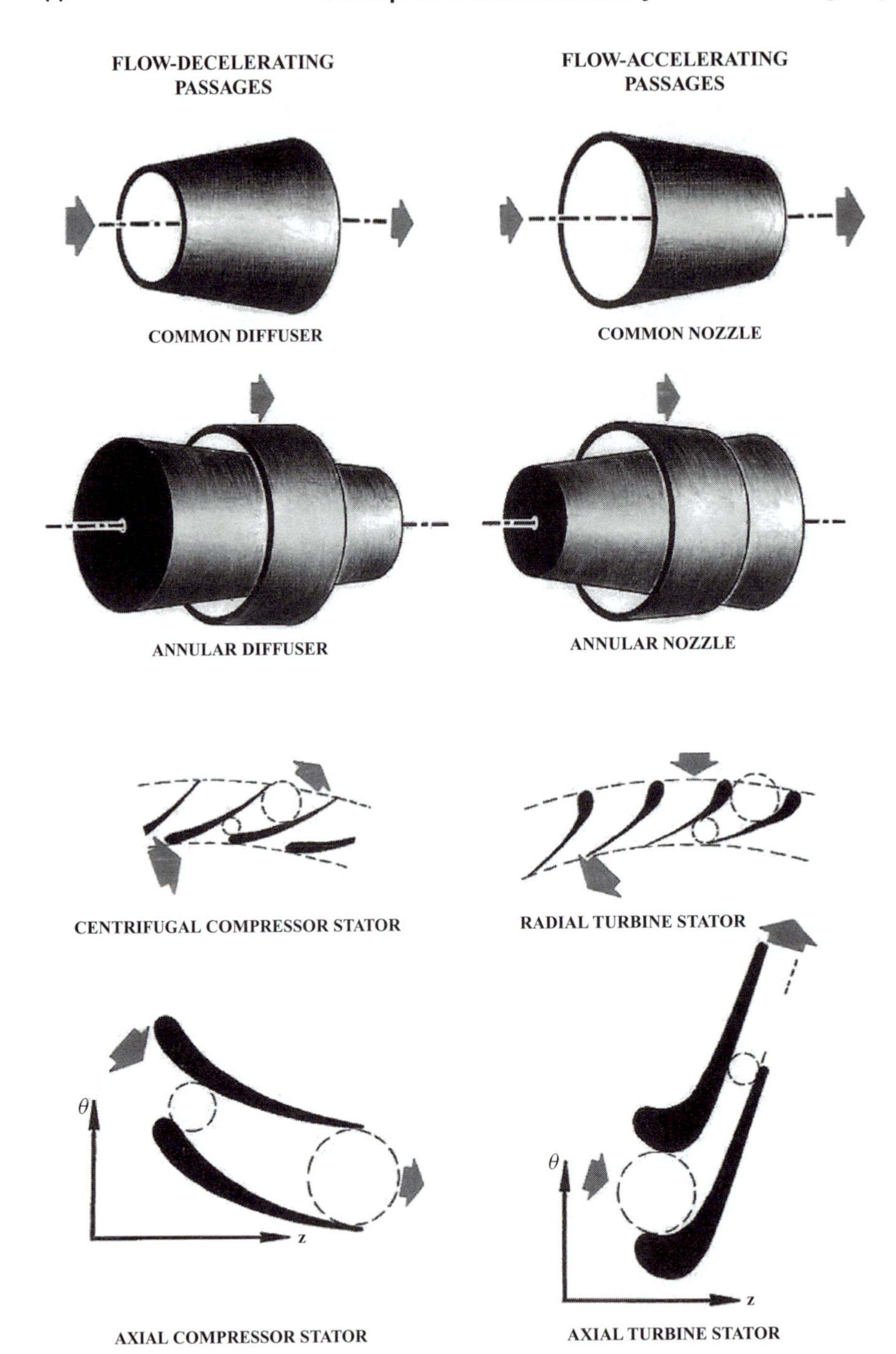

Figure 3.6. Examples of flow decelerating and accelerating passages.

velocity, namely being constant within an adiabatic stator, will no longer exist in a rotating subdomain. A rotor, by definition, is where shaft work is produced (or absorbed) at the expense of, in particular, the total temperature (3.2), the variable on which V_{cr} depends. Later in this chapter, another velocity nondimensionalizer, namely the *relative* critical velocity W_{cr}, will be defined to serve the same purpose within a rotor as V_{cr} does in the case of a stator. However, the advantage of W_{cr}, that of being constant throughout the rotor, will no longer exist in a nonaxial rotor, as the mere change of radius has its own effect of changing the relative total temperature T_{tr}, on which W_{cr} depends.

Figure 3.6 presents vaned and unvaned stationary components in a gas-turbine engine where the full benefit of using V_{cr} and M_{cr} is achieved, provided that the flow process is adiabatic. The components in this figure are categorized as flow-decelerating (diffuser-like) and flow-accelerating (nozzle-like) passages.

Definition of the Pitch Line in Turbomachines

The pitch (or mean-radius) line is a fictitious line in the meridional (z-r) projection of the turbomachine, along which the hub-to-casing average properties are assumed to prevail. The need to define this line would obviously arise in the preliminary design phase, where "bulk-type" one-dimensional flow calculations are carried out. The physical definition of this line is as controversial as the averaging method itself. Whereas some turbomachinists save themselves the trouble by choosing the arithmetic average of the local hub and tip radii to provide the pitch line, others differ in viewing where the average properties nearly exist. One of the averaging methods here is referred to as the "mass"-averaging method. The logic behind this method is that more weight should be placed on those radial locations in the flow stream with a higher concentration of fluid particles. Aside from being tedious, the process in this case requires prior knowledge of the radial distribution of static density, for it is part of the local mass flux. Another alternative in computing the mean radius (r_m) is loosely termed *volume averaging* and is discussed next.

Referring to Figure 3.7, the local annulus is divided into two equal-area annuli, and the dividing-circle radius is considered as the mean radius at this particular flow station. With this in mind, and referring to Figure 3.7, we can easily compute this mean radius (r_m) as

$$r_m = \sqrt{\frac{(r_h{}^2 + r_t{}^2)}{2}} \tag{3.29}$$

Construction of the entire pitch line therefore requires computing such a radius at all locations within the turbomachine. Between the preceding two options, we will adopt the simpler means of defining the mean radius, namely the arithmetic average of the hub and tip radii:

$$r_m = \frac{1}{2}(r_h + r_t) \tag{3.30}$$

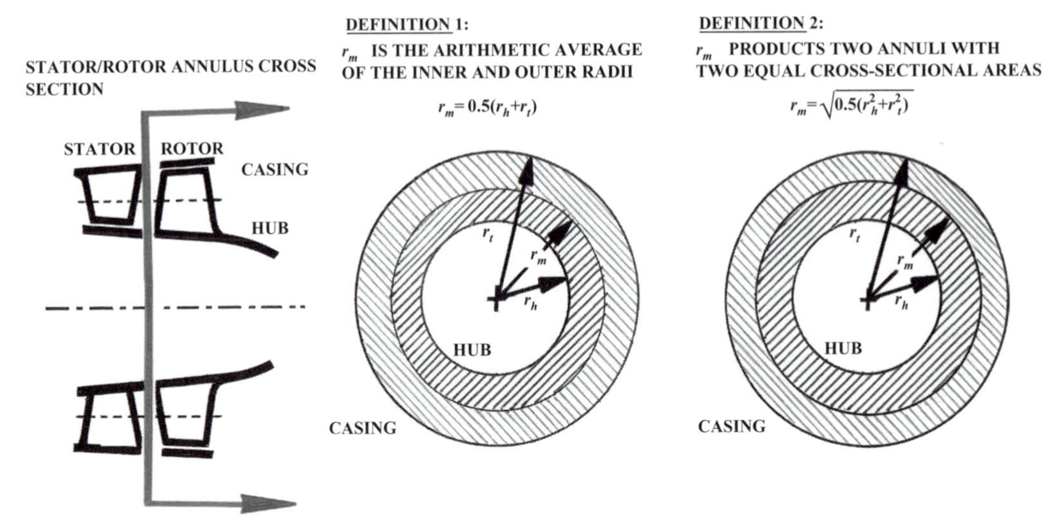

Figure 3.7. Two popular definitions of the pitch-line mean radius.

Continuity Equation in Terms of Total Properties

With no lack of generality, let us now focus on a stationary cascade, such as the turbine stator in Figure 3.4, for the purpose of simplicity. This has a nozzle-like vane-to-vane channel, and the only complexity here is that the passage not only accelerates the flow but turns it as well. To this end, we should expect the flow (swirl) angle to interfere with the continuity equation that we are about to produce. In a categorically subsonic-nozzle passage such as this, it makes sense to apply the continuity equation to the exit station, the only station where the flow can reach a critical Mach number of unity. This station is labeled "1" in Figure 3.4, and the starting-point form of the continuity equation is

$$\dot{m} = \rho_1 V_1 A_1 \tag{3.31}$$

where

ρ_1 is the stator-exit static density,
V_1 is the stator-exit velocity, and
A_1 is the cross-flow area at the exit station.

As can be seen in Figure 3.4, the vane-to-vane hub-to-casing cross-flow area A_1 can be expressed as

$$A_1 = (2\pi r_{m1} h_1)\cos\alpha_1 \tag{3.32}$$

where r_{m1}, h_1, and α_1 are the mean radius, the annulus height, and the flow swirl angle, respectively. Next, we introduce the critical Mach number in expressing ρ_1 as

$$\rho_1 = \rho_{t1}\left(1 - \frac{\gamma - 1}{\gamma + 1} M_{cr1}{}^2\right)^{\frac{1}{\gamma-1}}$$

Application of the equation of state to the total state will alter the preceding ρ_1 expression into the following form:

$$\rho_1 = \frac{p_{t1}}{RT_{t1}}\left(1 - \frac{\gamma - 1}{\gamma + 1}M_{cr1}{}^2\right)^{\frac{1}{\gamma-1}} \tag{3.33}$$

The stator's exit velocity V_1 can also be cast in terms of the critical Mach number as follows:

$$V_1 = M_{cr1}V_{cr1} = M_{cr1}\sqrt{\frac{2\gamma}{\gamma+1}RT_{t1}} \tag{3.34}$$

Substituting equations (3.33) and (3.34) in (3.31), and dropping all subscripts, we arrive at the following form of the continuity equation:

$$\dot{m} = \frac{p_t}{RT_t}\left(1 - \frac{\gamma - 1}{\gamma + 1}M_{cr}{}^2\right)^{\frac{1}{\gamma-1}} . M_{cr} . \sqrt{\frac{2\gamma}{\gamma+1}RT_t} . A$$

or

$$\dot{m} = \sqrt{\frac{2\gamma}{\gamma+1} \cdot \frac{p_t{}^2}{RT_t}}\, M_{cr}\left(1 - \frac{\gamma - 1}{\gamma + 1}M_{cr}{}^2\right)^{\frac{1}{\gamma-1}}$$

or

$$\dot{m} = \frac{p_t A}{\sqrt{T_t}}\sqrt{\frac{2\gamma}{(\gamma+1)R}}\, M_{cr}\left(1 - \frac{\gamma - 1}{\gamma + 1}M_{cr}{}^2\right)^{\frac{1}{\gamma-1}}$$

which, upon algebraic manipulation, assumes the following final form:

$$\frac{\dot{m}\sqrt{T_t}}{p_t A} = \sqrt{\frac{2\gamma}{(\gamma+1)R}}\, M_{cr}\left(1 - \frac{\gamma - 1}{\gamma + 1}M_{cr}{}^2\right)^{\frac{1}{\gamma-1}} \tag{3.35}$$

Equation 3.35 is the turbomachinery-adapted form of the continuity equation, where the left-hand side is commonly known as the "flow function." Note that the right-hand side is a function of the critical Mach number (M_{cr}), the chemical composition of the flowing medium, i.e., the gas constant (R), and the specific heat ratio (γ), with the latter being a function of the static temperature. In applying expression 3.35, we will always assume that the flowing medium is pure air and that the temperature changes between different flow stations are "sufficiently" small to justify a fixed γ magnitude.

Because some of the most important thermophysical relationships can be cast in terms of the traditional Mach number ($M = \frac{V}{a}$), it is more convenient to reexpress equation (3.35) using M instead. The procedure here is very similar to that leading to equation (3.35), with the static density expression (3.22) being utilized this time. The final outcome is

$$\frac{\dot{m}\sqrt{T_t}}{p_t A} = \sqrt{\frac{\gamma}{R}}M\left(1 + \frac{\gamma - 1}{2}M^2\right)^{\frac{1+\gamma}{2(1-\gamma)}} \tag{3.36}$$

A relatively challenging item in applying the continuity equation (3.35) is the composition of the cross-flow area (A). This area, as indicated earlier, will always be the projection of the local annulus area perpendicular to the local velocity direction, i.e.

$$A = \pi(r_t^2 - r_h^2)\cos\alpha$$

where

r_t is the local casing radius,
r_h is the local hub radius, and
α is the local flow swirl angle (Fig. 3.4).

The term *local* refers to the flow station (inlet, exit, or any station in between) where the continuity equation is to be applied.

Isentropic Flow in Varying-Area Passages

Despite the idealism involved in an isentropic flow (no viscosity effects, no secondary cross-flow streams, no lack of thermodynamic equilibrium, etc.), such a flow regime aids in identifying important compressible-flow characteristics. For instance, consideration of such a flow helps a great deal in such cases as the area transition from a station immediately upstream to one immediately downstream around the throat (where $M_{cr} = 1.0$) in a converging-diverging subsonic-supersonic nozzle (better known as De Laval nozzle).

To this end, the first task is to reexpress the flow-governing equations in their differential form while inserting the Mach number wherever it is appropriate. To achieve this, let us assume that the flow passage only accelerates (or decelerates) the flow without turning it. Now, let us begin with the continuity equation in its simplest form:

$$\dot{m} = \rho V A$$

Noting that $\dot{m}$ is constant, we can differentiate both sides of equation (3.41), with the result being

$$0 = (\rho V\, dA) + (\rho A\, dV) + (V A d\rho)$$

Dividing by $\rho V A$, we get the following differential form of the continuity equation

$$\frac{dA}{A} + \frac{dV}{V} + \frac{d\rho}{\rho} = 0 \tag{3.37}$$

Similarly, we could cast Euler's equation of motion in its differential form as

$$VdV + \frac{dp}{\rho} = 0 \tag{3.38}$$

Equation (3.37) is clearly based on a steady and inviscid flow, assumptions that constitute part of the larger isentropic-flow assumption. This very assumption defines the flow path through the following equation:

$$\frac{p}{\rho^{\gamma}} = \text{constant} \tag{3.39}$$

Differentiating both sides of equation (3.39), we get the following result:

$$\gamma \frac{d\rho}{\rho} - \frac{dp}{p} = 0 \tag{3.40}$$

Finally, combining the equation of state together with the Mach number definition, we get the following relationship:

$$M^2 = \frac{V^2}{\gamma\left(\frac{p}{\rho}\right)} \tag{3.41}$$

Equations (3.37) through (3.41) can be lumped up to yield the following two relationships:

$$\frac{dA}{A} = \frac{dp}{p}\left(\frac{1 - M^2}{\gamma M^2}\right) \tag{3.42}$$

and

$$\frac{dV}{V} = \left(\frac{-1}{\gamma M^2}\right)\frac{dp}{p} \tag{3.43}$$

The implications of these two equations are perhaps some of the most important in aerodynamics. Specifically speaking, the left-hand side of equation (3.42) will determine, by its sign, the shape of the flow passage that is required to produce a subsonic or a subsonic/supersonic flowfield. For instance, a positive sign of dA in equation (3.42) implies a diverging flow passage, and vice versa. Equation (3.43) on the other hand, spells out the fact that a flow passage that causes flow acceleration (i.e., a positive magnitude of dV) will have to simultaneously cause a decline in static pressure (i.e., a negative magnitude of dp), regardless of the Mach number value. Consequences of this, as well as the preceding set of differential equations, are discussed next. In so doing, we will distinguish subsonic flow fields from those which are supersonic. Onset of the transitional sonic state will also be discussed.

Subsonic Flow Fields ($M < 1.0$)

This is obviously the simplest and most common case, for it embraces most of the real-life flow applications with which we are familiar. Let us discuss the results of applying equations (3.42) and (3.43) to flow-accelerating and then flow-decelerating flow passages.

In a subsonic nozzle (or nozzle-like) flow passage, the condition $dV > 0$ is, by definition, fulfilled. As cited earlier, this will necessarily cause a pressure decline (i.e., $dp < 0$). With this in mind, equation (3.42) dictates a flow passage with shrinking cross-flow area in order for the task to be achieved. This is the all-familiar case of a

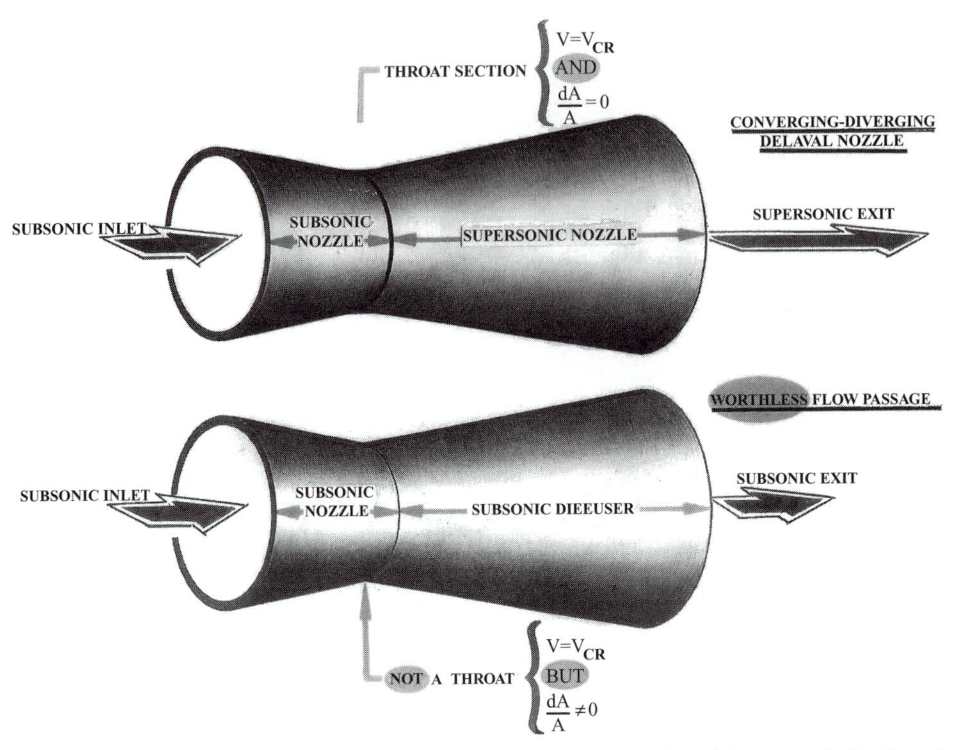

Figure 3.8. Satisfaction of the throat geometrical condition in a subsonic/supersonic De Laval nozzle.

subsonic nozzle, which is the upstream (converging) segment of the De Laval nozzle in Figure 3.8.

The process of creating a flow-decelerating passage is conceptually similar to the process just discussed, with the exception that we now have a velocity differential (dV) that is negative. According to equation (3.43) the pressure differential (dp) will have to be positive, indicating a streamwise rise in static pressure. In this case, equation (3.42) dictates a rising cross-flow area. This is the simple case of a diverging subsonic diffuser, examples of which appear on the left-hand side of Figure 3.6.

Supersonic Flow Fields ($M > 1.0$)

For the purpose of consistency, we begin (as we did earlier) with a flow-accelerating passage (i.e., $dV > 0$). Again, equation (3.43) reveals the fact that this very passage will simultaneously cause a streamwise decline in pressure (i.e., $dp < 0$). Because of its importance in this particular exercise, let us rewrite equation (3.42) as

$$\frac{dA}{A} = \frac{dp}{p}\left(\frac{1 - M^2}{\gamma M^2}\right) \tag{3.44}$$

Noting that $dp < 0$ and that the term $1 - M^2/\gamma M^2$ is negative for the supersonic flow at hand, we can now conclude that the area differential dA will have to be positive. This simply means that, contrary to the case of the subsonic nozzle, a supersonic nozzle will have to be diverging. An example of such a flow passage is the downstream segment, past the throat section, of the converging/diverging De Laval nozzle in Figure 3.8.

Aspects concerning the case of a supersonic diffuser can be identified in the same fashion. The final result is that such a flow passage will have to be of the converging type. This is clearly at odds with the subsonic diffuser passage discussed earlier. In fact, one can save a great deal of mental effort by keeping in mind the fact that supersonic flow passages will always have the opposite streamwise area variation in contrast with their subsonic counterparts.

The Sonic State

This is an interesting thermophysical state whereby the flow velocity becomes identical to the sonic speed. Such a state cannot be sustained over any finite length of the flow passage, for it is physically a transitional state. Occurrence of this state would be at the minimum-area location of a subsonic/supersonic nozzle (Fig. 3.8), a cross-flow section that is commonly referred to as the "throat."

An interesting feature of the throat section, which allows the subsonic-to-supersonic flow transition, is obtained by reference to equation (3.44). According to this equation, the throat section, in this type of nozzle, will have to satisfy the following condition:

$$\frac{dA}{A} = 0 \tag{3.45}$$

This means that for the throat to allow such a flow transition, it has to exist in a region where the converging-to-diverging area change is ideally smooth; in other words, an infinitesimally small passage length where the cross-flow area remains constant. With this in mind, and referring to Figure 3.8, we can tell that the subsonic-to-supersonic flow transition will never take place in the lower flow passage as a result of the abrupt area change at what may have been intended to be the throat section. What will happen in this flow passage is that the flow stream will indeed be brought to a Mach number of unity at this section. However, the next (diverging) segment will act as a subsonic diffuser, a situation that defeats the initially intended purpose.

Switching from the traditional Mach number (M) to the newly defined critical Mach number (M_{cr}) provides the easiest means of calculating the cross-flow area anywhere in the converging/diverging De Laval nozzle in Figure 3.8. This is possible through the modified version of the continuity equation in equation (3.35), except that the flow swirl is assumed not to exist. For example, substituting 1.0 for the critical Mach number in this expression will yield the throat's cross-flow area. Note that equation (3.35) is as applicable for supersonic flows as it is for subsonic flows. The supersonic-segment exit flow area, however, depends on

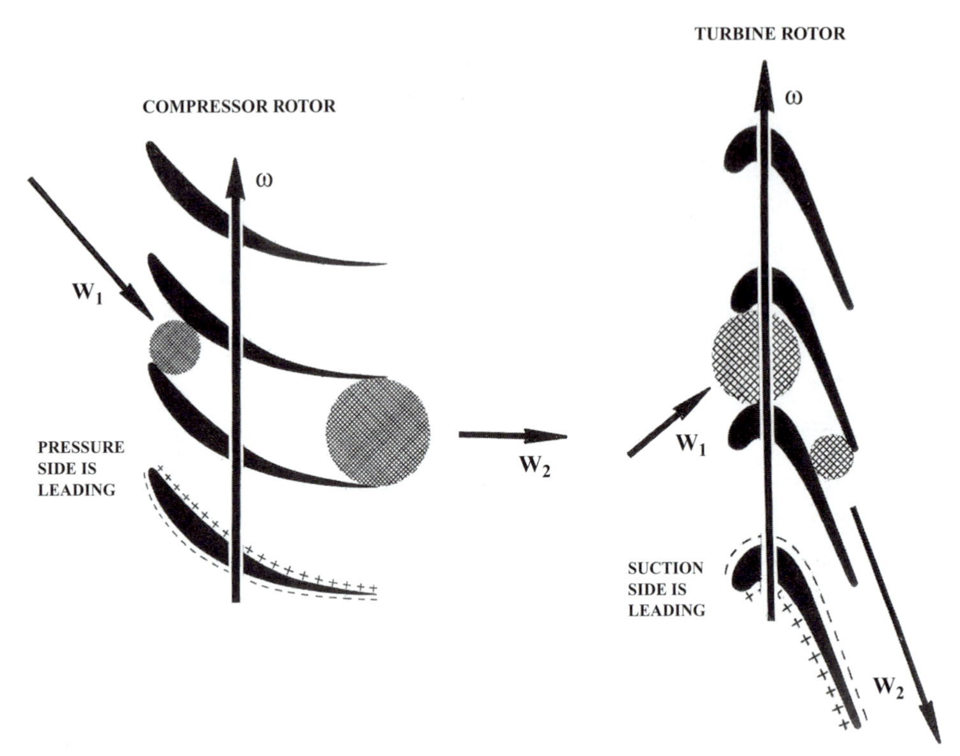

Figure 3.9. Cross-flow area variation in compressor and turbine rotors.

the desired exit Mach number and can also be obtained through application of equation (3.35).

Nozzle- and Diffuser-Like Airfoil Cascades

At this point, we confine our attention to subsonic flow passages. Unfortunately, such turbomachinery passages will hardly be of the simple type shown in the upper row of Figure 3.6. Instead, these will either be annular or radial, as seen in the remainder of this figure. Another category of such passages is the rotating airfoil cascades in Figure 3.9. Focusing on these airfoil cascades, we see that a compressor rotor's blade-to-blade passage is conceptually a diffuser, whereas that of a turbine is categorically a nozzle. Referring to Figure 3.9, another basic difference has to do with the direction of rotation. In the case of a compressor rotor, we see that the airfoil (concave) pressure side is leading the (convex) pressure side in the direction of rotation. The situation is quite the opposite in the turbine rotor.

Figure 3.10 shows a few fine details concerning a turbine stator, as contrasted with a typical compressor stator. Also shown in the figure is the static pressure distribution along the vane pressure and suction sides in both cases. In the turbine stator (on the left in Fig. 3.10), the overall pressure gradient is negative and, as such, is termed a favorable pressure gradient. The compressor stator, on the contrary,

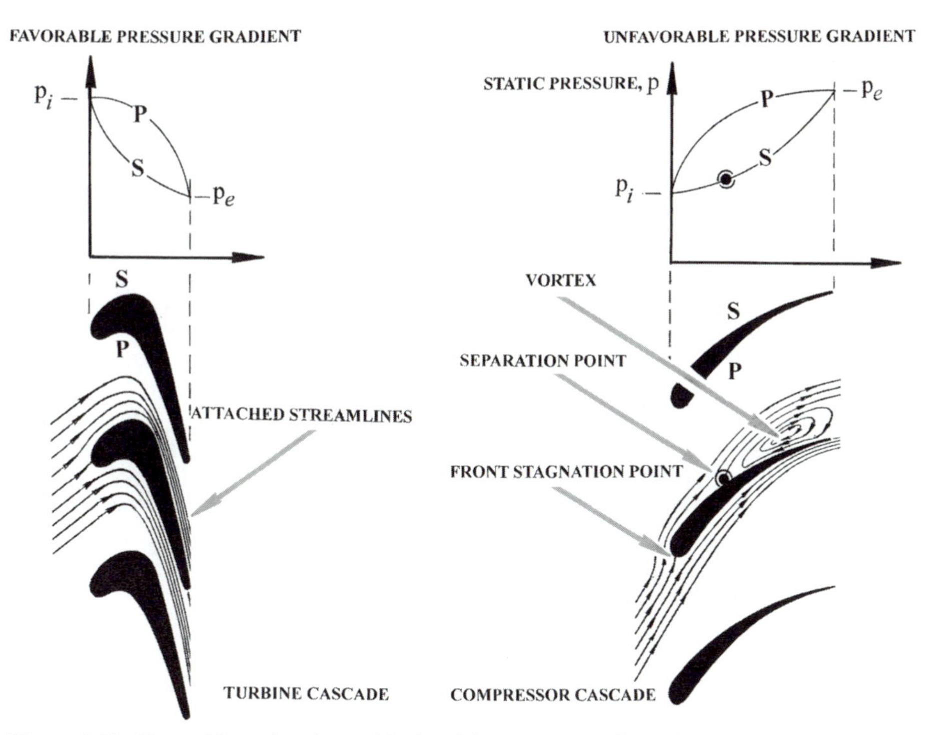

Figure 3.10. Favorable and unfavorable (static) pressure gradients in turbine and compressor cascades.

gives rise to a positive, or adverse, pressure gradient and is consistently referred to as an unfavorable pressure gradient. A continually accelerating flow is associated with a declining pressure, which, in turn, discourages the boundary layer buildup, preventing, in essence, any flow-separation tendencies. The environment, however, is quite the opposite in the compressor stator vane-to-vane passage. The rising pressure in this case aggravates the boundary-layer growth, especially over the airfoil suction side. In fact, the highly degrading viscosity/pressure-gradient interaction in this case could lead to a premature flow separation and zones of flow recirculation, as shown in Figure 3.10. As a result, a more significant streamwise total pressure loss will take place in this case.

Another example of an unfavorable pressure gradient is shown in Figure 3.11 and concerns the annular exhaust diffuser of a commercial turboprop engine. Viscous-flow analysis by Baskharone (1991) produced the velocity vector plot shown in the figure. The plot to the left in this figure corresponds to a turbine-exit/diffuser-inlet swirl angle of 6.3° (which is the design-point magnitude), whereas that on the right corresponds to a swirl angle of 25°. In the latter (high-swirl) case, the number of tangential "trips" of the flow particles adjacent to the endwalls is much larger by comparison. This has the effect of elongating the flow trajectory within the diffuser, leading (as seen in Figure 3.11) to a hub-region flow separation, with flow reversal downstream from the separation point.

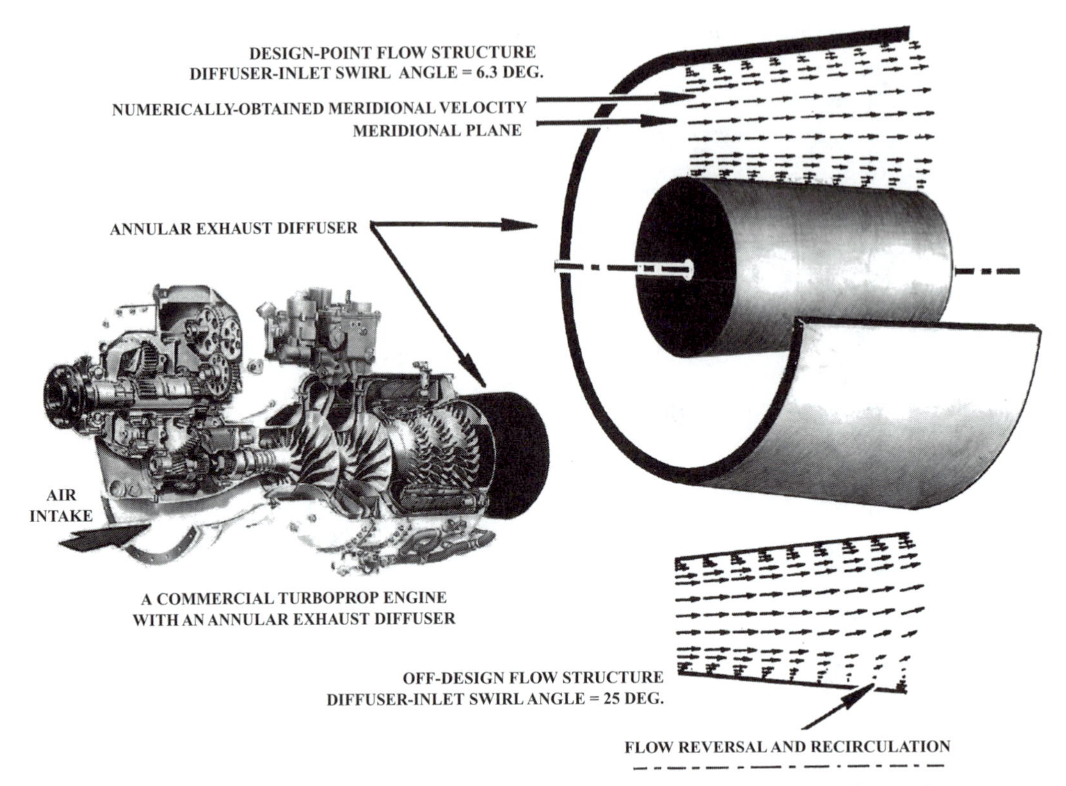

Figure 3.11. Effect of the inlet swirl angle on the exhaust-diffuser performance in a turboprop engine. Swirl velocity components is not shown.

Bernoulli's Equation: Applicability and Limitations

As discussed earlier in this chapter, an incompressible flow field does not necessarily have to be that of a liquid. For a flow in motion, the compressibility criterion is based on the Mach number magnitude. In fact, an air flow stream, say, will remain categorically incompressible, provided that the highest Mach number is below 0.3. However, a liquid flow stream will always constitute an incompressible flow field, no matter how high the velocity might get. The reason is that the sonic speed in liquids is a great deal higher than that in gases. This fact is clear by reference to equation (3.14), in the sense that no matter how elevated the applied pressure, the liquid density change will be much smaller compared with that in gases.

For an incompressible flow field, Bernoulli's equation offers a simple means of computing the total pressure. The equation itself can be written as

$$p_t = p + \frac{1}{2}\rho V^2 \tag{3.46}$$

The simplicity of this total-pressure definition stems from the fact that the total pressure here is breakable into two distinct components: the static component (p) and the dynamic pressure ($\frac{1}{2}\rho V^2$). Contrasted with expression (3.21) for a generally compressible flow field, the latter is hardly breakable into static and dynamic contributions.

In order to weigh the accuracy of Bernoulli's expression, let us assume that the working medium is a gas (e.g., air) that is proceeding under a Mach number that is sufficiently small from an engineering viewpoint. Furthermore, application of the equation of state, together with the Mach number definition, leads to the following result:

$$\frac{1}{2}\rho V^2 = \frac{\gamma}{2} p M^2 \tag{3.47}$$

Combining equations (3.47) and (3.21), we can recast the latter as

$$\frac{p_t - p}{(\rho V^2)/2} = \frac{2}{\gamma M^2}\left(1 + \frac{\gamma - 1}{2} M^2\right)^{\frac{\gamma}{\gamma - 1}} \tag{3.48}$$

Equation (3.48) is the accurate compressible-flow total/static pressure relationship and has (yet) nothing to do with Bernoulli's approximation. It is at this computational point that we attempt to bring Bernoulli's approximation into the picture. To this end, we will expand the bracketed term on the right-hand side as a binomial as follows:

$$\frac{p_t - p}{(\rho V^2)/2} = 1 + \frac{1}{4} M^2 + \frac{1}{40} M^4 + \frac{1}{1600} M^6 + \cdots \tag{3.49}$$

Ignoring higher-order terms in the binomial expression in eq. (3.49), we can now assess how the accuracy of Bernoulli's equation depends on the Mach number. Before we carry this out, it is important to point out that stopping at the first term on the right-hand side, namely 1, is equivalent to simply casting Bernoulli's equation. Now, referring to equation (3.49) (where four binomial terms are retained), we achieve the following results:

$M = 0.0$ Bernoulli's expression results in a zero error.

$M = 0.1$ The error is approximately 0.25%.
$M = 0.2$ The error is approximately 1.00%.
$M = 0.3$ The error is approximately 2.25%.
$M = 0.4$ The error is approximately 4.06%.
$M = 0.5$ The error is approximately 6.40%.

⋮

$M = 1.0$ The error is approximately 27.6%.

⋮

$M = 1.74$ The error is approximately 100%.

Reviewing these results with an engineering "mentality," it is reasonable to use Bernoulli's equation in computing the total pressure up to a Mach number that is somewhere between 0.3 and 0.4. Of course, the final judgment here is based on the required level of precision.

EXAMPLE 2

Referring to the turbine stator in Figure 3.4, the operating conditions are as follows:

- Inlet total temperature $(T_{t0}) = 1200$ K
- Inlet total pressure $(p_{t0}) = 1.8$ bars
- Mass-flow rate $(\dot{m}) = 6.2$ kg/s
- Hub-to-tip exit annulus area $(A_{annulus}) = 0.02\ \text{m}^2$
- Stator mean-radius exit angle $(\alpha_1) = 78^\circ$

Assuming an ideal hub-to-casing flow guidedness, an isentropic stator flow, and a stator average specific-heat ratio (γ) of 1.33, compute the stator-exit static pressure (p_2) in *two* different ways:

a) By computing the exit critical Mach number;
b) By applying Bernoulli's equation.

SOLUTION

Because the flow process is adiabatic (part of being isentropic),

$$T_{t1} = T_{t0}$$

Now, applying the continuity equation at the exit station,

$$\frac{\dot{m}\sqrt{T_{t1}}}{p_{t1}A_1} = \sqrt{\frac{2\gamma}{(\gamma+1)R}}M_{cr1}\left(1 - \frac{\gamma-1}{\gamma+1}M_{cr1}{}^2\right)^{\frac{1}{\gamma-1}}$$

or

$$\frac{7.8\sqrt{1200}}{1.8\times 10^8 \times 0.02 \times \cos 78^\circ} = 0.063M_{cr1}(1 - 0.1416M_{cr}{}^2)^{3.0303}$$

or

$$M_{cr1}(1 - 0.1416M_{cr1}{}^2)^{3.0303} = 0.361$$

Unfortunately, the preceding equation is nonlinear and has to be solved iteratively. Barring computer usage, a simple trial-and-error procedure would be adequate. The final result is

$$M_{cr1} = 0.385$$

a) Computing p_1 through knowledge of M_{cr1}:

$$p_1 = p_{t1}\left(1 - \frac{\gamma-1}{\gamma+1}M_{cr1}{}^2\right)^{\frac{\gamma}{\gamma-1}} = 1.652 \text{ bars}$$

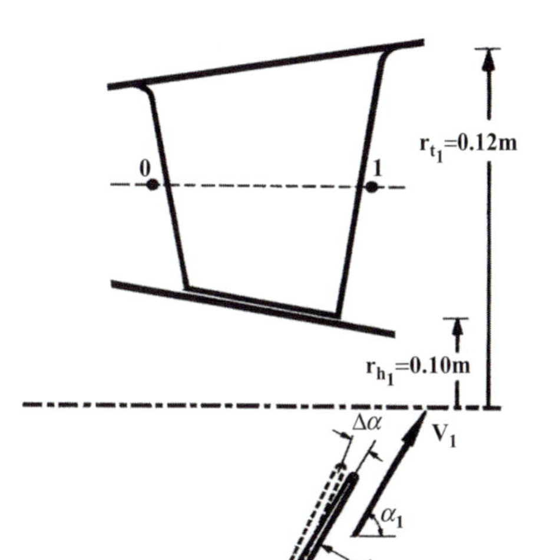

Figure 3.12. Stator-vane rotation in a variable-geometry stator.

b) Computing p_1 by applying Bernoulli's equation: First, we calculate both ρ_1 and V_1 as follows:

$$\rho_1 = \frac{p_{t1}}{RT_{t1}}\left(1 - \frac{\gamma - 1}{\gamma + 1} M_{cr1}{}^2\right)^{\frac{1}{\gamma-1}} = 0.490 \text{ kg/m}^3$$

$$V_1 = M_{cr1} V_{cr1} = 188.1 \text{ m/s}$$

Now, we substitute these magnitudes in Bernoulli's equation:

$$p_1 = p_{t1} - \frac{1}{2}\rho V_1{}^2 = 1.713 \text{ bars}$$

Finally, we calculate the error resulting from the application of Bernoulli's equation:

$$\text{error}\,(e) = \frac{(1.713 - 1.652)}{1.652} = 3.7\%$$

EXAMPLE 3

Consider the variable-geometry axial-flow turbine stator shown in Figure 3.12, where the stator vanes are rotatable. Initially, the vane setting was sufficient to choke the stator. In this and all other setting angles, the stator-exit total temperature T_{t1} was held constant at 1150 K.

Beginning at the choking state, the stator vanes were repeatedly rotated open, with the objective of reaching a final stator-exit critical Mach number ($M_{cr,1,f}$) of 0.5 in increments intended to reduce $M_{cr,1}$ by 0.1 every time. For each of these vane settings, calculate and tabulate the critical versus the traditional Mach number. In doing so, you may assume the following:

- Full flow guidedness by the stator vanes at all setting angles.
- Specific-heat ratio (γ) is 1.33.

SOLUTION

In the following, we will repeatedly reset the critical Mach number value. We will then proceed to calculate the "traditional" Mach number (M). Knowing that the critical velocity is

$$V_{cr1} = \sqrt{\frac{2\gamma}{\gamma + 1} R T_{t1}} = 613.8 \text{ m/s}$$

we can subsequently move to calculate the critical Mach number M_{cr1} as follows:

Set $M_{cr1} = 1.0$:

$$V_1 = M_{cr1} V_{cr1} = 613.8 \text{ m/s}$$

$$T_1 = T_{t1} - \frac{{V_{cr1}}^2}{2c_p} = 987.1 \text{ K}$$

$$a_1 = \sqrt{\gamma R T_1} = 613.8 \text{ m/s}$$

$$M_1 = \frac{V_1}{a_1} = 1.0 \text{ (just as expected)}$$

Set $M_{cr1} = 0.9$:

$$V_1 = 552.4 \text{ m/s}$$

$$T_1 = 1018.1 \text{ K}$$

$$a_1 = 623.4 \text{ m/s}$$

$$M_1 = 0.88$$

Repeating the same procedure, we obtain the following:

$$M_{cr1} = 0.80 \ldots M_1 \approx 0.78$$

$$M_{cr1} = 0.70 \ldots M_1 \approx 0.67$$

$$M_{cr1} = 0.60 \ldots M_1 \approx 0.57$$

$$M_{cr1} = 0.50 \ldots M_1 \approx 0.47$$

Favorable and Unfavorable Pressure Gradients

Axial-Flow Turbomachines

Turbine and compressor examples of this turbomachinery category are shown in Figure 3.5. Almost a "must" in propulsion applications, axial turbomachines offer an envelope size (in terms of the tip radius) that is sufficiently small to minimize the profile drag that is exerted on the engine as a whole. Drawbacks with this turbomachinery classification include high sensitivity to tip clearances as well as a smaller per-stage pressure ratio in comparison with radial turbomachines. In the following, the pressure-gradient effect is discussed as it pertains to bladed and then unbladed components. The latter are commonly known as axial "gaps." Of these, the stator/rotor gaps are particularly designed to be of a smaller length compared with the interstage axial gaps. This is because of the significant swirl-velocity components in the former gap category, particularly in axial-flow turbines.

Bladed components: One of the most crucial factors in designing turbomachinery components is the streamwise variation of static pressure to which the flow stream is exposed. Examples of these, as well as other simpler flow vessels, are shown in Figures 3.5 and 3.6. These are either diffuser-like (diverging) or nozzle-like (converging) flow passages. Of these, the former category produces a positive pressure gradient, whereas the latter gives rise to a negative streamwise gradient. The list, of course, also includes rotors, as shown in Figure 3.9, with the exception that it is the *relative* velocity component W in this case that is influenced by the flow-passage shape.

As is well known in classical boundary-layer theory (Schlichting 1979), the positive pressure gradient is potentially degrading in an aerodynamic sense. A rising pressure in the streamwise direction would aggravate the boundary-layer buildup. Depending on the local or overall pressure gradient, boundary-layer flow separation may occur over the solid walls under such circumstances. This, by reference to the compressor stator in Figure 3.10, may very well give rise to a sizable zone of flow recirculation within which there could be a massive loss of kinetic energy and total pressure, a real-life effect that gives rise to a high magnitude of the so-called *profile* losses. In the case of airfoil cascades, such a region of flow recirculation may also extend down to the trailing edge, thereby increasing the losses across the so-called airfoil *wake*, where the pressure and suction-side flow streams mix together (Fig. 3.1). Different from a boundary layer but equally a function of the fluid viscosity, the airfoil wake is categorically termed a free-shear layer. In addition, one could intuitively tell that in the exit flow stream in cases such as this, the circumferential velocity gradients can be severe (Fig. 2.16) to the point of substantially degrading the downstream airfoil cascade. As a matter of identification, the source of these challenges, namely the adverse pressure gradient, is termed, perhaps fittingly, an *unfavorable* pressure gradient.

Nozzle-like turbomachinery passages provide a *favorable* pressure gradient and are much more "forgiving" of changes in the operating conditions by comparison. In fact, the boundary-layer growth over the airfoil surface and endwalls would normally produce little to practically nonexistent design challenges in this case.

An interesting example of the facts just stated is the stage-count difference between the compressor and turbine sections in a typical gas-turbine engine. For instance, the auxiliary power unit in Figure 1.1 is equipped with a two-stage turbine that is driving a four-stage compressor as well as generating the required power

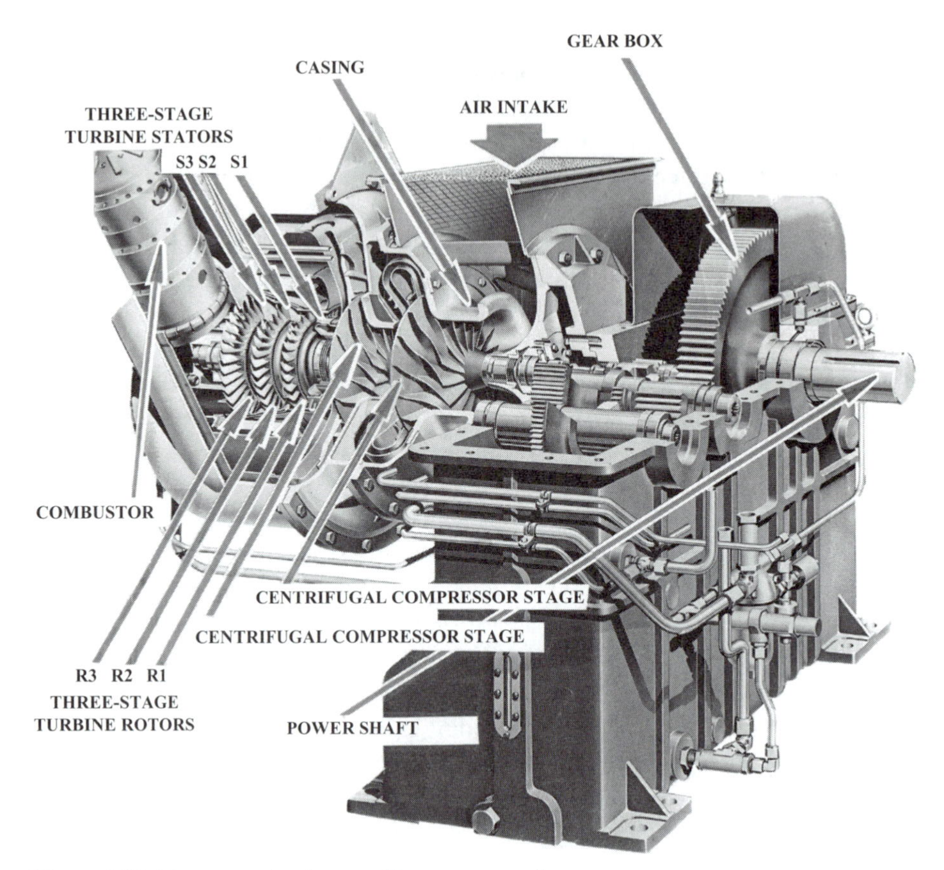

Figure 3.13. Components of an auxiliary power unit.

output. Note that the turbine's overall total-to-total pressure ratio is usually very much comparable to that of the compressor section. The reason behind the drastic stage-count difference here is that all the compressor components are exposed to unfavorable pressure gradients, a fact that directs the designer toward a large number of compressor stages, each with a small fraction of the overall pressure ratio.

The comparison is fair in the sense that all stages in the auxiliary power unit (Fig. 1.1) are of the axial-flow type. However, the situation becomes different as centrifugal compressor stages are utilized instead. Referring to Figure 3.13, utilization of such a stage type makes the compressor and turbine stage counts identical. The reason is that centrifugal compressor stages are typically capable of producing large pressure ratios at acceptable efficiency magnitudes. The same advantage is offered by radial-inflow turbines (Fig. 3.14). Nevertheless, this advantage is not a good excuse to resort to this type of large-envelope, bulky, and heavy turbomachines in propulsion applications, particularly in transoceanic turbofan engines. The engine configuration of the TPE331-14 turboprop engine in Figure 1.2 is an exception, for this is a small engine (550 shp), two of which propel a light airframe over missions that do not exceed a few hundred miles.

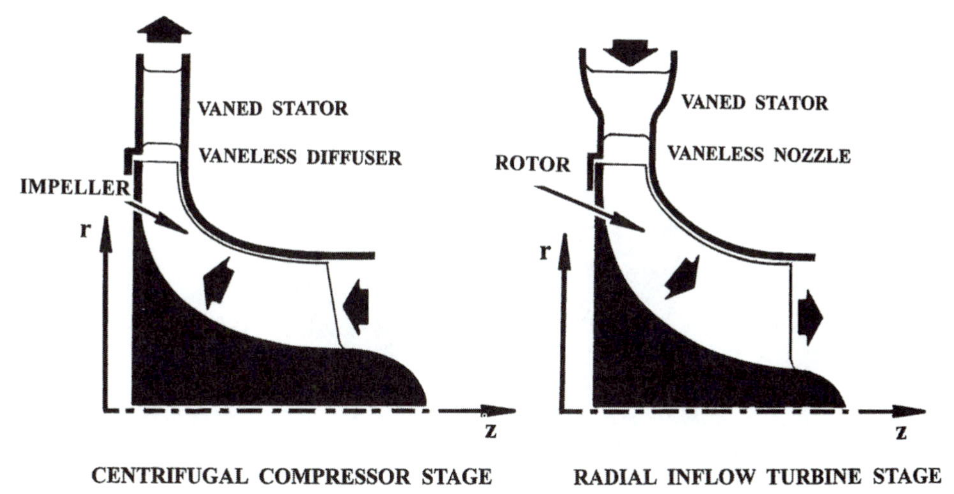

Figure 3.14. Meridional projections of a centrifugal-compressor stage versus a radial-turbine stage.

It is also important in this context to realize that the manner in which the cross-flow area is increased through diffuser-like passages is governed by strict guidelines. Such practice is aimed at suppressing the rate of boundary-layer growth and, in the process, avoiding premature flow separation. As seen in Figure 3.10, and despite almost identical vane-to-vane spacing, the cross-flow area increase in the compressor stator is much more conservative by comparison. The same observation applies to the compressor rotor cascade in Figure 3.9 as contrasted with that of the turbine rotor in the same figure.

Unbladed passages: These include spacings (or gaps) between the stationary and rotating cascades within the same stage, as well as the interstage gaps in multistage turbines and compressors. It is clearly the case that the cross-flow area is that of an annulus for this class of flow passages (Fig. 3.6). In some of these passages, the flow stream may (by design) be totally in the axial direction. Such an ideal situation is possible over the interstage gaps only. A zero-swirl situation is preferable from a flow-viscosity viewpoint. The reason is that the mere existence of a stage-exit swirl velocity component would elongate the flow path, particularly over the endwalls, for it adds its own tangential "trips" of the fluid particles across the gap. This has the effect of providing the boundary layer with an added length on which to grow. It should be emphasized, however, that the ideal situation of a zero swirl angle cannot continue to exist under off-design operating modes.

The performance degradation imparted by the swirl-velocity component is significantly large in unbladed passages with unfavorable pressure gradients, such as the turboprop exhaust diffuser in Figure 3.11, as well as axial-compressor interstage gaps. In the former example, and as indicated in Figure 3.11, elevation of the diffuser-inlet swirl angle is seen to give rise to a near-exit boundary-layer separation and flow

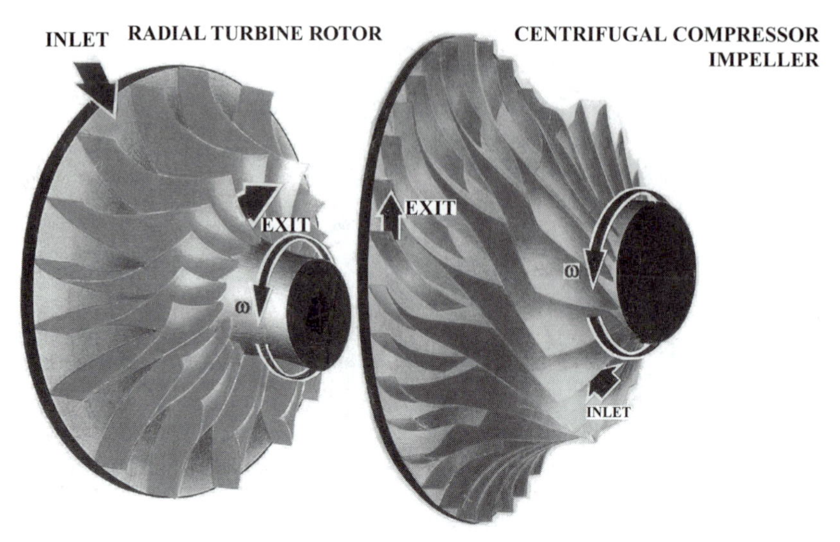

Figure 3.15. Relative flow direction in a radial rotor versus a centrifugal impeller.

reversal, both of which have negative impacts on, among others, the diffuser static-pressure recovery process.

Radial-Flow Turbomachines

In this class of turbomachines, the gaps separating rotating and stationary components are typically of the purely radial type. Although the radius change in a radial-turbine stage (Fig. 3.14) adds to the flow acceleration and therefore improves the stage performance, the rising radius in a centrifugal compressor worsens an already unfavorable pressure gradient. The same effect exists in both the stator and rotor components, emphasizing even more the adverse pressure gradient across each component.

The flow acceleration or deceleration in a rotating cascade of blades is experienced (only) by the relative velocity W and not the absolute velocity V. In fact, the latter velocity will typically experience the opposite effect, one that is hardly indicative of the blade-to-blade passage shape.

Bladed components: Examination of Figure 3.15 reemphasizes a previously noted fact in connection with the cross-flow area composition. The streamwise variation of this area is not only a function of the blade-to-blade passage shape but also the streamwise radius shift. The latter is particularly pronounced across rotors (Figs. 3.14 and 3.15), as they typically extend over a much larger radius change in comparison with the corresponding stators and vaneless gaps.

Radial gaps: In radial (or centrifugal) turbomachinery stages, the stator/rotor gaps are typically of the purely radial type, meaning a practically zero axial-velocity component everywhere within the gap (Fig. 3.14). In this case, the cross-flow area will always be of the type

$$A = 2\pi r b \cos\alpha \tag{3.50}$$

where

r is the local radius,
b is the sidewall spacing, and
α is the swirl angle, referenced to the local radial direction.

Away from the sidewalls, the flow structure is typically approximated to be that of the so-called free vortex type. The applicable equations that define the flow trajectory across the gap are the continuity equation and the conservation of angular momentum relationships. Assuming a negligible static density change across the radial gap, these relationships can be expressed as

$$\dot{m} = 2\pi r b \rho V_r$$

or

$$r V_r = \text{ constant}$$

and

$$r V_\theta = \text{ constant}$$

The last two equations can be combined to give the following important result

$$\frac{V_\theta}{V_r} = \tan\alpha = \text{ constant}$$

which means that the flow swirl angle is constant between the two radii defining the radial gap. The result is a flow trajectory that is a logarithmic spiral.

Design-Point and Off-Design Operation Modes

In designing a gas-turbine engine for, say, aircraft propulsion, the so-called design point is first defined. The thermophysical properties associated with this point (or state) are then utilized in the design of each of the engine components. It should therefore be obvious that the final engine design would give rise to the utmost performance only at this point of engine operation. Exceptions to this rule are rare, as the best performance, for instance, may intentionally correspond to, say, the 110% power setting. Note that there are many ways to assess the engine performance, such as the thermodynamic efficiency, specific fuel consumption, and, in this case, the propulsive efficiency. It is almost unimaginable to build an engine that ensures the optimum magnitudes of each and every one of these performance gauges, even at the design point. Yet, one can still speak of optimum performance, which in this case would be in reference to the designer's own performance-variable choice. Nevertheless, the fact is that the rest of the aircraft mission may, and does, deviate to varying degrees from the ideal design-point performance. Indeed, the engine may never achieve the design-point operation mode throughout the entire mission. An example of this matter's delicacy is the design of a turboprop engine, normally with a sea-level-takeoff design point, would take into account a hot-day and a normal-day takeoff. This is perhaps understandable in the sense that the design-point operation is the only mode where the engine is totally adapted to the ambient conditions.

Analysis of the design point in each engine is unique and entails such variables as ambient conditions, as well as externally imposed constraints on, for instance, the turbine-inlet total temperature and shaft speed. These variables are then used as input to a simple engine-wise "cycle analysis" whereby:

- The principal conservation laws are enforced, and the most basic turbine/compressor matching method is implemented (Chapter 12).
- The output of the process is then cast in the form of thermophysical properties at specific computational "nodes." These nodes span the entire engine, with each component being represented by two (inlet and exit) nodes.

Utilizing the zeroth-order results in the preliminary and then the detailed design phases will ultimately produce an engine that:

- Meets the optimum-performance criterion at the design point, and
- Suffers varied levels of performance deterioration as the engine enters off-design operation modes.

Note that the engine's operating condition at any given point in time is likely to belong to an off-design performance mode. Some of these operating modes are recognized ahead of time to constitute far-off-design modes in any mission. These include taxiing and ground idle, where the shaft speed may very well reach 40% of the design speed. Such a substantial deviation will cause equally tremendous changes in the velocity diagrams, producing, in the process, excessive flow underturning, which is very likely to produce a significant performance degradation. Of course, the startup and shutdown engine operations are even farther from the design-point performance, but the concern under these operating modes is mechanical in nature. Under such operating modes, the danger is a violently unsteady operating mode of the compressor, known as compressor "surge," a topic that is discussed in Chapter 5.

Choice of the Design Point

In the early design phase, a specific set of engine inlet conditions are set to define the engine design point. In power-system applications, such as gas-turbine power plants (Fig. 1.7) and auxiliary power units (Fig. 3.13), the design point is defined by the local ambient conditions. These do not always have to be the standard sea-level temperature and pressure, for auxiliary power units are also utilized in a typical airframe fuselage, where the inlet conditions would obviously depend on the altitude of operation. As for propulsion applications, it is usually a chosen point within the aircraft mission that defines the design point.

The choice of the engine's operating mode in the design-point definition is based either on concern about a critical phase within the aircraft mission or on economic factors. In the following examples, we apply these two criteria to turboprop and turbofan engines, respectively.

Example 1: Turboprop engines: The design point in this case is traditionally the sea-level takeoff. One may comprehend such a choice by recalling that a turboprop engine (Fig. 1.2) is categorically tailored for significantly short missions. In this case, the combined takeoff-climb phases would naturally constitute a "healthy" segment of the entire mission. In addition, these combined phases are those where the engine is assumed to produce the maximum thrust force. The carefulness with which this topic is handled typically leads the engine designers to account for what, in the bigger picture, would be considered minor changes in the ambient conditions.

Example 2: Transoceanic turbofan engines: These propulsion-efficient systems are normally assumed to face little to no problems in propelling the airframe within any of the mission phases. Recalling that the cruise-altitude engine operation is by far the longest of the mission phases and the most fuel-consuming, it is the steady-state cruise altitude that traditionally defines the engine design point. Supporting the choice as well is the fact that the substantial amount of fuel weight, which is predominantly consumed within this mission phase, curbs what would otherwise be a profitable payload. Again, it is the designer's responsibility to ensure nearly flawless engine operation during this cruise-operation phase, with the objective of minimizing the specific fuel consumption (the fuel mass per one unit of the thrust force).

Figure 1.3 shows an example of a turbofan engine with a design point that is at odds with the design-point choices just mentioned. This particular engine is designed to propel a subsonic trainer. As such, the engine has to support maneuvers at low to extremely low flight altitudes. As unique as it may appear, the engine is so designed to be dynamically similar over the entire range of zero to 15,000 ft altitudes. Defined in Chapter 5, the term *dynamic similarity* here involves the engine itself and limits all nondimensional variables (such as the Mach number) and flow angles to fixed magnitudes throughout the entire altitude range. The design process under such a constraint is one of the most challenging.

Variable-Geometry Turbomachines

For a gas-turbine engine that operates over a wide spectrum of engine operation modes, there are several means of improving its off-design performance by abandoning the fixed-geometry option in the design of bladed components. Turbomachines in this engine category are collectively referred to as "variable-geometry" turbomachines and are depicted in Figures 3.16 through 3.21. As seen in these figures, the

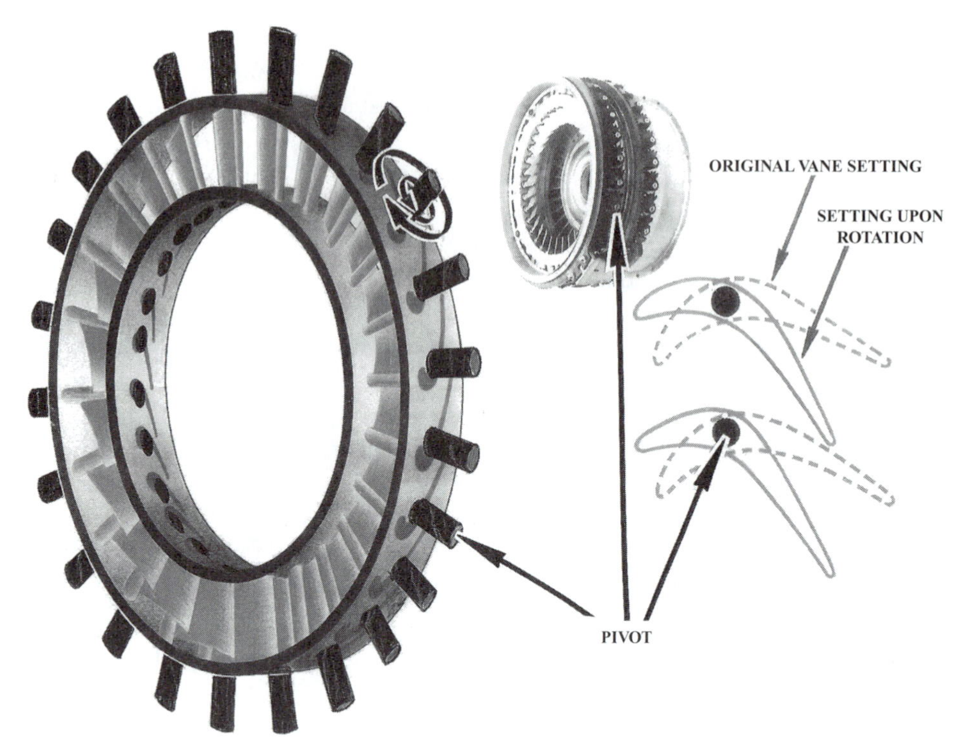

Figure 3.16. Mechanism of a rotating-vane configuration in an axial-turbine stator.

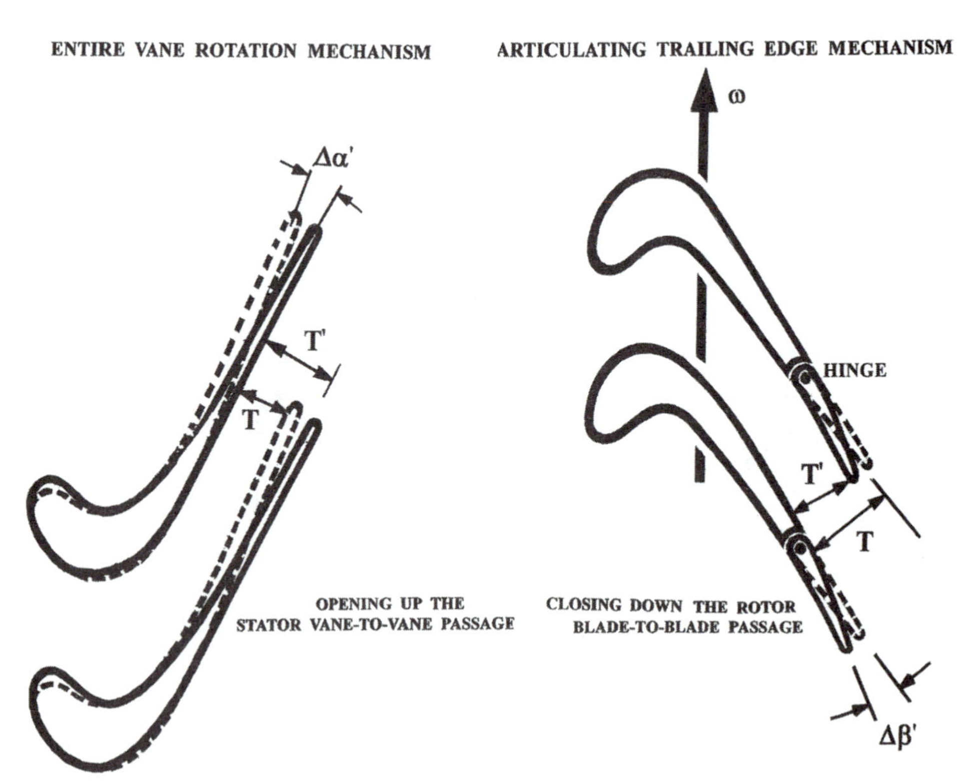

Figure 3.17. Whole-vane versus trailing-edge-segment rotation in variable-geometry axial-turbine cascades.

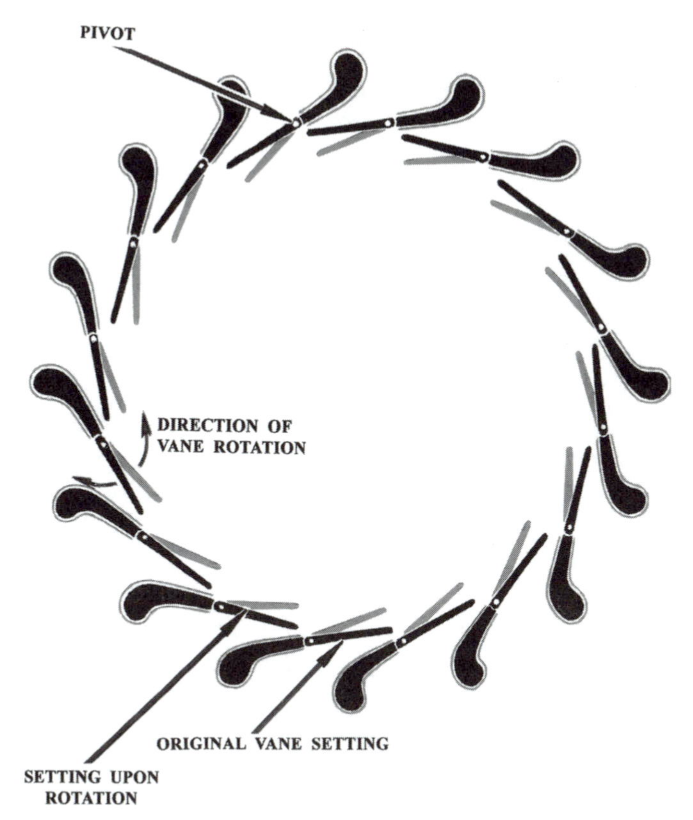

Figure 3.18. Rotating trailing edge in a variable-geometry radial stator.

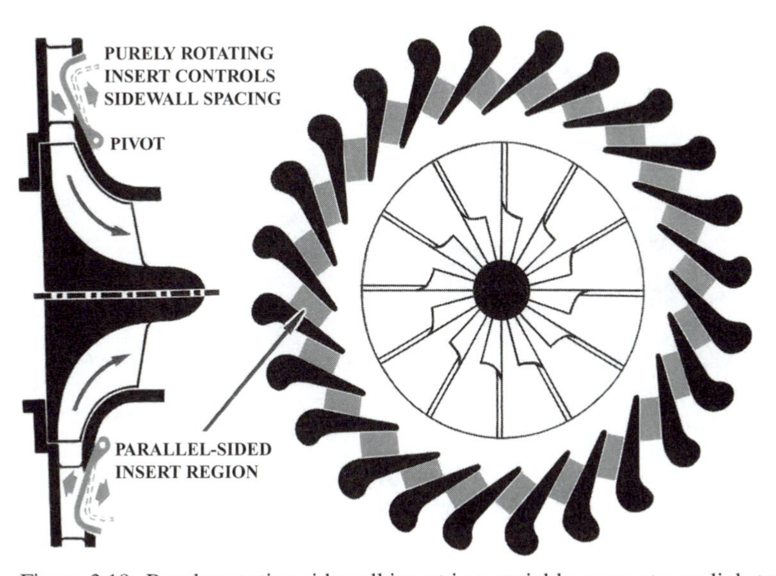

Figure 3.19. Purely rotating sidewall insert in a variable-geometry radial stator.

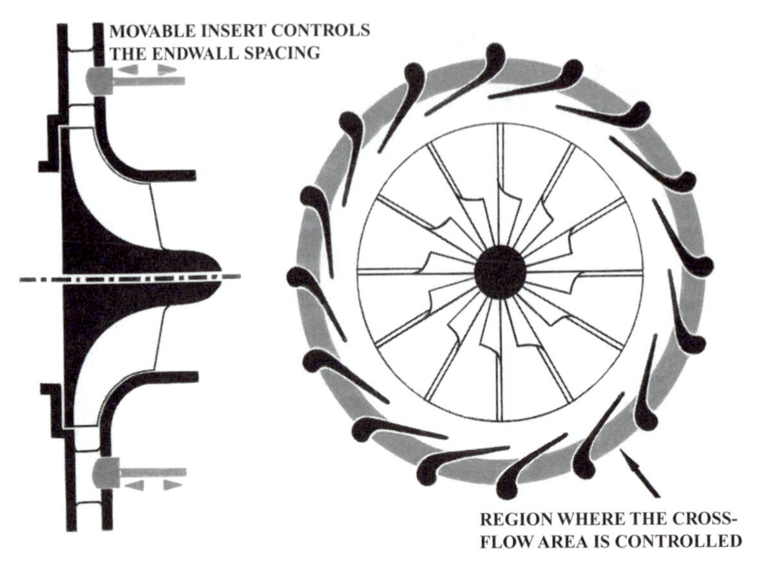

Figure 3.20. Purely translating sidewall insert in a variable-geometry radial stator.

change in geometry in such turbomachines corresponds to a process where the stator-exit cross-flow area of a stator cascade is either decreased or increased by rotating the entire airfoil, rotating its exit segment, or changing the shape or position of one of the endwalls. The common objective here is to efficiently respond to a change in the mass-flow rate and/or shaft speed. Figure 3.22 shows an example of the efficiency

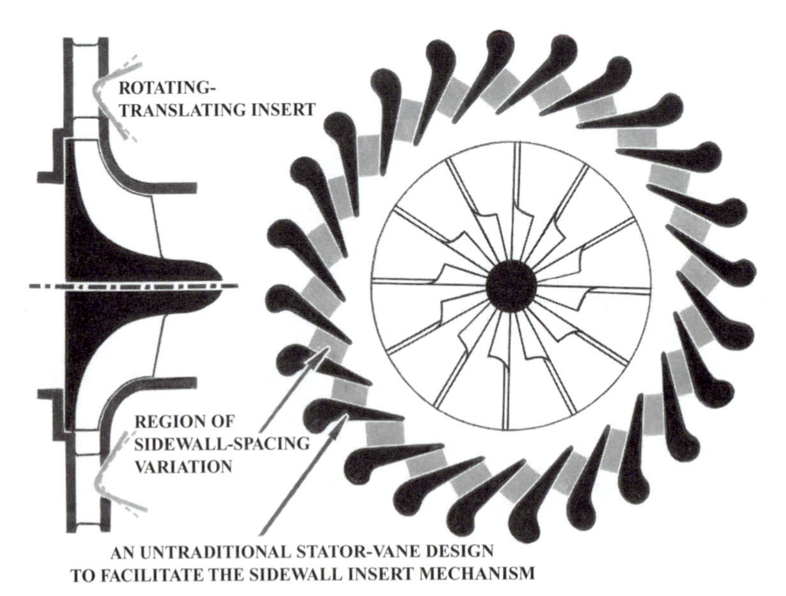

Figure 3.21. Rotating-translating sidewall insert in a variable-geometry radial stator.

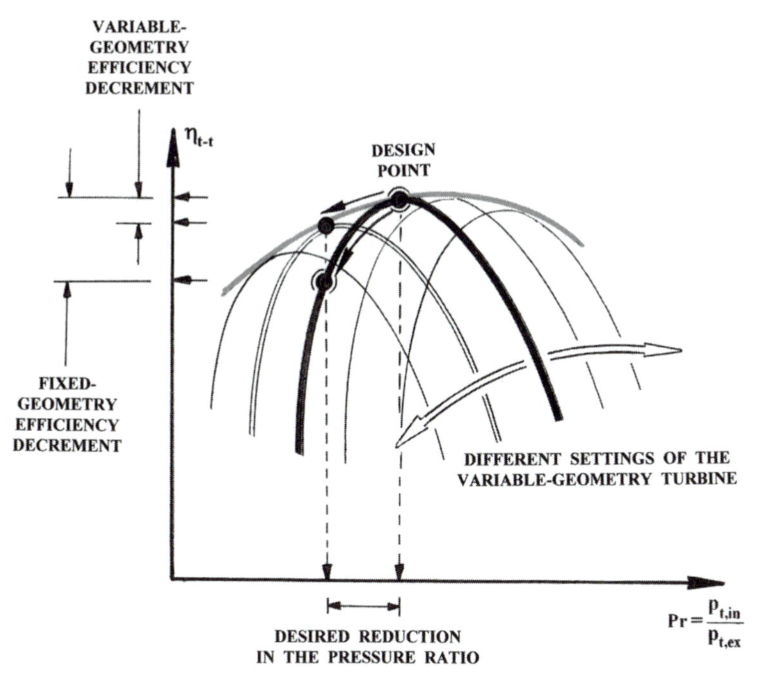

Figure 3.22. Benefit of variable-geometry stators in changing the pressure ratio.

gain as a result of using variable versus fixed-area radial turbines in a process where the shaft speed is reduced during operation.

In the area of axial-flow turbines, the stator or rotor cascades can be rotated open or closed during operation. This is normally effected through hydraulic means because of the airfoil aerodynamic load, which could be rather significant. Figure 3.16 shows the situation where the entire stator vane is rotated around a fixed pivot. Because it is only the exit cross-flow area that is targeted, the same effect can be achieved by rotating a finite exit segment of the airfoil as shown in Figure 3.17. Note that the airfoil-surface continuity, in a "slope" sense, should ideally be maintained during the process to minimize the possibility of local diffusion (meaning a sudden static pressure rise) at the pivot. This may prove to be geometrically hard, although maintaining the curvature continuity over the blade surface is even harder. In practice, however, there is always a surface discontinuity at this location, which will have a usually local impact on the airfoil-surface pressure distribution.

A similar geometrical arrangement can be extended to the stator of a radial turbine (Fig. 3.18). Alternatively, the geometry of a sidewall can be partially altered to meet a given cross-flow-area requirement. Figure 3.19 shows a rotating insert that affects part of one sidewall within the stator subdomain. This figure also shows the vane-to-vane regions that will be affected in the sense of sidewall spacing because of the insert rotation. The insert arrangement in Figure 3.20 differs in the way the insert is moved, with the case here being an axial translation of a similar segment of the sidewall. Finally, the insert in Figure 3.21 makes it possible to mix the former two arrangements. This particular insert has a carefully shaped surface, rotates around a displaced center, and is usually referred to as a rotating-translating sidewall insert. Among these three options, the last was chosen to effect the required cross-flow area variation in a NASA-sponsored Technology Demonstration Program entitled Variable-Area Radial Turbines (VART) early in the 1980s.

Means of Assessing Turbomachinery Performance

Ranging from the simple isentropic efficiencies to per-component losses in such variables as total pressure and kinetic energy, there are several ways of quantifying the performance level of turbomachines. In fact, and aside from universally known performance assessors, turbomachinists can, and do, define their own parameters to do the same should the design objective itself be nontraditional.

Total-to-Total efficiency

This is usually termed the *isentropic* efficiency, a well-known phrase in introductory thermodynamics texts, except that the properties at the end states in this case are perhaps left to imply static properties. In our case, the efficiency is defined on the basis of total properties at both the inlet and exit stations, as follows:

$$\eta_{t-t}/_{T} = \frac{\Delta h_{t\,act.}}{\Delta h_{t\,id.}} \text{ (for a turbine)} \tag{3.51}$$

$$\eta_{t-t}/_{C} = \frac{\Delta h_{t\,id.}}{\Delta h_{t\,act}} \text{ (for a compressor)} \tag{3.52}$$

Under an adiabatic-flow assumption, the turbine efficiency (3.51) is graphically represented and defined on the left-hand side of Figure 3.23. In the case of a compressor, it is the ideal (or isentropic) shaft work that constitutes the numerator, as it is the minimum magnitude with which the compressor would yield the same total-to-total pressure ratio. Assuming a sufficiently small *static* temperature change, we can simplify expressions (3.51) and (3.52) by considering the specific-heat ratio γ to be constant:

$$\eta_{t-t}/_{T} = \frac{1 - \frac{T_{t\,ex}}{T_{t\,in}}}{1 - \left(\frac{p_{t\,ex}}{p_{t\,in}}\right)^{\frac{\gamma-1}{\gamma}}} \text{ (for a turbine)} \tag{3.53}$$

$$\eta_{t-t}/_{C} = \frac{\left(\frac{p_{t\,ex}}{p_{t\,in}}\right)^{\frac{\gamma-1}{\gamma}} - 1}{\frac{T_{t\,ex}}{T_{t\,in}} - 1} \text{ (for a compressor)} \tag{3.54}$$

Referring to the left-hand side of Figure 3.23, a beginner may make a usually undetectable mistake by simply stating the phrase: "*assuming* the ideal exit pressure to be equal to the actual exit pressure, ..." The fact of the matter is that the actual

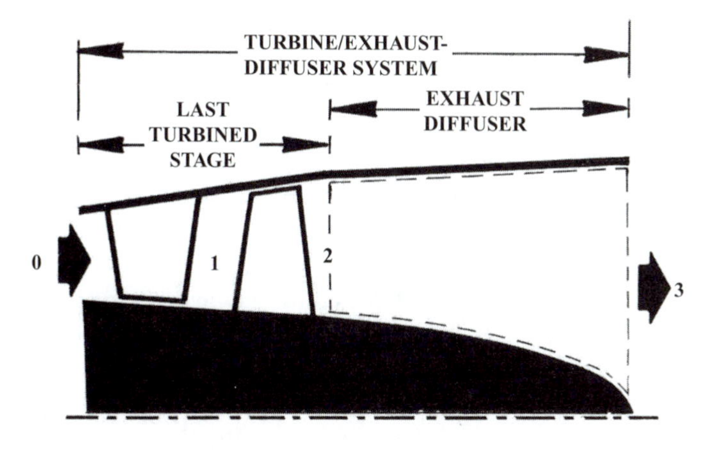

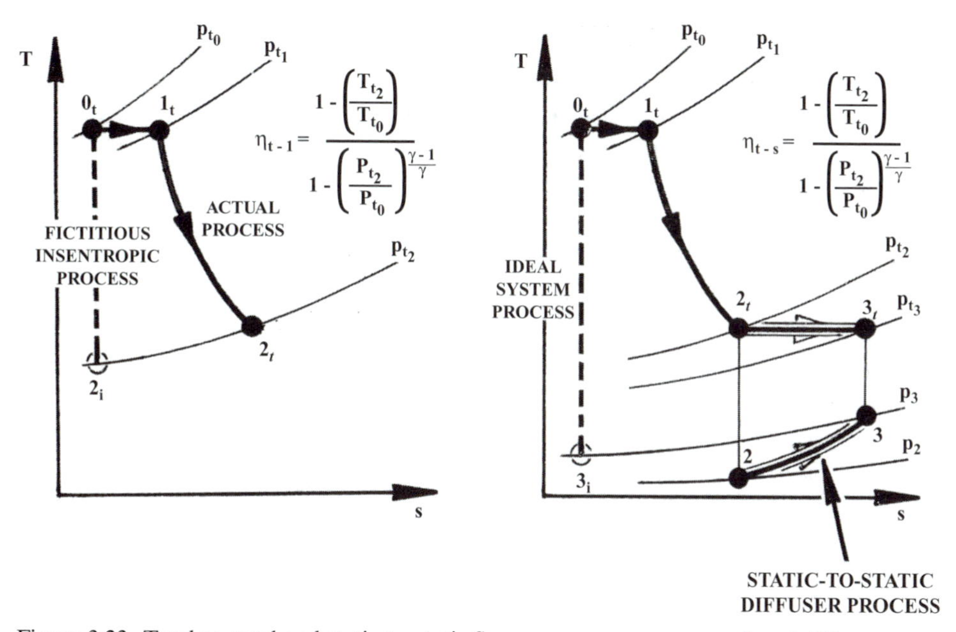

Figure 3.23. Total-to-total and static-to-static flow processes across an exhaust diffuser.

(entropy-producing) process will exist first. As a matter of *definition*, we next construct an isentropic process between the same inlet and exit total pressures for the purpose of comparison. There is accordingly no room here for any assumption, as this is simply a matter of implementing a definition.

Total-to-Static Efficiency

Restricting the discussion to the exit station of an entire turbine or compressor section, the mere existence of a finite (nonzero) swirl angle is indicative of inefficiency.

For one thing, a nonzero swirl-velocity component in this case simply means a waste of what could have been a shaft-work contributor, as will be explained in Chapter 4. Moreover, the nonzero swirl angle at such a location will give rise to higher magnitudes of friction-related losses. Whether that is the combustor-leading duct, in the case of a compressor, or an exhaust system in the case of a turbine, the nonzero swirl angle will elongate the flow trajectories over both endwalls. This may very well cause fast growth of the boundary layer and possibly flow separation, particularly in the case of an exhaust diffuser.

In any turbomachinery stage, it is both desirable and nearly possible to produce a zero exit swirl velocity, at least under the design-point operating mode. Although the nonzero exit swirl angle is itself a loss indicator, a separately defined variable termed the total-to-static efficiency is used to characterize, in particular, the turbine performance in turboshaft engines. These include virtually all power-system applications, as well as turboprop engines. Such engines typically end with exhaust diffusers, which, by definition, are unfavorable pressure-gradient passages. The static pressure recovery, in this case, is highly sensitive to the flow inlet swirl, as explained earlier. Note that the turbine-exit static pressure in this case is usually less than the ambient pressure, which would cause a highly damaging flow reversal to reach the turbine section should the diffuser-wise static-pressure rise be offset by a total pressure loss.

The right-hand side of Figure 3.23 shows a T-s representation of the total-to-static efficiency in the case of a turboprop engine turbine. The definition of this efficiency is

$$\eta_{t-s}/T = \frac{1 - \frac{T_{t\,ex}}{T_{t\,in}}}{1 - \left(\frac{p_{ex}}{p_{t\,in}}\right)^{\frac{\gamma-1}{\gamma}}} \quad \text{(for a turbine)} \tag{3.55}$$

A similar expression of the same variable can be obtained for a compressor by simply replacing $p_{t\,ex}$ by p_{ex}, the exit static pressure, in expression 3.54

Examination of expressions (3.53) and (3.55) reveals that η_{t-s} will always be less than η_{t-t}. However, these two efficiencies will be very close to one another in the event that the turbine-exit swirl angle α_{ex} is zero. In the case of a turboshaft engine, it is clear that η_{t-s} embraces the turbine performance as well as that of the exhaust diffuser (Fig. 3.11).

Kinetic-Energy Loss Coefficient

This is a popular loss coefficient, as it applies to vaned and unvaned turbomachinery components. Although the definition applies to rotating cascades (as based on the relative velocity, W), it is almost exclusively a primary performance assessor in stators. This coefficient (represented as $\bar{e}$ in Figure 3.24) is defined as

$$\bar{e} = 1 - \frac{V_{act.}^2}{V_{id.}^2} \tag{3.56}$$

where $V_{act.}$ is the actual exit velocity and $V_{id.}$ is the ideal velocity that is associated with a component-wise isentropic flow process.

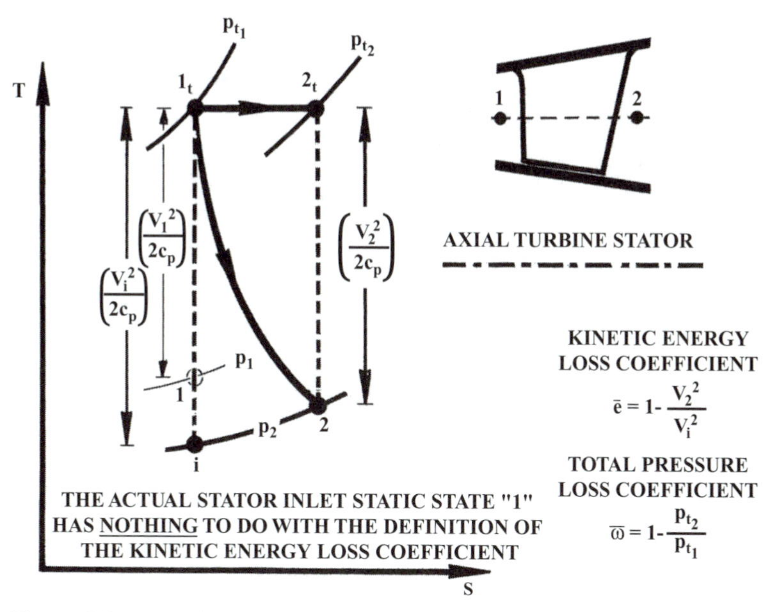

Figure 3.24. Definition of the kinetic-energy and total-pressure loss coefficients. Unsubscripted states are all *static* states.

Total-Pressure Loss Coefficient

Based on its name alone, one can detect the fact that this variable applies to stationary components. The reason is that the total pressure will certainly change across the rotor should any shaft-work interaction take place. The popularity of this coefficient is due to the fact that it is an explicit statement of the real-life decline in total pressure, one of the most physically meaningful properties and the most traceable on the *T-s* diagram (Fig. 3.24) because of irreversibilities (e.g., friction) within the component at hand. This coefficient (denoted $\bar{\omega}$) is defined as

$$\bar{\omega} = \frac{\Delta p_t}{p_{t\,in}} = 1 - \frac{p_{t\,ex}}{p_{t\,in}} \tag{3.57}$$

Later in this chapter, we will derive a similar variable that applies to a rotor subdomain. This will be based on the total *relative* pressure. Unfortunately, the definition will apply only to axial-flow turbomachines.

Entropy-Based Efficiency

As its name might imply, this particular variable is based on the Second Law of Thermodynamics. The variable is not particularly popular in turbomachinery applications despite its unique versatility, as it is applicable to rotors as it is to stators. In fact, this particular variable even applies to the entirety of a multistage compressor and turbine sections, as shown in Figure 3.25. Moreover, the variable remains applicable to any flow process that is defined (or electively viewed) on a total-to-total, total-to-static, or static-to-static basis, giving rise to the same magnitude every time. Let us

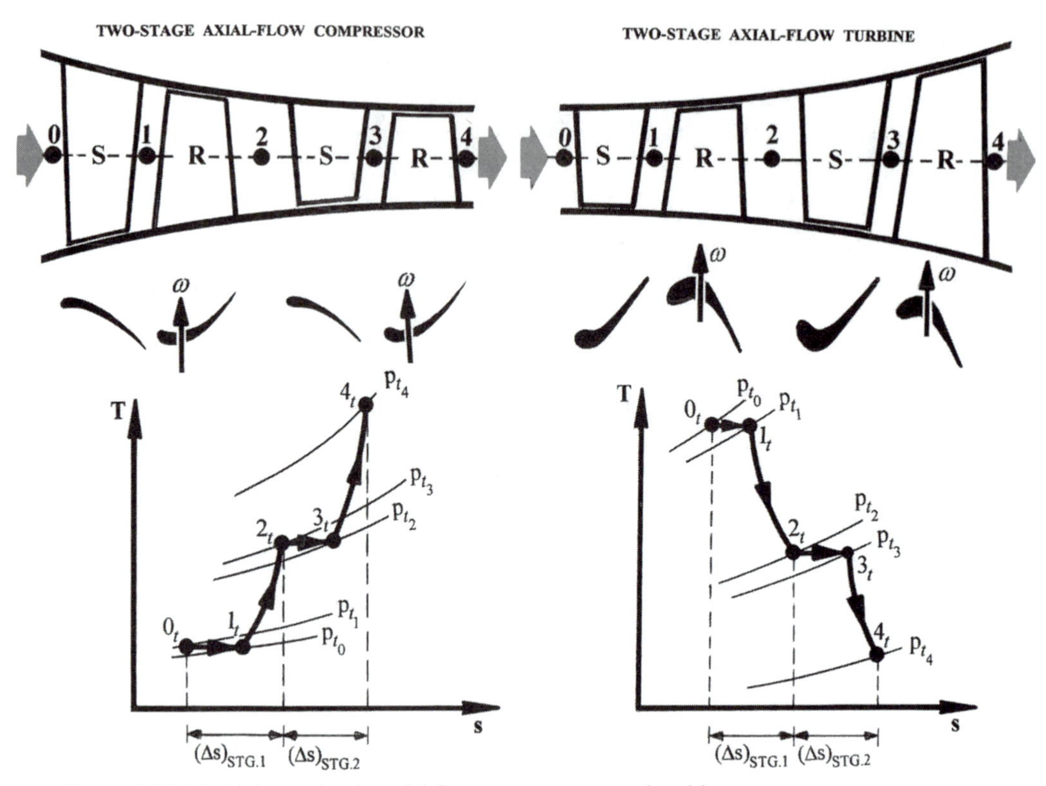

Figure 3.25. Multiple staging in axial-flow compressors and turbines.

begin with the loss coefficient ($\bar{q}$), which was defined earlier as

$$\bar{q} = 1 - e^{\frac{-\Delta s}{R}} \tag{3.58}$$

where Δs is the sum of all specific entropy productions (Fig. 3.25) for all components under consideration.

An efficiency-like parameter, η_s, is also common in the gas-turbine industry. This is derived from expression (3.57), with the exception of replacing the gas constant (R) by the specific heat (c_p) as follows:

$$\eta_s = e^{\frac{-\Delta s}{c_p}}$$

Note that the magnitude of η_s will always lie between zero and 100%, with the latter corresponding to the ideal case of zero entropy production (i.e., a perfectly isentropic flow process).

Total Relative Flow Properties

In this section, we aim to simplify the flow process through a rotor by defining the total properties on the basis of the relative velocity W. In the process, we will define the so-called relative critical Mach number, $M_{cr} = W/W_{cr}$. The thermophysical state that is defined by these properties will consistently be referred to as the "total relative" state, which is as legitimate as any other thermophysical state.

Let us begin with the definition of total relative temperature (T_{t_r}) as follows:

$$T_{t_r} = T + \frac{W^2}{2c_p} \tag{3.59}$$

This expression can be rewritten in terms of the usually known total temperature T_t as

$$T_{t_r} = T_t + \frac{W^2 - V^2}{2c_p} \tag{3.60}$$

On the basis of T_{t_r}, we can now define the total relative pressure p_{t_r} as

$$p_{t_r} = p_t \left(\frac{T_{t_r}}{T_t} \right)^{\frac{\gamma}{\gamma - 1}} \tag{3.61}$$

The total relative density (ρ_{t_r}) can also be obtained by applying a similar isentropic relationship:

$$\rho_{t_r} = \rho_t \left(\frac{T_{t_r}}{T_t} \right)^{\frac{1}{\gamma - 1}} \tag{3.62}$$

or, by applying the equation of state to the total relative state, as

$$\rho_{t_r} = \frac{p_{t_r}}{R T_{t_r}} \tag{3.63}$$

Having defined the total relative temperature (3.60), we next consider an adiabatic axial-flow rotor within which (as we will now prove) the total relative temperature remains constant. To this end, let us examine the change of total relative temperature in this case as follows:

$$\Delta T_{t_r} = (T_{t2} - T_{t1}) - \left[\left(\frac{V_2{}^2 - V_1{}^2}{2c_p} \right) + \left(\frac{W_1{}^2 - W_2{}^2}{2c_p} \right) \right] \tag{3.64}$$

Over the rotor defined previously, the two bracketed terms in equation (3.64) are proven to be identical in Chapter 4. We can therefore conclude that the total relative temperature will remain constant over an axial-flow rotor.

Introduction of the Relative Critical Mach Number

In a rotating frame of reference, let us begin by defining the so-called relative critical velocity (W_{cr}) as

$$W_{cr} = \sqrt{\left(\frac{2\gamma}{\gamma + 1} \right) R T_{t_r}} \tag{3.65}$$

which remains constant across an axial-flow adiabatic rotor. This particular velocity can be used to nondimensionalize the relative velocity (W), producing the so-called

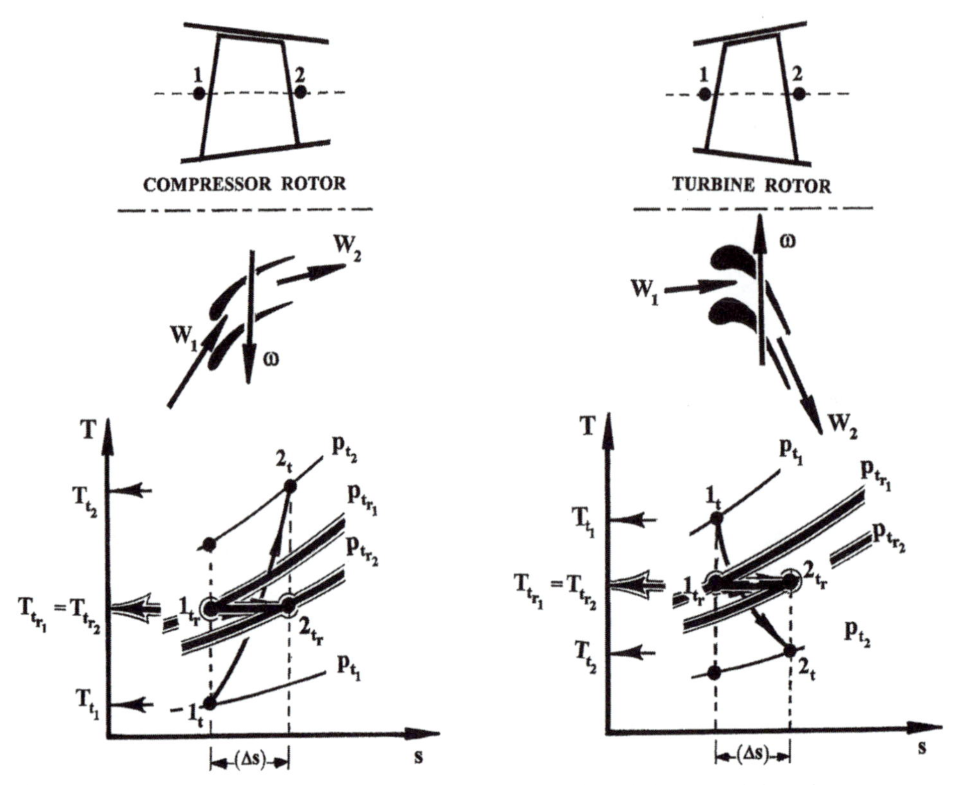

Figure 3.26. Introduction of the total relative properties in axial-rotor subdomains.

relative critical Mach number, M_{crr}, as follows:

$$M_{crr} = \frac{W}{W_{cr}} \tag{3.66}$$

Let us now redefine the total relative temperature and pressure in terms of the newly defined velocity ratio as follows:

$$T_{tr} = \frac{T}{\left[1 - \frac{\gamma-1}{\gamma+1}\left(\frac{W}{W_{cr}}\right)^2\right]} \tag{3.67}$$

and

$$p_{tr} = p\left(\frac{T_{tr}}{T}\right)^{\frac{\gamma}{\gamma-1}} \tag{3.68}$$

Figure 3.26 shows two T-s representations of the flow process through an axial-flow compressor rotor as contrasted with that in an axial-flow turbine rotor. These representations imply an adiabatic-flow assumption, under which the total relative temperature remains constant. The figure also illustrates the fact that the factors degrading the flow process (termed irreversibilities) will present themselves in the form of a decline in total relative pressure. This fact, added to expressions (3.65)

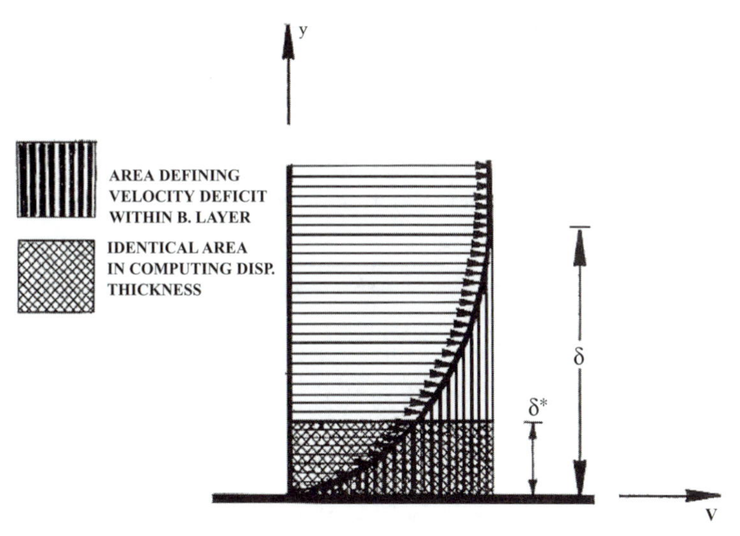

Figure 3.27. Definition of the boundary-layer displacement thickness.

through (3.68) supports an interesting conclusion that an axial-flow rotor will (in effect) act like a stator once relative (rather than absolute) flow variables are used instead.

Losses in Constant-Area Annular Ducts (Fanno Line)

This section highlights the effects of boundary-layer buildup over the endwalls in unbladed constant-area gaps. In order to quantify such effects, let us first define a relevant boundary layer variable that is referred to as the "displacement" thickness (δ^*). Referring to Figure 3.27, this variable is defined as

$$\delta^* = \int_0^\delta \left(1 - \frac{\rho V}{\rho_e V_e}\right) dy \qquad (3.69)$$

where

δ is the boundary-layer thickness,
ρ is the static density, and
V is the local velocity.

The subscript e refers to the boundary-layer edge. The displacement thickness represents the loss of mass flux caused by the loss in velocity within the boundary layer. According to the boundary-layer theory (Schlichting 1979), the flow stream will "sense" the existence of the solid wall as though it were displaced (toward the boundary-layer edge) through a distance that is numerically equal to δ^*.

Figure 3.29 shows the two displacement-thickness lines over the hub and casing surfaces. Based on the preceding remarks, the distance between these two lines defines the effective annulus height at any axial location, within this interstage duct, as

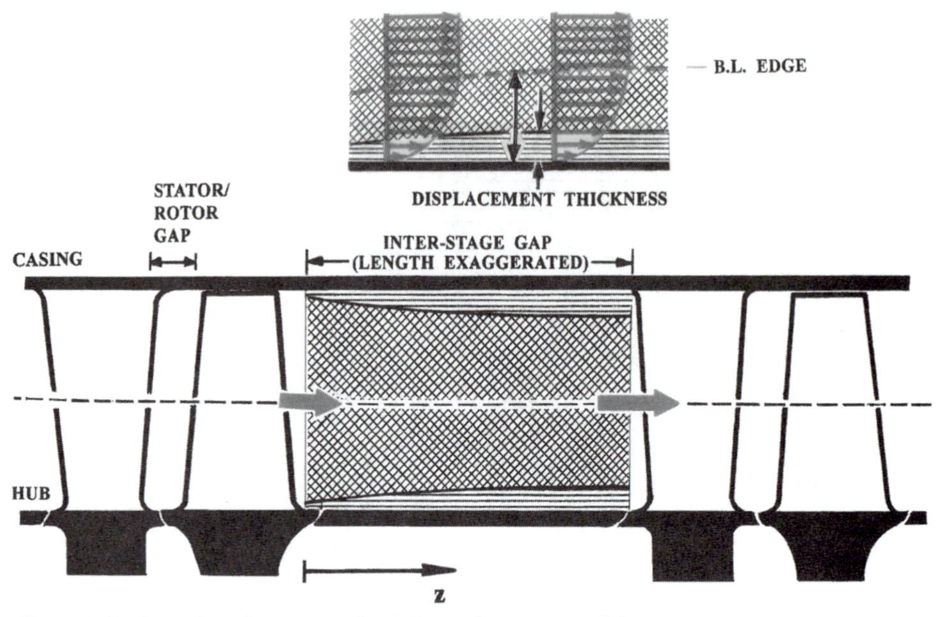

Figure 3.28. Boundary-layer growth within an interstage axial gap.

a result of the "blockage" effect of the boundary layer. In other words, the initially constant-area annular duct, shown in Figure 3.28, is now replaced by a converging subsonic nozzle. Going even further, phenomena such as choking can theoretically prevail at the flow-passage exit station, depending on the operating conditions, friction factor, and interstage duct length.

The preceding discussion is intended to pave the way to a well-known flow type that is referred to as the Fanno flow process. The process concerns a necessarily adiabatic flow and a constant-area duct configuration. Of these restrictions, the former is clear in Figure 3.29, in which the total-to-total flow process is represented by a horizontal line on the *T*-*s* diagram. In the following, we will proceed along the upper subsonic branch of the Fanno line discussing, in the process, the physical implications of key points on this branch.

Because of the irreversibility related to the flow viscosity, the sequence of events on the *T*-*s* diagram will have to progress from left to right over the subsonic branch in Figure 3.29, implying (as is truly the case) continuous entropy production throughout the entire process. Construction of this subsonic Fanno-line branch in this figure spells out the fact that the constant-area duct is now a nozzle-like flow-accelerating passage, a feature that is consistent with the converging shape of the "effective" annulus height in the same figure. At any point on this branch, the vertical distance away from the total temperature (horizontal) line is nothing but the local dynamic temperature $V^2/2c_p$. The subsonic branch ends at the maximum-entropy point (identified as the choking point). According to the entropy-production principle, there is no way for the process path to enter the supersonic branch at this point, as this will

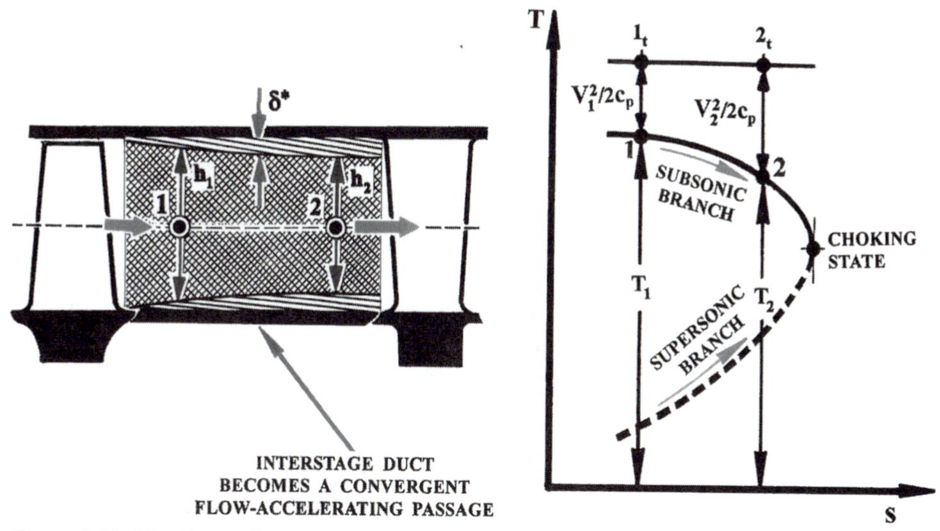

Figure 3.29. The Fanno-line process in a constant-area annular duct. Viscosity effects on interstage ducts.

imply an entropy detraction, which violates the Second Law of Thermodynamics. In fact, the only way for the process to exist on the supersonic branch is to *start* with a supersonic flow and then proceed toward a higher entropy magnitude. The process in this case is clearly one of flow deceleration up to, and terminating at, the maximum-entropy "choking" state.

Reverting back to the subsonic Fanno-line branch, the choking point (if reached) will have to correspond physically to the end of the converging "effective" duct in Figure 3.28.

The different property relationships covering the Fanno-line process can be found in virtually any basic book on the topic of compressible-flow aerodynamics (e.g., Zucker 1977) in both the functional and tabulated forms. Once the annular-duct inlet conditions and relative surface roughness are known, these relationships will yield such variables as the accumulated total pressure loss and the extra duct length that is needed for the choking state to be reached.

In the event of a swirling duct flow, the distance along the duct should be measured along the actual flow trajectory over the endwalls. The trajectory elongation in this case can alternatively be simulated in the form of an added surface roughness. This added roughness should be carefully calibrated in such a way as to yield a one-to-one equivalency of the axial distance (under such a roughness level) and the distance along the actual (swirling) flow trajectory. A simpler means of achieving the same purpose is to proceed with the actual (and not the axial) distance along the flow trajectory over the endwall of interest. Of course, each of the two endwalls will have to be considered separately because each endwall will be associated with its own flow-swirl angle.

Fanno-Flow Relationships

In texts belonging to the internal aerodynamics discipline, the Fanno-flow discussion is usually complemented by a simple one-dimensional analysis, where the different conservation principles (i.e., those of mass, momentum, and energy) are enforced. The result, as would be anticipated, is a set of expressions relating the thermophysical properties at a given axial location in the constant-area duct to those at a different location. Because the flow regime here is frictional, one should naturally expect the appearance of such variables as the friction coefficient f in most such relationships.

Of these relationships, two are very much relevant to turbomachinery applications. First, the following expression essentially provides the decline in total pressure $(p_{t1} - p_{t2})$ across a duct segment, with M_1 and M_2 being the segment's inlet and exit Mach numbers:

$$\frac{p_{t2}}{p_{t1}} = \left(\frac{M_1}{M_2}\right)\left[\frac{1+\left(\frac{\gamma-1}{2}\right)M_2{}^2}{1+\left(\frac{\gamma-1}{2}\right)M_1{}^2}\right]^{\frac{\gamma+1}{2(\gamma-1)}} \tag{3.70}$$

The second Fanno-flow expression relates the duct length to the Mach number rise across the duct, and is as follows:

$$\frac{fL}{D_h} = \left(\frac{\gamma+1}{2\gamma}\right)\ln\left[\frac{1+\left(\frac{\gamma-1}{2}\right)M_2{}^2}{1+\left(\frac{\gamma-1}{2}\right)M_1{}^2}\right] - \left(\frac{1}{\gamma}\right)\left(\frac{1}{M_2{}^2}-\frac{1}{M_1{}^2}\right) - \left(\frac{\gamma+1}{2\gamma}\right)\ln\left(\frac{M_2{}^2}{M_1{}^2}\right) \tag{3.71}$$

where

f is the friction coefficient,
L is the duct (or endwall-streamline) length (discussed later in this section),
M_1 and M_2 are the inlet and exit magnitudes of the Mach number, and
D_h is the duct hydraulic diameter.

The variable D_h needs to be defined in our particular case, meaning that of an annulus cross section. In general, the hydraulic diameter is defined as

$$D_h = \frac{4A}{P}$$

with A and P referring to the area and circumference of the cross section, respectively. Referring to the mean radius of the annulus by r_m and the annulus height by Δr, the hydraulic diameter of the annulus can be expressed as

$$D_h = \frac{4(2\pi r_m \Delta r)}{2\pi\left(r_m + \frac{1}{2}\Delta r + r_m - \frac{1}{2}\Delta r\right)} = 2\Delta r$$

Referring to equation (3.71), the length L represents that of the actual flow trajectory over the endwalls. In the event that the flow is nonswirling (i.e., $\alpha_1 = \alpha_2 = 0$), L would simply be the duct's axial extension, Δz. However, the trajectory length in

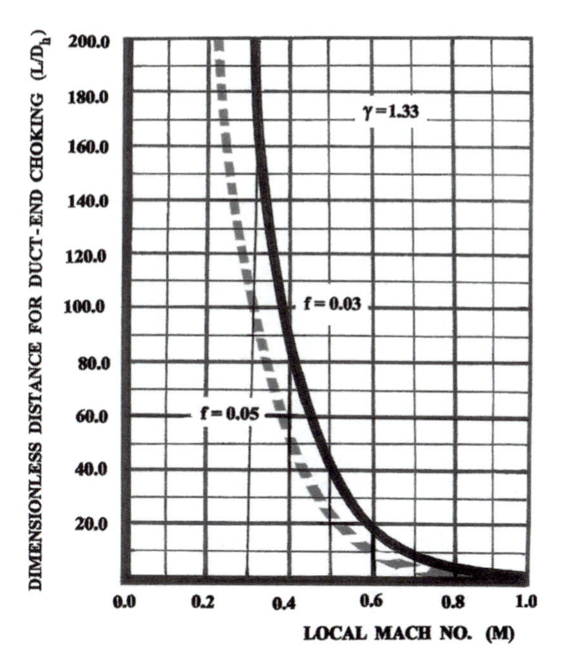

Figure 3.30. Effect of the friction coefficient on friction choking.

a swirling flow field, where the average swirl angle is α, can be expressed as

$$L = \frac{\Delta z}{\cos\alpha}$$

An interesting and useful manipulation of equation (3.71) makes it possible to compute the duct length, L_c, which is necessary to bring the flow stream from a given Mach number, M, to the state of "friction" choking (Fig. 3.30). Such a relationship is easily obtained by setting M_1 to M and M_2 to unity, as follows:

$$\frac{fL_c}{D_h} = \left(\frac{\gamma+1}{2\gamma}\right)\ln\left[\frac{\left(\frac{\gamma+1}{2}\right)}{1+\left(\frac{\gamma-1}{2}\right)M^2}\right] - \left(\frac{1}{\gamma}\right)\left(1-\frac{1}{M^2}\right) - \left(\frac{\gamma+1}{2\gamma}\right)\ln\left(\frac{1}{M^2}\right) \tag{3.72}$$

What gives equation 3.72 a special value is the fact that the physical duct length, with given inlet and exit Mach number magnitudes, can be viewed as the difference between the L_c magnitudes corresponding to each of these Mach numbers. The task becomes even easier with the help of Figure 3.30, which is a graphical representation of the left-hand side of equation (3.72) as a function of the Mach number M.

In applying equations (3.70) through (3.72), and contrary to what this text advocates, the flow kinematics appear in the form of the Mach number and not the critical Mach number. It is true that the latter hardly enjoys the same universality, particularly in external aerodynamics as well as most internal flow fields, and the preceding equations are a clear testimony to this fact. However, it has been the tradition in the turbomachinery discipline to utilize the critical Mach number instead, for reasons that were outlined earlier.

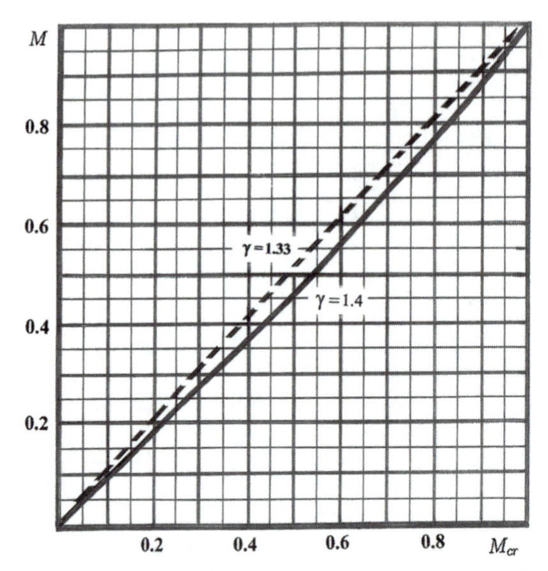

Figure 3.31. Relationship between the Mach number and the critical Mach number.

In view of the preceding discussion, it may now be appropriate to cast the Mach number in terms of its critical counterpart. Perhaps the easiest way to this end is to equate the two magnitudes of the static-to-total temperature ratios, one in terms of the Mach number and the other containing the critical Mach number instead:

$$T = \frac{T_t}{\left[1 + \left(\frac{\gamma-1}{2}\right) M^2\right]} = T_t \left[1 - \left(\frac{\gamma - 1}{\gamma + 1}\right) M_{cr}{}^2\right]$$

which can be rearranged to yield the expression

$$M = \sqrt{\left(\frac{2}{\gamma - 1}\right)\left[\frac{1}{1 - \left(\frac{\gamma-1}{\gamma+1}\right) M_{cr}{}^2} - 1\right]} \qquad (3.73)$$

Figure 3.31 provides a graphical representation of equation (3.73) for the two most common values of the specific-heat ratio, γ, namely 1.4 and 1.33. Examination of this figure reveals that the magnitude of M_{cr} is always greater than or equal to that of M. The two instances where these magnitudes are equal correspond to the Mach number magnitudes of zero and unity. The latter case was proven earlier in the context of defining the critical Mach number. The equality at the magnitude of zero is rather obvious.

EXAMPLE 4

Figure 3.32 shows the impeller of a "cheap" centrifugal compressor where the blades are made up of flat plates. The compressor's design point is defined as follows:

- Inlet total pressure $(p_{t1}) = 2.3$ bars
- Inlet total temperature $(T_{t1}) = 378.0$ K

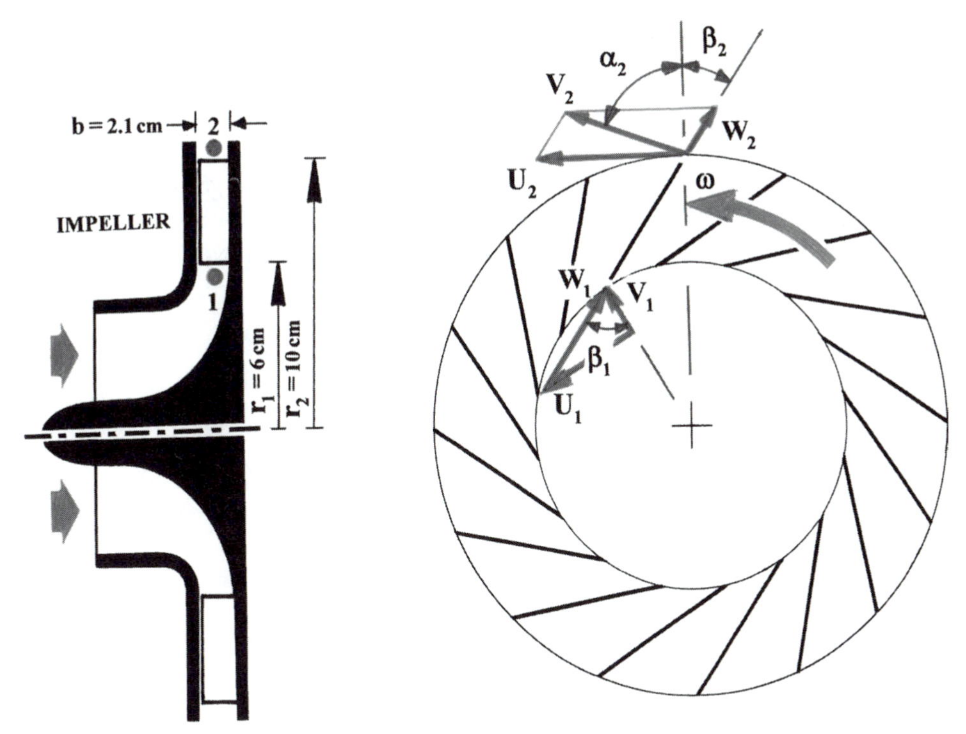

Figure 3.32. Input variables for Example 4.

- Inlet Mach number $(M_1) = 0.38$
- Inlet swirl angle $(\alpha_1) =$ zero
- Exit total pressure $(p_{t2}) = 6.1$ bars
- Exit total temperature $(T_{t2}) = 533.7$ K
- Exit Mach number $(M_2) = 0.5$
- Impeller speed $(N) = 42{,}000$ rpm

Assuming an adiabatic flow and a γ magnitude of 1.4:

a) Calculate the mass-flow rate $(\dot{m})$.
b) Sketch the impeller inlet and exit velocity diagrams.
c) Calculate ΔT_{t_r} across the impeller.

SOLUTION

Let us first focus on the "easy-to-compute" variables at the impeller's inlet and exit stations:

$$U_1 = \omega r_1 = N\left(\frac{2\pi}{60}\right) r_1 = 263.9 \text{ m/s}$$

$$U_2 = \omega r_2 = 439.8 \text{ m/s}$$

$$T_1 = \frac{T_{t1}}{\left(1 + \frac{\gamma-1}{2} {M_1}^2\right)} = 363.3 \text{ K}$$

$$a_1 = \sqrt{\gamma R T_1} = 382.1 \text{ m/s}$$
$$V_1 = M_1 a_1 = 145.2 \text{ m/s}$$
$$W_1 = \sqrt{U_1^2 - V_1^2} = 220.4 \text{ m/s}$$

In calculating W_1, note that we have made use of the fact that the vector $\mathbf{V}$ is totally in the radial direction (i.e., $\alpha_1 = 0$). With the solid-body rotational vector $\mathbf{U}_1$ being (as always) in the tangential direction, this creates a right-angle inlet velocity triangle.
Part a: Now, we apply the continuity equation at station 1:

$$\frac{\dot{m}\sqrt{T_{t1}}}{p_{t1}\{[2\pi r_1 b]\cos\alpha_1\}} = \sqrt{\frac{\gamma}{R}} M_1 \left(1 + \frac{\gamma - 1}{2} M_1^2\right)^{\frac{(1+\gamma)}{2(1-\gamma)}}$$

where $\alpha_1 = 0$

Direct substitution in this equation yields

$$\dot{m} = 2.29 \text{ kg/s}$$

Part b: Applying the continuity equation at the exit station, we get

$$\alpha_2 = 79.7^\circ$$

Now:

$$T_2 = 462.1 \text{ K}$$
$$a_2 = 430.9 \text{ m/s}$$
$$V_2 = 379.2 \text{ m/s}$$
$$V_{r2} = V_2 \cos\alpha_2 = 62.4 \text{ m/s}$$
$$V_{\theta 2} = V_2 \sin\alpha_2 = 374.0 \text{ m/s}$$
$$W_{r2} = V_{r2} = 62.4 \text{ m/s}$$
$$W_{\theta 2} = V_{\theta 2} - U_2 = -65.8 \text{ m/s}$$
$$W_2 = \sqrt{W_{\theta 2}^2 + W_{r2}^2} = 90.7 \text{ m/s}$$
$$\beta_2 = \arctan\left(\frac{W_{\theta 2}}{W_{r2}}\right) = -46.5^\circ$$

With these results, the impeller-exit velocity diagram will look like that in Figure 3.32.

Part c: We can now calculate both the inlet and exit total relative temperatures as follows:

$$T_{t,r1} = T_{t1} + \frac{W_1^2 - V_1^2}{2c_p} = 391.7 \text{ K}$$
$$T_{t,r2} = T_{t2} + \frac{W_2^2 - V_2^2}{2c_p} = 466.2 \text{ K}$$

Thus, the rise in total relative temperature across the impeller is

$$\Delta T_{tr} = T_{t,r2} - T_{t,r1} = 74.5 \text{ K}$$

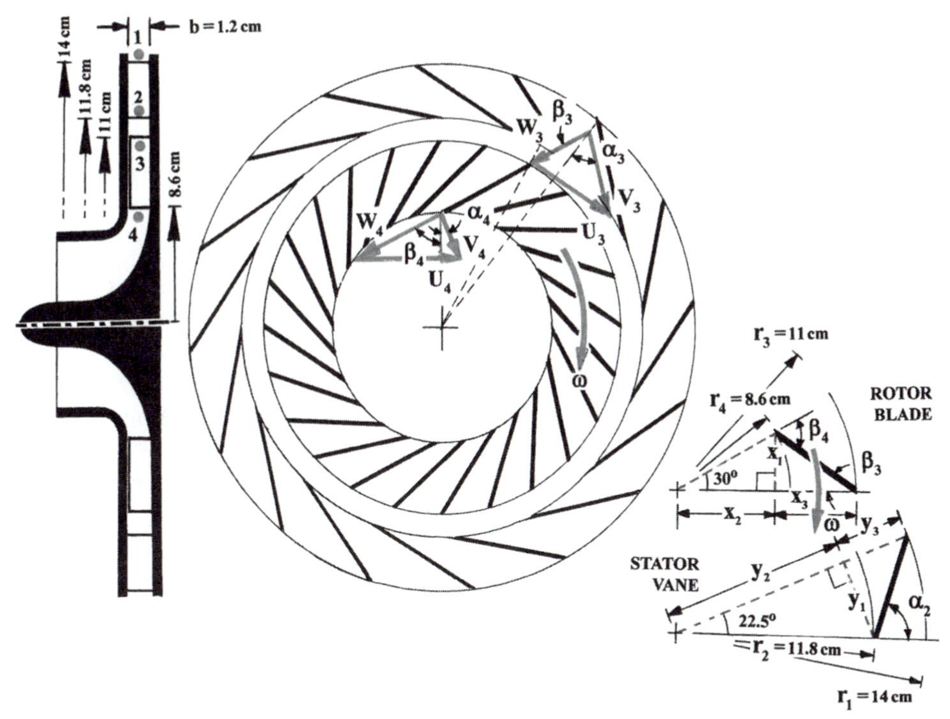

Figure 3.33. Input variables for Example 5.

EXAMPLE 5

Figure 3.33 shows a radial-turbine stage and its major dimensions. As seen in the figure, the stator vanes and rotor blades are all made of flat plates. The stage operating conditions are as follows:

- Inlet total pressure (p_{t1}) = 12.2 bars
- Inlet total temperature (T_{t1}) = 1180 K
- Stator-exit critical Mach number (M_{cr2}) = 0.92
- Rotor-exit radial-velocity component (V_{r4}) = 92.7 m/s
- Stage total-to-total efficiency (η_{t-t}) = 81%

The following information is also provided:

- Adiabatic flow throughout the stage
- Isentropic flow across the stator and stator/rotor gap
- Free-vortex flow structure in the stator/rotor gap
- Specific-heat ratio (γ) constant at 1.33

With this information:

(a) Calculate the mass-flow rate ($\dot{m}$).
(b) Calculate the rotor-inlet critical Mach number (M_{cr3}).
(c) Calculate the shaft speed in rpm.

(d) Calculate the stage-exit swirl angle (α_4).
(e) Using the energy-conversion relationship

$$T_{t4} = T_{t3} + \frac{1}{c_p}(U_3 V_{\theta 3} - U_4 V_{\theta 4})$$

which will be derived and discussed in Chapter 4, calculate:

(i) the rotor-wise change in relative critical Mach number (W/W_{cr});
(ii) the rotor-wise change in total relative pressure (p_{t_r}).

SOLUTION

Part a: Let us first compute the different variables on the isolated impeller blade and stator vane, which are sketched separately in Figure 3.33:

$$\beta_4 = 180^\circ - 60^\circ - \arctan\left(\frac{r_3 - r_4 \cos 30^\circ}{r_4 \sin 30^\circ}\right)$$

$$\beta_3 = -\arctan\left(\frac{r_4 \sin 30^\circ}{3.55}\right) = -50^\circ$$

$$y_2 = r_1 - r_2 \cos(22.5^\circ) = 3.10 \text{ cm}$$

$$\alpha_2 = 180^\circ - 67.5^\circ - \arctan\left[\frac{y_2}{r_2 \sin(22.5^\circ)}\right] = 78.0^\circ$$

$$\alpha_3 = \alpha_2 = 78.0^\circ \text{ (free-vortex gap flow)}$$

Applying the continuity equation at the stator exit station:

$$\frac{\dot{m}\sqrt{1180}}{12.2 \times 10^5 (2\pi 0.1180.012) \cos 78^\circ} = 0.0394$$

which yields

$$\dot{m} = 2.59 \text{ kg/s}$$

Part b: The procedure to calculate M_{cr3} is as follows:

$$V_2 = M_{cr2} V_{cr2} = 572.1 \text{ m/s}$$

$$V_{\theta 2} = V_2 \sin \alpha_2 = 559.6 \text{ m/s}$$

$$V_{r2} = V_2 \cos \alpha_2 = 118.9 \text{ m/s}$$

$$V_{\theta 3} = V_{\theta 2} \times \frac{r_2}{r_3} = 600.3 \text{ m/s}$$

$$V_{r3} = V_{r2} \times \frac{r_2}{r_3} = 127.5 \text{ m/s}$$

$$V_3 = \sqrt{V_{\theta 3}{}^2 + V_{r3}{}^2} = 613.7 \text{ m/s}$$

$$M_{cr3} = \frac{V_3}{V_{cr3}} = V_3 / \sqrt{\left(\frac{2\gamma}{\gamma + 1}\right) R T_{t3}} = 0.987$$

In the last equation, the flow adiabaticity was utilized in setting T_{t3}, T_{t2}, and T_{t1} to be identical.

Part c: Now we calculate the shaft speed (N) as follows:

$$W_{\theta 3} = V_{\theta 3} - U_3$$
$$W_{r3} = V_{r3} = 127.5 \text{ m/s}$$
$$\tan \beta_3 = W_{\theta 3}/W_{r3} = \tan -50.5^\circ$$

which yields

$$U_3 = 752.2 \text{ m/s} \ = \omega \times r_3$$
$$\omega = 6839 \text{ rad/s}$$
$$N = \omega \frac{60}{2\pi} = \ 65{,}300 \text{ rpm}$$

Part d: Referring to the rotor-exit velocity diagram in Figure 3.33, we have

$$U_4 = \omega r_4 = 588.2 \text{ m/s}$$
$$W_{\theta 4} = V_{r4} \tan \beta_4 = -548.1 \text{ m/s}$$
$$\alpha_4 = \arctan\left(\frac{U_4 + W_{\theta 4}}{V_{r4}}\right) = 23.4^\circ$$

Part e: In order to calculate the relative critical Mach numbers (W_3/W_{cr3}) and (W_4/W_{cr4}), we proceed as follows:

$$T_{t4} = T_{t3} - \frac{1}{c_p}(U_3 V_{\theta 3} - U_4 V_{\theta 4}) = 810.0 \text{ K}$$
$$V_4 = \sqrt{{V_{\theta 4}}^2 + {V_{r4}}^2} = 101.0 \text{ m/s}$$
$$W_4 = \sqrt{{W_{\theta 4}}^2 + {W_{r4}}^2} = 555.9 \text{ m/s}$$
$$W_3 = \sqrt{(V_{\theta 3} - U_3)^2 + {V_{r3}}^2} = 198.3 \text{ m/s}$$
$$T_{t,r3} = T_{t3} + \frac{{W_3}^2 - {V_3}^2}{2c_p} = 1034.2 \text{ K}$$
$$W_{cr3} = \sqrt{\left(\frac{2\gamma}{\gamma+1}\right) R T_{t_{r3}}} = 582.1 \text{ m/s}$$
$$W_3/W_{cr3} = 0.341$$

Similarly, we arrive at the following results:

$$T_{t4} = 939.2 \text{ K}$$
$$W_{cr4} = 556.0 \text{ m/s}$$
$$W_4/W_{cr4} = 0.998 (\approx 1.0)$$

The magnitude of the *relative* critical Mach number W_4/W_{cr4} suggests that the rotor passage is choked for all practical purposes. Note that the critical Mach number V_4/V_{cr4} may very well have a low subsonic magnitude. However, when judging the rotor-choking status, it is only the former velocity ratio that counts.

In order to compute the total relative pressures (p_{t,r_3}) and (p_{t,r_4}), the following procedure is perhaps the easiest:

$p_{t3} = p_{t2} = p_{t1}$ (stator and stator/rotor-gap flows are isentropic)

$$\eta_{t-t} = \frac{1 - T_{t4}/T_{t3}}{1 - (p_{t4}/p_{t3})^{\frac{\gamma-1}{\gamma}}}$$

These two equations produce the following results:

$$p_{t3} = 12.2 \text{ bars}$$
$$p_{t4} = 1.70 \text{ bars}$$

Recalling that the total, total relative, and even static states lie on the same vertical (constant-entropy) line on the T-s diagram, we can always relate the properties associated with any two of these states by using isentropic relationships. With this in mind, we can proceed as follows:

$$p_{t,r3} = p_{t3}\left(\frac{T_{t,r3}}{T_{t3}}\right)^{\frac{\gamma}{\gamma-1}} = 7.17 \text{ bars}$$

$$p_{t,r4} = p_{t4}\left(\frac{T_{t,r4}}{T_{t4}}\right)^{\frac{\gamma}{\gamma-1}} = 3.09 \text{ bars}$$

EXAMPLE 6

Figure 3.34 shows the stator/rotor gap in an axial-flow turbine stage, the station designation, and the major dimensions. The gap is a constant-area annular duct with an annulus height (Δr) of 2.2 cm and a mean radius of 11.0 cm. Across the gap, the flow is swirling at a constant swirl angle (α) of 78°, with a friction coefficient (f) of 0.05. The gap-inlet magnitudes of total pressure and total temperature are 7.6 bars and 782 K, respectively. Knowing that the gap inlet and exit Mach numbers are 0.70 and 0.72, respectively, calculate:

(a) The mass-flow rate $(\dot{m})$;
(b) The percentage of total pressure decline across the gap;
(c) The annulus-height shrinkage as a result of displacement-thickness growth.

SOLUTION

Part a: Using expression (3.72), we can calculate the variables L_{c1} and L_{c2} corresponding to the axial-gap inlet and exit Mach numbers as follows:

$$L_{c1} = L)_{M=0.70} = 0.223\frac{D_h}{f}$$

$$L_{c2} = L)_{M=0.72} = 0.184\frac{D_h}{f}$$

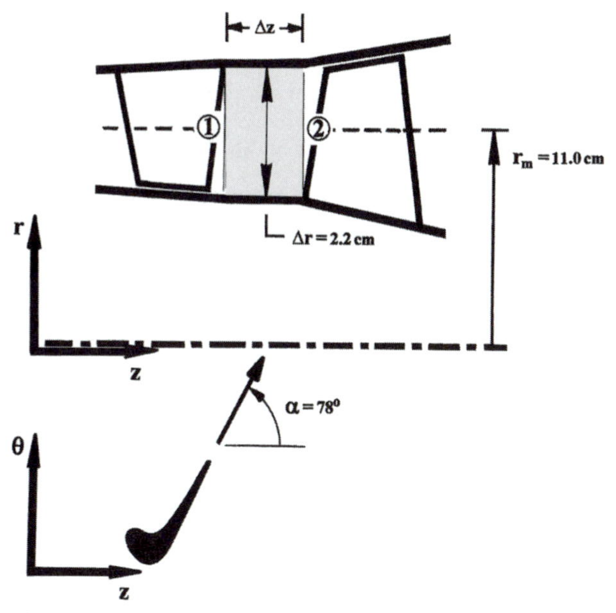

Figure 3.34. Input variables for Example 6.

where the hydraulic diameter (D_h) and friction coefficient (f) are

$$D_h = 2\Delta r = 4.4 \text{ cm}$$

$$f = 0.05 \text{ (given)}$$

This gives rise to the following:

$$L_{c1} = 0.1967 \text{ m}$$
$$L_{c2} = 0.1627 \text{ m}$$

Now the stator/rotor gap length (Δz) can be computed as follows:

$$\frac{\Delta z}{\cos\alpha} = L_{c1} - L_{c2} = 0.034 \text{ m, where } \alpha = 78^\circ$$

which yields

$$\Delta z = 0.71 \text{ cm}$$

Part b: Applying the continuity equation (3.41) at the gap inlet station, we get

$$\dot{m} = 2.87 \text{ kg/s}$$

Part c: The total-to-total pressure ratio across the gap is provided in expression (3.70), which gives

$$\frac{p_{t2}}{p_{t1}} = 0.987$$

Knowing that $p_{t1} = 7.6$ bars, we can calculate the decline in total pressure as follows:

$$\frac{\Delta p_t}{p_{t1}} = 1.3\%$$

Part d: Referring to Figure 3.32, let us refer to the gap-exit effective annulus height by h_2. This variable encompasses the blockage effect resulting from the displacement-thickness growth over the endwalls. The gap-exit cross-flow area with this definition can be expressed as

$$A_2 = 2\pi r_m h_2 \cos 78°$$

Substituting this into the continuity equation (3.36), we get

$$h_2 = 2.19 \text{ cm}$$

Now we can express the annulus-height shrinkage caused by the boundary layer buildup as follows:

$$\frac{\Delta h}{h_1} \approx 0.45\%$$

EXAMPLE 7

The meridional projection of an axial-flow compressor stator is shown in Figure 3.35 along with the major dimensions. The stator, which has a purely axial flow path, is operating under the following conditions:

- Mass-flow rate ($\dot{m}$) = 8.59 kg/s
- Inlet total pressure (p_{t1}) = 5.1 bars
- Inlet total temperature (T_{t1}) = 507.0 K
- Airfoil (or metal) exit angle ($\alpha_2{}'$) = 18°
- Flow deviation (or underturning) angle (ϵ) = 11°
- Kinetic-energy loss coefficient ($\bar{e}$) = 0.075

Assuming an adiabatic flow through the stator, and a constant specific-heat ratio (γ) of 1.4, calculate the streamwise rise or decline of the following variables:

(a) The total pressure (p_t);
(b) The static pressure (p);
(c) The static density (ρ);
(d) The axial velocity component (V_z).

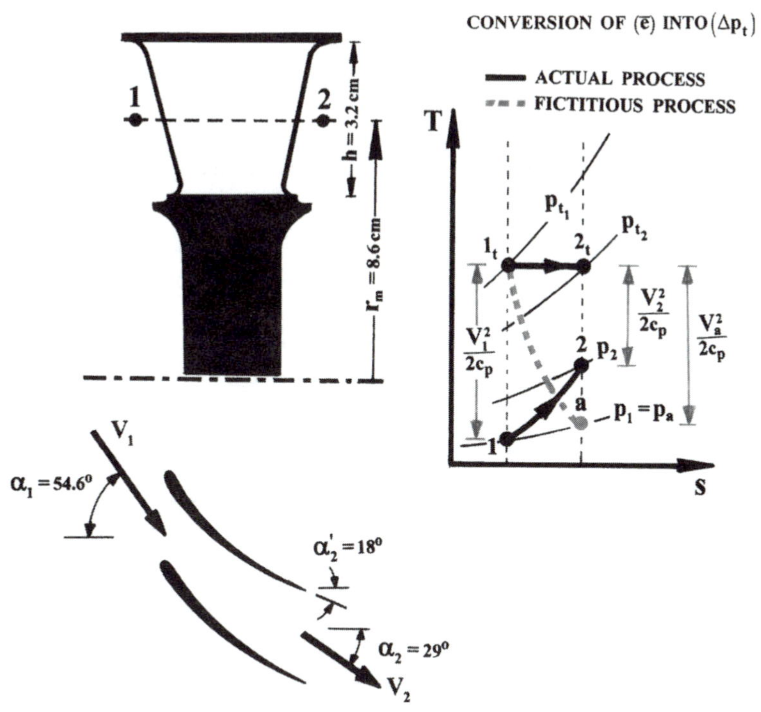

Figure 3.35. Input variables for Example 7.

SOLUTION

This particular problem underscores two important facts:

(1) A purely axial flow path (meaning horizontal hub and casing lines) will strongly (and undesirably) impact the through-flow (or axial) velocity component (V_z). This is at odds with a traditional design principle calling for a streamwise-constant V_z.

(2) Although a loss coefficient such as $\bar{e}$ can lead to such variables as the stator-wise decline in total pressure (Δp_t), it is hardly as critical when it comes to computing the static-pressure change (Δp). The latter is much more influenced by the streamwise vane-to-vane, hub-to-casing cross-flow area variation. This point is illustrated on the *T-s* diagram in Figure 3.35.

Let us first find the stator-inlet critical Mach number ($M_{cr,1} = V_1/V_{cr1}$) by applying the continuity equation at this station:

$$\frac{\dot{m}\sqrt{T_{t1}}}{p_{t1}A_1} = \sqrt{\frac{2\gamma}{(\gamma+1)R}}M_{cr1}\left(1 - \frac{\gamma-1}{\gamma+1}M_{cr,1}{}^2\right)^{\frac{1}{\gamma-1}}$$

where

$$A_1 = (2\pi r_m h_1)\ \cos\alpha_1$$

As we examine the two preceding expressions and the given data, we find that $M_{cr,1}$ is the only unknown for us to calculate. Unfortunately, the continuity equation is non-linear as far as $M_{cr,1}$ is concerned. A simple trial-and-error procedure is followed in this case, giving rise to the following final result:

$$M_{cr,1} \approx 0.77$$

It follows that

$$V_1 = M_{cr,1} V_{cr,1} = 317.3 \text{ m/s}$$
$$V_{z1} = V_1 \cos\alpha_1 = 183.8 \text{ m/s}$$

The next step is to calculate p_{t2} or, equivalently, the total pressure loss across the stator. In doing so, and by reference to Figure 3.35, note that the state a (defined by its static properties) is of our own creation and is clearly different from the static state 2, the actual stator-exit state. To explain this, and in conformity with the definition of $\bar{e}$, we are to visualize a *fictitious* process extending from state 1_t to another state that, by definition, shares the same inlet static pressure (p_1), with the horizontal intercept (i.e., Δs) being precisely identical to the entropy production associated with the actual process. To comprehend this rationale, it is probably beneficial to refer back to the $\bar{e}$ definition in Figure 3.24. Note that p_a will be considerably less than the actual static pressure p_2 (see Figure 3.35), for it is the cross-flow area divergence that dictates the exit/inlet static-pressure ratio.

By definition, we have

$$\bar{e} = 1 - \frac{{V_{act.}}^2}{{V_{id.}}^2} = 1 - \frac{{V_a}^2}{{V_1}^2}$$

which yields

$$V_a = 305.2 \text{ m/s}$$

Noting that $T_{t2} = T_{t1} = T_{ta}$ (adiabatic flow process), we have

$$T_a = T_{t1} - \frac{{V_a}^2}{2c_p} = 460.6 \text{ K}$$

In order to compute p_{t2}, we should first recognize the fact that states 2_t and a both lie on a vertical line (see the T-s diagram in Figure 3.35). Thus, we can indeed use the isentropic relationships to relate the properties, at these two states, to one another. Noting that $p_a = p_1$ (by definition), we can easily arrive at the following:

$$\frac{p_{t2}}{p_a} = \left(\frac{T_{t2}}{T_a}\right)^{\frac{\gamma}{\gamma-1}}$$
$$p_{t2} = 4.95 \text{ bars}$$

Now, we can calculate the total pressure loss across the stator:

$$\frac{\Delta p_t}{p_{t1}} = 2.94\%$$

Now we move to calculate the actual critical Mach number at the stator exit station as follows:

$$\frac{\dot{m}\sqrt{T_{t2}}}{p_{t2}[(2\pi r_m h)\cos\alpha_2]} = \sqrt{\frac{2\gamma}{(\gamma+1)R}} M_{cr2}\left(1 - \frac{\gamma-1}{\gamma+1} M_{cr2}{}^2\right)^{\frac{1}{\gamma-1}}$$

With ϵ being the vane-exit deviation angle, the exit flow angle, α_2, can be expressed as

$$\alpha_2 = \alpha_2{}' + \epsilon = 29^\circ$$

Direct substitution in the preceding continuity equation yields the following compact equation:

$$M_{cr2}(1 - 0.1667 M_{cr2}{}^2)^{2.5} = 0.4052$$

We are confronted, once again, with a nonlinear equation in M_{cr2}. A trial-and-error approach in this case gives rise to

$$M_{cr2} \approx 0.44$$

Now, we can calculate virtually all other properties at the stator exit station, as follows:

$$p_2 = p_{t2}\left(1 - \frac{\gamma-1}{\gamma+1} M_{cr2}{}^2\right)^{\frac{\gamma}{\gamma-1}} = 4.41 \text{ bars}$$

$$V_2 = M_{cr2} V_{cr2} = 181.3 \text{ m/s}$$

$$V_{z2} = V_2 \cos\alpha_2 = 158.6 \text{ m/s}$$

$$\rho_2 = \frac{p_{t2}}{RT_{t2}}\left(1 - \frac{\gamma-1}{\gamma+1} M_{cr2}{}^2\right)^{\frac{1}{\gamma-1}} = 3.13 \text{ kg/m}^3$$

Let us now calculate the same variables at the stator inlet station (recall that $M_{cr1} = 0.77$):

$$p_1 = 3.54 \text{ bars}$$

$$V_{z1} = 183.8 \text{ m/s}$$

$$\rho_1 = 2.70 \text{ kg/m}^3$$

Finally, we can calculate the streamwise change of properties across the stator, as follows:

$$\text{decline in total pressure} = \frac{p_{t1} - p_{t2}}{p_{t1}} = 2.94\%$$

$$\text{rise in static pressure} = \frac{p_2 - p_1}{p_1} = 24.6\%$$

$$\text{rise in static density} = \frac{\rho_2 - \rho_1}{\rho_1} = 15.9\%$$

$$\text{decline in axial-velocity component} = \frac{V_{z1} - V_{z2}}{V_{z1}} = 13.7\%$$

Exhaust Diffusers

A category of gas-turbine engines, namely turboshaft engines, will normally end with an annular exhaust diffuser, as opposed to a nozzle in pure propulsion applications. This engine category encompasses all power systems, as well as turboprop engines. The common feature here is that the engine output is in the form of shaft power. Figure 3.11 shows such a diffuser as the exhaust passage of a commercial turboprop engine. The computationally attained data provided in this figure prove but one important fact: that there is a performance penalty to pay for excessive flow swirl, particularly at the endwalls. Worse yet is the likelihood of boundary-layer separation, leading to the most damaging exit flow behavior, namely flow reversal, as shown in Figure 3.11.

It is not at all odd to see a turbine-exit static pressure that is less than the ambient pressure. It is therefore the critical task of the exhaust diffuser to efficiently convert the maximum amount of inlet kinetic energy into a static pressure rise. What this means is that a lossy diffuser flow field will lead to a total-pressure decrement and, subsequently, place a low ceiling on the exit static pressure. Quantifying the diffuser performance in this case is the so-called static-pressure recovery coefficient, C_p, which is defined next.

Static-Pressure Recovery Coefficient (C_p)

This is simply the ratio between the actual static pressure rise and the maximum possible magnitude, namely the inlet dynamic pressure,

$$C_p = \frac{(p_{ex} - p_{in})}{(p_{t\,in} - p_{in})} = \frac{\Delta p}{\frac{\gamma}{2} p_{in} {M_{in}}^2}$$

Figure 3.36 shows the variation of C_p with the inlet swirl angle, in a turbulent flow environment, for the same commercial turboprop engine as in Figure 3.11. Figure 3.36 reveals the fact that the onset of hub-surface boundary-layer separation occurs at an inlet swirl angle of approximately 18°, far from the design-point magnitude, which is roughly 6.5°.

Annular Diffuser Design

One of the most important tools in the preliminary design process for the diffuser is the graph shown in Figure 3.37, which is known as the Sovran and Klomp chart (Sovran and Klomp 1965). Clear on this chart are two straight lines: one corresponding to the maximum value of C_p for a given diffuser length, and the other yielding the minimum total pressure loss. In reality, however, a typical diffuser designer would construct a straight line midway between the two lines as the locus of "optimum" diffuser performance. The dark circle on Figure 3.36 represents the same exhaust diffuser in Figure 3.11.

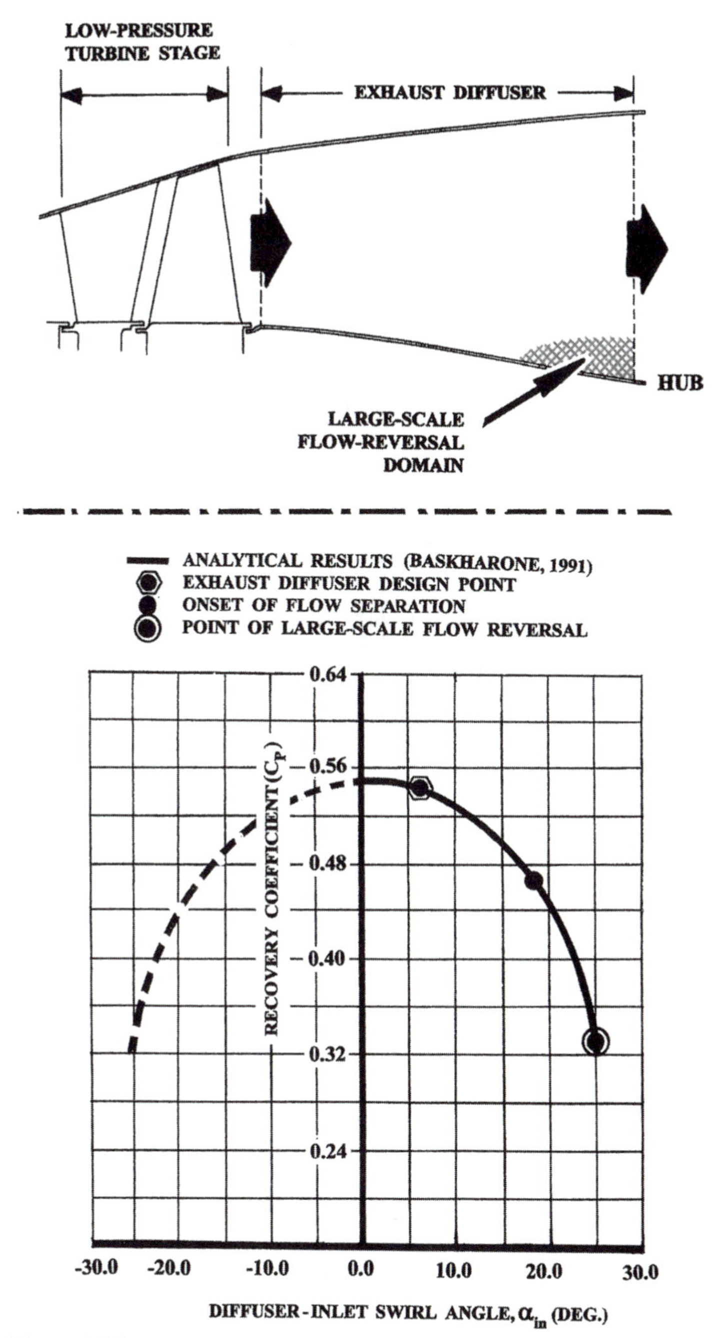

Figure 3.36. Dependence of the exhaust-diffuser recovery coefficient on the inlet swirl angle.

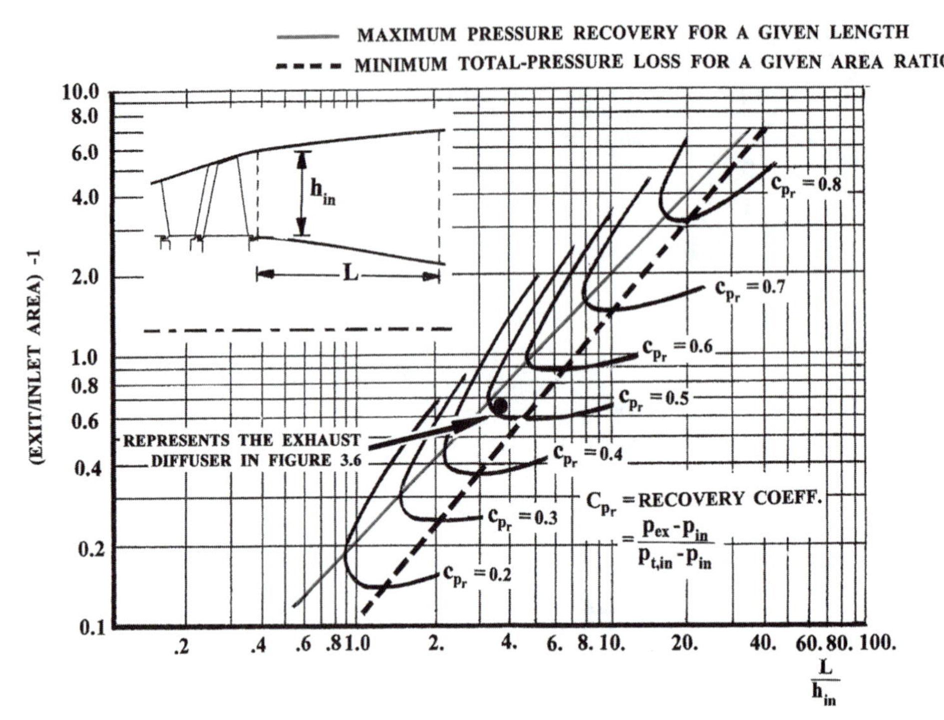

Figure 3.37. Sovran and Klomp annular-diffuser design chart.

A seldom-used definition of the total-to-static efficiency is based on the turbine/exhaust diffuser combination. The efficiency definition in this case is the same as that in expression 3.55, with the understanding that p_{ex} will now refer to the diffuser-exit static pressure. Figure 3.38 shows a more or less linear relationship between this

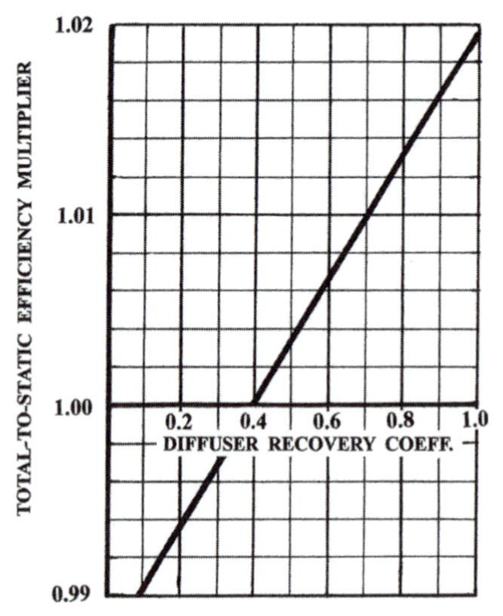

Figure 3.38. Effect of the exhaust-diffuser pressure recovery on the system total-to-static efficiency.

efficiency (as it relates to the low-pressure turbine/exhaust diffuser system) and the diffuser's recovery coefficient. Again, the figure pertains to the commercial turboprop engine in Figure 3.11.

Definition of the Momentum Thickness

Besides the displacement thickness (discussed within the context of the Fanno-flow process), there exists another boundary-layer variable that is equally useful, namely the momentum thickness. This variable quantifies the loss in linear momentum as a result of the loss in mass flux within the boundary layer. Represented by the letter θ, the momentum thickness is defined as

$$\theta = \int_0^\delta \frac{\rho V}{\rho_e V_e}\left(1 - \frac{V}{V_e}\right) dy \tag{3.74}$$

Added to the Reynolds number and geometry-related variables, the momentum thickness is frequently part of loss-correlating expressions (Chapters 8 and 10). These are commonly focused on the profile (or friction-related) loss in such variables as the total pressure over cascades of lifting bodies. In the following, some analytical and empirical expressions approximating different boundary-layer variables are provided. The need for these expressions arises, in particular, during the preliminary phases of the design procedure.

In the following, closed-form expressions for the boundary-layer variables are presented. These are commonly based on a flat-plate type of analogy. The expressions are "loosely" adapted to airfoils through, for instance, replacing the distance along the solid surface by that along the mean camber line. Recalling that a flat plate is aerodynamically one that is exposed to a zero pressure gradient, one should be careful not to use the results in the final design phase. Accurate calculations of these variables is possible through numerical modeling.

Laminar Boundary Layer

For a Reynolds number magnitude that is less than 2×10^5 (based on the distance along the mean camber line and the boundary-layer-edge velocity), the flow field is categorized as laminar. In this case, a simple solution of the viscous-flow governing equations by Blasius provides magnitudes for the boundary layer, displacement, and momentum thicknesses as follows:

$$\delta = 5.0\sqrt{\frac{\nu x}{V_e}} \tag{3.75}$$

$$\delta^* = 1.72\sqrt{\frac{\nu x}{V_e}} \tag{3.76}$$

$$\theta = 0.664\sqrt{\frac{\nu x}{V_e}} \tag{3.77}$$

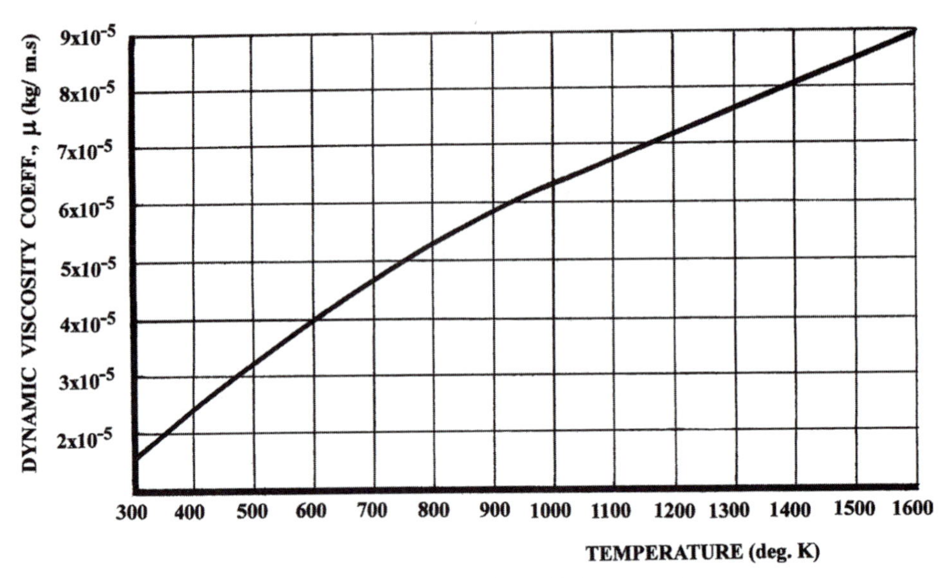

Figure 3.39. Variation of the air dynamic viscosity at high temperatures.

where the subscript e refers to the boundary-layer edge, x is the streamwise distance along the solid surface (mean camber line for airfoils), and ν is the kinematic viscosity coefficient. As is well-known, the latter can be expressed as

$$\nu = \frac{\mu}{\rho}$$

where μ is the dynamic viscosity coefficient. The problem with this coefficient is that its high-temperature magnitudes are virtually nonexistent in the traditional fluid dynamics textbooks. Figure 3.39 shows the variation of "μ" with temperature within a high (turbine-suited) temperature range. As for the boundary-layer velocity profile, the following expression is almost universal in the case of a laminar boundary layer:

$$\frac{V}{V_e} = \left(\frac{y}{\delta}\right)^{\frac{1}{5}} \tag{3.78}$$

where y is the distance perpendicular to the solid surface, and δ is the boundary-layer thickness. In expression (3.74), the boundary-layer edge is defined to exist at the y magnitude, where the nondimensional velocity ratio V/V_e reaches a value of 0.99.

Turbulent Boundary Layer

As the Reynolds number significantly exceeds 2×10^5 (based on the same preceding variables), the flow field is said to be turbulent. In this case, equations (3.71) through (3.74) are replaced by the following set of relationships:

$$\delta = 0.23x\left(\frac{V_e x}{\nu}\right)^{-\frac{1}{6}} \tag{3.79}$$

$$\delta^* = 0.057x\left(\frac{V_e x}{\nu}\right)^{-\frac{1}{6}} \tag{3.80}$$

$$\theta = 0.022x\left(\frac{V_e x}{\nu}\right)^{-\frac{1}{6}} \tag{3.81}$$

$$\frac{V}{V_e} = \left(\frac{y}{\delta}\right)^{\frac{1}{7}} \tag{3.82}$$

Expressions (3.79) through (3.82) are no more than rough approximations which, in latter chapters, will be beneficial in a first-order estimation of losses over airfoil cascades, in particular. Achieving exact closed-form expressions in this case is virtually unthinkable. Note that expressions (3.75) through (3.82) are also applicable to the (hub and casing) endwalls. In this case, the distance x will have to be that along the endwall flow trajectory, which is where the swirl angle (α) gets into the picture.

EXAMPLE 8

The following is a simplified version of a well-known kinetic-energy loss coefficient formula within a turbine stator:

$$\bar{e}_s = E\left(\frac{\theta_{tot.}}{S_m \cos\alpha_2 - t_{t.e.} - \delta^*_{tot.}}\right)\left(1 + \frac{S_m \cos\alpha_{av.}}{h_{av.}}\right)$$

where

- The energy factor $(E) = 1.8$
- S_m is the mean-radius vane-to-vane spacing
- $t_{t.e.}$ is the trailing-edge thickness
- h refers to the annulus height
- δ^* is the boundary-layer displacement thickness
- θ is the boundary-layer momentum thickness

In this expression, the subscript *av.* refers to the arithmetic average of the thermophysical property. The subscript *tot.*, however, stands for the combined (pressure- + suction-side) boundary-layer variables over the stator vane.

In this example, we recognize the fact (stated and discussed in Chapter 8) that the suction-side displacement and momentum thicknesses are much larger than those over the pressure side. Assuming the blade-surface boundary layer is laminar (on both sides of the airfoil), we will adopt expressions (3.80) and (3.81) as applicable to computing $\delta_{press.}{}^*$ and $\theta_{press.}$ (i.e., over the airfoil's pressure side). As for the suction side, we will multiply both variables by a factor of 3.5. In computing such variables, the distance x (in equations (3.76) and (3.77) will be that along the mean camber line, which is highlighted in Figure 3.40.

Referring to the axial-flow turbine stator in Figure 3.40, the following operating conditions are applicable:

- Inlet total pressure $(p_{t1}) = 11.3$ bars
- Inlet total temperature $(T_{t1}) = 1322.0$ K
- Inlet critical Mach number $(M_{cr1}) = 0.42$

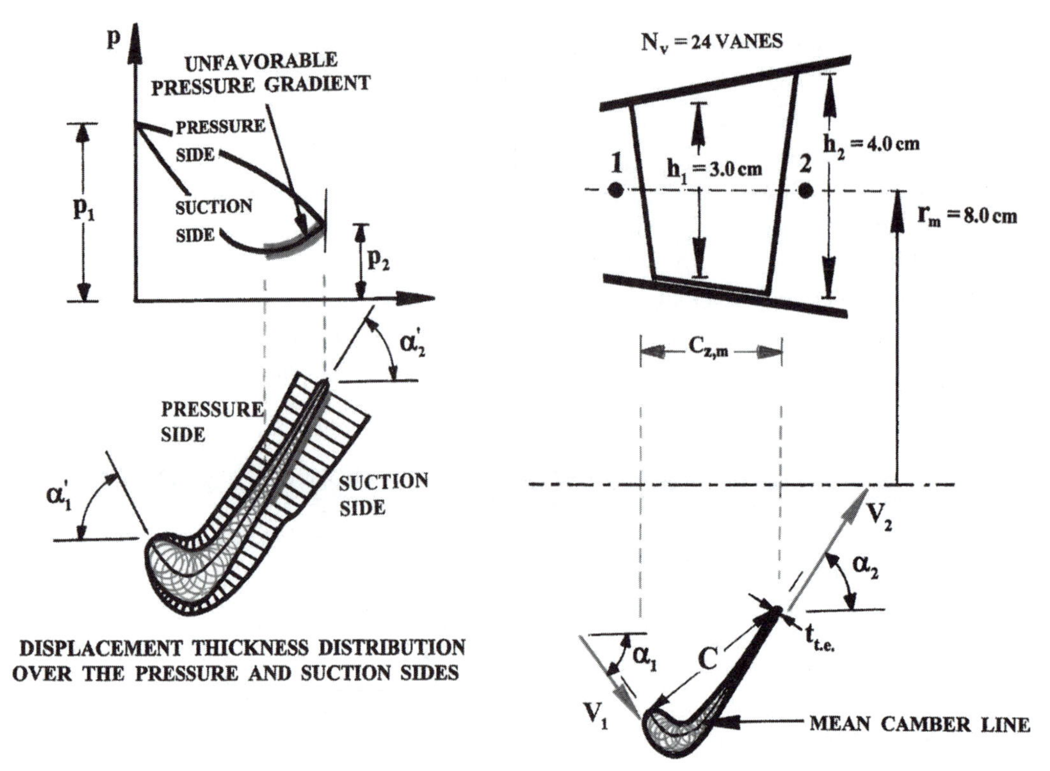

Figure 3.40. Input variables for Example 8.

- Inlet swirl angle $(\alpha_1) = 39.0$
- Stator is choked (i.e., $M_{cr2} = 1.0$)
- Flow process is adiabatic
- Average specific-heat ratio $(\gamma) = 1.33$

The following information also pertains to the airfoil cascade:

- Number of stator vanes $(N_v) = 24$
- Vane axial chord $(C_z) = 2.5$ cm
- Vane true chord $(C) = 4.6$ cm
- Mean camber-line length $(L) = 9.2$ cm
- Trailing-edge thickness $(t_{t.e.}) = 2.2$ mm

I) Starting with a total-pressure loss coefficient $(\frac{\Delta p_t}{p_{t1}})$ of 6.0%, calculate:
   Ia) The mass-flow rate $(\dot{m})$;
   Ib) The stator-exit swirl angle (α_2).
   Ic) The rise in axial velocity, ΔV_z, across the stator.
II) Using the expression for the kinetic-energy loss coefficient $\bar{e}$, calculate the total pressure loss across the stator.

SOLUTION

Part Ia: Let us apply the continuity equation at the stator inlet station:

$$\frac{\dot{m}\sqrt{T_{t1}}}{p_{t1}(2\pi r_m h_1 \cos\alpha_1)} = \sqrt{\frac{2\gamma}{(\gamma+1)R}} M_{cr1}\left(1 - \frac{\gamma-1}{\gamma+1}{M_{cr1}}^2\right)^{\frac{1}{\gamma-1}}$$

Direct substitution in this equation yields

$$\dot{m} = 9.03 \text{ kg/s}$$

Part Ib: As provided in the problem statement,

$$\frac{p_{t1} - p_{t2}}{p_{t1}} = 0.06$$

It follows that

$$p_{t2} = 10.62 \text{ bars}$$

$$T_{t2} = T_{t1} = 1322 \text{ K (adiabatic stator flow)}$$

Applying the same continuity-equation version at the stator exit station, we get

$$\alpha_2 = 67.8^\circ$$

Part Ic:

$$V_1 = M_{cr1} V_{cr1} = 276.4 \text{ m/s}$$

$$V_2 = V_{cr2} = V_{cr1} = 658.1 \text{ m/s} \quad \text{(choked stator)}$$

$$V_{z1} = V_1 \cos\alpha_1 = 214.8 \text{ m/s}$$

$$V_{z2} = V_2 \cos\alpha_2 = 248.7 \text{ m/s}$$

$$\frac{\Delta V_z}{V_{z1}} = 15.8\%$$

Part II: In order to compute $\bar{e}$ using the given expression, we first calculate all of the independent variables appearing in this expression:

$$\alpha_{av.} = 0.5(\alpha_1 + \alpha_2) = 14.3^\circ$$

$$S_m = \frac{2\pi r_m}{N_v} = 2.15 \text{ cm}$$

$$\alpha_2 = 67.8^\circ \text{ (computed earlier)}$$

$$h_{av.} = 0.5(h_1 + h_2)$$

$$T_1 = T_{t1} - \frac{{V_1}^2}{2c_p} = 1289 \text{ K}$$

$$\mu_1 = 7.6 \times 10^{-5} \text{ (Fig. 3.39)}$$

$$\rho_1 = \frac{p_{t1}}{RT_{t1}}\left(1 - \frac{\gamma-1}{\gamma+1}{M_{cr1}}^2\right)^{\frac{1}{\gamma-1}} = 2.75 \text{ kg/m}^3$$

$$\nu_1 = \mu_1/\rho_1 = 2.76 \times 10^{-5}\ \text{m}^2/\text{s}$$

$$T_2 = 1134.8\ \text{K}$$

$$\mu_2 = 6.85 \times 10^{-5}$$

$$\rho_2 = 1.76\ \text{kg/m}^3$$

$$\nu_2 = 3.89 \times 10^{-5}\ \text{m}^2/\text{s}$$

$$\nu_{av.} = 3.325 \times 10^{-5}\ \text{m}^2/\text{s}$$

$$\delta^*{}_{press.} = 1.72\sqrt{\frac{\nu_{av.} L}{V_2}} = 0.12\ \text{mm [expression (3.76)]}$$

$$\delta^*{}_{suc.} = 3.5\delta_{press.}{}^* = 0.41\ \text{mm}$$

$$\delta^*{}_{tot.} = 0.53\ \text{mm}$$

$$\theta_{press.} = 0.664\sqrt{\frac{\nu_{av.} L}{V_2}} = 0.045\ \text{mm [equation (3.77)]}$$

$$\theta_{suc.} = 3.5\theta_{press.} = 0.158\ \text{mm}$$

$$\theta_{tot.} = 0.203\ \text{mm}$$

$$t_{t.e.} = 2.2\ \text{mm (given)}$$

Substituting these variables into the expression provided for $\bar{e}_s$, we get

$$\bar{e}_s = 0.108$$

Conversion of $\bar{e}_s$ into a total pressure loss: By definition,

$$\bar{e}_s = 0.108 = 1 - \frac{V_{act,1}^2}{V_{id,1}^2}$$

where

$$V_{id.,1} = V_1 = 276.4\ \text{m/s}$$

Thus

$$V_{act.,1} = 261.0\ \text{m/s}$$

Also

$$T_{act.,1} = T_{t1} - \frac{V_{act.,1}{}^2}{2c_p} = 1292.5\ \text{K}$$

Referring to Figure 3.35, we observe that the mere definition of $\bar{e}$ requires that

$$p_{act.1} = p_1$$

and

$$T_{t2} = T_{t1}$$

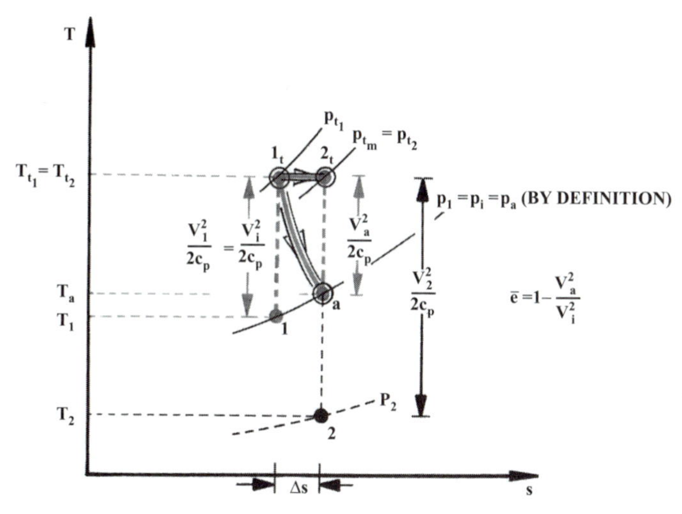

Figure 3.41. Relationship between the kinetic-energy loss coefficient and the stator-exit velocity.

Now, we can calculate the entropy production involved:

$$\Delta s = c_p \ \ln \frac{T_{act.,1}}{T_1} - R \ \ln \frac{p_{act.,1}}{p_1} = 3.175 \text{ J/(kg K)}$$

Referring to Figure 3.41, we can re-express the entropy production (Δs) on a total-to-total basis as

$$\Delta s = 5.997 \text{ J/(kg K)} = c_p \ \ln \frac{T_{t2}}{T_{t1}} - R \ \ln \frac{p_{t2}}{p_{t1}}$$

which, upon recognizing the fact that $T_{t2} = T_{t1}$, yields

$$p_{t2} = 11.17 \text{ bars}$$

Now, we can calculate the actual total pressure loss:

$$\frac{\Delta p_t}{p_{t1}} = \frac{p_{t1} - p_{t2}}{p_{t1}} = 1.10\%$$

Note that the computed magnitude happens to be much less than the value of 6%, which was assumed in Part (I) of the problem.

PROBLEMS

1) A simplified method of analyzing cascade flows away from solid surfaces is to cast the flow-governing equations in the s-θ frame of reference as shown in Figure 3.42, where s is the distance along the stream tube in the meridional view. In this case, the stream tube thickness [$b = b(s)$] and radius [$r = r(s)$] are introduced in the governing equations for the purpose of generality. The upper and lower sides of the stream tube are assumed to be surfaces of revolution.

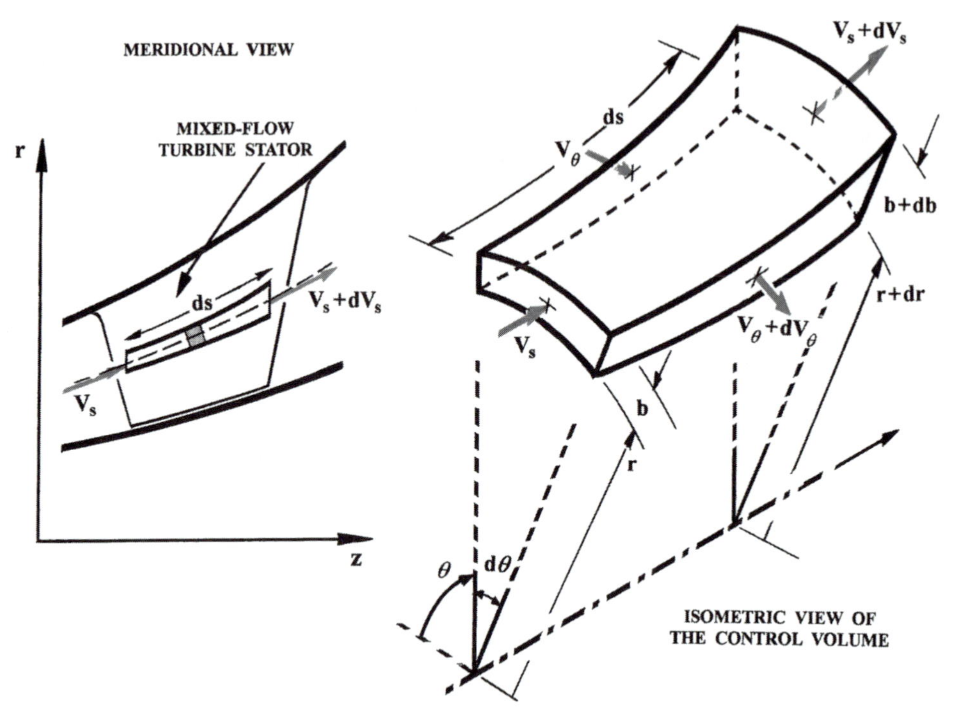

Figure 3.42. Geometry of a stream filament.

Considering the darkened fluid element that is magnified in the figure, and assuming a constant static density (ρ), prove that the differential form of the continuity equation in this case is

$$\frac{1}{b}\frac{\partial}{\partial s}(rbV_s) + \frac{\partial V_\theta}{\partial \theta} = 0$$

2) Figure 3.43 shows the vane-to-vane passage of a high-pressure-turbine first-stage stator, that has a mean radius (r_m) of 7.8 cm, and an exit annulus height (h_1) of 3.8 cm. The stator operating conditions are as follows:

- Total pressure (p_{t0}) = 10 bars
- Total temperature (T_{t0}) = 1433 K
- Stator-exit swirl angle (α_1) = 67°

In addition, the exit critical Mach number (V/V_{cr1}) = 0.75.

Assuming an isentropic stator flow:

(a) Calculate the mass-flow rate through the stator.

(b) If the stator vanes are rotated closed (as shown in the figure) in such a way as to choke the stator, calculate the angle of rotation ($\Delta\alpha$) assuming a fixed mass-flow rate and fixed inlet conditions.

3) Referring back to Example 6, *recalculate* the annulus-height shrinkage across the axial gap, with the difference being in the gap's friction coefficient (f).

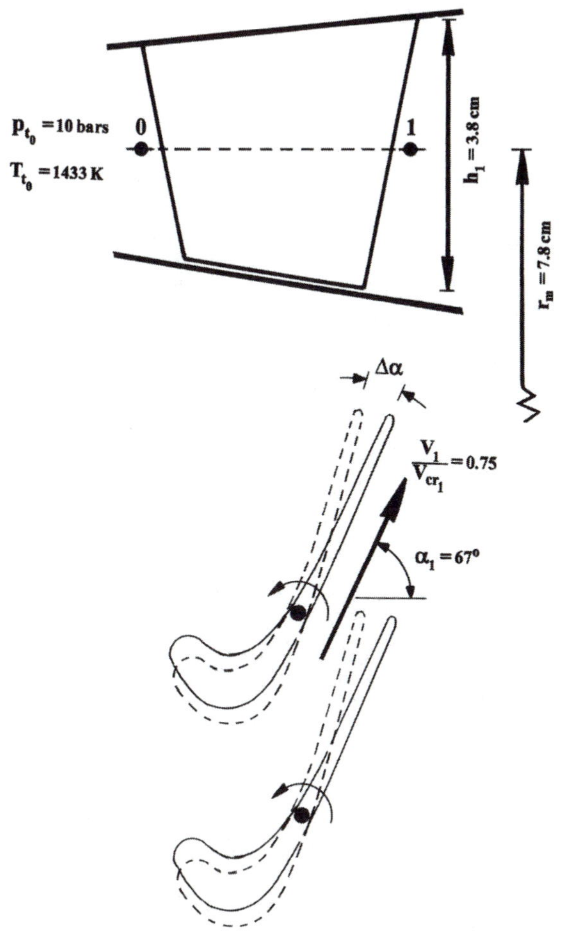

Figure 3.43. Input variables for Problem 2.

This coefficient is now 0.11 instead of 0.05. The gap axial length remains at the 0.71 cm magnitude previously calculated in Example 6 and so will the mass-flow rate ($\dot{m} = 2.87$ kg/s). Note that the gap-exit Mach number, being influenced by the new magnitude of f, will now be different.

4) Figure 3.44 shows the stator of a radial-inflow turbine. The flow across the darkened radial nozzle, downstream from the vaned stator, is assumed to be isentropic, incompressible, and axisymmetric. The flow angle (α), measured from the radial direction, is roughly constant over the entire unvaned nozzle. Considering the data provided in this figure, and applying the mass and energy conservation principles, calculate the following:

   a) The nozzle exit velocity (V_2);
   b) The nozzle-exit static temperature (T_2);
   c) The mass-flow rate ($\dot{m}$) through the stator.

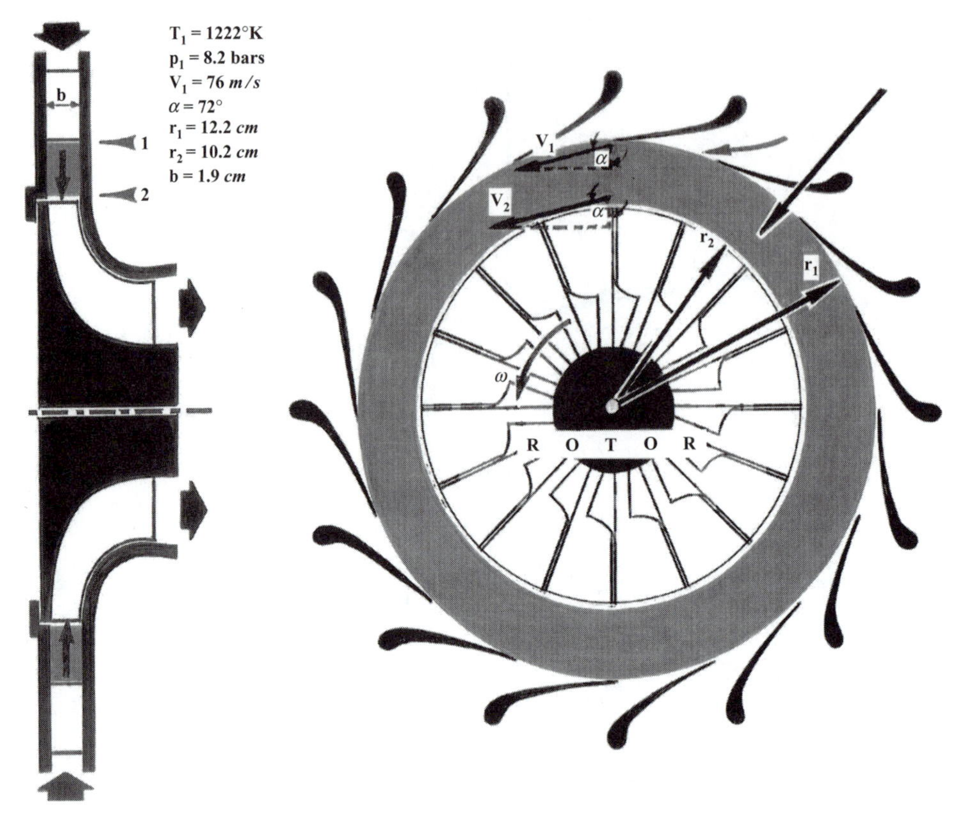

Figure 3.44. Stator/rotar vaneless gap and input variables.

5) Figure 3.45 shows an axial-flow compressor stage, which has a mean radius (r_m) of 10 cm, a stator-exit flow angle (α_1) of zero, and the rotor-inlet airfoil (or metal) angle that is shown in the figure. The following operating conditions also apply:

   - Stage-inlet total temperature (T_{t0}) = 446 K
   - Rotor incidence angle (i_R) = $-8°$
   - Stator-exit static temperature (T_1) = 410 K
   - Stator-exit static pressure (p_1) = 3.5 bars

   Assuming an average specific-heat ratio (γ) of 1.4,

   a) Calculate the shaft speed (N) in rpm; and
   b) Assuming a total pressure loss across the stator of 6%, calculate the stage-inlet total pressure (p_{t0}).

6) Recalculate all of the required variables in Example 8 upon considering the following changes:

   - h_1 = 2.6 cm (instead of 3.0 cm)
   - h_2 = 3.75 cm (instead of 4.0 cm)

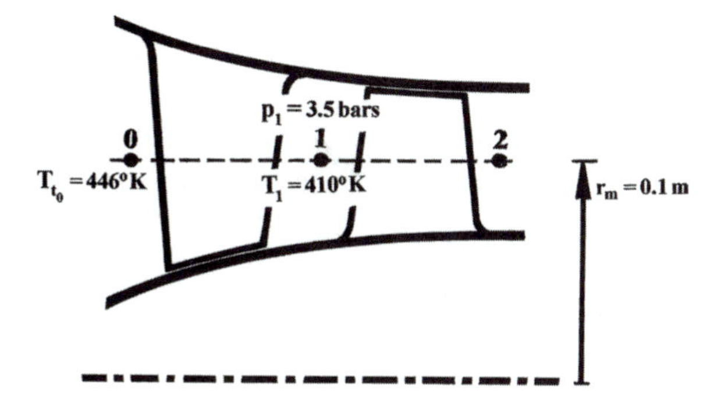

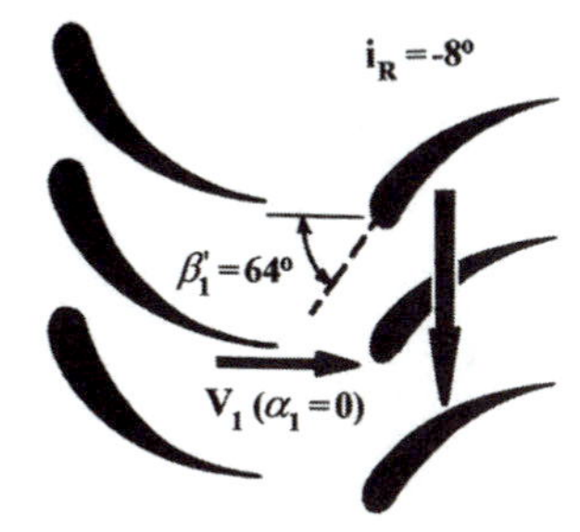

Figure 3.45. Axial-flow compressor stage and input variables.

7) Recalculate all three requirements in Example 6 upon making the following change:

   - Friction coefficient $(f) = 0.075$ (instead of 0.05)

   Compare your results with those obtained in Example 6 and Problem (3).

8) Recalculate all four requirements in Example 7 upon replacing the kinetic-energy loss-coefficient magnitude of 0.075 (last line of the input data) with the following line:

   - Total-pressure loss coefficient $(\bar{\omega}) = 6.2\%$

9) Figure 3.46 shows a radial-inflow turbine stage and its major dimensions. The stage operating conditions are as follows:

   - Rotor speed $(N) = 46{,}000$ rpm
   - Stage-inlet total pressure $(p_{t0}) = 12.8$ bars
   - Stage-inlet total temperature $(T_{t0}) = 1384$ K
   - Stator flow field is considered isentropic
   - Stator-exit critical Mach number $(M_{cr1}) = 0.91$
   - Stator kinetic-energy loss coefficient $(\bar{e}) = 0.06$
   - Rotor-inlet relative-flow angle $(\beta_1) = 0$
   - Rotor-exit total temperature $(T_{t2}) = 1065$ K
   - Rotor-exit critical Mach number $(M_{cr2}) = 0.46$
   - Total-to-static efficiency $(\eta_{t-s}) = 83\%$

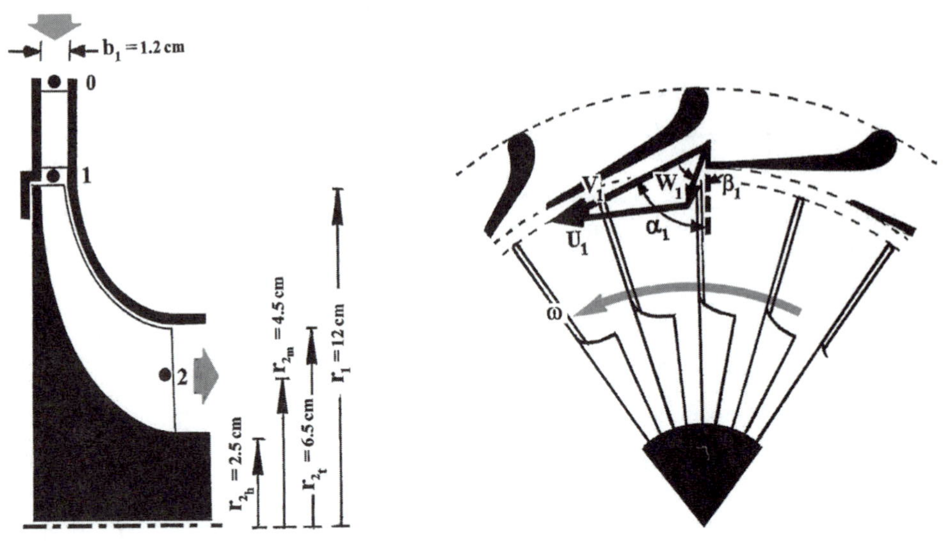

Figure 3.46. Input variables for Problem 9.

Assuming an adiabatic flow and a specific-heat ratio (γ) of 1.33, calculate:

a) The mass-flow rate ($\dot{m}$);
b) The change in total relative pressure caused by the rotor-flow irreversibility sources (e.g., friction).

Hint: In addressing item b, you may follow the following sequence of computational steps. First, compute the overall total relative pressure decline, which is caused by irreversibilities as well as the streamwise decrease in radius. Next, calculate the part of Δp_{t_r} that is caused solely by the radius decline by repeating the foregoing computational process but with a 100% rotor efficiency this time. The difference between these two magnitudes will represent the total relative pressure loss as a result of real-life irreversibilities.

10) Figure 3.47 shows the last stage of the turbine section in a turboprop engine. The stage is followed by an annular exhaust diffuser, which is designed in accordance with the Sovran and Klomp (1965) chart in Figure 3.37. The system operating conditions are as follows:

- Stage-inlet total pressure (p_{t0}) = 4.2 bars
- Stage-inlet total temperature (T_{t0}) = 860 K
- Stator is choked
- Stage-exit total temperature (T_{t2}) = 645 K
- Stage total-to-total efficiency (η_T) = 92%
- Mass-flow rate ($\dot{m}$) = 4.3 kg/s
- Rotor speed (N) = 52,000 rpm
- Ambient pressure (p_a) = 1.06 bars

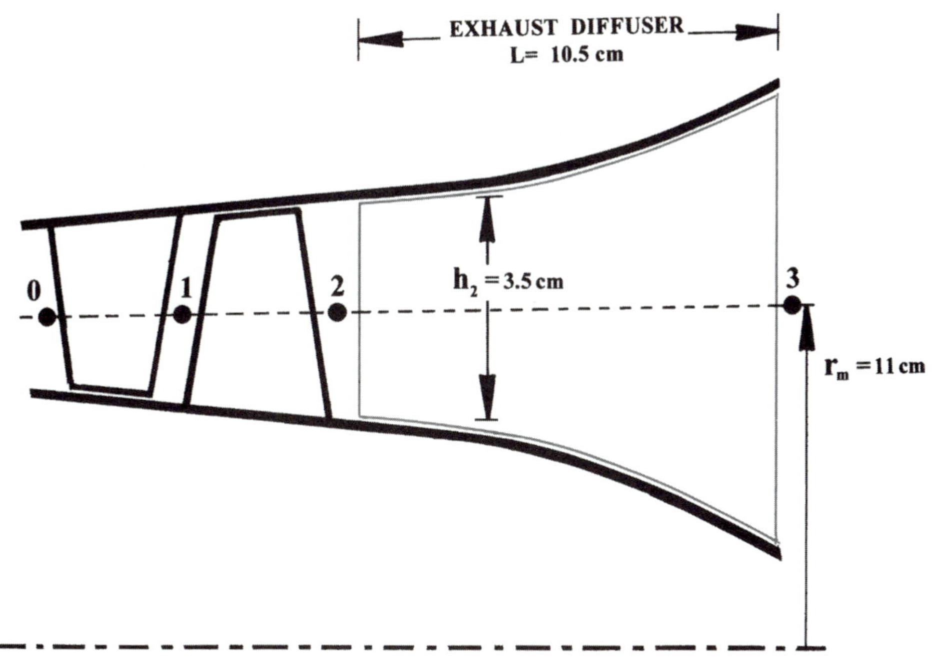

Figure 3.47. Geometry variables for Problem 10.

With an adiabatic flow throughout the rotor and exhaust diffuser, and assuming a γ average magnitude of 1.33, calculate:

a) The stator-exit absolute (or swirl) flow angle (α_1);
b) The maximum magnitude of rotor-exit critical Mach number (M_{cr2}), that would ensure no diffuser-exit flow reversal.

11) Figure 3.48 shows the interstage gap between two successive axial-flow stages in a compressor section. The operating conditions of the *first stage* are as follows:

- Inlet total pressure (p_{t0}) = 3.4 bars
- Inlet total temperature (T_{t0}) = 423 K
- Stator flow process is isentropic
- Stator-exit swirl angle (α_1) is positive
- Stator-exit critical Mach number (M_{cr1}) = 0.46
- Rotor total-to-total efficiency ($\eta_{rot.}$) = 82.5%
- Stage-exit total temperature (T_{t2}) = 531 K
- Mass-flow rate ($\dot{m}$) = 12.5 kg/s
- Shaft speed (N) = 32,500 rpm
- Axial-velocity component (V_z) is constant across the stage

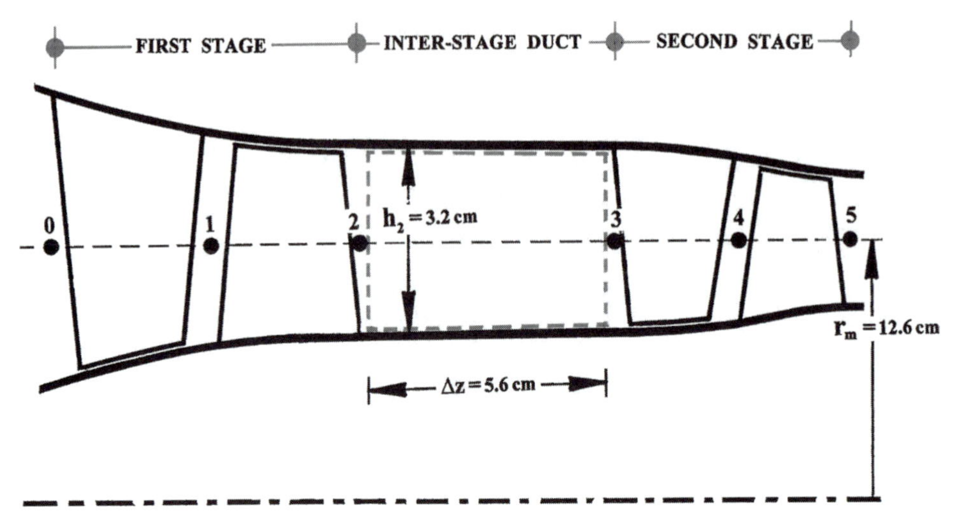

Figure 3.48. Geometry variables for Problem 11.

The flow is also choked at the interstage gap exit station. Considering an adiabatic flow process and a specific-heat ratio (γ) of 1.4, calculate:

a) The first-stage rotor-inlet critical Mach number (M_{cr1});
b) The percentage of total pressure loss ($\Delta p_t/p_{t2}$) over the interstage gap;
c) The endwall friction coefficient (f) across the gap.

12) Shown in Figure 3.49 are the major dimensions of a radial-inflow turbine stage. The stage operating conditions are as follows:

- Stator flow process is isentropic (by assumption)
- Rotor-inlet total pressure (p_{t1}) = 14 bars
- Rotor-inlet total temperature (T_{t1}) = 1365 K

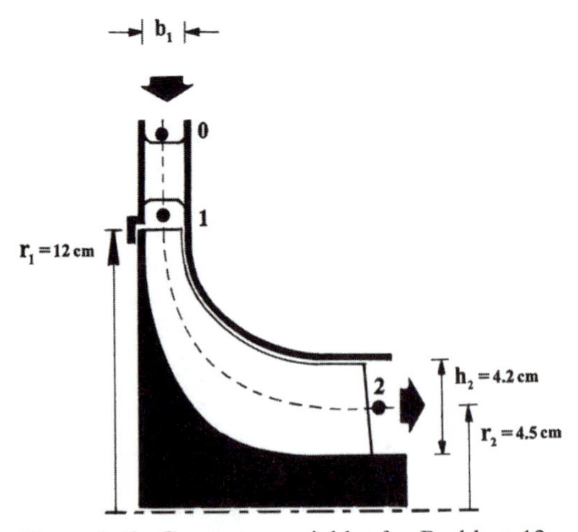

Figure 3.49. Geometry variables for Problem 12.

- Stator-exit critical Mach number (M_{cr1}) $= 0.94$
- Rotor-inlet relative flow angle (β_1) $= 0$
- Mass-flow rate ($\dot{m}$) $= 7.2$ kg/s
- Rotor-exit critical Mach number (M_{cr2}) $= 0.58$
- Rotor-exit static temperature (T_2) $= 1045$ K
- Stage total-to-total efficiency (η_T) $= 86\%$

Assuming an adiabatic flow process through the rotor, and a constant specific-heat ratio (γ) of 1.33:

a) Calculate the stator sidewall spacing (b_1).
b) Calculate the rotor efficiency on a:

   i) Total-to-static basis;
   ii) Total-relative to total-relative basis;
   iii) Static-to-static basis.

CHAPTER FOUR

Energy Transfer between a Fluid and a Rotor

Figure 4.1 shows a general-type mixed-flow compressor rotor. The thermophysical states 1 and 2 represent average conditions over the entire inlet and exit stations, respectively. With the rotor blade-to-blade hub-to-casing passage being the control volume, and with the continuity and energy equations (already covered in Chapter 3), we are now left with the momentum-conservation principle to implement.

The momentum equation is a vector relationship that can be resolved in the z, r, and θ coordinate directions. In the following, these three (scalar) relationships will be cast and their kinetic consequences discussed.

Axial-momentum equation:

$$F_z = \dot{m}(V_{z2} - V_{z1}) \tag{4.1}$$

The axial force F_z will be absorbed, in part, by a thrust bearing in most cases. Nevertheless, part of this force can be mechanically and/or aerodynamically damaging. Referring to Figure 4.2, for an axial-flow turbine rotor, the rotating blades will indeed move axially in response to this force.

Figure 4.2 shows the two different scenarios when a net axial force exists on the rotor blades. Movement to the left in this figure would close down the tip-to-casing clearance gap. Knowing that this clearance is normally less than 0.5 mm, particularly in high-pressure turbines, it is perhaps obvious that this rotor displacement can very well lead to the blades rubbing against the casing, potentially causing a catastrophic mechanical failure. Referring to the other rotor-displacement scenario in Figure 4.2, the rotor displacement to the right would open up the tip-clearance gap. This would encourage the flow to migrate even more from the pressure side to the suction side over the tip, which is one of the most aerodynamically degrading mechanisms. In cases of low aspect ratio (short) blades, this rotor motion will not only unload the tip region but will also render ineffective a good percentage of the blade height near the tip. This real-life effect will heavily influence the shaft-work extraction across the rotor.

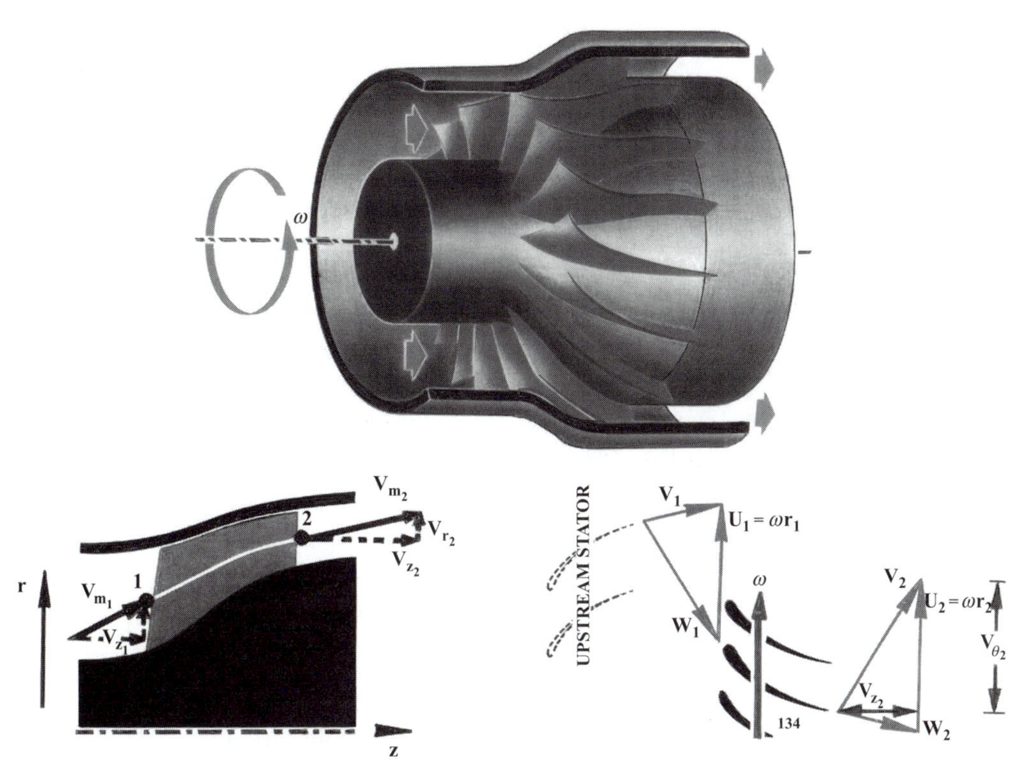

Figure 4.1. Velocity-vector relationships within a mixed-flow compressor rotor.

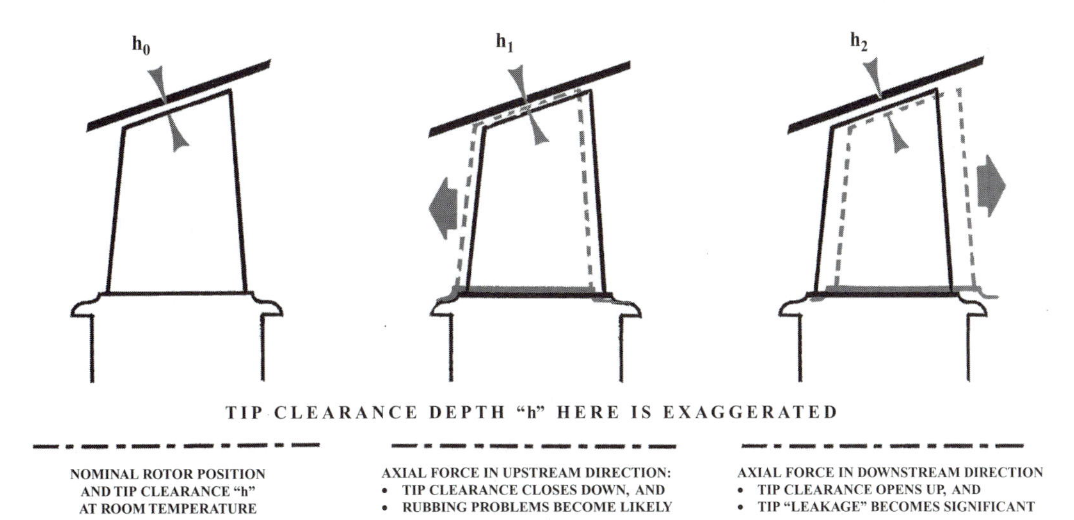

Figure 4.2. Rotor displacement caused by the change in axial momentum.

Radial-momentum equation:

$$F_r = \dot{m}(V_{r2} - V_{r1}) \tag{4.2}$$

The radial force component has little to do with the steady-state aerodynamic performance of a turbomachine. The force is normally absorbed as a journal-type load. However, depending on the lubricant's flow path and its properties, this force can aggravate a cyclic shaft motion known as "whirl." This off-center shaft motion can, and historically did, result in a catastrophic mechanical failure (Baskharone and Hensel, 1991). A famous example of this is the Space Shuttle Main Engine liquid-oxygen turbopumps, where an early design led to frequent premature incidents of mechanical failure. Also sharing the blame then was the tangential momentum in the secondary (or leakage) flow passages, for which remedies were later devised (e.g., Childs et al. 1991; Baskharone 1999).

Tangential-momentum equation: The change in tangential-velocity component from $V_{\theta 1}$ to $V_{\theta 2}$ (Figure 4.1) results in a net torque τ. This torque, together with the rotational speed, gives rise to the power that is produced or consumed by the rotor. The tangential-momentum equation can be written simply as

$$\tau = \dot{m}(r_2 V_{\theta 2} - r_1 V_{\theta 1}) \tag{4.3}$$

with $r_1 V_{\theta 1}$ and $r_2 V_{\theta 2}$ representing the specific (per unit mass) values of angular momentum at the inlet and exit stations, respectively. Now, multiplying through by the shaft speed (in radians per second), we get the power absorbed by the rotor in Figure 4.1:

$$P = \omega\tau = \omega\dot{m}(r_2 V_{\theta 2} - r_1 V_{\theta 1}) \tag{4.4}$$

The specific shaft work can be easily obtained by dividing equation (4.4) by the mass-flow rate as follows:

$$w_s = U_2 V_{\theta 2} - U_1 V_{\theta 1} \tag{4.5}$$

Equations (4.3) through (4.5) are different versions of what is historically known as Euler's "turbine" equation, although it does apply to the compressor subfamily as well. The latter includes such power-absorbing turbomachines as blowers, ducted fans, and full-scale compressors. Note that a propeller is not classified as a turbomachine because it is unbound.

The preceding discussion was focused on a work-absorbing turbomachine, thus excluding the turbine category as a work supplier. Should equations (4.3) through (4.5) be applied verbatim here, all the left-hand sides will come out negative. In order to remove this sign, we will reverse the angular-momentum terms, when it comes to the case of a turbine rotor, as follows:

$$\text{Torque}\,(\tau) = \dot{m}(r_1 V_{\theta 1} - r_2 V_{\theta 2}) \tag{4.6}$$

$$\text{Power}\,(P) = \omega\dot{m}(r_1 V_{\theta 1} - r_2 V_{\theta 2}) \tag{4.7}$$

$$\text{Specific shaft work}\,(w_s) = U_1 V_{\theta 1} - U_2 V_{\theta 2} \tag{4.8}$$

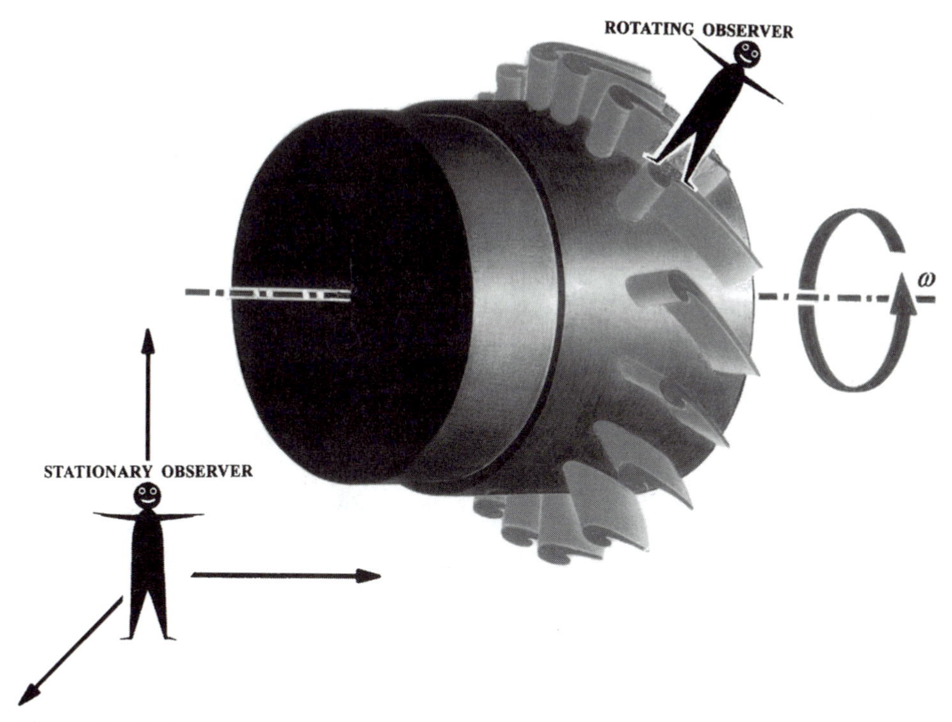

Figure 4.3. Stationary and rotating frames of reference.

Although this interchange of variables is optional, it is dislikable in an industry design setting to speak of, for instance, a negative power. Therefore, in subsequent chapters, it will be equations (4.6) through (4.8) that will be used when dealing with a turbine rotor, while holding applicable all of the velocity triangles' sign-convention rules cited in Chapter 3.

Used with the inlet and exit velocity diagrams, all derivatives of the angular-momentum equations are the most relevant to our (primary-passage) aerothermodynamic topics. Aside from the solid-body rotational velocity (U), the two other velocity-triangle contributors are the absolute (V) and relative (W) velocities, each of which is observed in two (stationary and rotating) frames of reference. With this in mind, we should perhaps turn our attention to the physical pictures of these two, emphasizing the need for both.

Stationary and Rotating Frames of Reference

Consider the two observers in Figure 4.3, one situated in the stationary and the other in the rotating frame of reference. The velocity difference between the two observers is clearly the vector **U**, the velocity vector of the blade-mounted observer. Otherwise, both observers will register the same axial and radial velocity components. In a vector form, the absolute/relative velocity relationship can be expressed as follows:

$$\mathbf{V} = \mathbf{W} + \mathbf{U} \tag{4.9}$$

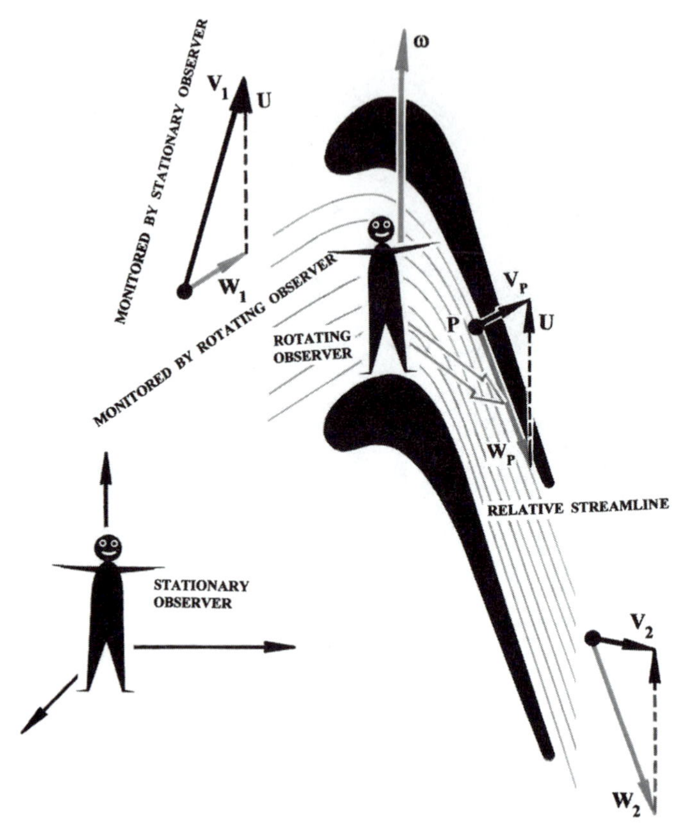

Figure 4.4. Relative streamlines in the rotating frame of reference.

Note that the relative velocity W is meaningless outside the rotor subdomain. In equation (4.9), the velocity of the spinning observer (**U**) can be separately expressed as

$$\mathbf{U} = \omega r \mathbf{e}_\theta \tag{4.10}$$

where e_θ is the unit vector in the circumferential direction.

Resolving the preceding two equations in the θ, z, and r directions, we obtain the following relationships:

$$V_\theta = W_\theta + \omega r \tag{4.11}$$

$$V_z = W_z \tag{4.12}$$

$$V_r = W_r \tag{4.13}$$

Figure 4.4 highlights the necessity of adopting a rotating frame of reference. The rotating observer, as shown in Figure 4.4, is intentionally placed inside

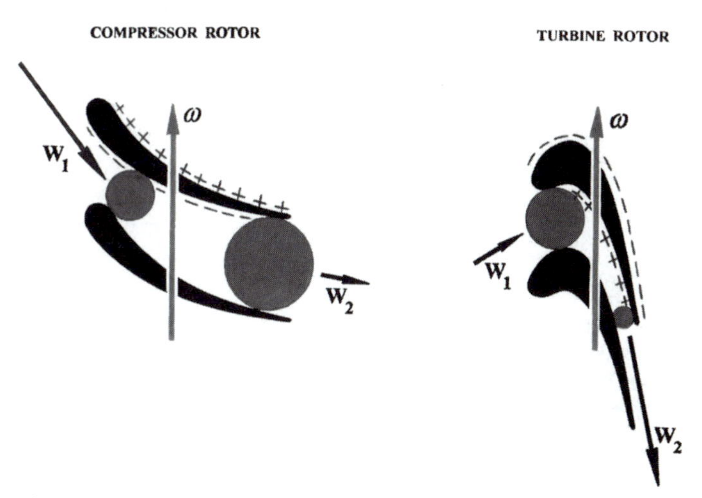

Figure 4.5. Direction of blade rotation in compressor and turbine airfoil cascades.

the rotor blade-to-blade passage in order to highlight the following important facts:

- The velocity diagram concept is not restricted to the flow inlet and exit stations but applies to any station in between.
- It is only through the reference frame attached to the rotor blade that a sensible set of streamlines (those associated with the relative flow field) can be obtained. The absolute flow field would comparatively give rise to incomprehensible streamlines.
- It is only the relative velocity that truly reflects the shape of the blade-to-blade passage. This fact is further clarified in Figure 4.5, in which the change in W is indicative of whether the passage is diffuser- or nozzle-like. In fact, the change in the absolute velocity V in Figure 4.4 is quite the opposite.
- Examination of Figure 4.4 makes it necessary to resort to the relative velocity W in dealing with many physical phenomena, such as choking of the blade-to-blade channels, especially in turbines, as well as the boundary layer profiles within the rotor subdomain.

To summarize, it is important to realize that both the absolute and relative velocities are legitimate rotor-domain variables from which to choose. In situations where variables such as torque are required, it is the absolute velocity V that is appropriate to use. On the other hand, real effects such as friction and choking are necessarily investigated on the relative flow-field basis. Figures 4.5 and 4.6 are intended to highlight some additional topics as follows:

- Because both the compressor and turbine sections are mounted on the same shaft, the spinning-speed magnitude and direction will be identical on both sides of the combustor. This is indicated in Figure 4.5, where the suction (convex) side of the turbine rotor blade is leading in the direction of rotation. The picture is quite the opposite in the case of a compressor rotor.

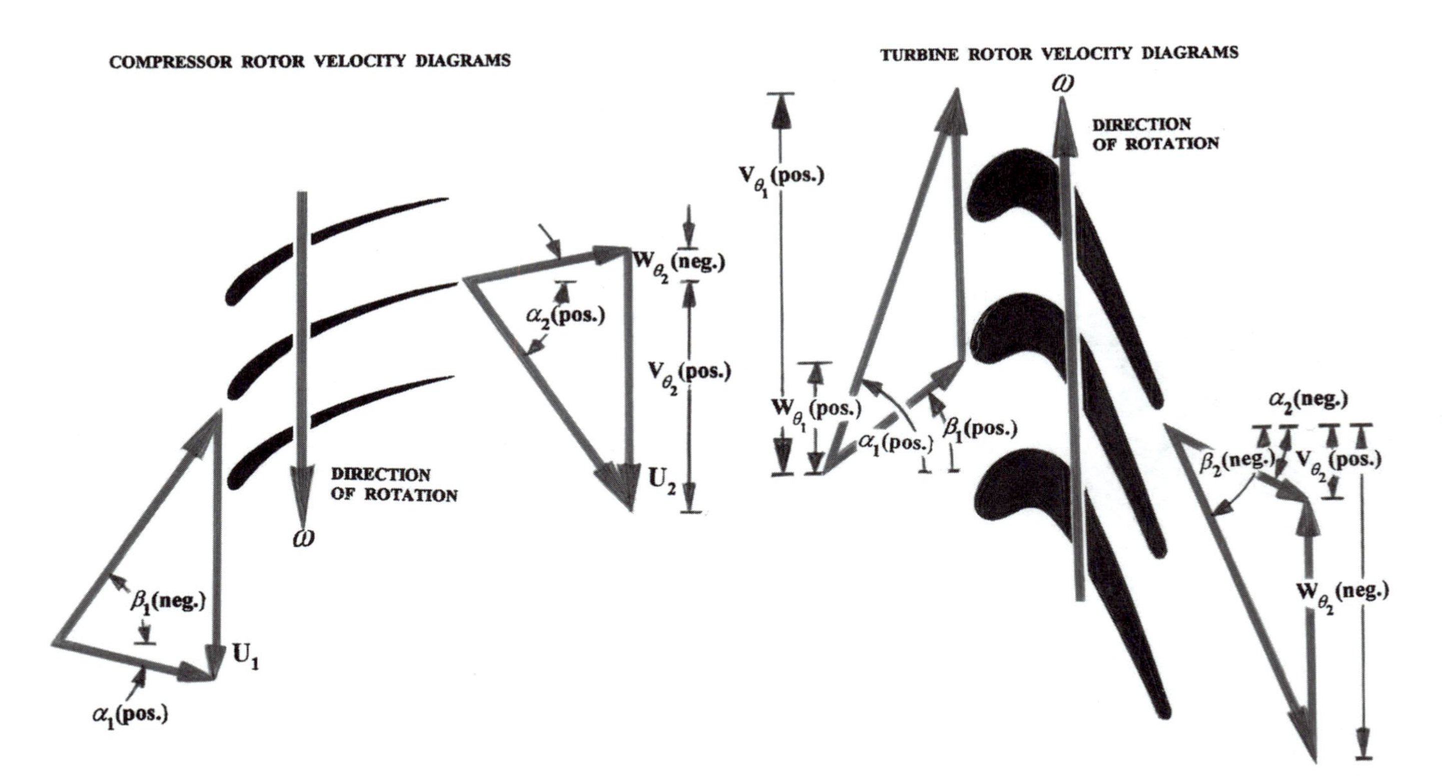

Figure 4.6. Typical velocity triangles and sign convention: all angles and velocity components in the direction of rotation are positive.
Notes:

- $U = \omega$ r is the rotor "linear" velocity, where "r" is the local radial location
- In a purely axial turbomachine, U_1 and U_2 are identical.

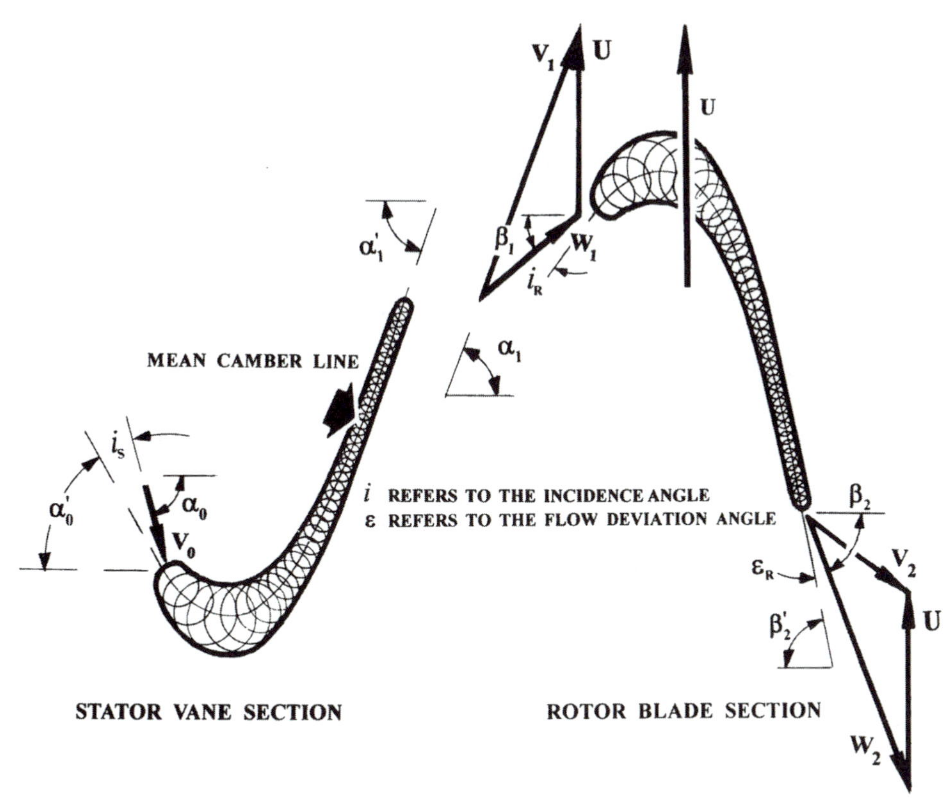

Figure 4.7. Definition of the stator and rotor incidence angles in axial-flow turbines.

- It only seems logical for the turbine-blade rotation to be in the direction shown in Figure 4.6. After all, we know that a flow passage that is shaped like the turbine blade-to-blade channel will experience an upward force and, if left free, will move in this same direction. As this rule is applied to the compressor blade-to-blade channel, one would initially think that an upward blade rotation is physically unattainable. However, we should always recall that a compressor rotor will always be driven by a turbine rotor that will dictate, among other variables, the direction of the compressor rotation. In other words, it is what the compressor rotor is compelled to do (in terms of the rotation direction) which facilitates the flow compression process to take place.

Flow and Airfoil Angles

Revisited in this section are the sign-convention rules, extending them to the airfoil (also called "metal") angles (Figure 4.7). Measured from the axial direction, the absolute and relative flow angles, as well as the tangential-velocity components, will always be positive in the direction of rotation and negative otherwise. This, by reference to Figure 4.6, is as applicable to compressors as it is to turbine cascades. In special cases such as a radial-turbine rotor inlet and a centrifugal-compressor impeller outlet (Chapters 10 and 11), the flow angles (both absolute and relative) are

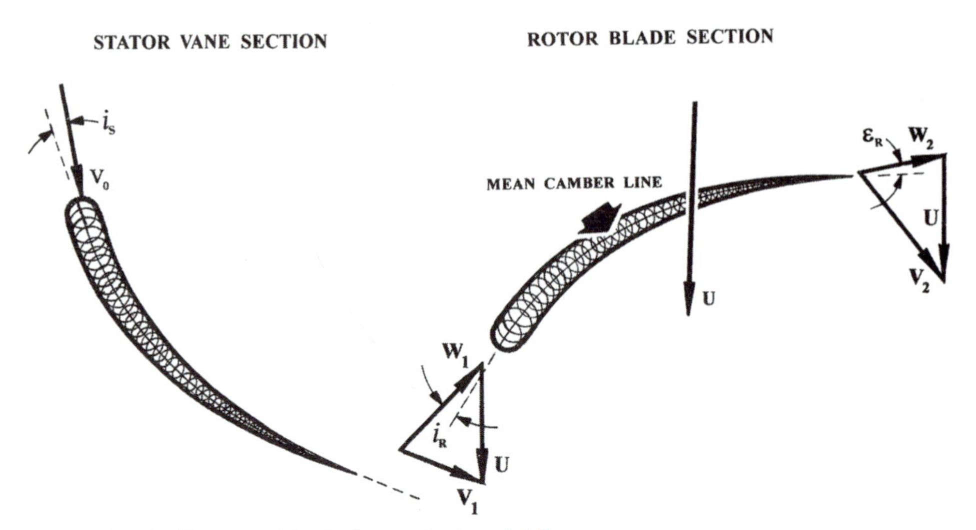

Figure 4.8. Incidence and deviation angles in axial-flow compressors.

measured from the local radial direction. They nevertheless abide by the same rule: positive in the direction of rotation and negative otherwise.

Figures 4.7 and 4.8 put in focus some detailed accounts of the flow/airfoil interaction for both the stator and rotor subdomains in axial-flow turbomachines. In both figures, each airfoil camber line is highlighted. This particular line is defined as the locus of all centers of circles of which the airfoil suction and pressure sides are tangents. At the leading and trailing edges, the tangents to the mean camber line define the airfoil (or metal) inlet and exit angles and, subsequently, the incidence and deviation angles, respectively. As is clear in Figures 4.7 and 4.8, the incidence angle is that between the incoming flow stream and the tangent to the mean camber line at the leading edge. The deviation angle is similarly defined as the angle between the departing flow stream and the camber line tangent at the trailing edge. The latter is much more significant in compressor cascades, as will be presented later in Chapter 9.

Components of Energy Transfer

In the following, we will rewrite Euler's energy-transfer equation using some trigonometric manipulations. This will enable us to identify and separate different components of the fluid/rotor energy transfer. In doing so, reference will continually be made to the general-type rotor in Figure 4.1. We begin with the exit velocity triangle in this figure, where the following relationship applies:

$$V_{z2}{}^2 = V_2{}^2 - V_{\theta 2}{}^2 = W_2{}^2 - W_{\theta 2}{}^2 = W_2{}^2 - (U_2 - V_{\theta 2})^2 \tag{4.14}$$

or

$$V_2{}^2 - V_{\theta 2}{}^2 = W_2{}^2 - U_2{}^2 + 2U_2 V_{\theta 2} - V_{\theta 2}{}^2 \tag{4.15}$$

which yields the following relationship:

$$U_2 V_{\theta 2} = \frac{1}{2}\left(V_2{}^2 + U_2{}^2 - W_2{}^2\right) \tag{4.16}$$

Following the same logic, a similar relationship that applies to the inlet velocity triangle can be obtained:

$$U_1 V_{\theta 1} = \frac{1}{2}\left(V_1{}^2 + U_1{}^2 - W_1{}^2\right) \tag{4.17}$$

Under an adiabatic-flow assumption, expressions (4.16) and (4.17) can be substituted in the specific shaft-work expression (4.5) for a general mixed-flow compressor stage, resulting in

$$w_s = h_{t2} - h_{t1} = U_2 V_{\theta 2} - U_1 V_{\theta 1} \tag{4.18}$$

which can be reexpressed as

$$w_s = \left(\frac{V_2{}^2 - V_1{}^2}{2}\right) + \left(\frac{W_1{}^2 - W_2{}^2}{2}\right) + \left(\frac{U_2{}^2 - U_1{}^2}{2}\right) \tag{4.19}$$

It is important to examine each term on the right-hand side of this expression and study its implications:

The term ($V_2^2 - V_1^2/2$): Recalling the way in which the total enthalpy was broken up into static and dynamic components, this term clearly represents a dynamic enthalpy change. It may seem at first that maximizing the absolute exit velocity (V_2) would be a rational means of ensuring the maximum contribution of this term to the overall total enthalpy rise. This is theoretically true, but with a major reservation. The drawback here is that the subsequent stator in this case will have the unfair responsibility of converting this large dynamic head into a static-pressure rise, a process that (under such circumstances) would be associated with a large-scale total pressure loss.

The term ($W_1^2 - W_2^2/2$): Having defined the dynamic enthalpy change, this term is clearly indicative of a static enthalpy change. For this term to be a positive contributor to the total enthalpy rise, W_2 will have to be much less than W_1. This, by itself, proves the fact that the rotor blade-to-blade passage must be diverging (i.e., diffuser-like).

The term ($U_2^2 - U_1^2/2$): Following the same rationale this term also represents a static enthalpy change. For the contribution of this term to be substantial, the flow exit station will have to be at a significantly higher radius in comparison with the rotor inlet radius. This very argument points to the subfamily of centrifugal compressors (Chapter 11), which is known for its high work-absorption capacity.

Definition of the Stage Reaction

For a general (turbine or compressor) stage, the stage reaction (R) is defined as

$$R = (\text{rotor static enthalpy change})/(\text{total enthalpy change})$$

In the preceding reaction definition, the denominator was purposely left without such a qualifier as "across the rotor" or "across the stage." The reason is that there would be no total enthalpy change across the stator (under an adiabatic flow condition), which means that either one of these qualifiers can be used. In terms of velocity components, the preceding reaction definitions can be reexpressed as

$$R_{comp.} = \frac{0.5\left[\left(W_1^2 - W_2^2\right) + \left(U_2^2 - U_1^2\right)\right]}{0.5\left[\left(W_1^2 - W_2^2\right) + \left(U_2^2 - U_1^2\right) + \left(V_2^2 - V_1^2\right)\right]} \tag{4.20}$$

$$R_{turb.} = \frac{0.5\left[\left(W_2^2 - W_1^2\right) + \left(U_1^2 - U_2^2\right)\right]}{0.5\left[\left(W_2^2 - W_1^2\right) + \left(U_1^2 - U_2^2\right) + \left(V_1^2 - V_2^2\right)\right]} \tag{4.21}$$

Reaction of Axial-Flow Stages

The preceding two expressions are the most general and apply to any (axial, radial, or mixed-flow) turbomachinery stage. Two similar expressions can be deduced to represent the axial-flow stages shown in Figure 4.5. Setting $U_2 = U_1$, these two expressions are as follows:

$$R_{comp.} = \frac{0.5\left(W_1^2 - W_2^2\right)}{0.5\left[\left(W_1^2 - W_2^2\right) + \left(V_2^2 - V_1^2\right)\right]} \tag{4.22}$$

$$R_{turb.} = \frac{0.5\left(W_2^2 - W_1^2\right)}{0.5\left[\left(W_2^2 - W_1^2\right) + \left(V_1^2 - V_2^2\right)\right]} \tag{4.23}$$

Examination of these two expressions in light of Figure 4.5 reveals the following:

- For a *compressor* stage, the magnitude of reaction is a reflection of how *diffuser-like* the rotor blade-to-blade passage is.
- For a *turbine* stage, the magnitude of reaction represents how *nozzle-like* the rotor passage is.

The optimum magnitude of reaction ranges between 50 and 60%, based on a mean-radius (or pitch-line) flow analysis. The danger, however, would come from the hub and/or tip reaction magnitudes. Note that at least one radius-dependent velocity, namely U, will vary along the blade radial extension, and so may other velocity magnitudes and/or angles, all of which do influence the local reaction magnitude. As a result, the blade-tip reaction value may be undesirably high, whereas the hub-section reaction might be extremely low or even negative. The latter would be a designer's nightmare should it go undetected up to the detailed design phase.

Figure 4.9 shows the rotor blade geometry and velocity triangles associated with three axial-flow turbine and compressor cascades for three popular reaction

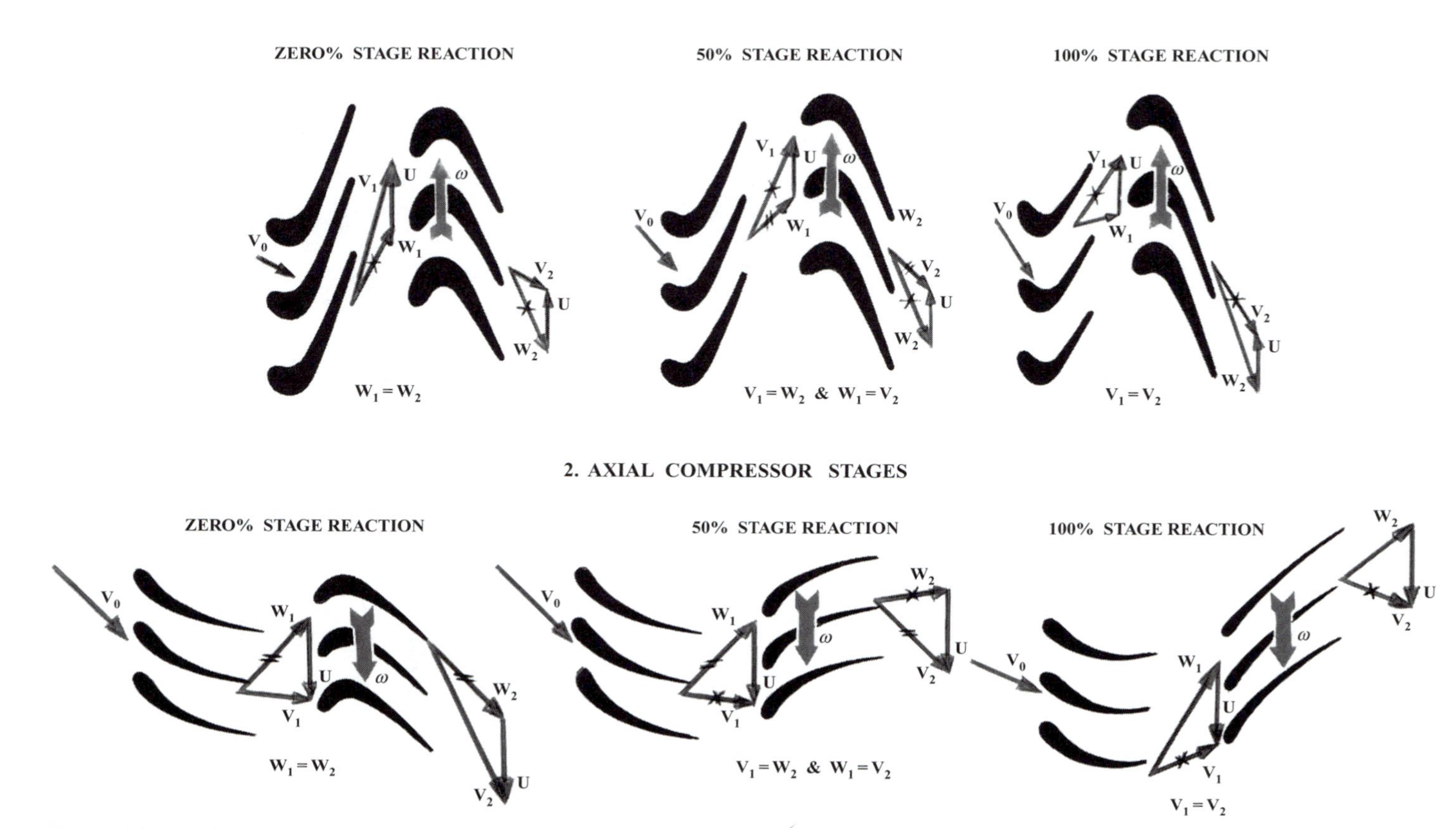

Figure 4.9. Special stage-reaction magnitudes in axial-flow turbomachines.

percentages, namely 0, 50%, and 100% reaction magnitudes. Focusing (in particular) on the 50% reaction, in both cases we clearly see that the inlet and exit velocity triangle and airfoil cascades are mirror images of one another. This appropriately adds the phrase "symmetric" to the 50% reaction stage. Also note that the zero-reaction (impulse) turbine blading is rather odd, with the rotor having only the objective of symmetrically turning the relative flow stream without accelerating it at all.

Invariant Thermophysical Properties

In the following, specific thermophysical-property groupings will be defined. These will remain constant across either the stator or rotor components within the same stage. In rotating blade cascades, the name and composition of the invariant property will depend on whether the rotor is of the axial, mixed-flow, or radial type.

The Total Enthalpy (h_t)

This, like any such total property, is composed of a thermal-energy part, namely the local static enthalpy, and a velocity-dependent dynamic part, as follows:

$$\text{Total enthalpy}\,(h_t) = h + \frac{V^2}{2} \tag{4.24}$$

In the absence of both heat transfer and shaft-work interaction, we know that the total enthalpy will be invariant across a stationary passage. This result applies to such flow passages as the stator subdomain and interstage ducts. Furthermore, should the specific-heat ratio (γ) be roughly constant across the flow passage, then equation (4.24) can be further simplified, leaving the total temperature to be the invariant property instead:

$$T_t = T + \frac{V^2}{2c_p} \tag{4.25}$$

The Total Relative Enthalpy (h_{t_r})

As the phrase relative would naturally imply, this variable would remain unaltered in an adiabatic flow process through an axial-flow rotor. The invariant property here is based on relative velocity and can be expressed as

$$\text{Total relative enthalpy}\,(h_{t_r}) = h + \frac{W^2}{2} \tag{4.26}$$

In order to prove that an adiabatic rotor of the axial-flow type will give rise to a constant magnitude of total relative enthalpy, let us first express the inlet and exit magnitudes of this property:

$$h_{t,r1} = h_1 + \frac{W_1{}^2}{2} = \left(h_{t1} - \frac{V_1{}^2}{2}\right) + \frac{W_1{}^2}{2}$$

$$h_{t,r2} = h_2 + \frac{W_2{}^2}{2} = \left(h_{t2} - \frac{V_2{}^2}{2}\right) + \frac{W_2{}^2}{2}$$

Subtracting the first equation from the second, we obtain the following result:

$$h_{t,r2} - h_{t,r1} = [h_{t2} - h_{t1}] - \left[\left(\frac{V_2{}^2 - V_1{}^2}{2}\right) + \left(\frac{W_1{}^2 - W_2{}^2}{2}\right)\right]$$

Noting that an axial-flow rotor will give rise to a constant solid-body rotational velocity U (based on a simple pitch-line analysis), and by reference to equation (4.19), we can easily conclude that both the bracketed terms are identical and that each of these represents the specific shaft work (w_s), meaning that

$$h_{t,r2} = h_{t,r1} \text{ (for an axial-flow rotor)}$$

For a sufficiently small change in the specific-heat ratio (γ), the invariant property becomes the total relative temperature (T_{tr}), where

$$T_{tr} = T + \frac{W^2}{2c_p} = \left(T_t - \frac{V^2}{2c_p}\right) + \frac{W^2}{2c_p} \tag{4.27}$$

Invariance of the total relative temperature, which is strictly limited to axial-flow rotors, should perhaps remind us of the fact that an axial-flow rotor can, in effect, be converted into a simple axial-flow stator, provided that the relative velocity vector is used instead.

The Rothalpy (I)

This is not a widely known thermophysical property and is not as widely used as others. The property is versatile enough to remain invariant throughout any rotor regardless of its meridional flow-path geometry (axial, radial, or mixed-flow type). The property rothalpy, termed I, is defined as

$$I = h_t - UV_\theta \tag{4.28}$$

In order to prove the invariance of rothalpy across any type of rotor, let us consider the flow process through the general mixed-flow compressor rotor in Figure 4.1:

$$I_{ex} - I_{in} = (h_{t\,ex} - h_{t\,in}) - (U_{ex}V_{\theta\,ex} - U_{in}V_{\theta\,in}) \tag{4.29}$$

Now, invoking Euler's energy equation in the right-hand side of equation (4.29), we see that the bracketed term is equal to the rest of the right-hand side in the same equation. With this result, the fact can now be confirmed that the rothalpy maintains a fixed magnitude across any rotor type. The rothalpy definition can equivalently be expressed as

$$I = c_pT_t - UV_\theta \tag{4.30}$$

Importance of the Invariant Properties

The example shown in Figure 4.10 is perhaps a testimony to the power of invariant properties in simplifying certain turbomachinery computations. The figure shows

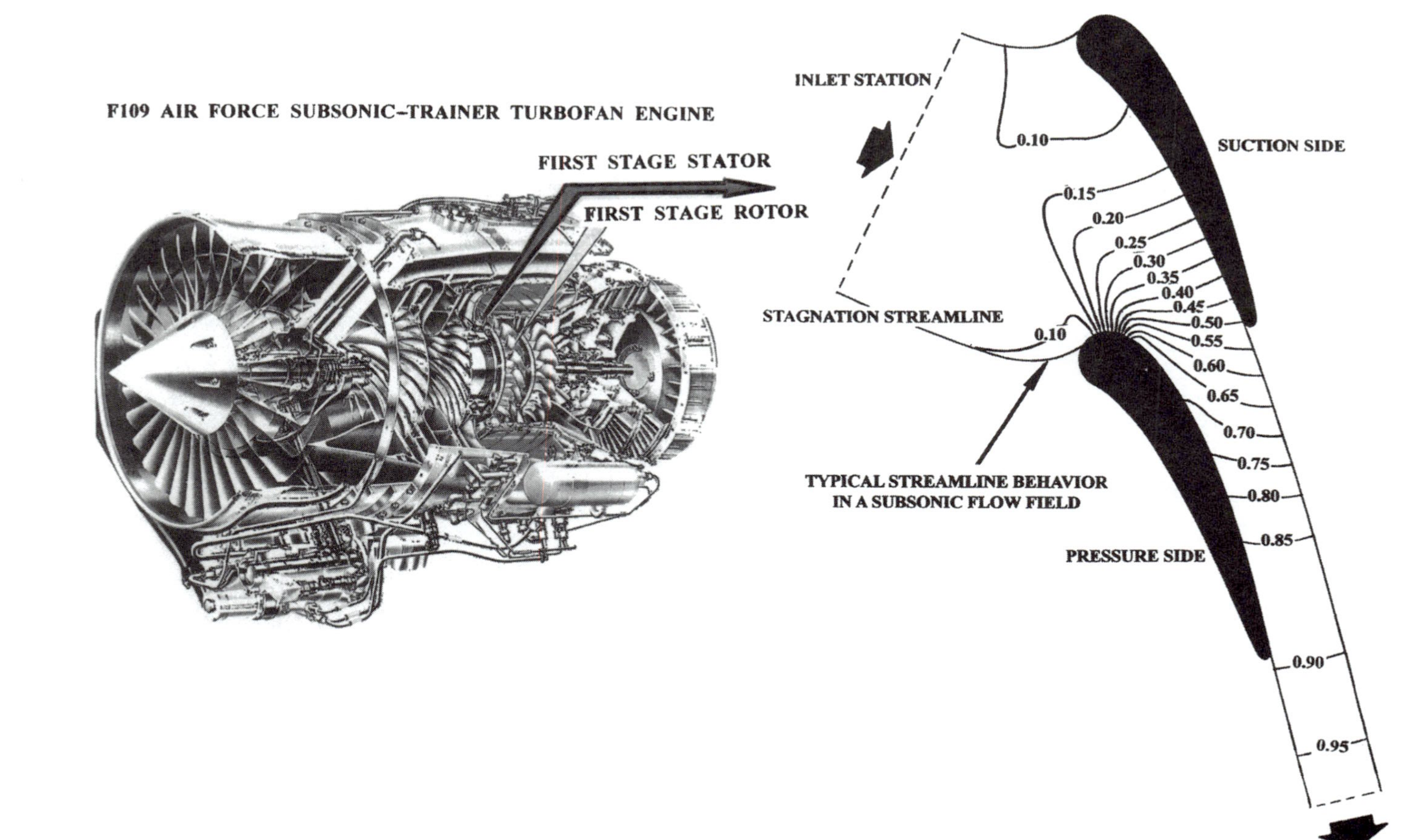

Figure 4.10. Critical Mach number contours within the first-stage stator in a turbofan engine.

contours of the critical Mach number in the vane-to-vane passage of the first-stage stator in the F109 turbofan engine. The critical Mach number distribution in this figure is the final outcome of an aerodynamic study that is described next.

In a parametric cooling effectiveness study of the stator in Figure 4.10, the static temperature field associated with the stator was readily available. This case required the corresponding velocity (or critical Mach number) distribution. The reason was to calculate the convection heat-transfer coefficient, which is a strong function of the local velocity.

The stator flow process was assumed to be adiabatic at this preliminary stage, an assumption that automatically identifies the total enthalpy h_t as the invariant property. Next, it was decided to proceed with a constant specific heat ratio on the basis that the maximum static temperature difference within the stator domain was approximately 120 K, which was judged to be sufficiently small. Next, the following version of the energy equation was utilized:

$$\Delta h_t = 0 = c_p(T_1 - T_2) + \frac{(V_1{}^2 - V_2{}^2)}{2} \tag{4.31}$$

where the thermophysical state 1 is a totally defined reference state (chosen to be the middle point on the flow inlet station), with the velocity magnitude known a priori. The procedure then was to proceed from one location to another, each time calculating a new velocity magnitude using equation (4.31) until the entire flow domain was acceptably covered. In the end, all of the computed velocities were converted into a critical Mach number, and the contour plot in Figure 4.10 was generated.

The preceding example illustrates the importance of the physical/thermal property groupings (the total temperature in this case) in computing unknown property magnitudes. Had the example involved an axial-turbine rotor, the invariant property would have been the relative total temperature (T_{t_r}) instead.

It may appear difficult to visualize a situation where the seemingly unattractive property rothalpy (I) in expression (4.30) could be practically useful. Part of the "unease" in this case is the fact that the property itself seems too abstract in the sense that it results from an algebraic manipulation of Euler's equation. However, the invariance of such a property can be beneficially used in many situations during a flow-modeling process within a general type of rotor subdomain where:

- The static temperature field throughout the entire rotating subdomain is known, and the relative velocity W is required at all corresponding points in the flow field.
- The absolute velocity V is known, and the corresponding total temperature is required.

In both cases, the solid-body rotational velocity U is calculable, as the shaft speed and local radial coordinate r are both known. Also note that there has to be a boundary condition involving I at a station (usually the rotor inlet) that is either supplied or calculable.

It is important to point out that solutions of the two examples just cited are by no means as easy as one may initially think, for they involve a great number of iterative steps and numerical interpolation over the rotor meridional projection. The unfortunate fact is that since its inception (at NASA-Lewis Research Center) in the mid-1970s, this particular property has virtually been written off, except in advanced turbomachinery flow-analysis settings.

Total Relative Properties

As previously defined, the total relative enthalpy h_{tr} will now be discussed in terms of its usefulness. Across an adiabatic axial-flow stage rotor and in the absence of significant γ changes, the invariant property becomes the total relative temperature:

$$T_{tr} = \text{constant}$$

It is perhaps appropriate at this point to define a reference velocity that is equivalent to V_{cr} for a rotating subdomain, on the basis of the total relative temperature T_{tr}. Referred to as the relative critical velocity, W_{cr}, the following definition applies:

$$W_{cr} = \sqrt{\left(\frac{2\gamma}{\gamma+1}\right) RT_{tr}} \tag{4.32}$$

This definition of W_{cr} paves the way to define the so-called relative critical Mach number, $M_{cr\,r}$, as follows:

$$M_{cr\,r} = \frac{W}{W_cr} \tag{4.33}$$

In a rotating blade-to-blade passage, the relative critical Mach number has the same implications as the absolute critical Mach number in a stationary passage. These include the subsonic/supersonic classification and the choking condition. Note that regardless of how total or total relative properties are defined, and whether the frame of reference is stationary or rotating, the static properties remain unaffected. In terms of $M_{cr\,r}$, we are now in a position to redefine the total relative temperature T_{tr} and the total relative pressure as follows:

$$T_{tr} = \frac{T}{\left[1 - \frac{\gamma-1}{\gamma+1}\left(\frac{W}{W_{cr}}\right)^2\right]} \tag{4.34}$$

$$p_{tr} = \frac{p}{\left[1 - \frac{\gamma-1}{\gamma+1}\left(\frac{W}{W_{cr}}\right)^2\right]^{\frac{\gamma}{\gamma-1}}}$$

or

$$p_{tr} = p\left(\frac{T_{tr}}{T}\right)^{\frac{\gamma}{\gamma-1}} \tag{4.35}$$

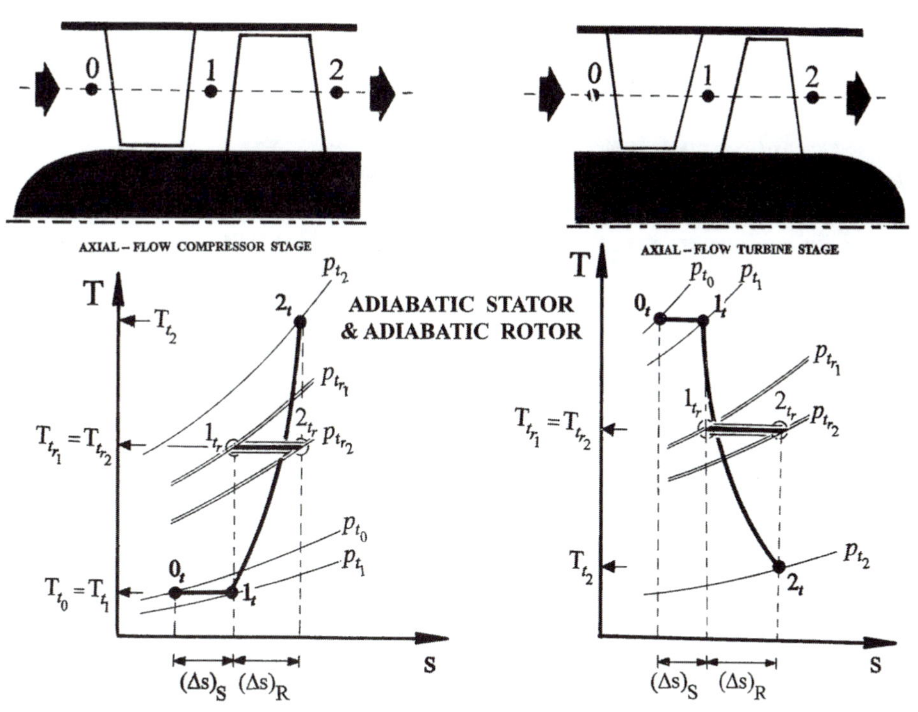

Figure 4.11. Invariance of the total relative temperature across an axial-flow adiabatic rotor.

A simpler alternative to this T_{tr} expression is

$$T_{tr} = \left(T_t - \frac{V^2}{2c_p}\right) + \frac{W^2}{2c_p} \tag{4.36}$$

$$p_{tr} = p_t \left(\frac{T_{tr}}{T_t}\right)^{\frac{\gamma}{\gamma-1}} \tag{4.37}$$

The variables T_{tr} and p_{tr} have the same implications as their p_t and T_t counterparts in stationary flow passage. For instance, an adiabatic-rotor flow process means that T_{tr} will have to be constant throughout the rotor. Furthermore, the total relative pressure p_{tr} in an axial-flow rotor can only decline as a reflection of the irreversibilities within the rotor, and it would remain constant only in an ideally isentropic axial-rotor flow process (Figures 4.11 and 4.12).

Expression (4.37) becomes clearer by reference to Figure 4.11, where the static, total, and total relative states on the T-s diagram are represented by points that are vertically above one another. The reason is that the total (or stagnation) state would be obtained at the end of a fictitious isentropic diffuser that is aimed at killing the absolute velocity V, which places the total (or stagnation) state vertically above the static state. As for the total relative state location on the T-s diagram, the fictitious process is conceptually the same (except that the isentropic diffuser here is rotating), which similarly places the total relative state vertically above the static state.

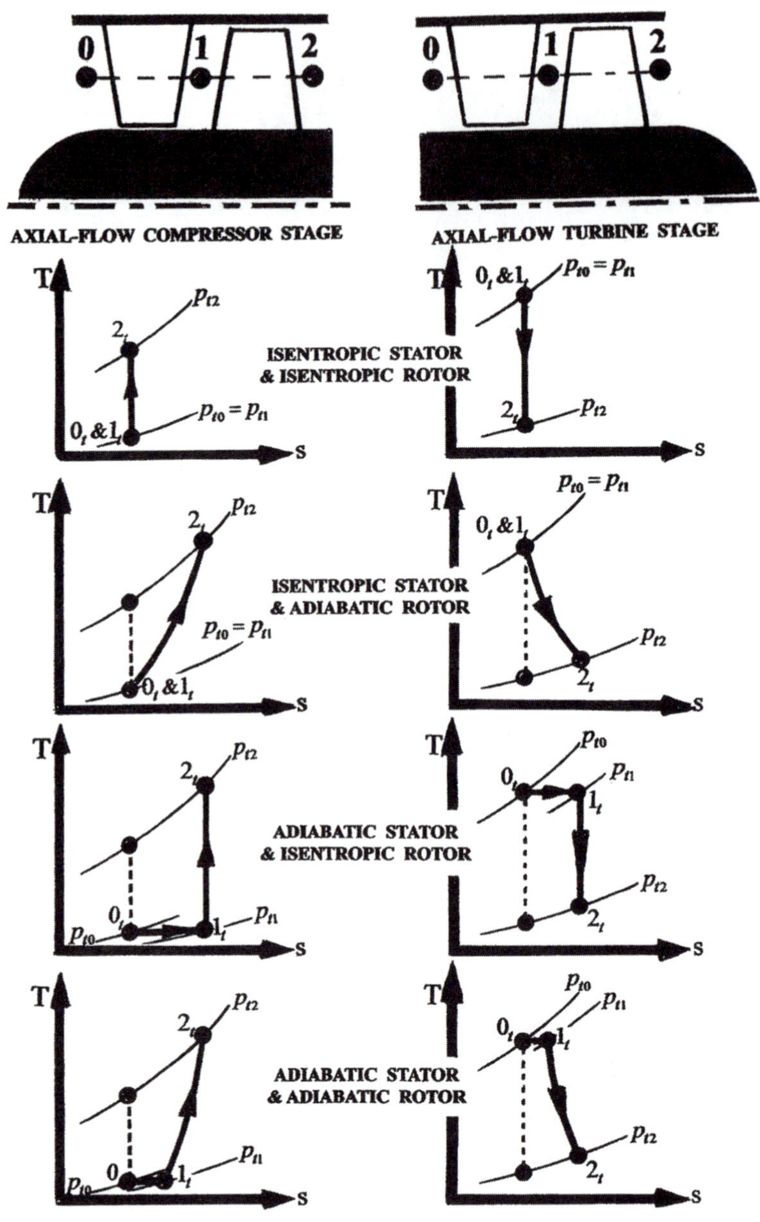

Figure 4.12. Different adiabatic/isentropic combinations in compressor and turbine stages.

Figures 4.11 and 4.12 illustrate the property-invariance principle in axial-flow stages. Figure 4.11 contrasts an axial-flow compressor stage with its turbine counterpart, with both permitting an adiabatic flow stream. Figure 4.13 widens the same thermodynamic facts to cover different stator/rotor combinations. The focus in Figure 4.12 is the adiabatic versus isentropic flow condition in connection with these combinations. Note that the second row of the *T-s* representations in

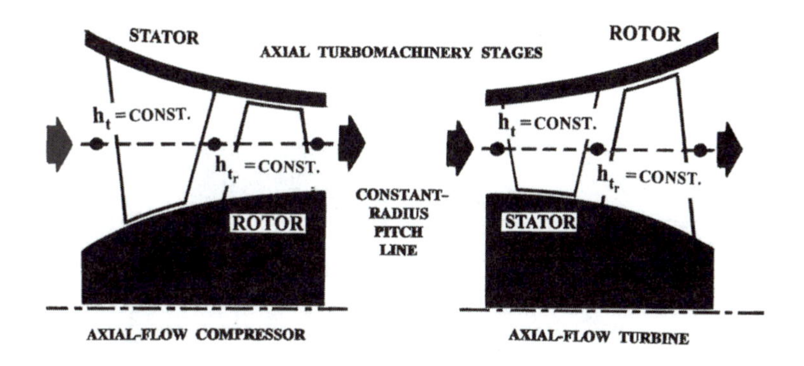

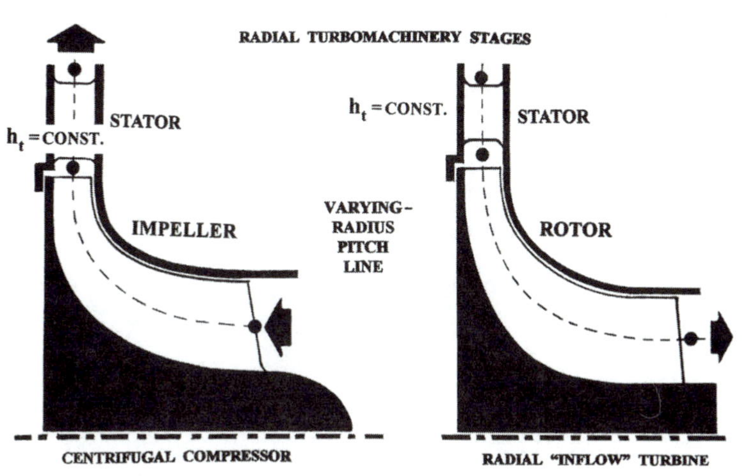

Figure 4.13. Invariant total properties in axial-flow compressor and turbine sttages.

Figure 4.12 (marked with arrowheads) is the customary means of representing the stage process in basic thermodynamics textbooks, although it is (in the bigger picture) just a special isentropic-stator case. Finally, Figure 4.13 presents a summary of the property-invariance topic as it applies to stators and rotors of axial- and radial-flow turbomachines.

Incidence and Deviation Angles

The first step here is to be familiar with the airfoil's mean camber line (Figures 4.7 and 4.8), as the inlet and exit tangents to this line act as datums for the incidence and deviation angles, respectively. Measured from the axial direction (in axial-flow turbomachines), the slope angles of these tangents are often referred to as the inlet and exit metal angles. In practice, these two angles can substantially differ from the inlet and exit flow angles. The metal angles will be distinguished from the flow angles by assigning a prime superscript to the former.

An angle of serious consequences is referred to as the incidence angle and stands as the cascade-theory equivalent to the angle of attack in the wing aerodynamics field. The incidence angle is defined (only in magnitude) as the difference between

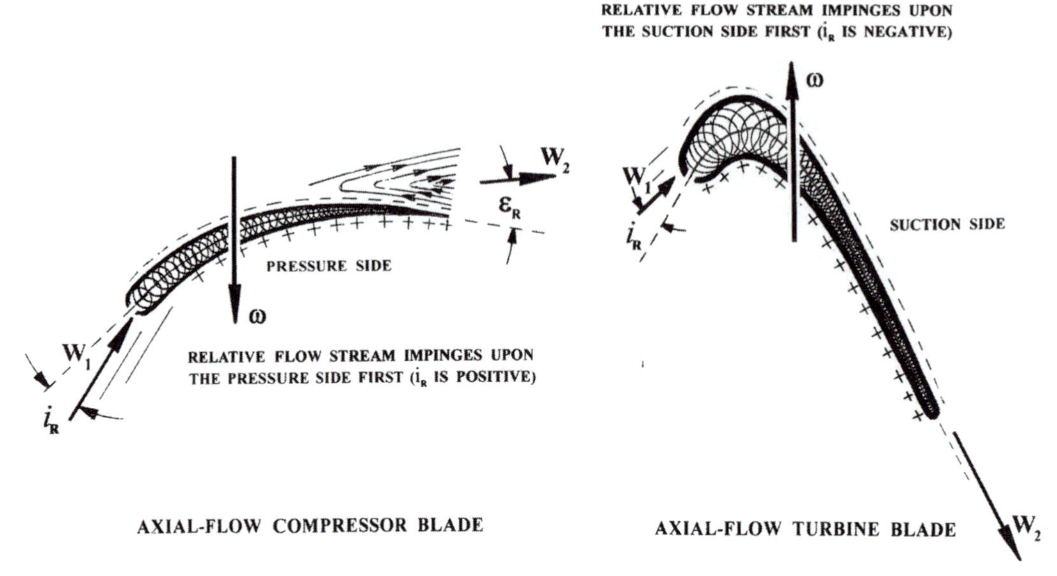

Figure 4.14. Signs of incidence and deviation angles in compressor and turbine rotors.

the incoming flow angle and the airfoil (metal) angle. Applied to stators and rotors, this statement can be broken up into two separate expressions as follows:

$$|i_{stator}| = |\alpha_1 - \alpha_1{}'| \tag{4.38}$$

$$|i_{rotor}| = |\beta_1 - \beta_1{}'| \tag{4.39}$$

As emphasized earlier, the preceding two expressions provide only the absolute value of the incidence angle. The sign of this angle, by reference to Figure 4.14, will always abide by the following rule: If extensions of the approaching streamlines impact (by simple inspection) the airfoil's pressure side first, the incidence angle is said to be positive. On the contrary, if these impinge upon the suction side first, the incidence angle is negative. The example in Figure 4.14 illustrates this, showing a turbine-blade incidence angle that is negative.

The deviation angle ϵ is computationally similar to the incidence angle, except that the flow and metal angles here correspond to the airfoil trailing edge instead (Figure 4.14). There is no need for, or use of, the sign of this particular angle, for it will always reflect a flow stream that is underturned. As seen in Figure 4.14, this flow unguidedness is the result of real-life degrading effects (e.g., viscosity-related consequences within the boundary layer, pressure-to-suction secondary-flow migration, flow separation and recirculation, etc.). Whereas the deviation angle in compressor cascades is considerably high, the same angle in turbine cascades is often so small that it can (intentionally) be overlooked. The reason, as pointed out in Chapter 3, is that compressor stator and rotor cascades provide, by definition, an adverse (or unfavorable) static pressure gradient in the streamwise direction. Such a pressure gradient is the primary factor in, for instance, the steep boundary-layer buildup (especially over the airfoil suction side) and an often premature flow separation. The favorable pressure gradient that prevails in turbine cascades, on the other hand, places the

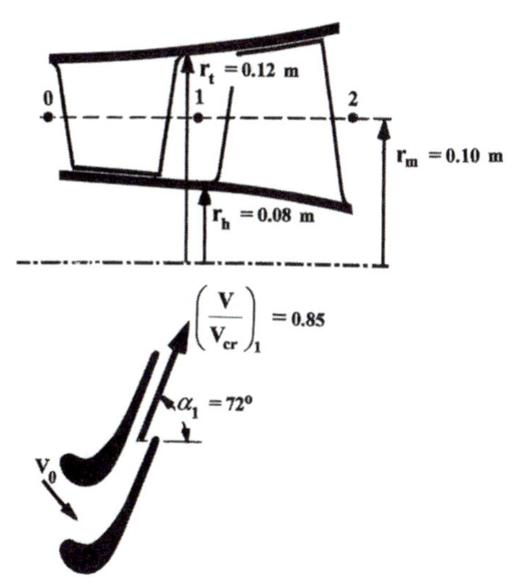

Figure 4.15. Input variables for Example 1.

deviation angle near zero unless:

- The cascade design itself is deficient, or
- The turbine stage is operating under a far-off-design mode, where an early flow separation takes place due to, say, excessively high incidence angles, particularly those that are positive (Chapters 8 and 10).

The compressor component unfavorable pressure gradient should perhaps explain why the number of compressor stages within the same engine can be double or even triple the number of turbine stages, in an attempt to minimize the per-stage adverse pressure gradient effects.

EXAMPLE 1

The turbine stage shown in Figure 4.15 is a symmetric (50% reaction) stage and is running at 45,000 rpm. The stage mean radius is 0.1 m, and the axial-velocity component is assumed to be constant throughout the entire stage. The flow conditions at station (1) are as follows:

- The stator-exit total pressure $p_{t1} = 7.4$ bars
- The stator-exit total temperature $T_{t1} = 1450$ K
- The stator-exit critical Mach number $M_{cr1} = 0.85$
- The stator-exit absolute flow angle $\alpha_1 = 72^\circ$

Assuming an adiabatic flow process and an average specific heat ratio (γ) of 1.33, calculate:

a) The static enthalpy drop across the rotor.
b) The rotor-exit absolute critical Mach number, M_{cr2};
c) The power (P) produced by the turbine stage.

SOLUTION

Part a:

$$V_{cr1} = \sqrt{\left(\frac{2\gamma}{\gamma+1}\right) R T_{t1}} = 689.3 \text{ m/s}$$

$$V_1 = 0.85 \times V_{cr,1} = 585.9 \text{ m/s}$$

$$V_{\theta 1} = V_1 \sin 72^\circ = 557.2 \text{ m/s}$$

$$V_{z1} = V_1 \cos 72^\circ = 181.1 \text{ m/s} = V_{z2} = V_z = W_{z1} = W_{z2}$$

$$U = \omega r_m = \frac{45,000 \times 2\pi}{60} \times 0.1 = 471.2 \text{ m/s}$$

$$W_{\theta 1} = V_{\theta 1} - U = 86.0 \text{ m/s}$$

$$W_{\theta 2} = -V_{\theta 1} = -557.2 \text{ m/s (50\% stage reaction, Figure 4.9)}$$

$$W_1 = \sqrt{{W_{\theta 1}}^2 + {W_{z1}}^2} = 200.4 \text{ m/s}$$

$$W_2 = \sqrt{{W_{\theta 2}}^2 + {W_{z2}}^2} = 585.9 \text{ m/s}$$

Now, we can calculate the static enthalpy drop across the rotor as follows:

$$\Delta h = \frac{{W_2}^2 - {W_1}^2}{2} = 151.5 \text{ kJ/kg}$$

Part b:

$$V_{\theta 2} = -W_{\theta 1} = -86.0 \text{ m/s (50\% reaction stage)}$$

$$w_s = U(V_{\theta 1} - V_{\theta 2}) = c_p(T_{t1} - T_{t2})$$

where

$$c_p = \frac{\gamma}{\gamma - 1} R \approx 1156.7 \text{ J/kg K}$$

Now, we can calculate the exit total temperature:

$$T_{t2} = 1188.0 \text{ K}$$

We now proceed to compute the exit critical Mach number, M_{cr2}:

$$V_{cr2} = 623.9 \text{ m/s}$$

$$V_2 = \sqrt{{V_{\theta 2}}^2 + {V_{z2}}^2} = 200.4 \text{ m/s } (= W_2, \text{ as would be expected})$$

$$M_{cr2} = V_2 / V_{cr2} = 0.32$$

Part c: Applying the continuity equation at the stator exit station,

$$\frac{\dot{m}\sqrt{T_{t1}}}{p_{t1}\left[\pi\left({r_{t1}}^2 - {r_{h1}}^2\right)\cos\alpha_1\right]} = \sqrt{\frac{2\gamma}{\gamma+1}} \times M_{cr1} \times \left(1 - \frac{\gamma-1}{\gamma+1}{M_{cr1}}^2\right)^{\frac{1}{\gamma-1}}$$

which, upon substitution, yields

$$\dot{m} = 5.83 \text{ kg/s}$$

Finally, we can calculate the power output:

$$P = \dot{m}w_s = \dot{m}U(V_{\theta 1} - V_{\theta 2}) = 1768.0 \text{ kW}$$

COMMENTS

1) Contrary to a compressor rotor, note that a turbine rotor will always produce a decline in the following variables:
   - The swirl (or tangential) velocity component (V_θ)
   - The total pressure (p_t)
   - The total temperature (T_t)
2) In *most* cases, however, the following thermophysical properties will decrease in the streamwise direction:
   - The static pressure (p)
   - The static temperature (T)
   - The absolute velocity (V) [An impulse (or 0% reaction) stage is the exception, in this case, as seen in Figure 4.9]
3) The stagewise equality of the axial-velocity component (V_z) is a design-related tradition but is not always the case. As will be explained in Chapters 8 and 9, dealing with axial-flow turbines and compressors, such a constraint (if implemented) will fix the magnitude of the so-called "flow coefficient" across the stage, and over a sequence of multiple stages, wherever applicable.

EXAMPLE 2

Shown in Figure 4.16 is the power-turbine stage in an auxiliary power unit (APU) and the stator-exit mean-radius absolute-velocity vector. The rotor speed is 17,900 rpm, and the specific shaft work, at the mean radius, is 75,000 J/kg. Assuming an average specific-heat ratio of 1.33, calculate:

a) The stage-exit flow angles α_2 and β_2;
b) The stage reaction;
c) The rotor incidence angle, knowing that $\beta_1' = 45°$.

SOLUTION

Part a: Let us assume that $V_{\theta 2}$ is positive, an assumption that can later be verified. Let us now proceed to compute the rotor-exit absolute and relative flow angles (α_2 and β_2, respectively):

$$U = \omega r_m = 150.0 \text{ m/s}$$

$$V_{\theta 1} = V_z \tan \alpha_1 = 321.3 \text{ m/s}$$

$$V_{\theta 2} = V_{\theta 1} - \tfrac{w_s}{U} = -178.7 \text{ m/s}$$

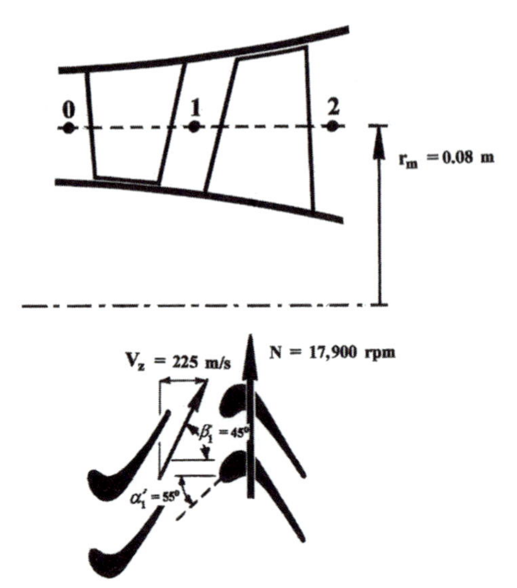

Figure 4.16. Input variables for Example 2.

This negative magnitude negates the assumption we made at the beginning. Now

$$\alpha_2 = \tan^{-1}\frac{V_{\theta 2}}{V_z} = -38.5°$$

$$\beta_2 = \tan^{-1}\frac{W_{\theta 2}}{V_z} = -55.6°$$

Part b:

$$V_1 = \sqrt{{V_{\theta 1}}^2 + V_z^2} = 392.3 \text{ m/s}$$

$$V_2 = \sqrt{{V_{\theta 2}}^2 + V_z^2} = 287.3 \text{ m/s}$$

$$W_1 = \sqrt{(V_{\theta 1} - U)^2 + V_z^2} = 282.8 \text{ m/s}$$

$$W_2 = \sqrt{(V_{\theta 2} - U)^2 + V_z^2} = 398.3 \text{ m/s}$$

The stage reaction (R) can now be calculated as

$$R = \frac{\left({W_2}^2 - {W_1}^2\right)}{\left({V_1}^2 - {V_2}^2\right) + \left({W_2}^2 - {W_1}^2\right)} = 52.4\%$$

Part c: In order to calculate the rotor-blade incidence angle (i_R), we first calculate the rotor-inlet relative flow angle (β_1) as follows:

$$\beta_1 = \tan^{-1}\frac{W_{\theta 1}}{V_z} = 37.3°$$

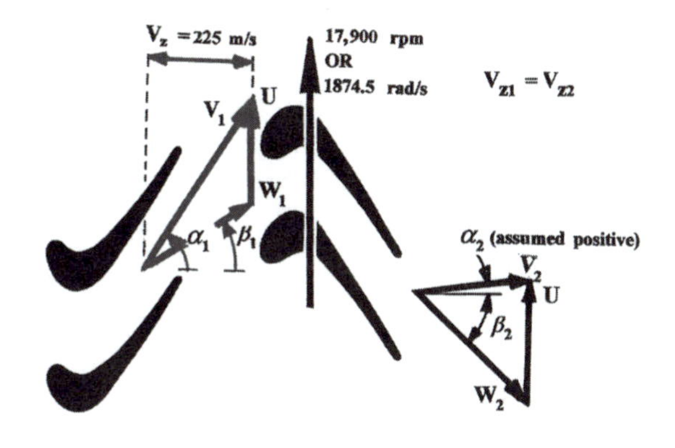

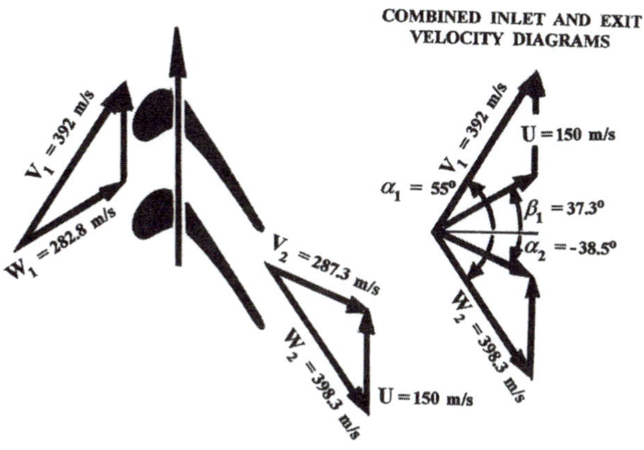

Figure 4.17. Details of the velocity triangles for Example 2.

Now, with the rotor-blade inlet (metal) angle β_1' being 45° (greater than β_1), we can conclude (by a simple sketch of the blade-inlet/flow interaction) that the blade incidence angle is negative:

$$i_R = \beta_1 - \beta_1' = -7.7^\circ$$

REMARK

Figure 4.17 provides sufficiently accurate velocity triangles at the rotor inlet and exit stations, including a pair that share the same vertex.

EXAMPLE 3

Figure 4.18 shows a single-stage turbine with a mean radius of 0.1 m. The turbine operating conditions are as follows:

- The mass-flow rate $\dot{m} = 4.3$ kg/s
- The turbine-inlet total temperature $T_{t0} = 1400$ K

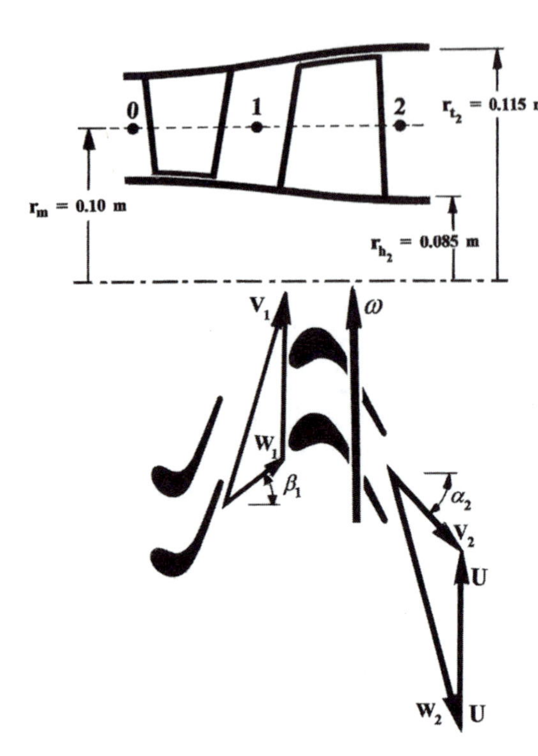

Figure 4.18. Velocity triangles for Example 3.

- The turbine-inlet total pressure $p_{t0} = 12.5$ bars
- The stator flow process is isentropic (by assumption)
- The rotational speed $(N) = 34{,}000$ rpm
- The axial velocity component is constant throughout the stage
- The stator-exit critical Mach number $= 0.85$
- The stator-exit absolute flow angle $\alpha_1 = 68^\circ$
- The rotor-exit relative critical Mach number $W/W_{cr2} = 0.82$
- The rotor relative-to-total pressure ratio $p_{tr2}/p_{tr1} = 0.822$

With this information:

(a) Calculate the torque τ delivered by the turbine stage.
(b) Calculate the stage total-to-total efficiency η_T.

SOLUTION

The importance of this problem stems from the fact that the continuity equation, this time, is applied in the rotating frame of reference. In order to do so, we will first have to compute the exit relative velocity (W_2), relative critical Mach number (W_2/W_{cr2}), the exit value of the relative flow angle (β_2), and the exit magnitudes of total relative pressure ($p_{t,r2}$) and temperature ($T_{t,r2}$). As a result of the flow adiabaticity, and because the turbine stage is of the axial-flow type, we know that the total relative temperature (T_{tr}) in particular will remain constant across the rotor. With this in mind, we can proceed as follows:

$$U = \omega r_m = 356.1 \text{ m/s}$$

$$V_{cr1} = \sqrt{\left(\frac{2\gamma}{\gamma+1}\right) R T_{t1}} = 677.3 \text{ m/s}$$

$$V_1 = M_{cr1} V_{cr1} = 575.7 \text{ m/s}$$

$$V_{\theta 1} = V_1 \sin\alpha_1 = 533.8 \text{ m/s}$$

$$W_{\theta 1} = V_{\theta 1} - U = 177.7 \text{ m/s}$$

$$V_{z1} = V_{z2} = V_1 \cos\alpha_1 = 215.7 \text{ m/s } = V_z$$

$$\beta_1 = \tan^{-1}\left(\frac{W_{\theta 1}}{V_z}\right) = 39.5^\circ$$

$$W_1 = \sqrt{{W_{\theta 1}}^2 + V_z^2} = 279.5 \text{ m/s}$$

$$T_{t,r1} = T_{t1} - \left(\frac{{V_1}^2 - {W_1}^2}{2c_p}\right) = 1290.5K = T_{t,r2}$$

$$W_{cr2} = \sqrt{\left(\frac{2\gamma}{\gamma+1}\right) R T_{t,r2}} = 650.3 \text{ m/s}$$

$$p_{t,r1} = p_{t1}\left(\frac{T_{t,r1}}{T_{t1}}\right)^{\frac{\gamma}{\gamma-1}} = p_{t0}\left(\frac{T_{t,r1}}{T_{t0}}\right)^{\frac{\gamma}{\gamma-1}} = 9.0 \text{ bars (isentropic stator)}$$

$$p_{t,r2} = 0.822 p_{t,r_1} = 7.4 \text{ bars (as listed in the problem statement)}$$

At this point, we are prepared to apply the continuity equation at the rotor exit station and in the rotating (rotor-attached) frame of reference:

$$\frac{\dot{m}\sqrt{T_{t,r_2}}}{p_{t,r_2}\left[\pi\left({r_{t2}}^2 - {r_{h2}}^2\right)\cos\beta_2\right]} = \sqrt{\frac{2\gamma}{(\gamma+1)R}} \times M_{r,cr,2} \times \left(1 - \frac{\gamma-1}{\gamma+1} M_{r,cr,2}^2\right)^{\frac{1}{\gamma-1}}$$

where

$$M_{r,cr} = \frac{W}{W_{cr}}$$

All variables in the preceding continuity equation are known, with the exception of the rotor-exit relative flow angle (β_2), which can now be calculated:

$$\beta_2 = -73.1^\circ \text{ (i.e., opposite to the direction of rotation)}$$

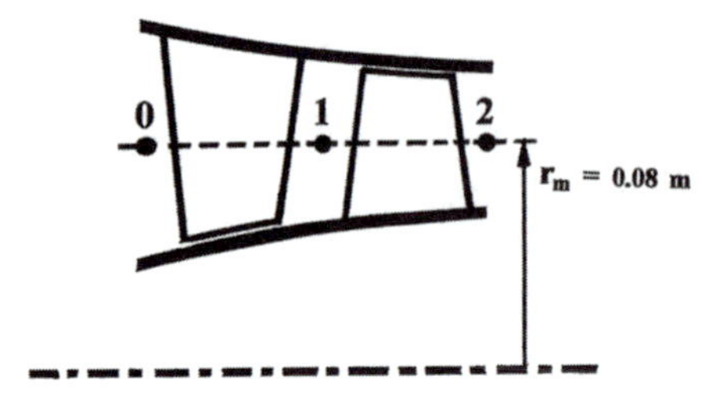

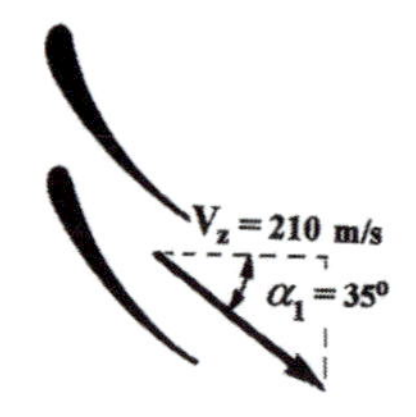

Figure 4.19. Input variables for Example 4.

Now, we proceed to calculate the shaft-exerted torque (τ) as follows:

$$W_{\theta 2} = V_z \tan \beta_2 = -711.6 \text{ m/s}$$

$$V_{\theta 2} = W_{\theta 2} + U = -355.6 \text{ m/s}$$

$$\tau = \dot{m} r_m (V_{\theta 1} - V_{\theta 2}) = 382.4 \text{ Nm}$$

In order to calculate the stage total-to-total efficiency (η_T), we should first calculate the stage-exit total temperature (T_{t2}) and pressure (p_{t2}) as follows:

$$T_{t2} = T_{t1} - \frac{U(V_{\theta 1} - V_{\theta 2})}{c_p} = 1126.3 \text{ K}$$

$$p_{t2} = p_{t,r,2} \left(\frac{T_{t2}}{T_{t,r,2}} \right)^{\frac{\gamma}{\gamma - 1}} = 4.28 \text{ bars}$$

Finally, we calculate the total-to-total stage efficiency (η_T) as follows:

$$\eta_T = \frac{1 - \left(\frac{T_{t2}}{T_{t1}} \right)}{1 - \left(\frac{p_{t2}}{p_{t1}} \right)^{\frac{\gamma - 1}{\gamma}}} = 83.6\%$$

EXAMPLE 4

Figure 4.19 shows an adiabatic compressor stage with a mean radius of 0.08 m. The stage is operating under the following conditions:

- The rotor-inlet absolute total pressure $p_{t1} = 1.85$ bars
- The rotor-inlet absolute total temperature $T_{t1} = 480$ K
- The rotational speed $(N) = 58{,}000$ rpm
- The axial-velocity component is constant throughout the stage

- The stator-exit absolute flow angle $\alpha_1 = +35^\circ$
- The rotor-exit absolute total pressure $p_{t2} = 3.1$ bars
- The rotor-exit relative flow angle (β_2) is negative
- The rotor-exit static temperature $T_2 = 496$ K

Assuming an average specific-heat ratio of 1.4, calculate:

1) The change in total relative pressure across the rotor;
2) The stage reaction;
3) The rotor-exit static pressure p_2.

SOLUTION

First, we calculate the magnitude of c_p knowing that $\gamma = 1.4$,

$$c_p = \frac{\gamma}{\gamma - 1} R \approx 1004.5 \text{ J/(kg K)}, \text{ where } R_{air} \approx 287.0 \text{ J/kg K}$$

Let us now calculate the relevant rotor-inlet and exit variables:

$$U = \omega r_m = 485.9 \text{ m/s}$$

$$V_1 = \frac{V_{z1}}{\cos \alpha_1} = 256.4 \text{ m/s (where } V_{z1} = V_{z2} = V_z)$$

$$V_{\theta 1} = V_{z1} \tan \alpha_1 = 147.0 \text{ m/s}$$

$$W_{\theta 1} = V_{\theta 1} - U = -338.9 \text{ m/s}$$

$$W_1 = \sqrt{{W_{\theta 1}}^2 + {V_{z1}}^2} = 398.7 \text{ m/s}$$

$$T_{(t,r,1)} = T_{t1} - \left(\frac{{V_1}^2 - {W_1}^2}{2c_p} \right) = 526.4 \text{ K}$$

$$= T_{t,r,2}$$

$$= T_2 + \frac{{W_2}^2}{2c_p} = 496.0 + \frac{{W_2}^2}{2c_p}$$

which yields

$$W_2 = 247.1 \text{ m/s}$$

Continuing the calculations at the rotor exit station, we have

$$\beta_2 = \cos^{-1}\left(\frac{V_z}{W_2}\right) = -31.8^\circ \text{ (sign is cited in the problem statement)}$$

$$W_{\theta 2} = V_z \tan \beta_2 = -130.2 \text{ m/s}$$

$$V_{\theta 2} = W_{\theta 2} + U = 355.7 \text{ m/s}$$

$$V_2 = \sqrt{V_{\theta 2}^2 + V_z^2} = 413.1 \text{ m/s}$$

$$T_{t2} = T_2 + \frac{V_2^2}{2c_p} = 580.9 \text{ K}$$

$$p_{(t,r,1)} = p_{t1}\left(\frac{T_{(t,r,1)}}{T_{t1}}\right)^{\frac{\gamma}{\gamma-1}} = 2.56 \text{ bars}$$

$$p_{t,r,2} = p_{t2}\left(\frac{T_{t,r,2}}{T_{t2}}\right)^{\frac{\gamma}{\gamma-1}} = 2.20 \text{ bars}$$

Indicative of the rotor total-to-total efficiency, the total relative pressure decline (across the rotor) can now be calculated:

$$\Delta p_{t_r} = p_{(t,r,1)} - p_{t,r,2} = 0.36 \text{ bars}$$

Part b: Having computed all absolute and relative velocities at the rotor inlet and exit stations, the stage reaction (R) is obtained by direct substitution in expression (4.22):

$$R = 48.3\%$$

Part c: Starting with the rotor-exit critical Mach number $M_{cr2} = V_2/V_{cr2}$, we can calculate the exit static pressure as follows:

$$M_{cr2} = \frac{V_2}{V_{cr2}} = \frac{V_2}{\sqrt{\left(\frac{2\gamma}{\gamma+1}\right)RT_{t2}}} = 0.937$$

$$p_2 = p_{t2}\left(1 - \frac{\gamma-1}{\gamma+1}M_{cr2}^2\right)^{\frac{\gamma}{\gamma-1}} = 1.78 \text{ bars}$$

EXAMPLE 5

The turbine-rotor blade cascade in Figure 4.20 is switched to operate as a compressor cascade by transmitting (to the blades) a torque that gives rise to the direction of rotation shown in the figure. (The pressure side is leading in the rotation direction.) If V_z is maintained constant across the rotor, indicate which of the following statements are factually true and which are false:

a) The reaction is greater than 100%. True or false
b) The change in Δh_t will be negative. True or false
c) α_2 will be greater than β_2. True or false

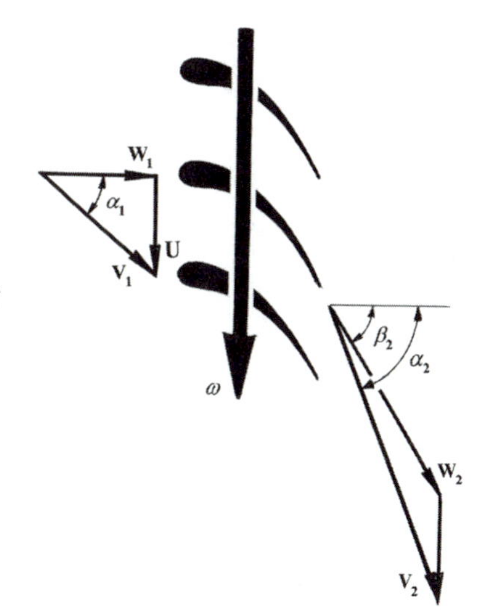

Figure 4.20. Inlet and exit velocity triangles for Example 5.

SOLUTION

Part a: Referring to Figure 4.20, and recalling the definition of the compressor-stage reaction in expression (4.22), we can conclude that:

- W_1 is less than W_2, or $(W_1{}^2 - W_2{}^2)$ is negative.
- On an absolute-magnitude basis, $(V_2{}^2 - V_1{}^2)$ is greater than $(W_1{}^2 - W_2{}^2)$, with V_2 being much greater than V_1, as shown in the figure.

Based on these two facts, and in light of equation (4.22), we now have

$$\text{Stage reaction}(R) = (\text{negative quantity})/(\text{positive quantity})$$

which means that the stage reaction in this case is negative. As a result, the first proposition in the problem statement is *false*.

Part b: Let us now recall the energy-conservation/Euler statement, namely

$$h_{t2} - h_{t1} = w_s = U(V_{\theta 2} - V_{\theta 1})$$

Referring to the velocity triangles in Figure 4.20, we clearly see that $V_{\theta 2}$ is much greater than $V_{\theta 1}$, with both components being in the direction of rotation (i.e., positive). With this in mind, we see that h_t will indeed rise across the rotor. This means that the second stated proposition is *false*.

Part c: Examination of Figure 4.20 reveals that the exit swirl angle (α_2) is much greater than the exit relative flow angle β_2, with both of them being positive. This simply means that the third proposed statement is *true*.

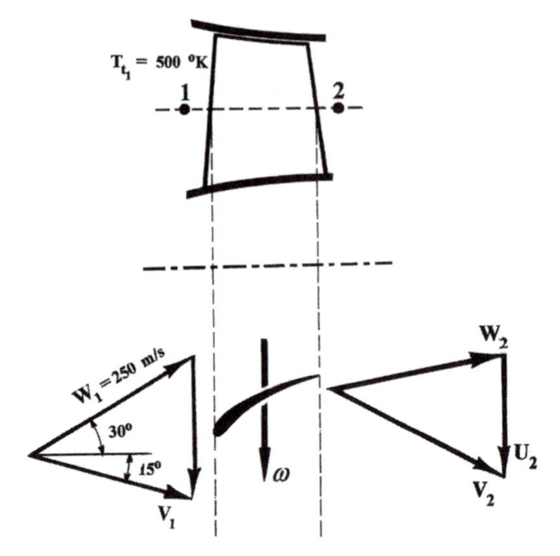

Figure 4.21. Input variables for Example 6.

EXAMPLE 6

Figure 4.21 shows the rotor-inlet velocity triangle in an axial-flow compressor rotor. The compressor-stage reaction is 50% at the mean radius. Also, the inlet absolute total temperature (T_{t1}) is 500 K, and the absolute velocity angle is 15°. Assuming a constant axial-velocity component, an isentropic stator, and a specific heat ratio (γ) of 1.4:

a) Sketch the rotor-exit velocity triangle.
b) At the rotor exit station, calculate T_t, T_{t_r}, and T.
c) Assuming a zero incidence angle and zero deviation angle, sketch the mean-radius rotor-blade airfoil.

SOLUTION

Part a: Noting that the stage reaction is 50%, and referring to Figure 4.21, the following relationships hold true:

$$W_2 = V_1$$

$$V_2 = W_1$$

$$\alpha_2 = -\beta_1$$

$$\beta_2 = -\alpha_1$$

With these facts, construction of the exit velocity triangle (shown in Figure 4.21) is straightforward.

Part b:

$$V_z = 250.0 \cos 30^\circ = 216.5 \text{ m/s}$$

$$V_1 = \frac{V_z}{\cos 15^\circ} = 224.1 \text{ m/s}$$

$$U = V_{\theta 1} - W_{\theta 1} = V_1 \sin(15^\circ) - W_1 \sin(-30^\circ) = 183.0 \text{ m/s}$$

$$T_{t,r,2} = T_{(t,r,1)} = T_{t1} - \frac{(V_1^2 - W_1^2)}{2c_p} = 506.1 \text{ K}$$

Combining fine the energy-conservation and Euler equations gives rise to the following well-known expression:

$$w_s = U(V_{\theta 2} - V_{\theta 1}) = c_p(T_{t2} - T_{t1})$$

Application of this expression enables us to proceed toward computing the rotor-exit total relative temperature and static temperature as follows:

$$T_{t2} = T_{t1} + \frac{U(V_{\theta 2} - V_{\theta 1})}{c_p} = 512.2 \text{ K}$$

$$T_2 = T_{t2} - \frac{V_2^2}{2c_p} = 475.0 \text{ K}$$

Part c: A sketch of the rotor blade, at the mean radius, is part of Figure 4.21.

EXAMPLE 7

Air enters the first stage of an axial-flow compressor stage at a total pressure of 2.0 bars and a total temperature of 350 K. The axial-velocity component is 200 m/s (constant throughout the stage), and the stator flow process is assumed to be isentropic. In addition, the stage total-to-total pressure ratio is 1.75, the total-to-total efficiency is 82%, the rotational speed is 32,000 rpm, the rotor-exit relative flow angle β_2 is zero, and the average specific-heat ratio is 1.4. Calculate the following variables:

a) Rotor-inlet and rotor-exit absolute flow angles;
b) Rotor-inlet and rotor-exit total relative temperature;
c) Stage reaction.

SOLUTION

In the following solution procedure, reference is made to the inlet and exit velocity triangles in Figure 4.21. Despite the differences in the magnitudes of the variables computed here, this figure conceptually depicts a similar picture in the current example.

Part a: By definition,

$$\eta_C = 0.82 = \frac{\left(\frac{p_{t2}}{p_{t1}}\right)^{\frac{\gamma-1}{\gamma}} - 1}{\left(\frac{T_{t2}}{T_{t1}}\right) - 1}$$

which yields

$$T_{t2} = 424.0 \text{ K}$$

Furthermore, we have

$$U = \omega r_m = 234.7 \text{ m/s}$$

$$V_{\theta 2} = U = 234.7 \text{ m/s } (\beta_2 = 0)$$

$$V_{\theta 1} = V_{\theta 2} - \frac{c_p(T_{t2} - T_{t1})}{U} = -82.3 \text{ m/s}$$

$$\alpha_1 = \tan^{-1}\frac{V_{\theta 1}}{V_z} = -22.4^0$$

Part b:

$$W_1 = \sqrt{(V_{\theta 1} - U)^2 + V_z^2} = 374.8 \text{ m/s}$$

$$T_{(t,r,1)} = T_{t1} - \left(\frac{V_1^2 - W_1^2}{2c_p}\right) = 396.6 \text{ K}$$

Part c:

$$V_1 = V_z / \cos\alpha_1 = 216.3 \text{ m/s}$$

$$W_1 = 374.8 \text{ m/s}$$

$$W_2 = V_z = 200 \text{ m/s (a result of } \beta_2 \text{ being zero)}$$

$$V_2 = \sqrt{V_{\theta 2}^2 + V_z^2} = \sqrt{U^2 + V_z^2} = 308.4 \text{ m/s}$$

Finally, we are now in a position to calculate the stage reaction (R) by direct substitution in expression (4.22):

$$R = 67.5\%$$

EXAMPLE 8

Figure 4.22 shows the rotor subdomain of an axial-flow turbine and its major dimensions. The rotor operating conditions are as follows:

- Inlet total pressure (p_{t1}) = 13.2 bars
- Inlet total temperature (T_{t1}) = 1405.0 K
- Inlet critical Mach number (V_1/V_{cr1}) = 0.92
- Exit total pressure (p_{t2}) = 5.4 bars
- Exit total temperature (T_{t2}) = 1162.0 K
- M_{cr2} is small enough to justify the equality of static and total densities at the rotor exit station
- Mass-flow rate ($\dot{m}$) = 11.4 kg/s
- Shaft speed (N) = 48,300 rpm

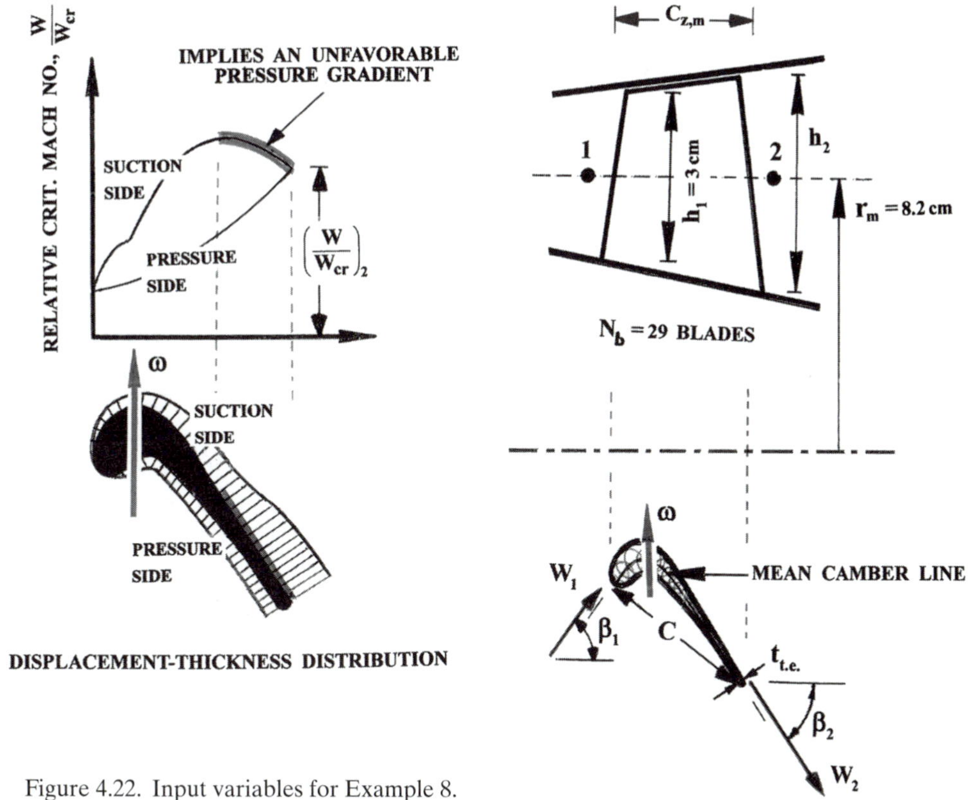

Figure 4.22. Input variables for Example 8.

The following geometrical data are also applicable:

- Number of blades (N_b) = 29
- Trailing-edge thickness ($t_{t.e.}$) = 2.5 mm
- Mean-radius axial chord length (C_z) = 2.5 cm
- Mean-radius true chord (C) = 4.6 cm
- Mean camber-line length (L) = 9.2 cm

SEGMENT 1

a) Beginning with the assumption of an unchoked rotor, calculate the blade-exit annulus height (h_2), that will ensure a rotorwise constant magnitude of V_z.
b) Now verify whether the rotor is actually choked.
c) In the event of rotor choking, recompute all variables that are pertinent to the exit velocity triangle.

SEGMENT 2

Consider the following expression for calculating the rotor kinetic-energy loss coefficient ($\bar{e}_R$):

$$\bar{e}_R = \left(\frac{\theta_{tot}}{S_m \cos\beta_2 - t_{t.e.} - \delta_{tot}{}^*}\right)\left(\frac{A_{3D}}{A_{2D}}\right)\left(\frac{Re}{Re_{ref}}\right)^{-0.2}$$

where

- $\bar{e}_R$ is the kinetic-energy loss coefficient
- θ_{tot} refers to the combined (pressure and suction sides) momentum thicknesses
- $\delta_{tot}{}^*$ refers to the combined displacement thicknesses
- S_m is the mean-radius blade-to-blade spacing
- β_2 is the rotor-exit relative flow angle
- $t_{t.e.}$ is the trailing-edge thickness
- A_{2D} is the blade surface area that is in contact with the flow stream
- A_{3D} is A_{2D} plus the blade-to-blade hub and casing surface areas
- Re is the Reynolds number, based on the true chord and the exit relative velocity (i.e., C and W_2, respectively)
- Re_{ref} is a reference Reynolds number and is equal to 7.57×10^6

Now, apply the $\bar{e}_R$ expression to calculate the total relative pressure loss (i.e., $p_{t r1} - p_{t r2}$) over the rotor.

In carrying out this task, you may assume turbulent boundary layers over the pressure and suction sides. Also, when using expressions (3.80) and (3.81), consider the outcome as belonging only to the pressure side. As for the suction side, use a magnifying multiplier of 3.5.

Clarification: The reason for this multiplier is shown in Figure 4.22. As seen in the figure, the boundary layer grows much faster on the suction side. Consequently, both the displacement and momentum thicknesses are typically much larger than those over the pressure side. Part of the reason behind such a substantial difference is the strong likelihood of the suction side being exposed to flow-deceleration regions (Figure 4.22), which causes a local rise in static pressure. As discussed in Chapter 3, this constitutes an adverse (or unfavorable) pressure gradient, a factor that aggravates the suction-side boundary-layer buildup.

COMMENT

The importance of this problem has to do with some of the most serious rotor-choking consequences. As seen in Figure 4.23 (discussed later), there can really be major changes in the rotor-exit velocity triangle as a result of choking, as opposed to "blindly" treating the rotor as unchoked. In reality, turbomachinists are not exactly immune from making the mistake of ignoring the process of verifying whether the passage (stationary or rotating) is choked. This is particularly the case in situations where the need to compute the critical Mach number (absolute or relative) is nonexistent.

Noteworthy as well is a late step in the problem solution where a loss assessor, namely the kinetic-energy loss coefficient ($\bar{e}$), is converted into a total-pressure loss coefficient ($\bar{\omega}$). Because the object in this problem is a rotor, a new definition of $\bar{e}$ will now be cast in terms of relative thermophysical properties. The conversion from $\bar{e}$ into $\bar{\omega}$ (the reader will notice) is made through an intermediate step, where the entropy production (Δs) will be executed.

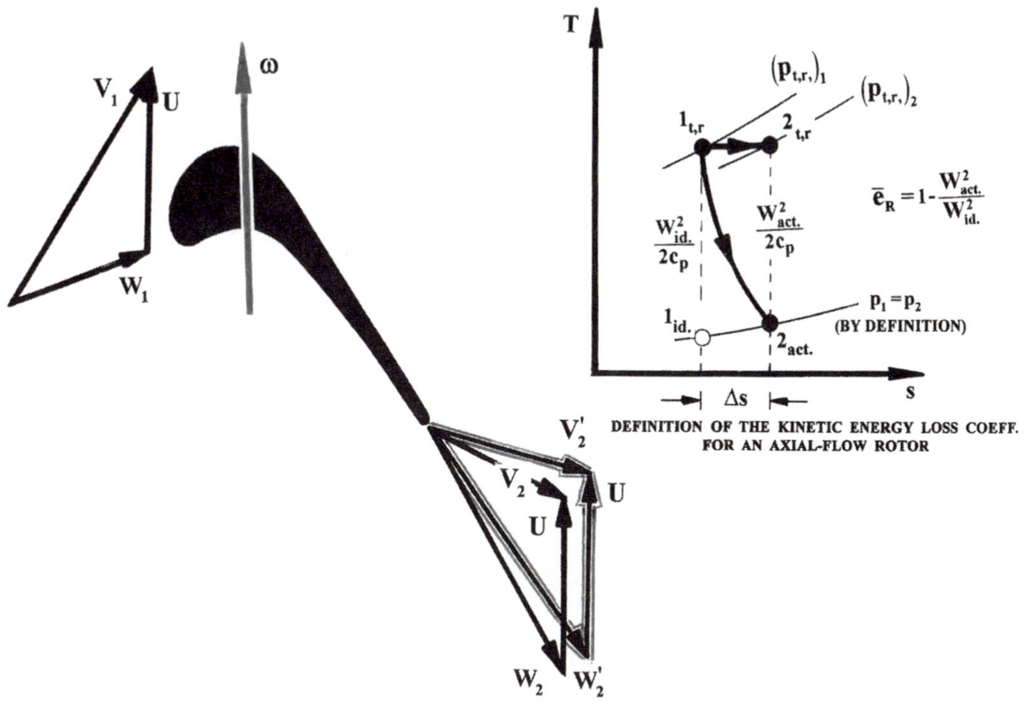

Figure 4.23. Rotor-choking effect on the exit velocity triangle.

SOLUTION

SEGMENT 1

Part 1a:

$$V_{cr1} = \sqrt{\left(\frac{2\gamma}{\gamma+1}\right)RT_{t1}} = 678.5 \text{ m/s}$$

$$V_1 = M_{cr1}V_{cr1} = 624.2 \text{ m/s}$$

$$\rho_1 = \left(\frac{p_{t1}}{RT_{t1}}\right)\left(1 - \frac{\gamma-1}{\gamma+1}M_{cr1}{}^2\right)^{\frac{1}{\gamma-1}} = 2.22 \text{ kg/m}^3$$

$$\dot{m} = \rho_1 V_{z1} A_1 = \rho_1 V_{z1}(2\pi r_m h_1)$$

which yields

$$V_{z1} = 332.2 \text{ m/s}$$

Now, we proceed toward eventually calculating the rotor-exit annulus height (h_2) as follows:

$$\alpha_1 = \cos^{-1}\left(\frac{V_{z1}}{V_1}\right) = 57.8^\circ$$

$$V_{\theta 1} = V_1 \sin\alpha_1 = 528.4 \text{ m/s}$$

$$U = \omega r_m = 414.8 \text{ m/s}$$

$$W_{\theta 1} = V_{\theta 1} - U = 113.6 \text{ m/s}$$

$$W_1 = \sqrt{{W_{\theta 1}}^2 + {V_{z1}}^2} = 351.1 \text{ m/s}$$

$$\beta_1 = \tan^{-1}\left(\frac{W_{\theta 1}}{V_{z1}}\right) = 18.9^\circ$$

$$T_{(t,r,1)} = T_{t1} - \left(\frac{{V_1}^2 - {W_1}^2}{2c_p}\right) = 1289.9 \text{ K}$$

$$p_{(t,r,1)} = p_{t1}\left(\frac{T_{(t,r,1)}}{T_{t1}}\right)^{\frac{\gamma}{\gamma-1}} = 9.35 \text{ bars}$$

Now, we set $V_{z2} = V_{z1} = 332.2$ m/s and the exit static density equal to the exit total magnitude, as an approximation,

$$\rho_2 \approx \rho_{t2} = \frac{p_{t2}}{RT_{t2}} = 1.62 \text{ kg/m}^3 \text{ (assuming } M_{cr2} \text{ to be negligible)}$$

These two simplifications enable us to compute the rotor-exit annulus height by applying the continuity equation as follows:

$$h_2 = \frac{\dot{m}}{\rho_2 V_{z2}(2\pi r_m)} = 4.11 \text{ cm}$$

Part 1b: In order to verify the choking status of the rotor passage, we have to calculate the *relative* critical Mach number at the passage exit station. To this end, we proceed as follows:

$$V_{\theta 2} = V_{\theta 1} - \frac{c_p(T_{t1} - T_{t2})}{U} = -149.2 \text{ m/s}$$

$$\alpha_2 = \tan^{-1}\left(\frac{V_{\theta 2}}{V_{z2}}\right) = -24.2^\circ$$

$$V_2 = \sqrt{{V_{\theta 2}}^2 + {V_{z2}}^2} = 364.2 \text{ m/s}$$

$$W_{\theta 2} = V_{\theta 2} - U = -564.0 \text{ m/s}$$

$$W_2 = \sqrt{{W_{\theta 2}}^2 + {V_{z2}}^2} = 654.6 \text{ m/s}$$

$$\beta_2 = \tan^{-1}\left(\frac{W_{\theta 2}}{V_{z2}}\right) = -59.5^\circ$$

$$T_{tr2} = T_{tr1} = 1289.9 \text{ K}$$

$$W_{cr2} = \sqrt{\frac{2\gamma}{\gamma+1} R T_{t2}} = 650.1 \text{ m/s}$$

$$\frac{W_2}{W_{cr2}} = 1.01 \text{ (\textit{impossible} for a convergent nozzle)}$$

What this means is that the rotor passage is *choked*.

Part 1c: Due to the newly realized rotor choking, we have to implement the following corrective actions:

- Set the rotor-exit relative velocity (W_2) equal to the relative critical velocity (W_{cr2}).
- Reapply the continuity equation at the rotor-exit station and in the rotating frame of reference, so that the actual value of the exit relative flow angle (β_2) can be obtained.
- Make all other changes in the variables comprising the rotor-exit velocity triangle. Note that the axial velocity component (V_z) will no longer remain constant across the rotor.

We now proceed to implement these changes, beginning with the calculation of rotor-exit relative properties:

$$p_{t,r,2} = p_{t2}\left(\frac{T_{t,r,2}}{T_{t2}}\right)^{\frac{\gamma}{\gamma-1}} = 8.23 \text{ bars}$$

$$T_{t,r,2} = 1289.9 \text{ K (computed earlier)}$$

Applying the continuity equation in the rotating frame of reference at the rotor-exit station, we have

$$\frac{\dot{m}\sqrt{T_{t,r,2}}}{p_{t,r,2}[(2\pi r_m h_2)\cos\beta_2]} = \sqrt{\frac{2\gamma}{(\gamma+1)R}} \times \frac{W_2}{W_{cr2}}\left[1 - \frac{\gamma-1}{\gamma+1}\left(\frac{W_2}{W_{cr2}}\right)^2\right]^{\frac{1}{\gamma-1}}$$

which yields

$$\beta_2 = 53.7^\circ$$

Referring to Figure 4.23, the sign of β_2 is indeed negative, namely

$$\beta_2 = -53.7^\circ$$

This value is certainly different from the previously computed value of -59.5°, with the difference being the recognition (at this point) of the rotor-passage choking status.

Pursuing this corrective procedure, we have

$$W_{\theta 2} = W_2 \sin \beta_2 = -523.9 \text{ m/s (instead of } -564.0 \text{ m/s)}$$

$$T_2 = T_{t,r,2} - \frac{W_2{}^2}{2c_p} = 1107.2K$$

$$V_{\theta 2} = W_{\theta 2} + U = -109.1 \text{ m/s (instead of } -149.2 \text{ m/s)}$$

$$V_{z2} = W_{z2} = W_2 \cos \beta_2 = 384.9 \text{ m/s (instead of 332.2 m/s)}$$

$$V_2 = \sqrt{V_{\theta 2}{}^2 + V_{z2}{}^2} = 400.1 \text{ m/s (instead of 364.2 m/s)}$$

$$\alpha_2 = \tan^{-1}(V_{\theta 2} V_{z2}) = -15.8^\circ \text{ (instead of } -24.2^\circ)$$

$$T_{t2} = T_2 + \frac{V_2{}^2}{2c_p} = 1176.4 \text{ K (instead of 1162.0 K)}$$

With the velocity components computed, we can easily generate the choking-altered rotor-exit velocity triangle. This is superimposed on the original triangle in Figure 4.23.

SEGMENT 2

A good first step is to list the magnitudes of different variables that are relevant to the calculation of the kinetic-energy loss coefficient, $\bar{e}_R$, as follows:

- $h_{av.} = \frac{1}{2}(h_1 + h_2) = 3.56$ cm
- $S_m = \frac{2\pi r_m}{N_b} = 1.78$ cm (blade-to-blade spacing)
- $\beta_2 = -59.7^\circ$
- $t_{t.e.} = 2.5$ mm
- $T_2 = 1107.2$ K
- $\mu_2 = 6.7 \times 10^{-5}$ kg/(m/s) (from Figure 3.39)
- $\rho_2 \approx \rho_{t2} = 1.62$ kg/m^3
- $\nu_2 = \frac{\mu_2}{\rho_2} = 4.14 \times 10^{-5}$m^2/s
- $W_2 = 650.1$ m/s
- mean camber-line length (L) = 9.2 cm
- true chord $(C) = 4.6$ cm
- $\delta_{press.}{}^* = 0.057L\,[\frac{W_2 L}{\nu_2}]^{-1/6} = 0.493$ mm
- $\delta_{tot.}{}^* = \delta_{press.}{}^* + \delta_{suc.}{}^* = \delta_{press.}{}^* + 3.5\delta_{press.}{}^* = 2.22$ mm
- $\theta_{press.} = 0.022L\,[\frac{W_2 L}{\nu_2}]^{-1/6} = 0.190$ mm
- $\theta_{tot.} = 0.857$ mm
- $A_{2D} \approx 2Lh_{av.} = 0.00655$ m^2
- $A_{3D} \approx A_{2D} + 2S_m C_{z,m} = 0.00744$ m^2
- $Re = \frac{W_2 C}{\nu_2} = 7.22 \times 10^{-5}$

Upon substituting these variables for $\bar{e}_R$ in the expression provided, we obtain

$$\bar{e}_R = 0.268$$

Noting that a rotor subdomain can be viewed as that of a stator, provided *relative* properties are utilized, we can legitimately adapt the $\bar{e}$ definition in equation (3.56) in Chapter 3 to suit our rotor subdomain as follows:

$$\bar{e}_R = 1 - \frac{W_{act.}{}^2}{W_{id.}{}^2} \text{ for a rotor subdomain}$$

In fact, Figure 3.24 may still be used as a graphical means here, except that the total pressure lines (in this figure) now represent total relative pressures. Note that the static states are common in both stationary and rotating frames of reference when dealing with a rotor (Figure 4.4). However, in the interest of clarifying the thermodynamic aspects behind the computational steps, the variables involved in computing $\bar{e}_{rotor}$ (or $\bar{e}_R$) are displayed on a T-s diagram as part of Figure 4.23.

Now, substituting in $\bar{e}_R$ expression, we get

$$W_{act.} = 300.4 \text{ m/s}$$

where the following substitution was made:

$$W_{id.} = W_1 = 351.1 \text{ m/s}$$

Also,

$$T_{id.} = T_{t_{r_1}} - \frac{W_1{}^2}{2c_p} = 1236.6 \text{ K}$$

$$T_{act.} = T_{(t,r,1)} - \frac{W_{act.}{}^2}{2c_p} = 1250.9 \text{ K}$$

By selecting the most convenient states (namely $1_{id.}$ and $2_{act.}$ in this case, both static states), we can calculate the rotor-generated entropy production as follows:

$$\Delta s_R = c_p \ln\left(\frac{T_{act.}}{T_{id.}}\right) - R \ln\left(\frac{p_{act.}}{T_{id.}}\right) = 13.29 \text{ J/(kg K)}$$

As explained in Chapter 3, we are free (in computing Δs) to select *any* two (inlet and exit) states. In other words, the rotor flow process (when we get to this point) can be viewed from any of the total-to-total, total-to-total relative or total relative-to-static standpoints (see Figure 4.23). In order to calculate $p_{t,r,2}$ (the next step), let us select the two states "1_{tr} and 2_{tr} (both being total relative states) for expressing $\Delta s_{rot.}$:

$$\Delta s_R = 13.29 \text{ J/(kg K)} = c_p \ln\left(\frac{T_{t,r,2}}{T_{(t,r,1)}}\right) - R\left(\frac{p_{t,r,2}}{T_{(t,r,1)}}\right)$$

Recalling that we are dealing with an *adiabatic axial-flow* rotor, we know that

$$T_{t,r,2} = T_{(t,r,1)}$$

which, together with the Δs_R expression, yields

$$p_{t,r,2} = 8.93 \text{ bars}$$

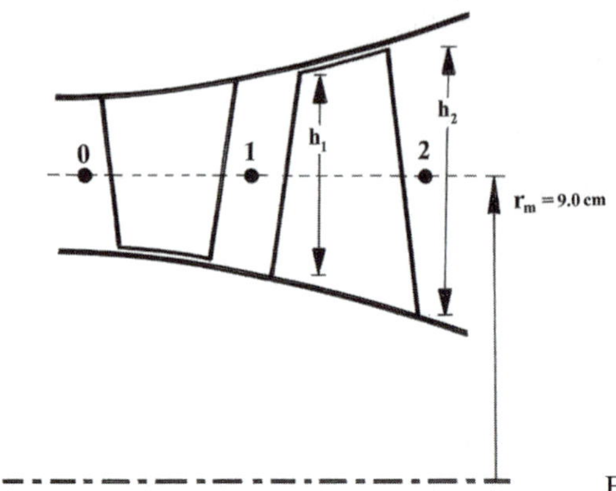

Figure 4.24. Input variables for Example 9.

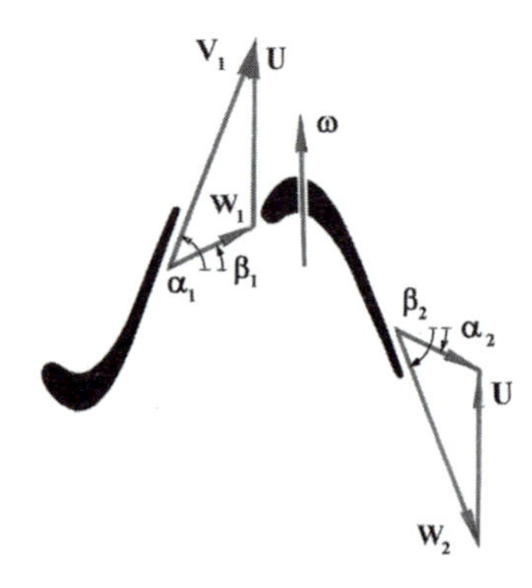

which is less than $p_{(t,r,1)}$, as would be anticipated. Finally, we can compute the total relative pressure loss as follows:

$$\frac{\Delta p_{t\,r}}{p_{(t,r,1)}} = 4.53\%$$

EXAMPLE 9

Figure 4.24 shows an axial-flow turbine stage, along with its major dimensions. The stage is operating under the following conditions:

- Rotor-inlet total pressure $(p_{t1}) = 12.5$ bars
- Rotor-inlet total temperature $(T_{t1}) = 1285.0$ K
- Rotor speed $(N) = 46{,}000$ rpm
- Rotor-inlet critical Mach number $(M_{cr\,1}) = 0.86$
- Rotor-inlet swirl angle $(\alpha_1) = 66.0^\circ$
- Rotor total-to-total efficiency $(\eta_{t-t}) = 89\%$
- Mass-flow rate $(\dot{m}) = 13.4$ kg/s
- Stage reaction $(R) = 50\%$
- Adiabatic flow throughout the stage
- A constant specific-heat ratio (γ) of 1.33
- Constant axial-velocity component (V_z)

Calculate the rotor-exit annulus height (h_2) by applying the continuity equation:

a) In the stationary frame of reference;
b) In the rotating frame of reference.

SOLUTION

Because the stage reaction is 50%, it then suffices to compute the rotor-inlet velocity-triangle variables, as the exit triangle is simply a mirror image of the former (this is the reason such a stage is often termed *symmetric*).

$$V_{cr1} = \sqrt{\left(\frac{2\gamma}{\gamma+1}\right) RT_{t1}} = 648.9 \text{ m/s}$$

$$V_1 = M_{cr1} V_{cr1} = 558.0 \text{ m/s}$$

$$V_{z1} = V_1 \cos\alpha_1 = 227.0 \text{ m/s}$$

$$V_{\theta 1} = V_1 \sin\alpha_1 = 509.8 \text{ m/s}$$

$$U = \omega r_m = 433.5 \text{ m/s}$$

$$W_{\theta 1} = V_{\theta 1} - U = 76.3 \text{ m/s}$$

$$W_1 = \sqrt{{W_{\theta 1}}^2 + {V_{z1}}^2} = 239.5 \text{ m/s}$$

$$\beta_1 = \tan^{-1}\left(\frac{W_{\theta 1}}{V_{z1}}\right) = 18.6^\circ$$

$$\rho_1 = \frac{p_{t1}}{RT_{t1}}\left[1 - \left(\frac{\gamma-1}{\gamma+1}\right) {M_{cr1}}^2\right]^{\frac{1}{\gamma-1}} = 2.42 \text{ kg/m}^3$$

Let us now apply the continuity equation at the stator-exit station in order to calculate h_1:

$$\dot{m} = \rho_1 V_{z1}(2\pi r_m h_1)$$

which yields

$$h_1 = 4.3 \text{ cm}$$

Moving to the rotor exit station, we have

$$V_{z2} = V_{z1} = 227.0 \text{ m/s (given in the problem statement)}$$

$$V_2 = W_1 = 239.5 \text{ m/s (50\% reaction stage)}$$

$$W_2 = V_1 = 558.0 \text{ m/s}$$

$$\alpha_2 = -\beta_1 = -18.6^\circ$$

$$\beta_2 = -\alpha_1 = -66.0^\circ$$

$$V_{\theta 2} = -W_{\theta 1} = -76.3 \text{ m/s}$$

Now, we apply the Euler/energy-conversion relationship, namely

$$U(V_{\theta 1} - V_{\theta 2}) = c_p(T_{t1} - T_{t2})$$

which yields

$$T_{t2} = 1065.3 \text{ K}$$

Also,

$$V_{cr2} = \sqrt{\left(\frac{2\gamma}{\gamma+1}\right) RT_{t2}} = 590.8 \text{ m/s}$$

$$M_{cr2} = \frac{V_2}{V_{cr2}} = 0.405$$

$$T_{t,r,2} = T_{(t,r,1)} = 1175.2 \text{ K (adiabatic flow)}$$

$$W_{cr2} = \sqrt{\left(\frac{2\gamma}{\gamma+1}\right) RT_{t,r,2}} = 620.5 \text{ m/s}$$

$$M_{cr2} = \frac{W_2}{W_{cr2}} = 0.899$$

Furthermore,

$$\eta_{t-t} = 0.89 = \frac{1 - (T_{t2}/T_{t0})}{1 - (p_{t2}/p_{t0})^{\frac{\gamma-1}{\gamma}}}$$

which yields

$$p_{t2} = 5.29 \text{ bars}$$

$$p_{t,r,2} = p_{t2}\left(\frac{T_{t,r,2}}{T_{t2}}\right)^{\frac{\gamma}{\gamma-1}} = 7.86 \text{ bars}$$

Part a: Let us now compute the rotor-exit annulus height h_2 by applying the continuity equation in the *stationary* frame of reference (i.e., using *absolute* thermophysical properties):

$$\frac{\dot{m}\sqrt{T_{t2}}}{p_{t2}[(2\pi r_m h_2)\cos\alpha_2]} = \sqrt{\frac{2\gamma}{(\gamma+1)R}} \times M_{cr2} \times \left(1 - \frac{\gamma-1}{\gamma+1} M_{cr2}{}^2\right)^{\frac{1}{\gamma-1}}$$

which yields:

$$h_2 = 6.48 \text{ cm}$$

Part b: We now apply the continuity equation in the *rotating* frame of reference (i.e., using *relative* flow properties):

$$\frac{\dot{m}\sqrt{T_{t,r,2}}}{p_{t,r,2}[(2\pi r_m h_2)\cos\beta_2]} = \sqrt{\frac{2\gamma}{(\gamma+1)R}} \times M_{r,cr2} \times \left(1 - \frac{\gamma-1}{\gamma+1} M_{cr2}{}^2\right)^{\frac{1}{\gamma-1}}$$

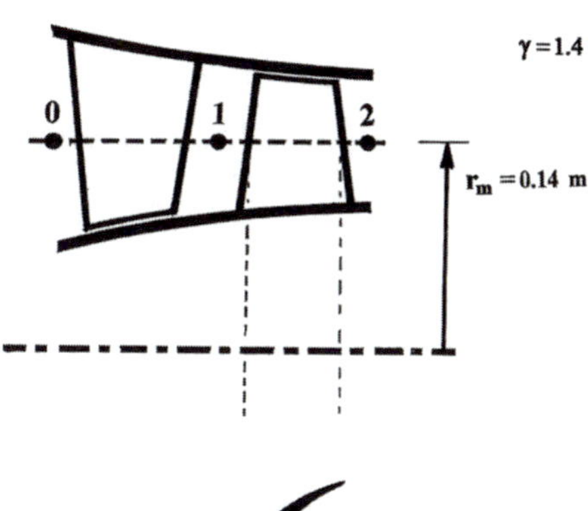

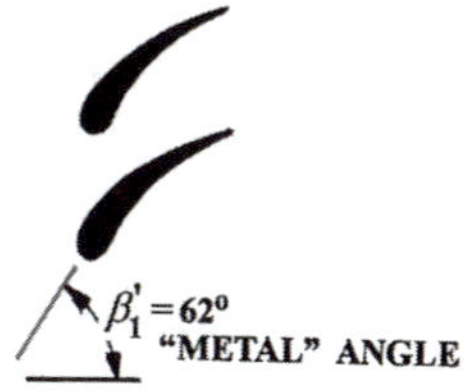

Figure 4.25. Input variables for Problem 1.

which yields

$$h_2 = 6.48 \text{ cm (same result as in Part a)}$$

PROBLEMS

1) Figure 4.25 shows a high-pressure axial-flow compressor stage with a constant mean radius of 0.14 m and the following operating conditions:

- The rotational speed (N) = 34,000 rpm
- The rotor-inlet total pressure = 10 bars
- The rotor-inlet total temperature = 580 K
- The stage axial-velocity component is constant at 216 m/s
- The rotor-blade inlet (metal) angle = 62°
- The rotor pressure ratio $\frac{p_{t2}}{p_{t1}} = 1.52$
- The stage total-to-total efficiency = 81%
- The rotor-blade incidence angle = −8°

   Assuming an isentropic stator and a specific-heat ratio of 1.38, calculate:

   a) The stage reaction;
   b) The change in total relative pressure p_{t_r}.

2) Figure 4.26 shows a turbine stage with a constant mean radius of 0.1 m. The stage operating conditions are as follows:

- The stage flow process is isentropic (by assumption)
- The stator-exit absolute critical Mach number = 0.88
- The stage-inlet total pressure = 11.5 bars
- The stage-inlet total temperature = 1200 K
- The mass-flow rate = 8.7 kg/s

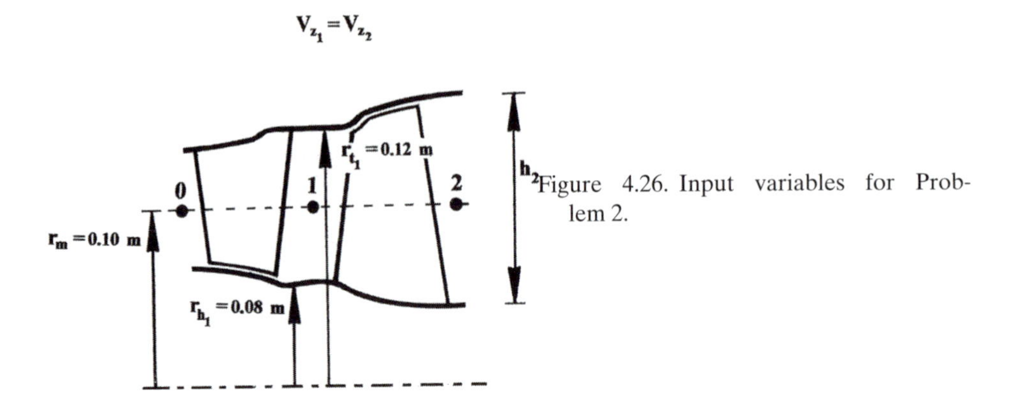

Figure 4.26. Input variables for Problem 2.

- The stage total-to-total efficiency = 86%
- The torque produced = 540 N/m
- The rotational speed (N) = 28,000 rpm

  Also, the stage-wise axial-velocity component is constant.

  a) Calculate the rotor-exit blade height h_2.

  b) Now consider the situation in which the rotor blades are rotated open to accommodate an increase in the mass-flow rate of 14% by altering the stage-exit total properties while maintaining constant both of the rotor-exit static properties and the rotor-exit relative critical Mach number W/W_{cr}. Calculate the rotor blade angle of rotation.

3) Figure 4.27 shows a single-stage turbine. The turbine design point is defined as follows:

- The rotor speed (N) = 46,800 rpm
- The axial-velocity component is constant throughout the stage
- The rotor blade-to-blade passage is choked
- The stage reaction is 50%
- The mass-flow rate = 3.6 kg/s
- The rotor-inlet static pressure (p_1) = 4.4 bars
- The rotor-exit static pressure (p_2) = 2.2 bars
- The rotor-exit static temperature (T_2) = 1020 K

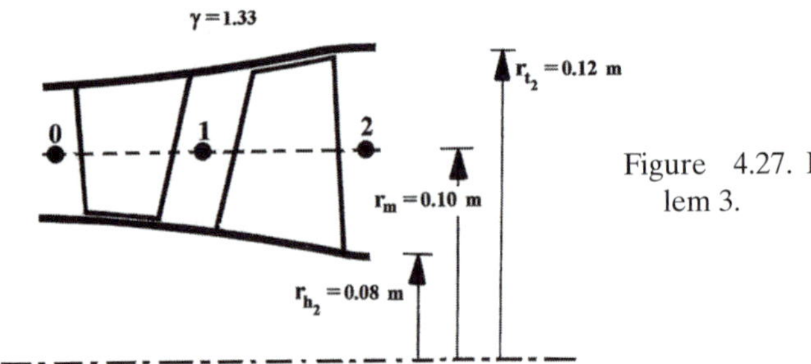

Figure 4.27. Input variables for Problem 3.

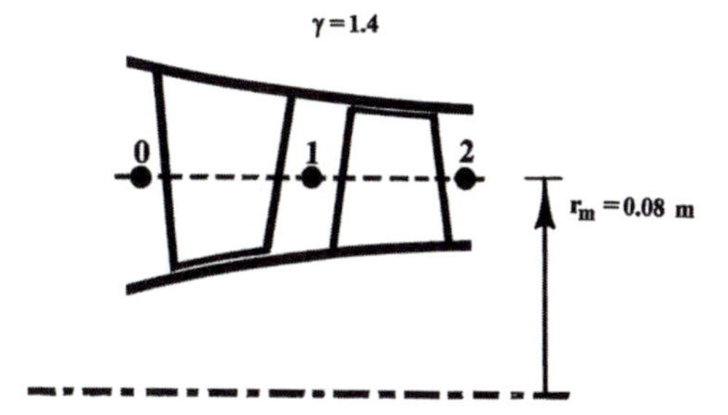

Figure 4.28. Input variables for Problem 4.

Assuming an average specific-heat ratio of 1.33, calculate the stage total-to-total efficiency.

4) Figure 4.28 shows a low-pressure compressor stage with a constant mean radius of 0.08 m. The stage operating conditions are as follows:

- The stage-inlet total pressure = 2.4 bars
- The stage-inlet total temperature = 380 K
- The rotor-inlet absolute velocity V_1 is totally axial
- The stator total pressure loss = 0.22 bars
- The entire stage is adiabatic
- The total relative pressure experiences a loss of 4% across the rotor
- The rotor-inlet static pressure $p_1 = 1.6$ bars
- The rotor-exit static pressure $p_2 = 2.1$ bars
- The shaft speed $(N) = 32{,}000$ rpm
- The rotor-exit relative flow angle β_2 is negative

Assuming an average specific-heat ratio of 1.4, calculate:

a) the static temperature rise across the rotor;
b) the stage total-to-total efficiency.

5) Figure 4.29 shows the first stage in a multistage compressor and the downstream stator. The stage operating conditions are as follows:

- The stage-inlet total pressure = 1.6 bars
- The stage-inlet total temperature = 350 K
- The stator-exit absolute flow angle $\alpha_1 = +30°$

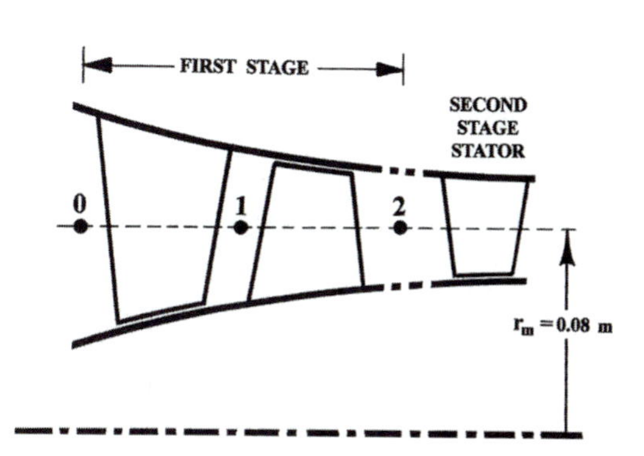

Figure 4.29. Input variables for Problem 5.

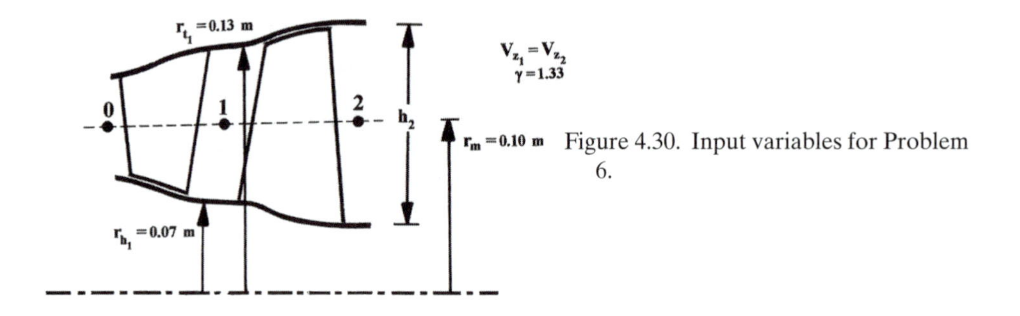

Figure 4.30. Input variables for Problem 6.

- The stator-exit static temperature $T_1 = 310$ K
- The rotor-inlet relative critical Mach number $W_1/W_{cr1} = 0.88$
- The rotor-exit relative flow angle $\beta_2 = +18^\circ$
- The power absorbed by the stage $(P) = 850{,}000$ W

Assuming an isentropic first-stage stator, calculate:

a) The shaft speed;
b) The mass-flow rate;
c) The stage reaction;
d) The second-stage stator airfoil (or metal) angle, knowing that this stator suffers a positive incidence angle of 8.5°.

6) Shown in Figure 4.30 is a single-stage turbine where the operating conditions are as follows:

- The stator passage is choked
- The mass-flow rate = 12.0 kg/s
- The stage-inlet total pressure $p_{t0} = 8.5$ bars
- The stage-inlet total temperature $T_{t0} = 1050$ K
- The rotational speed $(N) = 40{,}000$ rpm
- The stage-exit static pressure $p_2 = 3.6$ bars

Assuming an isentropic flow process throughout the entire stage and a fixed V_z magnitude, calculate:

a) The torque τ delivered by the turbine rotor;
b) The rotor-exit blade height h_2.

7) Figure 4.31 shows the last-stage rotor of a turboprop engine turbine. The stage is axial with a mean radius of 0.075 m. The operating conditions are as follows:

- The stator flow process is isentropic (by assumption)
- The stator-inlet total pressure = 2.35 bars
- The stator-inlet total temperature = 1033 K
- The rotor speed $(N) = 38{,}197$ rpm
- The axial-velocity component $V_z = 186$ m/s (constant)

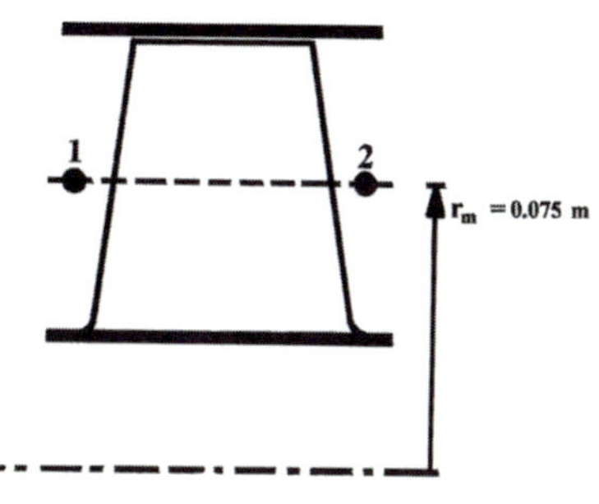

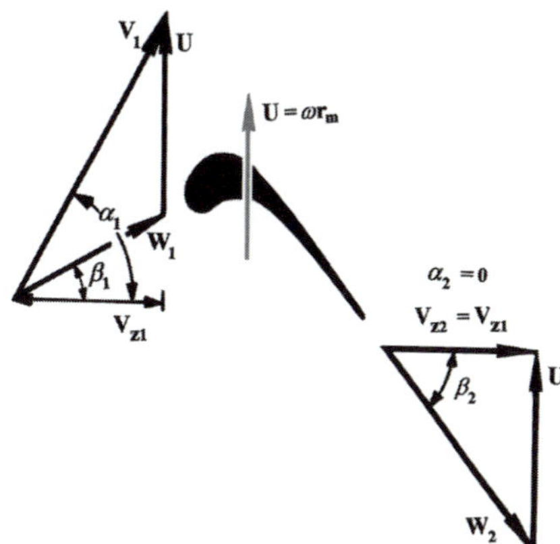

Figure 4.31. Input variables for Problem 7.

- The rotor-inlet absolute flow angle $(\alpha_1) = 65.5°$
- The rotor-inlet relative flow angle $(\beta_1) = 30.0°$
- The rotor-exit absolute flow angle $(\alpha_2) = 0$

Assuming an average specific heat ratio of 1.33:

a) Draw the rotor-exit velocity triangle;
b) Calculate the specific shaft work;
c) Calculate the rotor-exit total temperature T_{t2};
d) Calculate the rotor-exit relative flow angle β_2.

8) Figure 4.32 shows an axial-flow compressor stage, together with the rotor-inlet velocity triangle at the mean radius. If the rotor-inlet total temperature T_{t1} is 444 K and the rotor-exit relative velocity is perfectly axial (i.e., $\beta_2 = 0$), calculate the following variables:

a) The rotor-inlet total relative temperature T_{tr1};
b) The rotor-exit static temperature T_2;
c) The stage reaction.

You may assume that the stagewise axial-velocity component (V_z) is constant. Knowing that this is a normal stage (i.e., $\alpha_2 = \alpha_0$), sketch both the stator vane and the rotor blade, showing, in particular, the airfoil (metal) angles at the inlet and exit stations. In doing so, assume that the incidence and deviation angles are both zero.

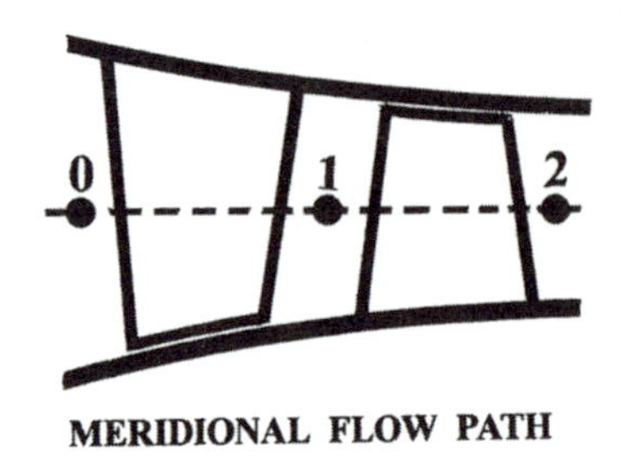

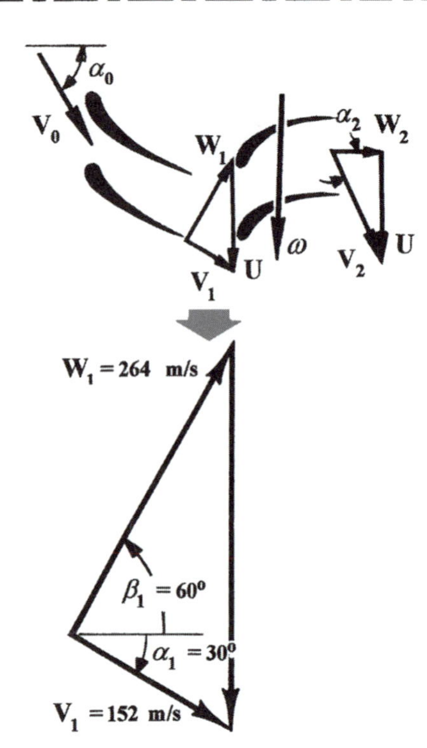

Figure 4.32. Input variables for Problem 8.

9) Figure 4.33 shows an axial-flow turbine stage where:

- Stator-exit critical Mach number $(V/V_{cr\,1}) = 0.88$
- Stator-inlet total temperature $(T_{t0}) = 1200$ K
- Stator-inlet total pressure $(p_{t0}) = 11.5$ bars
- Mass-flow rate $(\dot{m}) = 8.7$ kg/s
- Stage-produced torque $(\tau) = 540$ N m
- Rotor speed $(N) = 28{,}000$ rpm
- Isentropic stator flow field (by assumption)
- Rotor total-to-total efficiency $(\eta_{t-t}) = 91\%$
- Fixed magnitude of the axial velocity component (V_z)

Assuming a specific-heat ratio (γ) of 1.33, calculate the blade half height (h_2) at the rotor exit station.

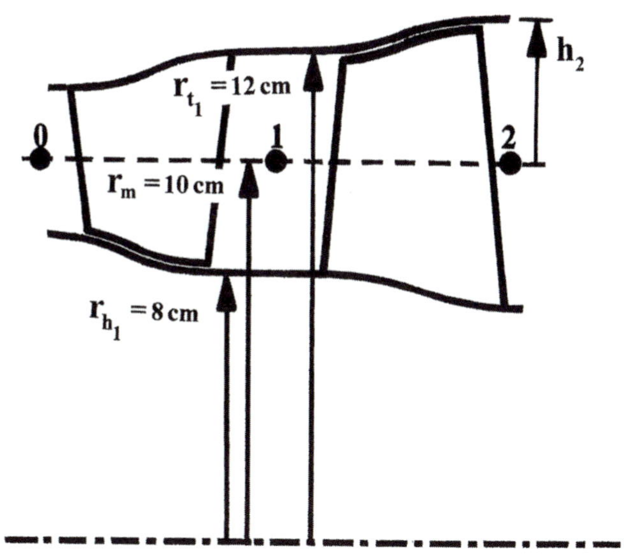

Figure 4.33. Stage geometry in Problem 9.

10) Figure 4.34 shows an axial-flow turbine stage where the entire flow process is adiabatic. The stage operating conditions are as follows:

- Inlet total pressure (p_{t0}) = 11.6 bars
- Inlet total temperature (T_{t0}) = 1287 K
- Mass-flow rate ($\dot{m}$) = 2.6 kg/s
- Stage-exit total temperature (T_{t2}) = 985.0 K
- Rotor speed (N) = 52,000 rpm
- V_z is constant throughout the stage
- Stator flow process is assumed isentropic
- Stator passage is choked
- Stage reaction (R) = 0
- Total-to-total efficiency (η_{t-t}) = 84%

Calculate the decline in total relative pressure (Δp_{t_r}) across the rotor.

11) Figure 4.35 shows an axial-flow compressor stage. The stage operating conditions are such that:

- Stator-exit absolute-velocity vector, $\mathbf{V}_1$, is as shown in the figure
- Rotor-inlet static temperature (T_1) = 560 K
- Rotor-inlet static pressure (p_1) = 8.6 bars
- Rotor-exit static temperature (T_2) = 581 K
- Rotational speed (N) = 32,000 rpm
- Stage-wide axial-velocity component is constant
- Total-to-total pressure ratio (p_{t2}/p_{t0}) = 1.24
- Stator flow process is assumed isentropic
- Rotor flow process is assumed adiabatic

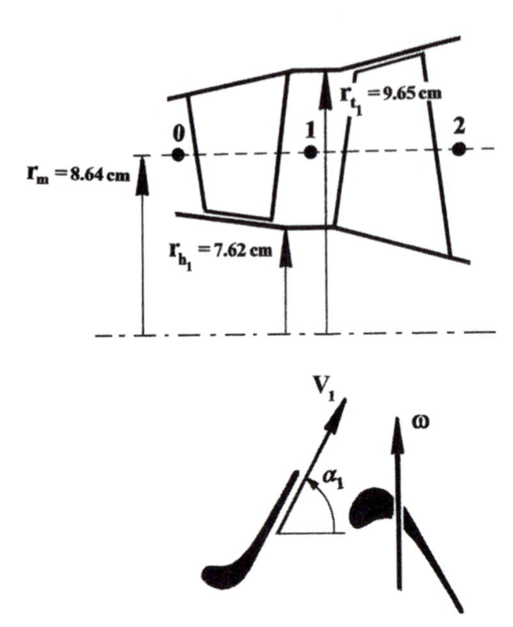

Figure 4.34. Input variables for Problem 10.

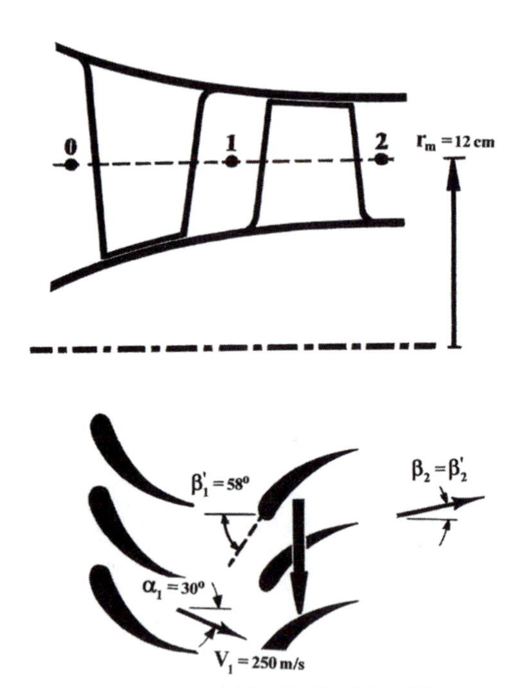

Figure 4.35. Input variables for Problem 11.

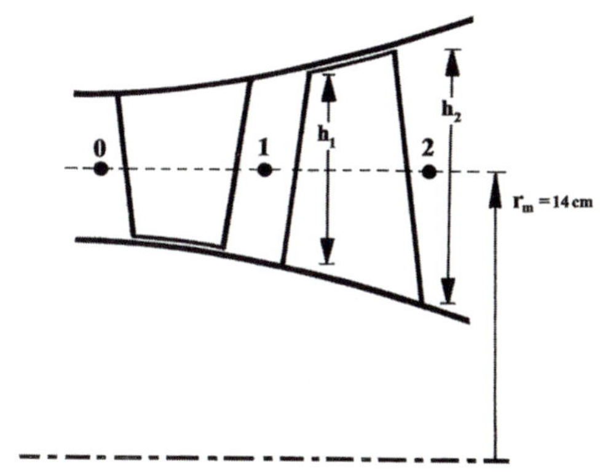

Figure 4.36. Input variables for Problem 12.

Also, the rotor-blade geometry is as shown in the figure, where the airfoil (metal) inlet angle is $-58°$. Calculate:

a) The rotor-blade incidence angle (i_R), indicating its sign;
b) The rotor-exit relative velocity (W_2);
c) The rotor-exit relative flow angle (β_2);
d) The stage total-to-total efficiency (η_C);
e) The stage reaction (R).

12) Figure 4.36 shows a single-stage axial-flow turbine that is operating under the following conditions:

- Mass-flow rate ($\dot{m}$) = 11.0 kg/s
- Rotor-inlet total pressure (p_{t1}) = 8.5 bars
- Rotor-inlet total temperature (T_{t1}) = 1100 K
- Stage total-to-total efficiency (η_T) = 88%
- Rotor-exit swirl angle (α_2) = 0
- Rotor flow passage is choked
- Rotor-exit static pressure (p_2) = 2.8 bars
- Stage-produced power (P) = 2762 kW
- Stagewide axial-velocity component is constant

Calculate the following variables:

a) The rotor speed (N).
b) The rotor-blade inlet and exit heights, h_1 and h_2.

13) Figure 4.37 shows an axial-flow turbine stage where the mean radius (r_m) is 8.5 cm. The stage operating conditions are as follows:

- Inlet total temperature (T_{t0}) = 1400 K
- Rotational speed (N) = 52,000 rpm
- Axial-velocity component (V_z) = 213.0 m/s (assumed constant)
- Stator-exit swirl angle (α_1) = 70°

If the stage-exit total temperature (T_{t2}) is 1088 K, calculate

a) The rotor inlet and exit relative flow angles (i.e., β_1 and β_2);
b) The stage reaction (R)

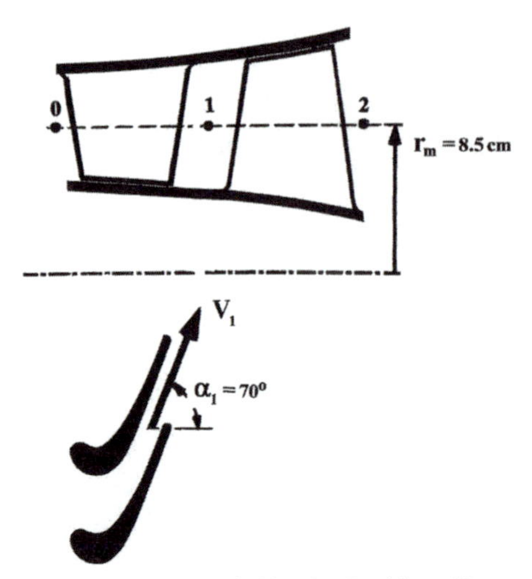

Figure 4.37. Input variables for Problem 13.

c) The rotor-exit absolute Mach number (i.e., V_2/a_2);
d) The change in dynamic enthalpy ($\Delta h_{dyn.}$) across the rotor.

14) Figure 4.38 shows an axial-flow compressor stage where the entire flow process is adiabatic and the stage-wide axial-velocity component (V_z) is constant. The stage operating conditions are as follows:

- Mass-flow rate ($\dot{m}$) = 2.2 kg/s
- Rotor speed (N) = 36,000 rpm
- Inlet total temperature (T_{t0}) = 450 K
- Inlet total pressure (p_{t0}) = 2.8 bars
- Change of total relative pressure $[(\Delta p_{t,r}/p_{t,r,in.})_{rotor}]$ = 13.28%

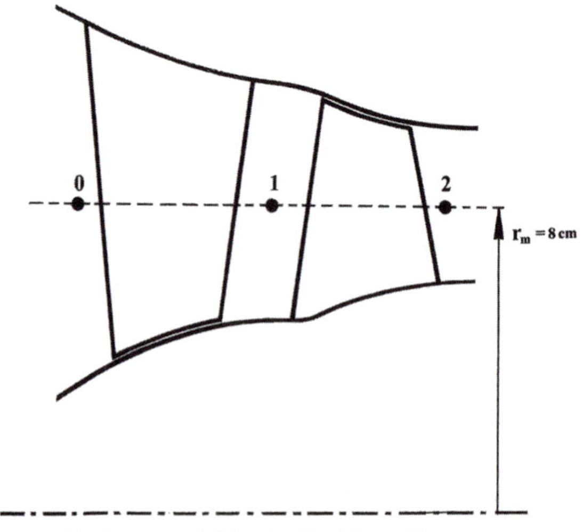

Figure 4.38. Input variables for Problem 14.

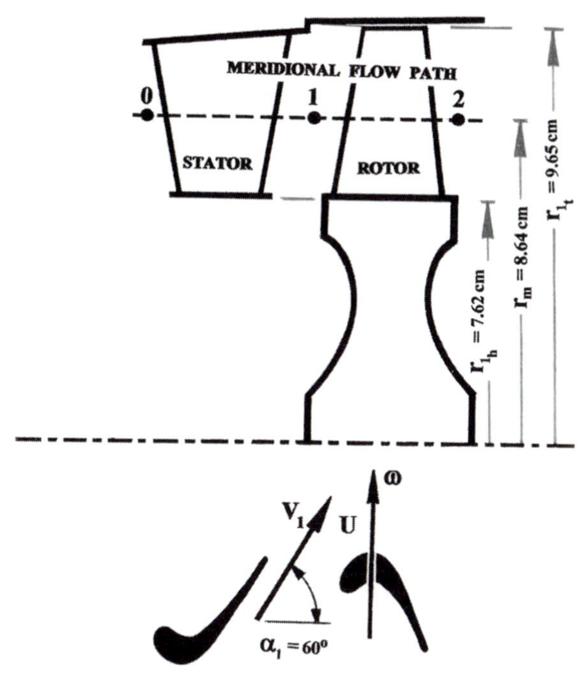

Figure 4.39. Input variables for Problem 15.

- Stator flow process is assumed isentropic
- Stator-exit critical Mach number $(V_1/V_{cr1}) = 0.68$
- Stator-exit swirl angle $(\alpha_1) = 0$
- Rotor-exit relative flow angle $(\beta_2) = -25°$

With this information, perform the following tasks:

a) Calculate the stage total-to-total efficiency (η_C).
b) Calculate the torque (τ) exerted on the shaft.
c) Calculate the rotor-exit static pressure (p_2).
d) If the rotor-blade incidence angle (i_R) is $-12°$, determine the blade-inlet (metal) angle (β_1').

15) Figure 4.39 shows a sketch of an uncooled axial-flow turbine stage where the stator passage is choked. The stage's other operating conditions are as follows:

- Stator-inlet total pressure $(p_{t0}) = 8.84$ bars
- Stator-inlet total temperature $(T_{t0}) = 1389$ K
- Stator mean-radius exit flow angle $(\alpha_1) = 60°$
- Rotor-inlet relative flow angle $(\beta_1) = 0$
- Stage total-to-total efficiency $(\eta_T) = 85\%$
- Rotor mean-radius exit swirl angle $(\alpha_2) = -20°$
- Constant axial-velocity component (V_z) across the stage
- Isentropic flow process across the stator (by assumption)

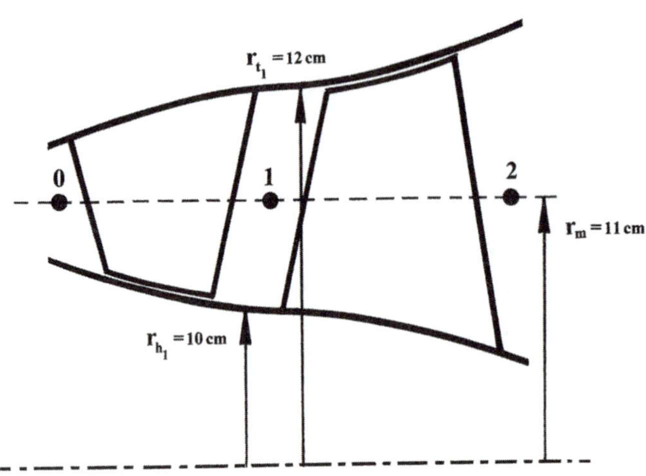

Figure 4.40. Input variables for Problem 16.

Calculate the following variables:

a) The mass-flow rate ($\dot{m}$) through the stage;
b) The rotor speed (N);
c) The rotor-inlet total relative temperature ($T_{t,r1}$);
d) The exit critical Mach number ($V_2/V_{cr,2}$);
e) The exit static pressure (p_2).

16) Figure 4.40 shows the meridional view of a single-stage axial-flow turbine. The turbine operating conditions are as follows:

- Stator passage is choked
- Mass-flow rate ($\dot{m}$) = 5.2 kg/s
- Stage-inlet total pressure (p_{t0}) = 12.0 bars
- Stage-inlet total temperature (T_{t0}) = 1385 K
- Stator flow process is assumed isentropic
- Rotor-inlet relative flow angle (β_1) = +52°
- Stage reaction is zero
- Stage total-to-total efficiency (η_T) = 81%
- V_z is constant across the entire stage

Considering the stage major dimensions in the figure, calculate:

a) The rotor speed (N);
b) The rotor-exit relative critical Mach number (W_2/W_{cr2});
c) The rotor-exit swirl angle (α_2);
d) The rotor-exit total temperature (T_{t2});
e) The rotor-exit total relative pressure ($p_{t,r2}$).

17) The first stage of a turboprop engine turbine (shown in Figure 4.41) has the following operating conditions:

- Mass-flow rate ($\dot{m}$) = 3.25 kg/s
- Rotor speed (N) = 41,730 rpm

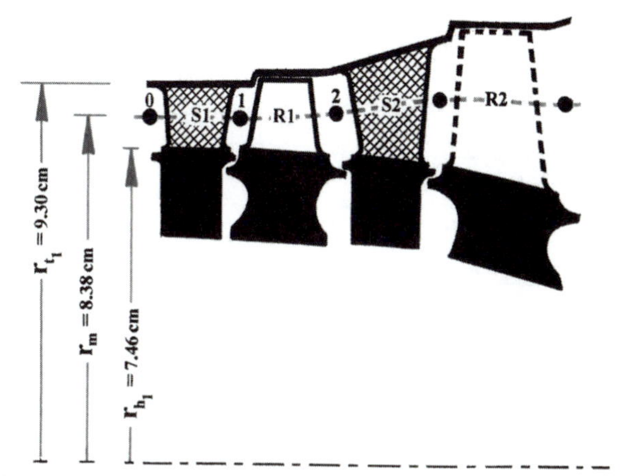

Figure 4.41. Input variables for Problem 17.

- Inlet total pressure (p_{t1}) = 10.57 bars
- Inlet total temperature (T_{t1}) = 1400 K
- Stator-inlet swirl angle (α_0) = 13.0°
- Stator-exit swirl angle (α_1) = 69.5°
- Rotor-exit swirl angle (α_2) = −22.5°
- Total-to-total pressure ratio (p_{t1}/p_{t2}) = 2.6

a) Draw the rotor inlet and exit velocity triangles.
b) Knowing that the stator and rotor incidence angles are −5.0° and 8.0°, respectively, and that the deviation angles are zero, sketch the stator-vane and rotor-blade airfoils at the mean radius.
c) Calculate the rotor-exit critical Mach number (V_2/V_{cr2}).
d) Calculate the stage total-to-total efficiency (η_T).
e) Calculate the entropy-based loss coefficient ($\bar{q}$).

18) Figure 4.42 shows an axial-flow turbine stage, along with some major dimensions. The stage operating conditions are as follows:

- Mass-flow rate ($\dot{m}$) = 4.3 kg/s
- Stage-inlet total pressure (p_{t0}) = 12.5 bars
- Stage-inlet total temperature (T_{t0}) = 1400 K
- Stator flow process is assumed isentropic
- Rotor speed (N) = 34,000 rpm
- $V_{z1} = V_{z2}$
- Stator-exit critical Mach number (V_1/V_{cr1}) = 0.85
- Rotor-exit relative critical Mach number (W_2/W_{cr2}) = 0.82
- Stator-exit swirl angle (α_1) = +68°
- Overall total-to-total pressure ratio (p_{t0}/p_{t2}) = 2.08

Calculate the following variables:

a) The stage-delivered torque (τ);
b) The stage total-to-total efficiency (η_{t-t}).

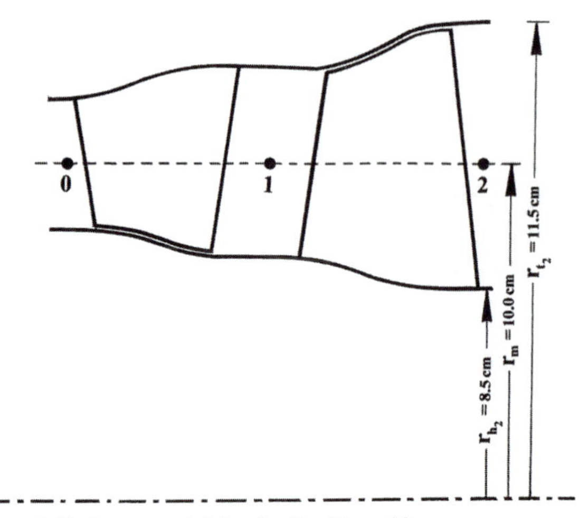

Figure 4.42. Input variables for Problem 18.

19) Figure 4.43 shows the major dimensions of an axial-flow turbine stage. The turbine design point provides the following data:

- Both the stator and rotor flow passages are choked
- The entire flow process is assumed isentropic
- Stage-inlet total pressure $(p_{t0}) = 8.85$ bars
- Stage-inlet total temperature $(T_{t0}) = 1388$ K
- Mass-flow rate $(\dot{m}) = 3.3$ kg/s
- Rotational speed $(N) = 56{,}000$ rpm
- $V_{z1} = V_{z2}$ (by assumption)

Calculate the following variables:

a) The rotor-inlet relative flow angle (β_1);
b) The rotor-exit swirl angle (α_2);

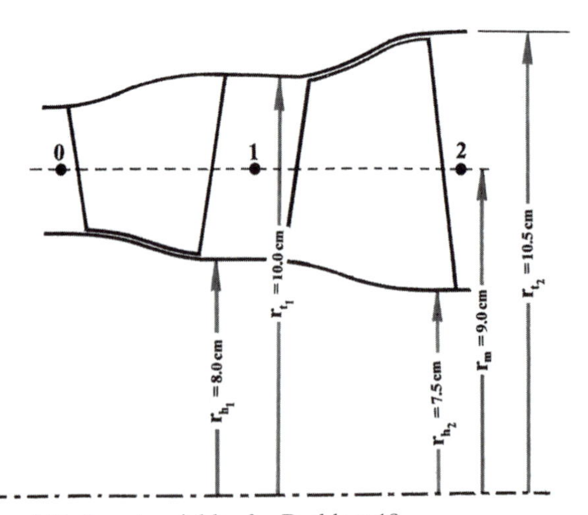

Figure 4.43. Input variables for Problem 19.

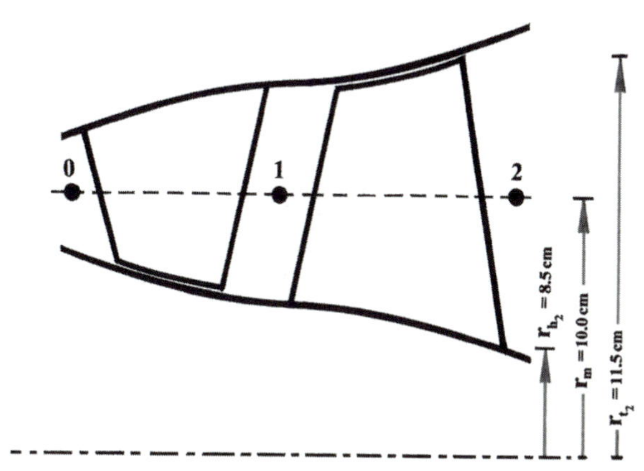

Figure 4.44. Input variables for Problem 20.

c) The rotor-exit total relative pressure ($p_{t,r2}$);
d) The power (P) delivered by the turbine stage;
e) The stage reaction (R).

20) Figure 4.44 shows an axial-flow turbine stage. The stage operating mode is defined as follows:

- Stage-inlet total pressure (p_{t0}) = 12.0 bars
- Stage-inlet total temperature (T_{t0}) = 1400 K
- Mass-flow rate ($\dot{m}$) = 4.88 kg/s
- Shaft speed (N) = 44,000 rpm
- Stator flow process is assumed isentropic
- Rotor-inlet relative flow angle (β_1) = 36.0°
- Rotor-exit static pressure (p_2) = 4.6 bars
- Rotor-exit static temperature (T_2) = 1127 K
- Rotor-exit total relative pressure ($p_{t,r2}$) = 7.02 bars
- $V_{z2} = V_{z1}$

Calculate the following variables:

a) The stage total-to-total efficiency (η_{t-t}),
b) The stage reaction (R),
c) The stator-exit Mach number ($M_1 = V_1/a_1$).

CHAPTER FIVE

Dimensional Analysis, Maps, and Specific Speed

Introduction

In this chapter, a turbomachinery-related nondimensional groupings of geometrical dimensions and thermodynamic properties will be derived. These will aid us in many tasks, such as:

- Investigating the full-size version of a turbomachine by testing (instead) a much smaller version of it (in terms of the total-to-total pressure ratio), an alternative that would require a much smaller torque and shaft speed, particularly in compressors;
- Alleviating the need for blade cooling in the component test rig of a high-pressure turbine section by reducing (in light of specific rules) the inlet total temperature;
- Predicting the consequences of the off-design operation by a turbomachine using the so-called turbine and compressor maps;
- Making a decision, at an early design phase, in regard to the flow path type (axial or radial) of a turbomachine for optimum performance.

A good starting point is to outline the so-called similitude principle, beginning with the definition of geometric and dynamic similarities as they pertain to turbomachines.

Geometrical Similarity

Two turbomachines are said to be geometrically similar if the corresponding dimensions are proportional to one another. In this case, one turbomachine is referred to as a scaled-up (or scaled-down) version of the other. An obvious (but not silly) case here is the turbomachine and itself.

Dynamic Similarity

Two geometrically similar turbomachines are said to be dynamically similar if the velocity vectors at all pairs of corresponding locations are parallel to one another and with proportional magnitudes. In this case, the fluid-exerted forces, at corresponding locations, will be proportional to each other. Under the dynamic similarity condition, all nondimensional variables (so-called π terms), such as the Mach number, pressure ratio, flow angles, and efficiency, will be identical.

Buckingham's π Theorem: Incompressible Flows

Consider the flow process of an incompressible fluid through a pipe. Now let it be our aim to express the pipe total pressure loss as a function of geometrical and thermodynamic properties, in terms of nondimensional π terms, following the guidelines of the well-known Buckingham's theorem. In other words, we are after the nondimensional form of the following expression:

$$\Delta p_t = f(V, \rho, \mu, L, D) \tag{5.1}$$

where

V is the inlet velocity,
ρ is the fluid density,
μ is the dynamic viscosity coefficient, and
L, D are the pipe length and diameter, respectively.

Without going through the details of the π theorem, we know (perhaps by mere inspection) that the nondimensional terms will come out as follows:

$$\pi_1 = \frac{\Delta p_t}{\rho V^2} \tag{5.2}$$

$$\pi_2 = \frac{\rho V D}{\mu} \tag{5.3}$$

$$\pi_3 = \frac{L}{D} \tag{5.4}$$

Of these terms, π_1 is clearly the sensible choice for a dependent variable. In other words, one can write the following expression to govern the pipe flow process:

$$\frac{\Delta p_t}{\rho V^2} = f(Re, L/d) \tag{5.5}$$

where the Reynolds number (Re) is normally defined as

$$Re = \frac{\rho V D}{\mu} \tag{5.6}$$

Application of Buckingham's Theorem to Compressible-Flow Turbomachines

Consider the flow process through the axial-flow compressor stage in Figure 5.1. We can probably expect the following variables to be interrelated:

- Total-to-total pressure ratio (Pr)
- Total-to-total efficiency (η_{t-t})

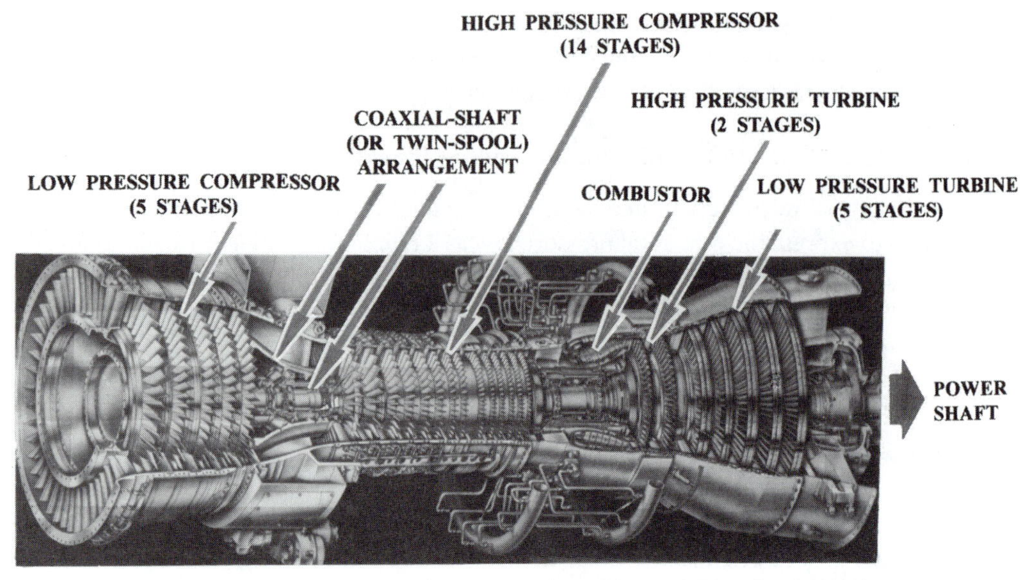

Figure 5.1. Example of a gas-turbine in a power plant. (General Electric LM6000)

- Inlet total pressure ($p_{t\,in}$)
- Inlet total temperature ($T_{t\,in}$)
- Mass-flow rate ($\dot{m}$)
- Shaft speed (N)
- Gas constant (R)
- Specific-heat ratio (γ)
- Characteristic length, chosen as an average diameter (D)

Avoiding once again the details of identifying the nondimensional terms, the following is a list of these terms:

$$\pi_1 = \text{The total-to-total pressure ratio}\,(Pr) \tag{5.7}$$

$$\pi_2 = \text{The total-to-total efficiency}\,(\eta_{t-t}) \tag{5.8}$$

$$\pi_3 = \frac{\dot{m}\sqrt{T_{t\,in}}}{D^2 p_{t\,in}} \tag{5.9}$$

$$\pi_4 = \frac{N}{\sqrt{RT_{t\,in}}} \tag{5.10}$$

$$\pi_5 = \text{The specific-heat ratio}\,(\gamma) \tag{5.11}$$

Of these, the last term (γ) would be relevant only when testing a version of the turbomachinery component under a vastly different inlet temperature. Moreover, the Reynolds number effect would be undeniable should one (or both) of the dynamically similar states give rise to a Reynolds number magnitude (based on an average airfoil chord) that is substantially less than 5×10^5. Should this be the case, the flow field is referred to as viscosity-dominated and the would-be considerable performance degradation becomes rather sensitive to the Reynolds number magnitude.

The situation just described is hard to come by and will therefore be eliminated from the remainder of the topic at hand.

In view of the preceding discussion, and taking (as we should) the first two of the dimensionless variables as being dependent on the others, we obtain the following functional relationship:

$$(Pr, \eta_{t-t}) = f\left(\frac{\dot{m}\sqrt{T_{t\,in}}}{D^2 p_{t\,in}}, \frac{N}{\sqrt{RT_{t\,in}}}\right) \tag{5.12}$$

Examination of the preceding functional relationship reveals that whatever the size and operation modes of two geometrically similar turbomachines, the rig model of the turbomachine will correctly simulate the actual engine-operation mode once the two non-dimensional arguments (on the right-hand side) are maintained the same in both turbomachines.

Compressor and Turbine Maps

As mentioned earlier, a special case of two turbomachines which are geometrically similar is the turbomachine and itself. In this case, the characteristic length's role fails to exist. Furthermore, in this and all subsequent chapters, the working medium will be (exactly or predominantly) air. This eliminates the gas constant, R, from the picture. With this in mind, the preceding functional relationship assumes the following simpler form:

$$(Pr, \eta_{t-t}) = f\left(\frac{\dot{m}\sqrt{T_{t\,in}}}{p_{t\,in}}, \frac{N}{\sqrt{T_{t\,in}}}\right) \tag{5.13}$$

Note that the two arguments on the right-hand side of this expression are no longer nondimensional. In fact, turbomachinists go even further by making nondimensional both the temperature and pressure in equation (5.5) using the standard sea-level magnitudes (T_{stp} and p_{stp}). The resulting nondimensional variables are referred to as θ and δ, which are defined as follows:

$$\theta = \frac{T_{t\,in}}{T_{stp}} \tag{5.14}$$

$$\delta = \frac{p_{t\,in}}{p_{stp}} \tag{5.15}$$

where T_{stp} is approximately 288 K, and p_{stp} is approximately 1 bar. Using θ and δ in place of T_t and p_t in expression (5.13) gives rise to the following functional relationship:

$$(Pr, \eta_{t-t}) = f\left(\frac{\dot{m}\sqrt{\theta_{in}}}{\delta_{in}}, \frac{N}{\sqrt{\theta_{in}}}\right) \tag{5.16}$$

Now, because θ_{in} and δ_{in} are dimensionless, we can clearly see that both of the two arguments on the right-hand side of expression (5.7) have the units of kg/s and rpm, respectively. As a matter of tradition, these two arguments are referred to as

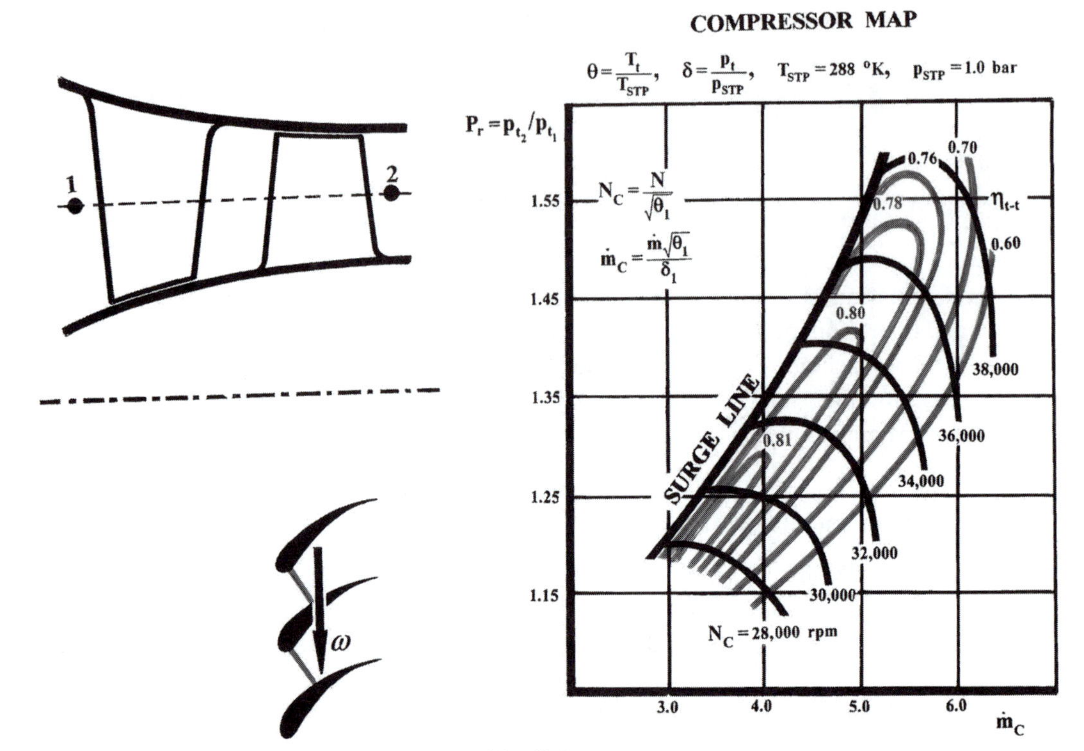

Figure 5.2. A typical compressor map with efficiency contours.

the corrected mass-flow rate ($\dot{m}_C$) and corrected speed (N_C), respectively, and are defined as follows:

$$\dot{m}_c = \frac{\dot{m}\sqrt{\theta_{in}}}{\delta_{in}} \tag{5.17}$$

$$N_c = \frac{N}{\sqrt{\theta_{in}}} \tag{5.18}$$

With these two variables, expression (5.7) can be compacted as follows:

$$(Pr, \eta_{t-t}) = f(\dot{m}_c, N_c) \tag{5.19}$$

Four important remarks are noteworthy at this point. First, the functional relationship in expression (5.19) is solely dependent on the turbomachine itself, for each one will have its own design and off-design features. Second, the expression is as applicable to work-producing turbomachines (turbines) as it is to those that are work-absorbing (compressors, fans, and blowers). Third, the expression is as applicable to a multistage turbomachine as it is to a single stage. Finally, it is often the practice in turbomachinery nomenclature to refer to $\dot{m}$ and N by the terms physical mass-flow rate and physical speed, respectively. The reason, of course, is to distinguish them from their corrected counterparts.

Compressor and turbine maps, shown in Figures 5.2 and 5.3, are graphical means of representing the functional relationship (5.19). Construction of these maps

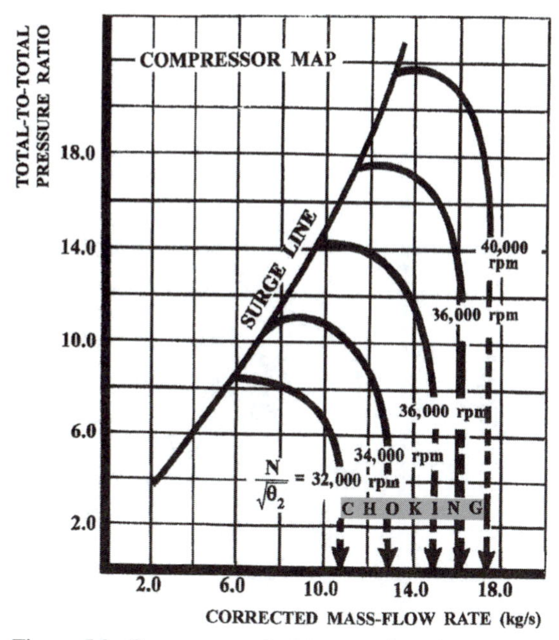

Figure 5.3. Compressor choking as a function of the corrected speed.

requires testing of either the same unit or one that is geometrically similar to it. Points on these maps will represent thermophysical states for which both versions of the turbomachine are dynamically similar. Either way, the map will be that of the real turbomachine, no matter what the rig inlet conditions may be.

The experimental work leading to a map consists of a repetitive flow-measurement procedure covering an acceptable range of off-design operation modes. Because the process is much more involved (perhaps tricky) in a compressor case, we should probably focus on this very case. Referring to Figure 5.3, the compressor map consists, for the major part, of constant corrected-speed curves extending from the so-called surge line down to the choking state, at which point the curves become vertical lines. As will be indicated in Chapter 9, compressor operation in the immediate vicinity of the surge line is characterized by instability that could cause mechanical failure, which makes it much wiser to begin the flow measurements from a point at or close to the choking state upon selecting a magnitude for the corrected speed and work the process up. Note that maintaining the corrected speed fixed does not necessarily involve the physical speed but can also be effected by changing the inlet total temperature. Regardless of the means, the corrected mass-flow rate in a rig situation is repeatedly reduced and the total-to-total pressure ratio obtained. As the compressor surge status is closely approached, a trained operator will sense it by simply monitoring the noise and subsequent shaking of the compressor. Referring to Figure 5.3, another sign of surge proximity is that the constant-speed line begins to level out regardless of the flow-rate reduction. It is at this point that the flow measurements involving this particular corrected-speed line must be terminated and another corrected-speed magnitude considered. Complementing the process each time flow measurements are made, it is highly preferable to also measure the total

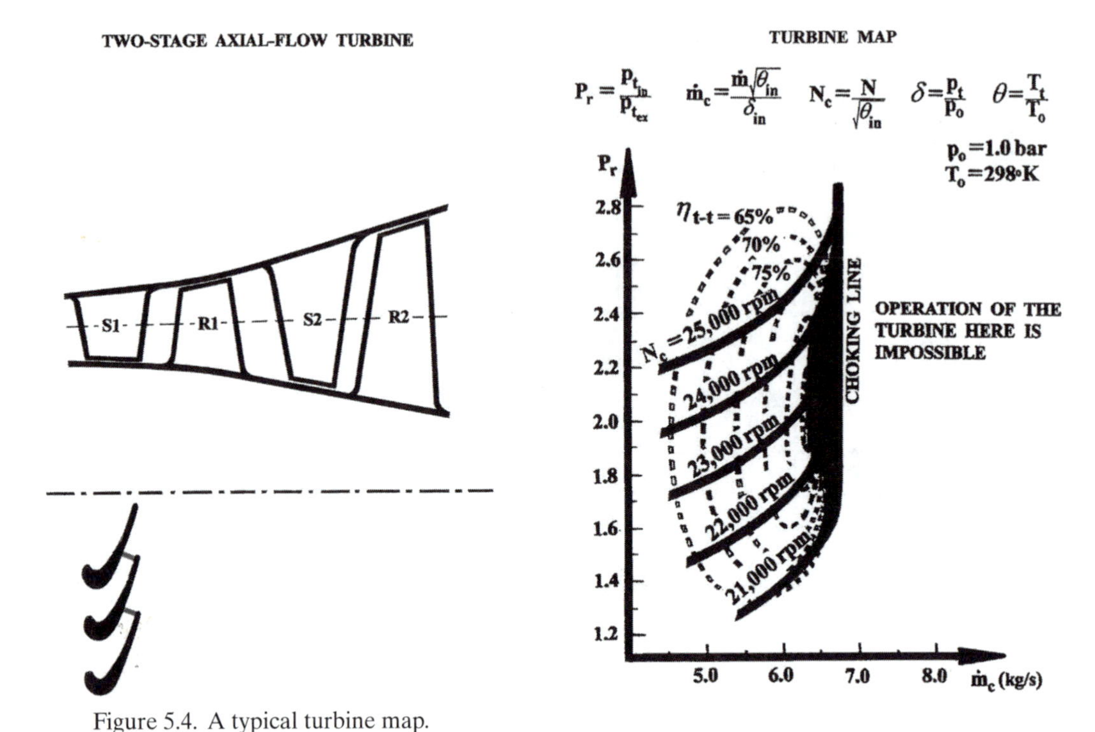

Figure 5.4. A typical turbine map.

exit temperature as well. This way, the total-to-total temperature ratio across the compressor can be calculated and later used, together with the corresponding pressure ratio, in order to compute the compressor total-to-total efficiency. In the end, points on the constant N_C lines with equal efficiency values can be joined together to obtain the constant-efficiency contours shown in Figure 5.2. Worth noting, and further discussed in Chapter 9, is the need to get sufficiently (hopefully safely) close to the surge line as possible. The reason is that the optimum efficiency point, which is close to or is itself the compressor design point, lies notably close to the surge line.

Choking of Compressors and Turbines

Reaching the sonic state, namely the choking condition, will take place at the minimum cross-flow area within a given airfoil cascade. Citing the computational differences between a stationary and a rotating cascade, in the sense of absolute versus relative flow properties, this condition can theoretically be met in either type of cascade. Judging by the typical shape of a turbine (as opposed to a compressor) cascade, the choking condition in the latter will be met at or close to the first cross-flow station (i.e., near the cascade front) (Fig. 5.3).

Figure 5.4 reemphasizes an important fact, namely that there is a unique magnitude of choking mass-flow rate that is associated with any given rotor-speed value. Examination of this very statement reveals one fact: that compressor choking, in most

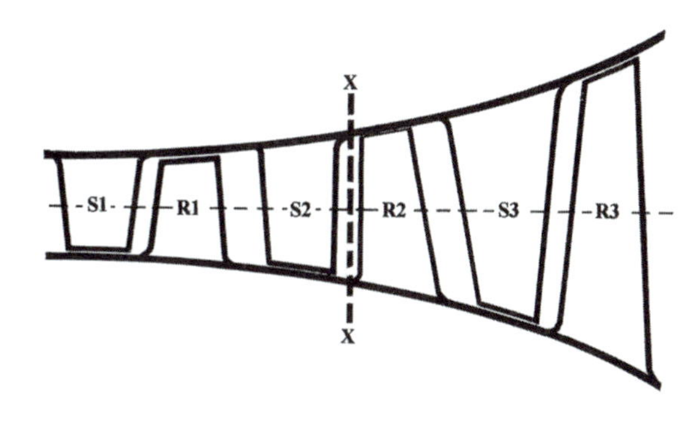

Figure 5.5. Choking within the interstage gap in a multistage turbine.

cases, will prevail within the rotor subdomain and at the minimum blade-to-blade passage area. Recalling the manner in which the relative properties are computed (Chapter 4), such variables as the relative velocity and total relative temperature for example, are functions of the rotor speed. Therefore, the association of the choking state and the shaft speed simply proves the above-cited rotor-choking claim. It is, at least, theoretically possible for a stator to be the choked component instead. In fact, choking could even occur in the stator-to-rotor or interstage gap. In all cases, however, the choking status that is speed-dependent. As stated earlier, it is often the rotor cascade that would, at inlet, reach this critical state. As the rotor physical speed is elevated (Fig. 5.3), so will the relative velocity and (at a higher rate) the relative critical velocity. This, in turn, will allow a higher mass-flow rate toward a higher choking magnitude.

Turbine choking is comparatively the norm rather than the exception. In fact, it is nearly a common practice, in a multistage turbine, to intentionally choke the first-stage stator so that the mass-flow rate is fixed, at least under design-point operation, throughout all succeeding stages. As is clear in Figure 5.4, and regardless of the running speed, there is a unique magnitude of corrected mass-flow rate above which a fixed-geometry turbine will not tolerate. Referring to the same figure, note that the turbine total-to-total pressure ratio is defined as the inlet-to-exit magnitude, which, aside from being the definition adopted throughout this text, happens to be the only way in which this ratio is communicated in the turbomachinery community.

Despite being a rare occurrence, choking of a multistage turbine can prevail in one of the stator/rotor or interstage gaps (station X in Figure 5.5). This situation would not materialize by design but could happen as a result of faulty endwall contouring and/or a tragic boundary-layer buildup in such a gap. Choosing station X to apply the continuity equation (3.35), we get

$$\frac{\dot{m}\sqrt{T_{t\,x}}}{p_{t\,x}A_x} = \left(\frac{2}{\gamma+1}\right)^{\frac{1}{\gamma-1}}$$

where a value of unity has already been substituted for the local critical Mach number, with A_x being the projected cross-flow area. This equation implies that the flow function (on the left-hand side) becomes a function of only the specific-heat ratio γ. Substitution of expressions (5.14) and (5.15) yields the following expression:

$$\frac{\dot{m}\sqrt{\theta_x}}{\delta_x} = \frac{A_x p_{stp}}{\sqrt{T_{stp}}}\left(\frac{2}{\gamma+1}\right)^{\frac{1}{\gamma-1}} \tag{5.20}$$

In a fixed-geometry turbine, and aside from the boundary-layer blockage effect (or displacement thickness), A_x would naturally be fixed. With this in mind, equation (5.20) can be rewritten as

$$\frac{\dot{m}\sqrt{\theta_x}}{\delta_x} = \left(\frac{A_x p_{stp}}{\sqrt{T_{stp}}}\right)\left(\frac{2}{\gamma+1}\right)^{\frac{1}{\gamma-1}} \tag{5.21}$$

The importance of equation (5.21) stems from the fact that the magnitudes of θ_x and δ_x can be tracked back to the turbine inlet station. These variables are in this sense unique and calculable. Recognizing the left-hand side as the corrected mass-flow rate, it is also true that the constant on the right-hand side is the maximum possible corrected mass-flow rate, for choking was assumed to take place at station X.

We can generalize the foregoing conclusions by perceiving station X as any streamwise station where the critical Mach number reaches a value of 1.0. Referring to Figure 5.2, let us refer to the left-hand side in equation (5.21) by $\dot{m}_{c,max}$ and refer to A_x by A_{min}. The former can then be expressed as

$$\dot{m}_{C,max} = \frac{A_{min} p_{stp}}{\sqrt{T_{stp}}}\left(\frac{2}{\gamma+1}\right)^{\frac{1}{\gamma-1}} \tag{5.22}$$

As an example, if we accept an average γ magnitude of, say, 1.33, the choking value of $\dot{m}_{C,max}$ will roughly be equal to $233.9 \times A_{min}$.

Specific Speed

At the earliest design phase, one is often confronted with two diametrically opposed alternatives, namely an axial versus radial (or centrifugal) turbomachine. Assuming that both options are on the table, a rule of thumb states that a radial turbomachine is best suited for those applications where a substantial shaft-work capacity, under a significantly low value of mass-flow rate, is required (Fig. 5.6). The single most helpful parameter in making the decision is a nondimensional term named the specific speed (Ns). Referring to the features of axial versus radial turbomachines (highlighted in Chapter 2), it is perhaps understandable why the latter is often barred from propulsion applications. Apart from the difficulty of multiple staging in the radial-turbomachinery category, the reasons also have to do with weight, size, and bulkiness considerations (e.g., the two radial turbine stages in the power system

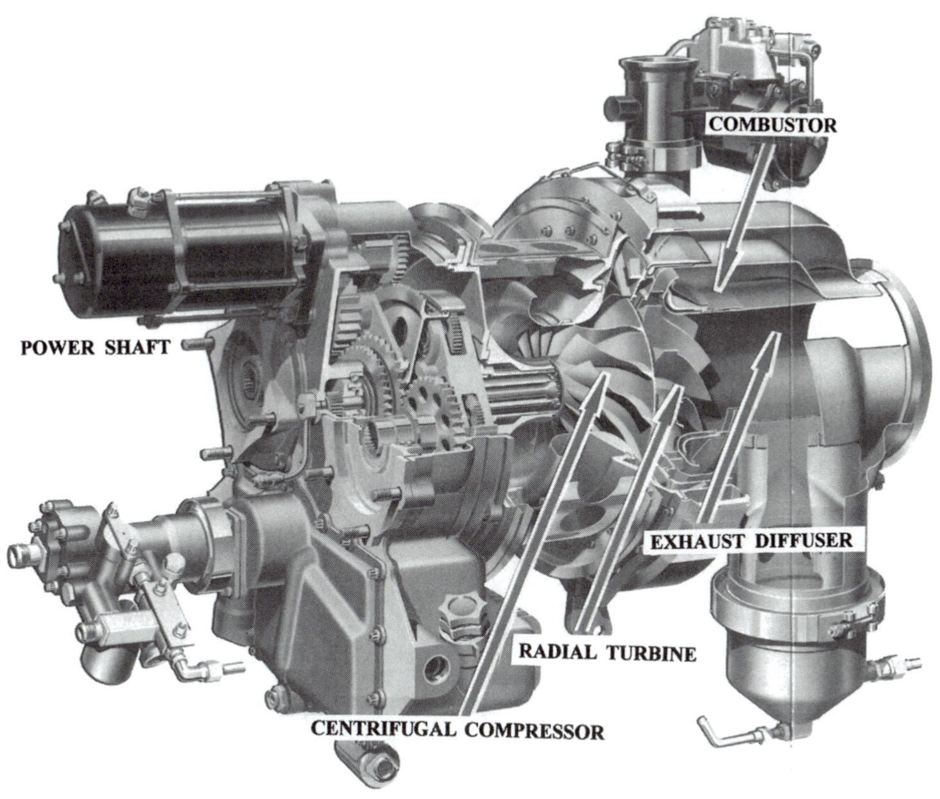

Figure 5.6. Example of a small auxiliary power unit. GT601 Starter (Garrett Turbine Engine Co.)

in Figure 5.6). Nevertheless, there are primary propulsion engines (e.g., the turbofan engine in Figure 1.3) where centrifugal compressor stages were indeed used. The decision in such a case is based on a well-known fact that a single centrifugal stage would possess a power capacity that may very well be equivalent to three or more axial-flow stages. Referring back to the turbofan engine in Figure 1.3, use of the centrifugal compressor stages may be somewhat justified. The fact is that replacing the two centrifugal stages here with, say, five or seven axial-flow stages would certainly elongate the engine. The outcome, with such replacement, would be higher levels of outer-surface skin friction, which adds to the overall drag force on the engine.

The term specific speed was first developed and used in the incompressible-flow turbomachinery area. The process, as explained next, was that of algebraically manipulating other nondimensional π terms, with the objective being a size-independent dimensionless term. The term, named specific speed, was later proven to possess some of the most important design implications in compressible-flow turbomachines as well. It is fitting, for more than just historical reasons, that we begin the discussion where it was initiated, namely in the incompressible-flow turbomachinery category.

Application of Specific Speed to Incompressible-Flow Turbomachines

With no lack of generality, let us consider the case of, say, a hydraulic pump as an example. Going through the dimensional-analysis process, the following dimensionless terms can be obtained:

1) *The flow coefficient*

$$\pi_1 = \frac{Q}{ND^3} = \frac{VA}{ND^3} \tag{5.23}$$

where

Q is the volumetric flow rate,
N is the shaft speed,
D is a characteristic diameter,
V is the local velocity at a pre-specified location in the flow path, and
A is the corresponding area that is perpendicular to V.

2) *The head coefficient*

$$\pi_2 = \frac{g_c H}{N^2 D^2} = \frac{\Delta p_t/\rho}{N^2 D^2} \tag{5.24}$$

where

g_c is the local gravitational acceleration,
H is the head of the turbomachine (in meters), and
Δp_t is the total pressure rise across the pump.

3) *The power coefficient*

$$\pi_3 = \frac{P}{\rho N^3 D^5} = \frac{\dot{m}\delta p_t/\rho}{N^2 D^2} \tag{5.25}$$

where

P is the pump power capacity, and
$\dot{m}$ is the mass-flow rate.

As previously indicated, the specific speed is but the outcome of the non-dimensional terms, whereby the size of the turbomachine (represented by D) is eliminated. This can be achieved by taking the square root of π_1, raising π_2 to the power $\frac{-3}{4}$, and multiplying the two together. The final outcome is the following expression for the specific speed N_s:

$$N_s = \frac{N\sqrt{Q}}{(g_c h)^{\frac{3}{4}}} \tag{5.26}$$

With the physical speed (N) being in radians per second, the specific speed comes out in radians, meaning dimensionless. The versatility of this term stems from the fact that it is size-independent. The significance of this feature is that it can be used, and its implications utilized, in the preliminary design phase, where the component dimensions are as yet unknown. The reason is that N_s is dependent on variables that are known before the design process even starts. Figure 5.7 shows the influence of this variable on the efficiency of a typical pump.

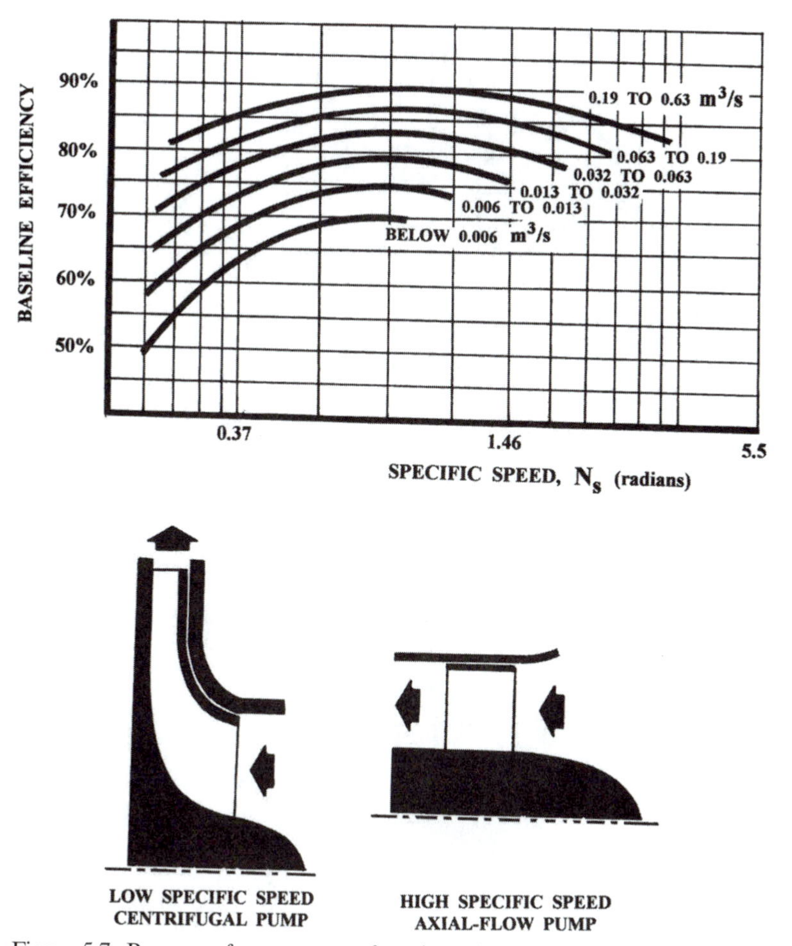

Figure 5.7. Pump performance as a function of specific speed and volumetric flow rate.

Application of the Specific Speed to Compressible-Flow Turbomachines

In making the transition to compressible-flow turbomachines, the specific speed definition (5.26) has to be modified. Take, for instance, the volumetric flow rate, which clearly varies in a compressible-flow turbomachine as a result of the streamwise density changes. Furthermore, the head of a turbomachine (in meters) is hardly a means of representing the shaft work associated with a gas turbine or compressor.

Adapted for use with compressible-flow turbomachines, the specific speed is defined as

$$N_s = \frac{N\sqrt{Q_{ex}}}{(\Delta h_{t\,id})^{\frac{3}{4}}} \tag{5.27}$$

or

$$N_s = \frac{N\sqrt{\frac{\dot{m}}{\rho_{ex}}}}{(\Delta h_{t\,id})^{\frac{3}{4}}} \tag{5.28}$$

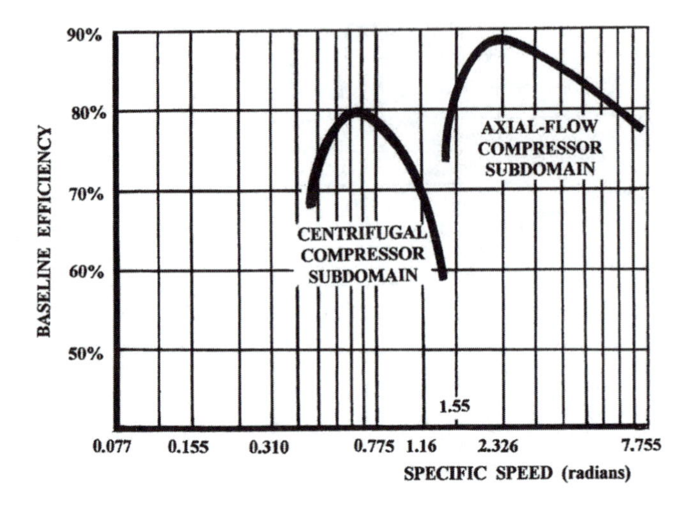

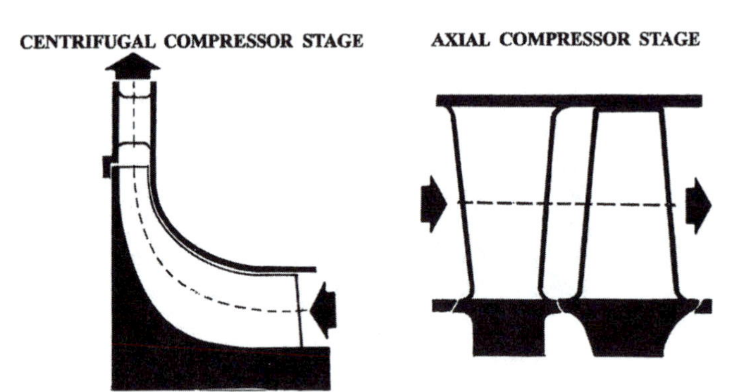

Figure 5.8. Performance of axial and centrifugal compressor stages as a function of specific speed.

In expression (5.28), the magnitude of the volumetric flow rate (nonfixed in compressible-flow turbomachines) is chosen to be that at the stage exit station. Noteworthy is the fact that the denominator is the ideal (or isentropic-process) total enthalpy change (or ideal shaft work) and not the actual magnitude. Perhaps there is a logical reason for this. The calculation of specific speed (which is the only step aimed at investigating the suitability of axial versus radial turbomachines) will occur at a computational point where only the inlet conditions and total-to-total pressure ratio are known, with the efficiency being one of the last variables to entertain.Regardless of this postulate's validity, the ideal total enthalpy change can be expressed as follows:

1) *Compressor stage*

$$\Delta h_{t\,id} = c_p T_{t\,in} \left[\left(\frac{p_{t\,ex}}{p_{t\,in}} \right)^{\frac{\gamma-1}{\gamma}} - 1 \right] \tag{5.29}$$

2) *Turbine stage*

$$\Delta h_{t\,id} = c_p T_{t\,in} \left[1 - \left(\frac{p_{t\,ex}}{p_{t\,in}} \right)^{\frac{\gamma-1}{\gamma}} \right] \tag{5.30}$$

Design Role of the Specific Speed

Figures 5.8 and 5.9 illustrate the baseline (or datum) efficiency versus the specific speed for both axial and radial (or centrifugal) stages in compressors and turbines, respectively. On each of these two graphs, two distinct domelike regions exist, one spanning a range of low specific speed and the other covering a range of high specific speed. The dome to the left in each figure represents the region where a centrifugal (or radial) turbomachine will give rise to the best possible performance. In fact, the two graphs reveal the fact that an axial-stage design in this region of low specific speed is hardly wise by reference to the dramatic efficiency deterioration for an axial stage in this region. On the other hand, Figures 5.8 and 5.9 both emphasize that turbomachines with high specific speed are better off being of the axial-flow type. In connection with compressor staging, Figure 5.8 shows a

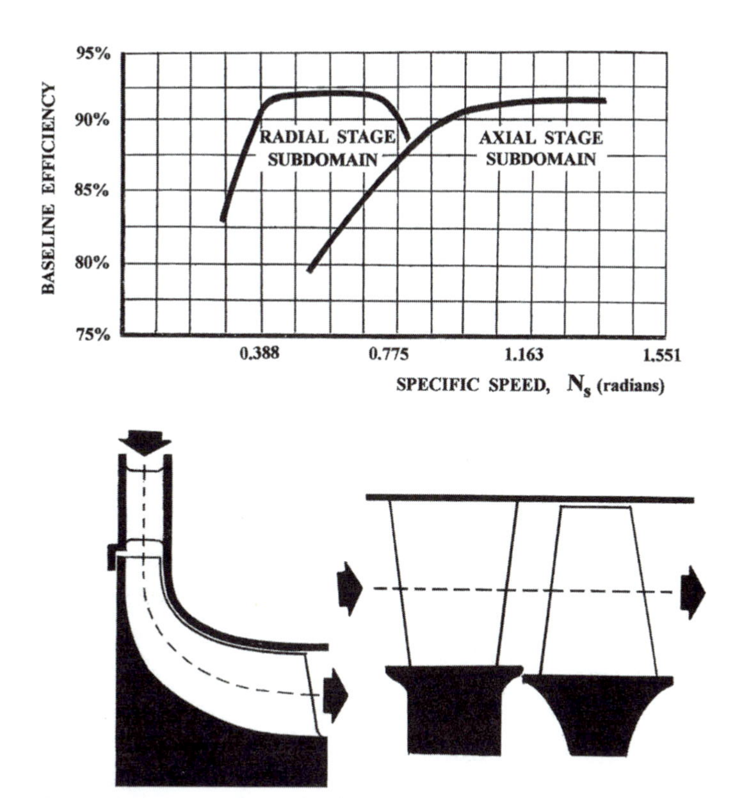

Figure 5.9. Performance of axial and radial turbine stages as a function of specific speed.

centrifugal-to-axial compressor-stage cutoff magnitude of the specific speed that is in the neighborhood of 1.4. For turbine staging (Fig. 5.9), the cutoff value seems to be around 0.776.

On many design occasions, a turbine stage will be associated with a set of design requirements that gives rise to a considerably low specific speed. In a case such as this, Figure 5.9 will not only suggest a radial-type stage but, as a consequence, a notably large tip radius as well. Should this stage be part of a propulsion system, the radial stage option will often be rejected. The usual remedy in this case is to break up the overall total-enthalpy drop (i.e., shaft work) into several increments, with an axial-flow stage being responsible for each of these increments. The specific speed associated with each stage should then be computed and Figure 5.9 used to justify the axial-flow stage choice. Note that such a strategy is more suited to turbines as opposed to compressors. The reason is that the turbine overwhelmingly favorable pressure gradient would make it possible to utilize a comparatively lower count of axial stages in the radial-to-axial stage-conversion process.

Charts such as those in Figures 5.8 and 5.9 are best used as multistaging tools. To explain this, let us assume a two-stage turbine section with a 70%: 30% "work split" ratio. With the higher first-stage work extraction, the denominator in the specific-speed definition (5.28) may very well place the stage in the radial-stage domain. Should this be undesirable, the designer may first set the work split at a different ratio (e.g., 60–40%). The outcome may still imply a radial stage or even two radial stages. It could be at this point that the designer will be forced to consider a three-stage arrangement.

Traditional Specific-Speed Approximations

It is obvious that the compressible-flow version of the specific speed N_s has "inherited" the volumetric-flow rate (Q) from its incompressible-flow counterpart. Referring to expression (5.27), it is the stage exit magnitude that is required, a task that is hardly easy. The reason is that the static density ρ_{ex} at the stage exit station is unknown, at least during the preliminary design procedure, where

$$Q_{ex} = \frac{\dot{m}}{\rho_{ex}} \tag{5.31}$$

The exit static density (ρ_{ex}) can be expressed in terms of the exit critical Mach number as

$$\rho_{ex} = \rho_{t\,ex}\left(1 - \frac{2\gamma}{\gamma+1} M_{cr\,ex}{}^2\right)^{\frac{1}{\gamma-1}} \tag{5.32}$$

The total density, however, can be expressed, using the equation of state, as

$$\rho_{t\,ex} = \frac{p_{t\,ex}}{RT_{t\,ex}} \tag{5.33}$$

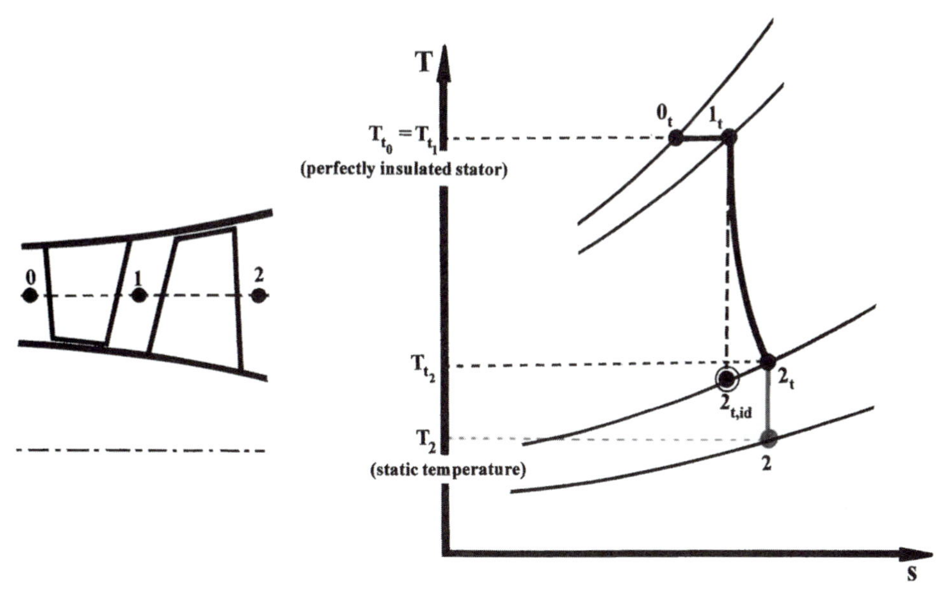

Figure 5.10. Exact versus approximate evaluation of the specific speed.

The exit total temperature can be computed by assuming a rough magnitude of the total-to-total efficiency, including an ideal 100% value. Referring to Figure 5.10, the following two simplifications are normally acceptable:

a) One can assume a substantially small exit critical Mach number, which is normally a turbine-stage design objective, or a reasonable Mach number in the case of a compressor stage. In the latter case, equation (5.25) can be employed to compute a rough magnitude of the exit static density. In the case of a turbine stage, the proposed low critical Mach number option paves the way to the following practical result:

$$\rho_{ex} \approx \rho_{t\,ex} = \frac{p_{t\,ex}}{RT_{t\,ex}} \tag{5.34}$$

b) For a substantial axial-velocity component, one can assume an ideally zero exit swirl, in which case the following expression can be utilized:

$$\rho_{ex} = \frac{p_{t\,ex}}{RT_{t\,ex}}\left[1 - \left(\frac{V_{z\,ex}}{V_{cr\,ex}}\right)^2\right]^{\frac{1}{\gamma-1}} \tag{5.35}$$

The preceding options are two other simplifications that can be employed at the preliminary design phase. In most cases, the axial/radial stage choice with such assumptions is so clear that there is no need to go back and verify the stage-type choice. However, should the stage-type selection be less than convincingly clear, the designer will usually get a more accurate picture once an upgraded magnitude of exit Mach number is known.

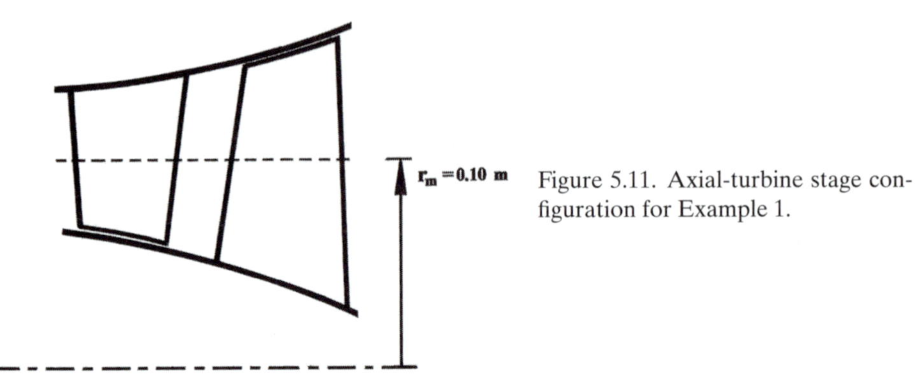

Figure 5.11. Axial-turbine stage configuration for Example 1.

EXAMPLE 1

The design-point operation of a single-stage axial-flow turbine (Fig. 5.11) is simulated in a cold rig. The two sets of operating conditions are as follows:

Design-point operation mode

- Stage-inlet total pressure $p_{t0} = 12$ bars
- Stage-inlet total temperature $T_{t0} = 1420$ K
- Stage-exit total pressure $p_{t2} = 5.4$ bars
- Rotational speed $(N) = 58,000$ rpm
- Mass-flow rate $\dot{m} = 3.8$ kg/s
- Rotor-inlet relative flow angle $(\beta_1) = 0.0°$
- Stage efficiency $(\eta_{t-t}) = 86\%$
- Stator-exit critical Mach number $(V/V_{cr})_1 = 0.96$

Cold-rig operation mode

- Stage-inlet total pressure = 2.6 bars
- Stage-inlet total temperature = 388 K

The stage has a mean radius r_m of 0.1 m, and the axial velocity component across the stage is assumed constant, which applies to both sets of operating conditions. With an isentropic stator assumption and an average specific-heat ratio γ of 1.365 (to apply to both sets of operating conditions), calculate:

a) The stage-exit total temperature T_{t2} in the cold rig;
b) The power produced by the turbine stage in the cold rig.

SOLUTION

Part a: According to the rules of dynamic similarity, we have

$$\left(\frac{p_{t0}}{p_{t2}}\right)_{rig} = \left(\frac{p_{t0}}{p_{t2}}\right)_{des.\ pt.} = 2.22$$

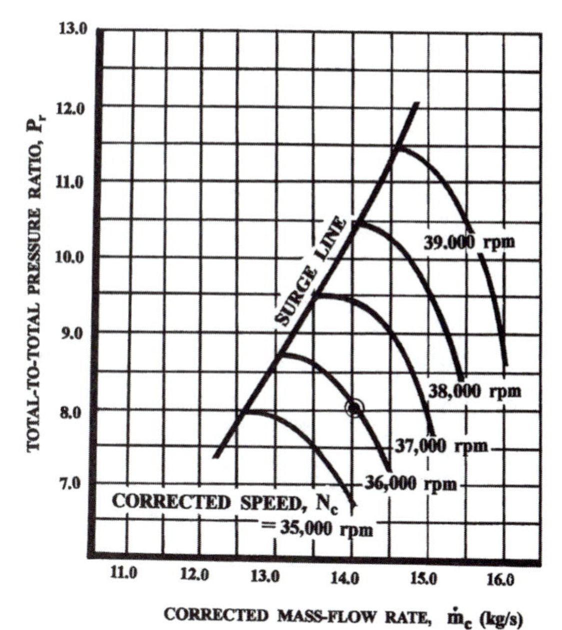

Figure 5.12. Compressor map for Example 2.

Also

$$\eta_{rig} = \eta_{des.\ pt.} = 0.86 = \frac{1 - (T_{t2}/T_{t0})}{1 - (p_{t2}/p_{t0})^{\frac{\gamma-1}{\gamma}}}$$

which yields

$$(T_{t2})_{rig} = 323.9\,\text{K}$$

Part b: The dynamic similarity rules also require equality of the corrected flow rate, meaning that

$$\left(\frac{\dot{m}\sqrt{\theta_0}}{\delta_0}\right)_{rig} = \left(\frac{\dot{m}\sqrt{\theta_0}}{\delta_0}\right)_{des.\ pt.}$$

which, upon substitution, gives the "physical" mass-flow rate in the rig as follows:

$$\dot{m}_{rig} = 1.58,\ \text{kg/s}$$

Finally, we can calculate the stage-supplied power (P) in the test rig as follows:

$$P_{rig} = \dot{m}_{rig} c_p (T_{t0} - T_{t2}) = 108.4\,\text{kW}$$

EXAMPLE 2

Figure 5.12 shows the map of a compressor in a turboprop engine. The cruise operation of the compressor is defined as follows:

- The power supplied $P_{alt} = 836.0$ kW
- The mass-flow rate $\dot{m}_{alt} = 3.8$ kg/s

- The inlet total pressure $(p_{t\,in})_{alt} = 0.235$ bars
- The inlet total temperature $(T_{t\,in})_{alt} = 216.0$ K
- Total-to-total efficiency $\eta_C = 80\%$

These operating conditions are simulated in a test rig utilizing air at inlet total pressure and temperature of 1.0 bar and 288 K, respectively. Assuming an average specific-heat ratio of 1.4 for both sets of operating conditions, calculate the following variables in the test rig:

a) The "physical" shaft speed (N) in rpm.
b) The torque (τ) transmitted to the compressor.

SOLUTION

Part a: Let us first calculate the cruise-operation total-to-total pressure ratio (Pr), starting with the supplied-power expression, namely

$$P = \frac{\dot{m} c_p T_{t\,in}}{\eta_C}\left[\left(\frac{p_{t\,ex}}{p_{t\,in}}\right)^{\frac{\gamma-1}{\gamma}} - 1\right]$$

which, upon substitution, yields

$$(Pr)_{cruise} = \frac{p_{t\,ex}}{p_{t\,in}} \approx 8.0$$

Also,

$$\frac{\dot{m}\sqrt{T_{t\,in}}}{\delta_{in}} \approx 14.0\,\text{kg/s}$$

These newly computed variables allow us to place the point signifying the cruise operation mode on the compressor map, as shown in Figure 5.12. Now we can read off the cruise-operation corrected-speed magnitude:

$$\left(\frac{N}{\sqrt{\theta_{in}}}\right)_{cruise} = 36{,}000\,\text{rpm}$$

The foregoing three variables correspond not only to the compressor cruise operation but also its rig operation because the two operating modes are dynamically similar. Let us now use the equality of corrected speeds:

$$\left(\frac{N}{\sqrt{\theta_{in}}}\right)_{rig} = \left(\frac{N}{\sqrt{\theta_{in}}}\right)_{cruise} = 36{,}000\,\text{rpm}$$

where

$$(\theta_{in})_{rig} = \frac{T_{t\,in}}{T_{STP}} = \frac{288.0}{288.0} = 1.0$$

which means that the "physical" speed in the test rig is

$$N_{rig} = 36{,}000\,\text{rpm}$$

Part b: Next, let us implement the equality of the corrected flow rate under both sets of operating conditions:

$$\left(\frac{\dot{m}\sqrt{\theta_{in}}}{\delta_{in}}\right)_{rig} = 14.0\,\text{kg/s}$$

where

$$(\delta_{in})_{rig} = \frac{p_{t\,in}}{p_{STP}} = \frac{1.0}{1.0} = 1.0$$

The rig-operation "physical" mass-flow rate can now be determined:

$$(\dot{m})_{rig} = 14.0\,\text{kg/s}$$

Finally, we can calculate the shaft-transmitted torque (τ) in the rig as follows:

$$\tau_{rig} = \frac{(Power)_{rig}}{\omega_{rig}} = 1089.6\,\text{N} \cdot \text{m}$$

EXAMPLE 3

Figure 5.13 shows a single-stage turbine and its map. The cruise-operation point of the turbine is defined as follows:

- Inlet total pressure = 8.5 bars
- Inlet total temperature = 1020 K
- Total-to-total (isentropic) efficiency = 81%
- Specific shaft work produced = 200 kJ/kg
- Rotor-inlet relative flow angle $\beta_1 = 0°$
- Rotor-exit absolute flow angle $\alpha_2 = 0°$
- Stator flow process is isentropic (by assumption)
- Axial-velocity component is constant
- Flow is assumed incompressible at the rotor exit station

I) Assuming an average specific-heat ratio of 1.365, and using the turbine map in Figure 5.13, calculate the following variables:

   a) The actual power P_{alt} delivered by the turbine stage;
   b) The loss in total relative pressure $(p_{t r_1} - p_{t r_2})$ across the rotor.

II) The preceding operating conditions are simulated in a test rig that utilizes air at inlet total pressure and temperature of 4.8 bars and 625 K, respectively. Considering the same assumptions (i.e., isentropic stator, stagewise constant V_z, rotor-exit incompressible flow, and an average specific-heat ratio of 1.365), calculate the following variables in the test rig:

   a) The torque (τ) delivered by the turbine stage;
   b) The rotor-exit absolute velocity (V_2).

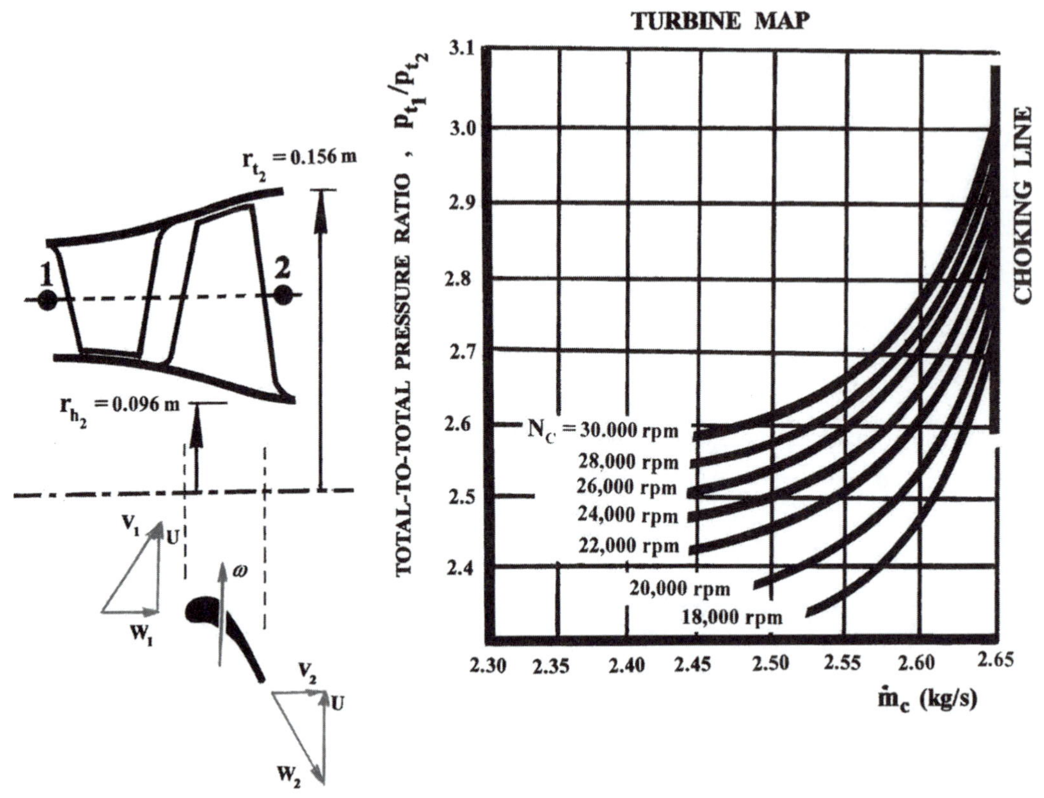

Figure 5.13. Input variables for Example 3.

SOLUTION

Part Ia: In order to calculate the cruise-operation power delivered, we proceed as follows:

$$w_s = U(V_{\theta 1} - V_{\theta 2}) = U^2 \text{ (because } V_{\theta 2} = 0)$$

$$U = 447.2\,\text{m/s}$$

$$N = \frac{U}{r_m}\left(\frac{60}{2\pi}\right) = 33,892\,\text{rpm}$$

$$\frac{N}{\sqrt{\theta_1}} \approx 18{,}000\,\text{rpm}$$

However,

$$w_s = \eta_T c_p T_{t1}\left[1 - \left(\frac{p_{t2}}{p_{t1}}\right)^{\frac{\gamma-1}{\gamma}}\right]$$

which yields

$$\frac{p_{t1}}{p_{t2}} = 2.6$$

We are now in a position to use the given turbine map to attain the following:

$$\frac{\dot{m}\sqrt{\theta_1}}{\delta_1} = 2.55\,\text{kg/s}$$

which, upon substitution, gives the physical mass-flow rate:

$$\dot{m}_{cruise} = 11.52\,\text{kg/s}$$

Now, the cruise-operation power (P) delivered by the turbine stage can be computed:

$$P = \dot{m}\eta_T c_p T_{t1}\left[1 - \left(\frac{p_{t2}}{p_{t1}}\right)^{\frac{\gamma-1}{\gamma}}\right] = 2303\,\text{kW}$$

Part Ib: With the magnitude of γ being 1.365, let us calculate the c_p magnitude:

$$c_p = \left(\frac{\gamma}{\gamma - 1}\right) R = 1073.3\,\text{J/(kg K)}$$

Now, we calculate the stage-exit total temperature and pressure:

$$p_{t2} = \frac{8.5}{2.6} = 3.27\,\text{bars}$$

$$T_{t2} = T_{t1} - \frac{w_s}{c_p} = 833.7\,\text{K}$$

Applying the continuity equation at the stage exit station, we have

$$\dot{m} = \rho_2 V_z (2\pi r_m h_2)$$

The simplification in the problem statement – that the stage-exit flow stream is incompressible – means that

$$\rho_2 \approx \rho_{t2} = \frac{p_{t2}}{RT_{t2}} = 1.37\,\text{kg/m}^3$$

which yields

$$V_z = 177.4\,\text{m/s}$$

In computing the inlet and exit total relative pressures, it is important to recall that in an axial-flow adiabatic rotor (such as the current rotor), the total relative temperature $T_{t,r}$ remains constant. Now, let us proceed with the aim being the inlet and exit total relative pressures:

$$V_1 = \sqrt{{V_{\theta 1}}^2 + {V_z}^2} = 481.1\,\text{m/s}$$

$$T_{t,r,1} = T_{t1} - \left(\frac{{V_1}^2 - {W_1}^2}{2c_p}\right) = 926.8\,\text{K} = T_{t,r,2}$$

$$p_{t,r,1} = p_{t1}\left(\frac{T_{t,r,1}}{T_{t1}}\right)^{\frac{\gamma}{\gamma-1}} = 5.94\,\text{bars}$$

$$p_{t,r,2} = p_{t1}\left(\frac{T_{t,r,2}}{T_{t2}}\right)^{\frac{\gamma}{\gamma-1}} = 4.86\,\text{bars}$$

$$\frac{\Delta p_{t_r}}{p_{t,r,1}} = 18.2\%$$

Part IIa: Equality of the corrected flow rate under the cruise and rig operating conditions will now enable us to compute the rig "physical" mass-flow rate. The rig speed, however, will be the result of equating the corrected speed, as follows:

$$\dot{m}_{rig} = \dot{m}_{cruise}\sqrt{\frac{T_{t1cruise}}{T_{t1rig}}} \times \frac{p_{t1rig}}{p_{t1cruise}} = 26.06\,\text{kg/s}$$

$$N_{rig} = N_{cruise}\sqrt{\frac{T_{t1rig}}{T_{t1cruise}}} = 26{,}530\,\text{rpm}$$

Thus

$$\tau_{rig} = \left(\frac{\dot{m}w_s}{\omega}\right)_{rig} = 1148.9\,\text{N}\cdot\text{m}$$

Part IIb:

$$\left(\frac{V_2}{V_{cr_2}}\right)_{rig} = \left(\frac{V_2}{V_{cr_2}}\right)_{cruise} = 0.338$$

$$(\eta_{t-t})_{rig} = (\eta_{t-t})_{cruise} = 0.81$$

With this total-to-total efficiency magnitude, and substituting in equation (3.59), we obtain the exit total temperature in the rig:

$$(T_{t2})_{rig} = 510.9\,\text{K}$$

The exit magnitude of critical velocity can now be calculated:

$$(V_{cr_2})_{rig} = \sqrt{\left(\frac{2\gamma}{\gamma+1}\right) RT_{t2rig}} = 411.4\,\text{m/s}$$

Finally, the exit velocity (V_2) in the test rig can now be calculated:

$$(V_2)_{rig} = \left(\frac{V_2}{V_{cr_2}}\right)_{rig} \times V_{cr_2} = 139.1\,\text{m/s}$$

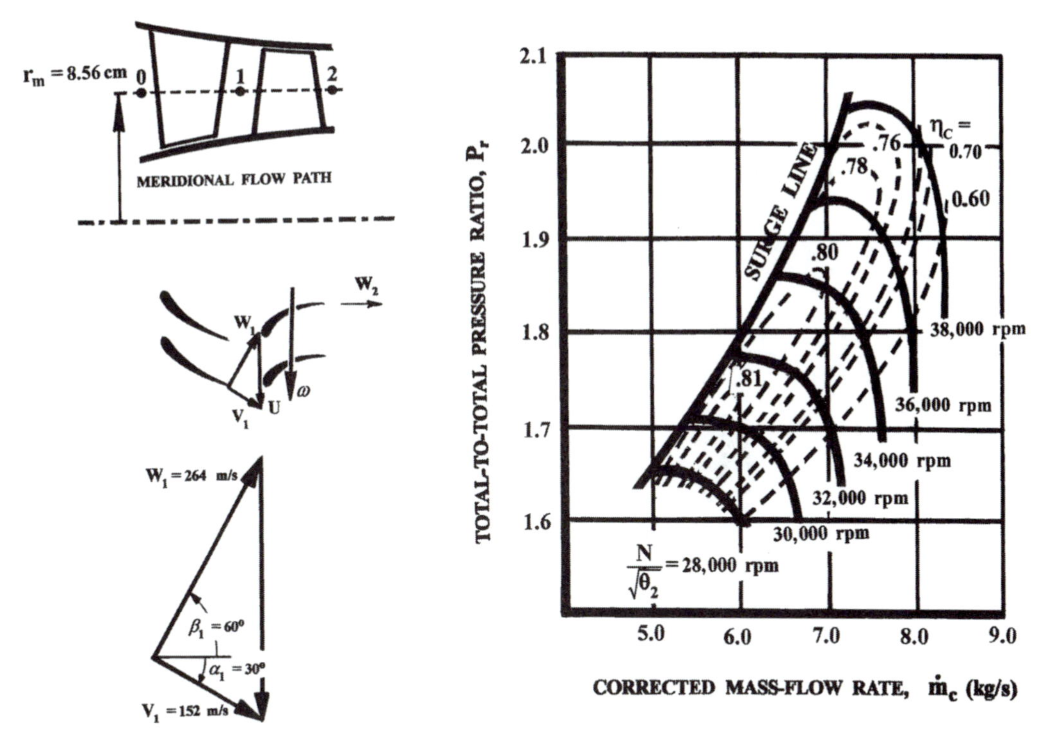

Figure 5.14. Input variables for Example 4.

EXAMPLE 4

Figure 5.14 shows the rotor-inlet velocity diagram for an adiabatic compressor stage that has a constant mean radius (r_m) of 8.56 cm, together with the corresponding compressor map. The following operating conditions also apply:

- A stagewise constant axial-velocity component
- Inlet total pressure = 1.0 bar
- Inlet total temperature = 288 K
- Mass-flow rate = 7.0 kg/s

Assuming an average specific-heat ratio of 1.4, justify (on the basis of Figure 5.8) the axial-flow stage choice. You may take the standard sea-level pressure and temperature to be 1.0 bar and 288 K, respectively.

SOLUTION

We begin by locating the stage operating conditions on the stage map:

$$\theta_{in} = \theta_0 = \frac{T_{t0}}{T_{STP}} = 1.0$$

$$\delta_{in} = \delta_0 = \frac{p_{t0}}{p_{STP}} = 1.0$$

It follows that

$$\dot{m}_C = \frac{\dot{m}\sqrt{\theta_{in}}}{\delta_{in}} = 7.0\,\text{kg/s}$$

Referring to the rotor-inlet velocity triangle in Figure 5.14, we get

$$U_m = W_1 \sin\beta_1 + V_1 \sin\alpha_1 = 304.6\,\text{m/s}$$

$$V_z = V_1 \cos\alpha_1 = 131.6\,\text{m/s}$$

$$V_{\theta 1} = V_1 \sin\alpha_1 = 76.0\,\text{m/s}$$

$$N = \frac{U_m}{r_m}\frac{60}{2\pi} \approx 34{,}000\,\text{rpm}$$

$$\frac{N}{\sqrt{\theta_{in}}} = 34{,}000\,\text{rpm}$$

Now, referring to the stage map, we get the following variables:

$$\frac{p_{t2}}{p_{t1}} = 1.85$$

$$\eta_{t-t} = 0.8$$

Using this efficiency magnitude, and referring to expression (3.60), we have

$$T_{t2} = 357.2\,\text{K}$$

Now, applying Euler's equation,

$$w_s = c_p(T_{t2} - T_{t1}) = U_m(V_{\theta 2} - V_{\theta 1})$$

which yields

$$V_{\theta 2} = 304.2\,\text{m/s}$$

Pursuing the solution procedure,

$$V_2 = \sqrt{{V_{\theta 2}}^2 + {V_z}^2} = 331.5\,\text{m/s}$$

$$V_{cr_2} = \sqrt{\left(\frac{2\gamma}{\gamma+1}\right)RT_{t2}} = 345.8\,\text{m/s}$$

$$M_{cr_2} = \frac{V_2}{V_{cr_2}} = 0.96$$

$$\rho_2 = \left(\frac{p_{t2}}{RT_{t2}}\right)\left[1 - \frac{\gamma-1}{\gamma+1}(M_{cr_2})^2\right]^{\frac{1}{\gamma-1}} = 1.19\,\text{kg/m}^3$$

At this point, we are prepared to compute the stage specific speed (N_s), by substituting the foregoing variables in expressions (5.28) and (5.29),

$$N_s = 2.38\ \text{radians}$$

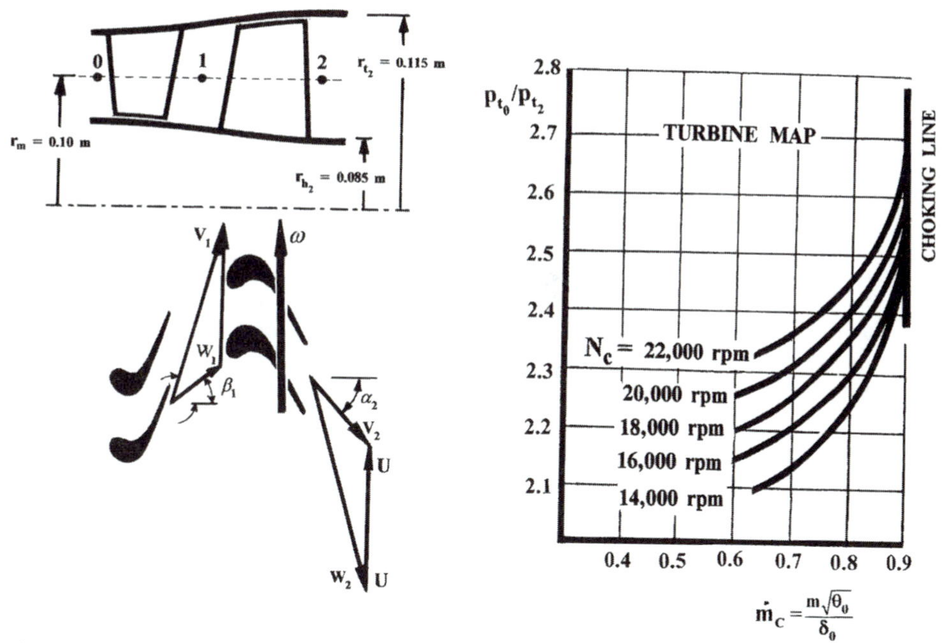

Figure 5.15. Input variables for Example 5.

Finally, we see (by reference to Figure 5.8) that this specific-speed magnitude places this stage well within the axial-flow compressor-stage "dome." This clearly justifies the axial-flow choice for this compressor stage.

EXAMPLE 5

Figure 5.15 shows a schematic of an adiabatic single-stage turbine with a mean radius of 0.1 m, as well as the turbine map. The turbine design point is defined as follows:

- Rotational speed (N) = 42,000 rpm
- Inlet total pressure = 11.6 bars
- Inlet total temperature = 1270 K
- Rotor-inlet critical Mach number = 0.82
- Rotor-inlet absolute flow angle = 73°
- Total relative pressure loss across the rotor = 6.2%
- Exit relative critical Mach number $(W_2/W_{cr2}) = 1.0$
- Axial-velocity component is constant throughout the stage
- Total pressure loss across the stator is negligible

Assuming an average specific-heat ratio of 1.33:

a) Calculate the stage total-to-total efficiency;
b) Discuss, in light of Figure 5.9, whether the axial-flow stage choice is justified.

SOLUTION

Part a:

$$V_1 = M_{cr_1} V_{cr_1} = 529.0 \text{ m/s}$$

$$U_m = \omega r_m = 356.0 \text{ m/s}$$

$$V_{\theta 1} = V_1 \sin\alpha_1 = 505.9 \text{ m/s}$$

$$V_z = V_1 \cos\alpha_1 = 154.7 \text{ m/s}$$

$$W_{\theta 1} = V_{\theta 1} - U_m = 149.9 \text{ m/s}$$

$$W_1 = \sqrt{W_{\theta 1}^2 + V_z^2} = 215.4 \text{ m/s}$$

$$T_{tr1} = T_{tr2} = T_{tr} = T_{t1} - \left(\frac{V_1^2 - W_1^2}{2c_p}\right) = 1169.1 \text{ K (axial stage)}$$

$$p_{tr1} = p_{t1}\left(\frac{T_{tr1}}{T_{t1}}\right)^{\frac{\gamma}{\gamma-1}} = 8.31 \text{ bars}$$

Counting the total relative pressure loss, we have

$$p_{tr2} = (1 - 0.062)p_{tr1} = 7.79 \text{ bars}$$

$$W_{cr_2} = \sqrt{\frac{2\gamma}{(\gamma+1)R} T_{tr2}} = 618.9 \text{ m/s} = W_2 \text{ (rotor is choked)}$$

$$\beta_2 = \cos^{-1}\left(\frac{V_z}{W_2}\right) = -75.5^\circ$$

The negative sign (above) was inserted in view of Figure 5.15. Note that any traditionally designed turbine stage will always give rise to a negative rotor-exit relative flow angle.

Now, applying the continuity equation at the rotor-exit station in the *rotating* frame of reference, using *relative* flow properties, we have

$$\frac{\dot{m}\sqrt{T_{tr2}}}{p_{tr2}\left[\pi\left(r_{t2}^2 - r_{h_2}^2\right)\cos\beta_2\right]} = \sqrt{\frac{2\gamma}{(\gamma+1)R}}\left(\frac{W_2}{W_{cr2}}\right)\left[1 - \frac{\gamma-1}{\gamma+1}\left(\frac{W_2}{W_{cr2}}\right)^2\right]^{\frac{1}{\gamma-1}}$$

Note that we could, at least theoretically, apply the continuity equation, at the same station, in the stationary frame of reference (using absolute flow properties). However, we first would have had to translate the rotor-choking statement in terms of absolute flow properties (i.e., V_2, α_2, etc.).

Substituting in the chosen version of the continuity equation (above), we get

$$\dot{m} = 4.27 \text{ kg/s}$$

The "corrected" temperature θ_0 and pressure δ_0 for the turbine stage can be calculated as

$$\theta_0 = T_{t0}/T_{STP} = 4.41$$
$$\delta_0 = p_{t0}/p_{STP} = 11.6$$

Now, the corrected mass-flow rate ($\dot{m}_C$) and corrected speed (N_C) can be calculated:

$$\dot{m}_C = \frac{\dot{m}\sqrt{\theta_0}}{\delta_0} = 0.773 \text{ kg/s}$$

$$N_C = \frac{N}{\sqrt{\theta_0}} \approx 20{,}000 \text{ rpm}$$

Using the provided turbine map, we get

$$\frac{p_{t0}}{p_{t2}} = 2.38$$

Note that this is on the high side of the total-to-total pressure ratio for an *axial*-flow turbine. Therefore, and despite the moderate-to-high mass-flow rate, the specific speed N_s (to be computed later) may be too low for an axial-flow turbine.

Let us now calculate the rotor-exit thermophysical properties:

$$W_{\theta 2} = W_2 \sin \beta_2 = -599.2 \text{ m/s}$$
$$V_{\theta 2} = W_{\theta 2} + U_m = -243.2 \text{ m/s}$$

Applying the Euler/energy-conservation relationship, we have

$$c_p(T_{t1} - T_{t2}) = w_s = U_m(V_{\theta 1} - V_{\theta 2})$$

which, upon substitution, yields

$$T_{t2} = 1039.5 \text{ K}$$

We are now in a position to calculate the stage total-to-total efficiency as follows:

$$\eta_T = \frac{1 - (T_{t2}/T_{t0})}{1 - (p_{t2}/p_{t0})^{\frac{\gamma-1}{\gamma}}} = 93.7\%$$

Part b: In order to calculate the specific speed, we first have to calculate the static magnitude of the rotor-exit density, as follows:

$$V_2 = \sqrt{V_{\theta 2}{}^2 + V_z{}^2} = 288.2$$

$$V_{cr_2} = \sqrt{\left(\frac{2\gamma}{\gamma+1}\right) R T_{t2}} = 583.6 \text{ m/s}$$

$$M_{cr_2} = \frac{V_2}{V_{cr_2}} = 0.494$$

$$\rho_2 = \left(\frac{p_{t2}}{R T_{t2}}\right)\left(1 - \frac{\gamma-1}{\gamma+1} M_{cr_2}{}^2\right)^{\frac{1}{\gamma-1}} = 1.467 \text{ kg/m}^3$$

Finally, we can calculate the stage specific speed (N_s) by direct substitution in expression (5.28) as follows:

$$N_s = 0.609\,\text{radians}$$

Referring to Figure 5.9, we see that this N_s magnitude qualifies the stage to be of the radial type.

PROBLEMS

1) It is desired to simulate the 7.62 km cruise operation of the high-pressure turbine (HPT) in a turbofan engine in a cold test rig using the full-scale turbine. The operating conditions to be simulated at this altitude are:

   - Turbine-inlet total temperature = 1350 K
   - Turbine-inlet total pressure = 8.5 bars
   - Physical speed (N) = 43,000 rpm
   - Physical mass-flow rate = 1.9 kg/s

   The turbine section of the rig receives air at a total inlet pressure and temperature of 1.2 bars and 300 K, respectively. Assuming the altitude specific-heat ratio to be 1.33 and the rig specific-heat ratio is 1.4, calculate the rig physical speed and physical mass-flow rate.

2) The design point of a compressor is defined as follows:

   - Inlet total pressure = 1.5 bars
   - Inlet total temperature = 345 K
   - Mass-flow rate = 3.8 kg/s
   - Physical speed = 45,000 rpm
   - Total temperature rise = 320 K

   These operating conditions are tested using air under standard sea-level conditions of 1.0 bars and 288 K, respectively, using the full-scale compressor. Calculate the mass-flow rate, speed, and total temperature rise during the test.

3) It is required to design a single-stage turbine according to the following specifications:

   - Physical speed = 75, 600 rpm
   - Inlet total pressure = 7.48 bars
   - Inlet total temperature = 1388.9 K
   - Total-to-total pressure ratio = 2.05

   Using the axial- and radial-stage efficiencies in Figure 5.9, and ignoring the turbine-exit compressibility effects, determine which turbine type (axial or radial) should be selected if:

   a) The mass-flow rate is 0.89 kg/s;
   b) The mass-flow rate is 5.33 kg/s.

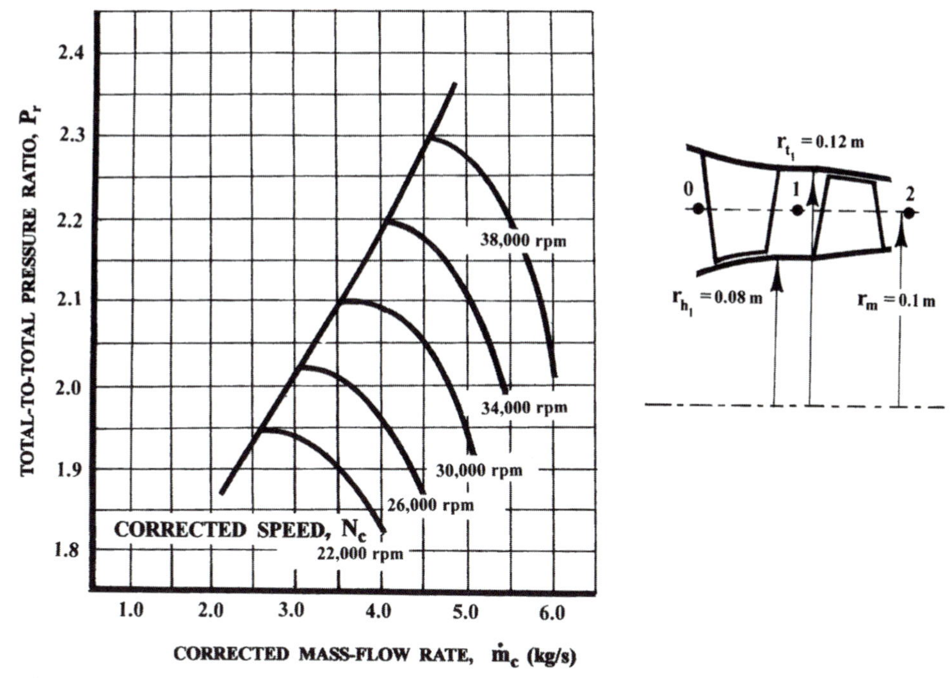

Figure 5.16. Input variables for Problem 5.

4) A single-stage high-pressure turbine is designed to be dynamically similar at both sea-level takeoff and 7.62 km altitude operations. The turbine inlet conditions at these two sets of turbine operation are:

   Sea level

   - Mass-flow rate = 2.89 kg/s
   - Inlet total pressure = 17 bars

   Altitude operation

   - Mass-flow rate = 1.91 kg/s
   - Inlet total pressure = 8.03 bars

   If the turbine physical speed (N) at sea-level takeoff is 45,000 rpm, calculate:

   a) The altitude-operation physical speed;
   b) The ratio of specific shaft work w_s between the sea-level and altitude operations, assuming a uniform specific heat ratio of 1.33 in both cases.

5) Figure 5.16 shows a compressor stage together with its map. The cruise operation of the stage is defined as follows:

   - The actual shaft work supplied = 54.05 kJ/kg
   - The rotor-inlet absolute flow angle = 0°
   - The rotor-exit relative angle = 0°
   - The stator flow process is isentropic (by assumption)
   - The inlet total pressure = 0.235 bars

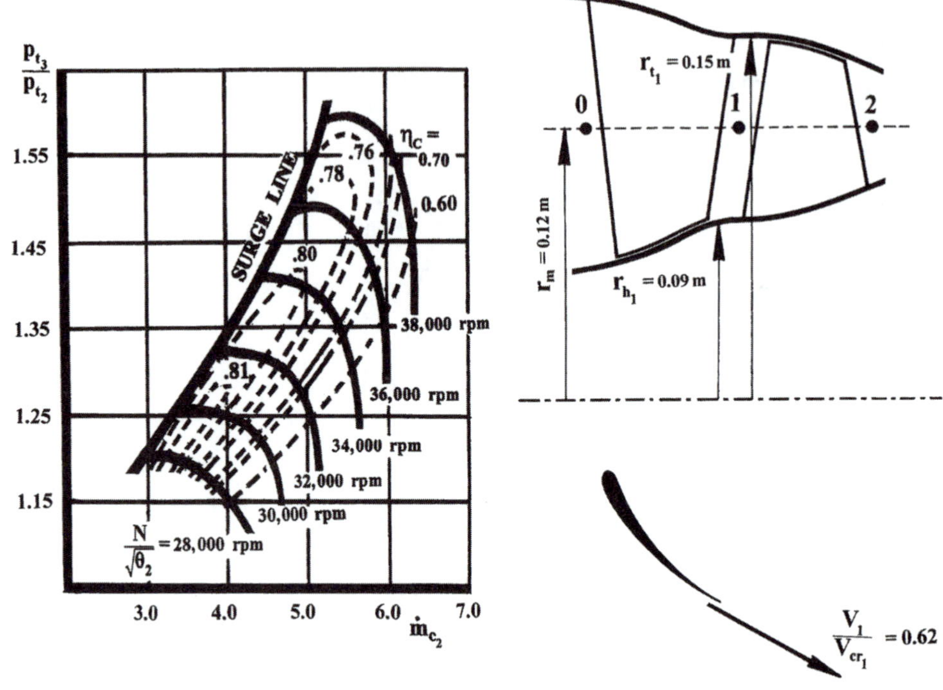

Figure 5.17. Input variables for Problem 8.

- The inlet total temperature = 210 K
- The total-to-total (isentropic) efficiency = 85.47%

These operating conditions are simulated in a test rig that utilizes air at a total inlet pressure and temperature of 1.0 bar and 288 K, respectively. The stator flow process is assumed isentropic under both sets of operating conditions. Also, the axial-velocity component is constant throughout the compressor.

a) Calculate the cruise-operation mass-flow rate.
b) Calculate the test-rig torque that is transmitted to the compressor.
c) Assuming an incompressible flow stream at the stator exit station, calculate the loss in total relative pressure across the rotor in the rig.

6) Repeat the solution procedure of Problem 4 upon altering the following item:
   - Mass-flow rate ($\dot{m}$) = 7.5 kg/s (instead of 7.0 kg/s)

7) Repeat the solution procedure of Problem 5 upon altering the following two items:
   - The stator flow passage is choked
   - The rotor-exit relative critical Mach number $(W_2/W_{cr2}) = 0.92$ (instead of 1.0)

8) Figure 5.17 shows a compressor stage in a propulsion system along with its map. Under the cruise operating mode, the stage operating conditions are as follows:

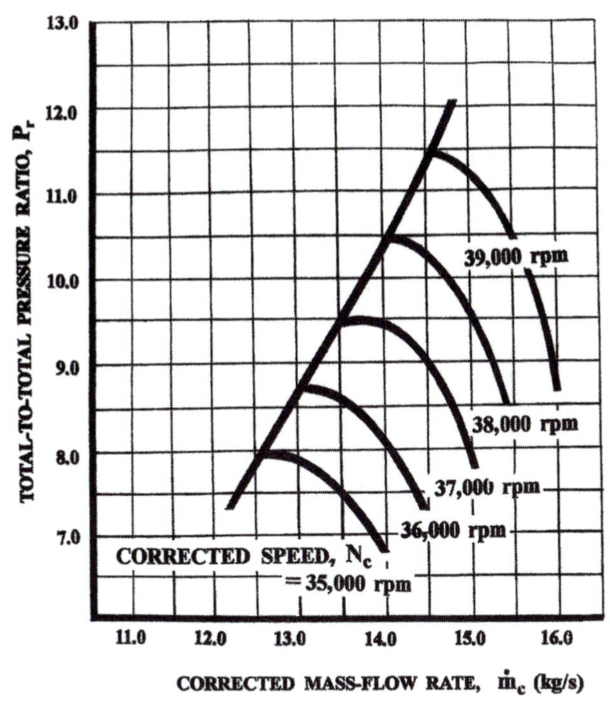

Figure 5.18. Input variables for Problem 9.

- Mass-flow rate ($\dot{m}$) = 1.5 kg/s
- Stage-inlet total pressure (p_{t_0}) = 0.25 bars
- Stage-inlet total temperature (T_{t_0}) = 200 K
- Stator flow process is assumed isentropic
- Stage-exit total pressure (p_{t2}) = 0.35 bars
- $V_{z_2} = V_{z_1}$
- Stator-exit critical Mach number (V_1/V_{cr_1}) = 0.62

Considering these operating conditions and referring to the map, calculate:

a) The rotor-inlet relative critical Mach number (W_1/W_{cr_1});
b) The stage-exit static pressure (p_2).

9) Figure 5.18 shows the map of an axial-flow compressor stage. The cruise-operation point of the stage is defined as follows:

- Supplied specific shaft work (w_s) = 146, 333 J/kg
- Rotor speed (N) = 31, 177 rpm
- Inlet total pressure ($p_{t\,in}$) = 0.235 bars
- Inlet total temperature ($T_{t\,in}$) = 216 K
- Total-to-total efficiency (η_C) = 80%

The foregoing operating conditions are simulated in a test rig utilizing air at an inlet total pressure of 1.0 bar and an inlet total temperature of 288 K. The specific-heat ratio (γ) is 1.4 under both sets of operating conditions.

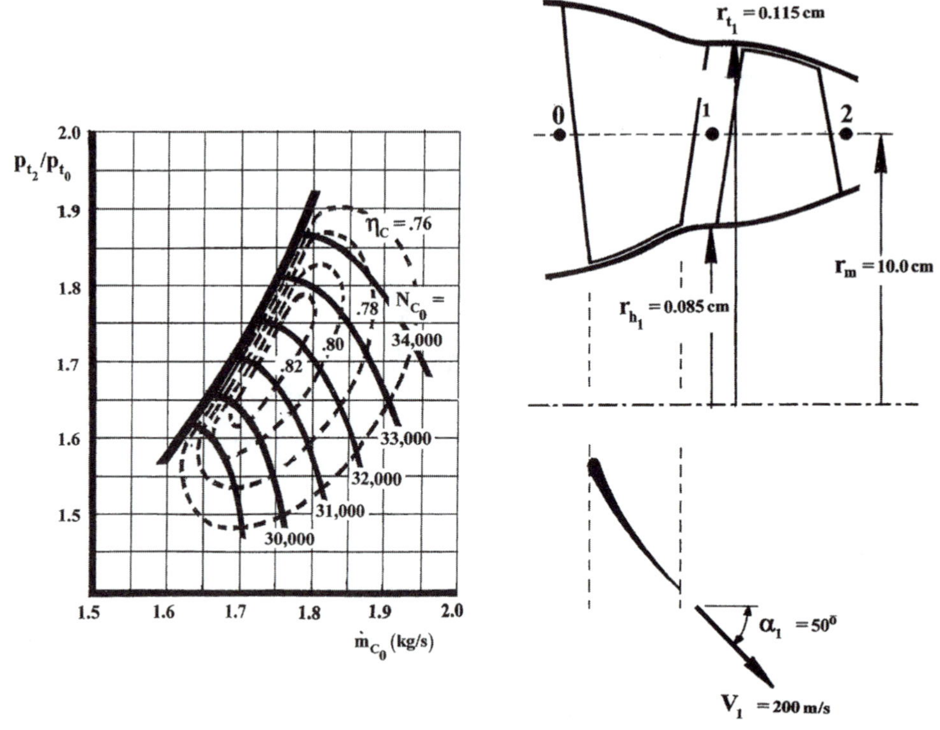

Figure 5.19. Input data for Problem 10.

Part I: Calculate the following variables:

a) The cruise-operation mass-flow rate;
b) The supplied power in the test rig.

Part II: Now consider the rig situation where a gradual increase in the inlet total pressure takes place while maintaining the inlet total temperature, mass-flow rate, and rotor speed at the same magnitudes as in Part 1. Calculate the critical value of inlet total pressure at which the compressor operation becomes unstable.

10) Figure 5.19 shows a single-stage high-pressure compressor stage together with its map. The stage operating conditions are as follows:

- Stage-inlet total temperature (T_{t0}) = 610 K
- Stage-inlet total pressure (p_{t0}) = 10.8 bars
- Stator total pressure loss coefficient ($\bar{\omega}_{st.}$) = 0.06
- Shaft speed (N) = 45,000 rpm
- Stage-wise axial-velocity component (V_z) is constant

Assuming an adiabatic flow process across the stage and a specific-heat ratio (γ) of 1.4, calculate:

a) The percentage of relative critical Mach number decline;
b) The percentage of total relative pressure decline.
c) The specific speed (N_s).

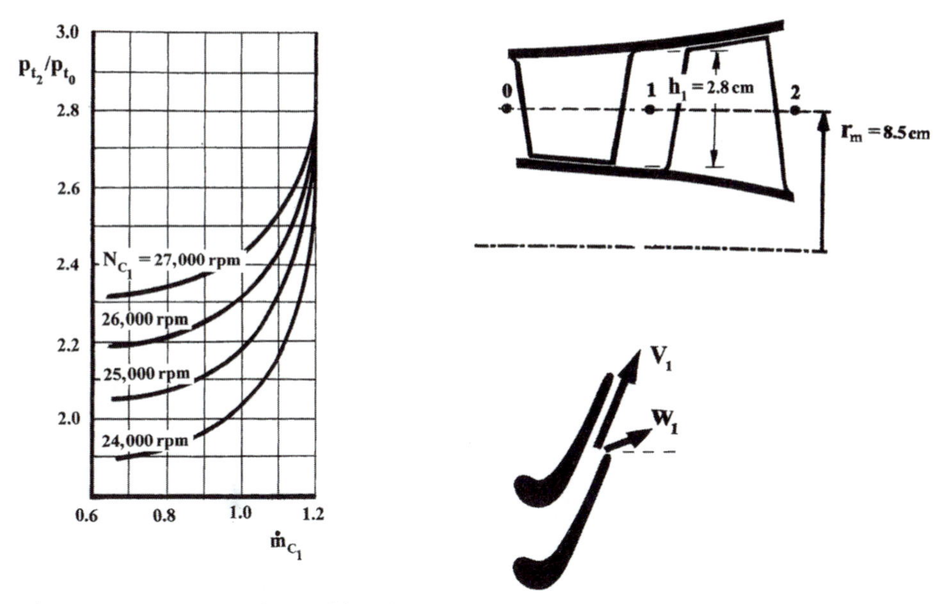

Figure 5.20. Input data for Problem 11.

11) Figure 5.20 shows an axial-flow turbine stage along with its map. The stage operating conditions are as follows:

- Inlet total temperature (T_{t0}) = 1198 K
- Inlet total pressure (p_{t0}) = 12.5 bars
- Rotor speed (N) = 51,000 rpm
- Rotor-inlet absolute flow angle (α_1) = 70°
- Rotor-inlet relative flow angle (β_1) = 18.8°
- Stator flow process is assumed isentropic
- Constant V_z magnitude across the stage
- Rotor passage is choked

Assuming an adiabatic rotor and a specific-heat ratio (γ) of 1.33, calculate:

a) The mass-flow rate ($\dot{m}$);
b) The exit critical Mach number (M_{cr_2});
c) The percentage of total relative pressure decline;
d) The specific speed (N_s). See whether this magnitude is within the axial-turbine range in Figure 5.9.

12) Figure 5.21 shows a centrifugal compressor and the corresponding map. The impeller operates under the following conditions:

- Impeller-inlet total pressure (p_{t1}) = 5.7 bars
- Impeller-inlet total temperature (T_{t1}) = 488 K
- Impeller-inlet swirl angle (α_1) = 0
- Impeller-inlet critical Mach number (M_{cr_1}) = 0.33
- Impeller total-to-total efficiency (η_{t-t}) = 79%
- Impeller-exit radial-velocity component (V_{r_2}) = 88.7 m/s

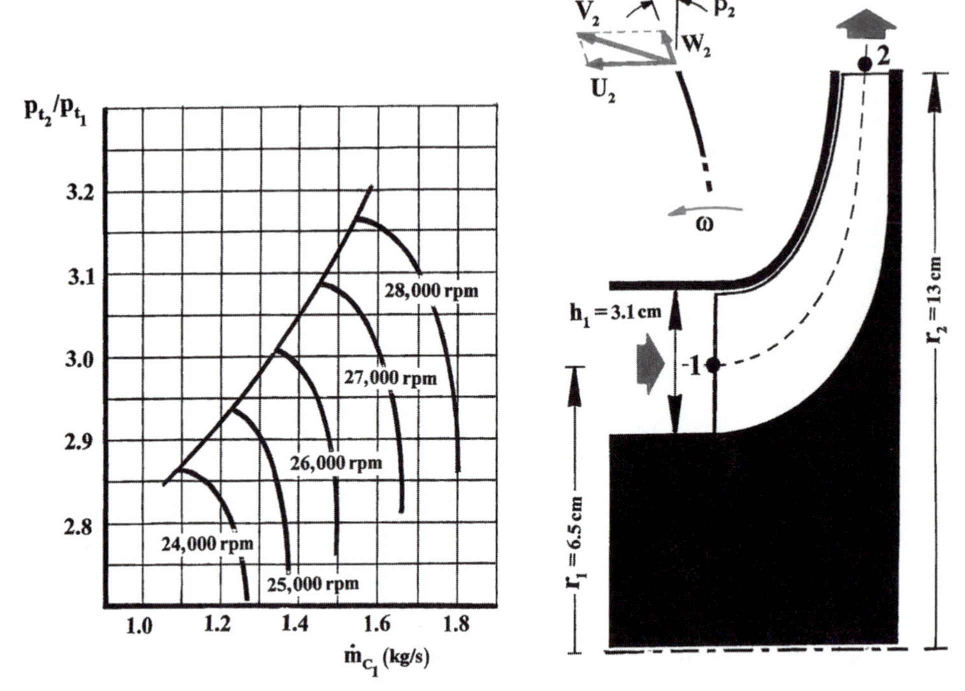

Figure 5.21. Input data for Problem 12.

With an adiabatic flow process and an average specific-heat ratio (γ) of 1.4, calculate

a) The impeller speed (N);
b) The impeller-exit relative flow angle (β_2);
c) The specific speed (N_s), stating whether the centrifugal-compressor choice is wise in this case.

13) Figure 5.22 shows a radial-inflow turbine stage along with its map. The operating conditions of the stage are as follows:

- Inlet total pressure (p_{t0}) = 12.5 bars
- Inlet total temperature (T_{t0}) = 1375 K
- Rotor-inlet critical Mach number (M_{cr_1}) = 0.82
- Rotor speed (N) = 30,590 rpm
- Mass-flow rate ($\dot{m}$) = 5.26 kg/s
- Rotor-inlet relative flow angle (β_1) = 0
- Rotor-exit critical Mach number (M_{cr_2}) = 0.38
- Stator flow process is isentropic (by assumption)
- Full guidedness of the flow in bladed components

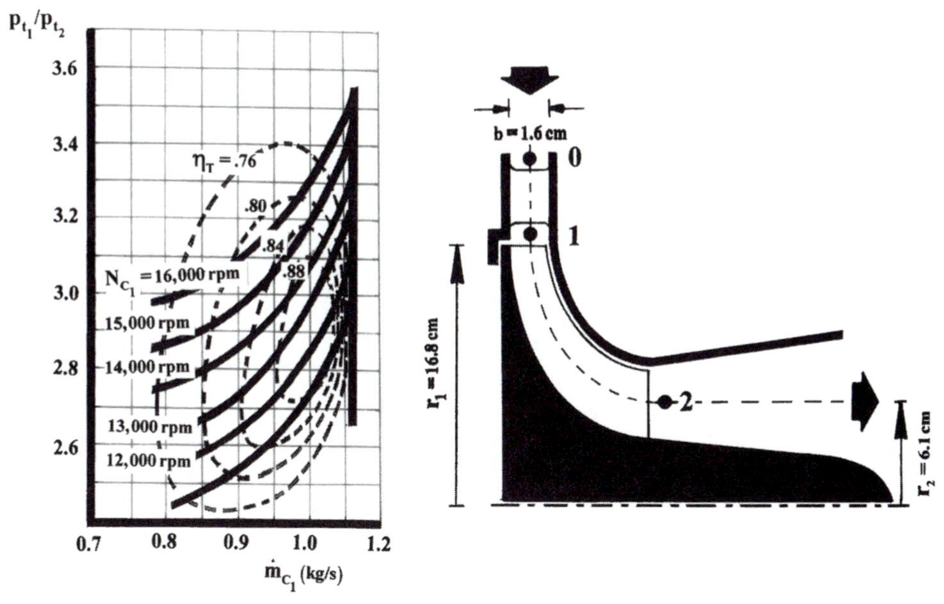

Figure 5.22. Input data for Problem 13.

Considering an adiabatic flow across the rotor and a γ magnitude of 1.33, calculate:

a) The specific speed (N_s) of the stage;
b) The total relative pressure decline that is *solely* produced by the rotor irreversibility sources (e.g., friction).

Hints: In addressing item *b*, you may want to follow the following steps:

1) Calculate the overall total relative pressure drop (i.e., due to the radius decline added to the irreversibily part).
2) Assuming a 100% efficient rotor, recalculate the total relative pressure drop. This drop is caused only by the rotorwise decrease in radius.
3) Subtract the latter total relative pressure drop from the former.
4) In carrying out step 2, note that

$$T_{tr2} = T_{t2} + \left(\frac{W_2^2 - V_2^2}{2c_p}\right) = T_{t2} + \left(\frac{W_{\theta 2}^2 - V_{\theta 2}^2}{2c_p}\right)$$

14) Referring to Problem 12, calculate the decline in total relative pressure across the impeller that is *exclusively* a result of the flow-process irreversibility sources.

CHAPTER SIX

Radial-Equilibrium Theory

In Chapters 3 and 4, we studied major changes in the thermophysical properties of a flow as it traverses a turbine or compressor stage. The analysis then was one-dimensional, with the underlying assumption that average flow properties will prevail midway between the endwalls. Categorized as a pitch-line flow model, this "bulk-flow" analysis proceeds along the "master" streamline (or pitch line), with no attention given to any lateral flow-property gradients.

However, we know of, at least, one radius-dependent variable, namely the tangential "solid-body" velocity vector (**U**). The question addressed in this chapter is how the other thermophysical properties vary along the local annulus height at any streamwise location. The stator and rotor inlet and exit stations, being important "control" locations, are particularly important in this context. In the following, the so-called radial equilibrium equation is derived and specific simple solutions offered. Despite the flow-model simplicity, the radial-equilibrium equation enables the designer early on to take a look at preliminary magnitudes of such important variables as the hub and tip reactions prior to the detailed design phase.

Assumptions

For any axial stator-to-rotor or interstage gap in Figure 6.1, the following assumptions are made:

1) The flow is under a steady-state condition.
2) The flow is inviscid as well as adiabatic.
3) The flow is axisymmetric (i.e., θ-independent).
4) There are no radial shifts of the meridional streamlines (Fig. 6.2).

Implications

Across each of the three gaps in Figure 6.1, viscosity-dominated flow-mixing subregions (termed *wakes*) of the upstream cascade exist. Beginning at the upstream trailing edges, these subregions are where the suction- and pressure-side boundary

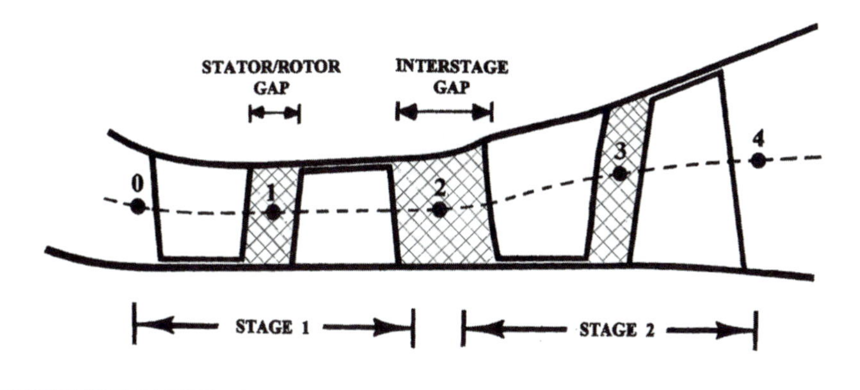

Figure 6.1. Stator/rotor and interstage gaps in a two-stage turbine.

layers will gradually mix toward a uniform flow stream, as shown in Figure 6.3. The problem is that such a uniform flow would theoretically exist at an infinite distance away from the trailing edge. When contrasted with the finite gap length, the situation would naturally spell mechanical and aerodynamic trouble for the downstream airfoil cascade.

The presence of these equidistant wakes (Fig. 6.3) destroys (by definition) the tangential uniformity (or axisymmetry) in the axial gap flow field. In addition, the existence of two airfoil cascades, one on each side of the gap, makes them both subject to cyclic stresses, which, in turn, may cause fatigue failure of the downstream cascade in particular. Note that the wakes themselves are viscosity-dominated in a flow field that is assumed inviscid.

These details of the flow in axial gaps would lead any theoretician to dismiss the very foundation of the radial equilibrium theory. Fortunately, turbomachinists have their own reasoning and interpretations. An example of this is the insertion of such terms as "sufficiently" far from the trailing edge in place of what is theoretically

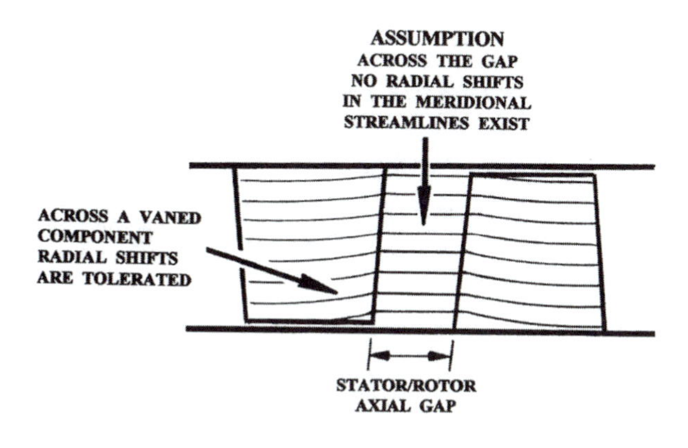

Figure 6.2. Radial shifts of the meridional streamlines within the stator and rotor subdomains.

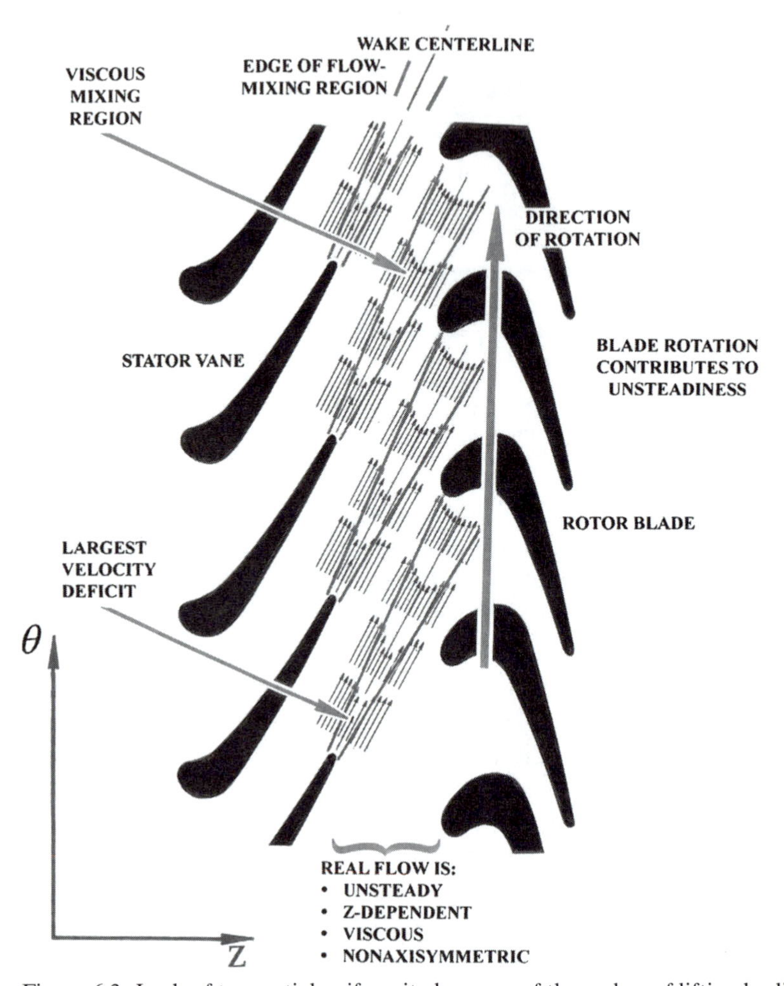

Figure 6.3. Lack of tangential uniformity because of the wakes of lifting bodies.

an infinite streamwise distance. Along these lines, the hypothesis is made that the suction- and pressure-side boundary-layer mixing will occur so suddenly that the mixing is in the immediate vicinity of the upstream trailing edges. Note that short of this and other engineering simplifications, the flow behavior in axial gaps would be attainable only through expensive large-scale computational models. These resource-consuming models will have to be capable of handling the flow field on a fully three-dimensional basis and with such real flow effects (e.g., turbulence) properly simulated.

Derivation of the Radial-Equilibrium Equation

Referring to Figure 6.4, consider the fluid element in the hub-to-casing annulus where no lifting bodies exist. The net radial force (F_p), caused by pressure acting on the

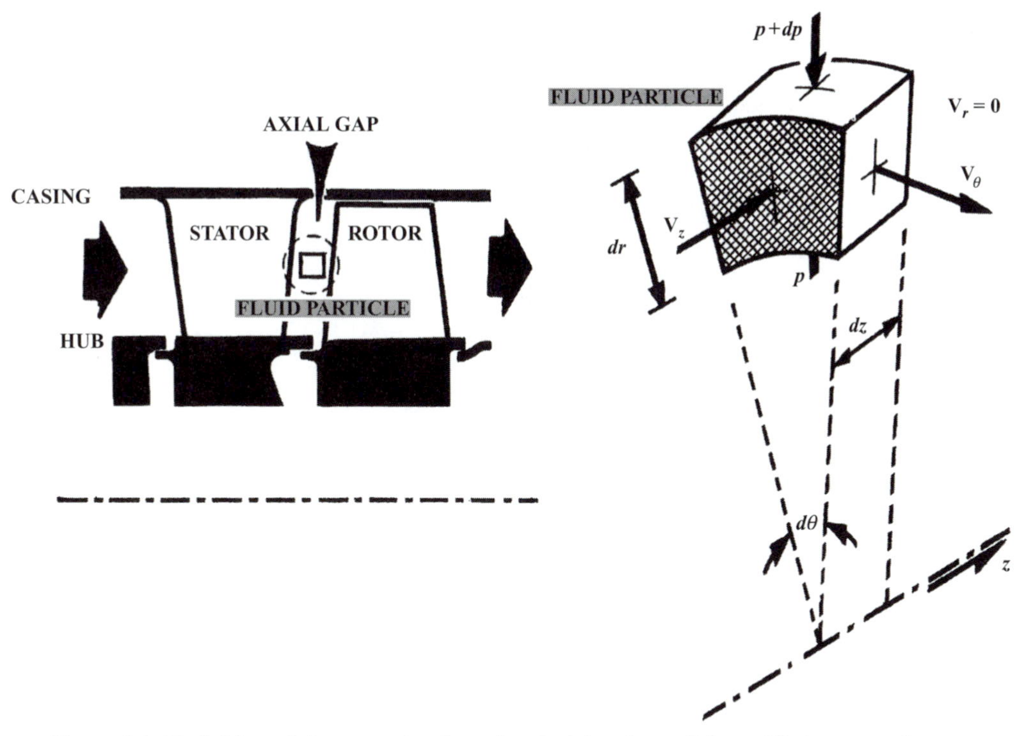

Figure 6.4. Definition of the control volume for deriving the radial-equilibrium equation.

control volume, can be expressed as follows:

$$F_p = (p + dp)(r + dr)d\theta - pr\,d\theta - 2\left(p + \frac{dp}{2}\right) dr \sin\left(\frac{d\theta}{2}\right) \tag{6.1}$$

With no lack of generality, the fluid element in Figure 6.4 is taken to have an axial extension of unity, as is probably evident in equation (6.1). Under the assumption of infinitesimally small $d\theta$, we can utilize the approximation

$$\sin\left(\frac{d\theta}{2}\right) = \frac{d\theta}{2} \tag{6.2}$$

Substituting (6.2) into (6.1), and ignoring high-order terms, we get the following F_p expression:

$$F_p = r\; dp\; d\theta \tag{6.3}$$

With an axial extension of unity, the fluid-element mass, where the net pressure force is acting, can be written as follows:

$$dm = \rho\left[\pi(r + dr)^2 - \pi r^2\right]\left(\frac{d\theta}{2\pi}\right) - \rho\, r\, dr\, d\theta \tag{6.4}$$

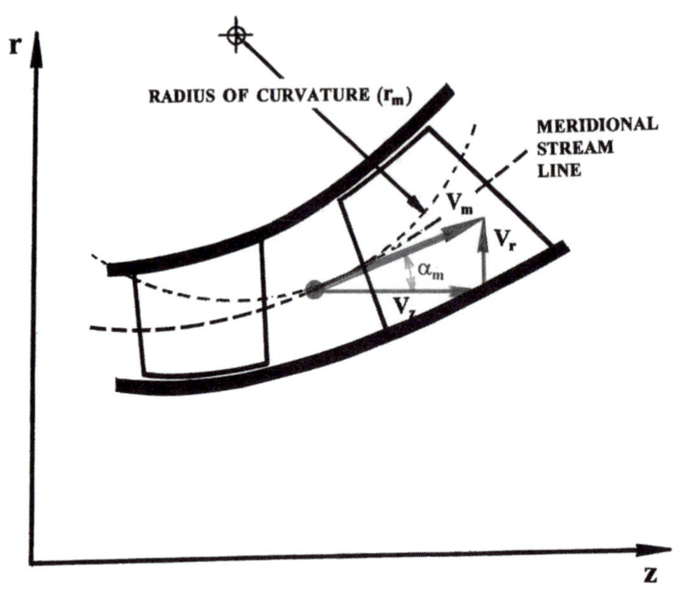

Figure 6.5. Fluid-particle acceleration as a result of the meridional streamline curvature.

The net pressure-exerted force F_p in equation (6.3) will have to account for the following acceleration components possessed by the particle:

1) The centripetal acceleration associated with the tangential velocity component V_θ, which is equal to ${V_\theta}^2/r$.

2) The acceleration that is associated with the curvature (if any) of the meridional streamlines. This acceleration component is locally perpendicular to the local meridional streamline.

3) Any radial component of a linear acceleration vector. It should be emphasized that this linear acceleration precludes that caused by the time dependency of the velocity vector (**V**), which is ignored here. Instead, it is simply the radial component of the inertia-produced acceleration that would naturally exist in an accelerating or decelerating flow passage.

The part of the pressure force that accounts for item 1 is

$$F_{p1} = dm\frac{{V_\theta}^2}{r} = \rho {V_\theta}^2 dr\, d\theta \tag{6.5}$$

The part that accounts for item 2 is illustrated in Figure 6.5 and can be expressed as follows:

$$F_{p2} = dm\left(\frac{{V_m}^2}{r_m}\right)\cos\alpha_m = \rho r\, dr\, d\theta\left(\frac{{V_m}^2}{r_m}\right)\cos\alpha_m \tag{6.6}$$

Finally, the part of the pressure force that accounts for item 3 is

$$F_{p3} = dm\dot{V}_m \sin\alpha_m = \rho r\, dr\, d\theta(\dot{V}_m \sin\alpha_m) \tag{6.7}$$

where

$$\dot{V}_m = V_m \frac{dV_m}{dm} \tag{6.8}$$

Equating the sum of F_{p1}, F_{p2}, and F_{p3} to F_p, we obtain

$$\frac{1}{\rho}\frac{dp}{dr} = \frac{V_\theta{}^2}{r} - \left(\frac{V_m{}^2}{r_m}\right)\cos\alpha_m - \dot{V}_m \sin\alpha_m \tag{6.9}$$

Equation (6.9) is the generalized form of the radial-equilibrium equation, which is as applicable to turbines as it is to compressors. It should be hinted that the same equation is applicable to incompressible-flow turbomachines, with the exception that the body force caused by gravity must be introduced as well should a considerable shift in elevation occur. However, such force in, say, a hydraulic pump or water turbine would normally be ignorably small.

Special Forms of the Radial-Equilibrium Equation

Under the assumption of straight, perfectly *axial* streamlines in the meridional projection, the following simple form of the radial-equilibrium equation will apply:

$$\frac{1}{\rho}\frac{dp}{dr} = \frac{V_\theta{}^2}{r} \tag{6.10}$$

This relationship is the simplest and most recognizable form of the radial-equilibrium equation. Under this simplification, a useful equation that provides the radial enthalpy distribution can also be obtained.

Let us begin with the total enthalpy definition, namely

$$h_t = h + \frac{V^2}{2} \tag{6.11}$$

Under the assumption of a purely axial gap, we can ignore the existence of the radial-velocity component (V_r) and rewrite equation (6.11) as

$$h_t = \left(\frac{\gamma}{\gamma - 1}\right)\frac{p}{\rho} + \frac{1}{2}\left(V_z{}^2 + V_\theta{}^2\right) \tag{6.12}$$

where the equation of state has already been utilized. Note that all of the thermophysical variables in (6.12) are (or actually are made to be) functions of a single independent variable, namely the radius (r). We are now in a position to differentiate both sides of (6.12) to obtain

$$\frac{dh_t}{dr} = V_z\frac{dV_z}{dr} + V_\theta\frac{dV_\theta}{dr} + \frac{\gamma}{\gamma - 1}\left(\frac{1}{\rho}\frac{dp}{dr} - \frac{p}{\rho^2}\frac{d\rho}{dr}\right) \tag{6.13}$$

Assuming a locally inviscid flow field, let us now go a step further by assuming an isentropic-flow gap. Now we can make use of the basic isentropic relationship, namely

$$\frac{p}{\rho^\gamma} = \text{constant} \tag{6.14}$$

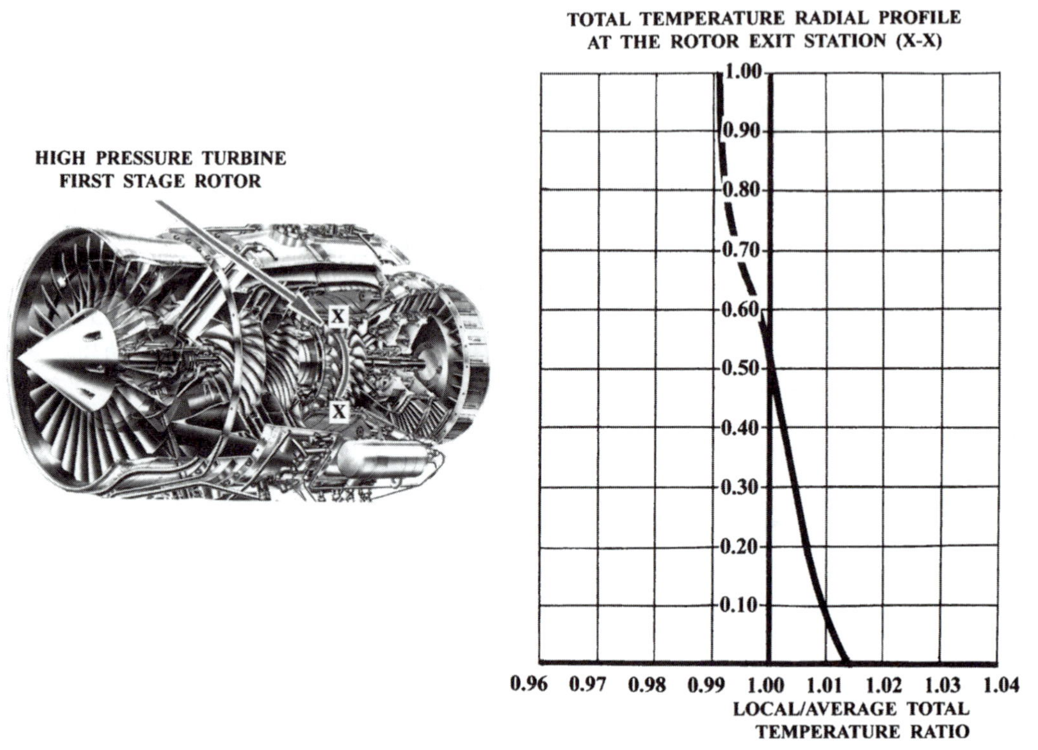

Figure 6.6. Example of a nearly constant total-temperature profile.

which, upon differentiation, yields

$$\frac{dp}{dr} = \left(\frac{\rho}{\gamma p}\right)\frac{dp}{d\rho} \tag{6.15}$$

Substituting (6.15) into (6.13), we get

$$\frac{dh_t}{dr} = V_z\frac{dV_z}{dr} + V_\theta\frac{dV_\theta}{dr} + \frac{V_\theta{}^2}{r} \tag{6.16}$$

Because the shaft work extracted (or consumed) appears in the form of a rotor-wise total-enthalpy drop (or rise), the solution of equation (6.16) across the upstream and downstream gaps will provide (upon subtraction) the shaft work at any given radial location.

Further Simplifications

Figure 6.6 shows the radial profile of the total temperature downstream from the high-pressure-turbine first-stage rotor that belongs to the turbofan engine in Figure 1.3. Examination of the relative deviations of the T_t magnitudes in Figure 6.6 from the average value reveals that this variable is practically constant at all radii. The figure is intended to allow the student to be "reasonably" comfortable when "assuming" the total temperature to be independent of the radial location, even downstream from a rotor-blade cascade. Such a simplifying assumption can be expressed as

$$\frac{dh_t}{dr} \approx 0 \tag{6.17}$$

Under this condition, equation (6.16) can be further compacted as

$$V_z \frac{dV_z}{dr} + V_\theta \frac{dV\theta}{dr} + \frac{V_\theta{}^2}{r} = 0 \tag{6.18}$$

A special case of the ordinary differential equation (6.18) is one where the axial-velocity component (V_z) is constant, and it can be expressed as

$$\frac{dV_\theta}{dr} = -\frac{V_\theta}{r}$$

or

$$\frac{dV_\theta}{V_\theta} = -\frac{dr}{r}$$

which, upon integration, provides a well-known relationship in the general turbomachinery area:

$$r V_\theta = \text{constant} \tag{6.19}$$

This relationship is commonly known as the free-vortex flow condition.

Despite some of the strong additional assumptions we have already made, the free-vortex condition is a frequently made simplification, particularly during the preliminary design phase. It may be appropriate at this point to summarize these assumptions:

- Ideally isentropic flow across the hub-to-casing gap
- Constant axial velocity (V_z) across the hub-to-casing gap
- A gapwise constant total enthalpy, h_t (or temperature, T_t)

Satisfaction of the free-vortex conditions upstream and downstream a rotor gives rise to what is referred to as a free-vortex blade design. Given the ideal circumstances involved in the free-vortex flow condition, it should perhaps be surprising to find that a great in any rotor blades are designed on this very basis (e.g., Fig. 6.7). Nevertheless, it should be emphasized that this in no way will guarantee a free-vortex flow structure in the upstream and downstream gaps. In fact, one could perhaps consider such a blade design as an aerodynamically naive attempt to lead the flow structure in these gaps to be close to such an ideal flow behavior. The reason is that it is under this ideal flow structure that the flow vorticity becomes nonexistent. The flow vortical motion, as is well-known, is a major source of kinetic-energy losses. Referring to Figure 6.7, note that a free-vortex blade typically suffers from a considerable amount of blade twist from hub to tip, a feature that could spell trouble from a mechanical standpoint.

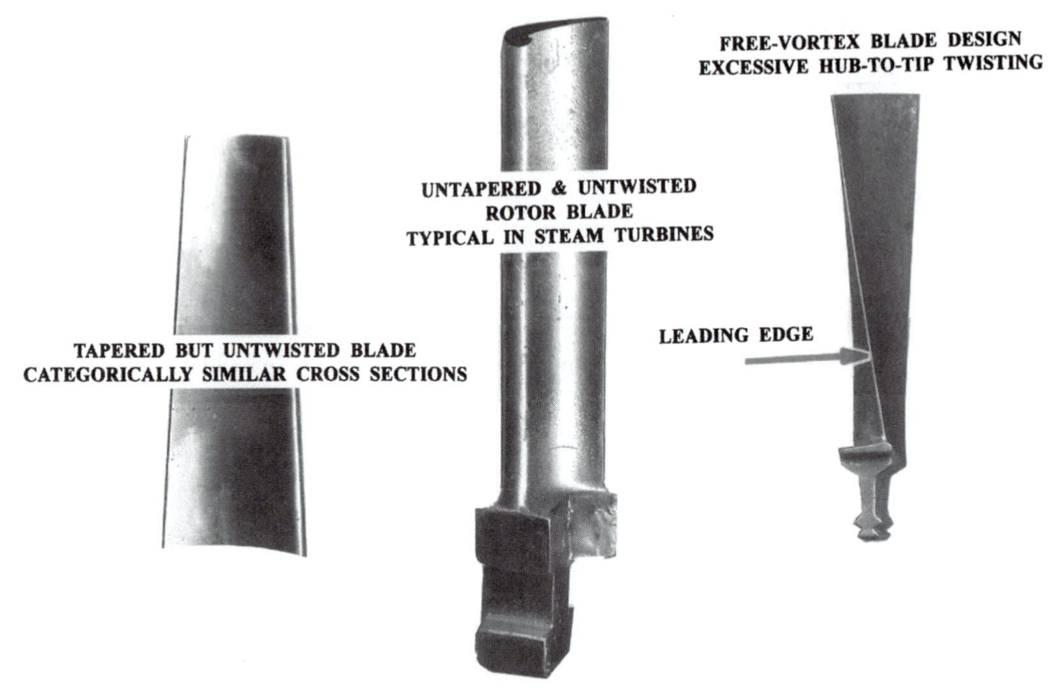

Figure 6.7. Excessive spanwise blade twist under a free-vortex design strategy.

EXAMPLE 1

Figure 6.8 shows the stator of an axial-flow turbine. The stator-inlet flow stream is totally in the axial direction at all radii. The exit flow pattern is that of the free-vortex type. At the hub radius, the flow conditions are as follows:

- The inlet static pressure $(p_{1_h}) = 7.0$ bars
- The exit static pressure $(p_{2_h}) = 6.0$ bars

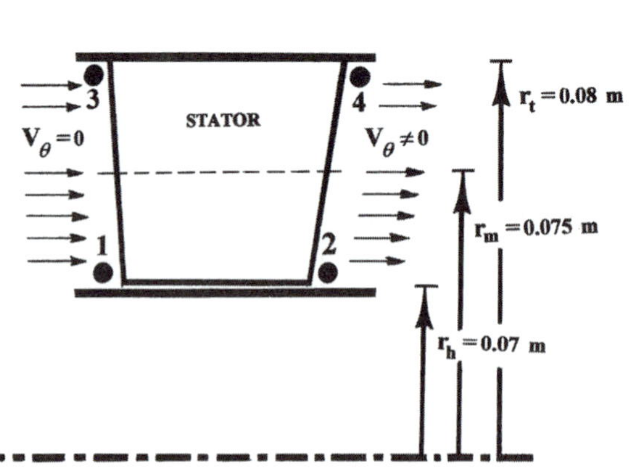

Figure 6.8. Input variables for Example 1.

- The exit tangential-velocity component ($V_{\theta 2h}$) = 220 m/s
- The axial-velocity component (V_z) over the inlet and exit stations is equal to 100 m/s
- The total temperature (T_t) over the inlet and exit stations is equal to 1400 K

Assuming a constant magnitude of the static density (ρ) over both the inlet and exit gaps of 1.5 kg/m^3, and an average specific-heat ratio (γ) of 1.33, calculate the following:

a) The stator-casing static-pressure difference ($p_3 - p_4$).
b) The casing-line stator-exit critical Mach number $(V/V_{cr})_4$.

SOLUTION

Part a:

$$p_3 - p_1 = 7 \times 10^5 \text{ N/m}^2$$

At the stator exit station, we have

$$V_\theta = \frac{K}{r}$$

The boundary condition is $[(V_\theta)_{r=0.07\text{ m}} = 220 \text{ m/s}]$, and we can evaluate the constant K as

$$K = 15.4$$

Now we can integrate the radial-equilibrium equation (6.10) as

$$p = \rho \int \frac{V_\theta{}^2}{r}\, dr = 1.5(15.4)^2 \left(\frac{-1}{2r^2}\right) + C = C - \frac{177.87}{r^2}$$

The integration constant (C) can be determined using the boundary condition $[(p)_{r=0.07\text{ m}}]$ as

$$C = 6.363 \times 10^5$$

which results in the following stator-exit static-pressure expression:

$$p = (3.363 \times 10^5) - \frac{177.87}{r^2}$$

Substituting the tip radius (0.08 m) in this expression yields

$$p_4 = 6.08 \times 10^5 \text{ N/m}^2 = 6.08 \text{ bars}$$

Thus,

$$p_3 - p_4 = 0.92 \text{ bars}$$

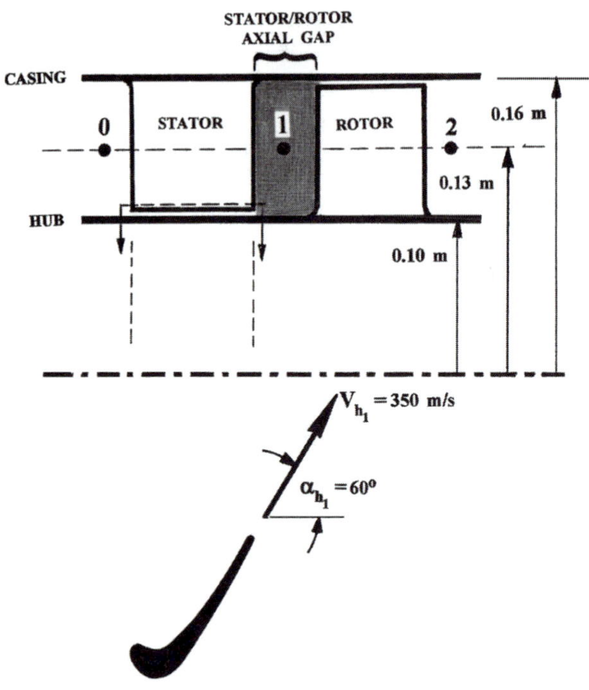

Figure 6.9. Input variables for Example 2.

Part b: Realizing that T_t is constant throughout the flow domain, we can calculate the critical velocity and critical Mach number as follows:

$$V_{cr} = \sqrt{\left(\frac{2\gamma}{\gamma + 1}\right) RT_t} = 677.3 \text{ m/s}$$

$$M_{cr\,4} = \frac{V_4}{V_{cr}} = 0.32 \text{ m/s}$$

where the V_θ expression given earlier was utilized in calculating V_4.

EXAMPLE 2

Figure 6.9 shows an axial-flow turbine stage, where the following conditions apply in the stator/rotor gap:

- The hub-radius velocity (V_h) = 350 m/s
- The absolute swirl angle (α_h) = 60°
- The mean-radius static pressure (p_m) = 8.0 bars

In addition, the flow behavior across the gap is that of the free-vortex pattern, and the rotor speed (N) is 13,000 rpm, the rotor blade is untapered and untwisted (i.e., the blade sections are identical at all radii), and the blade is designed to yield a zero incidence angle (i.e., $i_R = 0$) at the mean radius. Assuming an average specific-heat ratio of 1.33, calculate:

a) The rotor-blade hub and casing incidence angles.
b) The static pressure (in the gap) at the rotor tip radius, assuming a static density (ρ) of 3.25 kg/m^3 across the entire stator/rotor gap.

SOLUTION

Part a: At the mean radius, we have

$$(V_\theta)_m = V_{\theta h} \times \tfrac{r_h}{r_m} = 233.2 \text{ m/s}$$

$$U_m = \omega r_m = 177.0 \text{ m/s}$$

$$V_{z,m} = V_{z,h} = 175.0 \text{ m/s}$$

$$\beta_{1,m} = \arctan\left(\frac{W_{\theta,m}}{V_{z,m}}\right) = \arctan\left(\frac{V_{\theta,m} - U_m}{V_{z,m}}\right) = 17.8^\circ$$

where $\beta_{1,m}$ is the rotor-inlet relative flow angle. The mean-radius incidence angle $(i_R)_m$ is zero, so

$$(\beta_{1,m})' = 17.8^\circ$$

where $(\beta_{1,m})'$ is the mean-radius airfoil-inlet (metal) angle. This angle will be constant at all radii because the blade is untwisted. Let us now calculate the hub-radius incidence angle:

$$(i_R)_h = \beta_{1,h} - \beta_1{}' = \arctan\left(\frac{V_{\theta 1,h} - U_h}{V_{zh}}\right) = 25.8^\circ$$

Repeating the same computational process at the rotor-inlet tip radius, we get

$$(i_R)_t = -27.0^\circ$$

Figure 6.10 summarizes the results just obtained.

Part b: In order to obtain the tip-radius static pressure, we integrate the pressure version of the radial-equilibrium equation (6.10), which yields

$$(p)_{gap\ 1} = 8.88 \times 10^5 - \frac{1492.97}{r^2}$$

where the following data and boundary conditions were used:

$$(\rho)_{gap\ 1} = 3.25 \text{ kg/m}^3 \text{ (given)}$$

$$(V_{theta})_{gap\ 1} = \frac{K}{r}, \text{ with } K = r_h V_{\theta h} = 30.31 \text{ m}^2/\text{s}$$

$$(p_1)_{r=0.13 \text{ m}} = 8.0 \times 10^5 \text{ bars}$$

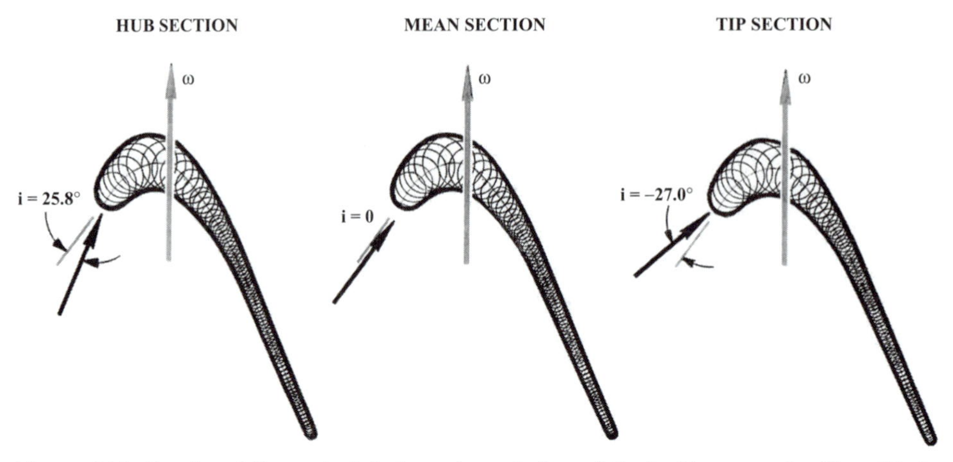

Figure 6.10. Results of Example 2: hub-to-tip variation of the incidence angle. *Note*: Blade sections are identical at all radii since the blade is untapered and untwisted.

Now, substituting the tip radius (0.16 m) in the pressure-radius relationship presented earlier, we get

$$(p_{tip})_{gap\ 1} = 8.30 \text{ bars}$$

EXAMPLE 3

Figure 6.11 shows the meridional flow path of a compressor stage. The rotor speed is 14,000 rpm, and the average specific-heat ratio (γ) is 1.4. Also, the following conditions apply over the rotor's inlet and exit gaps:

Rotor inlet gap

- T_t = constant
- $\alpha = 0°$
- $\rho = 2.25 \text{ kg/m}^3$
- $T_{hub} = 511$ K

Rotor exit gap

- T_t = constant
- Relative flow angle $\beta = 0$
- $\rho = 2.4 \text{ kg/m}^3$
- $T_{hub} = 552$ K
- $V_{z_{hub}} = 124$ m/s

Using any simple means of integration (such as the trapezoidal rule), and considering only the properties of the hub, mean, and tip sections (for simplicity), calculate the following variables:

a) The mass-flow rate ($\dot{m}$) through the compressor stage;
b) The exit total relative pressure at the tip section.

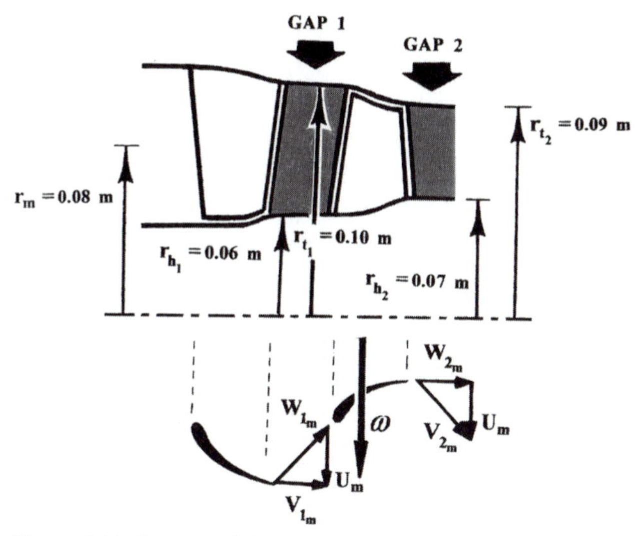

Figure 6.11. Input variables for Example 3.

SOLUTION

Part a: Let us at this point direct our attention to Gap 2 in Figure 6.11:

$$\omega = \frac{N \times 2\pi}{60} = 1466.0 \text{ radians/s}$$

Because the relative flow angle β_2 is zero at all radii, $W_{\theta 2}$ will also be zero at all radii. It follows that

$$(V_\theta)_{gap\ 2} = \omega r, \text{ where } \omega = 1466.0 \text{ radians/s}$$

Now, the radial-equilibrium equation (6.18) yields

$$V_z \frac{dV_z}{dr} = -2\omega^2 r$$

which, upon integration, gives

$$V_z = \sqrt{36{,}438 - 4{,}298{,}512 r^2}$$

where the following boundary condition was used:

$$(V_z)_{r=0.07\text{m}} = 124.0 \text{ m/s}$$

Now, we can calculate the mean and tip magnitudes of V_z as follows:

$$(V_z)_{mean} = (V_z)_{r=0.08\text{ m}} = 94.49 \text{ m/s}$$
$$(V_z)_{tip} = (V_z)_{r=0.09\text{ m}} = 40.27 \text{ m/s}$$

The mass-flow rate can be calculated (under the stated simplifications) as

$$\dot{m} = 2\pi\rho \int_{r_h}^{r_t} r V_z dr \approx 2\pi\rho \times \frac{1}{2}[r_h V_{zh} + r_m V_{zm} + r_t V_{zt}] \times \Delta r$$

where $\Delta r = 0.01$ m. Direct substitution in this expression yields

$$\dot{m} = 2.21 \text{ kg/s}$$

Part b: Remaining with the flow conditions in Gap 2, we have

$$V_h = \sqrt{V_{z,h}{}^2 + V_{\theta,h}{}^2} = 160.96 \text{ m/s}$$

$$(T_t)_{gap2} = (T_{t,h})_{gap2} = T_h + \frac{V_h{}^2}{2c_p} = 564.9 \text{ K}$$

$$(T_{t_r})_{tip} = T_t + \left(\frac{W_t{}^2 - V_t{}^2}{2c_p}\right) = T_t + \left(\frac{V_{\theta t}{}^2}{2c_p}\right) = 556.2 \text{ K}$$

$$(W_{cr})_{tip} = \sqrt{\left(\frac{2\gamma}{\gamma+1}\right) R(T_{tr})_{tip}} = 431.6 \text{ m/s}$$

$$\left(\frac{W}{W_{cr}}\right)_{tip} = \frac{(V_z)_{tip}}{(W_{cr})_{tip}} = 0.093$$

Let us now integrate the pressure version of the radial equilibrium equation (6.10):

$$\frac{dp}{dr} = \frac{\rho V_\theta{}^2}{r} = \frac{\rho(\omega r)^2}{r} = \rho\omega^2 r$$

which yields

$$(p)_{gap2} = \frac{1}{2}\rho\omega^2 r^2 + C$$

Substituting the boundary condition $[(p)_{r=0.07 \text{ m}} = \rho_h R T_h = 380{,}218 \text{ N/m}^2]$, we can calculate the integration constant (C):

$$C = 367{,}581 \text{ N/m}^2$$

which gives rise to the following static pressure expression:

$$(p)_{gap\,2} = \frac{\rho\omega^2}{2} r^2 + 367{,}581$$

Now, we can calculate the tip-radius pressure (where $r = 0.09$ m) as follows:

$$p_{tip} = 3.885 \text{ bars}$$

Finally, the tip-radius magnitude of the total relative pressure can be computed as follows:

$$(p_{t,r})_{tip} = \frac{p_{tip}}{\left[1 - \left(\frac{\gamma-1}{\gamma+1}\right)\left(\frac{W_t}{W_{cr,t}}\right)^2\right]^{\frac{\gamma}{\gamma-1}}} = 3.904 \text{ bars}$$

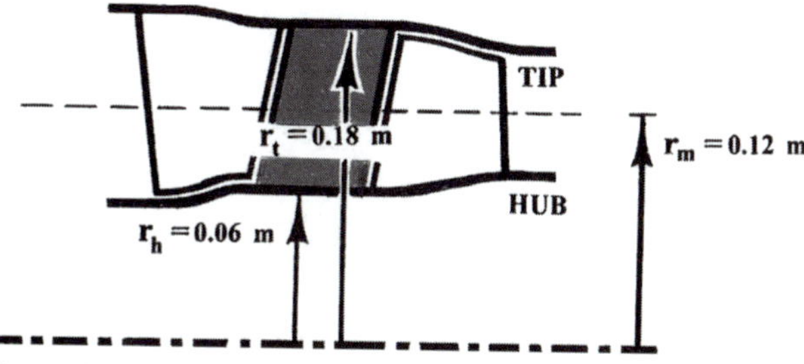

Figure 6.12. Input variables for Example 4.

EXAMPLE 4

Figure 6.12 shows the stator/rotor gap of an axial-flow compressor stage where the total temperature (T_t) is 550 K. The shaft speed (N) is 25,000 rpm. The following conditions also apply at the *mean* radius:

- Absolute velocity (V_{mean}) = 300 m/s
- Absolute flow angle (α_{mean}) = +30°
- Static density (ρ_{mean}) = 5.8 kg/m^3

The hub-to-tip tangential (or swirl) velocity component variation is

$$V_\theta \sqrt{r} = \text{constant}$$

The rotor-blade inlet (metal) angle is constant from hub to tip, and the rotor mean-radius incidence angle is zero. If the average specific-heat ratio γ is 1.4, calculate the following variables:

a) The hub and tip rotor-blade incidence angles.
b) The rotor-inlet total-pressure difference ($p_{t_{tip}} - p_{t_{hub}}$) between the hub and tip radii. In doing so, you may assume an average density magnitude to be that at the mean radius.

SOLUTION

Part a: Across the stator/rotor gap, we have

$$K = V_\theta \sqrt{r} = (V_\theta)_m \sqrt{r_m} = V_m \sin \alpha_m \sqrt{r_m} = 51.96$$

or

$$V_\theta = \frac{K}{\sqrt{r}} = \frac{51.96}{\sqrt{r}}$$

Substituting the hub and tip radii, we get

$$V_{\theta h} = 212.13 \text{ m/s}$$
$$V_{\theta m} = 150.0 \text{ m/s}$$
$$V_{\theta t} = 122.47 \text{ m/s}$$

Integrating the axial-velocity version of the radial equilibrium equation (6.18), we get

$$V_z = \sqrt{\frac{2700.0}{r} + 45{,}000}$$

where the following boundary condition was implemented:

$$(V_z)_{r=r_m} = V_m \cos \alpha_m = 259.81 \text{ m/s}$$

It follows that

$$V_{z,h} = 300.0 \text{ m/s}$$
$$V_{z,t} = 244.95 \text{ m/s}$$

We also have

$$U_h = 157.08 \text{ m/s}$$
$$U_m = 314.16 \text{ m/s}$$
$$U_t = 471.24 \text{ m/s}$$

In order to calculate the rotor-blade inlet (metal) angle, we make use of the stated item that the mean-radius incidence angle ($(i_R)_m$) is zero, meaning that the mean-radius flow and metal angles are identical:

$$\beta' = \beta_m' = \beta_m = \arctan\left(\frac{V_{\theta,m} - U_m}{V_{z,m}}\right) = -32.3^\circ$$

Now, we can calculate the hub and tip magnitudes (and signs) of the incidence angle as follows:

$$(i_R)_h = \beta_h - \beta' = \arctan\left(\frac{V_{\theta,h} - U_h}{V_{z,h}}\right) - (-32.3^\circ) = +42.3^\circ$$

$$(i_R)_t = \beta_t - \beta' = \arctan\left(\frac{V_{\theta,t} - U_t}{V_{z,t}}\right) - (-32.3^\circ) = -22.6^\circ$$

The foregoing results are graphically represented in Figure 6.13, which shows the spanwise variation of the rotor-blade incidence angle.

Part b:

$$T_m = T_t - \frac{V_m{}^2}{2c_p} = 505.3 \text{ K}$$

$$p_m = \rho R T_m = 8.41 \times 10^5 \text{ bars}$$

Now, we can integrate the pressure version of the radial-equilibrium equation, reaching the pressure expression

$$p = 9.715 \times 10^5 - \frac{15.66}{r}$$

where the value of the mean-radius static pressure (computed earlier) was used as a

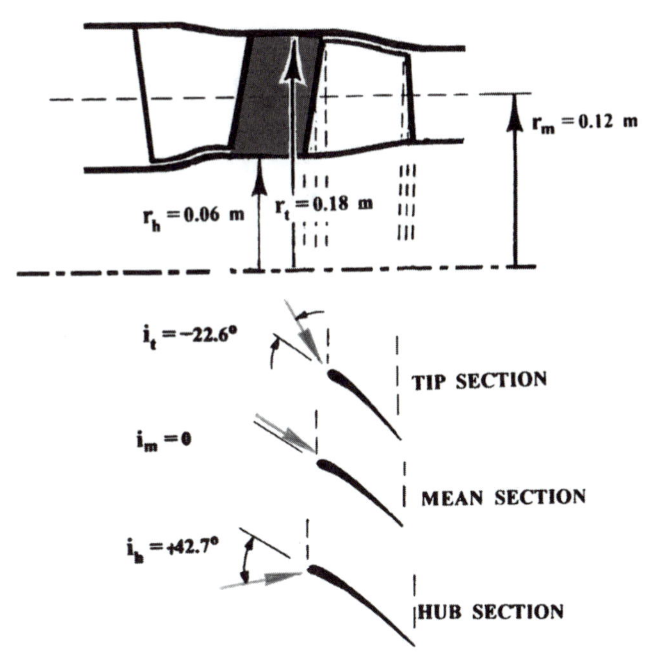

Figure 6.13. Results of Example 4.

boundary condition. Now we can calculate both the hub and tip pressure magnitudes as follows:

$$p_h = p_{r=0.06\text{ m}} = 7.11 \text{ bars}$$
$$p_t = p_{r=0.18\text{ m}} = 8.84 \text{ bars}$$

In order to calculate the hub and tip total pressures, we first calculate the corresponding critical Mach numbers as follows:

$$V_h = \sqrt{V_{\theta,h}{}^2 + V_{z,h}{}^2} = 367.4 \text{ m/s}$$
$$V_t = \sqrt{V_{\theta,t}{}^2 + V_{z,t}{}^2} = 273.9 \text{ m/s}$$
$$V_{cr} = \sqrt{\left(\frac{2\gamma}{\gamma+1}\right) RT_t} = 429.1 \text{ m/s (constant at all radii)}$$
$$M_{cr\,h} = \frac{V_h}{V_{cr}} = 0.856$$
$$M_{cr\,t} = \frac{V_t}{V_{cr}} = 0.638$$

Now, we can calculate the spanwise difference in total pressure as follows:

$$p_{t\,h} = \frac{p_h}{\left[1 - \left(\frac{\gamma-1}{\gamma+1}\right) M_{cr\,h}{}^2\right]^{\frac{\gamma}{\gamma-1}}} = 10.43 \text{ bars}$$
$$p_{t\,t} = \frac{p_t}{\left[1 - \left(\frac{\gamma-1}{\gamma+1}\right) M_{cr\,t}{}^2\right]^{\frac{\gamma}{\gamma-1}}} = 10.88 \text{ bars}$$
$$p_{t\,t} - p_{t\,h} = 0.45 \text{ bars}$$

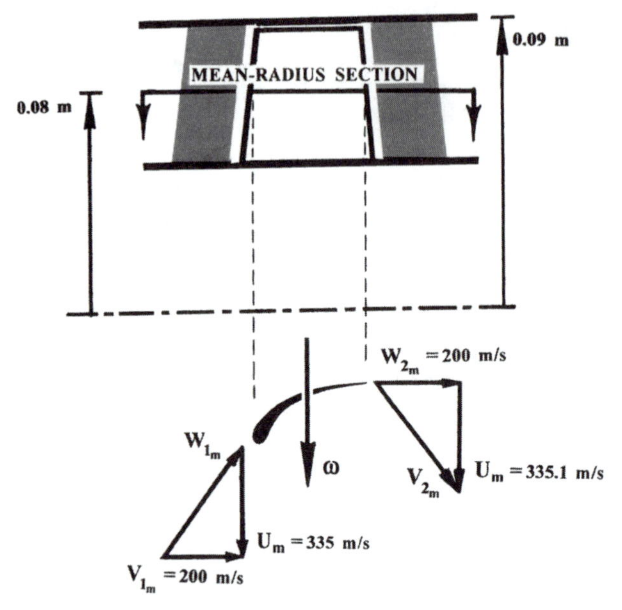

Figure 6.14. Input variables for Problem 1.

PROBLEMS

1) Figure 6.14 shows the upstream and downstream gaps of a high-pressure compressor rotor where the shaft speed (N) is 40,000 rpm. Referring to these gaps as "Gap 1" and "Gap 2," the flow conditions at the mean radius are as follows:

- $\alpha_{1m} = \beta_{2m} = 0$
- $V_{z1m} = V_{z2m} = 200$ m/s

The following average conditions also apply between the mean and tip radii:

- $T_{t1} = 650$ K
- $T_{t2} = 762$ K
- $(\rho_1)_{av.} = 1.24$ kg/m^3
- $(\rho_2)_{av.} = 1.35$ kg/m^3
- $V_{\theta 1} =$ constant
- $V_{\theta 2} =$ constant

The average static densities are provided for the sole purpose of integrating the radial-equilibrium equation. By integrating the proper version(s) of this equation:

a) Calculate the tip reaction (R_{tip}).
b) Calculate the change in static pressure $[(p_2)_{tip} - (p_2)_{mean}]$ between the mean and tip radii.
c) Sketch, with reasonable accuracy, the rotor-blade airfoil at the tip radius, assuming that the incidence and deviation angles are zero.

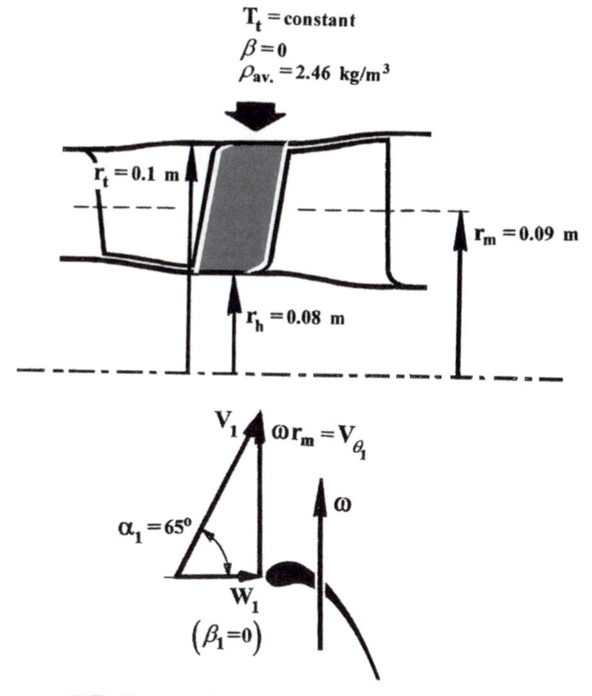

Figure 6.15. Input variables for Problem 2.

2) Figure 6.15 shows an axial-flow turbine stage where the shaft speed (N) is 36,000 rpm. The stage operating conditions in the stator/rotor gap are as follows:

- T_t = constant
- N = 52,000 rpm
- $\rho_{av.} = 2.46$ kg/m^3
- relative flow angle $\beta = 0$
- Hub magnitude of static temperature (T_{hub}) = 1142 K
- Tip-radius absolute flow angle (α_{tip}) = 62°
- Average value of the specific heat ratio $\gamma = 1.33$

Calculate the hub, mean-radius, and casing magnitudes of the axial velocity component.

3) The turbine stage in Figure 6.16 is designed in such a way that the rotor-inlet relative flow angle β_1 is zero everywhere in the stator/rotor gap between the hub and tip radii. In addition, the shaft speed is 46,000 rpm, and the following conditions apply at the mean radius in this gap:

- $(p_t)_m = 8.5$ bars
- $\alpha_m = 65°$
- $T_m = 1100$ K

Knowing that the total temperature (T_t) in this gap is 1220 K and is constant across the entire gap, and assuming a constant hub-to-tip density (ρ) that is

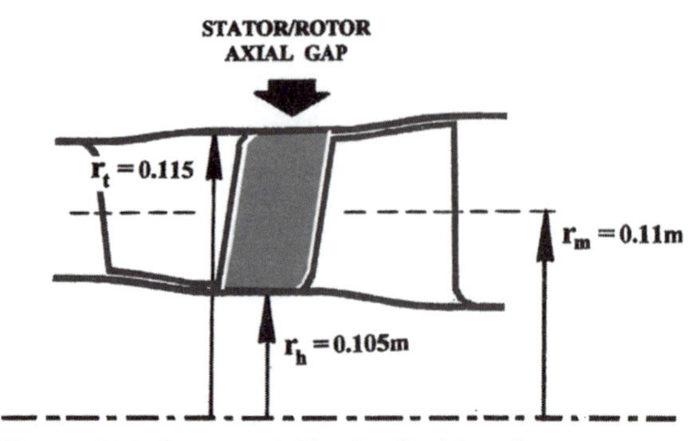

Figure 6.16. Input variables for Problem 3.

equal to the mean-radius magnitude only for the purpose of integrating the radial-equilibrium equation, calculate the hub-to-tip difference in total relative pressure $[(p_{t,r})_{tip} - (p_{t,r})_{hub}]$ over the stator/rotor gap.

4) Figure 6.17 shows a high-pressure compressor stage where the stator/rotor gap variables are as follows:

- $V_{hub} = 300$ m/s
- $\alpha_{hub} = 30°$
- $(V_\theta)_{tip} = 240$ m/s
- $V_\theta = a + br$, where a and b are constants
- The hub-to-tip total temperature $T_t = 550$ K
- The hub static pressure $p_h = 1.8$ bars
- The hub-to-tip static density is assumed constant and equal to the hub-radius magnitude for the purpose of integrating the radial-equilibrium equation

By computing the appropriate velocity component(s) at the hub, mean, and tip radii, calculate (through simple numerical-integration means) the mass-flow rate ($\dot{m}$) through the compressor stage.

5) Figure 6.18 shows the meridional view of a turbine rotor that is spinning at 46,000 rpm. The operating conditions of the rotor are as follows:

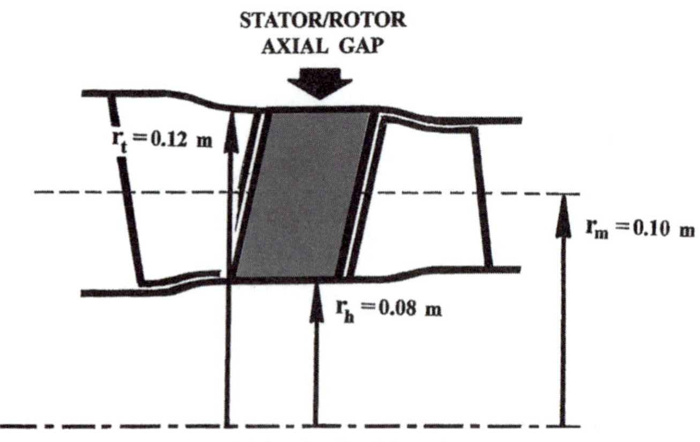

Figure 6.17. Input variables for Problem 4.

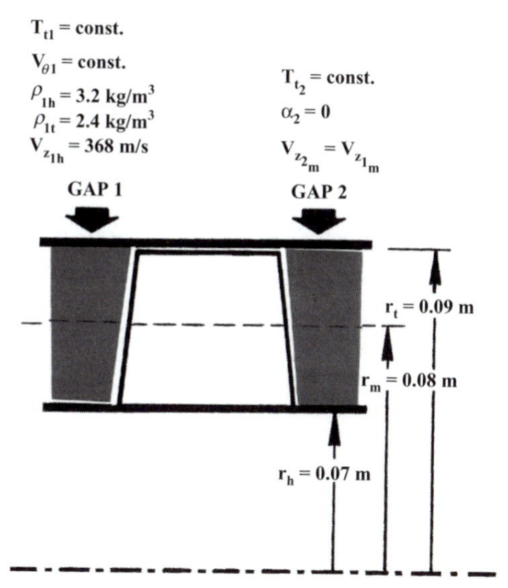

Figure 6.18. Input variables for Problem 5.

Rotor inlet gap

- T_{t1} = constant = 1200 K
- $V_{\theta 1}$ = 485 m/s and is constant
- $(\rho_1)_{hub} = 3.2\ \text{kg/m}^3$
- $(\rho_1)_{tip} = 2.4\ \text{kg/m}^3$

Rotor exit gap:

- $\alpha_2 = 0$ across the entire gap
- T_{t2} = constant

Assuming a gapwise linear density distribution, calculate the difference in static pressure between the hub and tip radii in the rotor-inlet gap.

6) Figure 6.19 shows a single-stage compressor where the shaft speed (N) is 34,000 rpm. The following rotor-exit flow conditions apply:

- T_t = 410 K (constant from hub to tip)
- $\rho_{av.} \approx 2.8\ \text{kg/m}^3$
- The tangential component of relative velocity W_θ has a hub-to-tip constant magnitude of −280 m/s

Also, the following rotor-exit condition prevails at the mean radius:

- p_{mean} = 3.4 bars

Calculate the tip-radius static pressure p_{tip} at the rotor exit station.

7) Figure 6.20 shows an axial-flow compressor stage. The stage operating conditions are as follows:

- Rotational speed (N) = 23,000 rpm
- Hub-to-casing stator-exit total temperature (T_{t1}) = 280 K
- *Average* density in the stator/rotor gap (ρ_1) = 1.2 kg/m^3

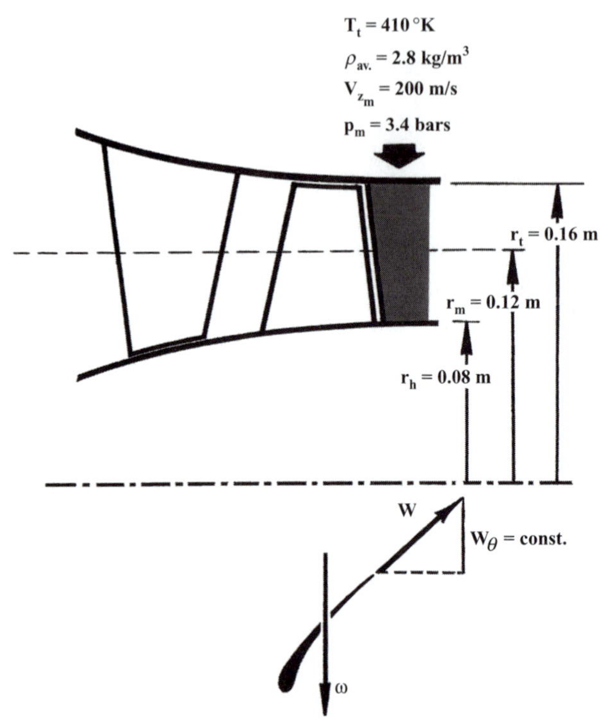

Figure 6.19. Input variables for Problem 6.

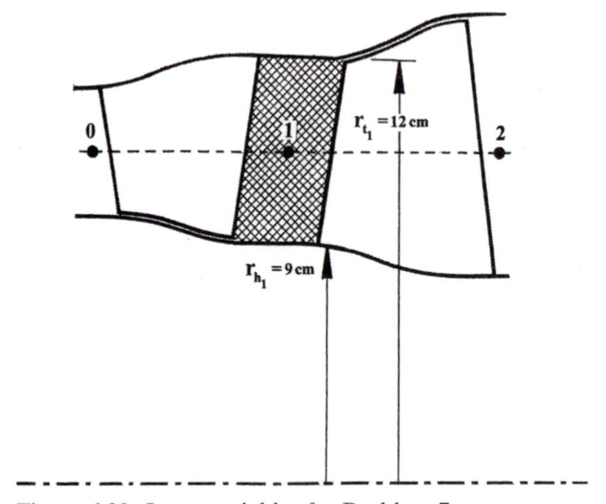

Figure 6.20. Input variables for Problem 7.

- Across the stator/rotor gap, the following conditions apply:

$$V_{hub} = 200\,\text{m/s}$$
$$\alpha_{hub} = +50^\circ$$
$$p_{mean} = 0.71\ \text{bars}$$

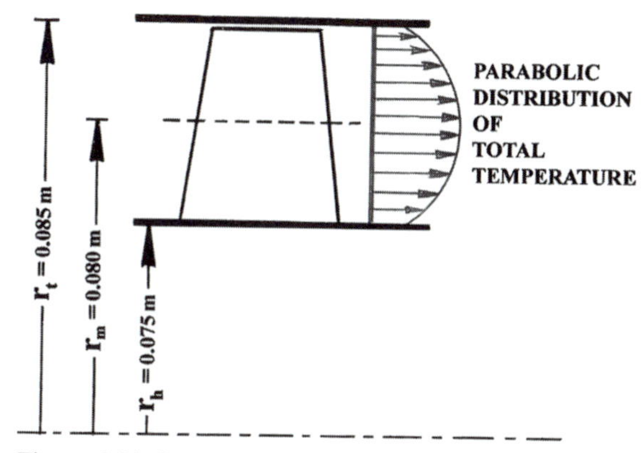

Figure 6.21. Input variables for Problem 8.

Assuming a free-vortex flow behavior in the stator/rotor gap, calculate the tip-radius values of the following variables within this gap:

a) The tip-radius relative flow angle (β_{tip});
b) The tip-radius static temperature (T_{tip}).

8) Figure 6.21 shows the total-temperature profile at the exit station of a compressor rotor. This is a symmetrical profile, which can be expressed as

$$T_t = a + b(r - r_m)^2$$

This relationship is subject to the following boundary conditions:

- $(T_t)_{hub} = 388$ K
- $(T_t)_{mean} = 582$ K

The following conditions also apply:

- $V_\theta = 188$ m/s and is constant from hub to tip
- $(V_z)_{mean} = 185.6$ m/s

Assuming a specific-heat ratio (γ) of 1.4, calculate the axial-velocity component (V_z) at both the hub and tip radii.

9) Figure 6.22 shows the rotor of a radial-inflow turbine stage together with graphical representations of the swirl velocity component and total temperature. The former is consistent with the so-called forced-vortex flow structure, whereby

$$V_\theta = a_1 + b_1 r$$

The total temperature profile is also linear, and can be represented as

$$T_t = a_2 + b_2 r$$

With these, the following boundary conditions apply:

$$(V_\theta)_{hub} = 200 \text{ m/s}$$
$$(V_\theta)_{tip} = 280 \text{ m/s}$$
$$(T_t)_{hub} = 750 \text{ K}$$
$$(T_t)_{tip} = 700 \text{ K}$$

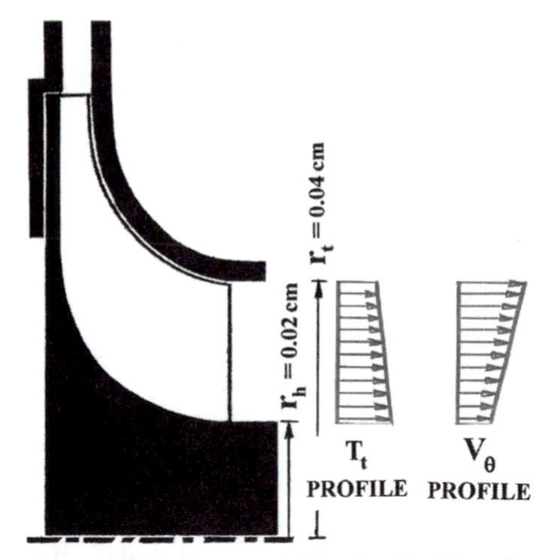

Figure 6.22. Input variables for Problem 9.

If the hub magnitude of the axial velocity component $(V_z)_{hub}$ is 370 m/s, calculate the same variable at the tip radius. You may assume an average γ magnitude of 1.36.

10) Figure 6.23 shows an axial-flow turbine rotor, together with a graphical representation of a parabolic and symmetric swirl velocity profile, which is subject to the following boundary conditions:
   - $(V_\theta)_{hub} = 360$ m/s
   - $(V_\theta)_{mean} = 216$ m/s

   Also, an incompressible rotor-exit flow field is assumed, with a uniform static density value of 1.15 kg/m^3, and the mean-radius static pressure is 2.5 bars.

   Assuming an average γ magnitude of 1.33, calculate the rotor-exit static-pressure differential $(p_{tip} - p_{hub})$.

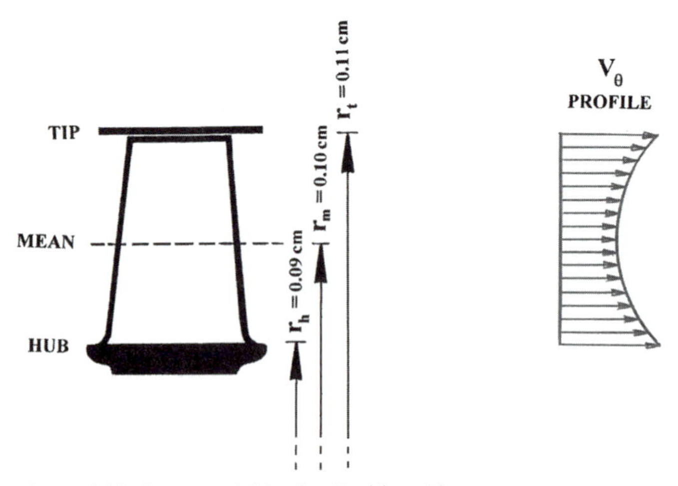

Figure 6.23. Input variables for Problem 10.

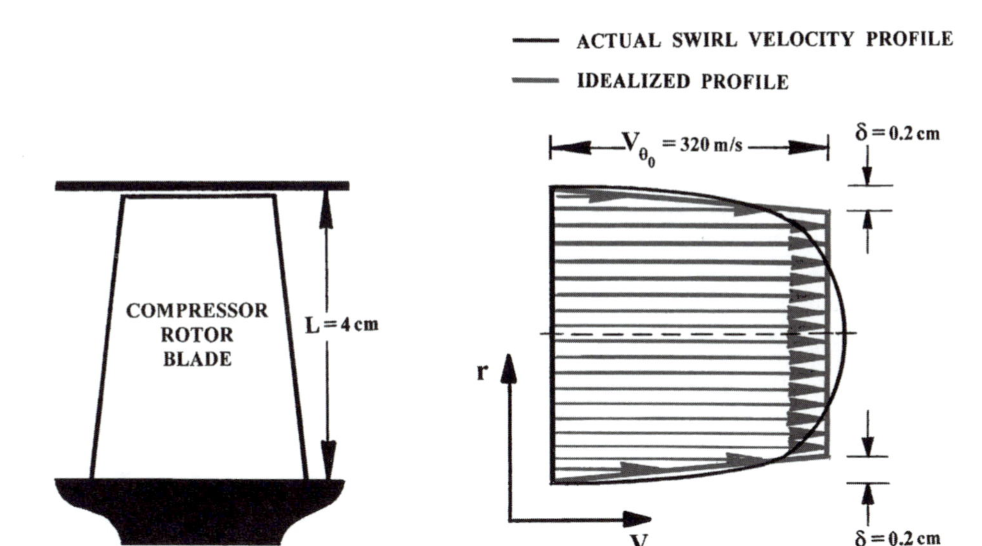

Figure 6.24. Input data for Problem 12.

11) Consider an axial-flow compressor rotor with constant hub and tip radii. The rotor, which is spinning at ω radians/s, operates under two free-vortex swirl-velocity patterns at its inlet and exit stations.

    Prove that the specific shaft-work consumption (w_s) is constant at all radii from hub to tip.

12) Figure 6.24 shows an idealized hub-to-casing tangential velocity profile at the exit station of an axial-flow compressor rotor. As seen in the figure, the profile is composed of simple straight lines. The hub-to-casing average magnitude of static density is 2.4 kg/m^3, and the hub value of static pressure (p_h) is 1.5 bars.

    Derive two expressions for the static-pressure distribution within the following two radial segments:

    - $(r_h) \leq r \leq (r_h + \delta)$
    - $(r_h + \delta) \leq r \leq (r_h + 2cm)$

    Hint: You will have to ensure the pressure "single-valuedness" at the interface of the above-defined radial segments. In fact, the pressure distribution along the first segment should provide a boundary condition for the pressure distribution along the second segment at the interface radius.

CHAPTER SEVEN

Polytropic (Small-Stage) Efficiency

In this chapter, we will get familiar with a new and unique performance gauge that is independent of the size of a turbomachine (in terms of the total-to-total pressure ratio). In addition, we will have a means of computing the overall efficiency of several stages, particularly those sharing the same total-to-total magnitudes of pressure ratio and efficiency without having to resort to the thermodynamics of each individual stage. The point is made that adding more stages to a multistage turbomachine will have drastic, but totally opposite, effects on turbines as contrasted with compressors. We will prove through this exercise that adding more turbine stages *enhances* the performance of the final turbine configuration. The effect in compressors, however, is that of performance *deterioration*.

Derivation of the Polytropic Efficiency

With no lack of generality, let us consider the stagewise compression process in Figure 7.1. Let us also view the process as a theoretically infinite sequence of compression processes over infinitesimally small compressor stages, as shown in the figure. The polytropic efficiency definition (to follow) applies to any of these stages. From such process increments, let us pick the segment that is highlighted in Figure 7.1. The polytropic efficiency is defined as

$$e_c = \frac{(dh_t)_{id.}}{(dh_t)_{act.}} \tag{7.1}$$

where the subscript c refers to "compressor," the case at hand. Noting that the infinitesimally small process is one that, in the limit, tends to collapse around the average state, it is perhaps clear why e_c is considered a state-dependent variable.

Applying the well-known Gibbs (or T-ds) equation for the ideal process, we obtain the following relationship:

$$(T_t ds)_{id.} = dh_t - \frac{1}{\rho_t} dp_t \tag{7.2}$$

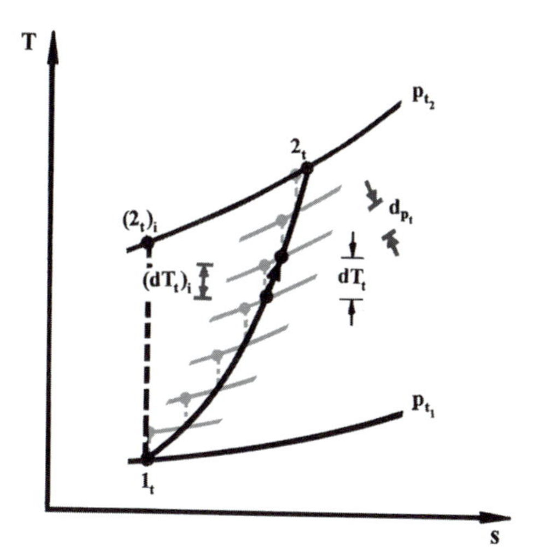

Figure 7.1. Breaking a compressor flow process into infinitesimally small processes.

The left-hand side of (7.2) is simply zero because an ideal (isentropic) process is one where $ds = 0$ by definition. It follows that

$$e_c = \frac{\left(\frac{1}{\rho_t}\right) dp_t}{c_p dT_t} = \frac{\left(\frac{RT_t}{p_t}\right) dp_t}{\left(\frac{\gamma}{\gamma-1}\right) R dT_t} \tag{7.3}$$

or

$$\frac{dT_t}{T_t} = \left(\frac{\gamma-1}{\gamma e_c}\right) \frac{dp_t}{p_t} \tag{7.4}$$

Integrating both sides of (7.4) between the end states 1 and 2, we obtain the following relationship:

$$\ln\left(\frac{T_{t2}}{T_{t1}}\right) = \ln\left(\frac{p_{t2}}{p_{t1}}\right)^{\frac{\gamma-1}{\gamma e_c}} \tag{7.5}$$

Referring to the temperature ratio (T_{t2}/T_{t1}) by τ and the pressure ratio (p_{t2}/p_{t1}) by π, we can obtain the following compact expression:

$$\tau_C = \pi_C^{\left(\frac{\gamma-1}{\gamma e_c}\right)} \tag{7.6}$$

Examination of equation (7.6) reveals that the equation reduces to a well-known isentropic relationship should the polytropic efficiency (e_c) be 100%, a result that is only sensible.

Let us now relate the polytropic efficiency (e_c) to the overall total-to-total (isentropic) efficiency (η_C) as follows:

$$\eta_C \equiv \frac{(\pi_C)^{\frac{\gamma-1}{\gamma}} - 1}{\tau_C - 1} = \frac{(\pi_C)^{\frac{\gamma-1}{\gamma}} - 1}{\pi_C^{\left(\frac{\gamma-1}{\gamma e_c}\right)} - 1} \tag{7.7}$$

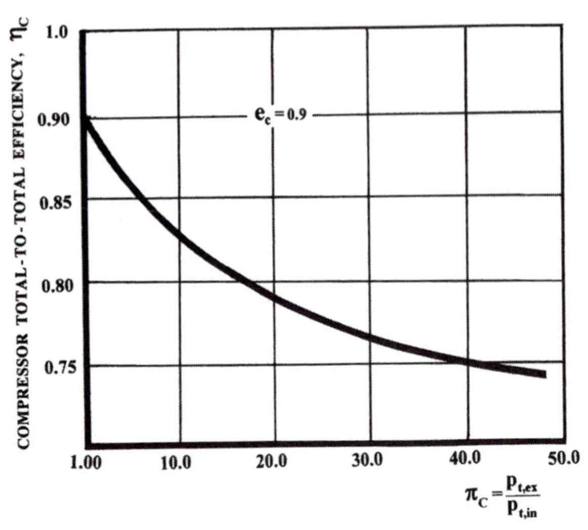

Figure 7.2. Dependence of the compressor efficiency on the pressure ratio.

Figure 7.2 shows a plot of equation (7.7) for a fixed e_c magnitude of 0.9. The figure clearly indicates that:

$$\eta_C \leq e_c \tag{7.8}$$

with the equality sign being applicable only to the case of a total-to-total pressure ratio of unity. Referring to Figure 7.2, one can conclude that as the pressure ratio goes up, the compressor isentropic efficiency (η_C) gets to be more and more penalized.

The same thermodynamic principles (above) can equally be extended to the case of a turbine (Fig. 7.3), where the polytropic efficiency is now labeled e_T. The final result in this case is

$$\eta_T = \frac{1 - (\pi_T)^{\left[e_T\left(\frac{\gamma-1}{\gamma}\right)\right]}}{1 - (\pi_T)^{\left[\frac{\gamma-1}{\gamma}\right]}} \tag{7.9}$$

Figure 7.4 is a plot of the isentropic versus polytropic efficiencies for a fixed e_T magnitude of 0.9. Contrary to the equivalent compressor plot (Fig. 7.2), examination of the figure reveals that:

$$\eta_T \geq e_T \tag{7.10}$$

with the equality sign being applicable only for a total-to-total pressure ratio of 1.0. Another important result, in the turbine case, is that the more the turbine pressure ratio (or turbine size) *increases*, the more *improved* the isentropic efficiency will become. This, again, is at odds with the equivalent compressor statement.

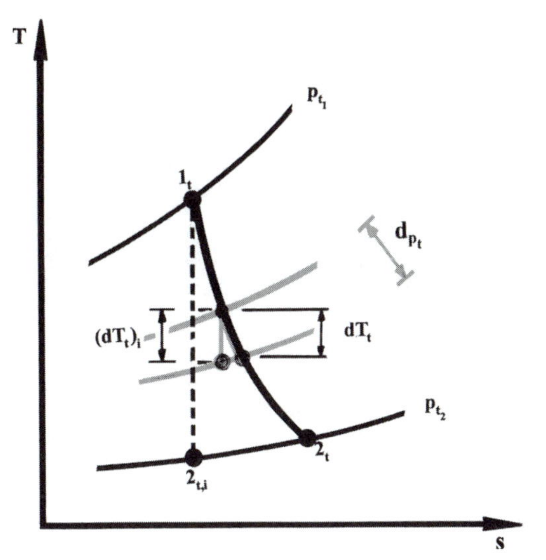

Figure 7.3. An infinitesimally small process within a turbine expansion process.

Multistage Compressors and Turbines

Suppose we have a multistage compressor (Fig. 7.5) that consists of N stages, each having its own total-to-total pressure ratio and isentropic efficiency. Required, in this case, is the overall total-to-total isentropic efficiency η_C. To address this issue, consider the jth stage, of which the end states are defined by $(T_{t\,j}, p_{t\,j})$ and $(T_{t\,j+1}, p_{t\,j+1})$,

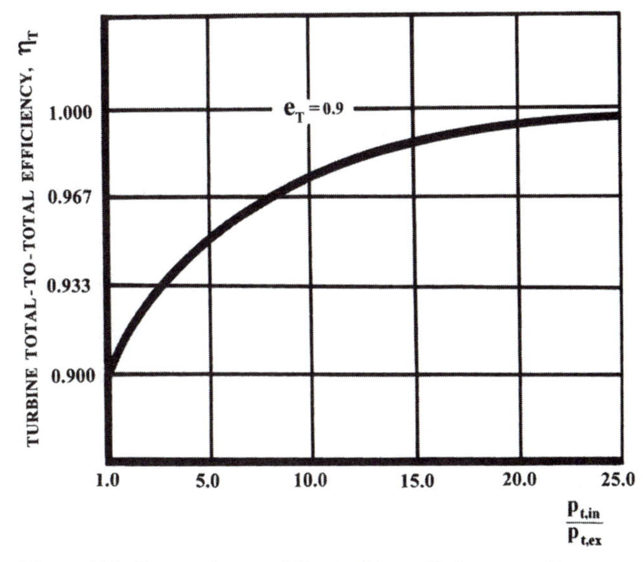

Figure 7.4. Dependence of the turbine efficiency on the pressure ratio.

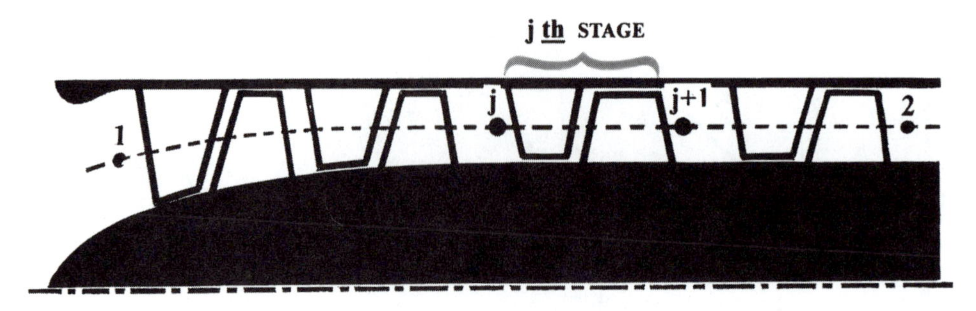

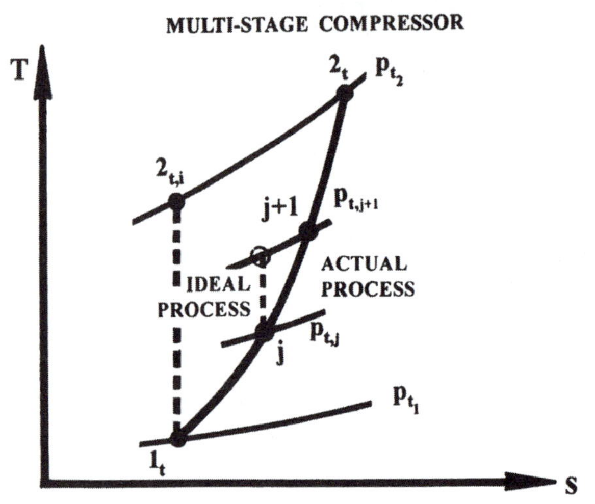

Figure 7.5. Overall-efficiency representation for a multistage compressor.

with the isentropic efficiency being η_{Cj}. The following expression is, in this case, applicable:

$$\eta_{Cj} = \frac{(\pi_{Cj})^{\frac{\gamma-1}{\gamma}} - 1}{\tau_{Cj} - 1} \tag{7.11}$$

where

$$\pi_{Cj} = \frac{p_{t\,j+1}}{p_{t\,j}} \tag{7.12}$$

and

$$\tau_{Cj} = \frac{T_{t\,j+1}}{T_{t\,j}} \tag{7.13}$$

Based on the isentropic efficiency definition, we can cast the following expression:

$$\frac{T_{t\,j+1}}{T_{t\,j}} = \tau_{Cj} = 1 + \frac{1}{\eta_{Cj}}\left[\pi_{Cj}{}^{\frac{\gamma-1}{\gamma}} - 1\right] \tag{7.14}$$

The overall total-to-total temperature ratio (τ_C) across the multistage compressor can be expressed as

$$\tau_C = \frac{T_{t(N_s+1)}}{T_{t1}} = \prod_{j=1}^{N_s}\left[1 + \frac{1}{\eta_{C\,j}}\left(\pi_{C\,j}^{\frac{\gamma-1}{\gamma}} - 1\right)\right] \quad (7.15)$$

where N_s is the number of compressor stages.

In addressing the total-to-total (isentropic) efficiency item, let us first substitute the relationship (7.15) in the general isentropic-efficiency expression, namely

$$\eta_C = \frac{\pi_C^{\left(\frac{\gamma-1}{\gamma}\right)} - 1}{\tau_C - 1} \quad (7.16)$$

This gives rise to the following overall isentropic-efficiency expression:

$$\eta_C = \frac{\pi_C^{\left(\frac{\gamma-1}{\gamma}\right)} - 1}{\prod_{j=1}^{N_s}\left[1 + \frac{1}{\eta_{C\,j}}\left(\pi_{C\,j}^{\frac{\gamma-1}{\gamma}} - 1\right)\right] - 1} \quad (7.17)$$

where the overall total-to-total pressure ratio (π_C) can be expressed as

$$\pi_C = \prod_{j=1}^{N_s} \pi_{C\,j} \quad (7.18)$$

As a special case, let us consider the situation where all of the compressor stages in Figure 7.5 possess the same total-to-total pressure ratio (π_S) and the same efficiency (η_S), where the subscript S refers to a typical compressor stage. Under these assumptions, the overall total-to-total (isentropic) efficiency can be expressed as

$$\eta_C = \frac{\pi_C^{\left(\frac{\gamma-1}{\gamma}\right)} - 1}{\left[1 + \frac{1}{\eta_S}\left(\pi_C^{\frac{\gamma-1}{\gamma N_s}} - 1\right)\right]^{N_s} - 1} \quad (7.19)$$

Similar expressions can systematically be derived for multistage turbines. Considering first the case where the turbine is composed of N_s stages with varying pressure ratios and isentropic efficiencies, the overall turbine efficiency (η_T) can be expressed as

$$\eta_T = \frac{1 - \tau_T}{1 - \pi_T^{\frac{\gamma-1}{\gamma}}} \quad (7.20)$$

In equation (7.20), the overall total-to-total temperature ratio (τ_T) can be expressed as

$$\tau_T = \prod_{j=1}^{N_s}\left\{1 - \eta_j\left[1 - (\pi_{T\,j})^{\frac{\gamma-1}{\gamma}}\right]\right\} \quad (7.21)$$

As for the situation where all the individual stages have identical magnitudes of the total-to-total pressure ratio and efficiency, the overall total-to-total N_s-stage turbine efficiency can be deduced as follows:

$$\eta_T = \frac{1 - \left[1 - \eta_S\left(1 - \pi_T^{\frac{\gamma-1}{\gamma N_s}}\right)\right]^{N_s}}{1 - \pi_T^{\frac{\gamma-1}{\gamma}}} \quad (7.22)$$

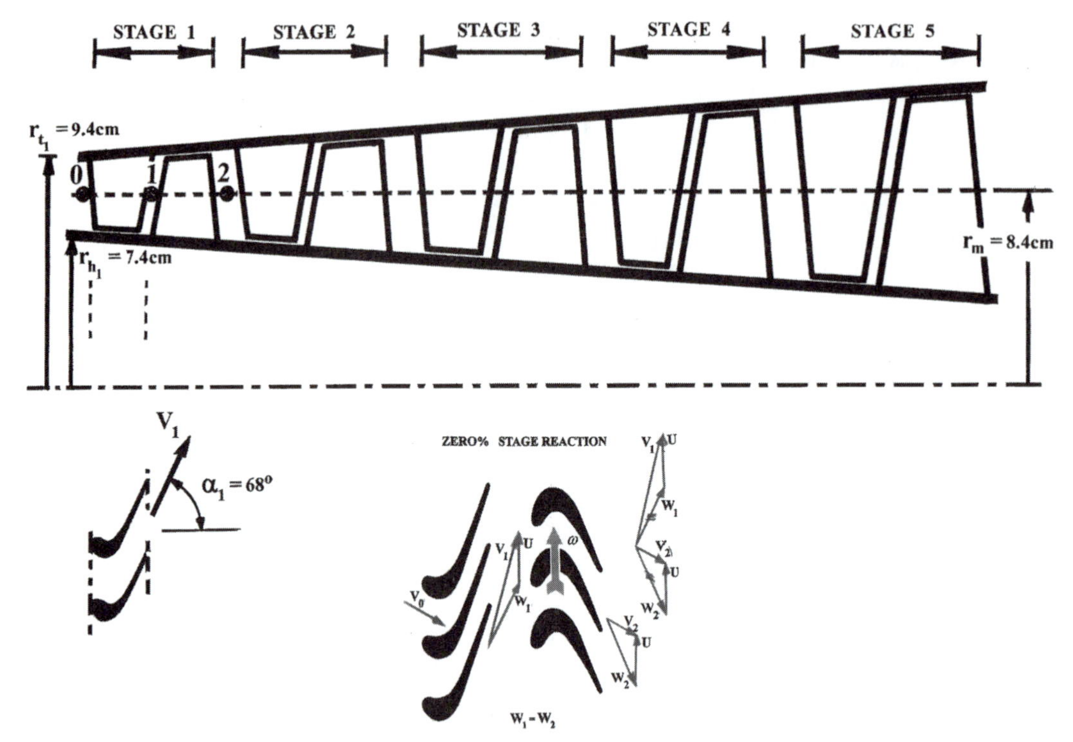

Figure 7.6. Input variables for Example 1.

where

$$\pi_T = {\pi_S}^{N_s} \tag{7.23}$$

In applying equation (7.21), note that both the stage and overall pressure ratios are defined as the exit-over-inlet pressure ratio, a ratio that is less than unity.

EXAMPLE 1

Figure 7.6 shows a schematic of a five-stage axial-flow turbine, where the stages' (inlet-to-exit) total-to-total magnitudes of pressure ratio and efficiency are 1.8 and 81%, respectively, and are the same for all five stages. The turbine inlet total pressure and temperature are 20 bars and 1400 K, respectively, and the rotational speed is 42,000 rpm. Throughout the turbine, the axial-velocity component is constant and the exit flow angle out of the first-stage stator (α_1) is 68°. This particular stator is choked, and the first-stage reaction is zero. Assuming an adiabatic flow throughout the entire turbine, an isentropic flow across the first-stage stator, and an average specific-heat ratio of 1.33, calculate:

a) The relative-flow angle (β_2) at the first rotor exit station;
b) The specific shaft work (w_s) produced by the first stage, and the specific shaft work $(w_s)_T$ that is produced by the entire turbine.

NOTE

Although it is possible to track down each stage-expansion process, a process that is clearly lengthy, the procedure in this problem, and those to follow, is to utilize the closed-form equations that directly lead to finding the overall efficiency.

SOLUTION

Part a:

$$V_{cr1} = \sqrt{\left(\frac{2\gamma}{\gamma+1}\right) RT_{t1}} = 684.7 \text{ m/s}$$

$$V_1 = V_{cr1} = 684.7 \text{ m/s (stator passage is choked)}$$

$$U = \omega r_m = 369.5 \text{ m/s}$$

$$V_{\theta 1} = V_1 \sin\alpha_1 = 634.8 \text{ m/s}$$

$$V_z = V_{z1} = V_1 \cos\alpha_1 = 256.5 \text{ m/s}$$

$$\beta_2 = -\beta_1 \text{ (zero stage reaction)}$$

$$\beta_2 = \arctan\left(\frac{W_{\theta 1}}{V_z}\right) = \arctan\left(\frac{V_{\theta 1} - U}{V_z}\right) = 46.0^\circ$$

$$W_{\theta 2} = -W_{\theta 1} = U - V_{\theta 1} = -265.4 \text{ m/s (see Figure 7.6)}$$

$$(w_s)_{stg.1} = U[V_{\theta 1} - (W_{\theta 2} + U)] = 196.1 \text{ kJ/kg}$$

Part b: The turbine overall exit/inlet total-to-total pressure ratio is obtained by substitution into expression (7.23):

$$\pi_T = \frac{1}{(1.8)^5} = 0.0529$$

The turbine overall total-to-total efficiency is obtained by substitution into expression (7.22):

$$\eta_T = 85.2\%$$

Finally, the turbine-produced specific shaft work is

$$(w_s)_T = c_p T_{t0} \eta_T \left[1 - \pi_T^{\frac{\gamma-1}{\gamma}}\right] = 714.3 \text{ kJ/kg}$$

EXAMPLE 2

The operating conditions of a four-stage axial-flow compressor are as follows:

- *First Stage:* $\pi_1 = 1.8$, and $w_{s1} = 78{,}400$ J/kg
- *Second Stage:* $\pi_2 = 2.1$, and $\eta_2 = 78\%$

- *Third Stage:* $\pi_3 = 2.3$, and $\eta_3 = 78\%$
- *Fourth Stage:* $\pi_4 = 2.6$, and $\eta_4 = 74\%$

where the terms π, w_s, and η are the total-to-total pressure ratio, the specific shaft work, and the total-to-total (isentropic) efficiency, respectively. In addition, the compressor inlet total pressure and temperature are 1.5 bars and 350 K, respectively.

a) Calculate the first-stage polytropic efficiency.
b) Determine and explain the effect of adding the next one, then two, and finally three stages to the first stage, in terms of the isentropic efficiency.

SOLUTION

The relationship between the total-to-total (isentropic) and polytropic efficiencies for a compressor is

$$\eta_C = \frac{\pi_C^{\left(\frac{\gamma-1}{\gamma}\right)} - 1}{\pi_C^{\left(\frac{\gamma-1}{\gamma e_C}\right)} - 1}$$

Part a: Let us apply the energy-conservation equation for the first stage:

$$(w_s)_{stg.1} = 78{,}400 \text{ J/kg} = c_p(T_{t2} - T_{t1})$$

which yields

$$T_{t2} = 428.0 \text{ K}$$

We are now able to compute the stage total-to-total (isentropic) efficiency as follows:

$$(\eta)_{stg.1} = \frac{\left(\pi_1^{\frac{\gamma-1}{\gamma}} - 1\right)}{(\tau_1 - 1)} = 82.05\%$$

Now, substituting in the isentropic-to-polytropic efficiency relationship we get

$$e_c = 84.0\%$$

Part b: The stage total-to-total temperature ratio (τ) can be expressed as

$$\tau = \frac{1}{\eta}\left(\pi^{\frac{\gamma-1}{\gamma}} - 1\right) + 1$$

The total-to-total efficiency, however, can be expressed as

$$\eta = \frac{\pi^{\frac{\gamma-1}{\gamma}} - 1}{\tau - 1}$$

Utilizing the former relationship, we get

$$(\tau)_{stg.1} = 1.223$$
$$(\tau)_{stg.2} = 1.303$$

$$(\tau)_{stg.3} = 1.344$$
$$(\tau)_{stg.4} = 1.424$$

For the two-stage configuration:

$$(\tau)_{config.1} = 1.593$$
$$(\pi)_{config.1} = 3.78$$
$$(\eta)_{config.1} = 77.9\%$$

For the three-stage configuration:

$$(\pi)_{config.2} = 8.694$$
$$(\tau)_{config.2} = 2.142$$
$$(\eta)_{config.2} = 74.9\%$$

For the four-stage configuration:

$$(\tau)_{config.3} = 3.05$$
$$(\pi)_{config.3} = 22.6$$
$$(\eta)_{config.3} = 70.1\%$$

Conclusion: Noting the deterioration in the total-to-total efficiency as more stages are added, we conclude that the compressor performance is penalized as the stage count is increased. This is contrary to the case of a multistage turbine, where the turbine performance is actually enhanced as more stages are added.

PROBLEMS

1) The inlet total temperature and pressure of a turbine stage are 1350 K and 11.5 bars, respectively. The stage inlet/exit total-to-total pressure ratio is 2.7, and the polytropic efficiency is 91%. Assuming an adiabatic flow and an average specific-heat ratio (γ) of 1.33, calculate:

   a) The stage total-to-total (isentropic) efficiency.
   b) The overall turbine-produced specific shaft work (w_s).

2) Figure 7.7 shows a four-stage axial-flow turbine, all with a total-to-total pressure ratio of 2.1, and an isentropic efficiency of 80%. The turbine is designed to yield the minimum absolute velocity (V_2) at the exit station. The turbine mean radius is 0.1 m, and the following operating conditions apply:

   - Inlet total temperature (T_{t_i}) = 1400 K
   - Inlet total pressure (p_{t_i}) = 22.5 bars
   - Shaft speed (N) = 30,000 rpm
   - Flow coefficient ($V_z/\omega r_m$) = 0.7 and is turbine-wise constant
   - Fourth-stage-stator exit swirl angle (α_1) = 64°

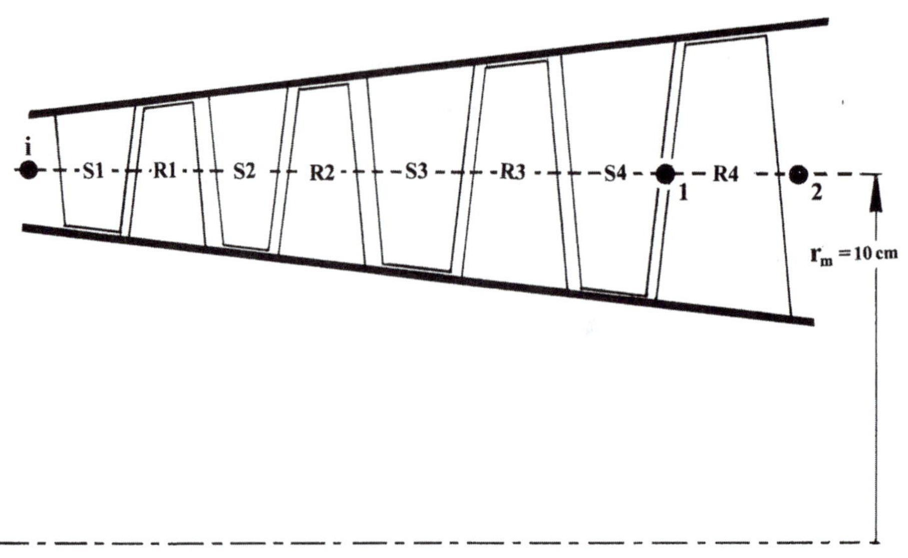

Figure 7.7. Input variables for Problem 2.

Assuming an adiabatic flow process and an average specific-heat ratio (γ) of 1.33, calculate:

a) The fourth-stage inlet total temperature;
b) The fourth-stage reaction;
c) The "appropriate" Mach number to verify the last-rotor choking status.

3) Consider separately a three-stage compressor versus a three-stage turbine. All six stages have a common total-to-total pressure ratio of 2.3 and a common stage efficiency of 81%. The flow in all stages is adiabatic, and an average specific-heat ratio (γ) of 1.37 is assumed.

   a) By computing the overall compressor and turbine efficiencies, prove (by comparison with the stage efficiency) that the efficiency of a multistage compressor will decline if more stages are added, whereas the turbine performance is, on the contrary, enhanced.
   b) Compute and compare the individual stage and the entire compressor poly tropic efficiencies. Repeat the same procedure for the three-stage turbine as well.

4) Figure 7.8 shows the meridional flow path of an eight-stage axial-flow compressor section. These stages have identical total-to-total magnitudes of the pressure ratio and efficiency. The compressor design point is defined as follows:

   - Stage total-to-total pressure ratio ($\pi_{stg.}$) = 1.36
   - Stage total-to-total efficiency ($\eta_{stg.}$) = 82%
   - Shaft speed (N) = 22,000 rpm
   - Compressor-inlet total pressure (p_{t_i}) = 1.4 bars
   - Compressor-inlet total temperature (T_{ti}) = 350 K

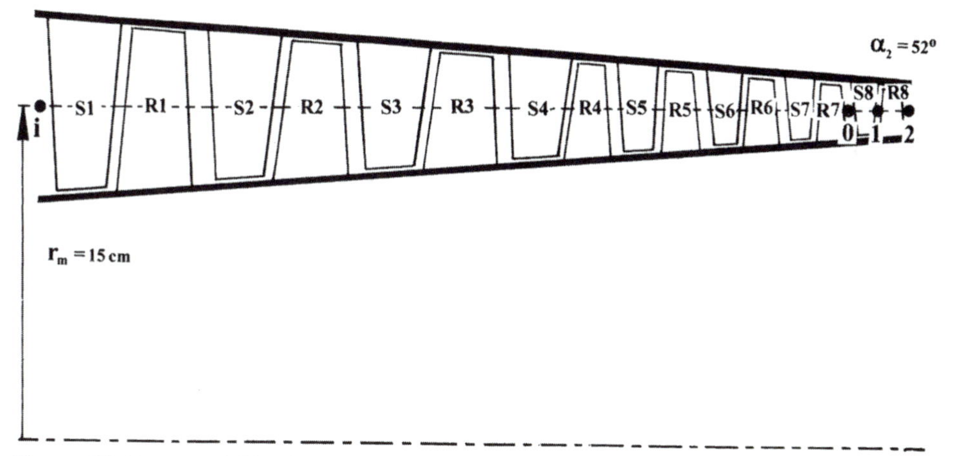

Figure 7.8. Input variables for Problem 4.

- A constant axial-velocity component (V_z) of 240 m/s
- Compressor-exit swirl angle (α_2) $= 52°$
- Adiabatic flow and a γ value of 1.4

Assuming zero incidence and deviation angles, sketch (with reasonable accuracy) the mean-radius rotor-blade section of the eighth stage, showing, in particular, the airfoil (or metal) inlet and exit angles.

5) Figure 7.9 shows a multistage compressor that is operating under the following conditions:

- Inlet total pressure ($p_{t\,in}$) $= 1.7$ bars
- Inlet total temperature ($T_{t\,in}$) $= 350$ K
- Compressor-exit total temperature ($T_{t\,ex}$) $= 870$ K
- Stage total-to-total pressure ratio (π_s) $= 1.88$
- Stage total-to-total efficiency (η_s) $= 85\%$

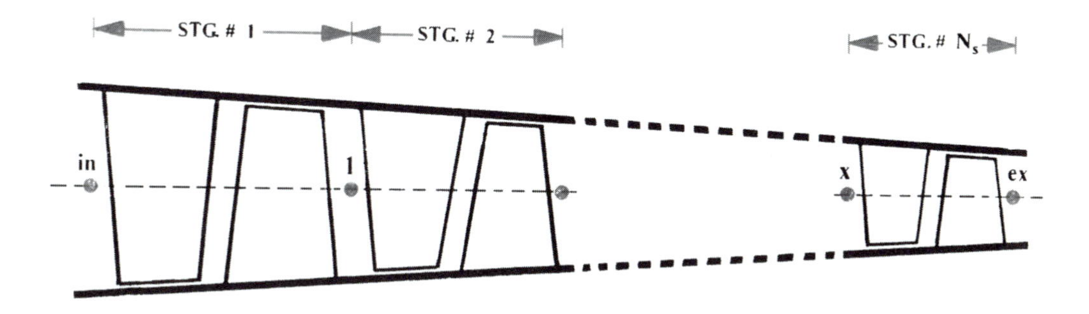

Figure 7.9. Illustration of the multistage compressor in Problem 5.

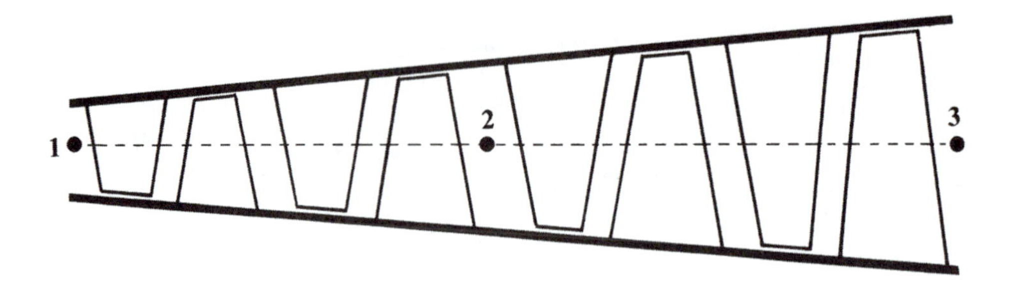

Figure 7.10. Illustration of the four-stage turbine in Problem 6.

- Mass-flow rate ($\dot{m}$) = 12.7 kg/s
- Shaft speed (N) = 52,000 rpm

Assuming an adiabatic flow throughout the entire compressor and a γ magnitude of 1.4, calculate:

a) The number of stages (N_s) that comprise the compressor;
b) The compressor-exit total pressure ($p_{t\,ex}$);
c) The specific speeds of both the first and last stage, knowing that each stage gives rise to an exit critical Mach number of 0.65.

6) Figure 7.10 shows a four-stage axial-flow turbine. All stages share the same total-to-total magnitudes of efficiency and pressure ratio. Referring to the figure, the following conditions also apply:

- $p_{t1} = 14.0$ bars
- $T_{t1} = 1280$ K
- $p_{t2} = 4.42$ bars
- $T_{t2} = 1010$ K

With a turbinewise adiabatic flow and a specific-heat ratio (γ) of 1.33, calculate the turbine-exit total pressure and temperature.

7) Figure 7.11 shows a five-stage axial-flow compressor. These stages have varying magnitudes of total-to-total pressure ratio, as indicated in the figure. All five stages, however, share a total-to-total efficiency of 84%. The flow process through the entire compressor is adiabatic, and the specific-heat ratio (γ) is assumed constant and equal to 1.4.

Considering the flow process over the entire compressor, calculate the average magnitude of polytropic efficiency (e_C).

8) A compressor is composed of nine stages, each with an identical pressure ratio and sharing a constant total-to-total efficiency of 83%. In addition, each of the nine stages gives rise to a total-to-total temperature ratio (τ_s) of 1.12.

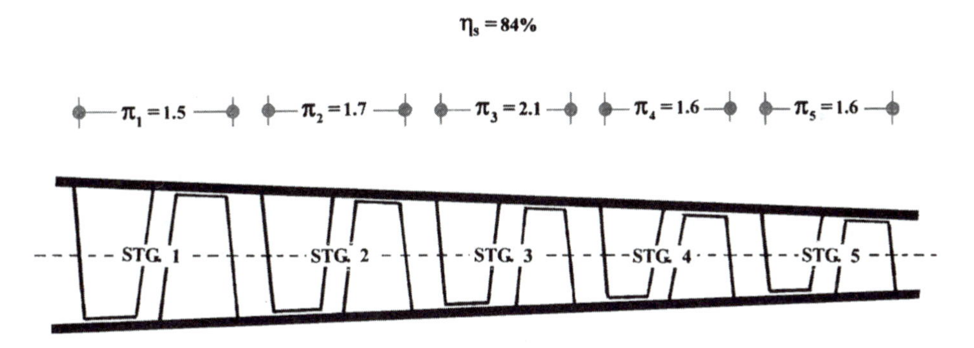

Figure 7.11. Input variables for Problem 7.

The compressor is then replaced by a six-stage compressor. Combined together, these six stages give rise to the same initial magnitudes of overall total-to-total efficiency and pressure ratio.

With a uniformly adiabatic flow and a specific-heat ratio (γ) of 1.4, calculate:

a) The initial magnitude of total-to-total efficiency;
b) The "replacement-compressor" magnitude of stage efficiency.

9) Two five-stage compressor configurations operate between the same magnitudes of inlet and exit total pressures. Details of each configuration are as follows:

Configuration 1: Five stages with pressure ratios of 1.5, 1.7, 1.8, 1.9, and 2.1. The corresponding stage efficiencies are 86, 85, 83, 80, and 76%, respectively.

Configuration 2: Five stages, each with an efficiency magnitude of 82%.

Compare the performances of the two compressor configurations in terms of:

a) The overall total-to-total efficiency (η_c);
b) The average magnitude of polytropic efficiency (e_c).

10) A four-stage axial-flow turbine is operating under the following conditions:

- Inlet total pressure ($p_{t\,in}$) = 15.4 bars
- Inlet total temperature ($T_{t\,in}$) = 1460 K
- Exit total pressure ($p_{t\,ex}$) = 1.13 bars
- Exit total temperature ($T_{t\,ex}$) = 813 K

All four stages have equal total-to-total values of pressure ratio and efficiency. Assuming a specific-heat ratio (γ) of 1.33:

a) Calculate the stage total-to-total efficiency (η_s).
b) Prove that the polytropic efficiencies of each stage and the entire turbine are identical.

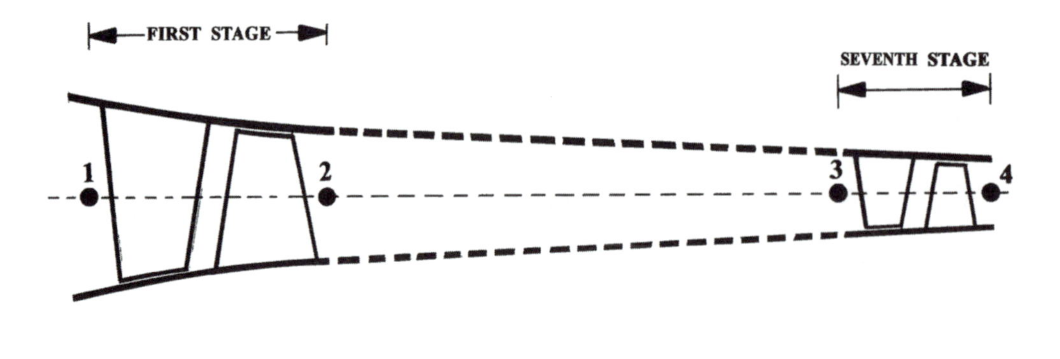

Figure 7.12. Compressor configuration for Problem 11.

11) It is required to build a multistage axial-flow compressor (Fig. 7.12) to meet the following specifications:

- Total-to-total pressure ratio (π_C) = 16.5
- Overall total-to-total efficiency (η_C) = 80%
- Shaft speed (N) = 38,000 rpm
- Mass-flow rate ($\dot{m}$) = 3.9 kg/s

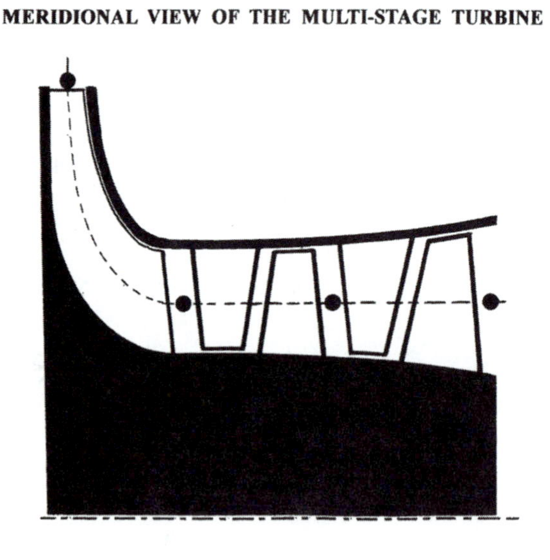

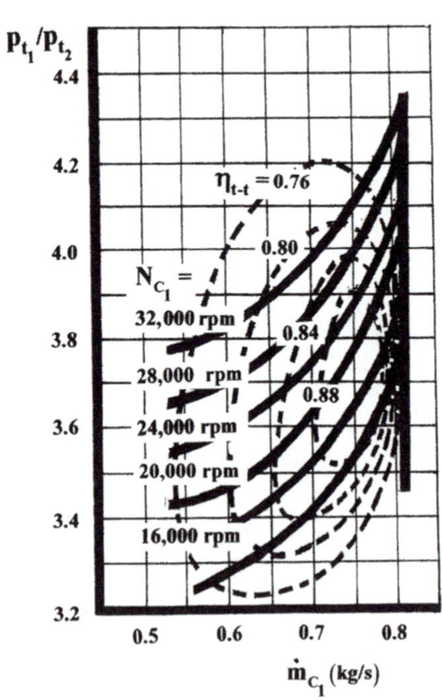

Figure 7.13. Input variables for Problem 12.

The compressor will be composed of a number of stages not to exceed seven. These stages share the same total-to-total magnitudes of pressure ratio and efficiency. The compressor receives air at the standard sea-level values of temperature and pressure, with a negligible value of inlet Mach number.

Assuming an adiabatic flow field throughout the compressor and a γ magnitude of 1.4, calculate:

a) The total-to-total efficiency (η_s) of the individual stages.
b) The specific speeds of the first and last stages. Comment on the difference between these two magnitudes.

12) Figure 7.13 shows a three-stage turbine of which the first is of the radial-inflow type and is governed by the map that is also provided in the figure. The following two axial-flow stages share the same values of total-to-total pressure ratio and efficiency. The turbine is operating under the following conditions:

- Mass-flow rate ($\dot{m}$) = 4.76 kg/s
- Rotational speed (N) = 48,000 rpm
- Turbine-inlet total pressure (p_{t1}) = 12.7 bars
- Turbine-inlet total temperature (T_{t1}) = 1152 K
- Overall inlet/exit total-to-total pressure ratio = 11.76

The first (radial) stage delivers a specific shaft work [$(w_s)_1$] of 294 kJ/kg. The two remaining stages share a total-to-total efficiency of 84%.

With an adiabatic flow process across the turbine and an average specific-heat ratio value of 1.33, calculate:

a) The turbine-exit total temperature (T_{t4}).
b) The radial-stage polytropic efficiency [$(e_T)_1$].

CHAPTER EIGHT

Axial-Flow Turbines

Historically, the first axial turbine utilizing a compressible fluid was a steam turbine. Gas turbines were later developed for engineering applications where compactness is as important as performance. However, the successful use of this turbine type had to wait for advances in the area of compressor performance. The viability of gas turbines was demonstrated upon developing special alloys that possess high strength capabilities at exceedingly high turbine-inlet temperatures.

In the history of axial turbines, most of the experience relating to the behavior of steam-turbine blading was put to use in gas-turbine blading and vice versa. This holds true as long as the steam remains in the superheated phase and not in the wet-mixture zone, for the latter constitutes a two-phase flow with its own problems (e.g., liquid impingement forces and corrosion).

Stage Definition

Figure 8.1 shows an axial-flow turbine stage consisting of a stator that is followed by a rotor. Figure 8.2 shows a hypothetical cylinder that cuts through the rotor blades at a radius that is midway between the hub and tip radii. The unwrapped version of the cylindrical surface in this figure is that where the stage inlet and exit velocity triangles will be required. As has been the terminology in preceding chapters, a stator airfoil will be termed a vane, and the rotor cascade consists of blades. An axial-flow turbine operating under a high (inlet-over-exit) pressure ratio would normally consist of several stages, each of the type shown in Figure 8.1, in an arrangement where the annulus height is rising in the through-flow direction (Fig. 8.3). In a typical industrial setting, a turbine design would undergo two steps: the preliminary design (or sizing) and the detailed design phases. Of these, and in contrast to common perception, the first design phase is comparatively more challenging, for it requires exploration of many turbine configurations as well as experience-based judgments.

The Preliminary Design Process

In this phase, general customer-specified or implied conditions and restrictions are known. Unless an initial decision of scaling up or down an existing turbine has been

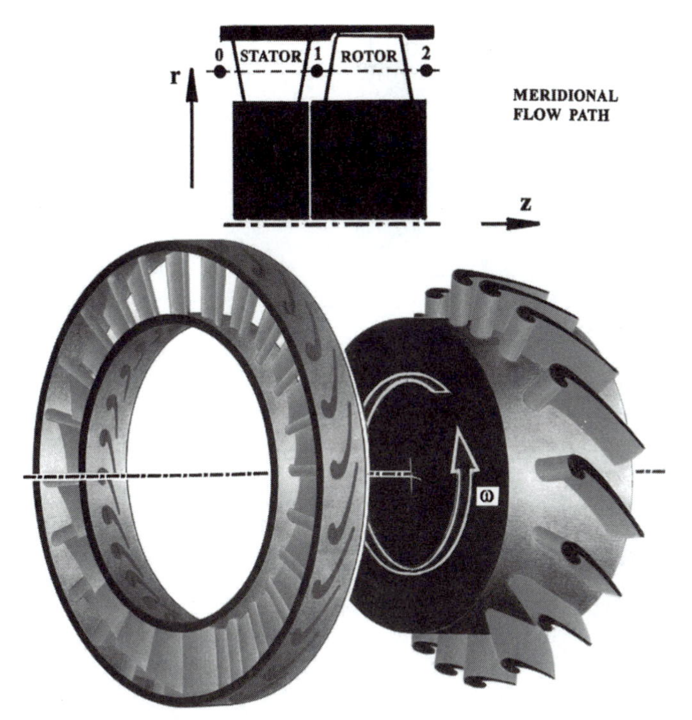

Figure 8.1. A purely axial turbine stage and station designation.

made, the designer will have to conduct a parametric study and examine the outcome carefully. In the general case of propulsion applications, considerations such as the turbine weight, length, and geometrical envelope become an indivisible part of the final choice. In the following, and in reference to Figure 8.3, some of the frequently given variables are listed:

Given data

- The turbine inlet conditions (e.g., p_{t0}, T_{t0}, and α_0)
- The rotational speed
- The total-to-total pressure ratio
- The mass-flow rate

If not explicitly provided, these variables are usually obtained in a preceding "cycle-analysis" phase, at which point they become practically unchangeable. Less restricting, but equally important, limitations can also be part of the problem definitions. Examples of these are:

- A minimum magnitude of clearance, depending on the manufacturing precision.
- A maximum tip Mach number value, for fear of tip-section supersonic pockets.
- Envelope constraints, meaning a limit on the last-stage tip radius.
- Interstage maximum value of temperature if the preceding stage is cooled and the following one is not planned to be.

Figure 8.2. Generating the axial-tangential plane of reference.

- A maximum value for the flow-path divergence angle (θ), for fear of endwall boundary-layer separation.
- A maximum magnitude for the turbine-exit Mach number. This restriction is often stated as one on the turbine-exit swirl angle.

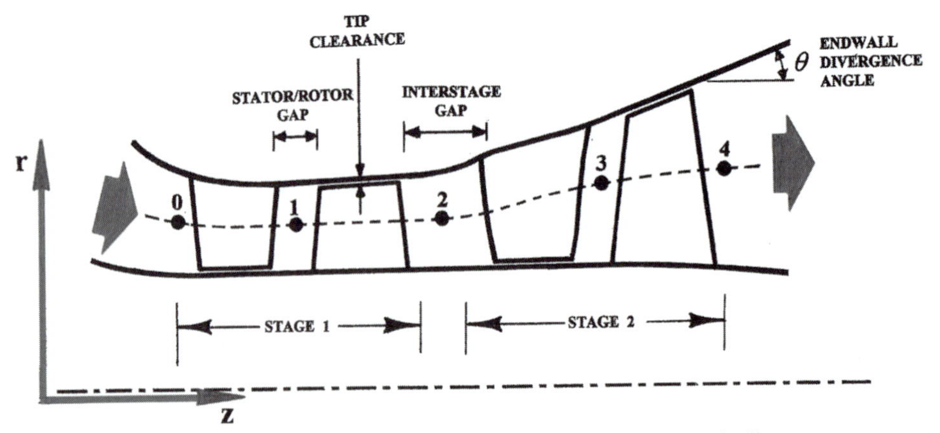

Figure 8.3. Endwall divergence to account for the streamwise density decline.

- Any relevant empirical relationship that would pose a design restriction. An example of such relationships is one that, in view of mechanical considerations, relates the magnitude of the turbine-exit area A_{ex} (in square meters) to the rotational speed (N) in rpm, as follows:

$$A_{ex}N^2 \leq 6 \times 10^5$$

Of course, the design conditions, limitations, and requirements are different for every turbine. However, the preliminary design requirements are essentially the same:

Requirements

- The turbine meridional flow path (sketched in Figure 8.3).
- The flow conditions along the pitch line at each component inlet and exit stations.
- The inlet and exit velocity triangles for each individual rotor.
- A zeroth-order prediction of each rotor hub and tip velocity triangle using such spanwise approximations as the free-vortex flow structure (Chapter 6). This is particularly important in the low-pressure (radially long) turbine section, where the large blade height may give rise to a negative hub reaction. Likewise, the tip reaction may be exceedingly high. (Recall that the optimum reaction magnitude is somewhere between 50 and 60%.)

In the following, the preliminary design procedure is presented in general terms. The procedure is based on a constant mean radius (horizontal pitch line) and begins with verifying, on a single-stage design basis, whether a single-stage turbine configuration is viable. After all, a multistage turbine would be undesirably longer, heavier, and more costly to fabricate.

First Step: Investigate a Single-Stage Configuration

Knowing the turbine-inlet total conditions, the total-to-total pressure ratio, make a reasonable assumption on the turbine efficiency so that you may calculate the turbine-exit temperature. With the exit total properties now known, you may elect to assume the exit critical Mach number or simply disregard it as zero, hoping (as you always should) for the minimum value in the final design. Assuming the exit critical Mach number enables you to compute the exit static density. With such simplifications, you should be able to calculate the specific speed for this single-stage turbine configuration. With this specific speed, you are now in a position to see whether such a configuration is viable by referring to Figure 5.9. In the event the process produces a radial stage instead, then the next step is to divide the shaft work among more than one stage.

Second Step: Define the Stage-to-Stage Work Split

Knowing Δh_t across the turbine, assume the share of each stage to be Δh_{t1}, Δh_{t2}, and so on, which is equivalent to defining each stage exit total temperature. In practice, the first stage is normally assigned a larger shaft-work share. There usually are several reasons for this, such as the short height (producing good hub and tip reactions) and the possibility (if cooled) to eject the coolant over the rear blade segment on the suction side. This will have the effect of suppressing the boundary-layer domination of this particular segment. The rear (low-pressure) turbine stages would have their own mechanically related reservations, as the large blade height will significantly elevate the bending stress at the root. In a two-stage turbine configuration, for instance, work splits such as 60:40 or 55:45 are frequent choices. Nevertheless, the designer can be limited, in making the choice, by several restrictions such as:

1) The lack of desire to cool, say, the second stage (assuming the first is cooled), for the simple reason that a cooled stage is much more expensive to fabricate. This is particularly true if there is internal cooling, as the cooling-flow passages are extremely narrow and difficult to produce. Such a cooling-related decision is, in effect, a restriction that is imposed on the second-stage inlet total temperature and can increase the first-stage share of shaft-work production.
2) An anticipated shift in the contribution of each stage to the overall amount of shaft work, as the turbine enters off-design operation modes. Perhaps a good example of this is an early design version of the turboprop three-stage turbine section in Figure 1.2. At 35,000 ft. operation, the last stage was found to be "windmilling," meaning that it was (operationally) performing as a compressor stage, contrary to its sea-level design-point operation. In general, drastic shifts in the off-design work split can be operationally degrading.
3) The specific speed of each individual stage, which will have to be within the axial-stage range according to Figure 5.9.

Third Step: Stage-by-Stage Turbine Design

Consider each stage separately, beginning with assumptions in connection with the stage-exit swirl angle and flow coefficient (to be defined later in this section). Proceed by calculating the mean-radius inlet and exit velocity triangles as well as the stage reaction. Should the latter be unsatisfactory, then change the exit swirl angle and/or the flow coefficient. Repeat the process until a satisfactory stage design is achieved. The streamwise annulus-height magnitudes can be computed at this point by repeatedly applying the continuity equation at key axial locations. To finally be able to construct the flow path, you need to know (or assume) the stator-to-rotor and interstage axial-gap lengths. In doing so, the following guidelines apply:

- Small axial gaps may trigger a strong and potentially dangerous stator/rotor unsteady flow interaction, for the rotor is normally spinning in the strong segments of the stator wakes, where the flow stream is far from being circumferentially uniform. This adds a cyclic stress pattern on top of the blade aerodynamic loading and thermal stresses, and may very well cause fatigue failure.
- Large axial gaps cause increased friction over the endwalls, which is a primary contributor to the stage overall losses. This is particularly true in the stator/rotor gaps due to the high magnitudes of swirl velocity across these gaps. Another

drawback is an increased skin friction drag on the engine, as the latter becomes longer in this case.

- The interstage axial gaps are typically set to be longer than the stator-to-rotor gap within each stage. Aside from the mechanical problems associated with the latter, a well-designed stage will produce a small exit-swirl angle. This, in effect, limits the number of tangential trips of the fluid particles as they proceed (along the hub and casing) from one stage to the next.

In computing the mean-radius axial chords of the different stator and rotor airfoil cascades, the so-called aspect ratio will prove to be beneficial. The aspect ratio λ is defined as

$$\lambda = \frac{h}{c_{zm}} \tag{8.1}$$

where h is the average annulus height and c_{zm} is the mean-radius axial chord of the airfoil.

The common magnitudes of λ range from 1.0 for the first (high-pressure) stage to magnitudes in excess of 3.0 for the last (low-pressure) stage(s). Generally speaking, an aspect ratio below 1.0 implies a rather short blade, to the point that the tip clearance becomes comparable with the blade height. This means a viscosity-dominated flow, which is associated with a large-scale total-pressure loss. A very large aspect ratio, however, leads to an unacceptably "tall" blade, causing high bending stresses at the root, which can possibly elevate the likelihood of mechanical failure.

In order to finalize the meridional flow-path construction, we also need to compute the stator/rotor as well as the interstage axial-gap lengths. Of these, the former is usually assumed to be one quarter to one half of the arithmetic average of both the stator and rotor axial chords. As for the interstage axial-gap length, an appropriate magnitude is typically one half of the average chords on both sides of the gap.

Stage Design: A Simplified Approach

In the following, a set of design-related variables will be defined. Next, simple relationships will be established among them. The purpose here is to simplify what is normally an iterative procedure to arrive at the optimum stage configurations:

The stage flow coefficient, ϕ:

$$\phi = \frac{V_z}{U} \tag{8.2}$$

In this definition, the axial-velocity component V_z is assumed to be streamwise constant. Within a mass-conservation framework, such an assumption, together with the stage-wise density decline, creates a geometrically divergent flow path as shown in Figure 8.3. The endwall divergence (justified in Figure 8.4 for a stator flow path, where

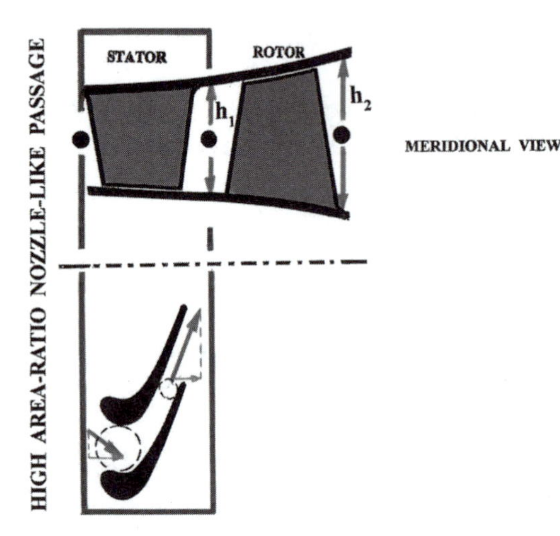

Figure 8.4. Divergence of the stator flow path is caused by the static density decline and the desire to maintain a constant axial velocity component.

the large exit Mach number significantly reduces the static density magnitude) should not be interpreted as a flow decelerator. The fact is that the nozzlelike vane-to-vane or blade-to-blade nature generates a sharp increase in velocity within each of the stator and rotor components. As for the common misconception that the endwalls of an axial-turbine stage should be horizontal, Figure 8.5 illustrates the fact that the axial-velocity component, in this case, will hardly remain constant. In a fixed axial-velocity configuration, however, a well-designed stage will normally have a flow coefficient that is somewhere between 0.4 and 0.6.

The stage work coefficient, ψ:

$$\psi = \frac{w_s}{U^2} = \frac{V_{\theta 1} - V_{\theta 2}}{U} = \frac{W_{\theta 1} - W_{\theta 2}}{U} \tag{8.3}$$

In terms of the relative-flow angles, the preceding definition can be rewritten as

$$\psi = \frac{V_z}{U}(\tan\beta_1 - \tan\beta_2) \tag{8.4}$$

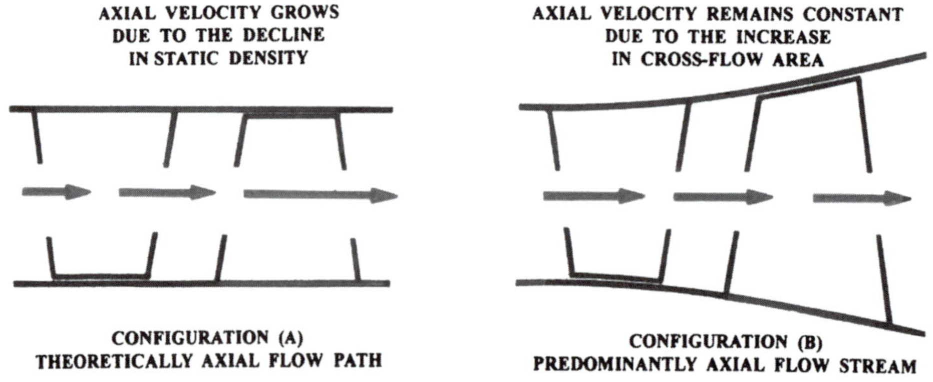

Figure 8.5. Purely axial versus diverging-endwall axial-turbine stage: note the change of axial-velocity component. Contoured-endwall stage configuration is often preferable.

where β_2 will always be negative. Normally a stage work coefficient around unity is considered optimum.

The stage reaction, *R*:

Our objective here is to redefine the reaction in terms of the inlet and exit relative-flow angles:

$$\begin{aligned} R &= \frac{(W_2^2 - W_1^2)/2}{U(V_{\theta 1} - V_{\theta 2})} \\ &= \frac{W_{\theta 1}^2 - W_{\theta 2}^2}{U(W_{\theta 1} - W_{\theta 2})} \\ &= \frac{W_{\theta 1} + W_{\theta 2}}{2U} \\ &= \frac{-V_z(\tan\beta_1 + \tan\beta_2)}{2U} \\ &= \frac{-\phi}{2}(\tan\beta_1 + \tan\beta_2) \end{aligned} \tag{8.5}$$

Equations (8.2), (8.4), and (8.5) can be further manipulated to achieve a simple set of relationships that are very workable (within an iterative optimization procedure) or programmable. These relationships are:

$$\tan\beta_1 = \frac{1}{2\phi}(\psi - 2R) \tag{8.6}$$

$$\tan\beta_2 = \frac{-1}{2\phi}(\psi + 2R) \tag{8.7}$$

$$\tan\alpha_2 = \tan\beta_2 + \frac{1}{\phi} \tag{8.8}$$

Of these, the relationship (8.8) is derived with reference to the exit velocity diagram in Figure 8.6 using simple trigonometry rules. Utilization of these relationships in executing the preliminary design phase is better illustrated by the following example.

EXAMPLE 1

Using a simple pitch-line flow analysis, design a turbine, preferably with one stage (Fig. 8.6). The turbine should comply with the following parameters:

- Flow inlet (swirl) angle, $(\alpha_0) = 0°$
- Mass-flow rate, $\dot{m} = 20$ kg/s
- Inlet total pressure, $p_{t0} = 4.0$ bars
- Inlet total temperature, $T_{t0} = 1100$ K
- Total-to-total pressure ratio, $p_{t\,in}/p_{t\,ex} = 1.873$
- Rotational speed, $N = 15{,}000$ rpm
- Blade mean-radius velocity (U) is not to exceed 340 m/s
- Stage work coefficient (ψ) is not to exceed 1.5

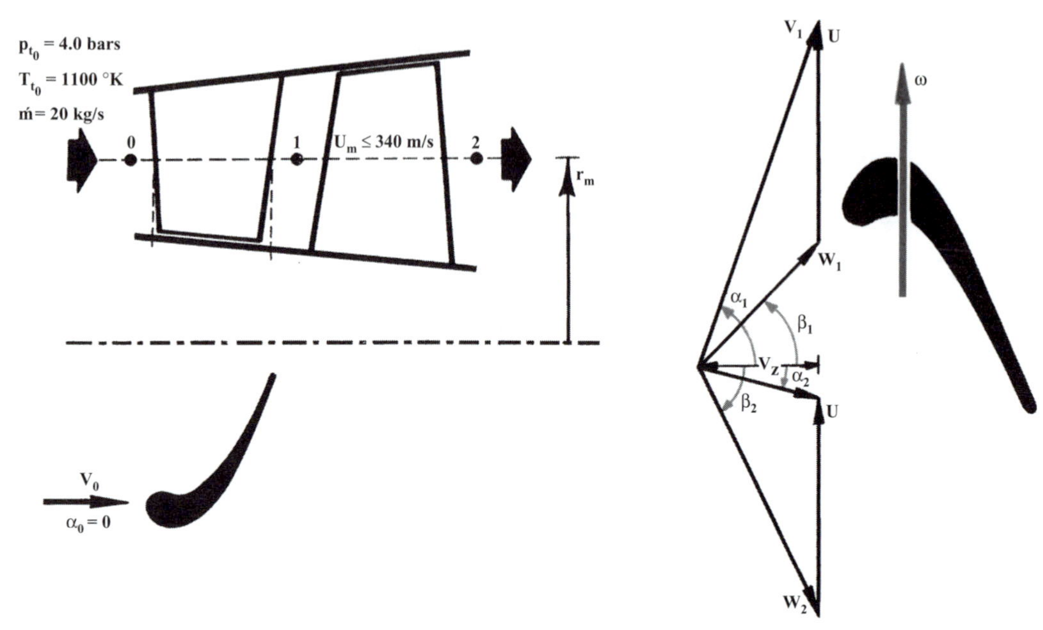

Figure 8.6. Input data for Example 1.

In addition, you may assume an initial total-to-total efficiency of 90%, and an average specific heat ratio γ of 1.33.

SOLUTION

DESIGN CHOICES AND VELOCITY TRIANGLES

A summary of the input data and the station-designation pattern are shown in Figure 8.6. Let us begin by assuming an ideal rotor-exit swirl angle α_2 of zero. Now, we calculate the stage work coefficient:

$$\psi = \frac{\Delta h_t}{U^2} = 1.427$$

In computing ψ, the following choices are made:

- U_m is set equal to the upper limit of 340 m/s.
- A total-to-total stage efficiency of 90% is assumed.

With ψ being less than the maximum allowable magnitude of 1.5, we conclude that the choice of a single-stage turbine is justified.

It should be reassuring, in addition, to arrive at the same conclusion using the specific-speed concept. Because we are not yet aware of the exit velocity diagram, we will resort to the simplification of zero rotor-exit swirl angle mentioned earlier, which should ensure a sufficiently small rotor-exit critical Mach number. In other words, this assumption allows us to treat the rotor-exit flow as incompressible. With this in mind, let us calculate the specific stage speed:

$$\Delta h_{t,id} = c_p T_{t1}\left[1 - \left(\frac{p_{t2}}{p_{t1}}\right)^{\frac{\gamma-1}{\gamma}}\right] = 183{,}448 \text{ J/kg}$$

$$T_{t2} = T_{t1}\left\{1 - \eta_{t-t}\left[1 - \left(\frac{p_{t2}}{p_{t1}}\right)^{\frac{\gamma-1}{\gamma}}\right]\right\} = 956.4 \text{ K}$$

$$p_{t2} = 2.136 \text{ bars}$$

$$\rho_2 \approx \rho_{t2} = \frac{p_{t2}}{RT_{t2}} = 0.778 \text{ kg/m}^3$$

$$N_s = \frac{\omega\sqrt{\frac{\dot{m}}{\rho_2}}}{\Delta h_{t,id}^{\frac{3}{4}}} = 0.898 \text{ radians}$$

Referring to Figure 5.9, we see that the proposed stage belongs to the axial-flow type. With the cutoff specific speed being approximately 0.776 radians, we see that the computed N_s does not represent a stage that is "solidly" axial. The reason, in part, is the high magnitude of shaft work (note the high value of the work coefficient, ψ), which influences the N_s denominator.

Because it is always desirable to minimize the rotor-exit swirl angle (α_2), for an exit swirl velocity is an unused valuable asset, we will begin by setting this angle to zero in the hope of obtaining an acceptable stage reaction with it. Furthermore, noting the high magnitude of the mass-flow rate (in the problem statement), let us set the flow coefficient (ϕ) to 0.8, which is definitely on the high side. In summary, we have just made the following choices:

$$\alpha_2 = 0$$

$$r_m = \frac{U_m}{\omega} = 21.6 \text{ cm}$$

$$\phi = \frac{V_z}{U_m} = 0.8$$

Now, direct substitution in equation (8.8) yields

$$\beta_2 = -51.3^\circ$$

Substitution in equation (8.4) gives rise to the rotor-inlet relative-flow angle (β_1) as follows:

$$\psi = 1.427 = \phi(\tan\beta_1 - \tan\beta_2)$$

which gives

$$\beta_1 = +28.2^\circ$$

In order to calculate the stage reaction, we substitute the preceding variables in equation (8.5):

$$R = \frac{-\phi}{2}(\tan\beta_1 + \tan\beta_2) = 0.285$$

Compared with the optimum magnitude range (0.5–0.6), this reaction magnitude is too small to accept.

One of the corrective actions we can implement is to change the rotor-exit swirl angle. Selecting an α_2 magnitude of -10° and repeating the preceding computational steps, we get the following results:

$$\beta_1 = +19.6^\circ$$
$$\beta_2 = -55.0^\circ$$
$$R = 0.428$$

This is perhaps sufficiently close to 0.5 (without having to increase the α_2 magnitude any further) and is probably an acceptable compromise.

Now, we can calculate other flow-kinematics variables as follows:

$$V_z = \phi U_m = 272.0 \text{ m/s}$$
$$\alpha_1 = 58.1^\circ$$

Note that the previously computed specific speed (N_s), with the chosen magnitude of α_2, is as accurate as an engineer would wish. The reason is that the 10° magnitude of this angle makes the exit velocity, and hence the exit-critical Mach number, significantly small, as the exit velocity becomes predominantly dictated by the axial-velocity component. Stated differently, equating the exit static density to its total counterpart (at exit) will have no significant impact on the magnitude of N_s.

At this point, we have attained all of the relevant variables to construct the inlet and exit velocity triangles, which are shown in Figure 8.7. Now, let us execute an important computational step, which is often ignored or forgotten.

VERIFICATION OF THE STATOR AND ROTOR CHOKING STATUS

Barring supersonic blading, the blade-to-blade passages will always resemble convergent subsonic nozzles. Should our computations thus far lead to a supersonic stator-exit critical Mach number or rotor-exit relative critical Mach number, then we have arrived at an aerodynamically impossible situation. The corrective action in this case is to equate such a number to unity, which will have the effect of reducing the stator-exit velocity (V_1) or rotor-exit relative velocity (W_2), whichever is the case. Note that such a change, should the need arise, will influence the flow angles to the point where a redesign is required.

Let us now execute this computational step:

$$V_1 = \frac{V_z}{\cos \alpha_1} = 514.7 \text{ m/s}$$

$$T_{t1} = T_{t0} = 1100 \text{ K (adiabatic stator flow)}$$

$$W_2 = \frac{V_z}{\cos \beta_2} = 474.2 \text{ m/s}$$

$$W_1 = \frac{V_z}{\cos \beta_1} = 288.7 \text{ m/s}$$

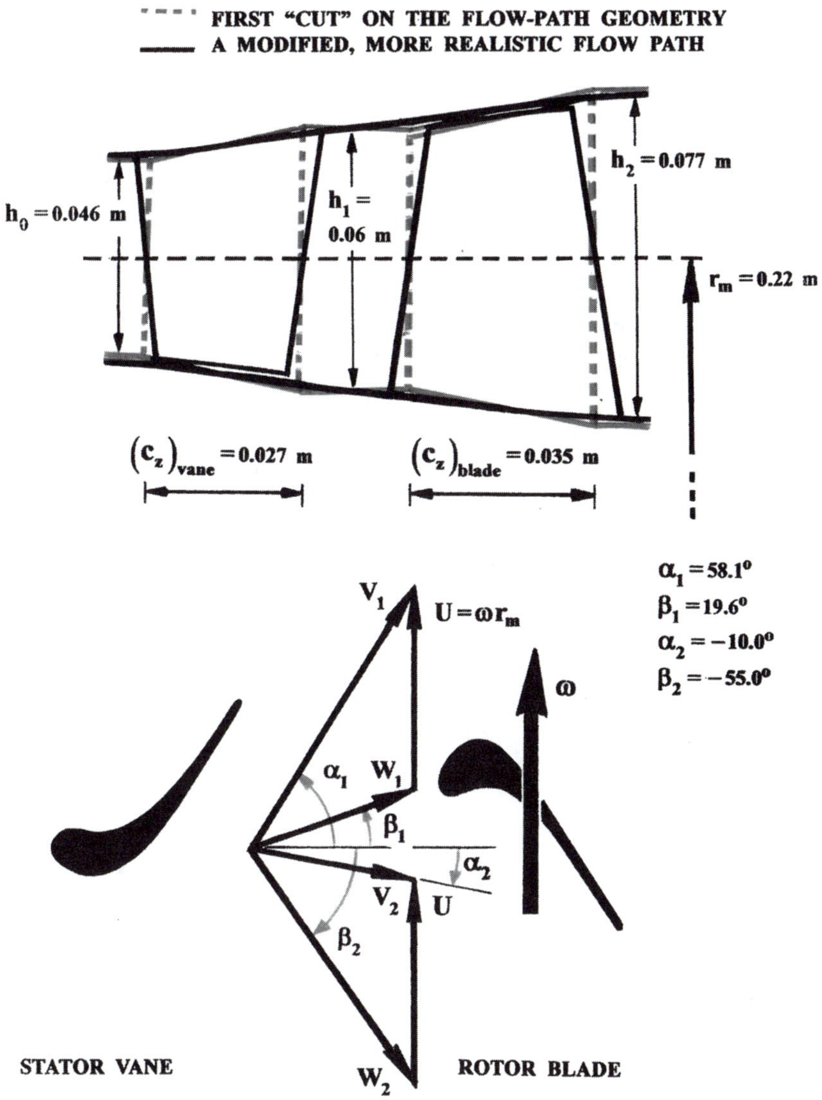

Figure 8.7. Results of the axial-stage preliminary design procedure.

$$T_{tr2} = T_{tr1} = T_{t1} - \left(\frac{V_1{}^2 - W_1{}^2}{2c_p}\right) = 1021.5 \text{ K (axial-flow stage)}$$

$$W_{cr2} = \sqrt{\left(\frac{2\gamma}{\gamma + 1}\right) RT_{tr2}} = 578.5 \text{ m/s}$$

$$M_{cr1} = \frac{V_1}{V_{cr1}} = 0.857$$

$$M_{r,cr2} = \frac{W_2}{W_{cr2}} = 0.820$$

With these two subsonic magnitudes, we have successfully executed the critical computational step at hand.

CALCULATION OF THE ANNULUS HEIGHTS

In this first step of what is the final computational phase, we will consider each flow station separately, calculating the local magnitudes of the critical Mach number and subsequently the static density. Knowing the V_z value, we will then apply the continuity equation, in its most "primitive" form, to calculate the local annulus height. At the stage inlet station "0," in particular, we are not aware of any swirl-velocity component. Therefore, we will proceed with the assumption that the stage-inlet velocity is identical to the axial-velocity component. The latter, incidentally, is assumed constant throughout the stage, a choice that enabled us to even speak of a stage flow coefficient, ϕ, as a fixed stage parameter.

$$M_{cr0} = \frac{V_z}{V_{cr0}} = 0.453$$

$$\rho_0 = \left(\frac{p_{t0}}{RT_{t0}}\right)\left[1 - \left(\frac{\gamma - 1}{\gamma + 1}\right) M_{cr0}{}^2\right]^{\frac{1}{\gamma-1}} = 1.16 \text{ kg/m}^3$$

$$h_0 = \frac{\dot{m}}{2\pi r_m \rho_0 V_z} = 4.6 \text{ cm}$$

$$M_{cr1} = 0.857 \text{ (computed earlier)}$$

$$\rho_1 = 0.908 \text{ kg/m}^3$$

$$h_1 = 6.0 \text{ cm}$$

$$M_{cr2} = 0.493$$

$$h_2 = 7.7 \text{ cm}$$

MEAN-RADIUS AXIAL CHORDS AND STATOR/ROTOR GAPS

Let us define the vane (or blade) aspect ratio (λ) on the basis of the mean-radius axial chord (which is typical in turbine blading) as follows:

$$\lambda = \frac{h_{av.}}{C_z}$$

where $h_{av.}$ is the average annulus height across the stator vane or rotor blade. As discussed earlier, the aspect ratio λ is hardly constant across a multistage turbine section. In this particular problem, we will proceed with the given upper limit of λ, namely 2.0. Thus

$$(\lambda)_{stator} = \frac{(h_0 + h_1)/2}{(C_{z,m})_{stator}}$$

which yields

$$(C_z)_{stator} = 2.7 \text{ m}$$

where the subscript m (signifying a mean-radius value) was dropped.

To calculate $(C_z)_{rotor}$, we follow the same logic, again maintaining an aspect ratio λ of 2.0. The final result is

$$(C_z)_{rotor} = 3.4 \text{ cm}$$

As for the stator/rotor axial-gap length, it is usually taken to be somewhere between 25 and 50% of the average of the stator and rotor axial chords. Selecting the upper limit (50%), we get

$$(\Delta z)_{gap} = 3.1 \text{ cm}$$

The preceding results are represented in Figure 8.7, which shows the stage meridional flow path. At this preliminary design stage, the stator and rotor mean-radius airfoils can only be sketched (as shown in the figure) on the basis of the already determined velocity diagrams.

EXAMPLE 2

The turbine section of a turboprop engine is to be designed according to the following specifications:

- Turbine-inlet swirl angle = 0°
- Mass-flow rate = 8.2 kg/s
- Turbine-inlet total pressure = 10 bars
- Turbine-inlet total temperature = 1380 K
- Overall total-to-total inlet/exit pressure ratio = 9.3
- Rotational speed = 35,000 rpm

The turbine design is also subject to the following set of constraints:

- Mean-radius solid-body velocity (U_m) is limited to 414 m/s
- Number of stages is not to exceed 3
- All stages will have the same total-to-total pressure ratio
- Aspect ratio of the last-stage rotor should not exceed 3.0

Assuming an initial stage efficiency (for all stages) of 91%, an adiabatic flow throughout the entire turbine, and an average specific-heat ratio of 1.33, design and draw (to scale) the meridional flow path, showing the major turbine dimensions as well as all velocity triangles.

SOLUTION

In the following, the preliminary step of determining the number of stages is executed. The idea is to rely on the specific-speed concept (Fig. 5.9) to make sure that any axial-flow stage has a specific-speed magnitude that is not in excess of approximately 0.776 radians. The problem statement gives us a maximum stage count of 3. However, it would be very beneficial if we could condense this number to one or two, as any of these would produce an engine with less weight and shorter length. Note that we are about to investigate stages with yet-unknown exit velocity triangles and therefore unknown exit static density. Because the latter appears in the definition of the specific speed in expression (5.28), it will always be assumed that the exit swirl-velocity component is sufficiently small to produce an exit critical Mach number that is less than, say, 0.4. Stated differently, we will always proceed with the exit total density instead.

INVESTIGATION OF THE STAGE COUNT

a) A single-stage turbine configuration: Let us assume that this "hypothetical" stage is, say, 91% efficient on a total-to-total basis. With the total-to-total pressure ratio being given as 9.3, this assumption will lead us to the exit total temperature

$$0.91 = \eta_{t-t} = \frac{1 - \left(\frac{T_{t\,ex}}{T_{t\,in}}\right)}{1 - \left(\frac{p_{t\,ex}}{p_{t\,in}}\right)^{\frac{\gamma-1}{\gamma}}}$$

which, upon substitution, yields

$$T_{t\,ex} = 846.4 \text{ K}$$

The exit total pressure can easily be found:

$$P_{t\,ex} = \frac{p_{t\,in}}{9.3} = 1.08 \text{ bars}$$

Now, implementing the density assumption, we have

$$\rho_{ex} \approx \rho_{t\,ex} = \frac{p_{t\,ex}}{RT_{t\,ex}} = 0.445 \text{ kg/m}^3$$

Next, the ideal magnitude of the total enthalpy drop (across the stage) is computed:

$$\Delta h_{t,id} = c_p T_{t\,in}\left[1 - \left(\frac{p_{t\,ex}}{p_{t\,in}}\right)^{\frac{\gamma-1}{\gamma}}\right] = 678{,}296 \text{ J/kg}$$

Finally, we calculate the stage specific speed (N_s) by substituting in expression 5.28:

$$N_s = 0.67 \text{ radians}$$

When compared with the radial/axial-stage cut-off value of 0.776 (Fig. 5.9), we are now in a position to perhaps settle for this specific speed and its implications, settling for a single-stage turbine. However, with an allowable stage count of three, we do not have to make such a borderline approximation.

b) A two-stage turbine configuration: In order to shorten this investigative step, we will make the following choices:

- The total-to-total efficiency, for each stage, will still be 91%.
- Among the 60:40, 55:45, and 50:50 traditional work-split ratios, let us pick the 55:45 option.

• Although the work-split definition applies to *actual* shaft-work magnitudes, we will consider the definition to apply to the ideal (or isentropic-path) magnitudes instead. After all, we have to find some simple noniterative means of computing the interstage total properties. Of course, this simplification will not yield a 55:45 work split. The deviation, however, is luckily minor in nature.

Implementing, in particular, the last two assumptions, we proceed as follows:

$$(\Delta h_{t,id})_{overall} = 678,296 \text{ J/kg (computed earlier)}$$

$$(\Delta h_{t,id})_{stg.1} = 0.55(\Delta h_{t,id})_{overall} = 373{,}063 \text{ J/kg}$$

$$= c_p T_{t\,in}\left[1 - \left(\frac{p_{t\,int}}{p_{t\,in}}\right)^{\frac{\gamma-1}{\gamma}}\right]$$

where the subscript "*int*" refers to the interstage gap. Substitution in the preceding equation gives rise to the following interstage total pressure:

$$(p_t)_{int} = 3.42 \text{ bars}$$

Substituting in the total-to-total efficiency definition, we get

$$(T_t)_{int} = 1086.5 \text{ K}$$

Now

$$\rho_{int} \approx (\rho_t)_{int} = \frac{(p_t)_{int}}{R(T_t)_{int}} = 1.097 \text{ kg/m}^3$$

The first-stage specific speed can now be computed, again by substituting in expression (5.29):

$$(N_s)_{stg.1} = 0.662 \text{ radians}$$

In regard to the second stage of this two-stage turbine configuration, we can similarly calculate the specific speed as follows:

$$(p_t)_{ex} = 1.11 \text{ bars}$$

$$(T_t)_{ex} = 845.7 \text{ K}$$

$$(\rho)_{ex} \approx (\rho_t)_{ex} = 0.457 \text{ kg/m}^3$$

$$(N_s)_{stg.2} = 1.20 \text{ radians}$$

Note that the computed value of $(p_t)_{ex}$ is not really consistent with what the given overall total-to-total pressure ratio implies. This is clearly a result of the last of our three declared assumptions.

Conclusion: At this point, we have a first stage with a specific speed of 0.662 radians (ineligible to be an axial stage) and a second stage with a specific speed of 1.20 radians (very much an axial-stage candidate). As a result, we will refuse this two-stage option. However, it is necessary to realize that a work-split ratio other than 55:45 may end up

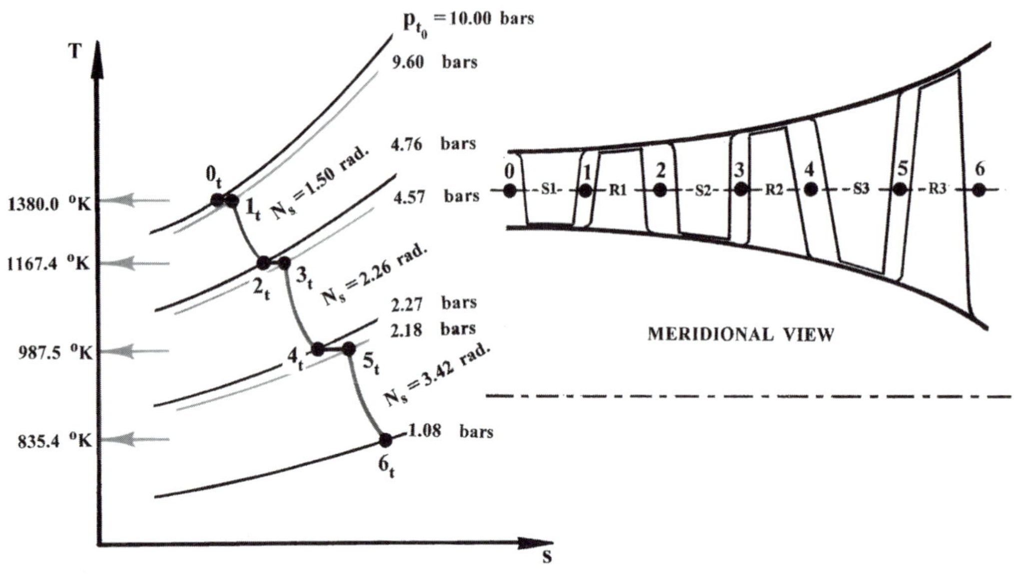

Figure 8.8. Therodynamics of a three-stage turbine for Example 2.

satisfying the specific-speed requirement for a turbine configuration with two axial stages. Examples of such work-split ratios would be 50:50 and 45:55. Nevertheless, it is equally important to recall that such ratios will result in a higher interstage total temperature, as the shaft-work percentage produced by the first stage gets lower (compared with the ratio of 55% chosen earlier). In such cases, and assuming that the first stage is cooled, the second stage may end up needing to be cooled as well. In fact, a first-stage shaft-work-extraction percentage that is less than 50% is hardly traditional.

c) Final choice: c three-stage turbine configuration: Assuming equal total-to-total pressure ratios, we can calculate the pressure ratio of each stage as follows:

$$(Pr)_{stg.} = (Pr)_{overall}{}^{\frac{1}{n}} = 2.1$$

where

Pr refers to the total-to-total pressure ratio, and
n is the number of stages (being three in this case)

Proceeding again with a per-stage total-to-total efficiency of 91%, we can represent the sequence of events on the T-s diagram, as explained next.

THERMODYNAMIC ANALYSIS

Figure 8.8 shows the flow processes across what is now a three-stage turbine configuration. The figure also shows the station-designation pattern. Each stage has a loss-free stator (i.e., one with an isentropic flow process) by assumption. With the stage efficiency($\eta_{stg.}$) being 91%, and the stage total-to-total pressure ratio $[(Pr)_{stg.}]$ being 2.1 (as computed earlier), we can calculate the first-stage exit properties as follows:

$$T_{t2} = T_{t0}\left\{1 - \eta_{stg.}\left[1 - (Pr)_{stg.}^{\frac{\gamma-1}{\gamma}}\right]\right\} = 1167.4 \text{ K}$$

$$p_{t2} = \frac{p_{t0}}{(Pr)_{stg.}} = 4.76 \text{ bars}$$

The same logic can be applied to the latter two stages, giving rise to the following results:

$$T_{t4} = 987.5 \text{ K}$$

$$p_{t4} = 2.27 \text{ bars}$$

$$T_{t6} = 835.4 \text{ K}$$

$$p_{t6} = 1.08 \text{ bars}$$

These results are shown in Figure 8.8 for clarity. The figure also shows the specific speed associated with each individual stage. Again, the stage-exit critical Mach number effect was ignored in computing the exit static density. The specific speed magnitudes, displayed in Figure 8.8, are all higher than the radial/axial-stage interface magnitude. This supports our choice of three axial-flow stages.

At this point, we can compute the work split among the three stages as follows:

$$(w_s)_1 : (w_s)_2 : (w_s)_3 = (\Delta T_t)_1 : (\Delta T_t)_2 : (\Delta T_t)_3 \approx 39 : 33 : 28$$

As stated earlier, it is always preferable (in designing a multistage turbine) to have the first stage deliver the largest amount of shaft work, which is clearly the case here.

The turbine-section mean radius can be calculated by selecting the upper limit of U_m (i.e., 414 m/s) as follows:

$$r_m = \frac{U_m}{\omega} = \frac{U_m}{N\left(\frac{2\pi}{60}\right)} = 11.3 \text{ cm}$$

FLOW KINEMATICS

The outcome of this computational phase is primarily the pair of velocity triangles at each rotor inlet and exit station. In proceeding from one stage to the next, we will begin (and, hopefully, end) with an exit swirl angle of zero. However, should this choice lead to an unacceptable stage reaction (i.e., far from 50%), we will then move to a negative exit-angle magnitude, repeating the process until the stage reaction is acceptable. In handling each stage, the set of equations (8.6) through (8.8) will be used repeatedly, just as they were in the preceding example.

First stage: Let us begin by making the following choices:

- exit swirl angle (α_2) = 0 (as mentioned earlier)
- stage flow coefficient (ϕ) = 0.6

Substituting these variables in equations (8.6) through (8.8), we get the following results:

$$\text{rotor-inlet relative flow angle } (\beta_1) = +36.3^\circ$$
$$\text{rotor-exit relative flow angle } (\beta_2) = -59.0^\circ$$
$$\text{stage reaction } (R) = 28\%$$

Of these, the stage reaction is far from the optimum reaction value of 50% and is therefore unacceptable. Now, let us reexecute the same procedure, with the following changes:

- $\alpha_2 = -18^\circ$
- $\phi = 0.65$

Substitution of these variables in equations (8.6) through (8.8) yields

$$\beta_1 = +18.5^\circ$$
$$\beta_2 = -61.8^\circ$$
$$\text{stage reaction } (R) = 49.7\%$$

The magnitude of the stage reaction is sufficiently close to the 50%, the optimum value. As a result, the foregoing choices and results are considered final as far as this stage is concerned.

The desire to maintain constant magnitudes of the mean radius (r_m) and axial velocity component (V_z) throughout the three-stage turbine means that the flow coefficient (ϕ) will have to be constant as we proceed to the second and third stages.

Second stage: Setting the stage-exit swirl angle (α_4) to -15°, the following results are obtained:

$$\beta_1 = +3.1^\circ$$
$$\beta_2 = -61.3^\circ$$
$$\text{stage reaction } (R) = 57.0\%$$

The stage reaction is clearly acceptable, and the other choices and results are therefore final.

Third stage: Because this is the last stage, any nonzero magnitude of the stage-exit swirl velocity component ($V_{\theta 6}$) will be viewed as a waste of work-producing angular momentum. With this in mind, let us assume the ideal situation, where α_6 is simply zero. The outcomes of this assumption are as follows:

$$\beta_1 = +2.6^\circ$$
$$\beta_2 = -57.0^\circ$$
$$\text{stage reaction } (R) = 49.0\%$$

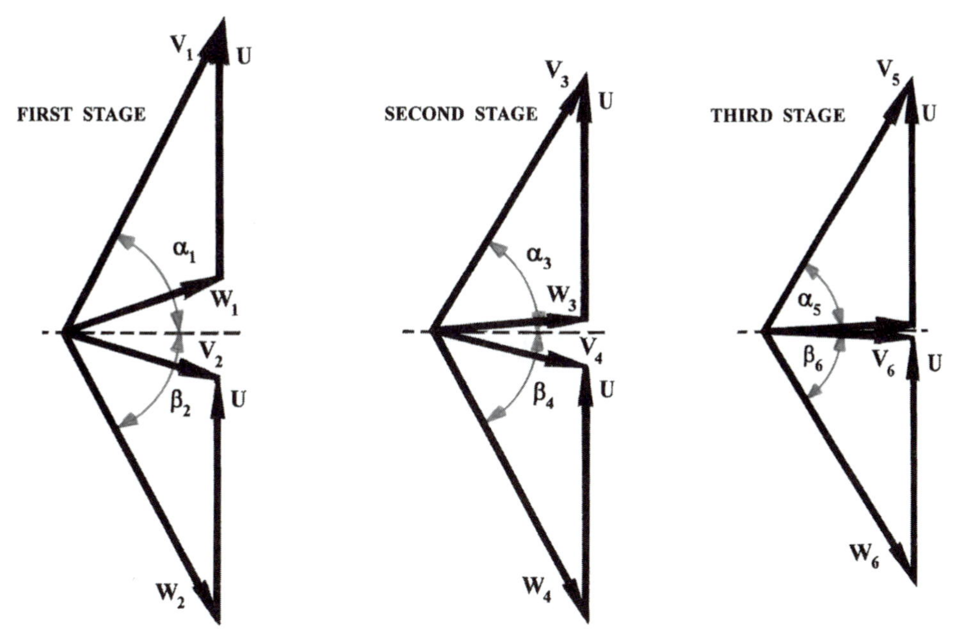

Figure 8.9. Velocity triangles associated with the chosen three-stage turbine configuration.

This reaction magnitude is clearly acceptable. In fact, we are with this result nothing less than fortunate, for we were able to achieve a no-swirl exit-flow stream, which is an ideal situation.

With the results just obtained, we can now construct the inlet and exit velocity triangles for each of the three rotors. Of course, the fixed-magnitude axial-velocity component is implied by our selection of the flow coefficient (ϕ), where

$$V_z = \phi U_m = 0.65 \times 414.0 = 269.1\text{m/s}$$

All of the velocity triangles associated with this three-stage turbine section are presented in Figure 8.9.

VERIFICATION OF NONSUPERSONIC MACH NUMBERS

In this computational step, we examine both the stator-exit critical Mach number (V/V_{cr}) and the rotor-exit relative critical Mach number W/W_{cr} for each individual stage. The objective is to confirm that each of these numbers is less than or equal to 1.0. This magnitude, by reference to Chapter 3, is the maximum any convergent (subsonic) blade-to-blade flow passage (stationary or rotating) can possibly reach.

First stage:

$$V_1 = \sqrt{V_z{}^2 + V_\theta{}^2} = 571.4 \text{ m/s}$$

$$T_{t1} = T_{t0} = 1380 \text{ K}$$

$$V_{cr1} = \sqrt{\left(\frac{2\gamma}{\gamma+1}\right) RT_{t1}} = 672.4 \text{ m/s}$$

$$V_1/V_{cr1} = 0.850$$

$$W_2 = \frac{V_z}{\cos\beta_2} = 560.4 \text{ m/s}$$

$$V_2 = \frac{V_z}{\cos\alpha_2} = 282.9 \text{ m/s}$$

$$T_{tr2} = T_{t2} - \left(\frac{V_2{}^2 + W_2{}^2}{2c_p}\right) = 1268.6 \text{ K}$$

$$W_{cr2} = \sqrt{\left(\frac{2\gamma}{\gamma+1}\right) RT_{tr2}} = 644.7 \text{ m/s}$$

$$M_{cr1} = V_1/V_{cr1} = 0.850 \text{ (i.e., subsonic)}$$

$$M_{r,cr2} = W_2/W_{cr2} = 0.870 \text{ (i.e., subsonic)}$$

Repeating the same computational procedure for the second and third stages allows similar conclusions to be drawn.

MERIDIONAL FLOW PATH CALCULATIONS

Calculation of the annulus heights: In the following, we will follow the same procedure as in Example 1. To be realistic, however, we will assume that the total pressure across each stator suffers a 5% loss, just as the "knee-type" segments of the process path intentionally suggests (Fig. 8.8).

First stage:

$$V_1 = 571.4 \text{ m/s (computed earlier)}$$

$$M_{cr1} = 0.85 \text{ (computed earlier)}$$

$$p_{t1} = 0.95 p_{t0} = 9.50 \text{ bars}$$

$$T_{t1} = T_{t0} = 1380.0 \text{ K (adiabatic stator flow)}$$

$$\rho_{t1} = \frac{p_{t1}}{RT_{t1}} = 2.40 \text{ kg/m}^3$$

$$\rho_1 = \rho_{t1}\left[1 - \left(\frac{\gamma-1}{\gamma+1}\right) M_{cr1}{}^2\right]^{\frac{1}{\gamma-1}} = 1.73 \text{ kg/m}^3$$

$$h_1 = \frac{\dot{m}}{\rho_1(2\pi r_m)V_z} = 2.48 \text{ cm}$$

$$V_0 = V_z = 269.1 \text{ m/s}$$

$$V_{cr0} = V_{cr1} = 672.4$$

$$M_{cr0} = 0.40$$

$$\rho_0 = 2.35 \text{ kg/m}^3$$

$$h_0 = 1.67 \text{ cm}$$

$$V_2 = 282.9 \text{ m/s}$$

$$V_{cr2} = \sqrt{\left(\frac{2\gamma}{\gamma + 1}\right) RT_{t2}} = 618.5 \text{ m/s}$$

$$M_{cr2} = 0.457$$

$$\rho_{t2} = 1.42 \text{ kg/m}^3$$

$$\rho_2 = 1.30 \text{ kg/m}^3$$

$$h_2 = 3.23 \text{ cm}$$

Second stage:

$$\rho_3 = 1.01 \text{ kg/m}^3$$

$$h_3 = 4.16 \text{ cm}$$

$$\rho_4 = 0.80 \text{ kg/m}^3$$

$$h_4 = 5.25 \text{ cm}$$

Third stage:

$$\rho_5 = 0.54 \text{ kg/m}^3$$

$$h_5 = 7.78 \text{ cm}$$

$$\rho_6 = 0.40 \text{ kg/m}^3$$

$$h_6 = 10.50 \text{ cm}$$

Axial chords and gap lengths: First, we recall the axial-chord-based aspect-ratio definition:

$$\lambda = \frac{h_{av.}}{c_z}$$

which is as valid for a stator vane as it is for a rotor blade.

In a multistage turbine, the aspect ratio (λ) grows from an average magnitude of nearly 1.0 over the first stage to a magnitude that is well above 3.0 over the last stage. Perhaps the turboprop-engine turbine section in Figure 1.2 demonstrates such a gradual rise in the aspect ratio, despite the fact that the figure only shows rotor blades. In fact, the current three-stage turbine resembles this particular turboprop engine to a good degree.

The high-pressure rotor(s), which will have significantly short blade heights, will typically suffer an aerodynamics-related problem. The reason is that the tip clearance here will be an unignorable percentage of such a short blade height. As will be explained later in this chapter, the so-called direct tip leakage, in this case, may very well "unload" a significant percentage of the blade height near the tip region. In fact, low-aspect-ratio blading is one of the toughest topics in this subdiscipline of internal aerodynamics.

High aspect ratio blades, however, would exist in the late (low-pressure) stage(s). The problem here is mechanical in nature. As discussed earlier, the long "span" of such blades may generate an excessive amount of bending stress at the root.

Starting with a first-stage-stator aspect ratio of, say, 1.5, let us proceed downstream, assuming the following sequence of aspect ratios:

$$(\lambda)_{stat.1} = 1.5$$
$$(\lambda)_{rot.1} = 1.8$$
$$(\lambda)_{stat.2} = 2.1$$
$$(\lambda)_{rot.2} = 2.4$$
$$(\lambda)_{stat.3} = 2.7$$
$$(\lambda)_{rot.3} = 3.0$$

Utilizing the formerly calculated annulus heights, and based on the average vane (or blade) spans, we are now in a position to compute the six axial-chord lengths as follows:

$$(C_z)_{stat.1} = 1.36 \text{ cm}$$
$$(C_z)_{rot.1} = 1.56 \text{ cm}$$
$$(C_z)_{stat.2} = 1.76 \text{ cm}$$
$$(C_z)_{rot.2} = 1.96 \text{ cm}$$
$$(C_z)_{stat.3} = 2.41 \text{ cm}$$
$$(C_z)_{rot.3} = 3.05 \text{ cm}$$

Turning our attention to the axial-gap length, we separate a stator-to-rotor gap from an interstage gap. The first will be taken as one-quarter of the average of the annulus heights on both sides of the gap. As for the interstage gap, the length will be taken as one-half of the arithmetic average of the annulus heights on both sides of the axial gap.

Now, referring to the station designation in Figure 8.8, and with the above-stated rules, we have:

$$(\Delta z)_1 = 0.73 \text{ cm}$$
$$(\Delta z)_2 = 1.66 \text{ cm}$$
$$(\Delta z)_3 = 0.93 \text{ cm}$$
$$(\Delta z)_4 = 2.19 \text{ cm}$$
$$(\Delta z)_5 = 1.30 \text{ cm}$$

Figure 8.10 offers a comparison between the stator-to-rotor and stage-to-stage axial-gap widths.

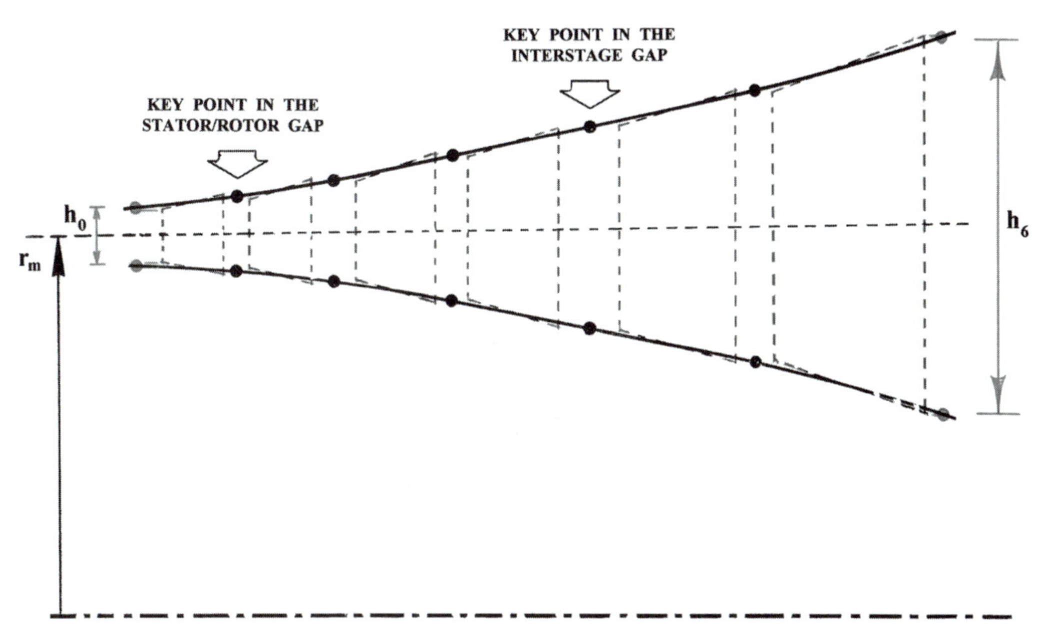

Figure 8.10. The process of creating the endwalls by simple interpolation.

CALCULATION OF THE TURBINE AXIAL LENGTH

Having computed all of the axial chords and gap widths, we can calculate the turbine axial length (L):

$$L = 18.91 \text{ cm}$$

CALCULATION OF THE SPECIFIC-SPEED

Now that the work split among the stages has been set and (more importantly) the stage-exit static densities are known, computing the stage specific speed is an easy task:

$$(N_s)_{stg.1} = \frac{\omega\sqrt{\frac{\dot{m}}{\rho_2}}}{(\Delta h_t)_{id}^{\frac{3}{4}}} = 0.777 \text{ radians}$$

$$(N_s)_{stg.2} = 1.12 \text{ radians}$$

$$(N_s)_{stg.3} = 1.80 \text{ radians}$$

All of these specific speeds are within the axial-turbine-stage zone in Figure 5.9.

CONSTRUCTION OF THE TURBINE FLOW PATH

Figure 8.11 shows a first approximation of the meridional flow path. This is symmetrical, and none of the stator vanes or rotor blades are tapered. The flow path in the

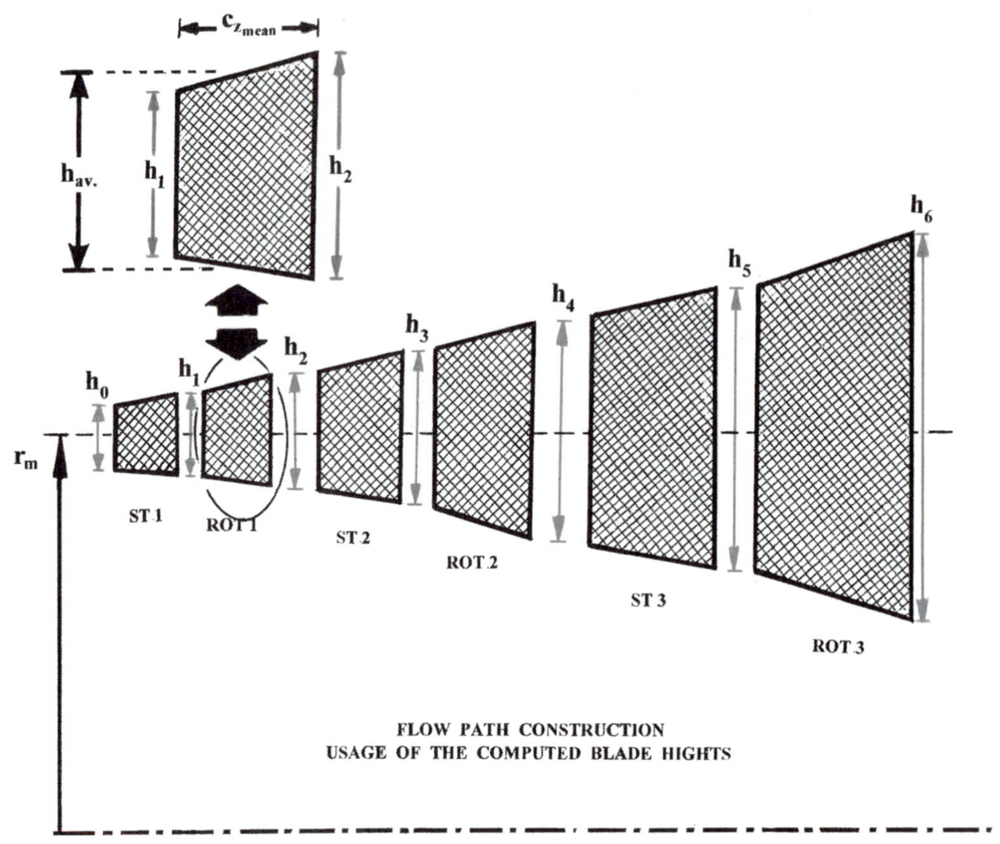

Figure 8.11. Axial extensions of stator and rotor components using the aspect-ratio criterion.

figure is simply based on the major dimensions computed previously, including those in the axial direction.

Superimposed on this "early-phase" flow path (Fig. 8.12) is the means to upgrade it. As Figure 8.12 shows, the process is started by locating key points, over both the hub and casing lines, midway in all axial gaps. These points are then used for interpolating smoother hub and casing lines. The process, however, turns out to be that of extrapolation over the first stator and last rotor. The resulting hub and casing lines are shown in Figure 8.13.

Because an excessive divergence angle (θ) of either the hub or casing endwall (above 15°) can lead to boundary-layer separation over the endwall(s), it is perhaps appropriate at this point to compute a rough magnitude of the turbinewide angle. Referring to Figure 8.13, an overall approximation of this angle is

$$\theta = \tan^{-1}\left[\frac{0.5(h_4 - h_0)}{L}\right] = 5.4^\circ < 15^\circ \text{ (i.e. acceptable)}$$

Proceeding further, the upgraded flow-path version in Figure 8.14 is where a taper angle of 10° is used for both sets of stator vanes and rotor blades. Experience has

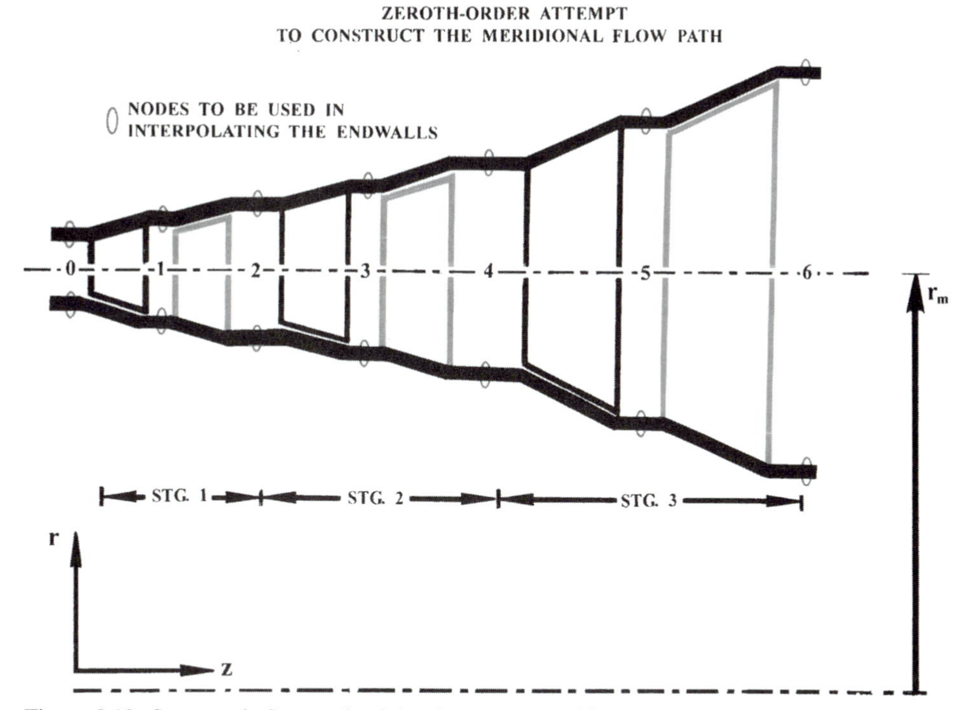

Figure 8.12. Symmetric flow path of the three-stage turbine configuration.

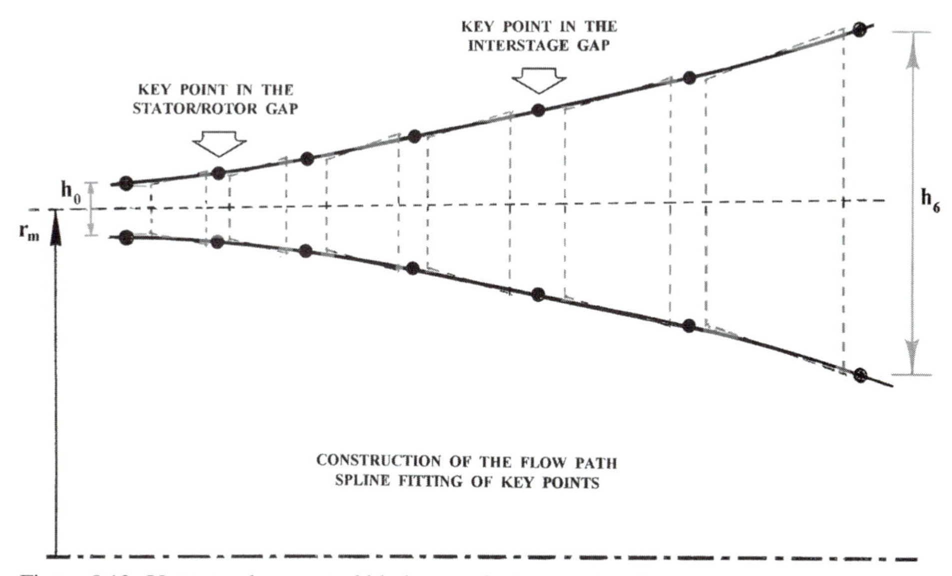

Figure 8.13. Untapered vanes and blades as a first approximation.

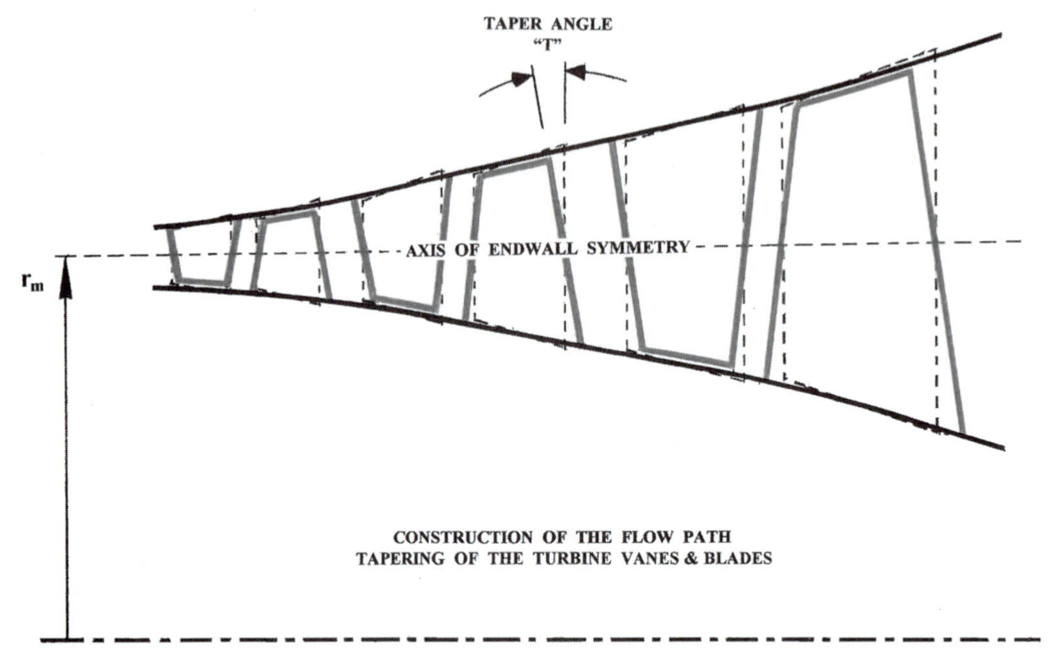

Figure 8.14. Tapering the rotor blades to avoid the consequences of bending stress.

shown that a stator/rotor gap that is constant at all radii, from hub to casing, will produce comparatively less total-pressure loss over the gap.

The meridional flow path is further modified in such a way as to ensure that all rotor tip lines and their corresponding casing segments have zero slope angles (Fig. 8.15). The reason behind such an adjustment is to accommodate axial displacements of the rotor blades without opening up the tip clearance or, on the contrary, avoid any blade/casing rubbing problems, two extremes that are illustrated in Figure 4.2. Figure 8.15 also shows what will later be discussed as a casing "recess," which in essence discourages the flow stream through the tip clearance. This stream would proceed undeflected and unavailable to produce shaft work.

The nonsymmetric flow path is regenerated in Figure 8.16 in order to show the different attachments of the stator vanes and rotor blades to what, in the end, constitutes the hub surface. Figure 8.16 also identifies the axial location of the point where the slope angle is the greatest, which happens to be on the hub line. This particular slope angle is, approximately 15°, which is right at the borderline of the allowable range.

Serving as a reminder, Figure 8.17 shows a sketch of what may become a costly consequence of a higher divergence-angle magnitude at the same point. As the figure suggests, the aerodynamic problem here stems from the fact that such a large angle may potentially cause flow separation and a large-scale recirculation zone. The final turbine flow path is shown in Figure 8.18.

Definitions of the Incidence and Deviation Angles

The magnitude of the incidence angle is the difference between the angle of the approaching flow stream and the airfoil-inlet (metal) angle. Figure 4.14 illustrates

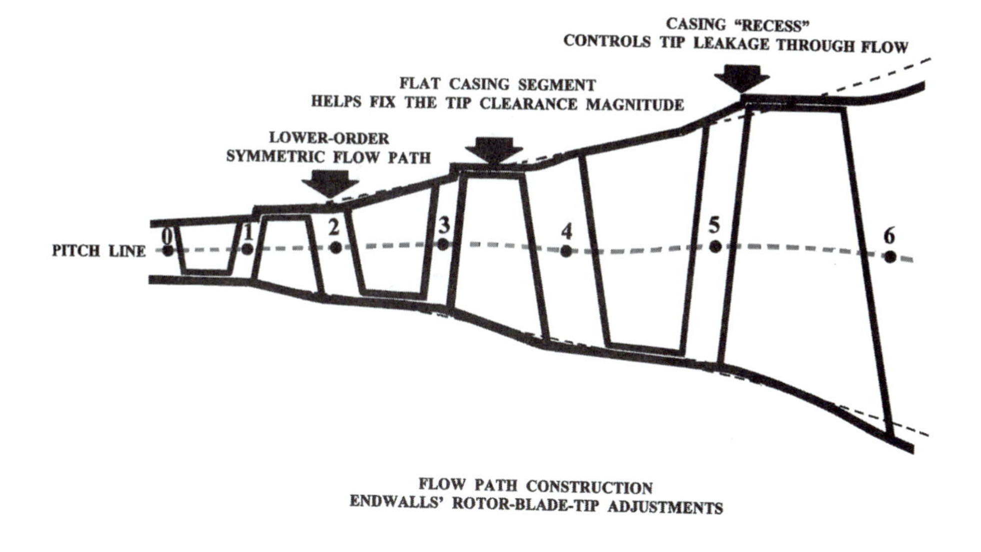

Figure 8.15. Flattening the rotor-blade tips to avoid the problem of rubbing against the casing and maintain constant the tip clearance.

the easiest means of determining the sign of this angle, which is as important as its magnitude.

The obvious external-aerodynamics equivalent to the incidence angle is the angle of attack in the general area of wing theory. The sign of the latter angle, however, is

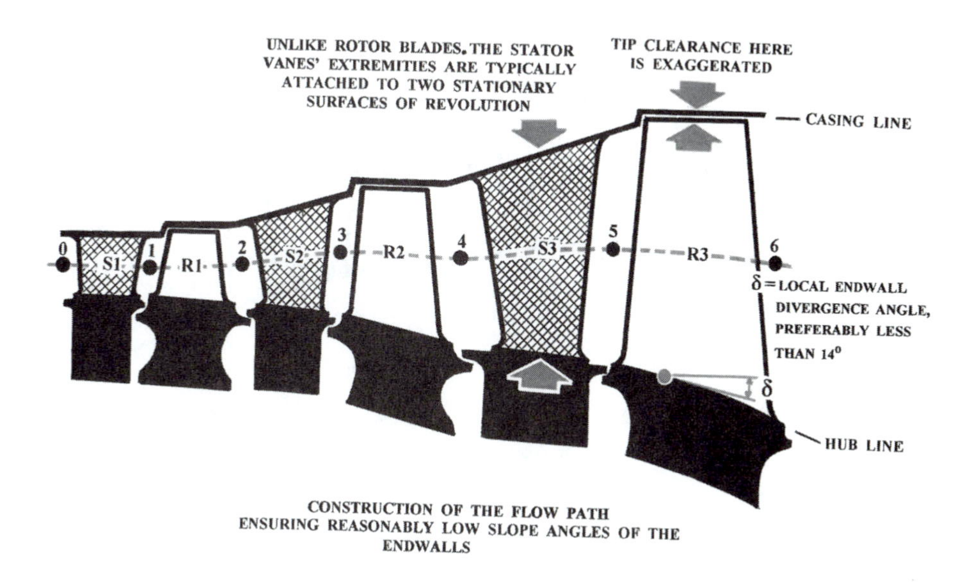

Figure 8.16. Final shape of the meridional flow path.

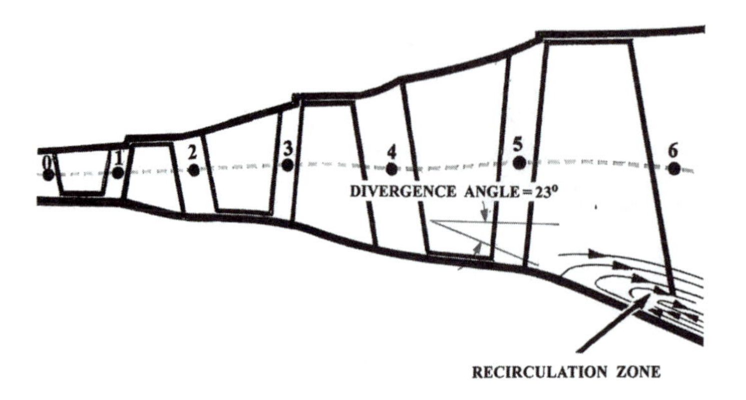

Figure 8.17. Effect of an excessive divergence angle of the flow path.

much easier by comparison. The reason is that a typical wing cross section is not as heavily "cambered" as a turbine airfoil normally is.

Of the two incidence-angle-sign possibilities, a positive angle is more damaging. As explained later in this chapter, a positive incidence may cause premature flow

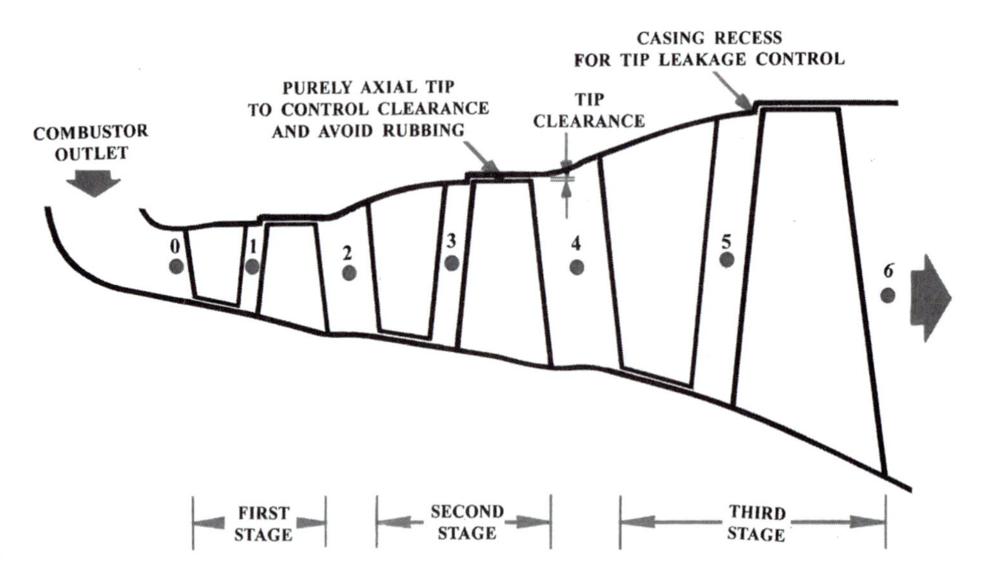

Figure 8.18. Adjusting the hub line to account for changes in the casing line.

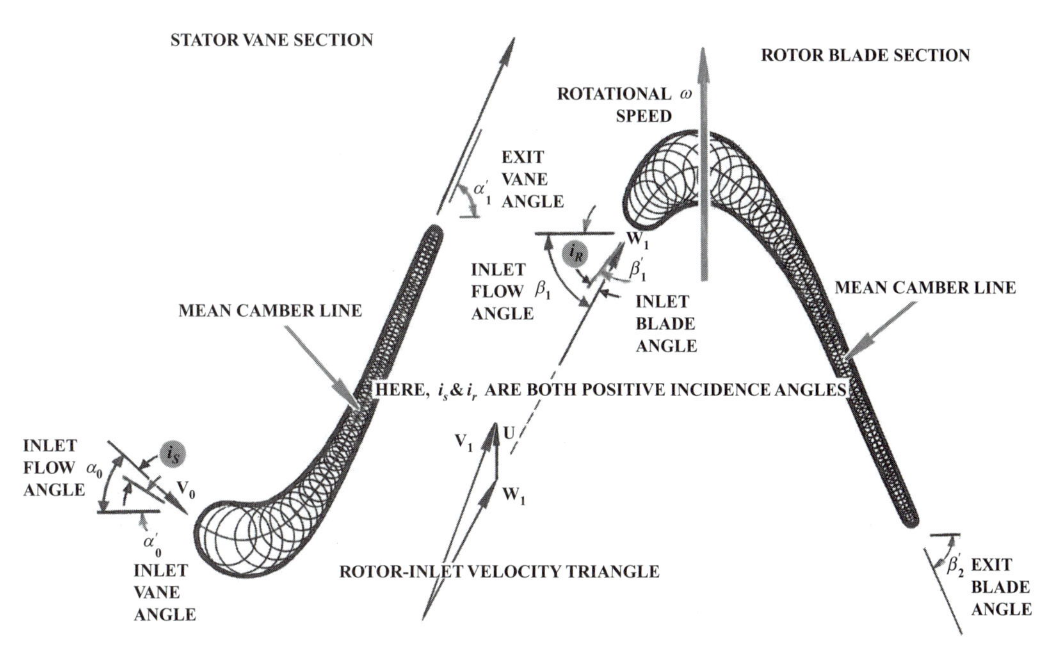

Figure 8.19. Stator-vane and rotor-blade incidence angles.

separation on the suction side. This, plus any other contributing factors (e.g., lack of reattachment, rear-segment flow unguidedness), could create a large-scale-loss environment. In fact, it is almost an axial-turbine blading fact that the optimum incidence angle is not even zero but is slightly negative (e.g. -5°) because of a justified concern about this situation to materialize.

It is important to distinguish between the stator and rotor incidence-angle definitions. Referring to Figure 8.19, the inlet flow angle for a stator vane is obviously that of the absolute velocity vector, namely α_0. As for a rotor blade, the flow angle that is performance-related is that of the relative-velocity vector (i.e., β_1). As a matter of terminology, the stator-vane and rotor-blade incidence angles in Figure 8.19 will be referred to as i_S and i_R, respectively.

Among the factors that influence the incidence angle impact on the total (or total relative) pressure is the leading-edge thickness. As a general rule, a thick leading edge is much more "forgiving" as far as incidence-angle variations are concerned. Such a leading edge is therefore adopted under circumstances where the turbine is expected to undergo a wide range of off-design operation modes.

The deviation-angle magnitude, however, is the difference between the flow exit angle and the trailing-edge mean camber-line angle (Fig. 8.19). The sign of this angle is immaterial, for it always reflects a flow-underturning situation. In Figure 8.19, this angle is intentionally set equal to zero. The reason is that a continually accelerating nozzle-like flow field (being the case here) will provide an environment where the boundary-layer buildup is significantly suppressed, and the likelihood of boundary-layer separation is practically nonexistent, at least under design-point operation. This is a great deal more than can be said in connection with compressor-airfoil

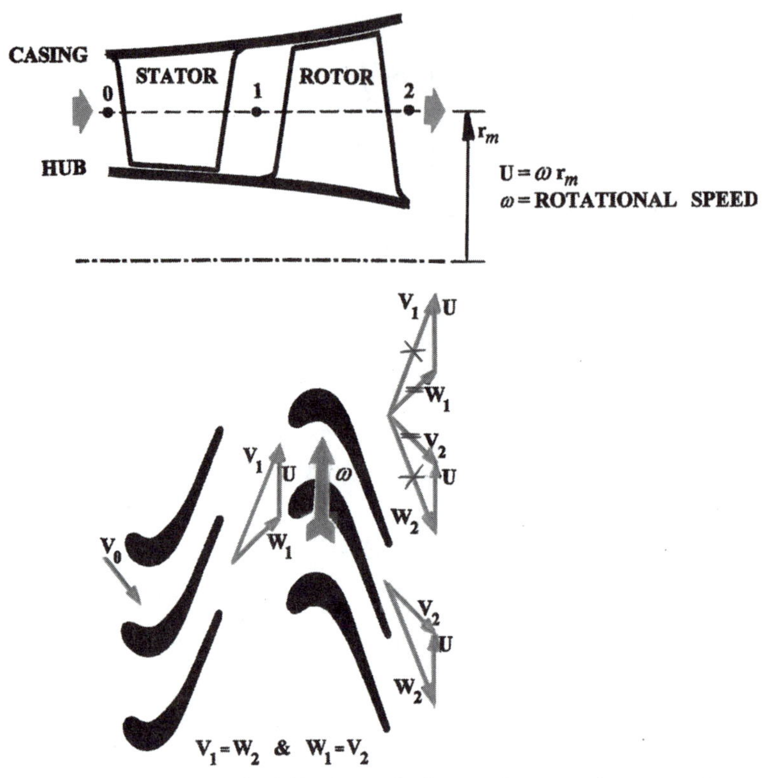

Figure 8.20. A symmetrical (50% reaction) stage for an optimum work-extraction process.

performance, where consequences of the adverse pressure gradient may cause a substantial deviation angle.

The blade reaction is another factor that influences the extent to which the incidence angle impacts the blade performance. Displayed in Figure 8.20 is the stator and rotor airfoil cascades and the velocity triangles associated with a 50% reaction stage. Although this reaction magnitude is traditionally recommended at the blade mean section, one should expect (more often than not) a lower to much lower hub reaction as well as a substantially higher tip reaction, depending on the blade radial extension.

Detailed Design of Airfoil Cascades

In contrast with the steam-turbine (Rankine) cycle, the gas-turbine (Brayton) cycle performance is highly sensitive to component efficiencies. It follows that gas-turbine blading, particularly for the axial-flow type, is a more tedious and thorough process toward the optimum airfoil-cascade configuration. It also follows that simple straight, untwisted (sometimes untapered) cascades, often utilized in steam turbines, are hardly employable here. Instead, modern axial-flow turbines, particularly those in propulsion applications, typically feature a substantial amount of blade twisting,

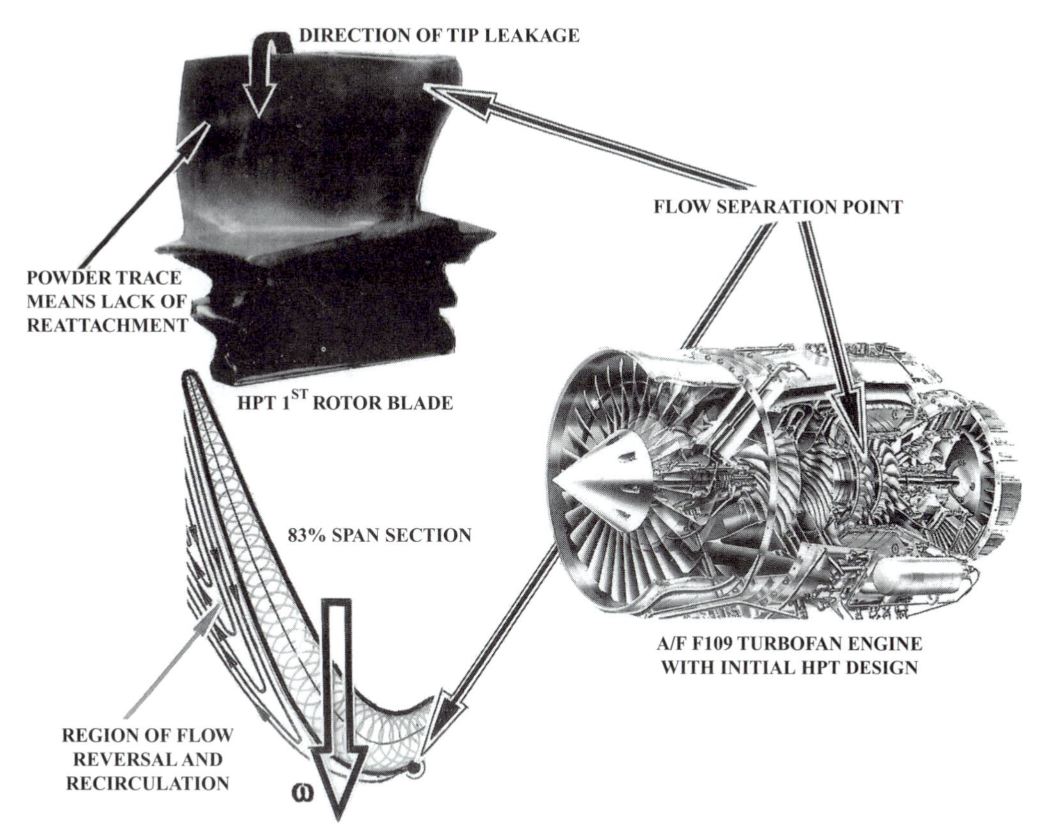

Figure 8.21. Flow separation caused by a "bump" on the blade suction side.

leaning (spanwise tangential shifting of the blade sections), and tilting (axial shifting of the blade sections) from hub to tip. Spanwise twisting of the blades accounts in part for the radial variation of the thermophysical properties of the flow as well as the velocity triangles. Figure 8.21 shows a typical high-pressure rotor blade, which is clearly three-dimensional and has a great deal of spanwise lean and twist.

Mechanical considerations are also of vital importance in turbine blading. Limitations in the form of engine-weight restrictions in propulsion applications, together with the persistent desire to elevate the inlet total temperature for a higher power output, create an environment of high mechanical/thermal stress levels. Aside from the high bending stress at the blade root, metallurgical means of strengthening the fabrication alloys in such a way to address the highly serious issue of stress concentration are needed. Of course, the bigger picture here, by reference to Figure 3.2, involves two airfoil cascades with tangential motion relative to one another and a finite axial distance in between. As mentioned earlier, the cyclic stresses generated within both of them worsen an already bad situation in the gas turbine. The only good news in all of this is that an aerothermo component design engineer does not have to worry about these issues, for they are "someone else's responsibility." The exception here is the need to respond designwise to such mechanical-component

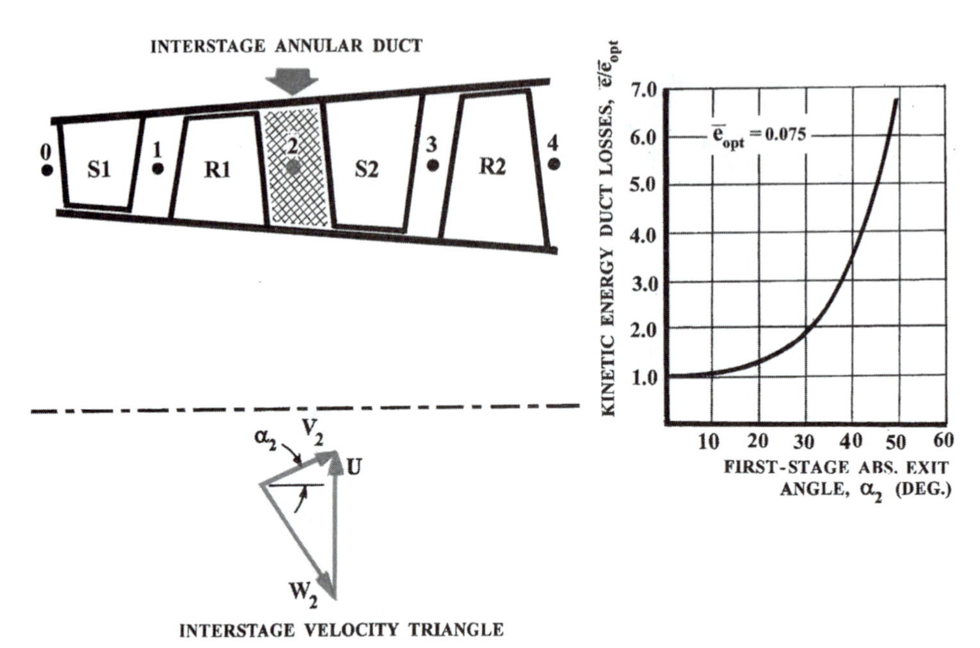

Figure 8.22. Kinetic-energy loss over an interstage duct.

design needs as thickening the root section, accommodating internal-cooling passages, and so forth.

Limiting the discussion to aerodynamic considerations, the stator-to-rotor and interstage axial gap lengths do abide by specific rules. As a result of the high stator-exit swirl velocity component, the stator/rotor gap is usually as small as can be mechanically tolerable. The reason is that the flow/endwall contact trajectory in this case is significantly elongated because of the many tangential "trips" that the fluid particle will undergo within this axial gap. As a result, the friction-related total pressure loss under such a situation can be unacceptably large. The interstage axial gap, by comparison, is one where the swirl-velocity level is considerably small, and the losses are typically low (Fig. 8.22). The designer in this case can elongate such a gap to the point of practically alleviating the rotor/stator dynamic interaction. In so doing, one of the very few constraints will be to abide by an engine-length restriction, if there is one.

Once the stage inlet and exit velocity triangles are established, the rotor-blade cascade is designed in such a way to efficiently give rise to nearly the same amount of flow deflection that is implied by these triangles. The stator-vane cascade is designed on the same basis.

On a two-dimensional basis, the designer would proceed from one radial location to the other, each time designing an airfoil-cascade section to suit the inlet and exit conditions. This is always possible through many commercial "canned" programs in existence today. However, the "stacking" of these airfoil sections on top of one another can in some cases have an unpleasant outcome. Figure 8.21 offers a dramatic picture of the flow behavior over a blade with a low height/chord ratio. As indicated in this figure, this blade was the outcome of the high-pressure turbine (HPT) initial design of the F109 turbofan engine. Regardless of what was a careful two-dimensional design procedure, at all radii, the blade came out with a clear "bump" on the suction side at approximately the 83% span section, as shown in Figure 8.21. In an engine test with a special powder injected in the main stream, this geometrical irregularity did lead to an early flow separation at this location, with no reattachment anywhere on the suction side. The unfortunate fact is that the aerodynamic penalty (in terms of the HPT efficiency) overwhelmingly met the requirement for a totally new blade redesign. Later in this chapter, the blading problem just outlined will be a topic of further discussion, particularly in the section covering the "implied stagger angle" topic. The example nevertheless underscores the need for reliable three-dimensional graphics to be part of the real design process so that geometrical irregularities can be detected prior to fabrication.

Existing computer codes (commercial or NASA-produced) are overwhelmingly tailored for the two-dimensional cascade design phase and will commonly produce an airfoil aerodynamic "loading." This will prompt the designer to either modify the input and try again or terminate the iterative procedure once the outcome is satisfactory. In the current turbine case, the aerodynamic loading is typically in the form of a velocity distribution over both the suction and pressure sides. There are specific loading features (discussed later in this chapter) that help evaluate the cascade design and make an informed judgment possible.

Airfoil-Cascade Geometry Variables

Figure 8.23 shows the major variables used to define the blade-cascade geometry. Each of these variables affects, to varying degrees, the airfoil's aerodynamic loading. As a matter of definition, an airfoil cascade is defined by the geometrical shape of a single airfoil plus the spacing between two successive airfoils. The latter is referred to as the cascade pitch "S," with a mean-radius magnitude S_m, where

$$S_m = \frac{2\pi r_m}{N_b} \tag{8.9}$$

with r_m being the mean radius and N_b the number of blades. The geometry variables in Figure 8.23 are particularly critical to the cascade aerodynamic performance. In the following, some of these variables with direct performance effects are identified and briefly discussed.

Leading-Edge Thickness

As discussed earlier, a thick leading edge would tolerate relatively substantial changes in the flow incidence angle. Nevertheless, a truly large thickness would cause the incoming flow to face a so-called stagnation flow process, whereby the leading edge would appear like a finite-width solid wall. Such a flow structure is widely known to be among those associated with a tremendous amount of kinetic energy and total-pressure losses. A thin leading edge, however, would give rise to a strong reliance

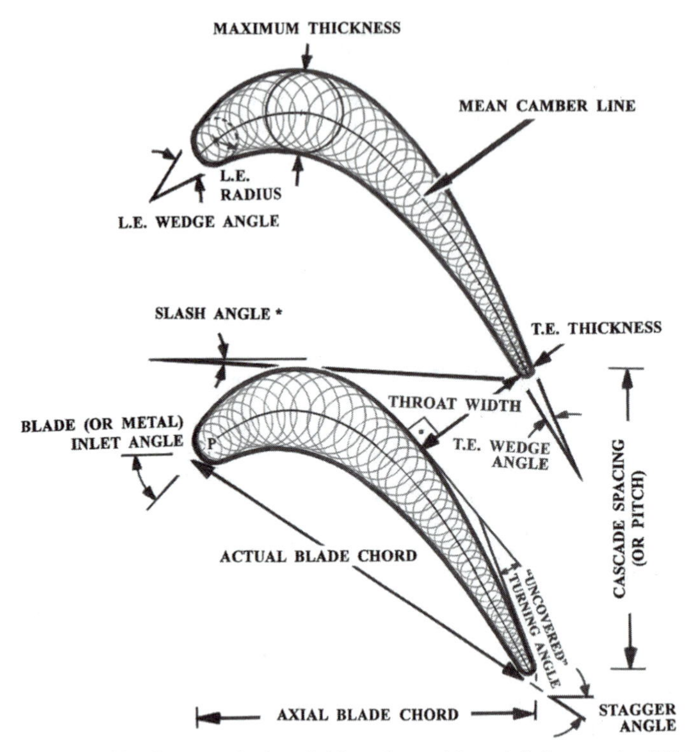

Figure 8.23. Geometrical variables of a turbine airfoil cascade. * This angle is relevant only to the hub section, as it controls the angle at which the base (or platform) is cut for those blades that are inserted, as separate units, around the circumference.

of the aerodynamic loading on the incidence angle. Under such circumstances, one should expect a significant level of airfoil performance deterioration as off-design operation modes are encountered.

Leading-Edge Wedge Angle

This is defined as the enclosed angle between two airfoil tangents at the two "merging" points, where the circular leading edge merges with the rest of the airfoil contour (Fig. 8.23). A large magnitude of this angle would give rise to a front-loaded airfoil. The result is typically a sudden expansion (or acceleration) that is unsustainable. In other words, such a large angle may produce a severe velocity jump followed by an immediate decline just downstream from the leading edge (usually at the merging points). This type of loading feature is normally referred to as a local diffusion, and is definitely undesirable.

Uncovered Turning Angle

Extended from the trailing edge and perpendicular to the adjacent airfoil surface is an intersection point (loosely) defined as the "throat" point (Fig. 8.23). Downstream from this suction-side point, the flow stream is referred to as uncovered (i.e., no longer a confined channel flow). This uncovered airfoil segment is typically designed to be a straight line in turbine airfoils. However, such a choice is incapable of preventing flow deceleration over this suction-side segment. The uncovered flow angle is that between two suction-side tangents, one at the throat point and the other at the trailing-edge merging point. Under a straight-line downstream segment, such an angle is clearly zero. However, a modest non-zero magnitude of this angle would facilitate an accelerating flow along the airfoil suction side. This, in some instances, may cause supersonic "pockets" to exist on the suction side. The concern here is that the flow within these pockets may experience oblique shocks, which are weak by category but will nevertheless produce their own total pressure losses.

Trailing-Edge Thickness

The ideal trailing-edge geometry is a zero-thickness cusp where both the suction and pressure sides share the same tangent. However, a round, finite-thickness trailing edge is the norm in the turbine bladed components. The reason has to do with the manufacturing precision, as well as the fear of rupture. A notably thick trailing edge may produce a region of vortical motion in the immediate vicinity of the trailing edge, worsening an already lossy flow-mixing zone, which is referred to as the airfoil wake. A thick trailing edge will also contribute to the creation of a sudden enlargement in the cross-flow area as the flow stream exits the trailing-edge plane. The performance penalty in this case is known as the "dump" effect and may indeed be substantial.

The flow field dependence on the cascade geometry is a highly involved and complex topic. The variables just discussed are merely intended to provide examples of the fluid/structure interaction effects. A more complicated set of loss-causing geometry-dependent flow mechanisms will be discussed later, and specific conclusions will be drawn.

Airfoil Aerodynamic Loading

A highly loaded vane (or blade) section is one where the pressure-to-suction static-pressure differentials at most of the chordwise locations are considerably high. These pressure differentials are naturally proportional to the corresponding suction-to-pressure-side velocity differentials at all locations. It is in this sense, by reference to the velocity distribution in Figure 8.24, that the enclosed area is indicative of the net torque-producing aerodynamic force, which is equivalent to the lift force in external aerodynamics. The figure also shows the computational domain boundary in a typical flow-analysis program, with the meridional projection being of the stream-filament type. The stream filament (or tube) in this figure has a streamwise thickness distribution that reflects the hub-to-casing blade-height distribution. This is the only means, within what is termed the "quasi-three-dimensional" flow-solver family, to account for the divergence of the flow path within what is conceptually an irrotational (potential) two-dimensional inviscid-flow program. Two famous and well-documented versions of such flow solvers were developed by Katsanis and

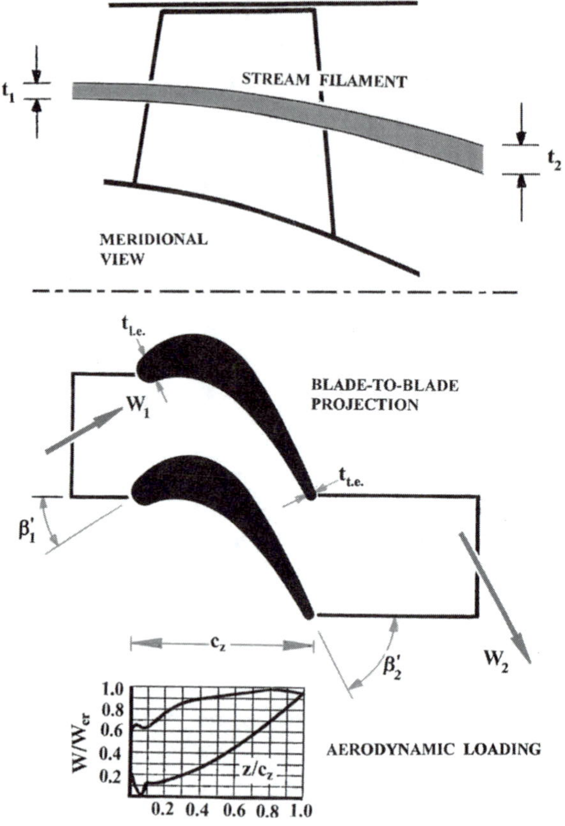

Figure 8.24. Solution domain for a quasi-three-dimensional potential-flow analysis and sample output.

McNally (1973) and McFarland (1984) at NASA-Lewis Research Center (Cleveland, Ohio) and are therefore in the public domain.

The use of simple potential flow-analysis codes is hardly intended to deny the existence or value of many three-dimensional viscous-flow programs existing today. For one thing, we know that a program that recognizes real-life loss-causing effects will be based on a much more complex set of flow-governing equations. Moreover, the fact that viscosity will present itself primarily within a very thin layer next to a solid wall makes it a necessity to work with a substantially refined computational grid right next to the solid walls (i.e., endwalls and airfoil surface). An example in this context is the work by Macgregor and Baskharone (1992) in which the flow viscosity was part of a time-dependent stator/rotor interaction model. In its final form, the computer program produced, in this study, consumed around two *days* of execution time on an IBM mainframe, occupying nearly its entire storage capacity. Yes, there are many equally sophisticated computer codes to do the job; the question, however, is whether they are needed, especially during the preliminary or midprocedure design phases. The answer is no, and the reason follows.

Regardless of the design phase, one can learn a great deal of performance information from the outcome of a simple inviscid-flow analysis code that would literally take seconds to execute. In fact, this particular point will be emphasized in the remainder of this chapter. Apart from mechanical considerations, it is the primary objective, in designing the turbine cascade, to produce a monotonically accelerating flow along the suction side, in particular. Such a velocity distribution will give rise to a continually decreasing pressure that substantially slows the boundary layer buildup and minimizes the possibility of flow separation. Of course, this cascade-loading information is nothing but an inviscid-flow computer-code outcome. Should the need arise, the blade-surface velocity distribution may very well be used in a separate, and inexpensive, boundary-layer analysis code as a postprocessor. There, the already-known surface velocity is treated as an imposed boundary layer-edge velocity. There are many simple (calculator-suited) models for doing the same, such as the flow model by Thwaites (1960).

The cascade design process is typically iterative and, sometimes, tedious. In spite of the cascade-geometry changes from one iteration to the next, the computational domain will always be characteristically similar to the blade-to-blade region in Figure 8.23. In its z-θ projection, the solution domain will always be tangentially bound by the so-called "periodic" boundaries. These are two pairs of straight lines that extend the solution domain upstream from the leading edge and downstream from the trailing edge. These periodic boundaries are where the flow behavior repeats itself in the tangential direction, a feature that establishes a solid distinction between the flow field around an isolated airfoil (such as an aircraft wing section) and that in a cascade. At any iterative step, the output, in one way or another, will consist of the blade aerodynamic loading in the form of the suction- and pressure-side velocity distributions. At this point, the designer should assess the numerical results in light of specific guidelines and eventually make the decision to investigate more cascade geometries or terminate the procedure. Note that the geometrical adjustments do not always concern the airfoil but could also involve the cascade pitch. Making such an adjustment is precisely equivalent to altering the number of blades.

Figure 8.25 shows the initial and final versions of the same airfoil cascade, with the difference being the result of recontouring the casing surface. The cascade in this figure is that of the second-stage stator in a three-stage turbine section of an existing turboprop engine. Referring to the initial and final velocity distributions (on the same figure), we find that the enclosed area associated with the latter is less by comparison. Such loading relief may go a long way in elongating the stator life. Moreover, the modified-casing velocity distribution features a suction-side segment of accelerated flow that was initially a region of adverse pressure gradient as a result of the surface velocity decline.

Geometrical Discontinuities

Local geometrical irregularities on the blade surface can produce a disproportionately massive flow-field distortion. Figure 8.26 shows the first-stage rotor of a

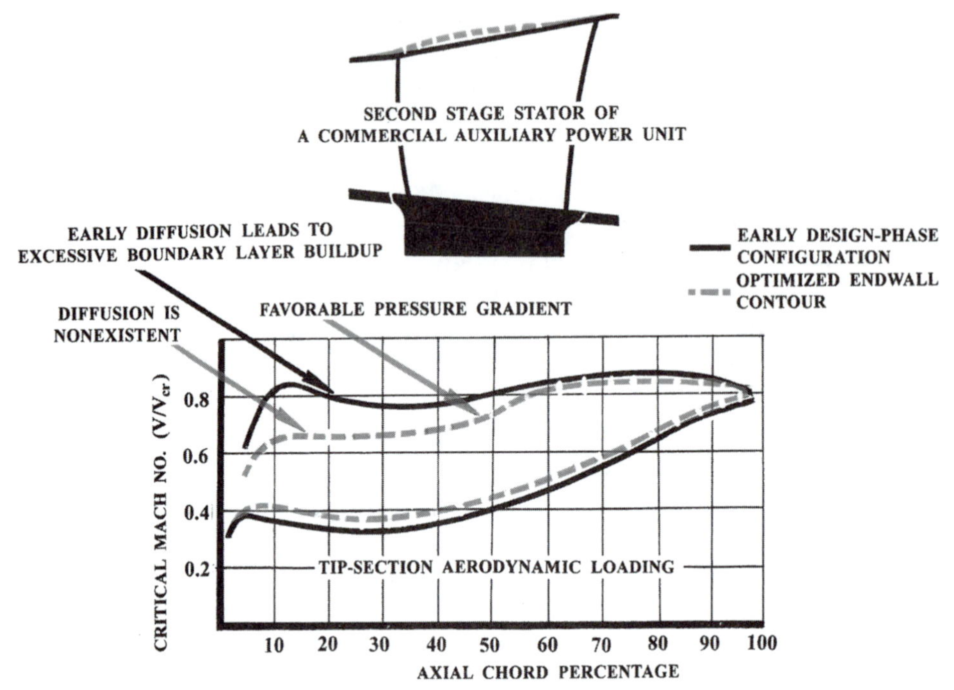

Figure 8.25. Effect of recontouring the casing line on the velocity distribution along a stator.

two-stage high-pressure turbine in an existing turbofan engine. The individual sections of this blade were carefully designed on a two-dimensional basis. However, because of incorrect "stacking" of these sections on top of one another, a small bump came to exist, as previously mentioned, on the suction side of the blade. This caused an estimated three-point loss in the overall turbine efficiency, and corrective action was required. To begin, the flow-field distortion in this case had to be comprehended first. To this end, a categorical flow-visualization process was employed during a cold-rig component test. The process in this case consisted of injecting a special adhesive powder within the incoming flow stream. The result, as shown in Figure 8.26, is different traces of powder at locations where the local velocity is small enough to allow the powder release. Upon component disassembly, the powder pattern on the blade surface was examined, and the following conclusions were reached:

- A premature boundary-layer separation did occur over the suction side at the approximate position of the bump.
- Downstream from the point of flow separation, the static-pressure gradient provided no environment for any flow reattachment, all the way down to the trailing edge.

Before embarking on a full-scale redesign process, one more diagnostic step was carried out. This involved a three-dimensional inviscid but vorticity-tolerating (rotational)-flow analysis code. Results of this computational process (also shown

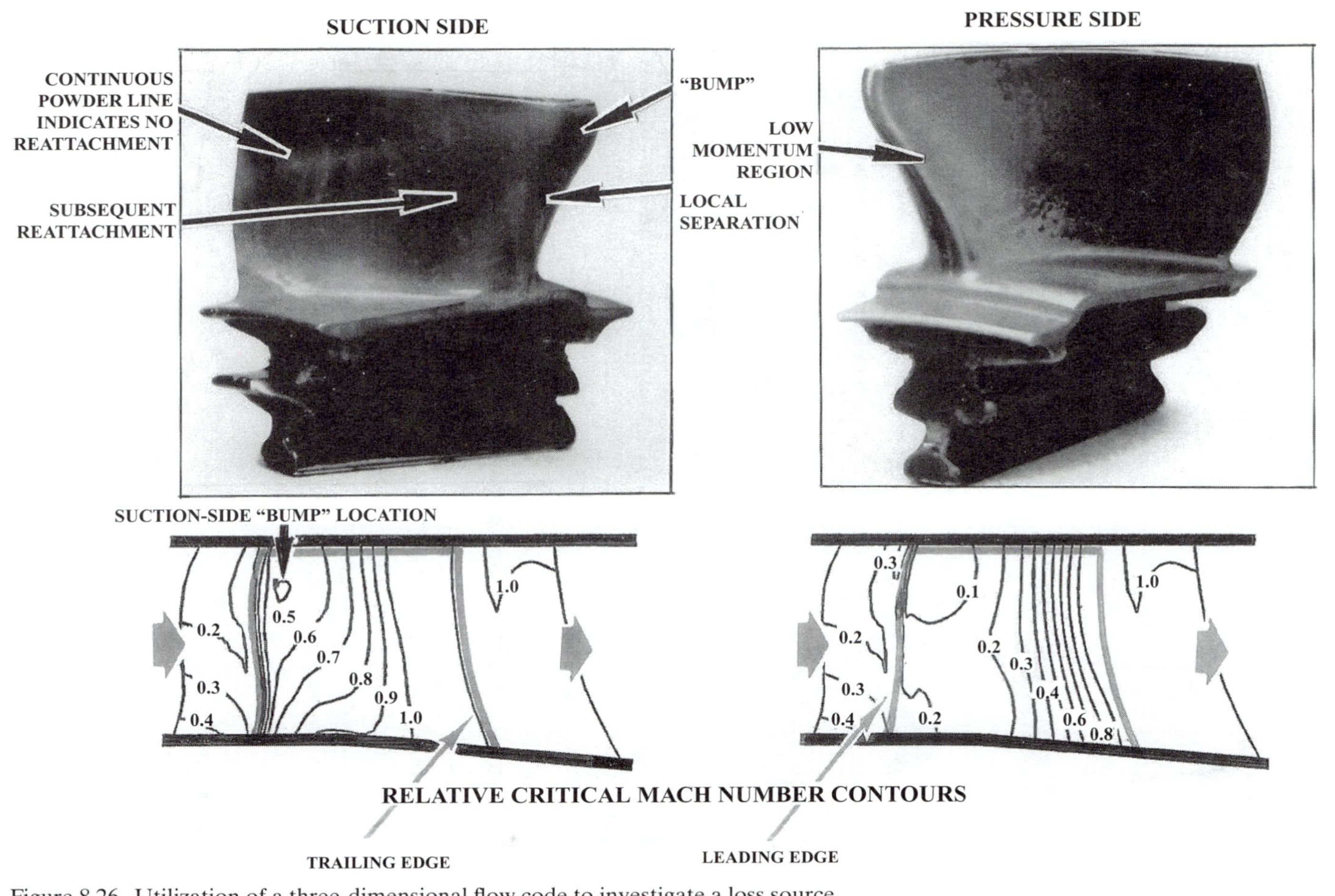

Figure 8.26. Utilization of a three-dimensional flow code to investigate a loss source.

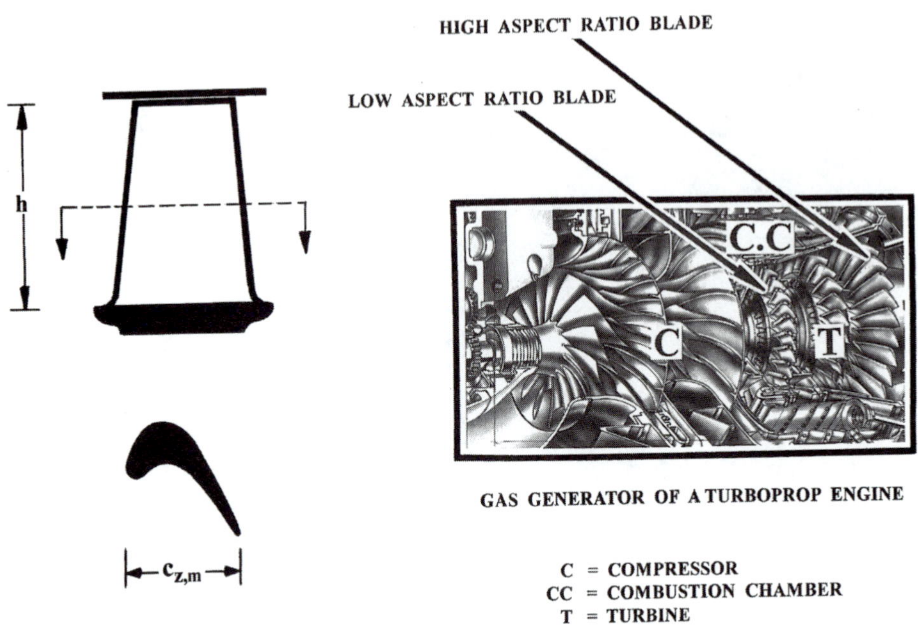

Figure 8.27. Definition and variation of the aspect ratio among turbine blades.

in Figure 8.26) were considered conclusive. As seen in this figure, the "island-like" velocity contours are observed nowhere in the figure except around the suction-side bump.

Performance-Controlling Variables

Several factors contribute to the total (or total-relative) pressure loss in a stator (or a rotor) turbine cascade. Chief among these are such geometrical aspects as the pitch/axial-chord and tip-clearance/span ratios as well as others. Last, the operating conditions themselves could give rise to an implied loss mechanism that is simply unremovable. In the following, some of these performance-related variables are discussed.

Aspect Ratio

As indicated earlier, the aspect ratio λ is defined as the ratio between the blade span (h) and the mean-radius magnitude of the axial chord (Fig. 8.27), namely

$$\lambda = \frac{h}{c_{zm}} \tag{8.10}$$

A value of this variable that is less than 1.0 is known to produce a blade-to-blade hub-to-casing cascade unit that is viscosity-dominated. In this case, the fast boundary layer buildup over the blade and endwalls as well as the potential for flow separation are worthy of monitoring. This is true in the sense that the boundary-layer thickness

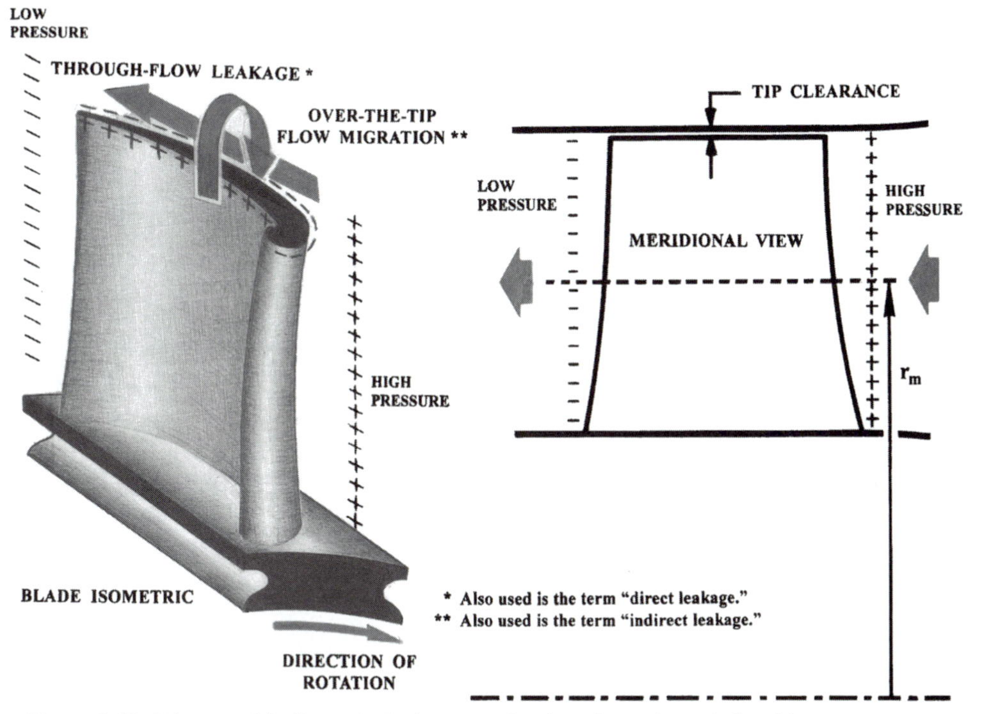

Figure 8.28. Direct and indirect tip-leakage mechanisms in unshrouded turbine rotors.

over the closely-located endwalls in this case could become comparable to the local annulus height, which is a bad sign, to say the least.

Turning to the low-pressure-turbine section (Fig. 8.27), the problem there is quite different. Because of the continuous rise in the annulus height, caused by the density decline, the blade span will have to grow, elevating the root bending-stress level.

Tip-Clearance Effects

In order to avoid rubbing between the rotor blades and the casing inner surface, a finite radial gap is allowed between the blade tip and the local casing segment. The gap height is typically as small as a fraction of a millimeter. However, it can very well have a disproportionately high influence on the stage efficiency. In unshrouded rotors, the flow migration over the tip from the pressure side to the suction surface (Fig. 8.28) would unload not only the tip section but also a significant percentage of the blade span. With the hub section having its own loss-producing mechanisms (e.g., the interaction between the hub and blade boundary layers), the hub section is also considered unreliable in the shaft-work production process. With these aspects in mind, and referring to Figure 8.22, the trend in modern turbines is to extract most of the shaft work at and around the mean radius. The resulting blade in this case is said to have a "reversed" lean geometry, which allows the flow deflection angle to be maximum at or near the mean radius.

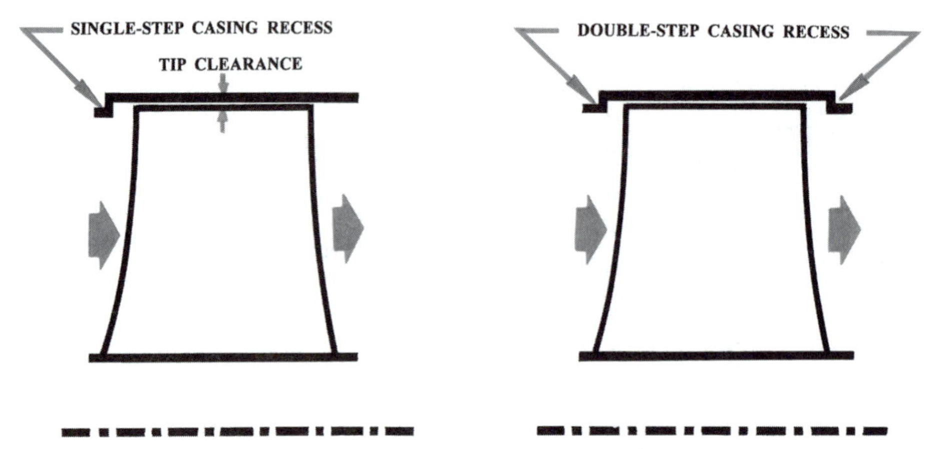

Figure 8.29. Casing recess as a means of direct-leakage control.

Direct Tip Leakage

This phrase describes the through-flow motion, in the tip clearance from the higher pressure point (upstream of the blade) to the downstream lower-pressure point (Fig. 8.28). This secondary flow stream will proceed through the tip clearance unguided, undeflected, and unavailable to participate in the shaft work production mechanism.

It is safe to say that direct leakage is a function of the blade tip reaction. This is true in the sense that a static enthalpy drop is convertible into a static pressure drop. For example, a zero tip reaction would virtually eliminate the streamwise static-pressure differential from the leading edge down to the trailing edge at the blade tip. This will have the effect of theoretically doing away with the direct tip-leakage stream.

There are many simple remedies for a persistent direct tip-leakage problem. Referring to Figure 8.29, a technique known as "casing recess," where the casing is shaped like a local abrupt step, is frequently used. The idea behind this and the double-step casing configuration (Fig. 8.29) is to place obstacles in the pathway of the direct-leakage flow stream.

Indirect Tip Leakage

This leakage mechanism is sustained by the pressure differential between the pressure and suction sides of the rotor tip section. As shown in Figure 8.28, the leakage flow stream is in the form of flow migration over the tip from the pressure side to the suction side. Obviously, the intensity of this leakage mechanism is dependent upon the tip-pressure differential between these two sides. In other words, it is the tip-section aerodynamic loading that dictates the magnitude of this particular leakage mechanism.

Suppression of this leakage stream requires physical isolation of the blade pressure and suction sides from one another. The term used in this context is tip "shrouding," of which Figure 8.30 shows two configurations. Of the two, the more common

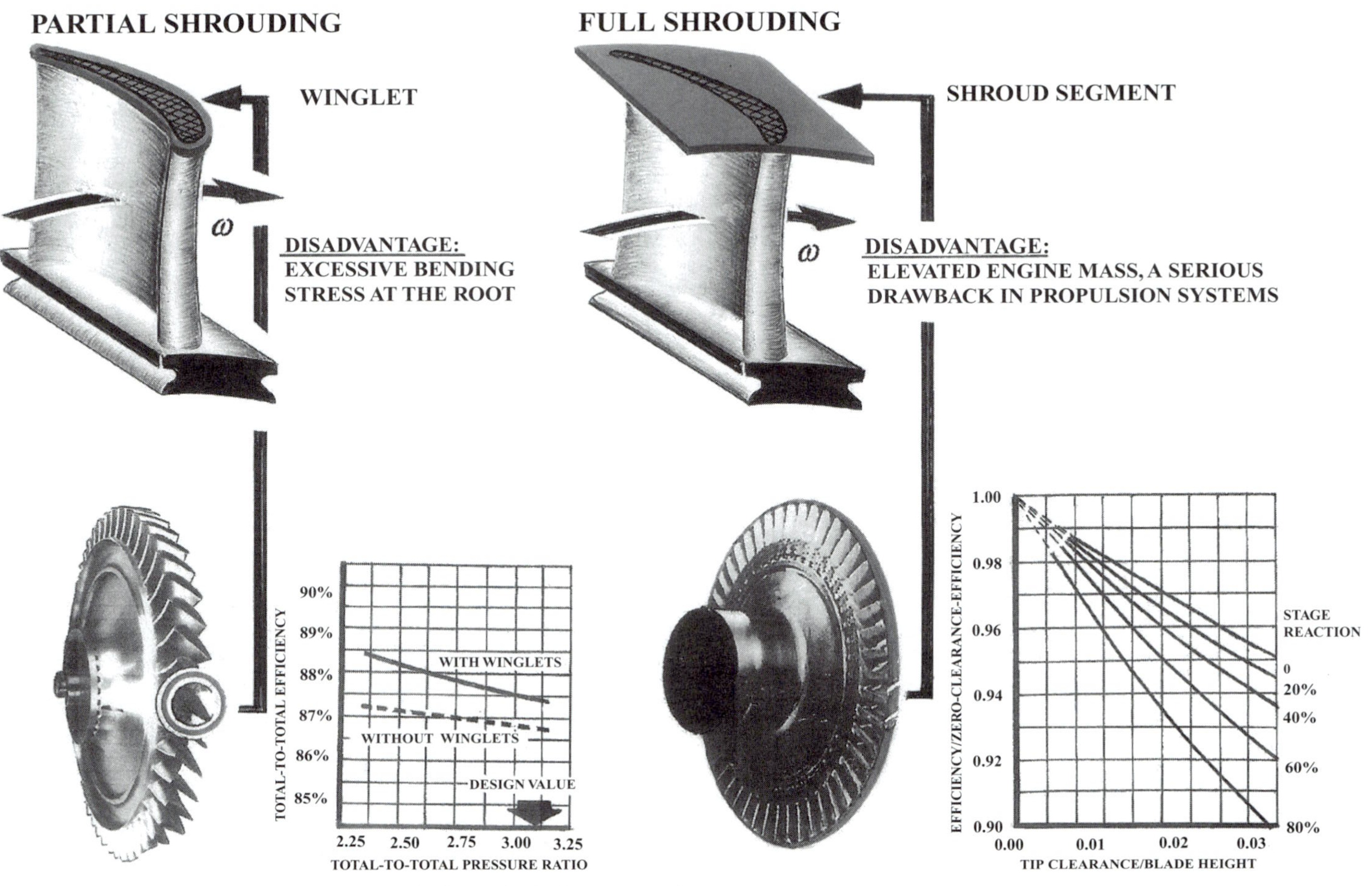

Figure 8.30. Partial and full shrouding for indirect tip-leakage control.

configuration is the full-shroud option in the figure. This is simply an ideally thin cylindrical sheet that is wrapped around all tips of the rotor blades. Partial shrouding, on the other hand, is where a smaller sheet of metal is attached to each blade tip, as seen in the same figure. This technique was devised specifically to reduce the weight of what would otherwise be a full shroud. Although the aerodynamic outcomes of both techniques might be very similar, the increase in the root bending stress with partial shrouding may come close to outweighing the weight-saving benefit. Yet many designers have resorted to this particular option, and the example provided in Figure 8.30 is one of many. The partially shrouded rotor in this Figure is the outcome of a technology demonstration program abbreviated as LART (Low Aspect Ratio Turbine). Funded by NASA-Lewis, this research, fabrication, and testing program was aimed at and indeed produced, a highly efficient single-stage turbine in the early and mid-1970s.

Empirical Correlation

Given a "baseline" turbine efficiency (η_{ref}), our objective here is to find a tip-leakage-related multiplier in order to arrive at a more realistic magnitude. Out of many existing means of doing so, the following expression is chosen for its simplicity:

$$\frac{\eta}{\eta_{ref}} = 1 - K\left(\frac{\delta}{h}\right)\left(\frac{r_t}{r_m}\right) \tag{8.11}$$

where the coefficient K is defined as

$$K = 1 + 0.586\left({\psi_{Ztip}}^{3.63}\right)$$

where

η is the rotor efficiency with the tip-clearance effects included,
δ is the tip-clearance height,
h is the blade span, and
ψ_{Ztip} is the tip Zweifel loading coefficient (defined later).

Reynolds Number Effect

Turbine blades are affected by the Reynolds number, based on the approach velocity and the length of the mean camber line (Fig. 8.23), in a fashion similar to that of flat plates. There is a critical Reynolds number magnitude of about 2×10^5 (based on the mean chord and approach velocity) above which the viscosity-related loss coefficient is approximately constant. Note that the Reynolds number is basically the ratio between the inertia and viscosity forces within the flow field. A low Reynolds number in this sense is indicative of viscosity domination. In this case, the flow field is highly sensitive to even slight Reynolds number changes. Fortunately, such a situation is virtually nonexistent under a turbine normal operating modes.

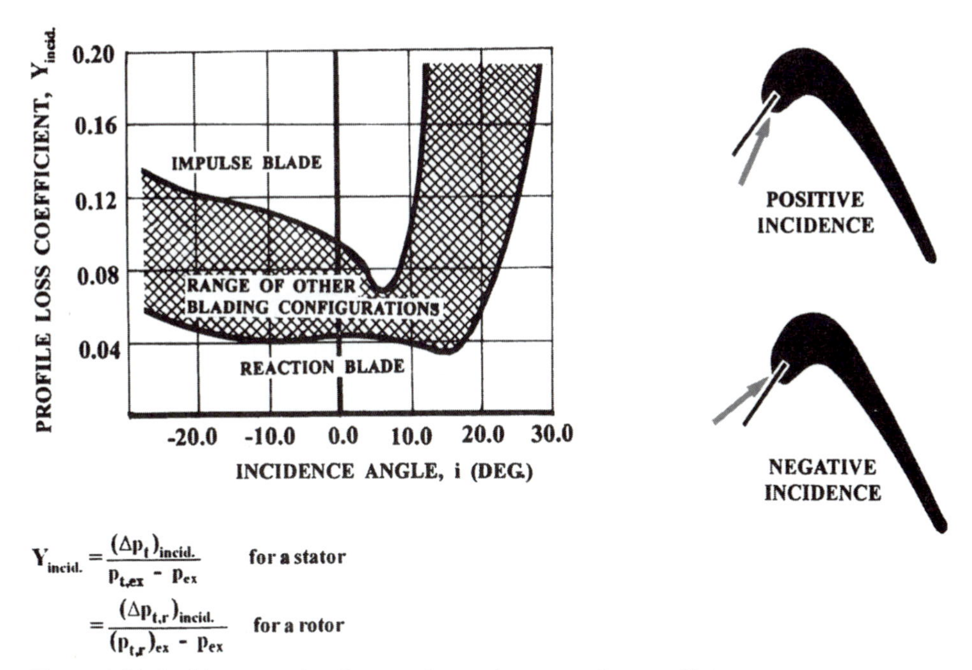

Figure 8.31. Incidence-angle effect on the total-pressure loss coefficient.

Incidence-Angle Effect

The incidence angle has a pronounced effect on the total (or total-relative) pressure in the vicinity of the stator vane (or rotor blade) leading edge and further downstream. Referring to the pure reaction-blading type in Figure 8.31, the incidence angle can vary from -15° to $+15^\circ$ with no appreciable increase in the total-pressure loss coefficient, Y_p (defined in the figure). The impulse (zero-reaction) blading, on the other hand, gives rise to a typically higher loss level within the same incidence range over the positive incidence-angle segment of this range. In general, negative incidence angles will normally produce moderate incidence losses. However, large positive incidence angles (by reference to Figure 8.31) are typically associated with large-scale losses.

The flow/leading-edge interaction mechanism is a strong function of the leading edge thickness. As previously indicated, a thick leading edge would yield a slightly high incidence loss at the design point but would be significantly tolerant to incidence-angle changes under off-design operating modes. A thin leading edge, by comparison, will be particularly suited for a given set of operating conditions, usually those associated with the design point. Deviations from the optimum magnitude, however, would give rise to greatly elevated losses.

Belonging, in effect, to the thin leading-edge category is a relatively rare type where the leading-edge is elliptic. An example utilizing this unconventional configuration is shown in Figure 8.32. The figure shows two spanwise locations within the same blade, where the 80% span section "had" to have an elliptic leading edge, as

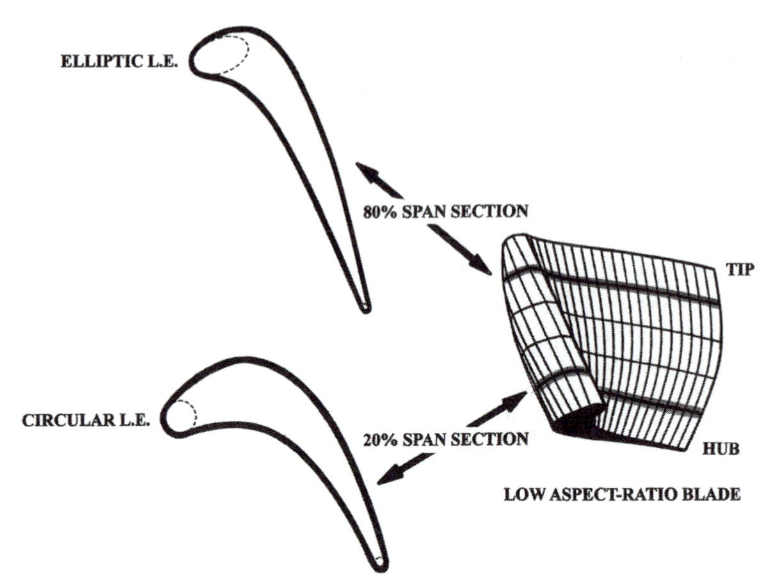

Figure 8.32. Circular versus elliptic leading-edge in axial-turbine blades.

several thickness/wedge-angle combinations were proved unsuccessful. Although an elliptic leading edge narrows down the "safe" incidence-angle range, it does offer a geometrical advantage, for it is, at least potentially, capable of improving the leading-edge loading characteristics. To explain this, note that a circular leading edge suffers from a curvature discontinuity at the two "tangency" points where the circular arc merges with the rest of the airfoil contour. Depending on other leading-edge variables (e.g., the wedge angle in Figure 8.23), this leads to a local "dip" in the velocity distribution over the suction side, which may cause a rather early flow separation. An ellipse, on the other hand, can be tailored to match the airfoil contour curvature at the two tangency points on the suction and pressure sides. A summary of a typical incidence/leading-edge interaction is shown in Figure 8.33, with the elliptic and thin leading edges lumped as one category. The figure emphasizes the sensitivity of such a category to relatively minor changes in the incidence angle.

Figure 8.34 shows two closely similar leading-edge regions obtained for the same blade section, at approximately the 70% span location, with the thinner-region configuration being that in the blade final design. As shown in the blade aerodynamic loading (in the same figure), the thinning adjustment produced a monotonically accelerating flow over the suction side. This has the effect of suppressing the boundary-layer buildup and minimizing the friction losses.

Suction-Side Flow Diffusion

Diffusion, in turbine-cascade terminology, is the effect associated with the flow deceleration over the airfoil suction side (Fig. 8.35). As indicated in Chapter 4,

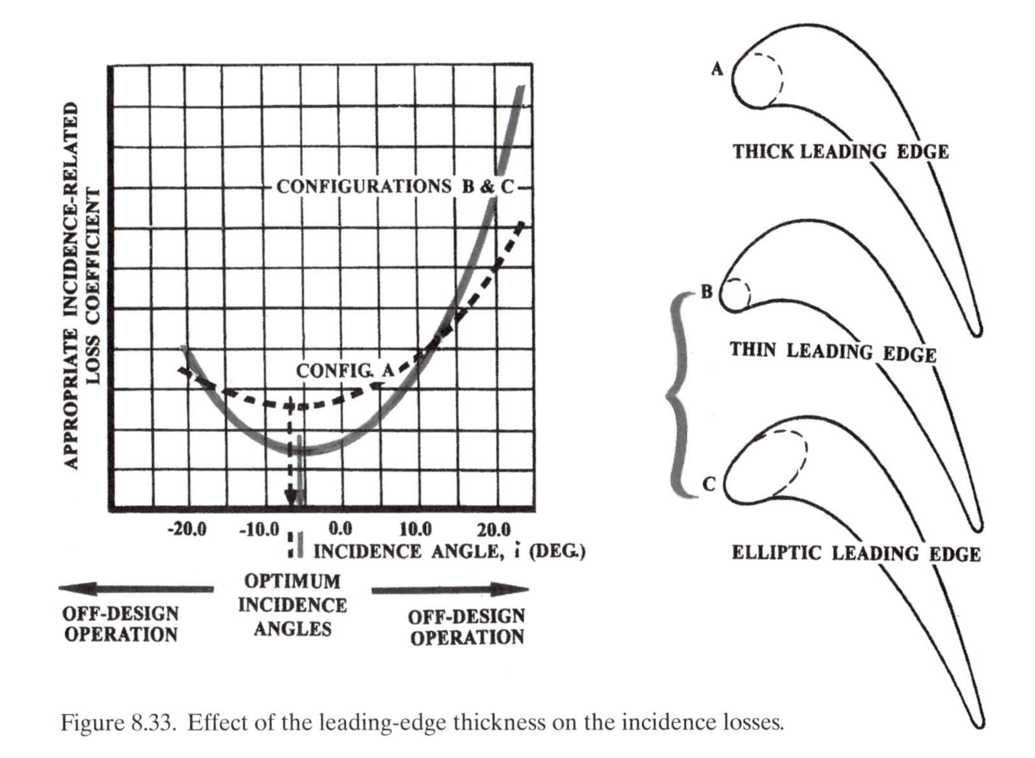

Figure 8.33. Effect of the leading-edge thickness on the incidence losses.

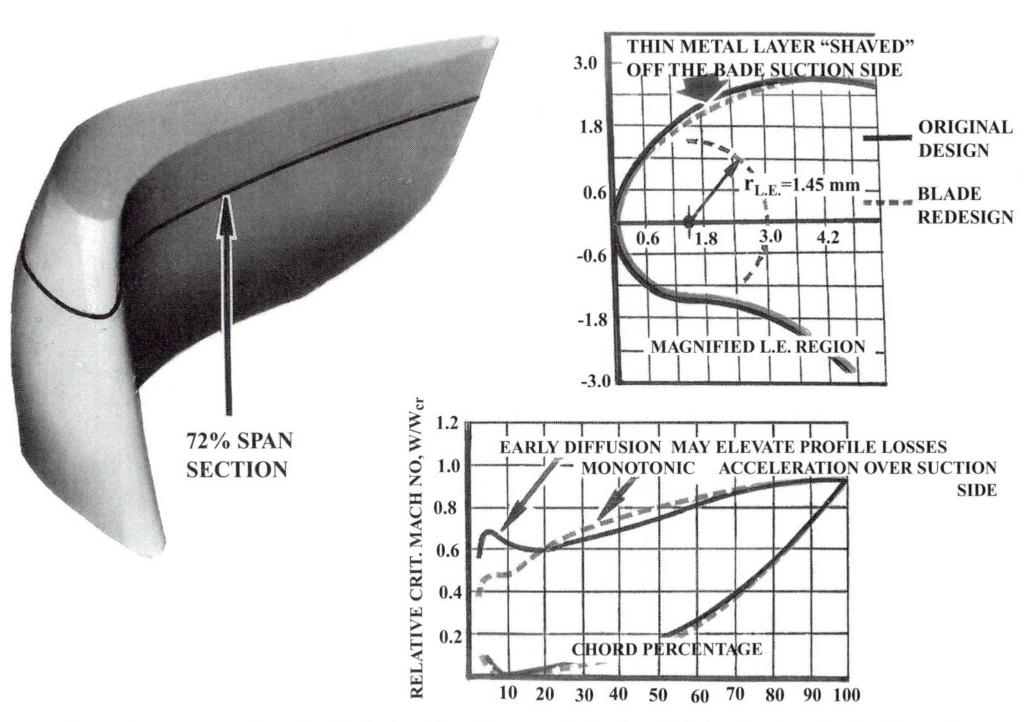

Figure 8.34. Improving the blade loading characteristics by slightly altering the leading-edge segment.

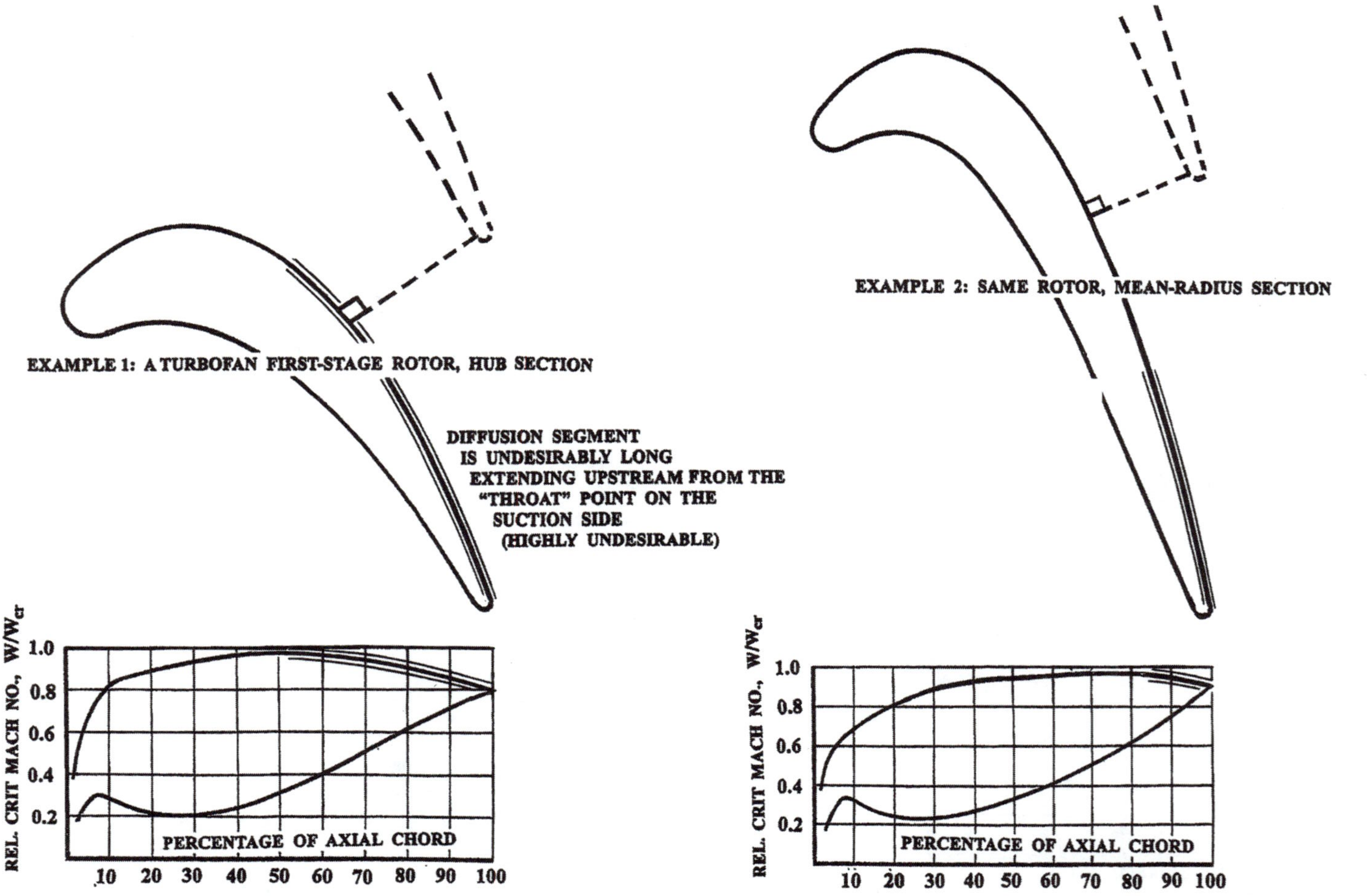

Figure 8.35. Diffusion over the suction side of a typical turbine airfoil.

this creates an unfavorable (or adverse) pressure gradient. Such an environment would locally aggravate the suction-side boundary-layer buildup and possibly lead to a costly flow separation. It is normally the uncovered-flow turning segment that is usually subject to such flow behavior. This segment, by reference to Figure 8.23, extends from the "geometric" throat point (on the suction side) down to the trailing edge. Depending on the blade geometry, however, the suction-side diffusion segment (marked in the figure) can start upstream or downstream from this point.

There is an upper limit on how much suction-side diffusion is tolerated for a good airfoil-cascade design. The so-called diffusion factor $\bar{\lambda}$ is defined as follows:

a) *The case of a stator*

$$\bar{\lambda} = \frac{(p/p_t)_{ex} - (p/p_t)_{maxV}}{1 - (p/p_t)_{maxV}} \tag{8.12}$$

b) *The case of a rotor*

$$\bar{\lambda} = \frac{(p/p_{t_r})_{ex} - (p/p_{t_r})_{maxW}}{1 - (p/p_{t_r})_{maxW}} \tag{8.13}$$

where p_t and p_{t_r} are the absolute and relative total pressures, respectively. The rule of thumb in the turbomachinery industry is for $\bar{\lambda}$ not to exceed 0.25.

Worth noting is the fact that local diffusion normally occurs just downstream from the leading edge on the pressure side as well, as a result of curvature discontinuity. However, this is typically followed by a strong and monotonic flow acceleration (Fig. 8.35). Therefore, even in the event of boundary-layer separation in this contour-merging region, reattachment can soon follow.

Location of the Front Stagnation Point

Depending on the incidence angle and the leading-edge geometry, the front stagnation point could be substantially far from its ideal location (point P in Figure 8.23). Figure 8.36 shows two spanwise locations on a turbine blade, where the front stagnation point associated with the section marked B is on the pressure side and far from its ideal location. In this case, Figure 8.36 shows a reversed-flow stream over a segment of the pressure side and around the leading edge on its way to the suction side. As the flow travels around the leading edge, the excessive boundary-layer growth, added to the high turning angle, may lead to early flow separation on the suction side. The resulting flow-recirculation region in this case is one where a significant amount of kinetic energy is wasted and total relative pressure suffers an early loss. Such a "lossy" flow structure could potentially extend to plague the entire suction side unless a flow reattachment soon occurs, which is the case in Figure 8.36.

In order to avoid such premature flow separation, the leading edge must be constructed in such a way as to give rise to a slightly negative incidence angle. Other remedies include a thicker leading edge and/or a larger leading-edge wedge angle.

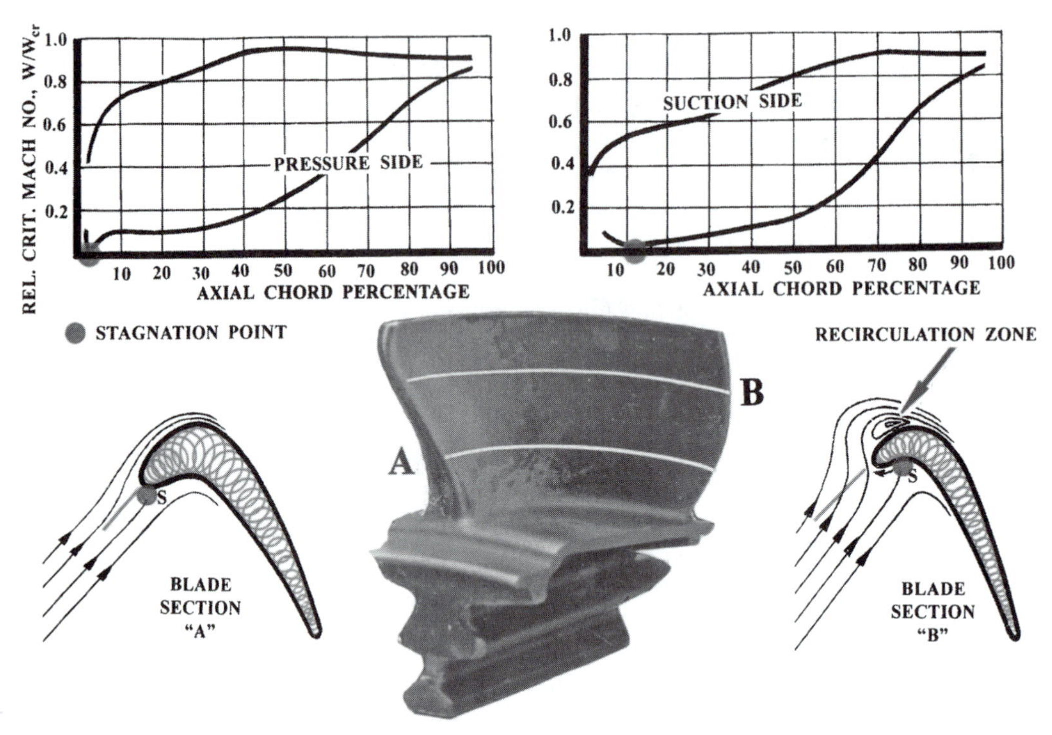

Figure 8.36. Aerodynamic consequences of the stagnation-point location.

Trailing-Edge Thickness

In the immediate vicinity of the trailing edge, two flow streams, with two different boundary layers, merge together. It is also at this location that a sudden flow-deceleration, due to a cross-flow area enlargement, takes place. Referring to Figure 8.37, the flow area just inside the trailing edge is comparatively smaller because of the trailing-edge thickness $t_{t.e.}$ and the so-called displacement thicknesses (δ_s and δ_p) on the suction and pressure sides, respectively. As represented in Figure 3.21, the latter two thicknesses reflect one effect of a typical boundary layer, namely the "viscous thickening" of the trailing edge. These passage-contraction elements obviously vanish immediately downstream from the trailing edge.

The trailing-edge "mixing" losses just outside the trailing-edge plane are in part a result of this sudden area enlargement. Quantitatively, the total pressure loss is inversely proportional to the factor f, where

$$f = \frac{S\cos\alpha_{ex} - t_{t.e.} - (\delta_s + \delta_p)}{S} \tag{8.14}$$

and where S is the cascade pitch (Fig. 8.23), α_{ex} is the airfoil exit angle, $t_{t.e.}$ is the trailing edge thickness, and δ_s and δ_p are the suction- and pressure-side displacement thicknesses, respectively as shown in Fig. 8.37.

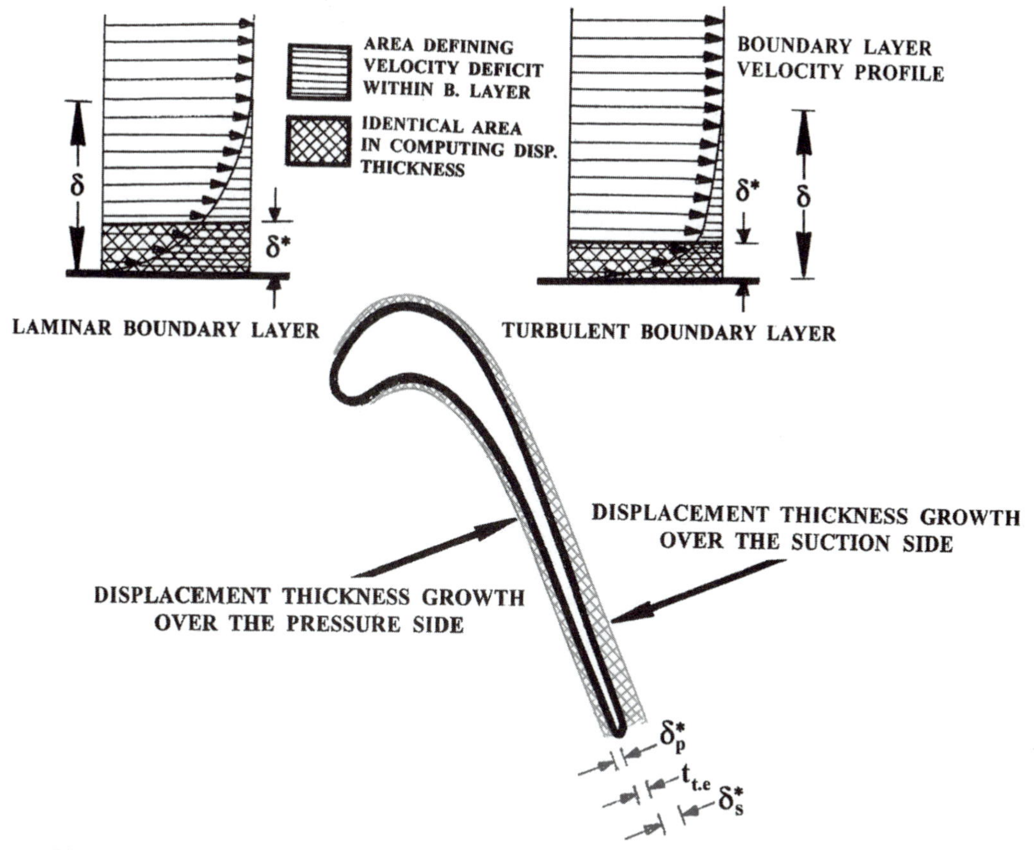

Figure 8.37. Displacement-thickness growth over the blade surface.

Of the preceding variables, one of the independent variables that can easily be altered during the design process is the physical trailing-edge thickness. As mentioned earlier, the ideal trailing-edge region is composed of coincident suction- and pressure-side segments, constituting what is termed a "cusp" trailing-edge configuration. Nevertheless, this zero-thickness trailing-edge segment would not withstand what would be an almost certain mechanical rupture. The next best option is to minimize the airfoil thickness as the trailing edge is approached. As for increasing the cascade pitch, by decreasing the blade count, the decision itself has to be in light of the bigger picture, namely the overall aerodynamic loading.

Design-Oriented Empirical Correlations

Implied Stagger Angle

In the process of creating the airfoil cascade, only the flow inlet and exit angles are initially known. These can be used to estimate the airfoil (or metal) angles, including

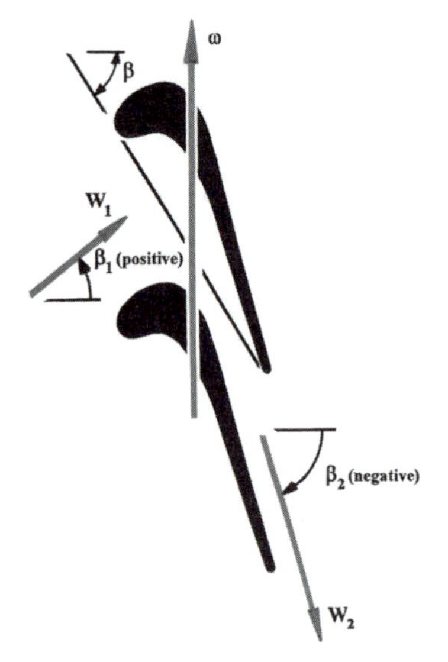

Figure 8.38. Variables needed to compute the implied stagger angle.

the option where the two sets of angles are simply equated. These airfoil angles, however, provide little help in constructing the airfoil, for there exists a theoretically infinite number of airfoils that will accommodate these airfoil-angle constraints.

A good starting point is to calculate what is referred to as the "implied" stagger angle. The term *stagger angle* is defined in Figure 8.23, and the qualifier "implied" refers to the fact that this angle is a function of (only) the flow inlet and exit angles and will in no way represent the final stagger angle, as the latter is a function of the final camberline, and its axial extension. Referring to Figure 8.38, for a rotor blade, a rough estimate of the implied stagger angle β is computed as

$$\beta = 0.95\left[\arctan\left(\frac{\tan\beta_1 - \tan\beta_2}{2}\right)\right] + 5.0^\circ \tag{8.15}$$

This angle makes the true chord angle unique and aids in constructing the mean camberline at the earliest detailed-design step.

The implied stagger angle β can also be used as a diagnostic tool in verifying the blade smoothness in the spanwise direction. This is illustrated in Figure 8.39, where two versions of the same rotor blade are evaluated in light of the spanwise distribution of β. The version of the blade with the suction-side bump in this figure has a β distribution that clearly suffers a slope discontinuity near the blade tip.

Optimum Blade Count (Zweifel's Loading Criterion)

Referring to the rotor cascade in Figure 8.40, let us define the airfoil-cascade "solidity" ratio, σ, as

$$\sigma = \frac{c_{zm}}{S_m} \tag{8.16}$$

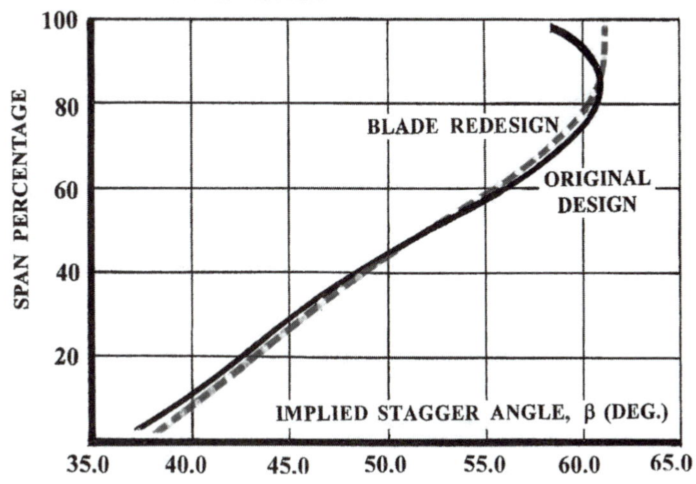

Figure 8.39. Implied stagger angle as a diagnostic tool to detect geometrical irregularities.

where the variables $c_{z,m}$ and S_m are the mean-radius magnitudes of the airfoil axial chord and cascade pitch, respectively. The term *solidity* here is perhaps understandable in the sense that the lower the pitch, the more the blade cascade gets to appear, in the limit, more like a solid wall to the incoming flow stream. Zweifel's loading coefficient ψ_Z (defined next) combines the cascade geometry (represented by σ) and the aerodynamic loading (represented by the flow-deflection angle) as

$$(\psi_Z)_{rotor} = \frac{2}{\sigma}\cos^2\beta_2[\tan\beta_1 - \tan\beta_2] \tag{8.17}$$

with β_1 and β_2 being the flow inlet and exit angles, respectively (note that β_2 will always be negative). Based on an optimum ψ_z value of approximately 0.8, the optimum solidity ratio σ_{opt} can be calculated. Once the mean value of the axial chord is known (or assumed), the definition of the solidity ratio will yield the optimum mean-radius blade-to-blade spacing S (Fig. 8.40). The corresponding optimum number of blades N_b can then be computed as

$$N_b = \frac{2\pi r_m}{S_m} \tag{8.18}$$

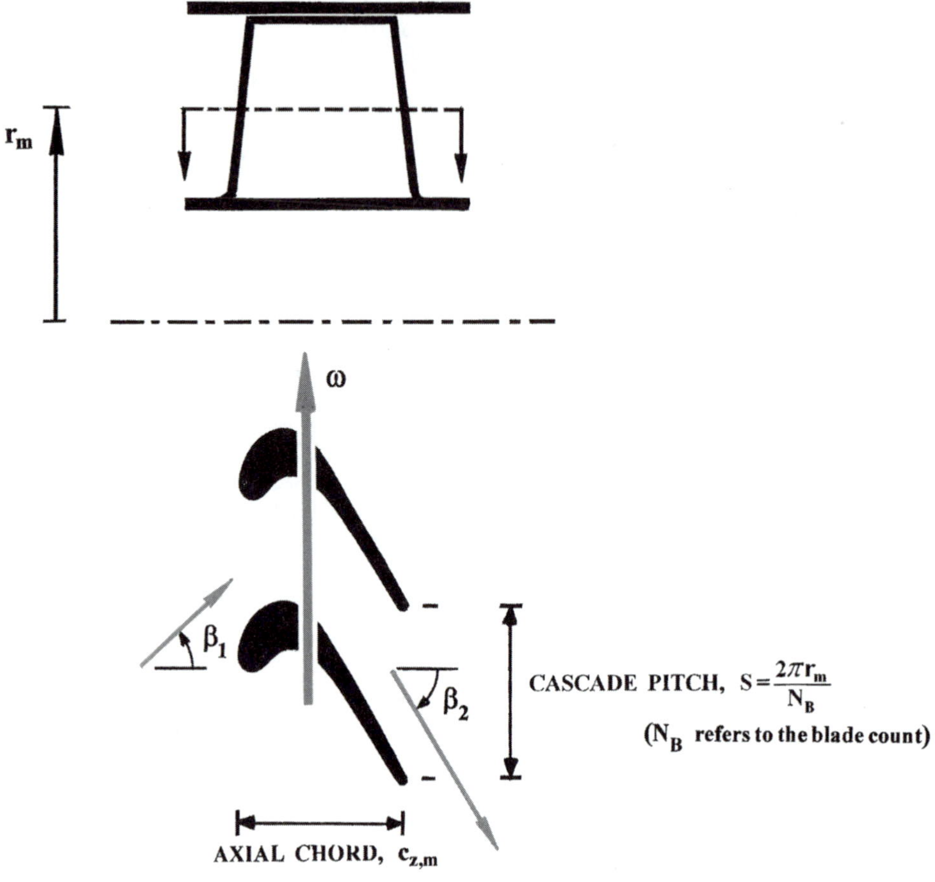

Figure 8.40. Variables needed to compute Zweifel's loading coefficient.

The same criterion can be utilized to calculate the optimum number of stator vanes. In this case, Zweifel's loading coefficient is a function of the inlet and exit "absolute" flow angles as follows:

$$(\psi_Z)_{stator} = \frac{2}{\sigma}\cos^2\alpha_1[\tan\alpha_1 - \tan\alpha_0] \tag{8.19}$$

with α_0 and α_1 being the stator inlet and exit flow angles, respectively. The optimum number of vanes can then be similarly computed.

Stacking of the Vane and Blade Airfoil Sections

As indicated earlier, each of the blade airfoil sections are designed on a two-dimensional basis. The result is a set of sections at an adequate number of hub-to-tip radial locations.

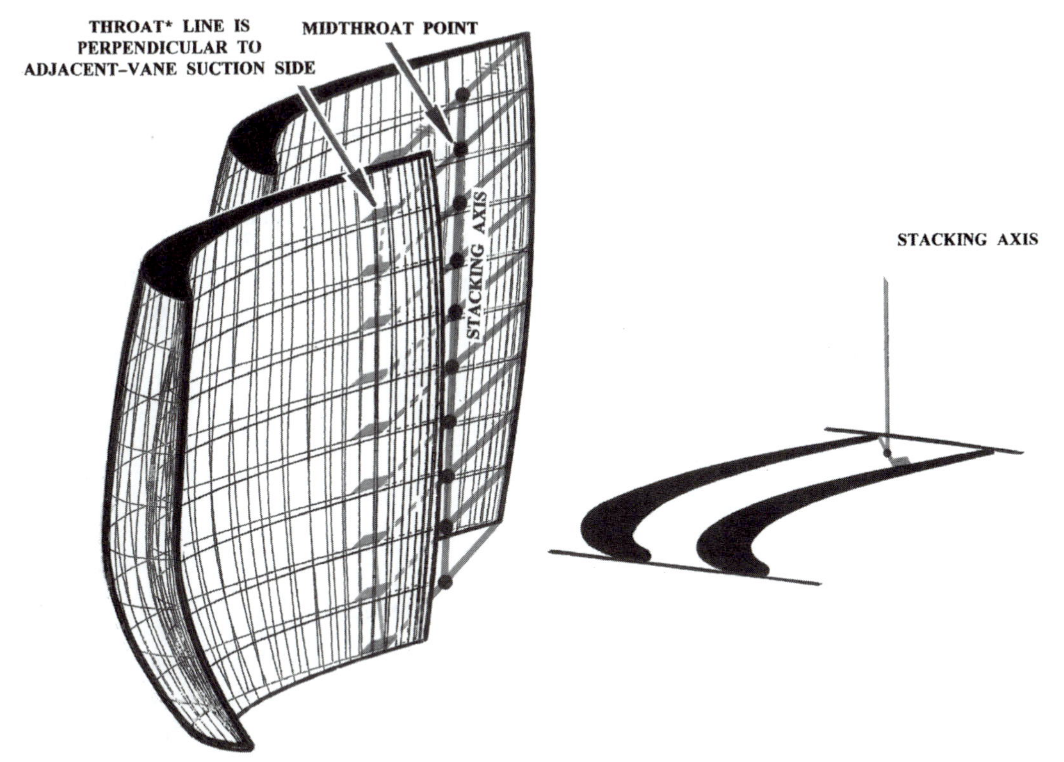

Figure 8.41. Midthroat stacking of the stator cross sections. *The term *throat* here does not embrace the aerodynamic condition, namely an M_{cr} of unity. It is rather the term referring to the position of minimum blade-to-blade cross-flow area in the gas-turbine industry community.

The next step is to "stack" these sections up, in order to create the stator vane or rotor blade, whichever is the case. Experience has shown that even with perfectly designed two-dimensional airfoils, the span-wise geometrical interaction among these sections may lead to an irregular three-dimensional body.

An important variable in this context is known as the stacking axis. By definition, this is a straight line that passes through the turbine axis of rotation as well as a set of corresponding airfoil points. Traditionally, the stacking axis of a stator vane is chosen to pass through the midthroat points associated with all of the cascade sections. Note that the term *throat* here may be aerodynamically wrong, for it implies a Mach number of unity, which might not be the case. Use of this term here merely identifies the shortest cross-flow line in the passage bound by two successive airfoils. This stacking pattern is shown in Figure 8.41, where the throat plane is seen to be symmetrical as a result. The symmetry of this plane has reportedly given rise to a minimum total pressure loss over the entire plane and slightly upstream of it.

In the case of a rotor blade, the stacking axis is typically chosen to pass through all centroids (centers of gravity) of the blade sections (Fig. 8.42). This stacking pattern is based solely on a mechanical-integrity foundation.

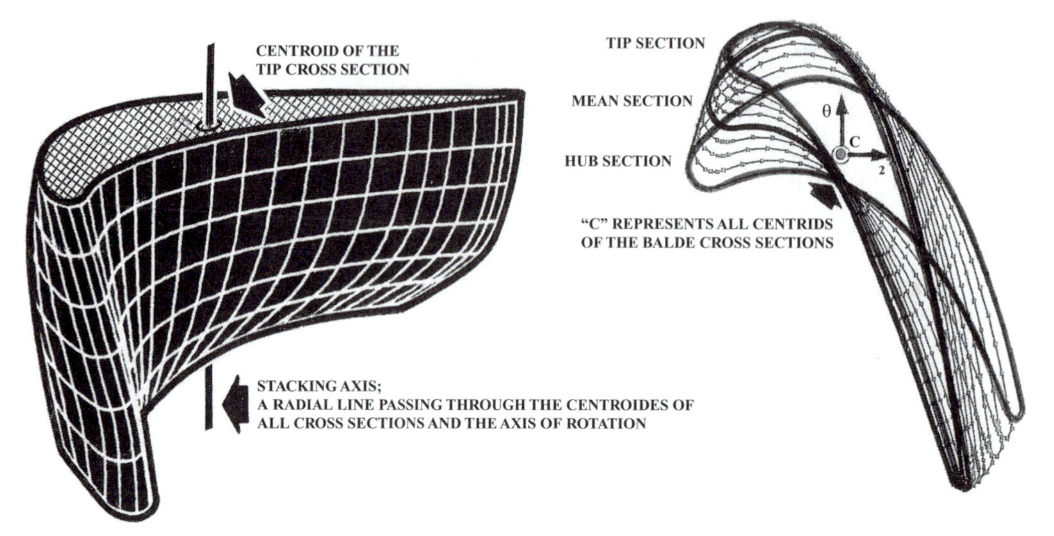

Figure 8.42. A typical rotor blade with c.g. stacking of the cross sections.

Shaft-Work Extraction in Low-Aspect-Ratio Blades

Based on its average dimensions, the aspect ratio $[(c_z/h)_{av.}]$ of the blade in Figure 8.43 is 1.01. Note that aspect ratios below unity are virtually intolerable because of what, in this case, would be excessive amounts of tip leakage and near-tip unloading. With the short span in a blade of this nature, the boundary-layer interaction over the blade and the hub surfaces adds to an already lossy flow field.

These loss spots in low aspect-ratio blades make it sensible to drop as reliable both the hub and tip sections in the shaft-work extraction process. The blade in Figure 8.43 offers a typical means of achieving just that. Examination of the mean section there reveals that the blade downstream segment is bent in such a way as to maximize the airfoil (or metal) exit angle. The result in this case is the largest magnitude of flow-deflection angle (δ) belonging to the mean section. This makes it the greatest contributor to the overall shaft work produced by all of the blade sections. In fact, the spanwise distribution of shaft work in this example is parabolic, with the maximum magnitude associated with the middle blade section.

The Supersonic Stator Option

One way of maximizing the turbine stage output power is to achieve a substantially high stator-exit kinetic energy. To this end, the stator vane-to-vane passage can be shaped like a converging-diverging De Laval nozzle, giving rise to a supersonic exit flow stream. An example of such a stator design is shown in Figure 8.44. The figure shows the two means of creating the minimum-area throat section across which the flow field proceeds to be supersonic. These include a special endwall contouring as well as a special vane-to-vane cross section that resembles a converging-diverging nozzle. Note the smooth area transition across the throat section, which is necessary for the subsonic-to-supersonic flow transition to occur.

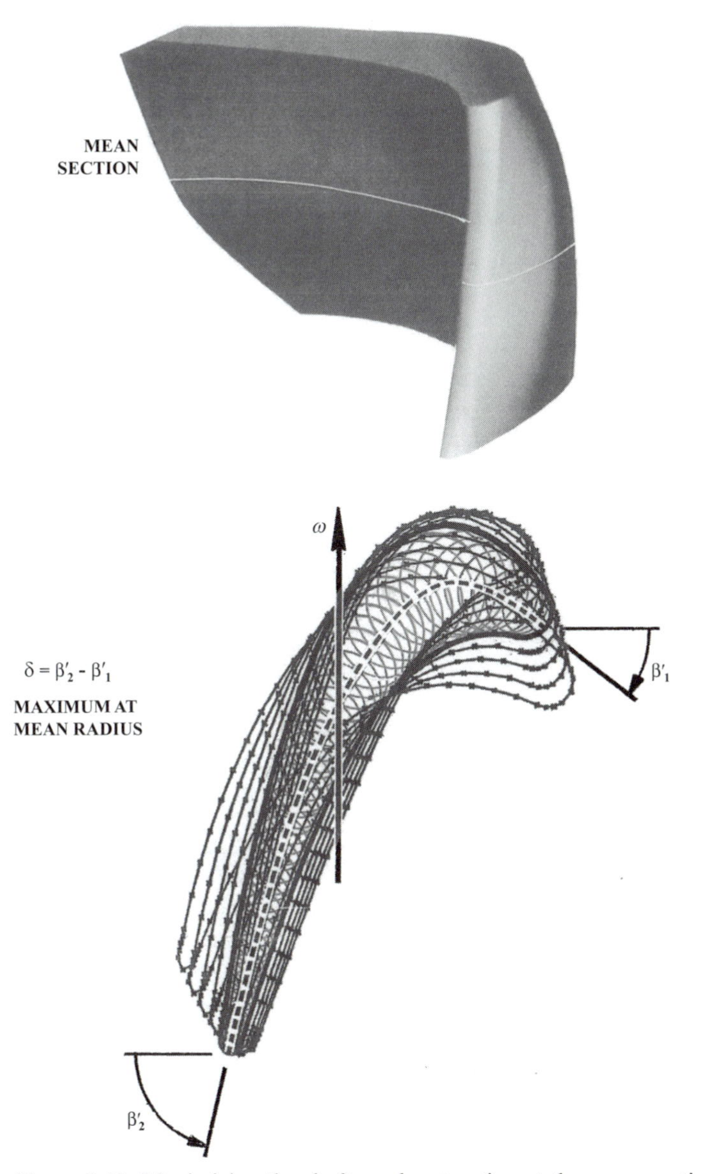

Figure 8.43. Maximizing the shaft-work extraction at the mean section.

An obvious drawback of the stator design in Figure 8.44 is the inability to produce a supersonic exit flow stream under some off-design operation modes. The different inlet total properties under such operation modes may very well require a different (smaller) throat-plane area. In a fixed-geometry stator, the inability to produce sonic conditions at the minimum-area location will result in a venturi-meter type of stator flow passage, with the diverging segment acting as a flow decelerator. The stator-exit flow under such circumstances would possess a significantly low Mach number, defeating the initial purpose of maximizing the stator exit velocity.

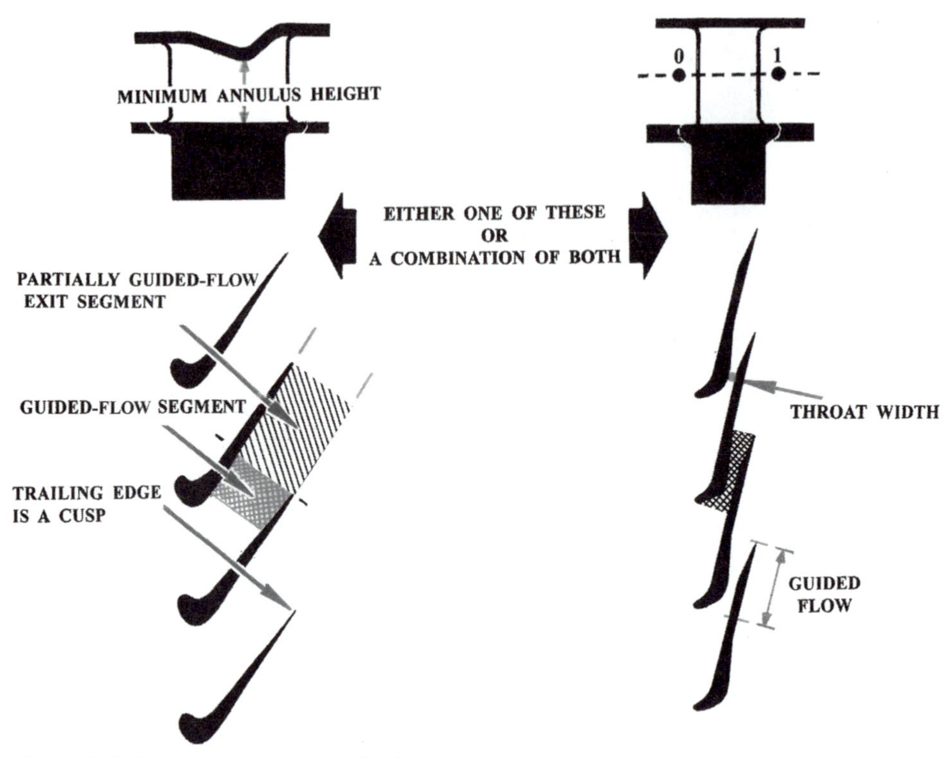

Figure 8.44. The supersonic stator-discharge option.

EXAMPLE 3

Consider the adiabatic subsonic-supersonic stator shown in Figure 8.45. At the design point, the stator operating conditions are as follows:

- Inlet total pressure $p_{t0} = 12.3$ bars
- Inlet total temperature $T_{t0} = 1452$ K
- Shaft speed $N = 46{,}000$ rpm
- Mass-flow rate $\dot{m} = 6.44$ kg/s
- Stator-exit critical Mach number $V_1/V_{cr1} = 2.35$
- Stator-exit (absolute) flow angle $\alpha_1 = 79°$

The stator mean radius r_m is 0.325 m. Furthermore, the following simplifications are applicable:

- Constant axial-velocity component (V_z)
- Isentropic flow between the throat and stator-exit plane
- An average specific heat ratio γ of 1.32

a) Calculate the total-pressure loss percentage.
b) Determine whether the rotor will receive a supersonic flow stream.
c) Repeat item "b" upon changing the shaft speed to 28,000 rpm.

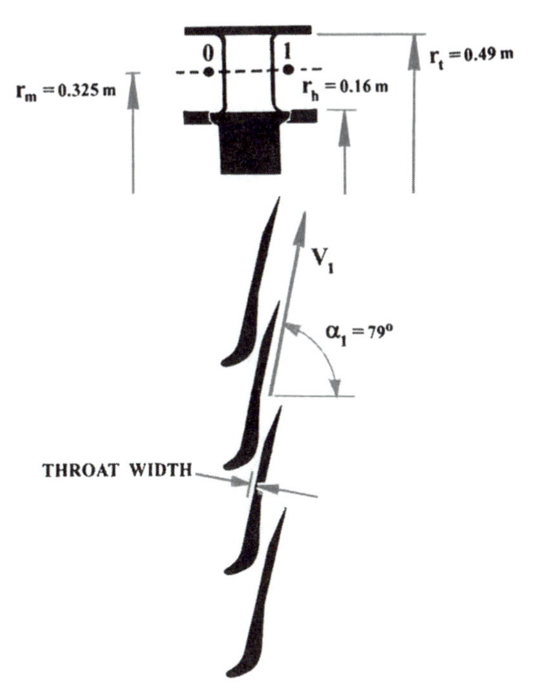

Figure 8.45. Input variables for Example 3.

SOLUTION

Proceeding with the station-designation pattern in Figure 8.45, let us add the superscript (*) to identify the thermophysical properties at the throat section.

Part a: Applying the continuity equation at the stator exit station, we have

$$\frac{\dot{m}\sqrt{T_{t1}}}{p_{t1}A_1} = \sqrt{\frac{2\gamma}{(\gamma+1)R}}M_{cr1}\left(1 - \frac{\gamma-1}{\gamma+1}{M_{cr1}}^2\right)^{\frac{1}{\gamma-1}}$$

where

$$\gamma = 1.32$$

$$T_{t1} = T_{t0} = 1452 \text{ K}$$

$$M_{cr1} = 2.35$$

$$\dot{m} = 6.44 \text{ kg/s}$$

$$A_1 = \pi\left({r_t}^2 - {r_h}^2\right)\cos\alpha_1 = 0.1286 \text{ m}^2$$

Direct substitution in the continuity equation (3.35) yields

$$p_{t1} = 11.4 \text{ bars}$$

Thus, the total-pressure loss percentage is

$$\frac{(p_{t0} - p_{t1})}{p_{t0}} = 7.3\%$$

Part b: At the stator exit station, we have

$$V_1 = M_{cr1} V_{cr1} = 1618.3 \text{ m/s}$$

$$V_{z1} = V_1 \cos\alpha_1 = 308.8 \text{ m/s}$$

$$V_{\theta 1} = V_1 \sin\alpha_1 = 1588.6 \text{ m/s}$$

$$U_m = \omega r_m = 1565.6 \text{ m/s}$$

$$W_{\theta 1} = V_{\theta 1} - U_m = 23.0 \text{ m/s}$$

$$W_1 = \sqrt{W_{\theta 1}{}^2 + V_{z1}{}^2} = 309.7 \text{ m/s}$$

$$T_{tr1} = T_{t1} - \frac{(V_1{}^2 - W_1{}^2)}{2c_p} = 386.5 \text{ K}$$

$$W_{cr1} = \sqrt{\left(\frac{2\gamma}{\gamma+1}\right) R T_{tr1}} = 355.3 \text{ m/s}$$

$$\frac{W_1}{W_{cr1}} = 0.87$$

In words, the last result simply means that an observer that is attached to the rotor blade will register an incoming flow stream that is subsonic, a statement that is relevant to the rotor flow field. A stationary observer, however, will monitor an "absolute" Mach number which is above 1.0 (actually 2.35), as the problem statement specifies.

In this case, the appropriate rotor-blade leading edge should be of the round (subsonic) type, as shown on the left-hand side of Figure 8.46.

Part c: Now we consider the lower shaft speed, namely 23,000 rpm:

$$U_m = 953.0 \text{ m/s}$$

$$W_{\theta 1} = V_{\theta 1} - U_m = 635.7 \text{ m/s}$$

$$W_1 = \sqrt{W_{\theta 1}{}^2 + V_z{}^2} = 706.7 \text{ m/s}$$

$$T_{tr1} = T_{t1} - \frac{(V_1{}^2 - W_1{}^2)}{2c_p} = 556.9 \text{ K}$$

$$W_{cr1} = \sqrt{\left(\frac{2\gamma}{\gamma+1}\right) R T_{tr1}} = 427.2 \text{ m/s}$$

Finally, we can compute the relative critical Mach number (W_1/W_{cr1}) at the rotor inlet station as follows:

$$\frac{W_1}{W_{cr1}} = 1.66$$

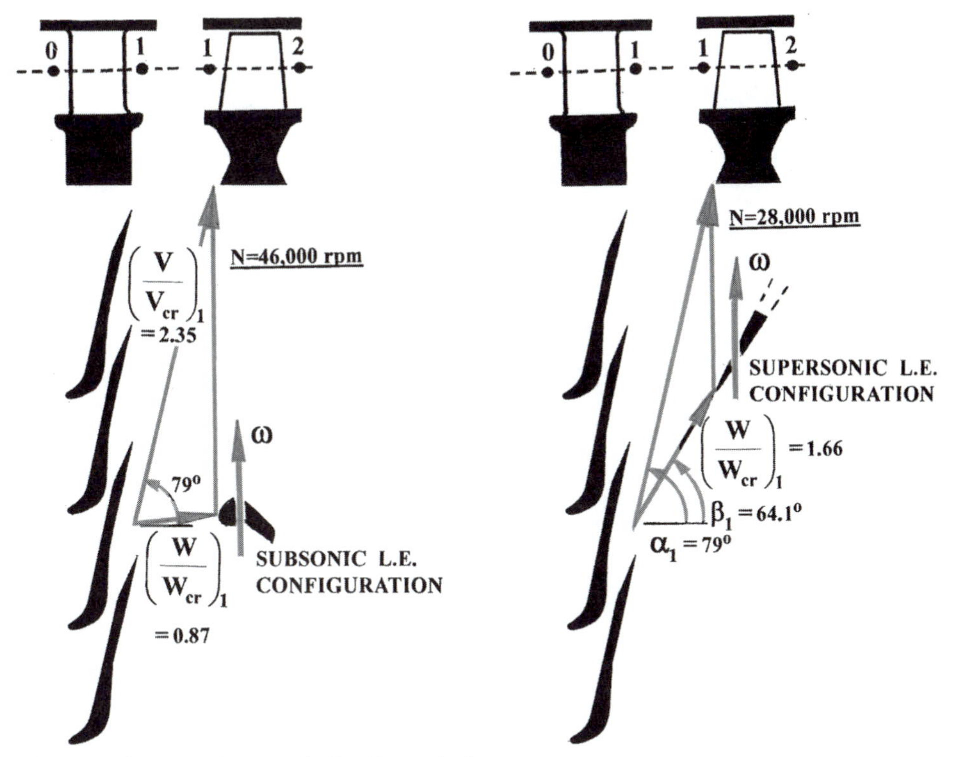

Figure 8.46. Numerical results for Example 3.

Again using the "observer" analogy, we can conclude that a rotating observer that is mounted on the blade, and spinning with it, will register a *relative* critical Mach number that is supersonic. The appropriate rotor-blade leading edge will therefore have to be of the wedge type, as shown on the right-hand side of Figure 8.46.

COMMENT

In solving the preceding problem, the point was made that despite the stator-exit critical Mach number, the rotor-inlet *relative* flow stream may be subsonic or supersonic, depending, in this case, on the shaft speed. The rotor leading-edge configurations that suit both relative-flow regimes are shown in Figure 8.46. Note, in particular, that a supersonic rotor blade will have to be wedge-shaped. A blunt circular-arc leading edge here would give rise to admission of a total relative pressure loss that is exceedingly high. In fact, such unfitting leading-edge configuration would convert the relative critical Mach number from supersonic to a subsonic magnitude, even before physically reaching the blunt leading edge. The reason is that a detached normal shock, under such circumstances, will prevail slightly upstream from the leading edge. This shock will abruptly reduce the initially supersonic relative flow stream into the subsonic range while causing a substantial total pressure loss.

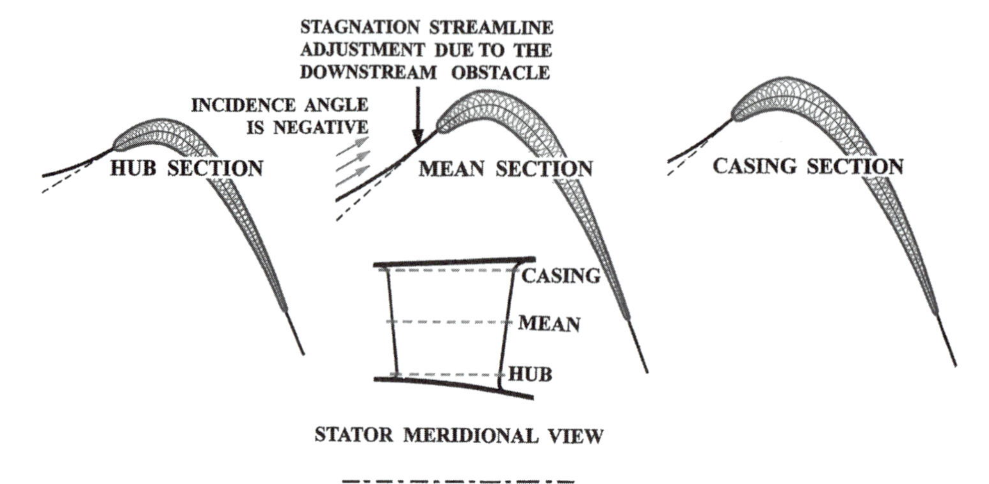

Figure 8.47. "Downstream effect" in a subsonic flow field.

Shape of the Stagnation Streamlines

Figure 8.47 shows the hub, mean, and casing sections of the second-stage stator in the auxiliary power unit GTCP-131, a product of Allied-Signal Aerospace Co. (Phoenix, Arizona). The incoming flow in this case is subsonic at all radial locations. The stagnation streamlines corresponding to the three vane sections are also shown as part of the output of a three-dimensional potential-flow computer code. These lines share one important characteristic as the airfoil leading edge is approached. In this and any subsonic flow field, the fluid particles will sense the presence of the airfoil prior to reaching it. This characteristic is simply nonexistent should the approaching flow stream be supersonic. In that case, a fluid particle will sense the presence of the lifting body only as it impacts it.

This stagnation streamlines' characteristic is important in a whole category of cascade-flow numerical modeling. Referring back to Figure 8.24, the tangentially bounding boundaries of the solution domain are horizontal (constant θ) straight lines. Such a choice has been traditional in numerical models which are based on an ideal potential flow behavior. With the advent of more realistic flow-analysis models, the boundary conditions imposed at these "periodic" boundaries became more complex. With this obvious choice of periodic boundaries, very fine resolution of the numerical model at and near these boundaries, especially close to the airfoil, became the norm. Under these circumstances, the curved stagnation streamlines upstream of the leading edge and downstream of the trailing edge (Fig. 8.47) have the effect of greatly reducing the numerical size of the model. With such a choice of periodic boundaries, we know that the velocity component perpendicular to the boundary is precisely zero, for the simple reason that the boundary itself is a streamline. This eliminates one of the "periodic" boundary conditions, leaving the analyst to equate the streamwise velocity component, as well as the pressure, at corresponding computational nodes on either pair of periodic boundaries. This is the reason that the execution of the full-blown numerical analysis is typically preceded by employing a usually

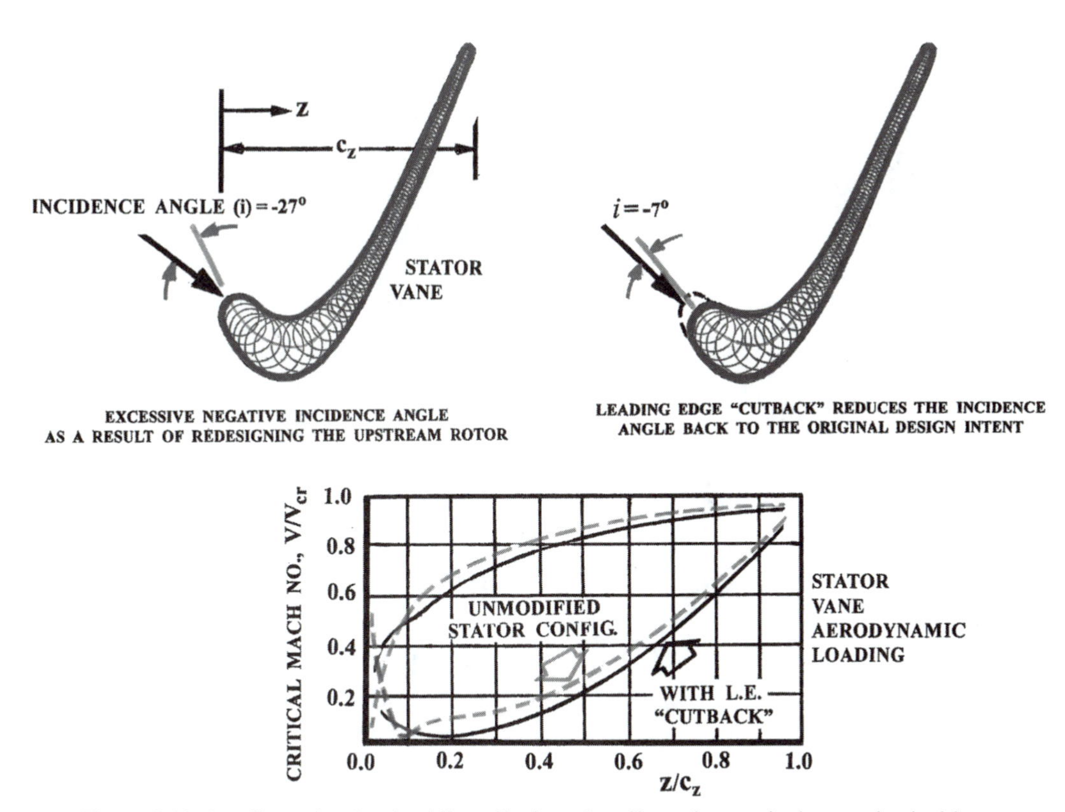

Figure 8.48. Leading-edge "cutback" to alleviate the effect of excessively negative incidence angles.

lower-order flow model for the sole purpose of being able to construct the stagnation streamlines. In fact, the stagnation streamlines in Figure 8.47 were numerically produced for a stator in an existing auxiliary power unit just for this very purpose.

Simple Component Adaptation Means

In the event of redesigning a poorly performing bladed component, it is natural to expect changes in the downstream component inlet conditions. Chief among these is the incidence angle, which, from experience, could reach real performance-degrading magnitudes. In most cases, the downstream component would not be under consideration for redesign because of any combination of excessive cost, extra effort, or time delay. Turbine designers nowadays are cleverly innovative in devising "quick fixes" to address problems of this nature, with virtually negligible additional costs. In the following, two of such methods are presented as examples.

Figure 8.48 shows one of the inexpensive solutions to alleviate a large negative incidence angle facing the second-stage stator because of redesigning the first-stage rotor. The stator-vane aerodynamic loading in the same figure displays the typical effects of an excessively high negative incidence, including early diffusion over the pressure side and virtually unloaded airfoil-front segment. The geometrical

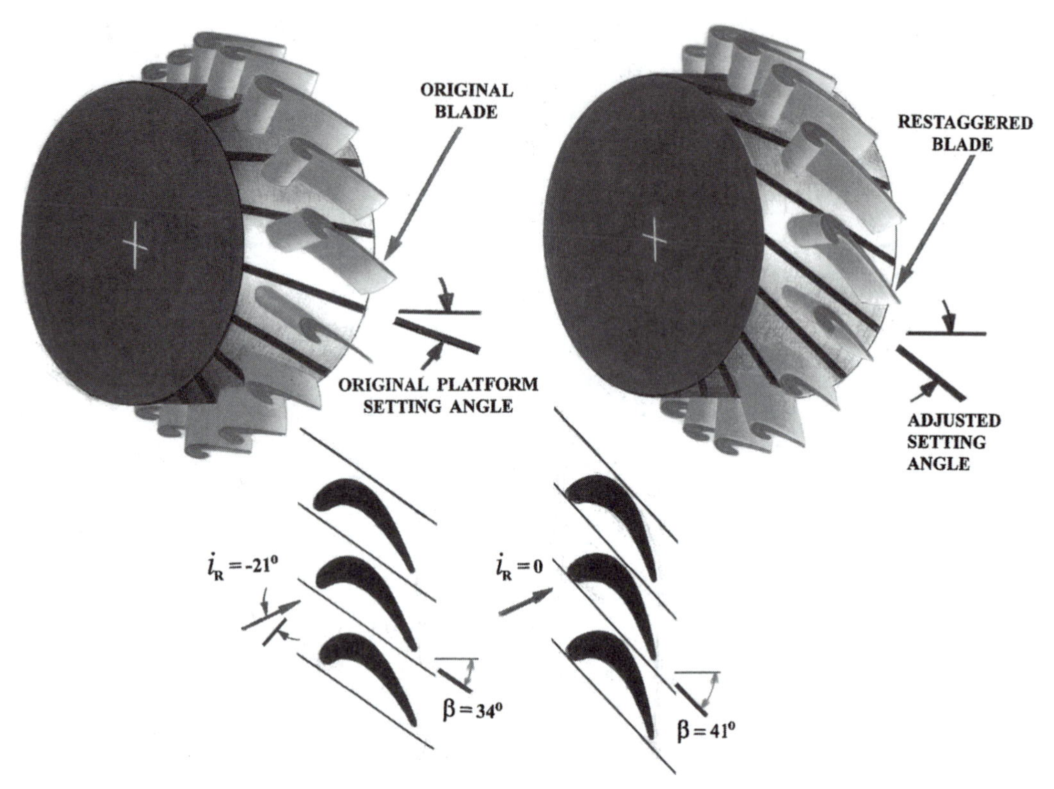

Figure 8.49. Restaggering the rotor blades by changing the "platform" setting angle.

second-stator modification in Figure 8.48 is commonly known as the leading-edge "cutback." In reality, this involves the creation of a different leading edge by "shaving off" a sufficient amount of the stator-vane material, to the point where the leading-edge mean camber-line tangent is acceptably close to the approaching-flow direction. Figure 8.48 also shows the favorable aerodynamic-loading effect that is associated with this airfoil modification.

Another way of controlling the damage arising from excessive magnitudes of incidence angle is termed airfoil *restaggering* (Fig. 8.49). The object here is a turbine-rotor cascade that is initially suffering from a large negative incidence as a result of redesigning the upstream stator. The blades of this rotor are of the individually inserted type, meaning blades which are manufactured as one unit with its "platform" segment. The process in the manufacturing phase is implemented by simply cutting the platforms at a different angle, as shown in Figure 8.49. Once assembled, the rotor blades will all be resituated in a process that is equivalent to rotating them in such a way to avoid what would have otherwise been an excessive incidence angle impact.

Hot-to-Cold Dimensions' Conversion

It is safe to assume that assembly of the turbine components takes place at room temperature. These components, however, are designed at the actual temperatures during real turbine operation, including an inlet magnitude which may very well

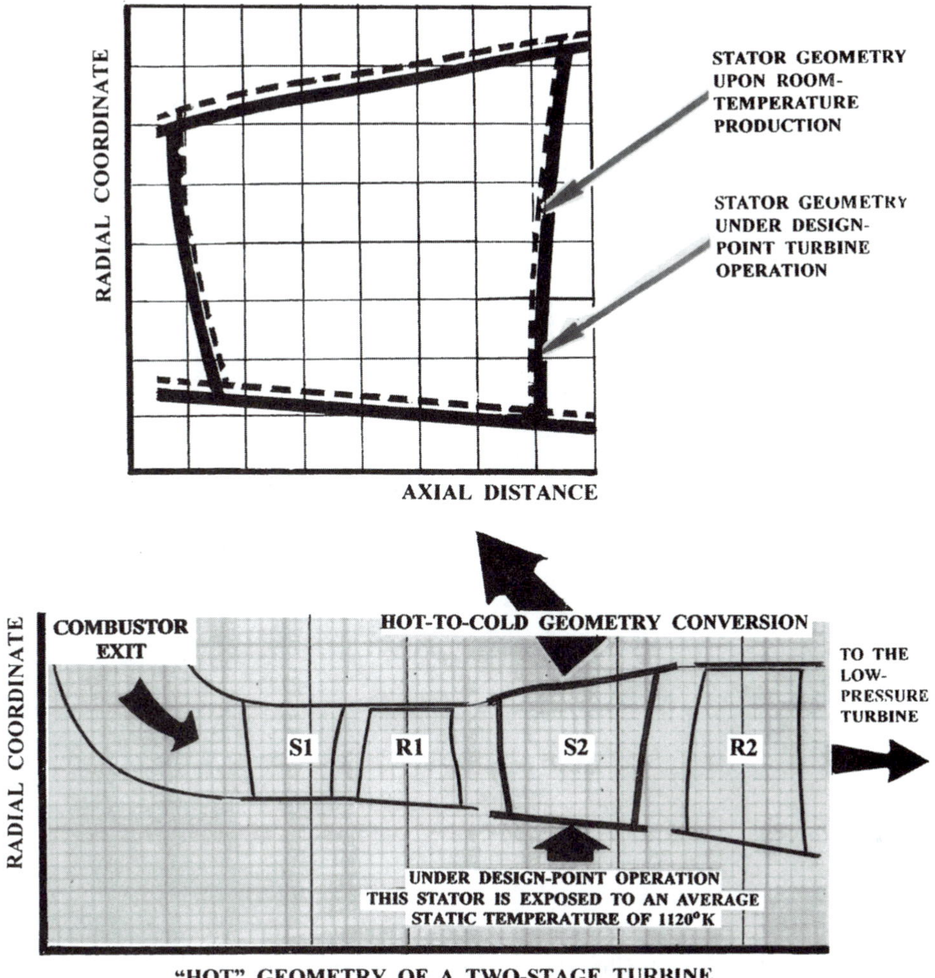

Figure 8.50. Hot and cold configurations of a stator vane.

exceed 1400 K. In view of this, the task of converting the "hot" into "cold" dimensions is unavoidable. Depending on the thermal properties of the metal and the flowing medium, there are computational means of identifying the changes of major dimensions during turbine operation.

Figures 8.50 and 8.51 present examples of the changes concerning a stator vane and a rotor blade, respectively. The two figures underscore the fact that the dimensions and shape changes in reality are much more than just fractions of a millimeter here and there. The changes in Figure 8.50 concern the second-stage stator in an existing auxiliary power unit (APU). While the leading- and trailing-edge lines in the meridional view remain virtually unchanged, the endwalls clearly undergo changes that are too significant to be ignored. The rotor blade in Figure 8.51 belongs to the low-pressure turbine section in a different auxiliary power unit. The numerical

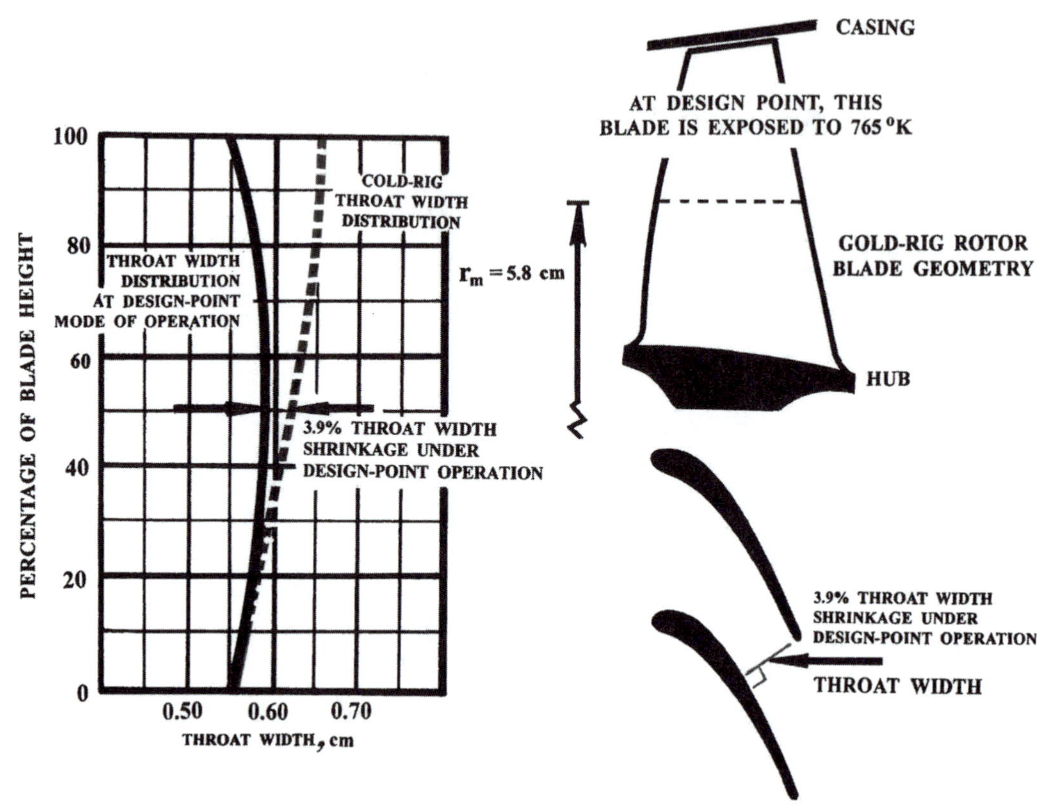

Figure 8.51. Effects of manufacturing and operational temperature change on the "geometric" throat width.

results shown in this figure prove that the dimension changes do include the critical dimension of the (geometric) throat width. Because an unshrouded blade is structurally a cantilever, the largest hot-to-cold throat-width change is at the tip section. Note that the change will also affect the uncovered segment of the blade suction side by changing the location of the suction-side "geometric" throat point on the suction side.

Cooling Flow Extraction and Path of Delivery

With the high-temperature environment of gas turbines, substantial thermal (in addition to mechanical and aerodynamic) loads can be dangerous contributors to a premature mechanical failure, particularly in the turbine early stages. In view of this, effective blade cooling becomes much more than an option, despite the large expense of creating the narrow coolant passages inside the blades.

Figure 8.52 shows an internally cooled first-stage rotor in the high-pressure turbine section of a turbofan engine. The cooling air, in this case, is ejected at the blade tip, forming, in effect, a barrier in the face of pressure-to-suction side flow leakage. This in itself has the effect of enhancing the aerodynamic performance of this unshrouded rotor. The cooling air is drawn, as is always the case, at the compressor-exit station. It may appear, at first, that a better location for drawing this air is

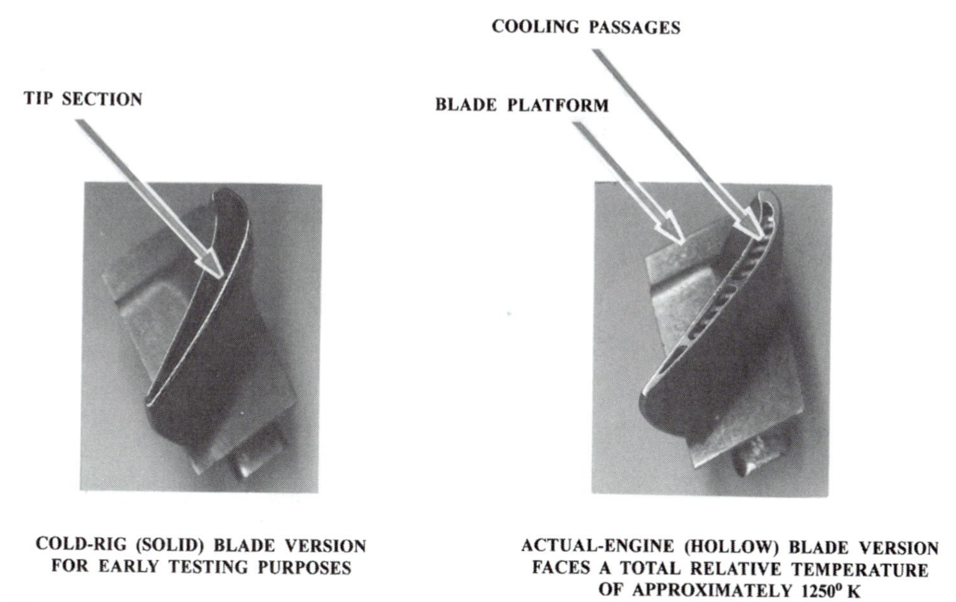

Figure 8.52. Cooled and uncooled versions of a high-pressure turbine blade.

somewhere close to the compressor inlet, because the temperature there is much lower by comparison. The mistake in doing so is that the cooling air will be at a very low pressure, much lower than that at the point of delivery within the turbine section. Theoretically speaking, the coolant extraction can be near the compressors inlet station, except that a significant amount of pumping must be applied to the cooling air, which makes such an option simply unwise.

Figure 8.53 shows the cooling-flow path in an industrial gas-turbine engine. Although the first (higher-temperature) stage in this figure is radial, the stage is bypassed in the cooling scheme, which is highly nontraditional. As seen in Figure 8.53, the cooling-flow stream is directed to the second (axial-flow) stage. The coolant pathway in this figure is composed of several irregularly shaped segments and is typical in gas-turbine cooling. This secondary flow passage consists, in part, of the teeth-to-wall cavities of labyrinth seals. Despite its commonality, the latter is hardly a good region in the cooling-flow passage because these seal cavities will, by design, cause a sequence of substantial total pressure losses. However, exclusion of the first higher-pressure stage (Fig. 8.53) helps in maintaining the cooling flow at a comfortable pressure above that at the coolant delivery station.

EXAMPLE 4

Figure 8.54 shows an adiabatic axial-flow turbine stage, where the rotor blade is untapered and untwisted. The stage operating conditions are as follows:

- Mass-flow rate ($\dot{m}$) = 12.6 kg/s
- Inlet total pressure (p_{t0}) = 10 bars

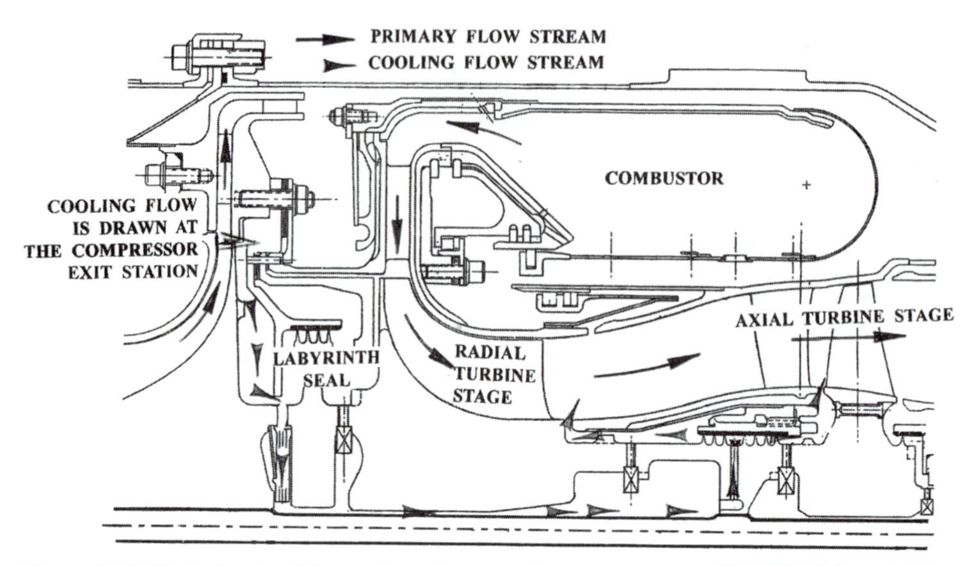

Figure 8.53. Typical path of the cooling air across the gas generator. (GTCP305-2 model for aircraft auxiliary power system by Garrett Turbine Engine Co.)

- Inlet total temperature (T_{t0}) = 1250 K
- Stator flow process is assumed isentropic
- Stator-exit static pressure (p_1) = 6.1 bars
- Rotor-exit static pressure (p_2) = 3.6 bars
- Total relative pressure loss across the rotor (Δp_{t_r}) = 0.8 bars
- Shaft speed (N) = 22,000 rpm

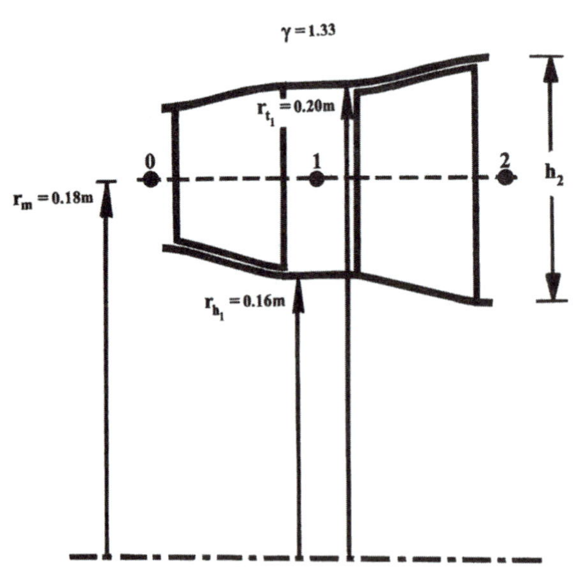

Figure 8.54. Input variables for Example 4.

Assuming a specific-heat ratio (γ) of 1.33:

a) Calculate the stage work coefficient (ψ).
b) Calculate the stage specific speed (N_s).
c) Assuming a free-vortex flow structure in the stator/rotor gap, and knowing that the mean-radius incidence angle is zero, calculate the rotor-tip incidence angle.
d) Knowing that the optimum number of rotor blades, according to Zweifel's loading criterion and based on the mean-radius flow conditions, is 28, calculate the solidity ratio at the average tip radius.

SOLUTION

Part a: The process to obtain the stage work coefficient in this problem is relatively lengthy. We begin by computing the rotor-inlet variables as follows:

$$p_1 = p_{t1}\left[1 - \frac{\gamma - 1}{\gamma + 1}\left(\frac{V_1}{V_{cr1}}\right)^2\right]^{\frac{\gamma}{\gamma-1}}$$

where

$$p_{t1} = p_{t0} = 10.0 \text{ bars (isentropic stator flow)}$$

Upon substitution, we get

$$\frac{V_1}{V_{cr1}} = M_{cr1} = 0.9027$$

But

$$V_{cr1} = \sqrt{\left(\frac{2\gamma}{\gamma + 1}\right) R T_{t1}} = 640.0 \text{ m/s}$$

Then

$$V_1 = 577.7 \text{ m/s}$$

Now, applying the continuity equation at the stator-exit station, we have

$$\frac{12.6\sqrt{1250.0}}{(10.0 \times 10^5)\pi[(0.2)^2 - (0.16)^2]\cos\alpha_1} = \sqrt{\frac{2\gamma}{(\gamma + 1)R}}\, M_{cr1}\left[1 - \left(\frac{\gamma - 1}{\gamma + 1}\right) M_{cr1}{}^2\right]^{\frac{1}{\gamma-1}}$$

which, upon substitution, yields

$$\alpha_1 = 75.47^\circ$$

We now proceed to calculate the variables of the rotor-inlet velocity diagram as follows:

$$V_z = V_{z1} = V_1 \cos\alpha_1 = 144.89 \text{ m/s}$$

$$V_{\theta 1} = V_1 \sin\alpha_1 = 559.23 \text{ m/s}$$

$$U_m = \omega r_m = 414.69 \text{ m/s}$$

$$W_{\theta 1} = V_{\theta 1} - U_m = 144.54 \text{ m/s}$$

$$W_1 = \sqrt{W_{\theta 1}^2 + V_z^2} = 204.66 \text{ m/s}$$

$$\beta_1 = \tan^{-1}\left(\frac{W_{\theta 1}}{V_z}\right) = +44.93^\circ$$

$$T_{tr1} = T_{t1} - \frac{V_1^2 - W_1^2}{2c_p} = 1123.84 \text{ K}$$

$$p_{tr1} = p_{t1}\left(\frac{T_{tr1}}{T_{t1}}\right)^{\frac{\gamma}{\gamma-1}} = 6.513 \text{ bars}$$

$$p_{tr2} = p_{tr1} - (\Delta p_{tr})_{rotor} = 5.713 \text{ bars}$$

In this problem, the junction, between the rotor-inlet and rotor-exit properties is the equality of total relative temperature (T_{tr}), which is a previously emphasized feature of an axial-flow turbomachine (Chapters 3 and 4). Also helpful is the newly computed p_{tr2}. The first of these conditions can be expressed as

$$T_{tr2} = T_{tr1} = 1123.84 \text{ K}$$

Now we can proceed to find all of the relevant thermophysical properties at the rotor-exit station as follows:

$$W_{cr2} = \sqrt{\left(\frac{2\gamma}{\gamma+1}\right) R T_{tr2}} = 606.81 \text{ m/s}$$

$$p_2 = p_{tr2}\left[1 - \frac{\gamma-1}{\gamma+1}\left(\frac{W_2}{W_{cr2}}\right)^2\right]^{\frac{\gamma}{\gamma-1}}$$

where

$$p_2 = 3.6 \text{ bars (given)}$$

Direct substitution yields

$$\frac{W_2}{W_{cr2}} = 0.874$$

The rest of the rotor-exit kinematical properties can be computed as follows:

$$W_2 = 530.53 \text{ m/s}$$

$$\beta_2 = \cos^{-1}\left(\frac{V_z}{W_2}\right) = -74.15^\circ \text{ (negative sign inserted)}$$

$$W_{\theta 2} = W_2 \sin\beta_2 = -510.37 \text{ m/s}$$

$$V_{\theta 2} = W_{\theta 2} + U_m = -95.68 \text{ m/s}$$

$$\alpha_2 = \tan^{-1}\left(\frac{V_{\theta 2}}{V_z}\right) = -33.44^\circ$$

$$V_2 = \sqrt{V_{\theta 2}^2 + V_z^2} = 173.63 \text{ m/s}$$

At this point, we are prepared to calculate the stage work coefficient (ψ) as follows:

$$\psi = \frac{U_m(V_{\theta 1} - V_{\theta 2})}{U_m{}^2} = \frac{(V_{\theta 1} - V_{\theta 2})}{U_m} = 1.579 \text{ (which is on the high side)}$$

Part b: In order to calculate an accurate value of the specific speed (N_s), we first calculate the rotor-exit static density (ρ_2) as follows:

$$T_{t2} = T_{t1} - \frac{U_m(V_{\theta 1} - V_{\theta 2})}{c_p} = 1015.21 \text{ K}$$

$$V_{cr2} = \sqrt{\left(\frac{2\gamma}{\gamma+1}\right) R T_{t2}} = 576.74 \text{ m/s}$$

$$\left(\frac{V}{V_{cr}}\right)_2 = 0.301$$

$$p_{t2} = p_{tr2}\left(\frac{T_{t2}}{T_{tr2}}\right)^{\frac{\gamma}{\gamma-1}} = 3.793 \text{ bars}$$

$$\rho_{t2} = \frac{p_{t2}}{R T_{t2}} = 1.302 \text{ kg/m}^3$$

$$\rho_2 = \rho_{t2}\left[1 - \frac{\gamma-1}{\gamma+1}\left(\frac{V}{V_{cr}}\right)_2^2\right]^{\frac{1}{\gamma-1}} = 1.252 \text{ kg/m}^3$$

The ideal total enthalpy drop across the rotor can be calculated as follows:

$$\Delta h_{tid} = c_p T_{t1}\left[1 - \left(\frac{p_{t2}}{p_{t1}}\right)^{\frac{\gamma-1}{\gamma}}\right] = 309{,}117 \text{ J/kg}$$

Now, the specific speed (N_s) can be calculated as follows:

$$N_s = \frac{\left(N \times \frac{2\pi}{60}\right)\sqrt{\frac{\dot{m}}{\rho_2}}}{(\Delta h_{tid})^{\frac{3}{4}}} = 0.557 \text{ radians}$$

This specific-speed magnitude is, by reference to Figure 5.9, too small to suggest the axial-flow stage design, being the subject of this example.

Part c: In order to compute the blade-tip incidence angle, we first need to compute the relative flow angle $[(\beta_1)_{tip}]$ as follows:

$$(V_{\theta 1})_{tip} = (V_{\theta 1})_{mean}\left(\frac{r_m}{r_t}\right) = 503.31 \text{ m/s (free-vortex flow structure)}$$

$$(W_{\theta 1})_{tip} = (V_{\theta 1})_{tip} - U_t = 42.54 \text{ m/s}$$

$$(\beta_1)_{tip} = \tan^{-1}\left[\frac{(W_{\theta 1})_{tip}}{V_z}\right] = 16.36^\circ$$

In identifying the tip-radius axial-velocity component (above) note that we made use of the fact that this component remains constant at all radii (from hub to tip) as a condition that comes under the umbrella of a free-vortex flow pattern. Now, we pursue our calculations at the rotor-inlet tip radius. In doing so, note that the rotor blade is untwisted (from hub to tip), which means that the blade tip inlet (metal) angle is identical to that at the mean radius. The latter is, in turn, equal to the mean-radius relative flow angle $[(\beta_1)_{mean}]$, for the mean-radius incidence angle is zero. With this in mind, we can pursue our calculations as follows:

$$(\beta_1{}')_{tip} = (\beta_1{}')_{mean} = (\beta_1)_{mean} = 44.93^\circ \text{ (computed earlier)}$$

Finally, we can compute the rotor-tip incidence angle $[(i_R)_{tip}]$ as follows:

$$(i_R)_{tip} = (\beta_1)_{tip} - (\beta_1{}')_{tip} = -28.6^\circ$$

Part d: In order to use Zweifel's loading criterion, we first need to calculate the average "linear" cascade pitch (S_m) using the number of blades (N_b), which is given to be 28. The starting point, however, is to calculate the average tip radius by calculating the blade-exit height (h_2) as follows:

$$h_2 = \frac{\dot{m}}{2\pi r_m \rho_2 V_z} = 6.14 \text{ cm}$$

$$r_{t,av.} = r_m + \frac{1}{2}\left(\frac{h_1}{2} + \frac{h_2}{2}\right) = 20.54 \text{ cm}$$

$$(S)_{tip,av.} = \frac{2\pi r_{t,av.}}{N_b} = 4.61 \text{ cm}$$

$$(S)_{mean} = \frac{2\pi r_m}{N_b} = 4.04 \text{ cm}$$

Noting that Zweifel's loading criterion utilizes average flow properties, and that these properties are assumed to prevail at the mean radius, we are now prepared to apply this criterion as follows:

$$(\psi_Z)_{opt.} = 0.8 = 2\left(\frac{S_m}{C_z}\right)\cos^2(\beta_2)_{mean}[\tan(\beta_1)_{mean} - \tan(\beta_2)_{mean}]$$

which yields

$$C_z = 3.4 \text{ cm}$$

Because the rotor blade is untapered, the magnitude of the axial chord C_z applies at all radii from hub to tip. Finally, we calculate the solidity ratio ($\sigma_{t,av.}$) corresponding to the tip-section average radius as follows:

$$\sigma_{t,av.} = \frac{C_z}{S_{t,av.}} = 0.738$$

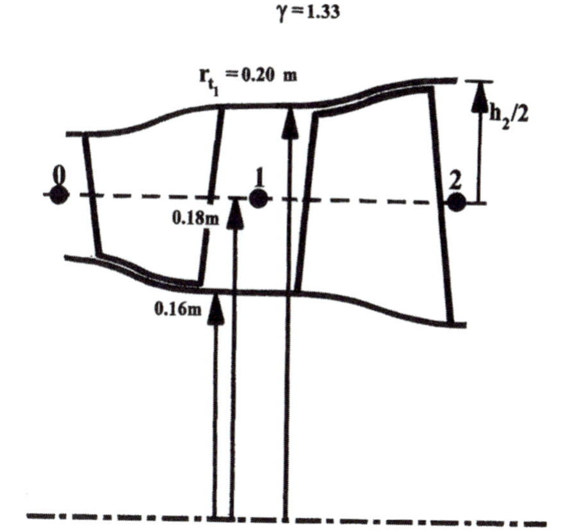

Figure 8.55. Input variables for Example 5.

EXAMPLE 5

Figure 8.55 shows an axial-flow turbine stage. The rotor blade is designed to have the optimum magnitude (i.e., 0.8) of Zweifel's loading coefficient at the mean radius. At this radius, both the inlet relative flow angle and exit absolute flow angle are zero. Besides, the rotor mean-radius and tip average-radius axial chords are 1.97 and 1.75 cm, respectively. The rotational speed is 48,000 rpm, and the axial-velocity component (V_z) is 280 m/s and is assumed constant across the rotor and at all radii from hub to tip. The following conditions also apply:

- Rotor-exit static density (ρ_2) = 1.24 kg/m^3
- Rotor-blade aspect ratio (λ) = 2.57

Calculate the rotor-cascade solidity ratio at the tip radius.

SOLUTION

The procedure leading to the tip-radius solidity ratio (σ_t) involves the Zweifel-criterion-based number of rotor blades (N_b). Note that this criterion applies almost exclusively at the mean radius (r_m). Also note that zero magnitudes of the mean-radius rotor-inlet relative angle ($\beta_{1,m}$) and rotor-exit swirl angle ($\alpha_{2,m}$) imply that both the inlet relative velocity ($W_{1,m}$) and exit absolute velocity ($V_{2,m}$), respectively, are entirely in the axial direction.

$$U_m = \omega r_m = 502.7 \text{ m/s}$$

$$\beta_{2,m} = -\tan^{-1}\left(\frac{U_m}{V_z}\right) = -60.9^\circ$$

Now, we apply Zweifel's loading criterion at the mean radius as follows:

$$(\psi)_{opt.} = 0.8 = 2\left(\frac{S_m}{C_{z,m}}\right)\cos^2\beta_{2,m}[\tan(\beta_{1,m}) - \tan(\beta_{2,m})]$$

which, upon substitution, yields

$$S_m = 1.85 \text{ cm}$$

Thus, the optimum number of rotor blades is

$$(N_b)_{opt.} = \frac{2\pi r_m}{S_m} = 34 \text{ blades}$$

Making use of the additional data given, we have

$$h_2 = \frac{\dot{m}}{(2\pi r_m)\rho_2 V_z} = 5.0 \text{ cm}$$

$$r_{t,2} = r_m + \frac{h_2}{2} = 20.5 \text{ cm}$$

$$(r)_{t,av.} = \frac{1}{2}(r_{t1} + r_{t2}) = 20.25 \text{ cm}$$

$$h_{av.} = \frac{1}{2}(h_1 + h_2) = 4.5 \text{ cm}$$

$$(c_z)_{tip} = \frac{h_{av.}}{\lambda} = 1.75 \text{ cm}$$

$$(S)_{tip,av.} = \frac{2\pi(r_t)_{av.}}{N_b} = 3.74 \text{ cm}$$

Finally, we calculate the tip-radius solidity ratio as follows:

$$(\sigma)_{tip} = \frac{(C_z)_{tip}}{(S)_{tip}} = 0.468$$

EXAMPLE 6

Figure 8.56 shows an off-design inlet velocity diagram for an axial-flow rotor. Use the loss versus incidence angle graph in the figure to calculate the total relative pressure at station *X*, which is just inside the airfoil leading edge, assuming a specific heat ratio (γ) of 1.33.

SOLUTION

First, we calculate the rotor-blade incidence angle:

$$i_R = \beta_1 - \beta_1{}' = -20^\circ$$

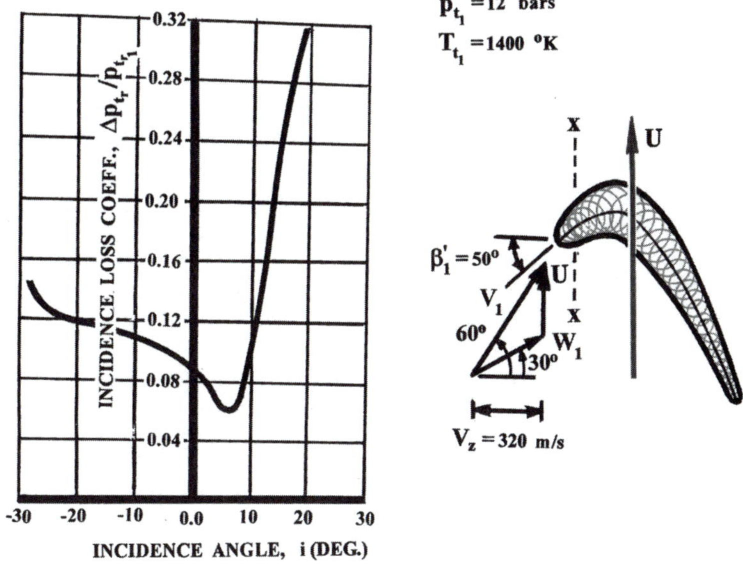

Figure 8.56. Input variables for Example 6.

Now, using the given graph, we get

$$\frac{(\Delta p_{t_r})_{incid.}}{(p_{t_r})_1} = 0.12$$

Let us now calculate the rotor-inlet total relative pressure $(p_{t_r})_1$:

$$V_1 = \frac{V_z}{\cos\alpha_1} = 640.0 \text{ m/s}$$

$$W_1 = \frac{V_z}{\cos\beta_1} = 369.5 \text{ m/s}$$

$$T_{tr1} = T_{t1} - \frac{(V_1{}^2 - W_1{}^2)}{2c_p} = 1281.9 \text{ K}$$

$$p_{tr1} = p_{t1}\left(\frac{T_{tr1}}{T_{t1}}\right)^{\frac{\gamma}{\gamma-1}} = 8.414 \text{ bars}$$

Using the fraction of total relative pressure loss computed earlier, we proceed to compute the total relative pressure $(p_{t_r})_X$ as follows:

$$(p_{t_r})_X = (p_{t_r})_1 - \left[\frac{(\Delta p_{t_r})_{incid.}}{(p_{t_r})_1}(p_{t_r})_1\right] = 7.4 \text{ bars}$$

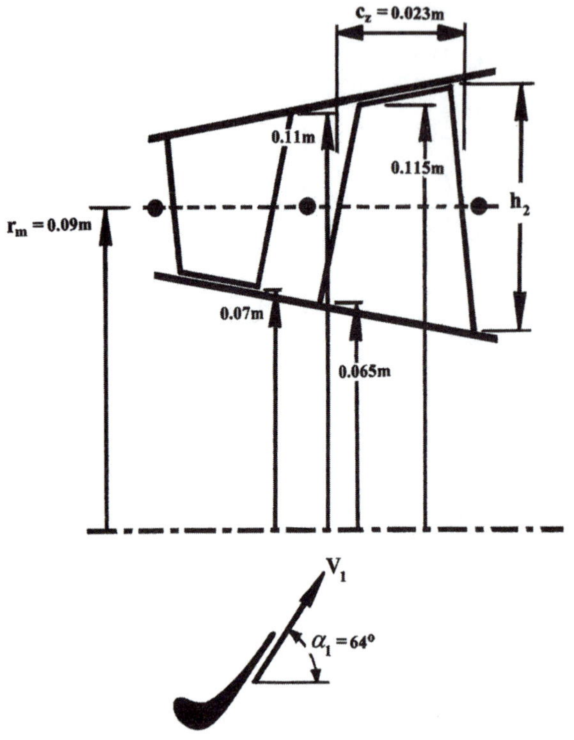

Figure 8.57. Input variables for Example 7.

EXAMPLE 7

The design point of the single-stage adiabatic turbine in Fig. 8.57 is defined as follows:

- Stator flow (assumed isentropic) is choked
- Stator-exit flow angle $(\alpha_1) = 64°$
- Stage-inlet total pressure $(p_{t0}) = 12$ bars
- Stage-inlet total temperature $(T_{t0}) = 1400$ K
- Stage-exit total pressure $(p_{t2}) = 3.6$ bars
- Stage work coefficient $(\psi) = 1.5$
- Rotational speed $(N) = 51{,}000$ rpm
- Axial velocity component (V_z) is constant throughout the stage

Assuming a specific-heat ratio (γ) of 1.33, calculate:

a) The mass-flow rate $(\dot{m})$ through the stage;
b) The stage-exit absolute flow angle (α_2);
c) The exit blade height (h_2);
d) The stage specific speed (N_s);
e) The optimum number of rotor blades (N_B) by applying Zweifel's loading criterion at the mean radius.

SOLUTION

Part a: Applying the continuity equation (3.35) at the stator-exit station, knowing that the stator is choked (i.e., $M_{cr1} = 1.0$), we have

$$\dot{m} = 12.63 \text{ kg/s}$$

Part b:

$$U_m = \omega r_m = 480.66 \text{ m/s}$$

$$\Delta h_t = (\psi)_{stg.} U_m{}^2 = 346{,}551 \text{ J/kg}$$

$$V_1 = V_{cr1} = \sqrt{\left(\frac{2\gamma}{\gamma + 1}\right) R T_{t1}} = 677.28 \text{ m/s}$$

$$V_{z1} = V_{z2} = V_z = V_1 \cos\alpha_1 = 296.9 \text{ m/s}$$

$$V_{\theta 1} = V_1 \sin\alpha_1 = 608.74 \text{ m/s}$$

$$V_{\theta 2} = V_{\theta 1} - \psi U_m = -111.26 \text{ m/s}$$

$$\alpha_2 = \tan^{-1}\left(\frac{V_{\theta 2}}{V_z}\right) = -20.5^\circ$$

Part c:

$$T_{t2} = T_{t1} - \left[\frac{U_m(V_{\theta 1} - V_{\theta 2})}{c_p}\right] = 1100.8 \text{ K}$$

$$V_{cr2} = \sqrt{\left(\frac{2\gamma}{\gamma + 1}\right) R T_{t2}} = 600.56 \text{ m/s}$$

$$M_{cr2} = \frac{V_2}{V_{cr2}} = 0.528$$

$$\rho_{t2} = \frac{p_{t2}}{R T_{t2}} = 1.139 \text{ kg/m}^3$$

$$\rho_2 = \rho_{t2}\left[1 - \left(\frac{\gamma - 1}{\gamma + 1}\right) M_{cr2}{}^2\right]^{\frac{1}{\gamma - 1}} = 1.008 \text{ kg/m}^3$$

Finally, we calculate the rotor-exit blade (or annulus) height as follows:

$$h_2 = \frac{\dot{m}}{\rho_2 (2\pi r_m) V_z} = 7.5 \text{ cm}$$

Part d:

$$(\Delta h_t)_{id.} = c_p T_{t1}\left[1 - \left(\frac{p_{t2}}{p_{t1}}\right)^{\frac{\gamma - 1}{\gamma}}\right] = 418{,}160 \text{ J/kg}$$

$$N_s = \frac{\omega\sqrt{\frac{\dot{m}}{\rho_2}}}{[(\Delta h_t)_{id.}]^{\frac{3}{4}}} = 1.15 \text{ radians}$$

Part e: Now, we apply Zweifel's loading criterion to the rotor, with the objective being the optimum number of blades:

$$\beta_1 = \tan^{-1}\left(\frac{V_{\theta 1} - U_m}{V_z}\right) = +23.3^\circ$$

$$\beta_2 = \tan^{-1}\left(\frac{V_{\theta 2} - U_m}{V_z}\right) = -63.4^\circ$$

$$0.8 = (\psi_Z)_{opt.} = 2\left(\frac{S_m}{C_{zm}}\right)\cos^2\beta_2[\tan\beta_1 - \tan\beta_2]$$

which, upon substitution, yields

$$S_m = 1.89 \text{ cm}$$

Finally, the optimum number of blades is calculated as follows:

$$(N_b)_{opt.} = \frac{2\pi r_m}{S_m} = 30 \text{ blades}$$

EXAMPLE 8

Figure 8.58 shows the "hot" section of an auxiliary power unit (APU) which consists of a single-stage turbine and the exhaust diffuser. The operating conditions are as follows:

- Mass-flow rate ($\dot{m}$) = 10.48 kg/s
- Inlet total pressure (p_{t0}) = 14.8 bars
- Inlet total temperature (T_{t0}) = 1406.0 K
- Stator passage is choked
- Stage total-to-total efficiency (η_T) = 88%
- Stage reaction = 50%
- Rotor speed (N) = 51,415 rpm
- Diffuser length (Δz) = 12.7 cm
- Diffuser endwalls' friction coefficient (f) = 0.14

Besides, the following simplifications are valid:

- The entire flow process is adiabatic
- The stator flow process is exclusively isentropic
- The turbinewise axial-velocity component (V_z) is constant
- The average specific-heat ratio (γ) is 1.33

a) Determine the turbine-exit critical Mach no. (M_{cr2}).
b) Calculate the overall total-to-static efficiency in two ways:
   1) Using the Fanno-flow relationships across the diffuser.
   2) Using Sovran and Klomp chart (Fig. 3.36).
c) Calculate the diffuser-exit annulus height (h_3).

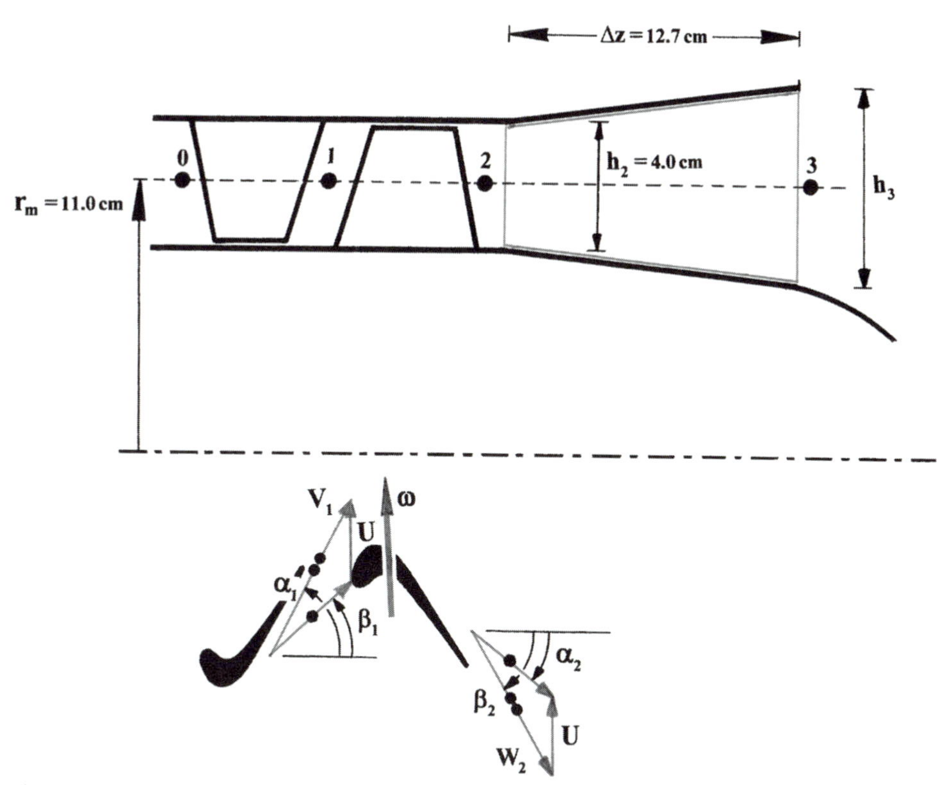

Figure 8.58. Input variables for Example 8.

Note: In addressing item b1, replace the diffuser by a constant-area duct, with the constant cross-sectional area being that at the diffuser inlet station as an approximation.

SOLUTION

Part a: With the stator being choked (i.e., $M_{cr1} = 1.0$), let us apply the continuity equation at the stator-exit station:

$$\frac{\dot{m}\sqrt{T_{t1}}}{p_{t1}(2\pi r_m h_1)\cos\alpha_1} = \sqrt{\frac{2\gamma}{(\gamma+1)R}}M_{cr1}\left[1-\left(\frac{\gamma-1}{\gamma+1}\right)M_{cr1}{}^2\right]^{\frac{1}{\gamma-1}}$$

Substitution in this equation gives rise to the stator-exit absolute (or swirl) angle:

$$\alpha_1 = 76^\circ$$

Let us now calculate the rotor-inlet velocity-triangle variables:

$$V_{\theta 1} = V_1 \sin\alpha_1 = V_{cr1}\sin\alpha_1 = \sqrt{\left(\frac{2\gamma}{\gamma+1}\right)RT_{t1}}\sin\alpha_1 = 658.6 \text{ m/s}$$

$$V_z = V_{z1} = V_1\cos\alpha_1 = V_{cr1}\cos\alpha_1 = 164.2 \text{ m/s}$$

$$U_m = \omega r_m = 592.3 \text{ m/s}$$

$$W_{\theta 1} = V_{\theta 1} - U_m = 66.3 \text{ m/s}$$

$$\beta_1 = \tan^{-1}\left(\frac{W_{\theta 1}}{V_z}\right) = +22.0^\circ$$

$$V_1 = \sqrt{V_{\theta 1}^2 + V_z^2} = 678.8 \text{ m/s}$$

$$W_1 = \sqrt{W_{\theta 1}^2 + V_z^2} = 177.1 \text{ m/s}$$

The following computational segment (marked at its beginning and end) could very well be part of a correct solution. But again, a great deal of it could be flatly wrong. The reason is that we (intentionally but mistakingly) will not compute the rotor-exit relative critical Mach number (W_2/W_{cr2}) to see whether it reflects a supersonic *relative* flow stream, which would be impossible with a subsonic nozzle-like blade-to-blade rotor passage. One may be surprised at how often this mistake is made in real life, particularly in rotating cascades, for it is unfortunately often the impression that the results will be correct once the exit *absolute* critical Mach number is less than unity.

At the end of the solution procedure, we will go back to calculate W_2/W_{cr2} and, if necessary, make the appropriate changes.

* * * * *

Let us now compute the rotor-exit flow properties (except, of course, for the variable W_2/W_{cr2}) as follows:

$$V_{z2} = V_{z1} = V_z = 164.2 \text{ m/s}$$

$$\alpha_2 = -\beta_1 = -22.0^\circ \text{ (stage reaction is 50\%)}$$

$$\beta_2 = -\alpha_1 = -76.0^\circ \text{ (stage reaction is 50\%)}$$

$$V_{\theta 2} = -W_{\theta 1} = -66.3 \text{ m/s}$$

$$V_2 = \sqrt{V_{\theta 2}^2 + V_z^2} = 177.1 \text{ m/s}$$

$$T_{t2} = T_{t1} - \frac{U_m}{c_p}(V_{\theta 1} - V_{\theta 2}) = 1034.8 \text{ m/s}$$

$$M_{cr2} = \frac{V_2}{V_{cr2}} = \frac{V_2}{\sqrt{\left(\frac{2\gamma}{\gamma+1}\right) R T_{t2}}} = 0.304$$

Because we are about to apply one of the Fanno-flow relationships, we should convert the critical Mach number (above) into the "traditional" Mach number (i.e., M_2) as follows:

$$M_2 = \sqrt{\left(\frac{2}{\gamma - 1}\right)\left\{\left[1 - \left(\frac{\gamma - 1}{\gamma + 1}\right) M_{cr2}^2\right]^{-1} - 1\right\}} = 0.281$$

Part b

i) System efficiency using Fanno-flow relationships: Recalling that the component we are about to consider is the exhaust *diffuser*, it is perhaps obvious that the mere phrasing of this part subtitle is aerodynamically inaccurate, for we know that the Fanno-flow relationships apply only to a duct with a constant cross-flow-area. We have not, yet, computed the diffuser-exit annulus height h_3, but we certainly expect it to be greater than h_2.

The implied assumption, then, is that h_3 will come out to be "sufficiently" close to h_2 from an engineering standpoint. In other words, we will proceed with the assumption that the exhaust-diffuser divergence angle is approximately zero. One of course could (in the end) go back to see whether this approximation is acceptable.

We now use one of the Fanno-flow relationships (Chapter 3) to calculate the diffuser-exit Mach number (M_3) as follows:

$$\frac{f\left(\frac{(\Delta z)_{diff.}}{\cos\alpha_2}\right)}{D_h} = \left(\frac{\gamma+1}{2\gamma}\right)\left\{\ln\left[\frac{1+\left(\frac{\gamma-1}{2}\right)M_3^2}{1+\left(\frac{\gamma-1}{2}\right)M_2^2}\right]\right\}$$

$$-\left(\frac{1}{\gamma}\right)\left(\frac{1}{M_3^2}-\frac{1}{M_2^2}\right)-\left(\frac{\gamma+1}{2\gamma}\right)\left[\ln\left(\frac{M_3^2}{M_2^2}\right)\right]$$

Upon substitution, we end up with a highly nonlinear equation in M_3. As a result, determination of this variable requires an iterative or a trial-and-error procedure. The final outcome of the latter is

$$M_3 = 0.285$$

Proceeding further with the Fanno-flow applicability assumption, we can calculate the (diffuser-exit/diffuser-inlet) total-to-total pressure ratio as

$$\frac{p_{t3}}{p_{t2}} = \left(\frac{M_2}{M_3}\right)\left[\frac{1+\left(\frac{\gamma-1}{2}\right)M_3^2}{1+\left(\frac{\gamma-1}{2}\right)M_2^2}\right]^{\frac{\gamma+1}{2(\gamma-1)}} = 0.987$$

But

$$p_{t2} = p_2\left[1+\left(\frac{\gamma-1}{2}\right)M_2^2\right]^{\frac{\gamma}{\gamma-1}} = 3.515 \text{ bars}$$

Thus

$$p_{t3} = 3.47 \text{ bars}$$

Note that the total pressure has dropped across the flow passage, which is to be expected as a result of friction over the endwalls. Now, the passage-exit static pressure can be calculated as follows:

$$p_3 = \frac{p_{t3}}{\left[1+\left(\frac{\gamma-1}{2}\right)M_3^2\right]^{\frac{\gamma}{\gamma-1}}} = 3.311 \text{ bars}$$

Now, the overall (stage + diffuser) total-to-static efficiency, defined in Chapter 3, can be computed as

$$\eta_{t-s})_{sys.} = \frac{1 - (T_{t3}/T_{t2})}{1 - (p_3/p_{t2})^{\frac{\gamma-1}{\gamma}}} = 85.1\%$$

ii) System efficiency using Sovrant and Klomp chart: This chart is included in Figure 3.37 and requires knowledge of the nondimensionalized diffuser length, which we will now calculate:

$$\frac{L}{h_{in}} = \frac{(\Delta z)_{diff.}}{h_2} = \frac{12.7}{4.0} = 3.175$$

Referring to Figure 3.37, we get

$$\text{Diffuser Recovery Coefficient } (C_{p_r}) = 0.465 = \frac{p_3 - p_2}{p_{t2} - p_2}$$

which, upon substitution, yields

$$p_3 = 3.418 \text{ bars}$$

It follows that:

$$\eta_{t-s})_{sys.} = \frac{1 - (T_{t3}/T_{t2})}{1 - (p_3/p_{t2})^{\frac{\gamma-1}{\gamma}}} = 86.6\%$$

Part c: Again, using Sovrant and Klomp chart (Fig. 3.37), we get:

$$\frac{A_3}{A_2} - 1 = 0.51$$

meaning that

$$\frac{A_3}{A_2} = 1.51 = \frac{(2\pi r_m)h_3}{(2\pi r_m)h_2}$$

which yields

$$h_3 = 1.51 \times 4.0 = 6.04 \text{ cm}$$

COMMENT

If we believe the newly computed empirically achieved h_3, then we do have an unignorable exhaust-diffuser divergence angle, where the exit annulus height is as high as 51% larger than that at the diffuser inlet station. The reason this point is being raised is that we have previously assumed that the annulus height was roughly constant across the exhaust diffuser. This was the case as we were about to compute the system total-to-static efficiency using a Fanno-flow relationship. A corrective action in this case, would be to go back and take the average of the diffuser-inlet and exit annulus heights to better represent the annulus height of what is simplified to be a constant-area duct.

* * * * *

Let us now go back and calculate (W_2/W_{cr2}) to make sure that the rotor passage is not choked. This will be the case if the rotor-exit relative critical Mach number is equal to or greater than unity, with the latter being impossible. It is this possibility, however, that is troubling. The reason is that we will have, in this case, to go back and set (W_2/W_{cr2}) to 1.0 and make the required corrections throughout the computational procedure.

The process for computing the rotor-exit relative critical Mach number (W_2/W_{cr2}) is as follows:

$$W_2 = \sqrt{W_{\theta 2}{}^2 + V_z{}^2} = 526.0 \text{ m/s}$$

$$V_2 = \sqrt{V_{\theta 2}{}^2 + V_z{}^2} = 177.1 \text{ m/s}$$

$$T_{tr2} = T_{t2} - \left(\frac{V_2{}^2 - W_2{}^2}{2c_p}\right) = 1220.4 \text{ K}$$

$$W_{cr2} = \sqrt{\left(\frac{2\gamma}{\gamma+1}\right) R T_{tr2}}$$

Thus

$$\frac{W_2}{W_{cr2}} = \frac{526.0}{632.4} = 0.83 \text{ (i.e., \textit{subsonic})}$$

In view of this result, we should perhaps feel fortunate in the sense that no adjustments in the numerical procedure are now needed.

EXAMPLE 9

Referring to Example 8, replace the 50% reaction item by the following:

- Rotor-exit total temperature $(T_{t2}) = 1031.5$ K
  - a) Verify whether the rotor is choked.
  - b) Determine whether V_z can actually remain constant across the rotor.
  - c) Recalculate the diffuser-inlet swirl angle (α_2).
  - d) Recalculate the diffuser-inlet critical Mach number (V_2/V_{cr2}).

SOLUTION

In solving this problem, we will be careful to compute the rotor-exit relative critical Mach number (W_2/W_{cr2}) and make the necessary changes should the rotor passage turn out to be choked.

Many variables in the following solution procedure are not computed. Instead, the formerly computed magnitudes (in Example 8) will simply be used.

$$V_{\theta 2} = V_{\theta 1} - \frac{c_p(T_{t1} - T_{t2})}{U_m} = -72.8 \text{ m/s}$$

$$W_{\theta 2} = V_{\theta 2} - U_m = -665.1 \text{ m/s}$$

$$W_2 = \sqrt{W_{\theta 2}{}^2 + V_z{}^2} = 685.1 \text{ m/s}$$

$$T_{tr2} = T_{tr1} = T_{t1} - \left(\frac{V_1{}^2 - W_1{}^2}{2c_p}\right) = 1220.4 \text{ K}$$

$$W_{cr2} = \sqrt{\left(\frac{2\gamma}{\gamma+1}\right) RT_{tr2}} = 632.3 \text{ m/s}$$

Now, we calculate the rotor-exit relative critical Mach number:

$$\frac{W_2}{W_{cr2}} = \frac{685.1}{632.3} = 1.08 \text{ (i.e., supersonic)}$$

The rotor-exit status implied by this result *cannot* possibly occur. As previously indicated, the sequence of blade-to-blade hub-to-casing flow channels is subsonic- and nozzle-like (as far as the relative flow stream is concerned) and will therefore be conceptually incapable of delivering a supersonic magnitude of W_2/W_{cr2}. The corrective action therefore will be to consider the rotor as choked, setting the rotor-exit relative critical Mach number to unity. One of the variables which will remain unchanged during this process is the rotor-exit total relative temperature (T_{tr2}), for it is determined by (and equal to) the magnitude of an *upstream* variable, namely T_{tr1}.

In order to compute the rotor-exit total relative pressure (p_{tr2}), we first turn to the definition of the total-to-total efficiency:

$$\eta_{t-t} = 0.88 = \frac{1 - \left(\frac{T_{t2}}{T_{t1}}\right)}{1 - \left(\frac{p_{t2}}{p_{t1}}\right)^{\frac{\gamma-1}{\gamma}}}$$

which, upon substitution, yields

$$p_{t2} = 3.515 \text{ bars}$$

The rotor-exit total relative pressure (p_{tr2}) can be computed as

$$p_{tr2} = p_{t2}\left(\frac{T_{tr2}}{T_{t2}}\right)^{\frac{\gamma}{\gamma-1}} = 6.92 \text{ bars}$$

At this point, we are prepared to apply the continuity equation using the rotor-exit relative properties, including the rotor-exit relative flow angle (β_2), which happens to be the variable we are after:

$$\frac{\dot{m}\sqrt{T_{tr2}}}{p_{tr2}(2\pi r_m h_2)\cos\beta_2} = \sqrt{\frac{2\gamma}{(\gamma+1)R}}\left(\frac{W_2}{W_{cr2}}\right)\left[1 - \frac{\gamma-1}{\gamma+1}\left(\frac{W_2}{W_{cr2}}\right)\right]^{\frac{1}{\gamma-1}}$$

where

$$\frac{W_2}{W_{cr2}} = 1.0$$

Substitution in the preceding continuity equation yields

$$\beta_2 = -61.2^\circ \text{ (negative sign inserted)}$$

With this result, we can proceed as follows:

$$W_2 = W_{cr2} = 632.3 \text{ m/s}$$

$$V_{z2} = W_2 \cos\beta_2 = 304.8 \text{ m/s} \neq V_{z1}$$

$$W_{\theta 2} = W_2 \sin\beta_2 = -554.0 \text{ m/s}$$

$$V_{\theta 2} = W_{\theta 2} + U_m = +38.3 \text{ m/s}$$

$$\alpha_2 = \tan^{-1}\left(\frac{V_{\theta 2}}{V_{z2}}\right) = +7.2^\circ$$

$$V_2 = \sqrt{V_{\theta 2}{}^2 + V_{z2}{}^2} = 307.2 \text{ m/s}$$

$$V_{cr2} = \sqrt{\left(\frac{2\gamma}{\gamma+1}\right) R T_{t2}} = 581.4 \text{ m/s}$$

Finally, the rotor-exit critical Mach number can be computed as

$$M_{cr2} = \frac{V_2}{V_{cr2}} = 0.528$$

PROBLEMS

1) Figure 8.59 shows the meridional flow path of a single-stage axial turbine. The design point of the stage is defined as follows:

   - Rotor speed (N) = 48,000 rpm
   - Stator-exit swirl angle (α_1) = 72°
   - The stator flow process is isentropic (by assumption)
   - The stage-inlet total pressure (p_{t0}) = 15 bars
   - The stage-inlet total temperature (T_{t0}) = 1450 K
   - The stator-exit critical Mach number (V_1/V_{cr1}) = 0.84
   - The axial-velocity component (V_z) is constant
   - The rotor-exit relative critical Mach number (W_2/W_{cr2}) = 0.92
   - The total-to-total stage efficiency (η_{t-t}) = 86%

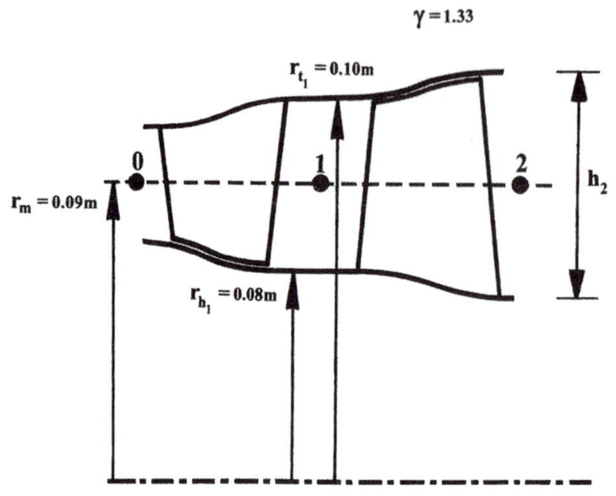

Figure 8.59. Input variables for Problem 1.

Taking an average specific-heat ratio (γ) of 1.33:

a) Calculate the stage flow coefficient (ϕ).
b) Calculate the stage work coefficient (ψ).
c) Calculate the stage specific speed (N_s).
d) Knowing that the optimum number of rotor blades according to Zweifel's loading criterion is 32, calculate the rotor-blade aspect ratio (based on the blade average dimensions).

2) Figure 8.60 shows an adiabatic single-stage axial-flow turbine that is operating under the following conditions:

- Mass-flow rate ($\dot{m}$) = 11.0 kg/s
- Rotor-inlet total pressure (p_{t1}) = 8.5 bars

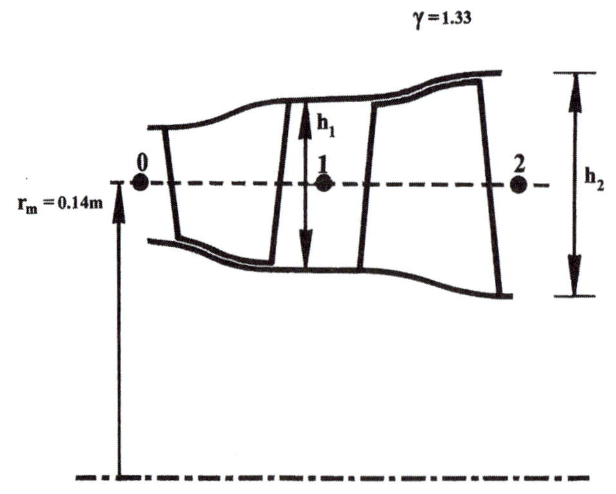

Figure 8.60. Input variables for Problem 2.

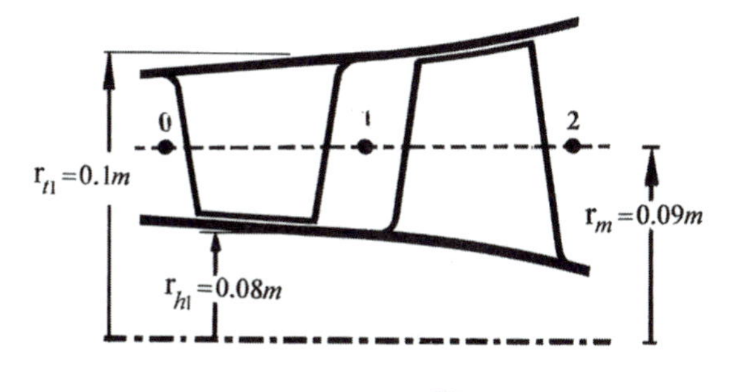

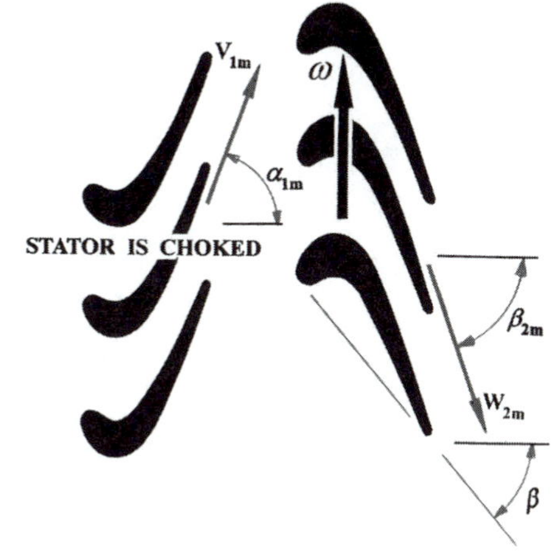

Figure 8.61. Input variables for Problem 3.

- Rotor-inlet total temperature (T_{t1}) = 1100 K
- Stage total-to-total efficiency (η_{t-t}) = 88%
- Rotor-exit absolute flow angle (α_2) = 0
- Rotor flow passage is choked (i.e., $W_2/W_{cr2} = 1.0$)
- Rotor-exit static pressure (p_2) = 2.8 bars
- Produced power (P) = 2,762 kW
- Axial-velocity component is constant across the rotor

Assuming an average specific heat ratio (γ) of 1.33, calculate:

a) The shaft speed;
b) The rotor-blade inlet and exit heights (h_1 and h_2, respectively).

3) Figure 8.61 shows the major dimensions of an axial-flow turbine stage. The operating conditions of the stage are as follows:

- Stator component is choked
- Stator-exit flow angle (α_1) = 66°
- Stator flow process is assumed isentropic
- Stage-inlet total pressure (p_{t0}) = 12.6 bars
- Stage-inlet total temperature (T_{t0}) = 1142 K
- Stage-exit total temperature (T_{t2}) = 930 K
- Stage total-to-total efficiency (η_{t-t}) = 91%

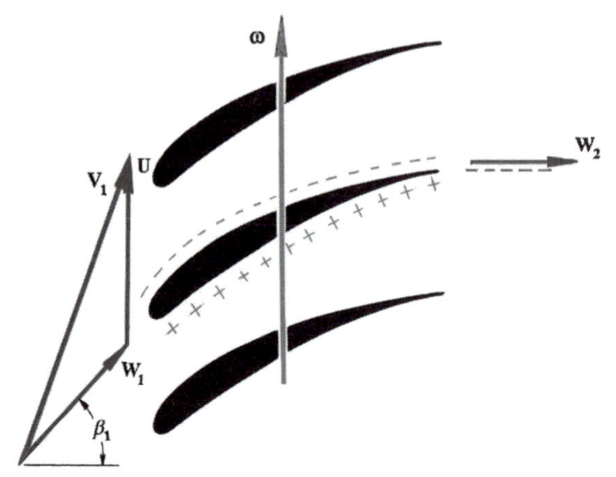

Figure 8.62. Input variables for Problem 4.

- Shaft speed $(N) = 38{,}000$ rpm
- Stator-inlet flow angle $(\alpha_0) = -28°$

Assuming an adiabatic-rotor flow process, a constant axial-velocity component, and an average specific-heat ratio (γ) of 1.33, calculate:

a) The rotor-exit absolute and relative flow angles (i.e., α_2 and β_2, respectively),
b) The rotor-exit annulus height by applying the continuity equation at the rotor exit station:

   i) Using the static properties (i.e., p_2, T_2, etc.),
   ii) Using the total properties (i.e., p_{t2}, T_{t2}, α_2, etc.),
   iii) Using the total-relative properties (i.e., p_{tr2}, T_{tr2}, β_2, etc.),

c) The specific speed (N_s) of the stage;
d) The rotor-blade implied stagger angle (β).
e) The optimum number of stator vanes (N_v) and rotor blades (N_b) by applying Zweifel's optimum loading criterion.

In part e, you may assume a stator-vane mean-radius axial chord (c_{zv}) of 2.2 cm and a rotor-blade mean-radius axial chord (c_{zb}) of 2.8 cm.

4) Figure 8.62 shows a compressor airfoil cascade that is intentionally placed downstream from the combustor and is therefore spinning in the same direction a turbine cascade would (i.e., with the suction side leading the pressure side in the direction of rotation). Assuming an adiabatic flow process, a fixed magnitude of V_z, and an exit relative flow angle, β_2, of zero, answer the following true or false questions:

   a) The rotor total temperature will rise (true or false)
   b) The reaction will be between 0 and 100% (true or false)

5) Figure 8.63 shows an axial-flow turbine stage, where the rotor blades are untapered (i.e., $c_{zhub} = c_{zmean} = c_{ztip}$). The stage operating conditions are as follows:

   - Inlet total pressure $(p_{t0}) = 10.0$ bars
   - Inlet total temperature $(T_{t0}) = 1250$ K

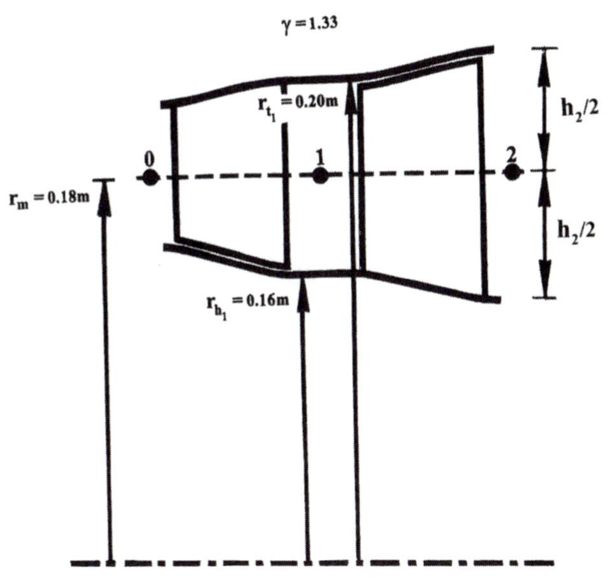

Figure 8.63. Input variables for Problem 5.

- Stator-exit swirl angle $(\alpha_1) = 68^\circ$
- Rotational speed $(N) = 22{,}000$ rpm
- Stage flow coefficient $(\phi) = 0.42$
- Stage work coefficient $(\psi) = 1.4$
- Stage-exit total pressure $(p_{t2}) = 4.1$ bars
- Stator flow process is assumed isentropic
- $V_{z1} = V_{z2}$

 a) Calculate the mass-flow rate $(\dot{m})$.
 b) Calculate the "exact" value of the specific speed (N_s).
 c) Knowing that the optimum number of rotor blades, based on the application of Zweifel's loading criterion along the mean radius, is 28 blades, calculate the rotor-cascade solidity ratio (σ) at the rotor-tip average radius.

6) The tapered turbine-blade row shown in Figure 8.64 is designed to have the optimum solidity ratio (σ) at the mean radius according to Zweifel's criterion of optimum loading. Along the mean radius, the rotor cascade is operating under the following conditions:

- Inlet relative flow angle $(\beta_1) = 0$
- Exit swirl angle $(\alpha_2) = 0$
- $V_{z1} = V_{z2} = 200$ m/s
- Rotational speed $(N) = 27{,}000$ rpm
- Free-vortex flow structures both at the inlet and exit stations

Calculate the following variables:

a) The number of rotor blades (N_b);
b) The solidity ratio (σ_{tip}) at the tip radius;

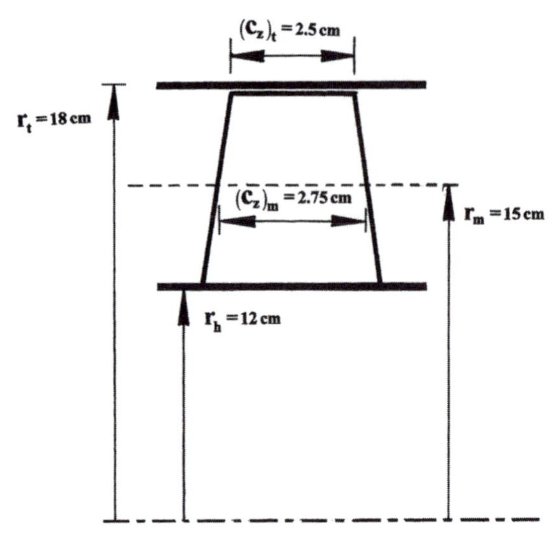

Figure 8.64. Input variables for Problem 6.

c) The Zweifel loading coefficient and the specific shaft work at the tip radius;
d) The reaction at the tip radius.

7) Shown in Figure 8.65 is an axial-flow turbine stage where the rotor blades are untapered and untwisted (i.e., the blade has identical cross sections at all radii). The stage operating conditions are as follows:

- Mass-flow rate $(\dot{m}) = 12.6$ kg/s
- Inlet total pressure $(p_{t0}) = 10.0$ bars
- Inlet total temperature $(T_{t0}) = 1250$ K
- Stator flow process is assumed isentropic
- Stator-exit static pressure $(p_1) = 6.1$ bars
- Rotor-exit static pressure $(p_2) = 3.6$ bars
- Rotor total relative pressure loss $(\Delta P_{tr}) = 0.8$ bars
- Rotational speed $(N) = 22{,}000$ rpm

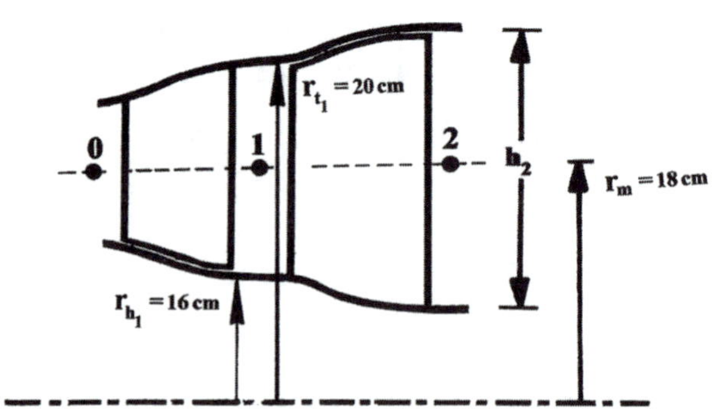

Figure 8.65. Input variables for Problem 7.

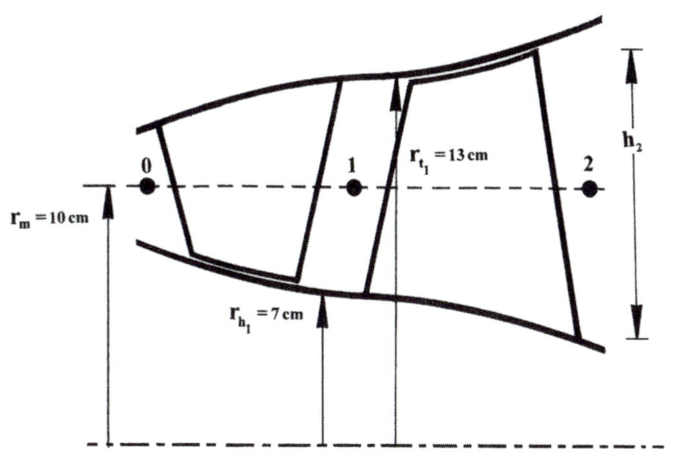

Figure 8.66. Input variables for Problem 8.

Assuming an adiabatic flow field throughout the stage and a specific-heat ratio (γ) of 1.33:

a) Calculate the stage work coefficient (ψ).
b) Calculate the stage specific speed (N_s).
c) Assuming a free-vortex flow pattern over the stator/rotor gap, and knowing that the rotor mean-radius incidence angle is zero, calculate the incidence angle at the rotor-inlet tip radius.

8) Figure 8.66 shows an axial-flow turbine stage that is operating under the following conditions:

- Flow process is assumed isentropic across the entire stage
- Stator flow passage is choked
- Mass-flow rate ($\dot{m}$) = 12.0 kg/s
- Stage-inlet total pressure (p_{t0}) = 8.5 bars
- Stage-inlet total temperature (T_{t0}) = 1050 K
- Rotor speed (N) = 40,000 rpm
- Stage-exit static pressure (p_2) = 3.6 bars
- Specific-heat ratio (γ) = 1.33

Applying the foregoing data and referring to the stage major dimensions in the figure, calculate:

a) The torque (τ) delivered by the turbine stage.
b) The rotor-exit blade height (h_2).

9) Figure 8.67 shows an axial-flow turbine stage where the blade spanwise taper and twist are nonexistent. The stage operating conditions are as follows:

- Inlet total pressure (p_{t0}) = 10 bars
- Inlet total temperature (T_{t0}) = 1400 K
- Rotational speed (N) = 26,000 rpm
- Mass-flow rate ($\dot{m}$) = 12.8 kg/s

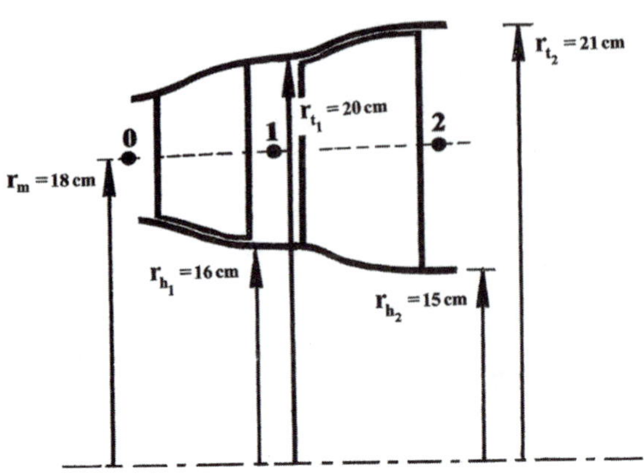

Figure 8.67. Input variables for Problem 9.

- Stator-exit critical Mach number $(V_1/V_{cr1}) = 0.8$
- Rotor flow passage is choked
  a) Assuming an isentropic flow field throughout the stage, calculate the stage work coefficient (ψ).
  b) Knowing that the rotor mean-radius incidence angle is zero, and assuming a spanwise free-vortex flow structure in the stator/rotor gap, calculate the rotor-blade incidence angle at the hub-section average radius.

10) Figure 8.68 shows the meridional view of the first in a multistage turbine section. The stage operating conditions are as follows:
- Rotational speed $(N) = 43{,}000$ rpm
- Inlet total pressure $(p_{t0}) = 10.0$ bars
- Inlet total temperature $(T_{t0}) = 1400$ K
- Flow coefficient $(\phi) = 0.555$
- Stator-exit flow angle $(\alpha_1) = 67.3°$

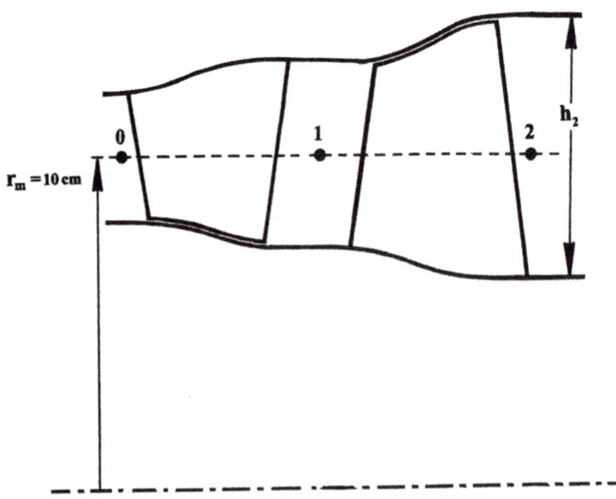

Figure 8.68. Input variables for Problem 10.

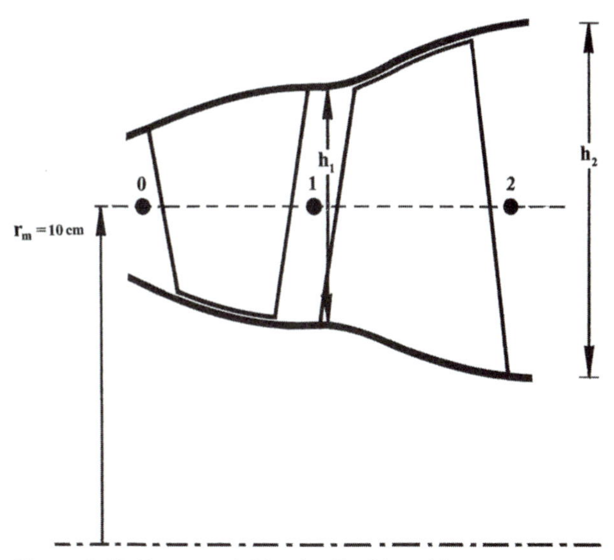

Figure 8.69. Input variables for Problem 11.

- Stage "actual" power output $(P) = 885$ kW
- Stage total-to-total efficiency $(\eta_{t-t}) = 88\%$
- Stage-exit swirl angle $(\alpha_2) = 18°$

The stator flow process is assumed isentropic, and the axial-velocity component (V_z) is taken as constant throught the stage. Considering the data and the stage major dimensions in the figure:

a) Calculate the mass-flow rate through the stage.
b) Calculate the total-to-total pressure ratio (p_{t0}/p_{t2}).
c) Calculate the stage work coefficient (ψ).
d) Calculate the stage specific speed (N_s).
e) Calculate the rotor-exit annulus height (h_2).

11) Figure 8.69 shows the meridional projection of an axial-flow turbine stage along with its major dimensions. The stage operating conditions are as follows:

- Inlet total pressure $(p_{t0}) = 8.2$ bars
- Inlet total temperature $(T_{t0}) = 1100$ K
- Stage entire flow process is assumed isentropic
- Rotor passage is choked
- Mass-flow rate $(\dot{m}) = 5.75$ kg/s
- Rotor speed $(N) = 42{,}000$ rpm
- Stator-exit swirl angle $(\alpha_1) = 68°$
- Stage flow coefficient $(\phi) = 0.46$
- Zweifel's optimum-loading criterion is applicable
- Rotor-blade aspect ratio $(\lambda) = 1.8$
- $V_{z1} = V_{z2}$

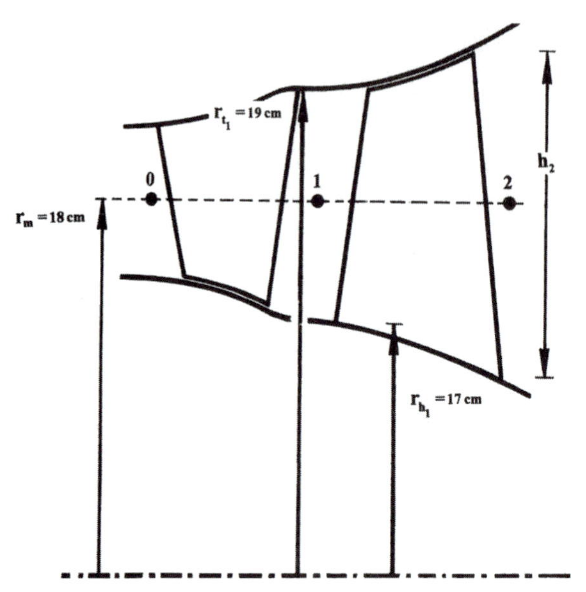

Figure 8.70. Input variables for Problem 12.

Calculate the following variables:

a) The rotor-exit swirl angle (α_2).
b) The stage work coefficient (ψ).
c) The rotor optimum number of blades (N_b).

12) Figure 8.70 shows the meridional flow path of an axial-flow turbine stage, together with the stage major dimensions. The stage design point is defined as follows:

- Stator flow is nearly isentropic and is assumed as such
- Inlet total pressure (p_{t0}) = 12.5 bars
- Inlet total temperature (T_{t0}) = 1450 K
- Stator-exit critical Mach number (V_1/V_{cr1}) = 0.84
- Mass-flow rate ($\dot{m}$) = 6.8 kg/s
- Rotational speed (N) = 29,000 rpm
- $V_{z1} = V_{z2}$
- Rotor-exit relative critical Mach number (W_2/W_{cr2}) = 0.92
- Stage total-to-total efficiency (η_{t-t}) = 86%

  a) Calculate the stage flow coefficient (ϕ)
  b) Calculate the stage work coefficient (ψ)
  c) Calculate the stage specific speed (N_s)
  d) Knowing that the optimum number of blades (N_b) is 32 (based on Zweifel's optimum loading criterion), calculate the rotor-blade aspect ratio (λ) based on the blade average dimensions.

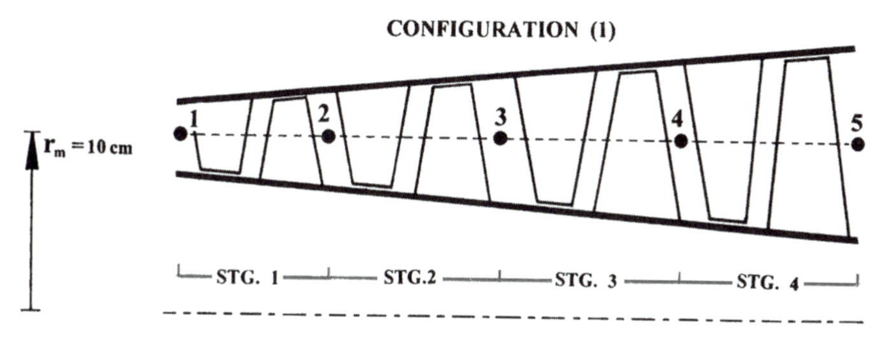

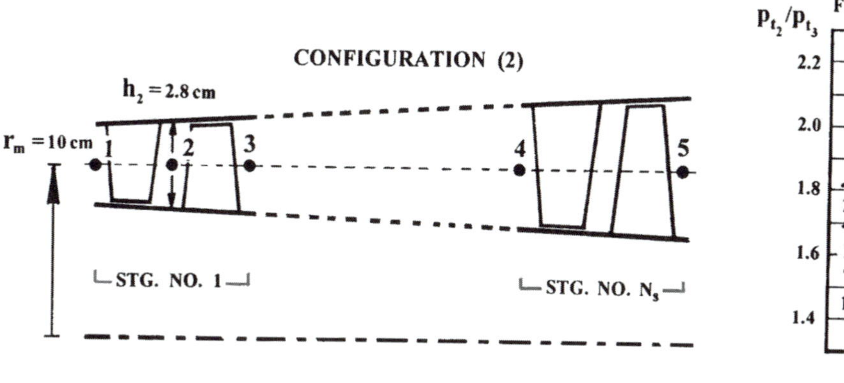

Figure 8.71. Input data for Problem 13.

13) Figure 8.71 shows an existing four-stage axial-flow turbine configuration, which is termed configuration 1. Each of these stages operates under a different set of operating conditions as follows:

- $(Pr)_{stg.1} = 2.1$ and $(\eta_{t-t})_{stg.1} = 84\%$
- $(Pr)_{stg.2} = 1.6$ and $(\eta_{t-t})_{stg.2} = 90\%$
- $(Pr)_{stg.3} = 1.9$ and $(\eta_{t-t})_{stg.3} = 88\%$
- $(Pr)_{stg.4} = 2.7$ and $(\eta_{t-t})_{stg.4} = 81\%$

where the symbol (Pr) refers to the inlet/exit total-to-total pressure ratio.

The arrangement just described is to be replaced by configuration 2, which is composed of a number of stages with identical total-to-total magnitudes of pressure ratio (1.8) and efficiency (82.1%). In both cases, the inlet conditions are:

- Inlet total pressure $(p_{t1}) = 17.8$ bars
- Inlet total temperature $(T_{t1}) = 1460$ K

In both configurations, isentropic flow processes are assumed to prevail across all stators.

a) Considering the first turbine configuration, calculate the overall total-to-total magnitudes of pressure ratio and efficiency.

b) Considering the second configuration, calculate the total number of stages.
c) Staying with the second configuration, the map corresponding to the first stage is also provided in Figure 8.71. Using this map and the inlet total properties, calculate the stage flow coefficient (ϕ), the stage work coefficient (ψ), and specific speed (N_s). In doing so, the following conditions are applicable:

- The stator-exit critical Mach number (M_{cr2}) = 0.92
- The stator-exit swirl angle (α_2) = 68.5°

14) The high-pressure turbine section of a turbofan engine is to be designed according to the following specifications:

- Turbine-inlet flow angle = 16°
- Mass-flow rate ($\dot{m}$) = 3.18 kg/s
- Total-to-total efficiency of each stage (η_S) is estimated at 89%
- Turbine-inlet total pressure (p_{t1}) = 13.3 bars
- Turbine-inlet total temperature (T_{t1}) = 1495 K
- Turbine overall total-to-total pressure ratio = 3.78
- Rotational speed (N) = 43,800 rpm

The turbine design is also subject to the following set of constraints:

- Number of stages should not exceed two
- Mean-radius blade velocity (U_m) is 395 m/s
- Maximum allowable first-stage work coefficient ($\psi_{stg.1}$) is 1.55
- Flow coefficient (ϕ) for each stage should not be less than 0.3
- Aspect ratio of the last-stage rotor should not exceed 2.0

Assuming an adiabatic flow throughout the turbine and an average specific heat ratio (γ) of 1.33:

a) Design and draw (to scale) the meridional flow path, showing the meridional views of the stator vanes and rotor blades.

b) Assuming a spanwise free-vortex flow conditions and zero incidence and deviation angles of zero, sketch (with reasonable accuracy) the blade airfoil sections of the first-stage rotor at the hub, mean, and tip locations, showing the "metal" inlet and exit angles. Note that the free-vortex simplification applies at the rotor exit station just as it does at the rotor inlet station.

c) Justify, on the basis of preliminary specific-speed magnitudes, the choice of each stage to be of the axial-flow type.

d) Calculate the turbine overall total-to-total efficiency (η_T), comparing it with the above supplied stage efficiency.

CHAPTER NINE

Axial-Flow Compressors

Introduction

The utilization of axial-flow compressors (Fig. 9.1) in gas-turbine engines has been relatively recent. The history of this compressor type began following an era when centrifugal compressors (Fig. 9.2) were dominant. It was later confirmed, on an experimental basis, that axial-flow compressors can run much more efficiently. Earlier attempts to build multistage axial-flow compressors entailed running multistage axial-flow turbines in the reverse direction. As presented in Chapter 4, a compressor-stage reaction in this case will be negative, a situation that has its own performance-degradation effect. Today, carefully designed axial-flow compressor stages can very well have efficiencies in excess of 80%. A good part of this advancement is because of the standardization of thoughtfully devised compressor-cascade blading rules.

Comparison with Axial-Flow Turbines

In passing through a reaction-turbine blade row, the flow stream will continually lose static pressure and enthalpy. The result is a corresponding rise in kinetic energy, making the process one of the flow-acceleration type. In axial compressors, by contrast, an unfavorable *static* pressure gradient prevails under which large-scale losses become more than likely. It is therefore sensible to take greater care in the compressor-blading phase. Another major difference, by reference to Figure 9.3, is that the compressor meridional flow path is geometrically converging as opposed to the typically diverging flow path of a turbine. This is a direct result of the streamwise density rise in this case, as will be discussed later. Referring to Figure 9.4, another difference in this context is a substantially greater blade count compared wth axial-turbine rotors, where the number of blades is typically in the low twenties. A similar observation can be made in reference to the ducted fan in Figure 9.5. Used primarily in turbofan engines, ducted fans are basically compressors, except that they operate under substantially low total-to-total pressure ratios. In these two figures, note that

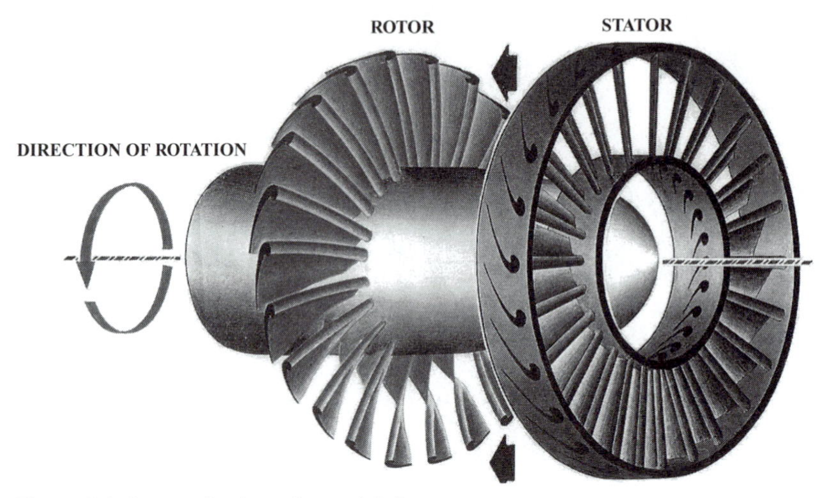

Figure 9.1. Isometric view of an axial-flow compressor stage.

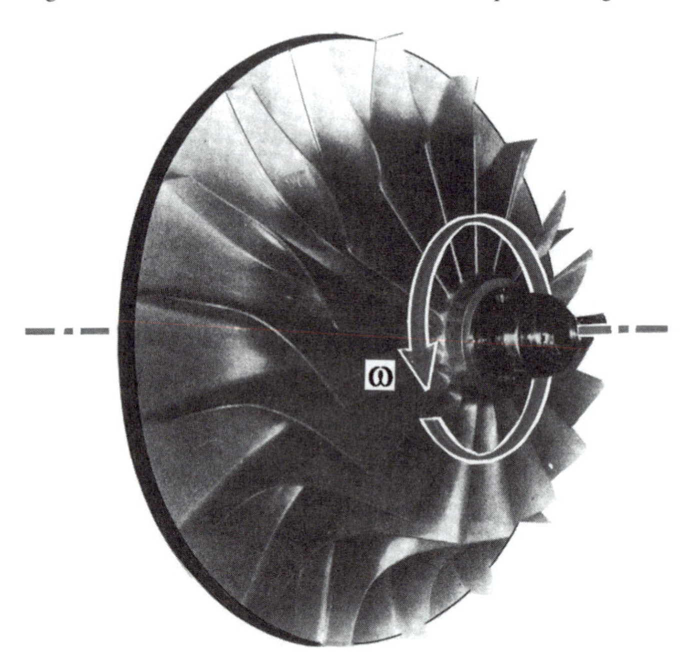

Figure 9.2. Centrifugal-compressor stage.

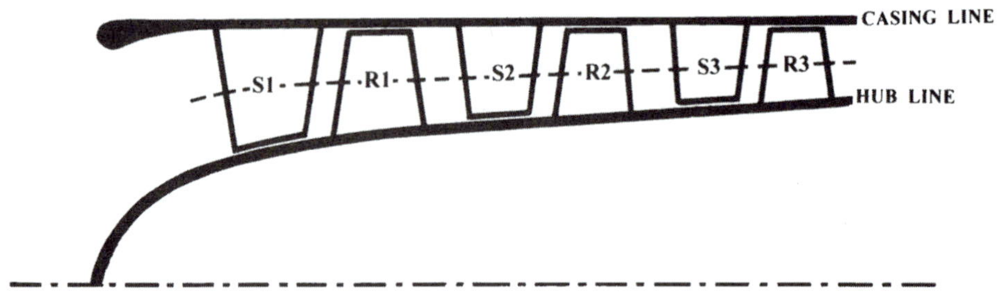

Figure 9.3. Multiple staging and meridional flow path.

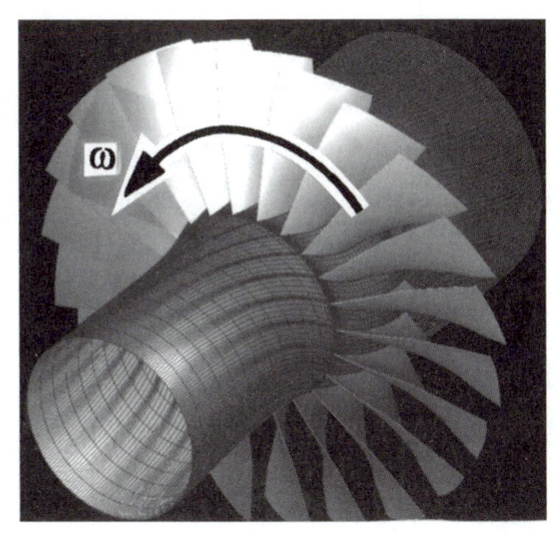

Figure 9.4. High-solidity-ratio rotor of an axial compressor.

the blade pressure side is leading the suction side in the direction of rotation, which is opposite the rotation direction of turbine rotors.

Whereas the turbine operating region is limited by the choking line, which represents a state that is mechanically harmless, the corresponding allowable/forbidden operation interface in a compressor is the surge line, where the compressor is facing mechanical failure. This is the reason behind the added emphasis in this chapter on the compressor off-design operation, including the phase of startup. However, such variables as the flow coefficient, stage reaction, and work coefficient will have the same definitions and similar implications as presented for axial-flow turbines in Chapter 8.

Figure 9.5. Ducted fan for use in turbofan engines.

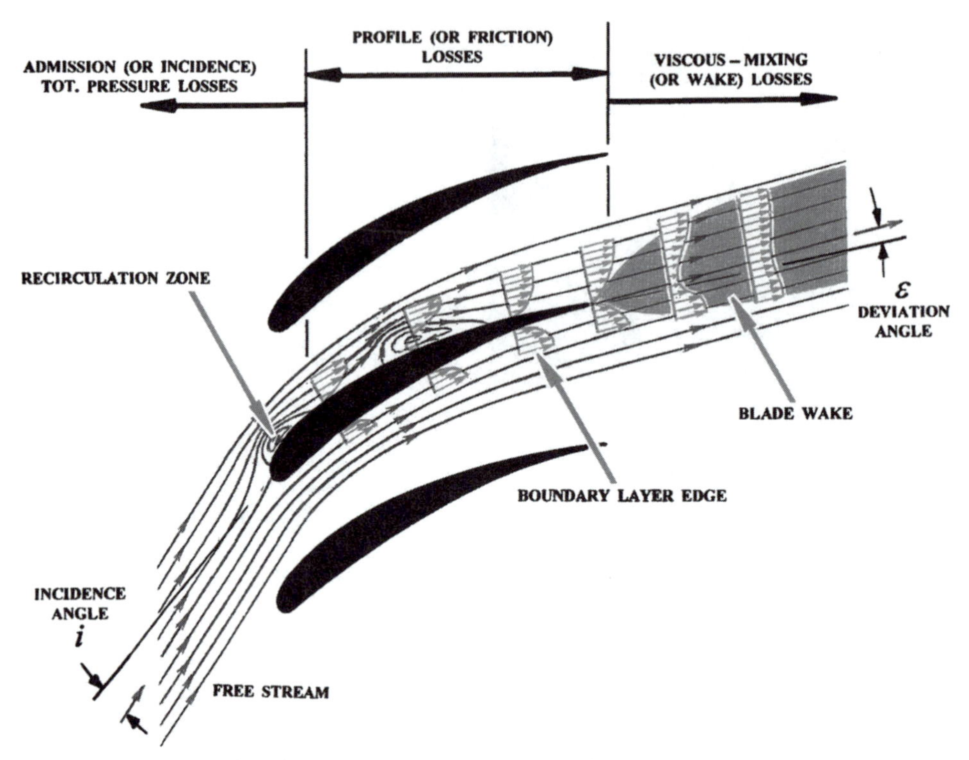

Figure 9.6. Loss classification over an axial-compressor airfoil cascade.

Stage Definition and Multiple Staging

Traditionally, a compressor stage is exclusively defined as a rotor followed by a stator, with a finite axial gap in between. In this chapter, however, we will consider the stage to be composed of a stator followed by a rotor, an option that simplifies the construction of the rotor-inlet velocity triangle, with no violation of the general rules.

Multiple staging is customary in axial-flow compressor sections. This is due to some strict limitations that are typically imposed on the stage total-to-total pressure ratio. A large stagewise pressure ratio will naturally worsen the already unfavorable pressure gradient, which in turn causes different loss mechanisms, threatening to offset most of the intended total-pressure rise. Figure 9.6 shows a sketch of the major loss mechanisms in this case, including flow separation and recirculation over the suction side in particular. Downstream from the trailing edge, however, is the airfoil "wake" (discussed in Chapter 4), where more losses are encountered. These real-life effects give rise to an exit-flow underturning, which in Figure 9.6 is represented by a finite deviation angle. Although elimination of these adverse flow effects is virtually impossible, it is possible, nevertheless, to suppress them by dividing the total power absorbed over a number of stages with smaller pressure ratios.

In addition to the familiar velocity triangles in Figure 9.7, a T-s representation of the stage flow process is provided as well. Note that the entropy production in this

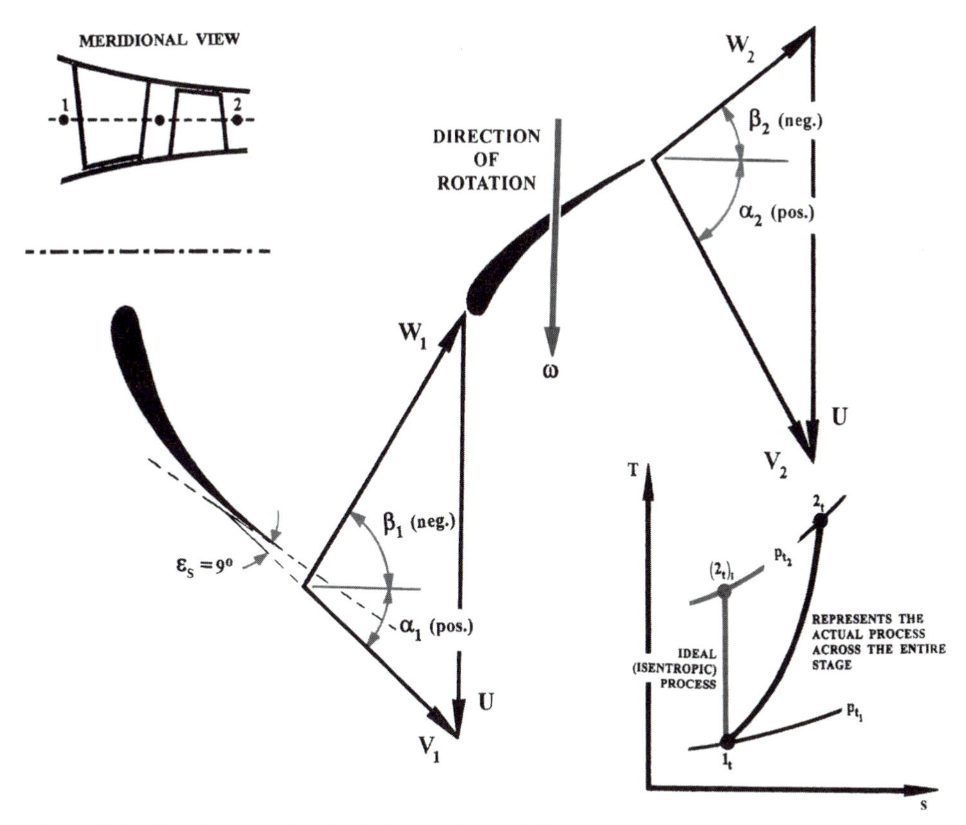

Figure 9.7. Velocity triangles, deviation angle, and T-s representation.

case will be far greater than that of a comparable-size turbine stage. Referring to the stator subdomain in Figure 9.7, the exit swirl angle (α_1) is seen to deviate from the airfoil (or metal) angle, with an indicated deviation angle (ϵ_S) of 9°.

The term *loss* here may be equally viewed as a decline in kinetic energy, as is clearly visible within the two recirculation zones in Figure 9.6. Better viewed in the rotating frame of reference, the rotor subdomain will suffer the same decline. The exception, however, is in the relative velocity (W), which will display this type of wasteful motion instead.

A perfectly axial-flow compressor stage, in the sense of horizontal endwalls, is simply a bad idea. Contrary to a turbine flow process, the (static) density will continually rise in the streamwise direction, a feature that is as true within a rotor as it is through a stator cascade. From a mass-conservation standpoint, in this case, the most direct means to obtain a much-desired constant through flow velocity is to monotonically decrease the cross-flow area. The convergent meridional flow path, being the net outcome, is schematically shown in Figure 9.7.

Figure 9.8 presents a typical rotor-flow process from the static-to-static, total-to-total, and total-relative-to-total-relative standpoints. Of these, the last is seen to be a horizontal process, implying a constant total relative temperature across the

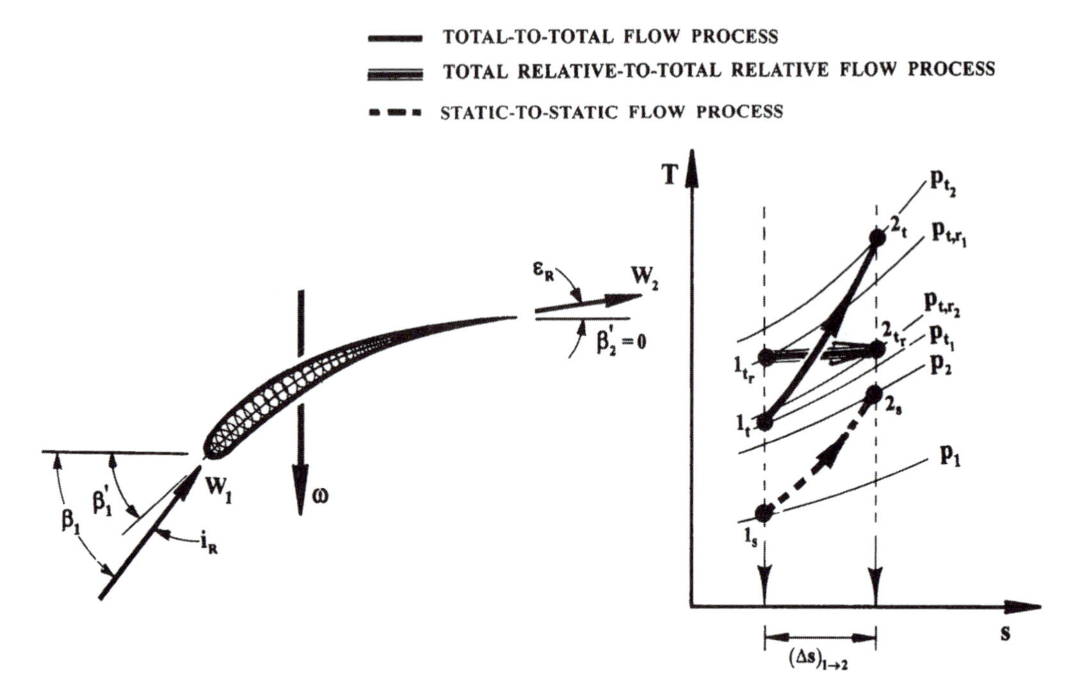

Figure 9.8. Introduction of total relative properties.

rotor. As discussed in earlier chapters, and assuming an adiabatic flow process, this result is precisely equivalent to the equality of total temperature across an adiabatic stator under the same assumption. As is seen in Figure 9.8, the static pressure and temperature will always be less than their total and total relative counterparts. The reason is that a total property is, by definition, composed of a static contribution plus a nonzero dynamic part. As for the total relative properties, Figure 9.8 shows that $(p_{t_r})_1$ and $(T_{t_r})_1$ at the inlet station are less than their total-property counterparts. Figure 9.8 also shows that the relationship is quite the opposite at the rotor exit station. Regardless of how the rotor-flow process is viewed, as seen in Figure 9.8, the amount of entropy production remains exactly the same, as would be expected.

Normal Stage Definition

A normal stage is one where the stage inlet and exit absolute-velocity vectors $(\mathbf{V}_{in})$ and $(\mathbf{V}_{ex})$ are identical in magnitude and direction. A multistage compressor section, which is composed of normal stages, will produce identical velocity triangles over each and every stage. It is worth emphasizing, however, that such properties as total temperature and pressure will continue to rise in the streamwise direction if any shaft work is to be transmitted to the compressor. Figure 9.9 shows the velocity triangles associated with a typical stage in a multistage compressor section. An important feature in this figure is that the stage geometry itself is the same throughout the entire compressor. The design of a compressor such as this is easy, as it involves

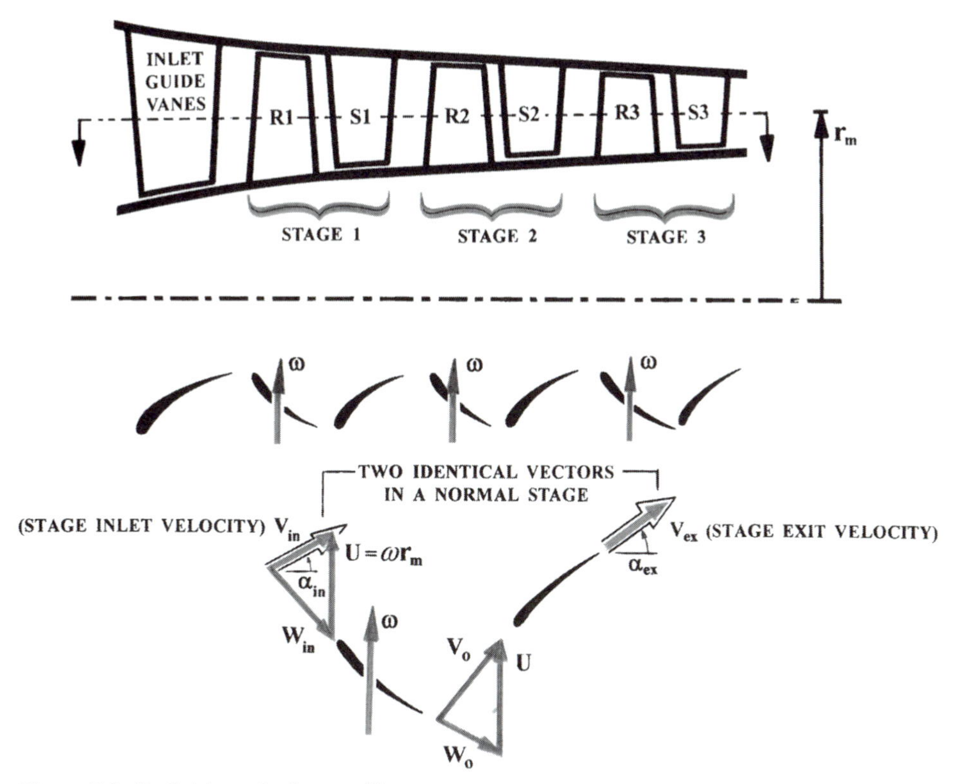

Figure 9.9. Definition of a "normal" stage.

designing and then repeating a single stage, saving (in the process) a great deal of design effort. What may seem odd in this context is the commonality of this particular multistage compressor configuration in gas-turbine applications. Figure 9.10 is an example of this fact, as it relates to the high-pressure compressor section of a turbofan engine.

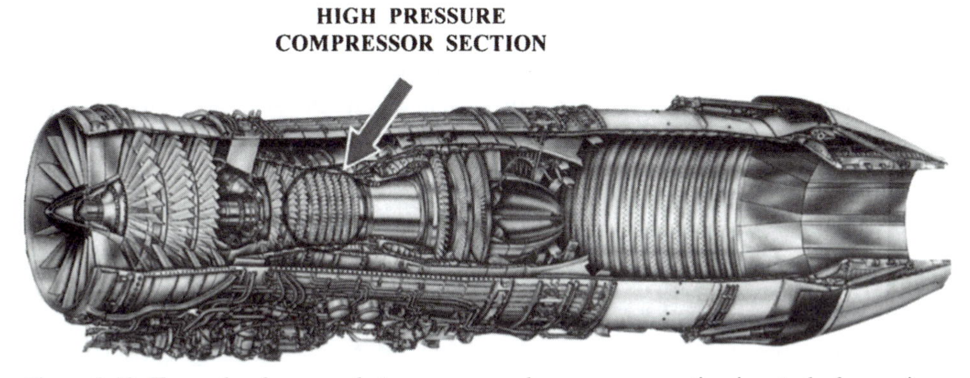

Figure 9.10. Example of a normal-stage-composed compressor section in a turbofan engine.

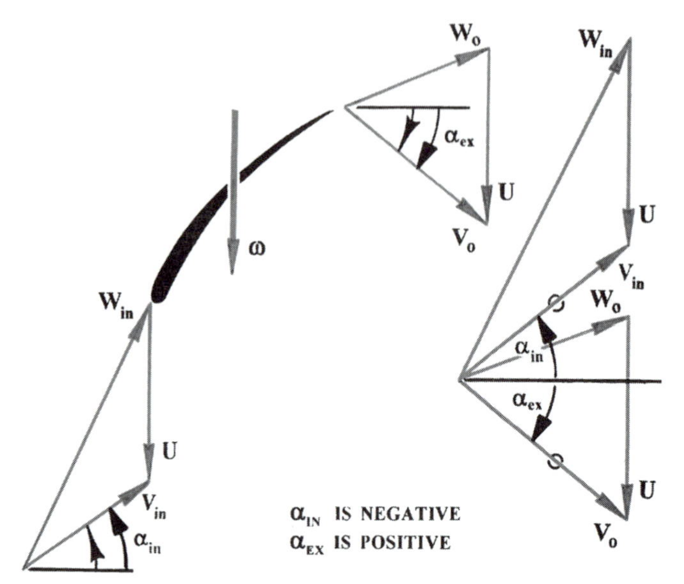

Figure 9.11. Example of a nonnormal stage despite the equality of inlet and exit absolute velocities.

Despite some kinematical similarities between the compressor stages in Figures 9.9 and 9.11, the latter is not a normal stage. The velocity triangles in Figure 9.11 satisfy the equality of inlet- and exit-velocity magnitudes. The triangles also show that the absolute value of the exit (swirl) angle is equal to its inlet counterpart. However, these two angles have different signs, where the stage-inlet flow angle is opposite to the direction of rotation (i.e., negative) and the exit angle is positive. The result is a nonparallel pair of inlet- and exit-velocity vectors.

Standard Airfoil Profiles

Compressor blade profiles are relatively simple by comparison with those of axial-flow turbines, where a virtually unlimited geometrical arbitrariness is overwhelmingly the rule. The compressor relative geometrical simplicity, however, should in no way mask the challenges stated earlier, which are exclusive to compressors. In the following, two airfoil subgroups, namely the British C4 and the American NACA series airfoil families, are presented and discussed.

The C4 Airfoil Family

In the British system, the two elements making up an airfoil are the base profile, being symmetrical, and the camber line. In its totality, the design process is executed with a predetermined airfoil-surface pressure distribution (i.e., aerodynamic loading) as the ultimate objective. A repetitive trial-and-error cascade design process to meet the given aerodynamic task, as in this case, is categorically known as the "indirect" design problem and is, by classification, tedious.

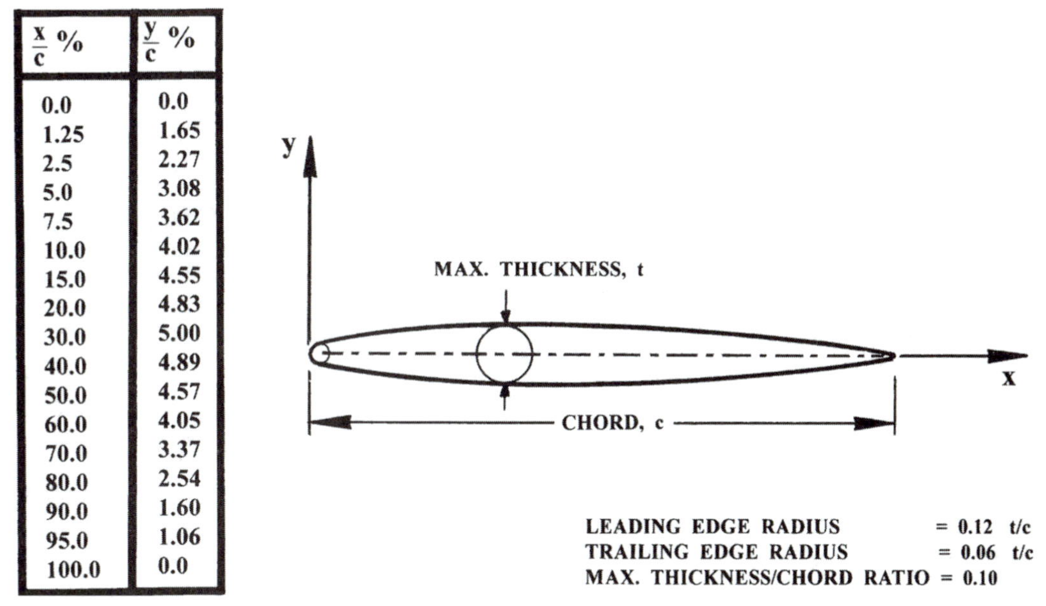

$\frac{x}{c}$ %	$\frac{y}{c}$ %
0.0	0.0
1.25	1.65
2.5	2.27
5.0	3.08
7.5	3.62
10.0	4.02
15.0	4.55
20.0	4.83
30.0	5.00
40.0	4.89
50.0	4.57
60.0	4.05
70.0	3.37
80.0	2.54
90.0	1.60
95.0	1.06
100.0	0.0

Figure 9.12. A typical base profile of the C4 airfoil family.

As shown in Figure 9.12, the definition of the base profile involves the ordinates of selected points on the airfoil surface, the maximum thickness magnitude, and where it is located. In a transonic Mach number environment, the t/c magnitude in Figure 9.12 is recommended to be as small as possible. In general, a maximum t/c magnitude of roughly 10% was shown to be acceptable. The chord-wise location of this point may vary from 30 to 50% of the chord length, with a nearly optimum location being around 40%.

The airfoil leading edge has to be round for reasons involving mechanical integrity as well as tolerance to changes in the incidence angle under off-design operating modes. Typical magnitudes of the leading-edge radius range between 8 and 12% of the airfoil maximum thickness. The optimum suction- and pressure-side stream mixing, in the immediate vicinity of the trailing edge, is attainable through a theoretical "cusp" trailing-edge configuration. This not only requires a zero trailing-edge thickness but also suction- and compressor-side tangents which are on top of one another. From both stress and fabrication viewpoints, such a trailing-edge configuration is practically out of consideration. In reality, a trailing-edge thickness of between 2% and 6% of the maximum thickness is considered a reasonable compromise. A minimum "included" angle between the tangents of the suction and pressure sides at the trailing edge is also recommended.

The other major step in defining the compressor airfoil is the camberline definition. This is normally a circular or parabolic arc, as shown in Figure 9.13. In defining this arc, the position and magnitude of maximum camber, a/c and b/c, are normally subject to optimization and compromise. Another critical camberline feature is the airfoil deflection angle, θ in Figure 9.13. To prevent flow separation, particularly over

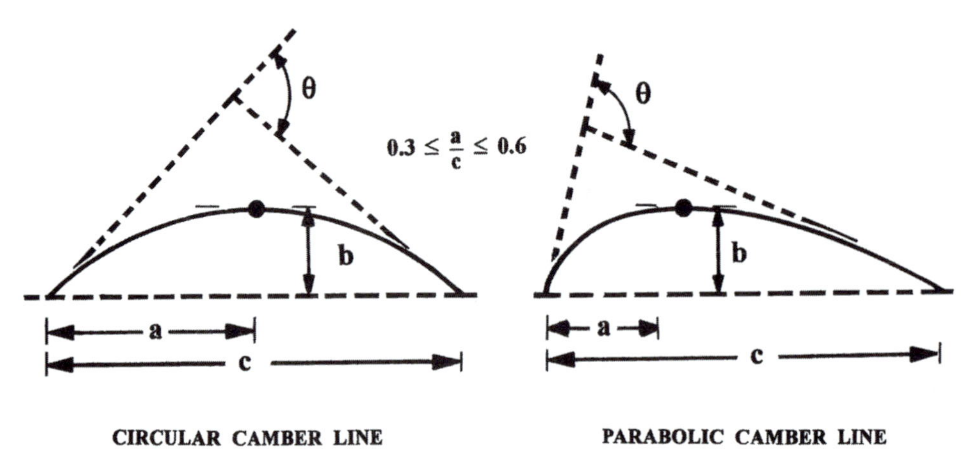

Figure 9.13. Circular versus parabolic mean camber line for use with the C4 family of airfoils.

the suction side, this angle should be as small as possible, which is in direct contrast with turbine airfoils.

A convincing amount of airfoil tests have established that an a/c ratio in the range of 0.3 to 0.6 would give rise to a high work-absorption capacity as well as a high blading efficiency. On the other hand, smaller magnitudes of this ratio give rise to a wider working range. In situations where the maximum Mach number is expected to be small, an a/c ratio in the neighborhood of 0.5 is generally acceptable.

The nomenclature of this airfoil-definition method is important to comprehend. For example, consider the following nomenclature:

$$\underline{10\ \text{C4}/30\ \text{P}\ 40}$$

This identifies a C4 airfoil with a t/c magnitude of 10%, a θ value of 30°, and a parabolic camber line with an a/c value of 40%. The different geometrical variables just cited are all defined in Figures 9.12 and 9.13.

The NACA Cascade Designation

The NACA approach to cascade data is somewhat different. Figure 9.14 shows the NACA nomenclature, where the tangent to the airfoil concave (or pressure) side is used as a datum.

In addition to the different symbols in Figure 9.14, the following variables are also used in the airfoil-cascade definition:

- The angle θ is referred to as the camber angle.
- The c/S ratio is recognized as the cascade "solidity" ratio.

Of these, the latter variable carries the same interpretation as its turbine counterpart in Chapter 8. In NACA nomenclature, the amount of airfoil camber is represented by the lift coefficient (C_{L0}) of a single isolated airfoil in a free air stream, as opposed to a member of an airfoil cascade. Note that the "lift coefficient" phrase is

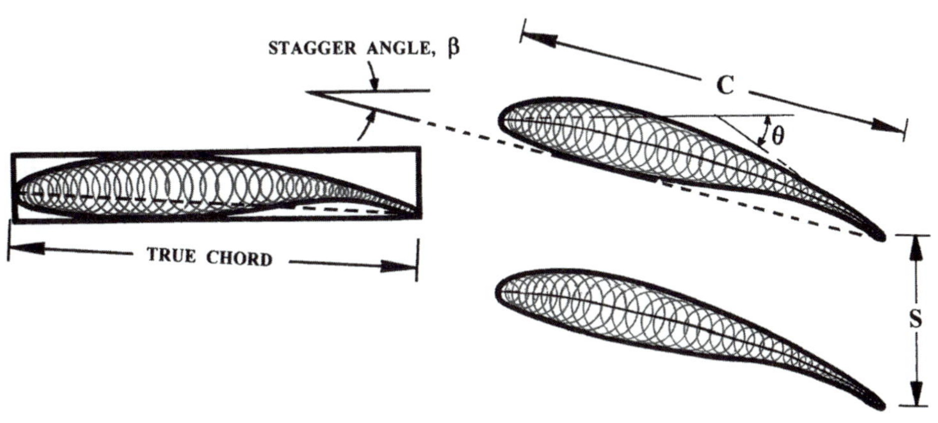

Figure 9.14. NACA cascade nomenclature.

exclusively an external-aerodynamics term that is particularly famous in the area of wing theory.

Definition of the blade profile within the NACA framework is much more complex by comparison with the C4 family. The reason, in part, is the existence of many configurations, namely the 4-, 5-, 6-, and 7-digit series. Fortunately, the airfoil configuration most used in cascade work is specifically one of the 6-digit series. The corresponding subset of airfoils within this category is commonly referred to as the 65-series, with the digit "5" denoting the chordwise position of minimum pressure (in tenths of the chord) in the basic symmetrical airfoil version at the zero-lift position. An example of such an airfoil designation method is

$$\underline{65 \ - \ (12)\,10}$$

which refers to a "NACA 65" airfoil type with a lift coefficient of 1.2 and a t/c ratio of 10%. Note that the lift coefficient (C_L) of a given airfoil versus the angle of attack is easy to obtain in a simple wind tunnel, a process that is more like routine work in external aerodynamics.

Real Flow Effects

Effect of the Incidence Angle

Figure 9.15 shows the variation of the total-pressure loss coefficient ($\bar{\omega}$) and the flow deflection angle ϵ (different from that of the camberline) versus the incidence angle. Note that the latter graph in no way suggests that $\bar{\omega}$ and ϵ are functions only of the incidence angle i. Other contributing factors include such aspects as the growth of the boundary layer and the onset of flow separation, if any. As was the case in axial-turbine cascades, the loss coefficient ($\bar{\omega}$) is much more damaging under positive incidence angles.

Referring to Figure 9.15, stalling is defined as the state at which the total-pressure loss coefficient ($\bar{\omega}$) is twice its minimum value. The flow-deviation angle (ϵ) in the figure can logically be thought of as depending on such variables as the amount of

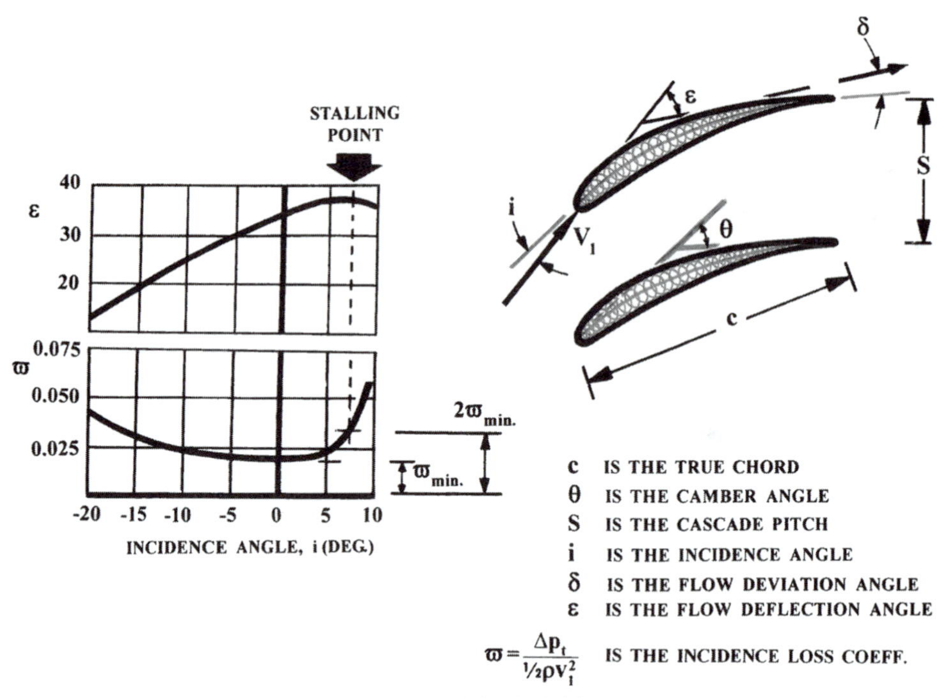

Figure 9.15. The effect of incidence angle and the definition of stalling.

flow turning, which is proportional to the camber angle (θ), as well as the level of flow guidedness and the blade profile itself. An empirical expression of this angle is

$$\delta = m\theta\sqrt{\frac{S}{C}}$$

where S is the mean-radius cascade pitch and C is the true airfoil chord. Figure 9.16 shows the functional dependency of the factor m on the airfoil stagger angle β.

Effect of the Reynolds Number

The Reynolds number (Re) is defined as

$$Re = \frac{V_{in}C}{\nu}$$

where V_{in} is the approach velocity. The same definition applies to a rotating cascade, provided that the inlet relative velocity W_{in} is used instead. A Reynolds number magnitude that is below the critical value of 2×10^5 leads to high profile losses as the flow field in this case is categorized as being dominated by viscosity (rather than inertia). Enhancement of the cascade performance through Reynolds number elevation, becomes minor once the critical Reynolds number magnitude is exceeded.

The critical Reynolds number magnitude, in reality, is partially influenced by the level of turbulence at the cascade inlet station. In general, low profile losses are obtained at a high Reynolds number and a low turbulence intensity, for the latter

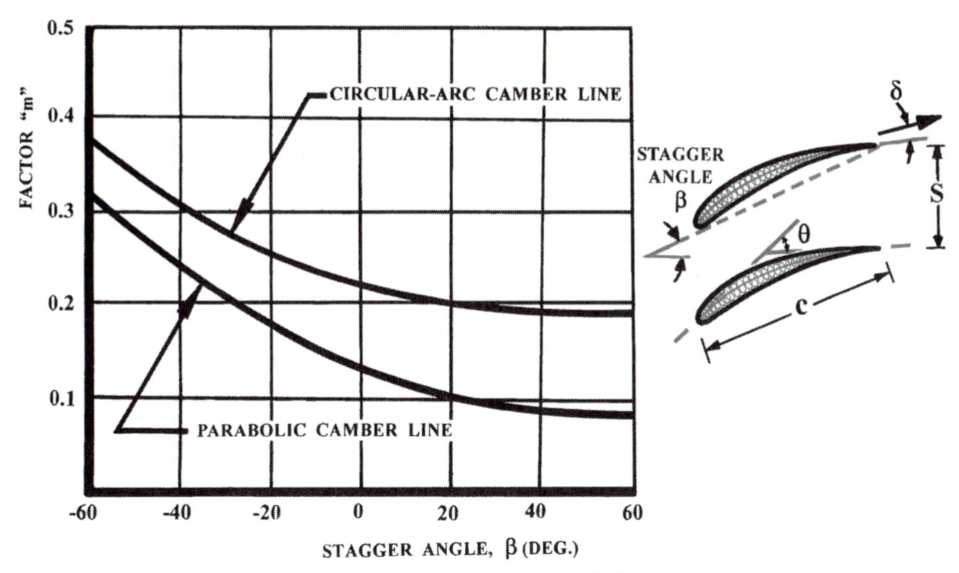

Figure 9.16. Determination of the factor m for loss calculations.

delays the laminar-to-turbulent boundary-layer transition. A different argument is true at low Reynolds numbers, where a high turbulence level will have the effect of delaying boundary-layer separation along the blade suction side.

Effect of the Mach Number

As the cascade-inlet Mach number is increased, the local velocity at some point on the airfoil surface may reach the sonic magnitude. The magnitude of the *inlet* Mach number at this point is termed the "critical" Mach number (M_C), not to be confused with M_{cr}, the local-velocity/critical-velocity ratio. The most important consequence of this state is a faster increase of the drag coefficient (C_D), which is indicative of the friction forces over the airfoil surface. As C_D reaches a magnitude that is 1.5 times its minimum value, the incoming Mach number is referred to as the drag-critical Mach number (M_{DC}). A further increase in the inlet Mach number will cause a notably rapid increase in the drag coefficient and therefore a fast decline in the pressure ratio toward a state where the rise in total pressure is totally offset by the total pressure loss. As this state prevails, the inlet Mach number is said to have its "maximum" magnitude ($M_{max.}$) because the situation at this point is that of choking. A compressor blade should preferably have high magnitudes of M_{DC} and $M_{max.}$, particularly the latter, for the reason that the interval ($M_{max.} - M_{DC}$) determines what is called the airfoil "working" range.

Tip Clearance Effect

In unshrouded rotors, being the norm in axial-flow compressors, both direct and indirect tip leakage (Fig. 8.28) will coexist. Of course, the difference here (in comparison with a turbine rotor) is that the direct-leakage flow stream is now in the direction opposite to that of the primary flow. On top of being unavailable for shaft-work absorption, this secondary flow stream is capable of producing more aerodynamic damage by comparison. One example (fully explored in Chapter 11) is the distortion of the rotor-inlet flow profile at and near the tip, as the two streams mix together.

A design-oriented quantification of the axial-compressor tip-clearance impact is that of Lakshiminarayana (1971). Systematically utilized in the preliminary design phase in particular, Lakshiminarayana's relationship accounts not only for the tip clearance but also the blade aspect ratio, the flow and work coefficients, and the tip-section average swirl angle. Relative to a baseline efficiency (one that does not include the tip-clearance effect), the relationship provides the efficiency decrement ($\Delta\eta$) caused by tip leakage in the following form:

$$\Delta\eta = \frac{0.07\bar{c}\psi}{\cos\alpha_{t,av}}\left(1.0 + 10.0\sqrt{\frac{\phi}{\psi}\frac{\bar{c}\lambda}{\cos\alpha_{t,av}}}\right)$$

where

$\bar{c}$ is the ratio of the tip clearance to the average blade height,
λ is the blade aspect ratio (height/average true chord),
$\alpha_{t,av}$ is the tip-radius average swirl angle,
ψ is the stage work coefficient,
ϕ is the flow coefficient.

Figure 9.17 shows a graphical illustration of the preceding relationship, where ψ and ϕ are fixed at the magnitudes of 0.8 and 0.6, respectively, and the aspect ratio (λ) at 1.5. Note the significant impact of the tip average swirl angle ($\alpha_{t,av}$) on the loss level in this figure.

Compressor Off-Design Characteristics

Upon development of a new compressor, the product is normally subjected to a sequence of tests in order to establish its performance characteristics. As presented in Chapter 5, the compressor pressure ratio ($p_{t\,ex}/p_{t\,in}$) is treated as a function of the "pseudo-dimensionless" variables $N/\sqrt{\theta_{in}}$ (the corrected speed) and $\dot{m}\sqrt{\theta_{in}}/\delta_{in}$ (the corrected mass-flow rate). A sketch of a typical compressor test facility and a typical compressor map are both shown in Figure 9.18. In the following, we focus on critical operating modes that, in the absence of a quick remedy, may cause a great deal of mechanical problems.

Rotating Stall and Total Surge

The line identified as the "surge line" in Figure 9.18 represents the limit of the pressure ratio that is attainable for any given corrected-speed magnitude. Signs of this critical state begin to appear when the compressor pressure rise is so extreme

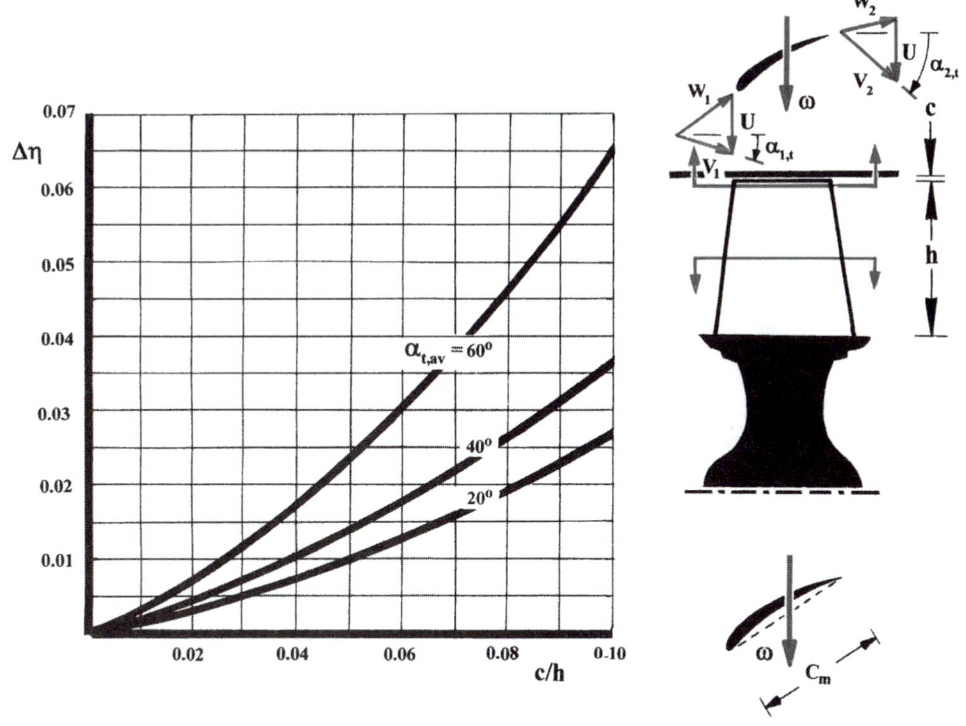

Figure 9.17. Effect of primary design variables on the tip-leakage phenomenon.

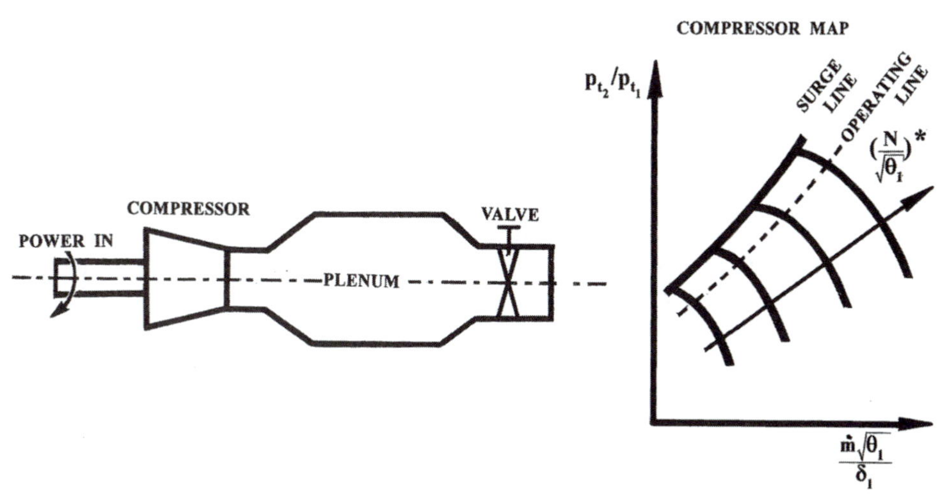

Figure 9.18. Rig simulation of the compressor behavior close to the surge line. * The arrowhead indicates the direction in which the corrected speed is increasing.

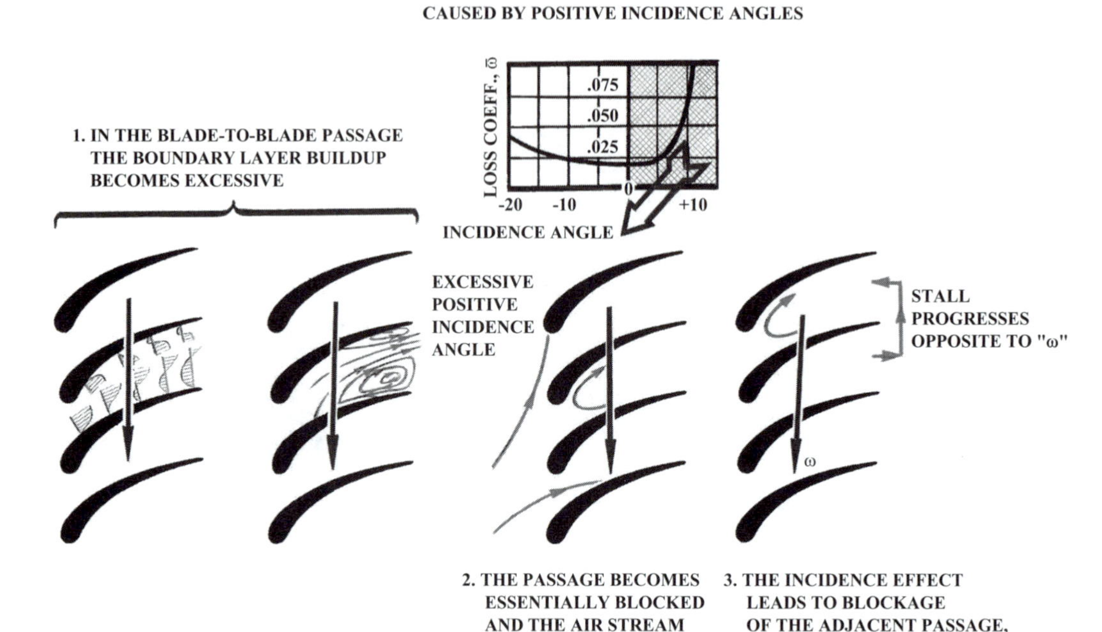

Figure 9.19. Sequence of events leading to the onset of rotating stall.

that the resulting adverse pressure gradient causes boundary-layer separation over a substantial blade segment. At this point, several forms of flow instability can occur. Principal among these is a state that is referred to as "rotating stall" and can very well be followed by a total compressor "surge."

The mechanism of rotating stall is rather complicated. Figure 9.19 spells out a sequence of events leading up to this unstable phase of compressor operation. The figure shows a packet of fluid that, under a large adverse pressure gradient, undergoes a severe flow reversal. To the incoming flow stream, the flow reversal in this particular passage essentially makes the passage appear like a totally blocked blade-to-blade channel. Under a subsonic inlet-flow situation, the so-called downstream effect will make it possible for the incoming flow stream to sense the existence of such an obstacle prior to physically getting there. This flow capability leads the incoming flow stream to branch out (as shown in Figure 9.19) so as to bypass the blocked flow channel. Such flow behavior will alter the incidence angles in the two passages on both sides of the blocked region. The adjustment, by reference to Figure. 9.19, is in the form of excessive negative incidence for the passage that is leading the blocked region, in the rotation direction, and a damagingly high positive incidence for the other passage. This will have the effect of "unstalling" the former while stalling the latter. As this continues to happen, the situation will be one where each flow passage will stall and then quickly unstall in a direction that is opposite to that of rotation. The inception of rotating stall is highly relevant to the rotor mechanical integrity. In fact, the amplitude of compressor vibration, as a result of this unsteady operation, may very well cause an abrupt mechanical failure unless quick action (e.g., speed reduction) is taken.

Compressor surge is the state reached as a substantial fraction of the rotor blades simultaneously reach their load-carrying capacity. As a result, a flow breakdown occurs and the compressor becomes incapable of supporting the overall pressure rise, with massive flow-reversal subregions appearing in the blade-to-blade passages. The frequency of such flow reversals is a function of the "storage volume" following the compressor, as well as the compressor behavioral characteristics themselves. In physical terms, the flow reversal will continue for a small fraction of a second until the back pressure is sufficiently small to bring the flow stream back to its forward direction. The duration of each such cycle is extremely short, giving rise to a violent sequence of vibrations, which are distinctly loud. Short of a quick remedial response (e.g., an immediate increase in the mass-flow rate or decrease in the shaft speed), a catastrophic mechanical failure can, and does, occur.

One of the "horror" stories in connection with the surge phenomenon has to do with a small turbofan-propelled airplane during its ground idle operating mode in the late 1980s. The compressor surge in this incident occurred in the compressor section of an auxiliary power unit (APU) placed in the fuselage, normally for such tasks as cockpit air conditioning. The violent APU vibrations in this case sent the APU flying, with a hazardous amount of momentum, to the outside, easily penetrating the skin of the airframe. The bottom line here is that compressor surge is a highly hazardous operating mode. Monitoring the events leading to such a compressor operating mode is therefore critical so that timely corrective action can be externally imposed.

The smallest distance between the compressor operating line (Chapter 12) on the compressor map (Fig. 9.18) and the surge line is referred to as the surge "margin." The ultimate dilemma here is that the most efficient compressor performance is in a region that is very close to nothing but the surge line itself. A careful balance must therefore be struck between selecting an overly large surge margin with poor steady-state performance, and a small margin that may compromise the compressor very issue of existence.

Compressor Behavior during Start-up

Figure 9.20 shows a schematic of a compressor section. As stated earlier, the overall contraction in the meridional flow path has the effect of ensuring a streamwise-fixed axial-velocity component. Under the design-point operating mode, the combination of this velocity component and the geometrical features of airfoil cascades collaborates to produce the optimum flow angles. However, the continuity of the axial-velocity component will be destroyed as the compressor is operated under an off-design pressure ratio. The reason is that the exit-over-inlet axial-velocity ratio will depend on the density ratio. This in turn is a function of the pressure ratio; namely the variable that is being altered.

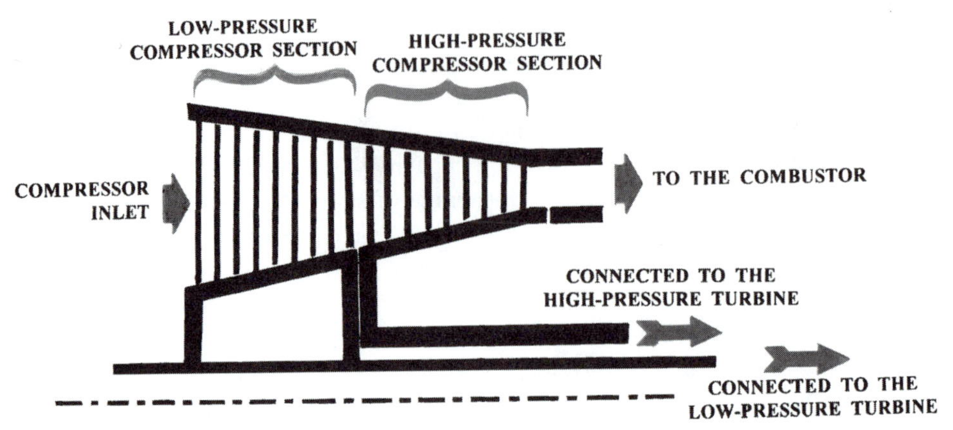

Figure 9.20. Symbolic representation of a coaxial-shaft (twin-spool) multistage compressor.

During the start-up process, the compressor will be running at substantially low speeds. With this being a major contributor in the energy-exchange process across each stage, the low speed will, in effect, reduce the stage shaft-work intake and hence the overall pressure ratio. Under such circumstances, the compressor will give rise to a significantly low density ratio, which, in turn, will raise the axial velocity component. This may lead the rearmost stage(s) producing negative total-pressure changes (i.e., acting like turbine stages in a pressure-ratio sense). In severe cases, this could lead to flow choking in this compressor segment. To summarize, the compressor start-up process is characterized by two undesirable features:

- The early blade rows will operate under extremely high positive incidence angles, leading to blade stalling and, potentially, total surge.
- The rearmost blade rows will operate under excessively negative incidence angles, providing an environment for turbine-like performance.

Means of Suppressing Start-up Problems

Between the viable threat of surge and the likelihood of rear stages causing flow expansion (as opposed to compression), the compressor start-up problems vary from mechanically threatening to aerodynamically degrading. Luckily, there are specific ways to more or less reduce, or totally extract, the most damaging features surrounding the start-up process. In the following, we discuss three different techniques toward this very task.

Utilization of Bleed Valves

Bleed valves can be employed to release air from the appropriate stage(s) in order to reduce the axial-velocity component. In the rearmost stages, this will have the effect of increasing the flow angles suppressing, in effect, the excessively negative incidence angles in these high-pressure stages. The net outcome in this case is to reduce or eliminate the likelihood that these stages will provide flow-expansion processes.

Utilization of Variable-Geometry Stators

Discussed conceptually in Chapter 3, this particular option is exclusively capable of suppressing the start-up problems on both ends of a multistage compressor. By varying the stator-vane setting angle, the downstream rotor can be guaranteed a practically acceptable incidence angle. Applied to the front stages, the large positive incidence angles, being the vehicle for the unstable rotor operation, can be controlled or even reversed. As for the rearmost stage(s), the aerodynamically degrading incidence angles, being excessively negative, will be favorably altered as well.

Utilization of Multiple Spools

Illustrated in Figure 9.20, the idea behind this option is to run different compressor segments at different appropriately chosen rotational speeds. The component arrangement in this figure is one where two coaxial shafts are used within the gas generator. First, a low-speed shaft connects the low-pressure (front) compressor section to the low-pressure (rear) turbine segment. From a compressor standpoint, the low-pressure, low-speed compressor segment will now be associated with moderately positive, or even negative, incidence angles. Under such circumstances, this particular compressor section may no longer be subject to stall or surge. As for the high-pressure compressor section, the shaft connecting it to the high-pressure turbine runs at a higher speed. The high-pressure, high-speed compressor section in this case is now sure to be the recipient of moderately negative, or even slightly positive, incidence angles. The likelihood of rear stage(s) expanding (rather than compressing) the flow stream may now be significantly reduced or even alleviated altogether.

EXAMPLE 1

Figure 9.21 shows an axial-compressor stage, where the flow process is adiabatic and the axial-velocity component (V_z) is constant. The stage operating conditions are as follows:

- Mass-flow rate ($\dot{m}$) = 2.2 kg/s
- Rotational speed (N) = 36,000 rpm
- Inlet total temperature (T_{t0}) = 450 K
- Inlet total pressure (p_{t0}) = 2.8 bars
- Rotor-wise loss in total relative pressure ($\frac{\Delta p_{tr}}{p_{tr1}}$) = 13.28%
- Stator flow process is assumed isentropic
- Stator-exit critical Mach number ($\frac{V_1}{V_{cr1}}$) = 0.68
- Stator-exit absolute flow angle (α_1) = 0
- Rotor-exit relative flow angle (β_2) = $-25°$.

Assuming an average specific-heat ratio (γ) of 1.4:

a) Calculate the stage total-to-total efficiency (η_C).
b) Calculate the torque transmitted to the compressor stage (τ).
c) Calculate the rotor-exit static pressure (p_2).
d) If the rotor-blade incidence angle is $-12°$, determine the blade-inlet (metal) angle.

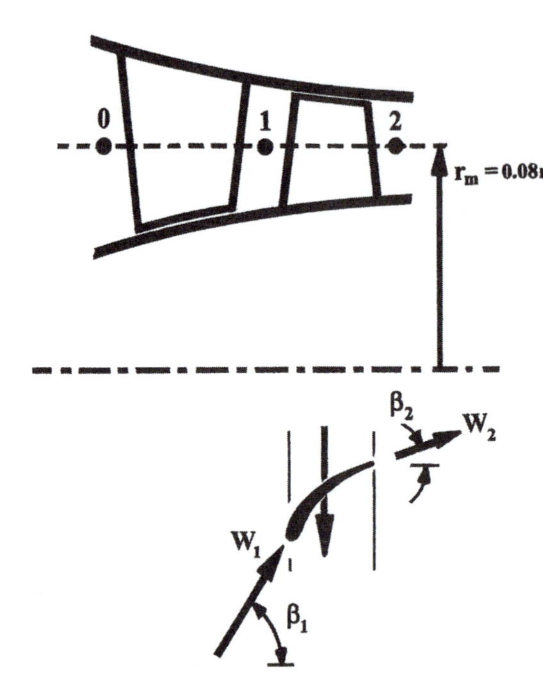

Figure 9.21. Input variables for Example 1.

SOLUTION

Part a:

$$V_{cr1} = \sqrt{\left(\frac{2\gamma}{\gamma+1}\right) RT_{t1}} = 388.17 \text{ m/s}$$

$$V_1 = M_{cr1} V_{cr1} = 263.95 \text{ m/s} = V_z \ (\alpha_1 = 0)$$

$$U_m = \omega r_m = 301.0 \text{ m/s}$$

$$\beta_1 = tan^{-1}\left(\frac{U_m}{V_z}\right) = +48.81^\circ$$

$$W_{\theta 2} = V_z \tan \beta_2 = 123.08 \text{ m/s}$$

$$W_2 = \sqrt{W_{\theta 2}{}^2 + V_z^2}$$

$$T_{tr1} = T_{t1} + \left(\frac{W_1{}^2 - V_1{}^2}{2c_p}\right) = 495.27 \text{ K}$$

$$p_{tr1} = p_{t1}\left(\frac{T_{tr1}}{T_{t1}}\right)^{\frac{\gamma}{\gamma-1}} = 3.916 \text{ bars}$$

$$p_{tr2} = p_{tr1} - (0.1328 p_{tr1}) = 3.396 \text{ bars}$$

$$c_p(T_{t2} - T_{t1}) = U_m(V_{\theta 2} - V_{\theta 1})$$

which, upon substitution, yields

$$T_{t2} = 503.6\ \mathrm{K}$$

$$T_{tr2} = T_{t2} + \left(\frac{W_2{}^2 - V_2{}^2}{2c_p}\right) = 495.27\ \mathrm{K} = T_{tr1}$$

$$p_{t2} = p_{tr2}\left(\frac{T_{t2}}{T_{tr2}}\right)^{\frac{\gamma}{\gamma-1}} = 3.60\ \text{bars}$$

$$\eta_C = \frac{(p_{t2}/p_{t1})^{\frac{\gamma-1}{\gamma}} - 1}{(T_{t2}/T_{t1}) - 1} = 62.5\%$$

Part b: We can now calculate the exerted torque as follows:

$$\tau = \dot{m} r_m (V_{\theta 2} - V_{\theta 1})\ (\text{where}\, V_{\theta 1} = 0)$$

or

$$\tau = 31.42\ \mathrm{N \cdot m}$$

Part c: Having computed the rotor-exit total pressure (p_{t2}), we can now calculate the static magnitude there:

$$V_{cr2} = \sqrt{\left(\frac{2\gamma}{\gamma+1}\right) R T_{t2}} = 410.64\ \mathrm{m/s}$$

$$V_2 = 318.65\ \mathrm{m/s}\ (\text{computed earlier})$$

$$M_{cr2} = \frac{V_2}{V_{cr2}} = 0.776$$

$$p_2 = p_{t2}\left[1 - \left(\frac{\gamma-1}{\gamma+1}\right) M_{cr2}{}^2\right]^{\frac{\gamma}{\gamma-1}} = 2.486\ \text{bars}$$

Part d: Referring to the relative-flow/blade interaction picture in Figure 9.21, we can calculate the blade-inlet (metal) angle as follows:

$$\beta_1{}' = \beta_1 + 12^\circ = 60.81^\circ$$

EXAMPLE 2

Figure 9.22 shows a single-stage axial-flow compressor and its map. The compressor cruise-operation conditions are as follows:

- Mass-flow rate ($\dot{m}$) = 1.5 kg/s
- Stage-inlet total pressure (p_{t0}) = 0.25 bars
- Stage-inlet total temperature (T_{t0}) = 200 K
- Stator flow process is assumed isentropic

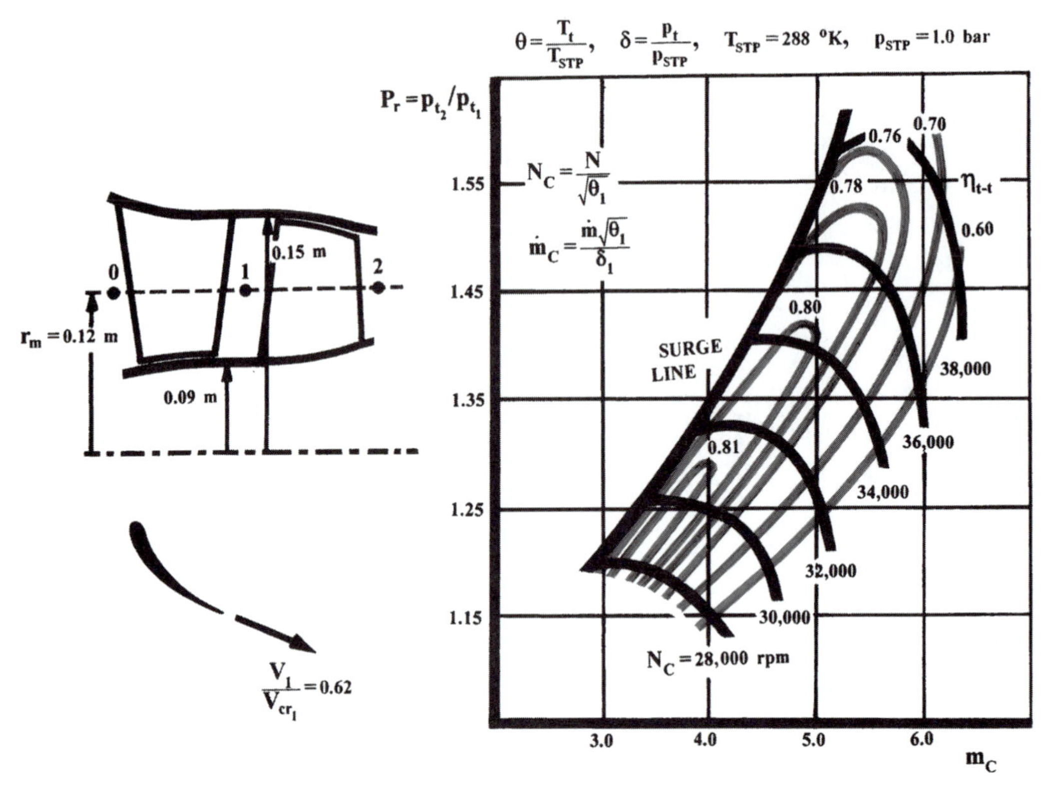

Figure 9.22. Input variables for Example 2.

- Stage-exit total pressure $(p_{t2}) = 0.35$
- Axial-velocity component (V_z) is constant across the stage
- Stator-exit critical Mach number $(V_1/V_{cr1}) = 0.62$

I) Considering these conditions and using the map, calculate:

a) The rotor-inlet relative critical Mach number (W_1/W_{cr1});
b) The stage-exit static pressure (p_2).

II) The compressor operating conditions are simulated in a test rig where the "actual" torque transmitted to the compressor is 82.3 N.m. Calculate:

a) The "physical" shaft speed in the rig (N_{rig});
b) The inlet total pressure and temperature in the rig.

SOLUTION

Part Ia: In preparation for using the given compressor map, some specific variables will have to be determined:

$$\theta_1 = \frac{T_{t1}}{288.0} = 0.6944$$

$$\delta_1 = \frac{p_{t1}}{1.0} = 0.25$$

$$\dot{m}_{C,1} = \frac{\dot{m}\sqrt{\theta_1}}{\delta_1} = 5.0 \text{ kg/s}$$

$$Pr = \frac{p_{t2}}{p_{t1}} = 1.40$$

where the subscript C refers to a "corrected" variable.

Now we can locate the point representing the stage operating conditions on the map and read off some other stage-performance variables as follows:

$$\eta_C = 80\%$$

$$N_{C1} = \frac{N}{\sqrt{\theta_1}} = 34{,}000 \text{ rpm}$$

Thus

$$N = 28{,}332 \text{ rpm}$$

$$U_m = N\left(\frac{2\pi}{60}\right)r_m = 356.03 \text{ m/s}$$

Now, applying the continuity equation at the stator exit station, we have

$$\frac{\dot{m}\sqrt{T_{t1}}}{p_{t1}[\pi(r_t^2 - r_h^2)\cos\alpha_1]} = \sqrt{\frac{2\gamma}{(\gamma+1)R}} M_{cr1}\left[1 - \left(\frac{\gamma-1}{\gamma+1}\right)M_{cr1}{}^2\right]^{\frac{1}{\gamma-1}}$$

which, upon substitution, yields

$$\alpha_1 = 55.95^\circ$$

Pursuing the calculations at the rotor inlet station, we have

$$V_1 = M_{cr1}V_{cr1} = 160.44$$

$$V_z = V_1\cos\alpha_1 = 89.83 \text{ m/s}$$

$$V_{\theta1} = V_1\sin\alpha_1 = 132.93 \text{ m/s}$$

$$W_{\theta1} = V_{\theta1} - U_m = -223.10 \text{ m/s}$$

$$W_1 = \sqrt{W_{\theta1}{}^2 + V_z^2} = 240.50 \text{ m/s}$$

$$T_{tr1} = T_{t1} + \left(\frac{W_1{}^2 - V_1{}^2}{2c_p}\right) = 215.99\text{K}$$

$$W_{cr1} = \sqrt{\left(\frac{2\gamma}{\gamma+1}\right)RT_{tr1}} = 268.9 \text{ m/s}$$

$$\frac{W_1}{W_{cr1}} = 0.894$$

Part Ib:

$$\eta_C = 0.80 = \frac{\left(\frac{p_{t2}}{p_{t1}}\right)^{\frac{\gamma-1}{\gamma}} - 1}{\left(\frac{T_{t2}}{T_{t1}}\right) - 1}$$

which yields

$$T_{t2} = 225.23 \text{ K}$$

We are now in a position to calculate $V_{\theta 2}$ as follows:

$$V_{\theta 2} = V_{\theta 1} + \frac{c_p}{U_m}(T_{t2} - T_{t1}) = 204.11 \text{ m/s}$$

In order to calculate the stage-exit static pressure (p_2), we proceed as follows:

$$V_2 = \sqrt{V_{\theta 2}{}^2 + V_z{}^2} = 223.0 \text{ m/s}$$

$$M_{cr2} = \frac{V_2}{V_{cr2}} = 0.812$$

$$p_2 = p_{t2}\left[1 - \left(\frac{\gamma - 1}{\gamma + 1}\right) M_{cr2}{}^2\right]^{\frac{\gamma}{\gamma-1}} = 0.233 \text{ bars}$$

Part IIa: In the following, we will construct two equations, which we will solve simultaneously to find the stage-inlet total temperature ($T_{t1_{rig}}$). Of these, the first equation will be based on the rig-supplied torque magnitude, which is given in the problem statement. The second equation, however, is based on the stage map provided and the fact that the two sets of operating conditions are dynamically similar and share (in particular) the same magnitude of corrected speed:

$$82.3 \text{ N} \cdot \text{m} = \tau_{rig} = \left\{\left(\frac{\dot{m} c_p T_{t1}}{\omega \eta_C}\right)\left[\left(\frac{p_{t2}}{p_{t1}}\right)^{\frac{\gamma-1}{\gamma}} - 1\right]\right\}_{rig}$$

where both the rig-operation total-to-total pressure ratio and efficiency are identical to those of the cruise-operation mode.

Upon substitution, the preceding relationship can be compacted as

$$\left(\frac{T_{t1}}{\omega}\right)_{rig} = 0.0866 \quad \text{(equation 1)}$$

As for the equality of corrected speeds, we have

$$\left(\frac{N}{\sqrt{\theta_1}}\right)_{rig} = \left(\frac{N}{\sqrt{\theta_1}}\right)_{altitude} = 34,000 \text{ rpm}$$

or

$$\left(\frac{\omega}{\sqrt{T_{t1}}}\right)_{rig} = 209.8 \quad \text{(equation 2)}$$

Simultaneous solution of equations 1 and 2 yields

$$(T_{t1})_{rig} = 330.11 \text{ K}$$

Note, by reference to the station designation in Figure 9.22, that the stator-inlet and stator-exit total properties are identical, as the stator flow is described as isentropic in the problem statement.

Now, substituting in equation 1, we get

$$(\omega)_{rig} = 3811.9 \text{ radians/s}$$

or

$$(N)_{rig} = 36{,}401 \text{ rpm}$$

Part IIb: In regard to the rig-inlet total temperature, we have obtained a magnitude of 330.11 K for it. As for the inlet total pressure, we proceed as follows:

$$\left(\frac{\dot{m}\sqrt{\theta_1}}{\delta_1}\right)_{rig} = \left(\frac{\dot{m}\sqrt{\theta_1}}{\delta_1}\right)_{altitude} = 5.0 \text{ kg/s}$$

which, upon substitution, yields

$$(p_{t1})_{rig} = 1.606 \text{ bars}$$

EXAMPLE 3

An adiabatic axial-flow compressor stage has a mean radius (r_m) of 0.12 m. The stage operating conditions are as follows:

- Stage-inlet total pressure (p_{t0}) = 1.72 bars
- Stage-inlet total temperature (T_{t0}) = 340 K
- Stator total-pressure loss ($\Delta p_{t\,stat.}$) = 6.2%
- Mass-flow rate ($\dot{m}$) = 4.7 kg/s
- Rotor-inlet absolute flow angle (α_1) = +25°
- Rotor-inlet relative flow angle (β_1) = −66°
- Shaft speed = 28,000 rpm
- Axial-velocity component (V_z) is constant throughout the stage
- Rotor-airfoil inlet (metal) angle ($\beta_1{}'$) = 58°
- Rotor-airfoil exit (metal) angle ($\beta_2{}'$) = 0
- Stage-exit total pressure (p_{t2}) = 3.64 bars
- Stage total-to-total efficiency (η_C) = 81%

Assuming an average specific-heat ratio (γ) of 1.38, calculate:

a) The stage reaction (R);
b) The rotor inlet and exit annulus heights (h_1 and h_2);
c) The rotor-blade incidence (i_R) and deviation (ϵ_R) angles;

d) The rotor total relative pressure-loss ratio $[(\Delta p_{tr})_{rot.}/p_{tr1}]$;
e) The specific entropy production (Δs) across the stage;
f) The stage specific speed (N_s).

SOLUTION

Part a:

$$p_{t1} = p_{t0} - (\Delta p_t)_{stator} = p_{t0} - (0.062 \times p_{t0}) = 1.613 \text{ bars}$$
$$U_m = V_{\theta 1} - W_{\theta 1} = V_1 \sin\alpha_1 - W_1 \sin\beta_1$$

Upon substitution, the preceding equation can be reduced to the following form:

$$0.423 V_1 + 0.913 W_1 = 351.9 \quad \text{(equation 1)}$$

On the other hand, the axial-velocity component (V_z) can be expressed as follows:

$$V_1 \cos\alpha_1 = V_z = W_1 \cos\beta_1$$

Substitution in this equation yields

$$0.906 V_1 - 0.407 W_1 = 0 \quad \text{(equation 2)}$$

Simultaneous solution of equations 1 and 2 yields

$$V_1 = 143.2 \text{ m/s}$$
$$W_1 = 319.0 \text{ m/s}$$

We can also compute the absolute-velocity components, as follows:

$$V_{z1} = V_1 \cos\alpha_1 = 129.8 \text{ m/s} = V_z$$
$$V_{\theta 1} = V_1 \sin\alpha_1 = 60.5 \text{ m/s}$$

The stage-exit total temperature (a key variable) can be computed by substituting in the compressor total-to-total efficiency, namely

$$\eta_C = 0.81 = \frac{\left(\frac{p_{t2}}{p_{t1}}\right)^{\frac{\gamma-1}{\gamma}} - 1}{\left(\frac{T_{t2}}{T_{t1}}\right) - 1}$$

Direct substitution in this equation yields

$$T_{t2} = 436.0 \text{ K}$$

To calculate the stage reaction, we proceed as follows:

$$V_{\theta 2} = V_{\theta 1} + \left(\frac{c_p}{U_m}\right)[T_{t2} - T_{t1}] = 344.8 \text{ m/s}$$

$$V_2 = \sqrt{V_{\theta 2}{}^2 + V_z^2} = 368.4 \text{ m/s}$$

$$W_2 = \sqrt{(V_{\theta 2} - U_m)^2 + V_z^2} = 130.0 \text{ m/s}$$

$$\alpha_2 = \tan^{-1}\left(\frac{V_{\theta 2}}{V_z}\right) = 69.4^\circ$$

$$\text{stage reaction}\,(R) = \frac{(W_1{}^2 - W_2{}^2)}{(V_2{}^2 - V_1{}^2) + (W_1{}^2 - W_2{}^2)} = 42.4\%$$

Part b: First, let us calculate the rotor inlet and exit critical Mach numbers:

$$M_{cr1} = \frac{V_1}{V_{cr1}} = \frac{V_1}{\sqrt{\left(\frac{2\gamma}{\gamma+1}\right) R T_{t1}}} = 0.43$$

$$M_{cr2} = \frac{V_2}{V_{cr2}} = \frac{V_2}{\sqrt{\left(\frac{2\gamma}{\gamma+1}\right) R T_{t2}}} = 0.97$$

Applying the continuity equation at both the rotor-inlet and exit stations, we can calculate the inlet and exit annulus heights as follows:

$$h_1 = 3.13 \text{ cm}$$

$$h_2 = 2.55 \text{ cm}$$

Part c: In order to compute the rotor-blade incidence (i_R) and deviation (ϵ_R) angles, the inlet and exit *relative* flow angles will be under focus and should first be identified:

$$\beta_1 = -66.0^\circ \text{ (given)}$$

$$\beta_2 = \tan^{-1}\left(\frac{W_{\theta 2}}{V_z}\right) = -3.1^\circ$$

Now, the required angles can be calculated:

$$i_R = -8.0^\circ$$

$$\epsilon_R = 3.1^\circ$$

Note that the deviation angle (ϵ) is among a few angles of which the sign is irrelevant. In fact, the mere implication of its name implies a flow (or relative-flow) stream underturning. As indicated in Chapter 3, the major contributor of this phenomenon is the suction-side boundary-layer separation.

The total relative pressure loss across the axial-flow rotor is indicative of a performance degradation, just as for the total pressure loss across the stator. In the following two chapters, the point is emphasized that the change in radius along the "master" streamline in a centrifugal-compressor impeller, or a radial turbine rotor will (itself) produce its own change in total relative properties.

Now, we calculate the rotor-inlet and exit total relative pressures:

$$p_{tr1} = p_{t1}\left(\frac{T_{tr1}}{T_{t1}}\right)^{\frac{\gamma}{\gamma-1}}$$

$$p_{tr2} = p_{t2}\left(\frac{T_{tr2}}{T_{t2}}\right)^{\frac{\gamma}{\gamma-1}}$$

Now, the percentage of total relative pressure loss across the rotor can easily be calculated:

$$\frac{\Delta p_{tr}}{p_{tr1}} = 14.2\%$$

REMINDER

In computing the total relative pressures p_{tr1} and p_{tr2}, we have used what is thermodynamically referred to as "isentropic" relationships, knowing fully well that any isentropic process in a turbomachine (or anywhere else, for that matter) is impossible, for it would negate the irreversibilities (internal and external), which we can reduce but never eliminate. Such relationships are therefore better viewed as simply (total relative)-to-total property conversion relationships at the same exact thermophysical state. In fact, a quick means of verifying this in any of such relationships is to ensure that it is the same subscript, say "1," that appears throughout the relationship. Recall that a relationship such as

$$\frac{p_1}{p_{t1}} = \left(\frac{T_1}{T_{t1}}\right)^{\frac{\gamma}{\gamma-1}}$$

clearly falls under the same category and is certainly valid.

Part e: Noting that the statorwise total temperature is constant (adiabatic flow), we can compute the entropy production as

$$(\Delta s)_{stator} = -R\ln\left(\frac{p_{t1}}{p_{t0}}\right) = 18.4 \text{ J/(kg K)}$$

As for the rotor, we have

$$(\Delta s)_{rotor} = c_p \ln\left(\frac{T_{t2}}{T_{t1}}\right) - R\ln\left(\frac{p_{t2}}{p_{t1}}\right) = 44.4 \text{ J/(kg K)}$$

Perhaps a "smarter" approach to calculating the rotor-produced entropy is to base it on a total relative-to-total relative basis, for we know that the total relative temperature (T_{tr}) remains constant across this *axial*-flow rotor. Implementing this approach, we have

$$(\Delta s)_{rotor} = -R\ln\left(\frac{p_{tr2}}{p_{tr1}}\right)$$

which, subject to slight numerical errors, should produce the same previously obtained result.

Part f: In order to calculate the stage specific speed (N_s), let us first calculate the "ideal" total-enthalpy rise across the rotor and the exit static density as follows:

$$(\Delta h_t)_{id.} = \frac{c_p}{\eta_C}(T_{t2} - T_{t1}) = 119,052 \text{ J/kg}$$

$$T_2 = T_{t2} - \left(\frac{V_2{}^2}{2c_p}\right) = 370.9\text{K}$$

$$p_2 = p_{t2}\left(\frac{T_2}{T_{t2}}\right)^{\frac{\gamma}{\gamma-1}} = 2.02 \text{ bars}$$

$$\rho_2 = \frac{p_2}{RT_2} = 1.90\,\text{kg/m}^3$$

Finally

$$N_s = \frac{N\left(\frac{2\pi}{60}\right)\sqrt{\frac{\dot{m}}{\rho_2}}}{(\Delta h_t)_{id.}{}^{\frac{3}{4}}} = 0.96 \text{ radians}$$

Reviewing Figure 5.8, we now know that this particular compressor stage should have been designed as a centrifugal stage for better performance. However, the magnitude of the specific speed is close to the axial-stage dome in Figure 5.8.

EXAMPLE 4

Figure 9.23 shows the interstage gap in a multistage axial-flow compressor where the mean radius is 12 cm. Also shown is the station-designation pattern in this particular problem. The compressor operating conditions are such that:

- Rotor speed (N) $= 31{,}500$ rpm
- Rotor-exit total temperature (T_{t_1}) $= 476.0$ K
- Rotor-exit total pressure (p_{t_1}) $= 4.2$ bars
- Mass-flow rate ($\dot{m}$) $= 8.0$ kg/s
- Rotor-exit absolute flow angle (α_1) $= 71.0°$
- Rotor-exit relative flow angle (β_1) $= -14.2°$.
- Friction coefficient over the endwalls (f) $= 0.01$

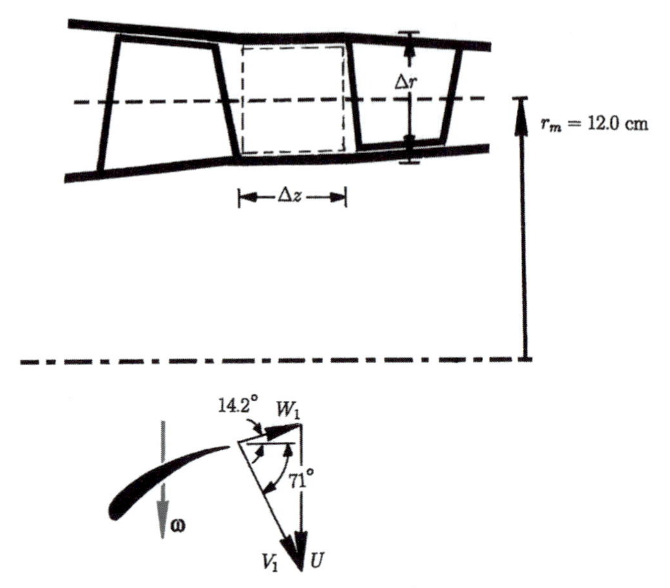

Figure 9.23. Input variables for Example 4.

Assuming an adiabatic flow, a constant magnitude of V_z, and a γ value of 1.4, calculate:

a) The rotor-exit annulus height (Δr);
b) The axial-gap length (Δz), which leads to a choked flow at the gap exit station.

SOLUTION

Anticipating the use of Fanno-flow relationships, specifically equation (3.71), we will proceed with the solution using the "traditional" Mach number (M), which is the common variable in practically all Fanno-flow relationships.

Part a: First, we focus on the rotor-exit velocity-triangle variables. Using simple trigonometry rules, we have

$$\omega r_m = U_m = V_{\theta 1} - W_{\theta 1} = V_z \tan \alpha_1 - V_z \tan \beta_1$$

which gives

$$V_z = 125.4 \text{ m/s (constant across the stage)}$$

In the following, we calculate the rotor-inlet Mach number:

$$\text{Rotor-exit absolute velocity } V_1 = \frac{V_z}{\cos \alpha_1} = 385.1 \text{ m/s}$$

$$T_1 = T_{t1} - \left(\frac{V_1^2}{2c_p}\right) = 402.2\,\text{K}$$

$$\text{Sonic speed } (a_1) = \sqrt{\gamma R T_1} = 402.1 \text{ m/s}$$

$$\text{Rotor-exit absolute Mach number } (M_1) = \frac{V_1}{a_1} = 0.958$$

Keeping the "traditional" Mach number (M) as our chosen nondimensional velocity ratio, let us now apply the continuity equation (in the stationary frame of reference) at the rotor exit station:

$$\frac{\dot{m}\sqrt{T_{t1}}}{p_{t1}[(2\pi r_m \Delta r)\cos\alpha_1]} = \sqrt{\left(\frac{\gamma}{R}\right)} \times M \times \left[1 + \left(\frac{\gamma - 1}{2}\right) M_1^2\right]^{\frac{1+\gamma}{2(1-\gamma)}}$$

which, upon substitution, yields

$$\Delta r = 4.2 \text{ cm}$$

Part b: Knowing that the flow is choked at the rotor/stator-gap exit station (i.e., $M_2 = 1.0$), and applying equation (3.71), we have

$$\frac{fL}{D_h} = \frac{f\left(\frac{\Delta z}{\cos\alpha_1}\right)}{D_h} = \left(\frac{\gamma+1}{2\gamma}\right)\ln\left[\frac{1 + \left(\frac{\gamma-1}{2}\right)M_2^2}{1 + \left(\frac{\gamma-1}{2}\right)M_1^2}\right]$$

$$-\left(\frac{1}{\gamma}\right)\left(\frac{1}{M_2^2} - \frac{1}{M_1^2}\right) - \left(\frac{\gamma+1}{2\gamma}\right)\ln\left(\frac{M_2^2}{M_1^2}\right)$$

where

$M_1 = 0.958$,
$M_2 = 1.0$, and the
hydraulic diameter $(D_h) = 2\Delta r = 8.4$ cm.

Direct substitution in the Fanno-flow relationship (above) gives the rotor-to-stator axial-gap length (Δz) as follows:

$$(\Delta z)_{gap} = 0.65 \text{ cm}$$

EXAMPLE 5

Figure 9.24 shows a purely axial compressor stage together with its major dimensions. Also shown is an incidence/total-pressure-loss correlation chart that specifically applies to the stator vanes. The stage design point is defined as follows:

- Mass-flow rate ($\dot{m}$) = 7.61 kg/s
- Inlet total pressure (p_{t0}) = 1.4 bars
- Inlet total temperature (T_{t0}) = 322 K
- Inlet Mach number (M_0) = 0.81
- Stator-exit Mach number (M_1) = 0.67
- Rotor-inlet (absolute) Mach number (M_2) = 0.681
- Stator-vane inlet (metal) angle ($\alpha_0{}'$) = +38.0°

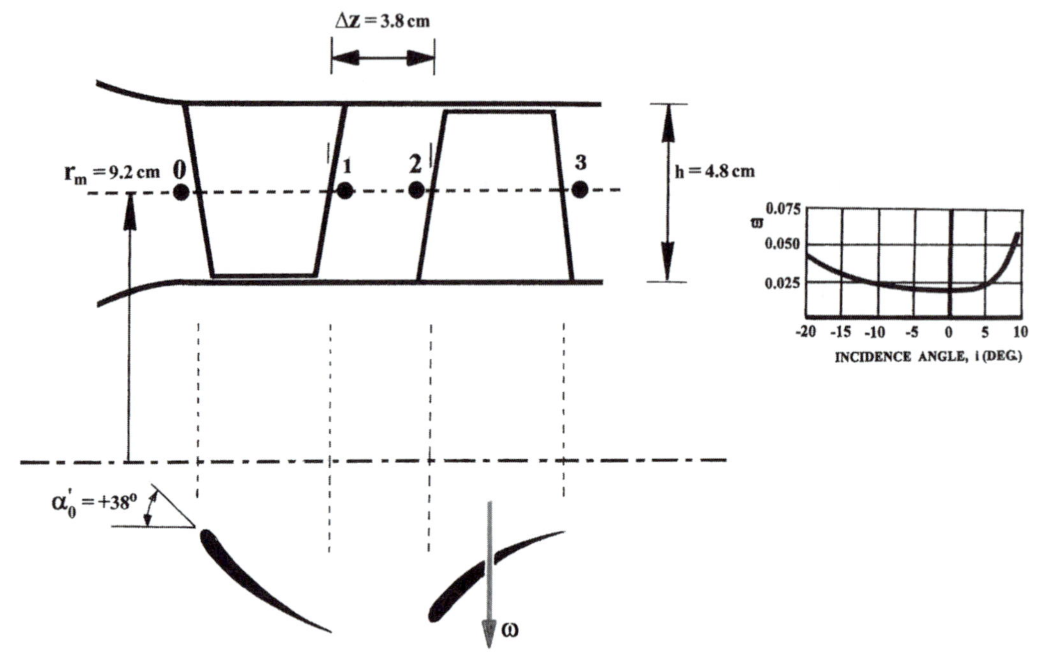

Figure 9.24. Input variables for Example 5.

It is also assumed that the stator incidence loss is the only degrading mechanism throughout the stator subdomain.

Assuming an adiabatic flow and a specific-heat ratio (γ) of 1.4, calculate:

a) The friction coefficient (f) across the stator/rotor axial gap;
b) The percentage of total pressure loss across the gap.

SOLUTION

Just as in the preceding example, and in anticipation of utilizing Fanno-flow relationship(s), we will continue to adopt the "traditional" (as opposed to the critical) Mach number as the nondimensional velocity ratio.

Part a: First, we apply the continuity equation to the stator exit station:

$$\frac{\dot{m}\sqrt{T_{t0}}}{p_{t0}[(2\pi r_m h_0)\cos\alpha_0]} = \sqrt{\left(\frac{\gamma}{R}\right)} \times M_0 \times \left[1 + \left(\frac{\gamma - 1}{2}\right) M_0{}^2\right]^{\frac{(1+\gamma)}{2(1-\gamma)}}$$

Upon substitution, we get

$$\alpha_0 = 26^\circ$$

Accordingly, the stator-vane incidence angle (i_S) is

$$i_S = \alpha_0 - \alpha_0{}' = -12^\circ$$

where $\alpha_0{}'$ is the stator-vane inlet (metal) angle, which is given as $+38°$. Now, using the incidence-loss graph provided in Figure. 9.24, we get

$$(\bar{\omega})_{stator} = 0.025 = \frac{(p_{t0} - p_{t1})}{p_{t0}}$$

which yields

$$p_{t1} = 1.365 \text{ bars}$$

Noting that the total temperature remains constant across the adiabatic stator, we now proceed to calculate the stator-exit static density (ρ_1) as follows:

$$T_1 = \frac{T_{t1}}{\left[1 + \left(\frac{\gamma-1}{2}\right)M_1{}^2\right]} = 295.5 \text{ K}$$

$$a_1 = \sqrt{\gamma R T_1} = 344.6 \text{ m/s}$$

$$V_1 = M_1 a_1 = 230.9 \text{ m/s}$$

$$p_1 = \frac{p_{t1}}{\left[1 + \left(\frac{\gamma-1}{2}\right)M_1{}^2\right]^{\frac{\gamma}{\gamma-1}}} = 1.01 \text{ bars}$$

$$\rho_1 = \frac{p_1}{RT_1} = 1.19 \text{ kg/m}^3$$

Applying the continuity equation at the stator-exit station, we get

$$V_{z1} = \frac{\dot{m}}{\rho_1(2\pi r_m h_1)} = 230.5 \text{ m/s}$$

Critical in the application of equation (3.71), we now calculate the stator-exit swirl angle:

$$\alpha_1 = \cos^{-1}\left(\frac{V_{z1}}{V_1}\right) = 3.5°$$

At this point, we are prepared to apply the Fanno-flow relationship (3.71), a step that will produce the friction coefficient (f), as follows:

$$f = 0.064$$

Part b: Let us now apply equation (3.70) to the stator/rotor axial gap:

$$\frac{p_{t2}}{p_{t1}} = \left(\frac{M_1}{M_2}\right)\left[\frac{1 + \left(\frac{\gamma-1}{2}\right)M_2{}^2}{1 + \left(\frac{\gamma-1}{2}\right)M_1{}^2}\right]^{\frac{\gamma+1}{2(\gamma-1)}} = 0.976$$

which yields

$$p_{t2} = 1.33 \text{ bars}$$

Thus

$$\frac{(\Delta p_t)_{gap}}{p_{t1}} = \frac{(p_{t1} - p_{t2})}{p_{t1}} = 2.6\%$$

EXAMPLE 6

Figure 9.25 shows an axial-flow compressor stage in which the stator is identical, both geometrically and operationally, to that in Example 7 of Chapter 3. For convenience, the stator data (geometrical and thermophysical) are reproduced in Figure 9.24. The rotor flow path, in the figure, is intended to produce a constant magnitude of the axial-velocity component (V_z) across the rotor. The stator/rotor gap is a constant-area annular duct with a 2.4 cm axial extension. The following operating conditions also apply:

- Rotor speed $(N) = 24,200$ rpm
- Mass-flow rate $(\dot{m}) = 8.59$ kg/s
- Duct friction coefficient $(f) = 0.269$
- Stage work coefficient $(\psi) = 0.92$
- Stage-exit total pressure $(p_{t4}) = 6.45$ bars

Assuming an adiabatic flow and a γ magnitude of 1.4:

a) Calculate the duct total pressure loss.
b) Calculate the rotor-exit swirl angle (α_4).
c) Calculate the stage total-to-total efficiency (η_C).
d) Calculate the rotor-exit annulus height (h_4).
e) Calculate the stage specific speed (N_s).
f) Sketch the rotor blade mean-radius airfoil section, assuming zero incidence and deviation angles

SOLUTION

Part a: Calculations regarding the rotor-inlet and exit velocity triangles are straightforward, and the triangles are sketched (to scale) in Figure 9.25. Now we move to apply the Fanno-flow relationships (3.70 and 3.71) to the stator/rotor axial gap. Because these relationships are cast in terms of the "traditional" Mach number (M), we begin by converting $(M_{cr})_2$ into M_2:

$$M_2 = \sqrt{\left(\frac{2}{\gamma - 1}\right)\left\{\left[1 - \left(\frac{\gamma - 1}{\gamma + 1}\right)M_{cr2}^{\ 2}\right]^{-1} - 1\right\}} = 0.408$$

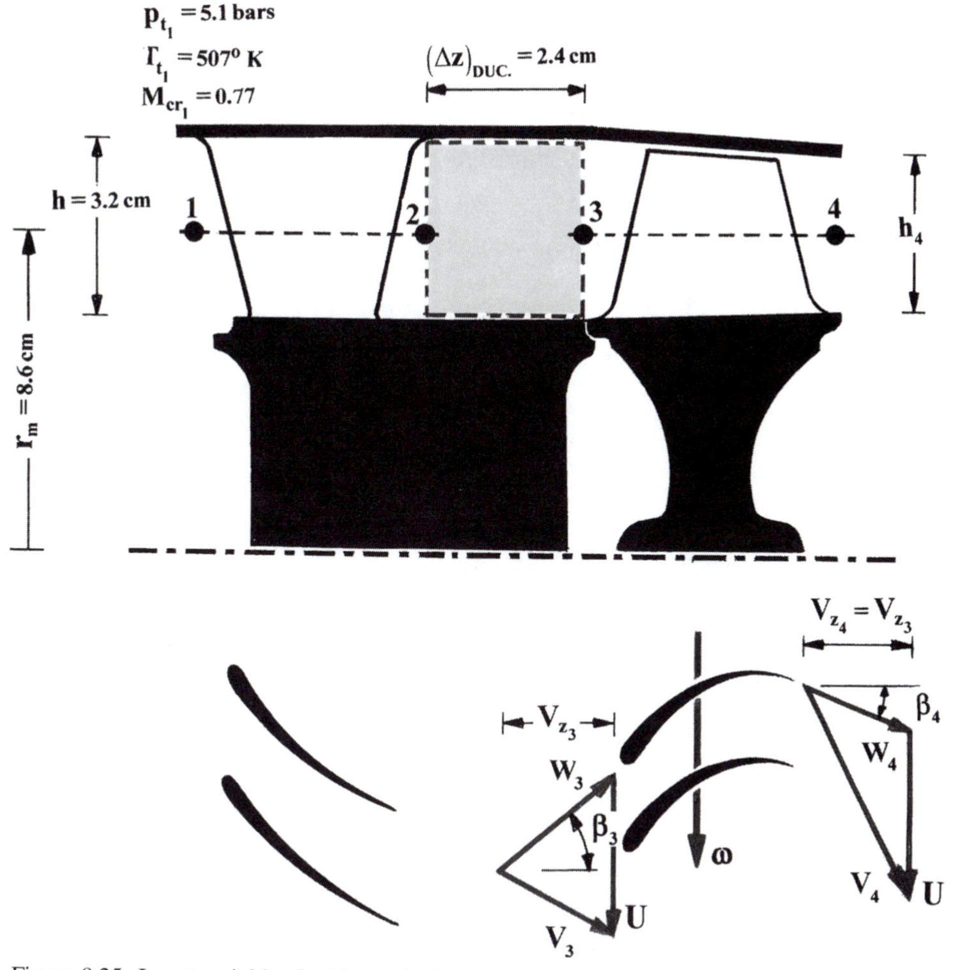

Figure 9.25. Input variables for Example 6.

Applying the Fanno-flow relationship (3.71) to the stator/rotor gap (i.e., between stations 2 and 3), we obtain

$$M_3 = 0.415$$

The Mach number M_3 can be converted into its critical counterpart as follows:

$$M_{cr3} = \sqrt{\left(\frac{\gamma+1}{\gamma-1}\right)\left\{1-\left[1+\left(\frac{\gamma-1}{2}\right)M_3^{\,2}\right]^{-1}\right\}} = 0.447$$

However, the application of equation (3.70) to the same gap gives rise to the gap-exit total pressure as follows:

$$p_{t3} = 4.87 \text{ bars}$$

As a result, the total pressure loss across this axial gap is

$$\frac{\Delta p_t}{P_{t2}} = 1.6\%$$

Note that this small percentage is consistent with the fact that this gap is notably short.

Part b: Within the rotor subdomain, we have

$$U_m = \omega r_m = 217.9\,\text{m/s}$$

$$V_{\theta 3} = V_3 \sin\alpha_3 = (M_{cr3} V_{cr3}) \sin\alpha_3 = 89.3\,\text{m/s}$$

$$V_{\theta 4} = V_{\theta 3} + (\psi U_m) = 289.8\,\text{m/s}$$

$$T_{t4} = T_{t3} + (\frac{\psi U_m^{\,2}}{c_p}) = 550.5\,\text{K}$$

$$V_{cr4} = \sqrt{\left(\frac{2\gamma}{\gamma+1}\right) R T_{t4}} = 429.3\,\text{m/s}$$

Referring to Figure 9.25, note that the annulus-height reduction across the rotor is intended to keep the axial-velocity component constant throughout the rotor; that is,

$$V_{z4} = V_{z3} = V_z = V_3 \cos\alpha_3 = 161.1\,\text{m/s}$$

Now

$$\alpha_4 = \tan^{-1}\left(\frac{V_{\theta 4}}{V_z}\right) = 60.9^\circ$$

Part c: At this point, we are prepared to calculate the stage total-to-total efficiency (η_C)

$$\eta_C = \frac{\left(\frac{p_{t4}}{p_{t3}}\right)^{\frac{\gamma-1}{\gamma}} - 1}{\left(\frac{T_{t4}}{T_{t3}}\right) - 1} = 81.0\%$$

Part d: Now, we will implement a numerical procedure with the objective being the rotor-exit annulus height (h_4):

$$M_{cr4} = \frac{\sqrt{V_{\theta 4}^{\,2} + V_z^2}}{V_{cr4}} = 0.772$$

$$\rho_4 = \left(\frac{p_{t4}}{R T_{t4}}\right)\left[1 - \left(\frac{\gamma-1}{\gamma+1}\right) M_{cr4}^{\,2}\right]^{\frac{1}{\gamma-1}} = 3.143\,\text{kg/m}^3$$

$$h_4 = \frac{\dot{m}}{\rho_4 V_{z4} (2\pi r_m)} = 3.1\,\text{cm}$$

Part e: Calculation of the stage specific speed (N_s) is straightforward for we have already computed its most critical contributor, namely ρ_4. The final result is

$$N_s = 1.413 \text{ radians}$$

When carried to Figure 5.8, this N_s magnitude places the stage within the axial-stage dome but also close to the centrifugal/axial stage interface.

Part f: In order to sketch the rotor-blade mean section, we need to compute the rotor inlet and exit relative flow angles (β_3 and β_4):

$$\beta_3 = \tan^{-1}\left(\frac{V_{\theta 3} - U_m}{V_z}\right) = -38.6^\circ$$

$$\beta_4 = \tan^{-1}\left(\frac{V_{\theta 4} - U_m}{V_z}\right) = +24.1^\circ$$

With these two angles, and the assumption of zero incidence and deviation angles, the rotor-blade mean-radius cross section should look like that in Figure 9.25.

EXAMPLE 7

Figure 9.26 shows the blade-to-blade passage in an axial-flow compressor rotor. Also shown in the figure is the midchannel distribution of the relative-velocity tangential component at the mean radius. The rotor mean radius (r_m) is 8.0 cm, and the shaft speed (N) is 39,170 rpm.

By calculating the net magnitude of the radial acceleration, as a rough measure of the relative-streamline radial shift, sketch the streamline at the midchannel, midspan location.

SOLUTION

Let us consider the four locations that are identified on the W_θ versus nondimensional axial distance. Because the mean radius is constant, the centripetal acceleration component will remain constant as well:

$$(a)_{cent.} = r_m \omega^2 = 0.08\left[N\left(\frac{2\pi}{60}\right)\right]^2 = 1.35 \times 10^6 \text{ m/s}^2$$

As will be derived in Chapter 11 (expression 11.17), the Coriolis-acceleration radial component can be expressed as

$$(a_r)_{Coriolis} = 2\omega W_\theta$$

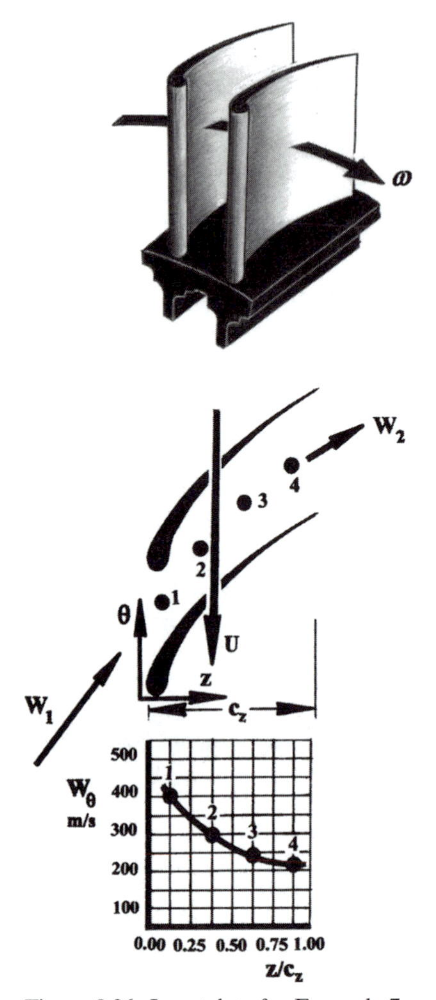

Figure 9.26. Input data for Example 7.

Referring to Figure 9.26, note that $\omega = -\omega \mathbf{e}_z$, where $\mathbf{e}_z$ is the unit vector in the z-direction. The figure also shows that the relative-velocity tangential component (W_θ) is positive at all axial locations (by reference to the direction of the θ-axis). As a result, the acceleration radial component (above) will be positive (i.e., in the radially outward direction). Let us now compute this component at the four selected points in the figure:

$$[(a_r)_{Coriolis}]_1 = 2\omega W_{\theta 1} = 1.44 \times 10^6 \text{ m/s}^2$$

$$[(a_r)_{Coriolis}]_2 = 2\omega W_{\theta 2} = 1.03 \times 10^6 \text{ m/s}^2$$

$$[(a_r)_{Coriolis}]_3 = 2\omega W_{\theta 3} = 0.79 \times 10^6 \text{ m/s}^2$$

$$[(a_r)_{Coriolis}]_4 = 2\omega W_{\theta 4} = 0.66 \times 10^6 \text{ m/s}^2$$

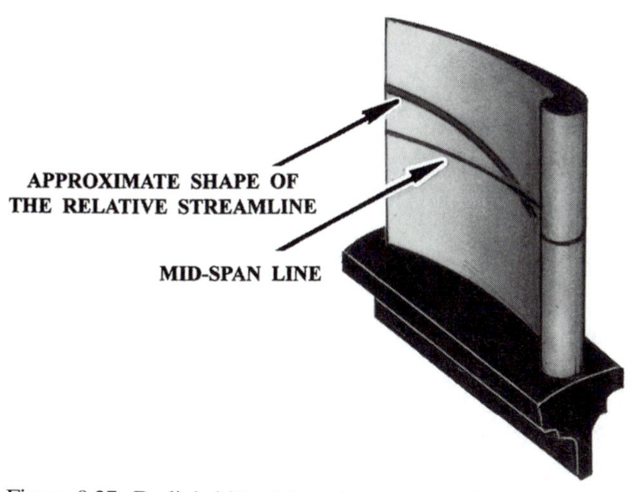

Figure 9.27. Radial shift of the relative streamline.

The net magnitudes of the radial acceleration component at these four points can be obtained by simply adding the centripetal component:

$$[(a_r)_{net}]_1 = 2.79 \times 10^6$$
$$[(a_r)_{net}]_2 = 2.38 \times 10^6$$
$$[(a_r)_{net}]_3 = 2.14 \times 10^6$$
$$[(a_r)_{net}]_4 = 2.01 \times 10^6$$

Taking these magnitudes to be indicative of the radial shift, the relative streamline should look like that in Figure 9.27. Due to the continuous streamwise decline in W_θ (Fig. 9.26), note that the Coriolis contribution, and, therefore, the net radial acceleration component continue to decline as well. The streamline shape in Figure 9.27 reflects this fact in the form of a continually decreasing rate of radius gain along the streamline.

PROBLEMS

1) Figure 9.28 shows the compressor map in a turboprop engine. The compressor cruise-operation point is defined as follows:

- Specific shaft work $(w_s) = 220$ kJ/kg
- Physical rotation speed $(N) = 31{,}177$ rpm
- Inlet total pressure $(p_{t\,in}) = 0.235$ bars
- Inlet total temperature $(T_{t\,in}) = 216$ K
- Total-to-total (isentropic) efficiency $(\eta_C) = 80\%$

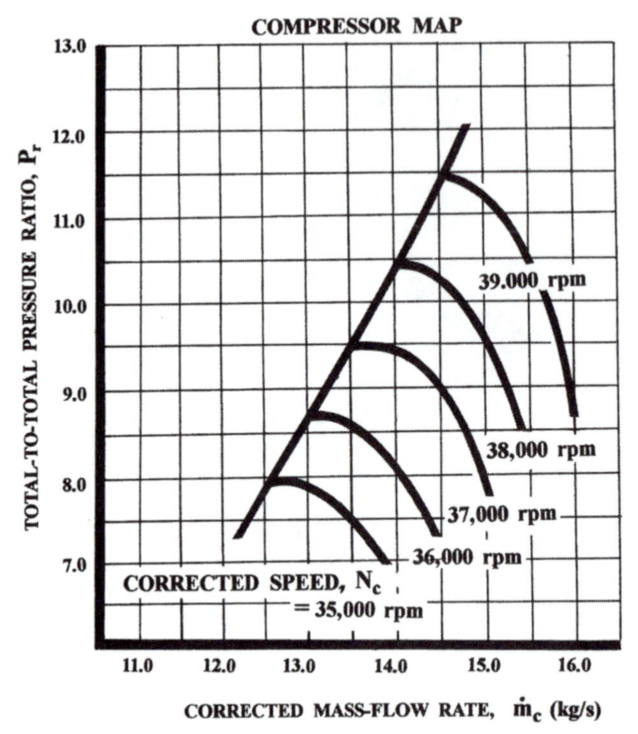

Figure 9.28. Input variables for Problem 1.

These operating conditions are simulated in a test rig utilizing air at inlet total pressure and temperature of 1.0 bars and 288 K, respectively. Assuming an adiabatic flow and an average specific heat ratio (γ) of 1.4:

I) Calculate the following variables:

Ia) The cruise-operation "physical" mass-flow rate ($\dot{m}$);

Ib) The test-rig compressor-supplied power (P).

II) Now consider the process where a gradual *increase* in the inlet total pressure is effected while the rig mass-flow rate ($\dot{m}_{rig}$), rotation speed (N_{rig}), and inlet total temperature ($T_{t\,rig}$) are all held constant at the same magnitudes given earlier. Considering this operation shift, calculate the value of inlet total pressure at which the compressor reaches the surge state.

2) A 50% reaction axial-flow compressor stage has a constant mean radius (r_m) of 8.5 cm. The stage design point is defined as follows:

- The flow process is adiabatic
- Inlet total pressure (p_{t0}) = 1.05 bars
- Inlet total temperature (T_{t0}) = 304 K
- Axial-velocity component is constant across the stage
- Stage total-to-total (isentropic) efficiency (η_C) = 79%
- Shaft speed (N) = 34,000 rpm

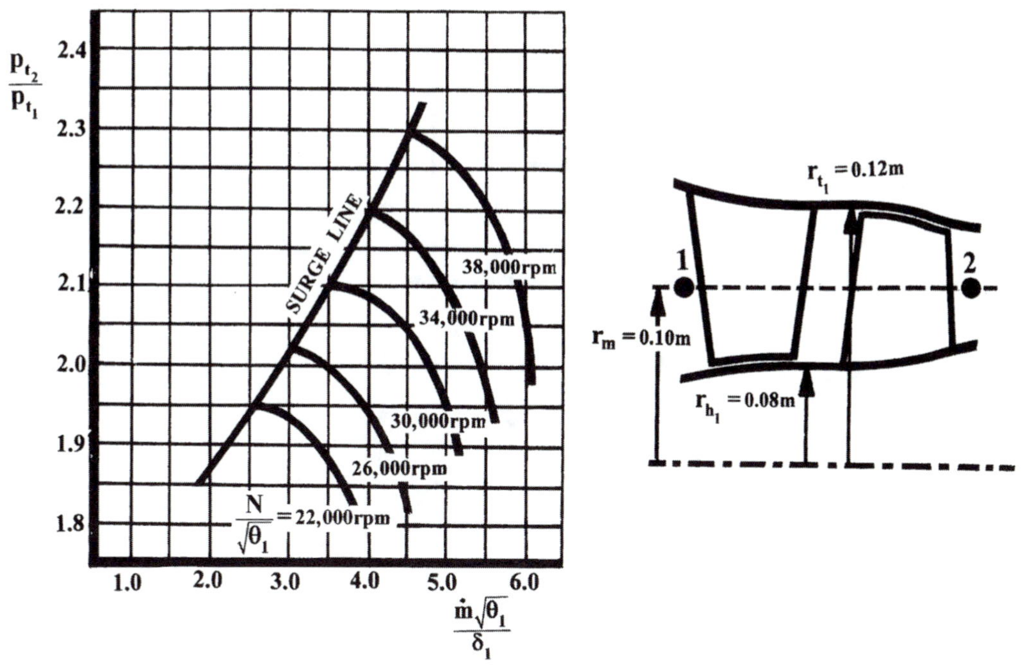

Figure 9.29. Input variables for Problem 3.

- Rotor average hub and tip radii are 7.3 and 9.7 cm, respectively
- Stator-exit static pressure (p_1) = 0.97 bars
- Rotor-inlet absolute flow angle (α_1) = +28°

Assuming an isentropic stator and an average specific heat ratio (γ) of 1.4, calculate:

a) The rotor-exit relative critical Mach number. ($W_2/W_{cr\,2}$);
b) The rotor-exit static temperature and pressure (T_2 and p_2);
c) The change in total-relative pressure (Δp_{t_r});
d) The stage hub and casing reaction magnitudes (R_h and R_t).

In item "d" you may assume a free-vortex flow pattern (i.e., $r V_\theta$ = constant = $r_m V_{\theta m}$), which implies a uniform axial-velocity component within the hub-to-casing gaps upstream and downstream from the compressor rotor.

3) Figure 9.29 show a single compressor stage and its map. The cruise operation point of the stage is defined as follows:

- "Actual" shaft work supplied (w_s) = 54.05 kJ/kg
- $\alpha_1 = \beta_2 = 0$
- Stage-inlet total pressure (p_{t0}) = 0.235 bars
- Stage-inlet total temperature (T_{t0}) = 210 K
- Stage total-to-total efficiency (η_C) = 85.47%

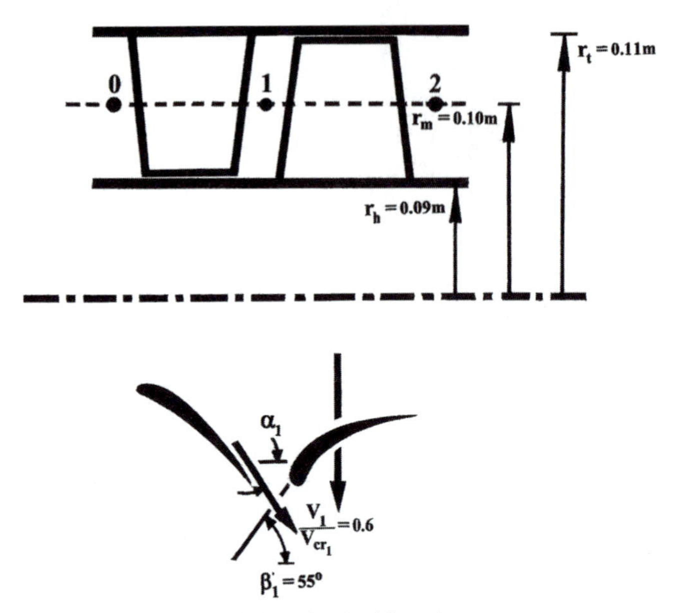

Figure 9.30. Input variables for Problem 4.

- Shaft speed $(N) = 22{,}200$ rpm
- Axial-velocity component is constant across the stage

These operating conditions are simulated in a test rig utilizing air at inlet total pressure and temperature (p_{t0}) and (T_{t0}) of 1.0 bar and 288 K, respectively. The stator flow process is assumed isentropic, and the average specific-heat ratio (γ) is fixed at 1.4 for both sets of operating conditions.

a) Calculate the rotor-inlet relative critical Mach number $(W_1/W_{cr\,1})$.
b) Calculate the test-rig torque transmitted to the stage.
c) Calculate the loss in total relative pressure $(p_{t_{r1}} - p_{t_{r2}})_{rig}$ in the test rig.

4) Referring to Figure 9.30, the cruise operation of a "purely" axial compressor stage is defined as follows:

- Inlet total pressure $(p_{t0})_{alt.} = 0.106$ bars
- Inlet total temperature $(T_{t0})_{alt.} = 210$ K
- Exit total temperature $(T_{t2})_{alt.} = 256$ K
- Shaft speed $(N)_{alt.} = 22{,}200$ rpm
- Mass-flow rate $(\dot{m})_{alt.} = 0.26$ kg/s
- Rotor-blade inlet (metal) angle $(\beta_1{}') = -55°$
- Stator-wise loss in total pressure $(\Delta p_{t\,stator}) = 6\%$

These conditions are simulated in a test rig where the following conditions apply:

- Inlet total pressure $(p_{t0})_{rig} = 1.0$ bar
- Inlet total temperature $(T_{t0})_{rig} = 288$ K

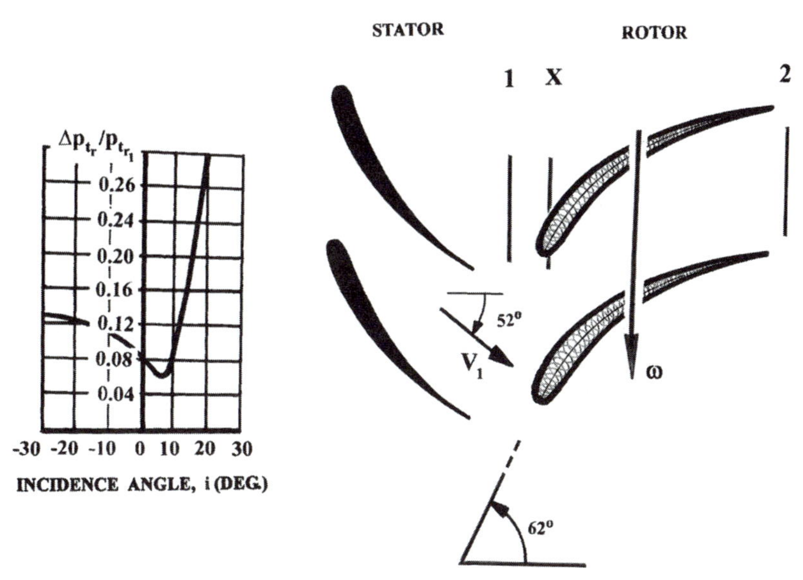

Figure 9.31. Input variables for Problem 5.

Considering the rig operating conditions, and assuming an average specific heat ratio (γ) of 1.4, calculate the following variables in the rig:

a) The mass-flow rate, speed, and exit total temperature;
b) The stator exit (swirl) angle (α_1);
c) The dynamic enthalpy rise across the rotor;
d) The rotor-blade incidence angle.

5) Figure 9.31 shows an axial compressor stage together with the stator-exit velocity vector (V_1) and the rotor-blade "metal" angle. The figure also shows a plot of the total relative pressure loss as a function of the rotor incidence angle. The station "X" in the figure is just inside the blade leading edge, where the only total relative pressure loss (Δp_{t_r}) is that due to the incidence angle. The stage operating conditions are:

- Stator-exit total pressure (p_{t1}) = 1.2 bars
- Stator-exit total temperature (T_{t1}) = 320 K
- Axial-velocity component (V_z) = 168 m/s
- Solid-body rotational velocity (U_m) = 366.3 m/s

Assuming a specific-heat ratio (γ) of 1.4, calculate the total relative pressure ($p_{t_r X}$) at station "X" in the figure.

6) *True or False:* A single-stage axial-flow compressor is about to enter a rotating-stall mode of operation. At this point, the inlet total pressure is reduced while the inlet total temperature; the physical mass-flow rate, and the physical speed remain fixed.

This leads to a stable compressor operation. (True or False)

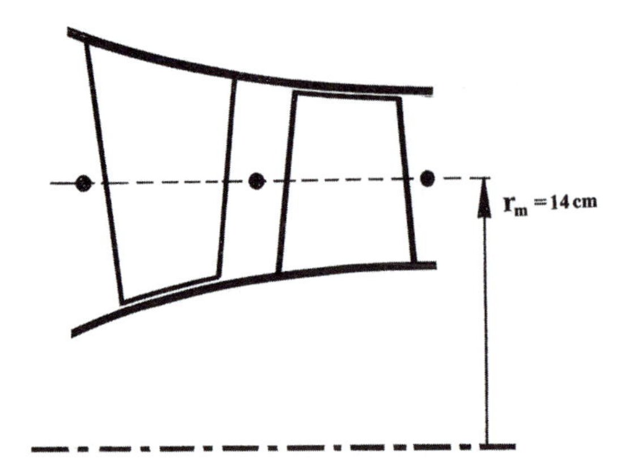

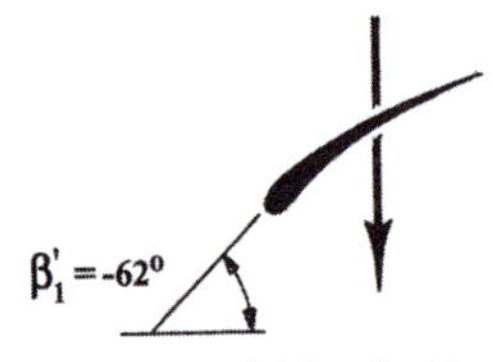

Figure 9.32. Input variables for Problem 8.

7) *Multiple Choice:* A single-stage axial-flow compressor is operating at the point of maximum efficiency. An off-design operation mode is then effected by gradually decreasing the "physical" speed (N), with the mass-flow rate and inlet conditions held constant. As a result:

   a) The compressor may enter an unstable operation mode.
   b) The rotor incidence angle will move toward more positive values.
   c) The surge margin will gradually decrease.
   d) All of the above.
   e) None of the above.

8) Figure 9.32 shows a high-pressure compressor stage where the operating conditions are as follows:

   - Rotational speed (N) = 34,000 rpm
   - Rotor-inlet total pressure (p_{t_1}) = 10 bars
   - Rotor-inlet total temperature (T_{t_1}) = 580 K
   - $V_{z_1} = V_{z2} = 216$ m/s
   - Rotor-blade inlet (metal) angle ($\beta_1{}'$) = 62°
   - Total-to-total pressure ratio (p_{t_2}/p_{t_1}) = 1.52
   - Total-to-total efficiency (η_C) = 81%

   If the rotor-blade incidence angle (i_R) is −8°, calculate:

   a) The stage reaction (R);
   b) The change in total relative pressure ($p_{t,r2} - p_{t,r1}$) across the rotor.

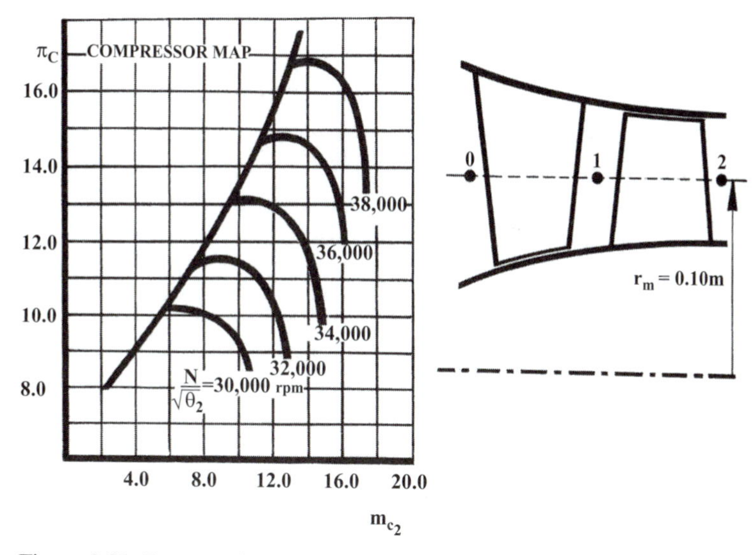

Figure 9.33. Input variables for Problem 9.

9) Figure 9.33 shows a turbofan-engine compressor stage and its map. The engine cruise-operation mode gives rise to the stage operating conditions

- Physical mass-flow rate ($\dot{m}$) = 6.4 kg/s
- Physical speed (N) = 29,033 rpm
- Inlet total pressure (p_{t1}) = 0.42 bars
- Inlet total temperature (T_{t1}) = 210 K
- Exit total temperature (T_{t2}) = 492 K

a) Using the stage map, calculate the compressor total-to-total efficiency.
b) Consider the off-design stage operation (as a result of a temporary loss of altitude) where the inlet total pressure rises. Assuming that the inlet total temperature change is negligible and that the physical speed and the physical mass-flow rate are both fixed, calculate the stage exit pressure (p_{t2}) just before the stage enters an unstable operation mode.
c) The stage operating conditions in item a (above) is simulated in a test rig using air at pressure and temperature of 1.0 bar and 288 K, respectively. Calculate the power (P) that is needed to drive the compressor in this case.

10) Figure 9.34 shows the meridional view of an axial-flow compressor stage. The stage operating conditions are as follows:

- Rotor speed (N) = 35,000 rpm
- Stator-inlet total temperature (T_{t0}) = 468 K
- Stator flow process is assumed isentropic
- $V_{z1} = V_{z2} = 200$ m
- Total-to-total pressure ratio (p_{t2}/p_{t1}) = 1.586
- The rotor-inlet absolute-velocity and rotor-exit relative-velocity vectors (i.e., $\mathbf{V}_1$ and $\mathbf{W}_2$) are both in the axial direction

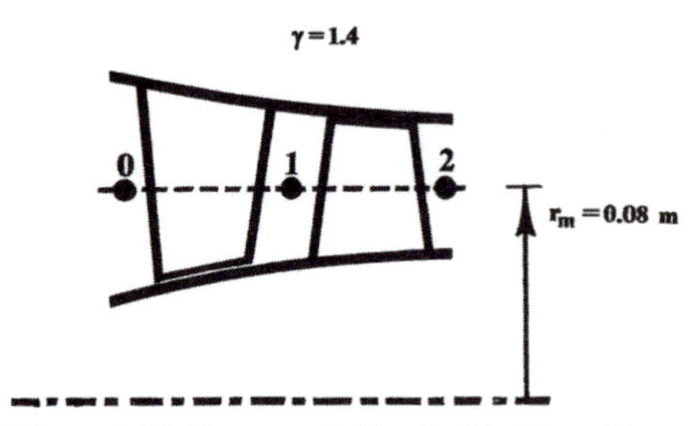

Figure 9.34. Input variables for Problem 10.

Part 1: Assuming an adiabatic flow process throughout the stage and a specific heat ratio (γ) of 1.4, calculate:

a) The rotor-exit static temperature (T_2);
b) The stage-consumed specific shaft work (w_s);
c) The rotor-exit relative critical Mach number (W_2/W_{cr2});
d) The stage total-to-total efficiency (η_C).

Part 2: Knowing that the rotor-blade incidence angle (i_R) is $-8.0°$, and assuming a deviation angle of zero, sketch the mean-radius blade section, showing both the inlet and exit airfoil (or metal) angles.

11) Figure 9.35 shows the meridional view of an axial-flow compressor stage. The stage operating conditions are as follows:

- Rotor-exit relative flow angle (β_2) $= -25°$
- Rotational speed (N) $=$ 32,000 rpm

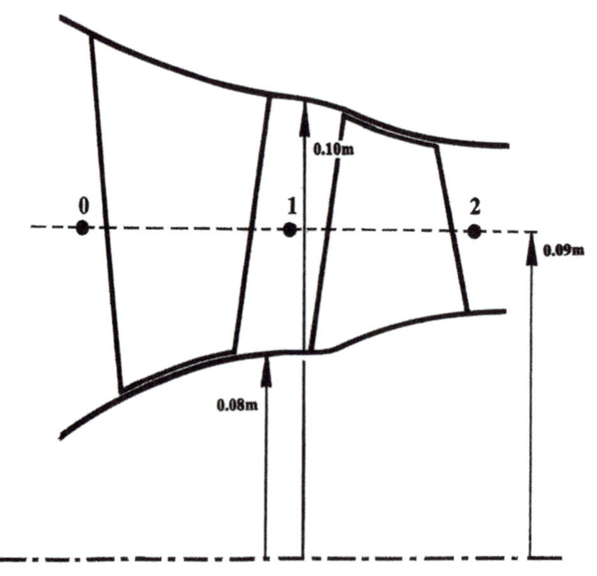

Figure 9.35. Input variables for Problem 11.

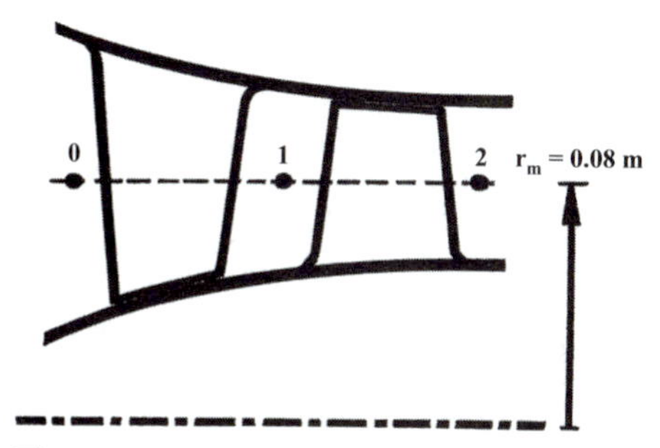

Figure 9.36. Input variables for Problem 12.

- $V_{z1} = V_{z2} = 200$ m/s
- Stator flow process is assumed isentropic
- Stator-inlet total temperature $(T_{t0}) = 468$ K
- Stage reaction $(R) = 50\%$
- Stage-inlet total pressure $(p_{t0}) = 3.2$ bars
- Stage-exit total pressure $(p_{t2}) = 4.02$ bars

Assuming an adiabatic stage flow and a specific-heat ratio (γ) of 1.4, calculate:

a) The stator-exit absolute flow angle (α_1);
b) The stator-exit critical Mach number (V_1/V_{cr1});
c) The mass-flow rate $(\dot{m})$ through the stage;
d) The power (P) supplied to the stage;
e) The rotor-exit relative-critical Mach number (W_2/W_{cr2});
f) The rotor-exit (traditional) Mach number (V_2/a_2);
g) The stage total-to-total efficiency (η_C).

12) Figure 9.36 shows a low-pressure compressor stage that is operating under the following conditions:

- Inlet total pressure $(p_{t0}) = 2.4$ bars
- Inlet total temperature $(T_{t0}) = 380$ K
- Stator-exit velocity vector $(\mathbf{V}_1)$ is totally axial
- Stator-wise loss in total pressure $(\Delta p_{t\,st.}) = 0.22$ bars
- Rotor-wise loss in total relative pressure $= 4\%$
- Rotor-inlet static pressure $(p_1) = 1.6$ bars
- Rotor-exit static pressure $(p_2) = 2.1$ bars
- Rotor speed $(N) = 32{,}000$ rpm
- Exit relative tangential-velocity component $(W_{\theta 2})$ is negative

Assuming an adiabatic flow field throughout the stage and a specific heat ratio (γ) of 1.4, calculate:

a) The rotorwise static temperature change $(T_2 - T_1)$;
b) The stage total-to-total efficiency (η_C).

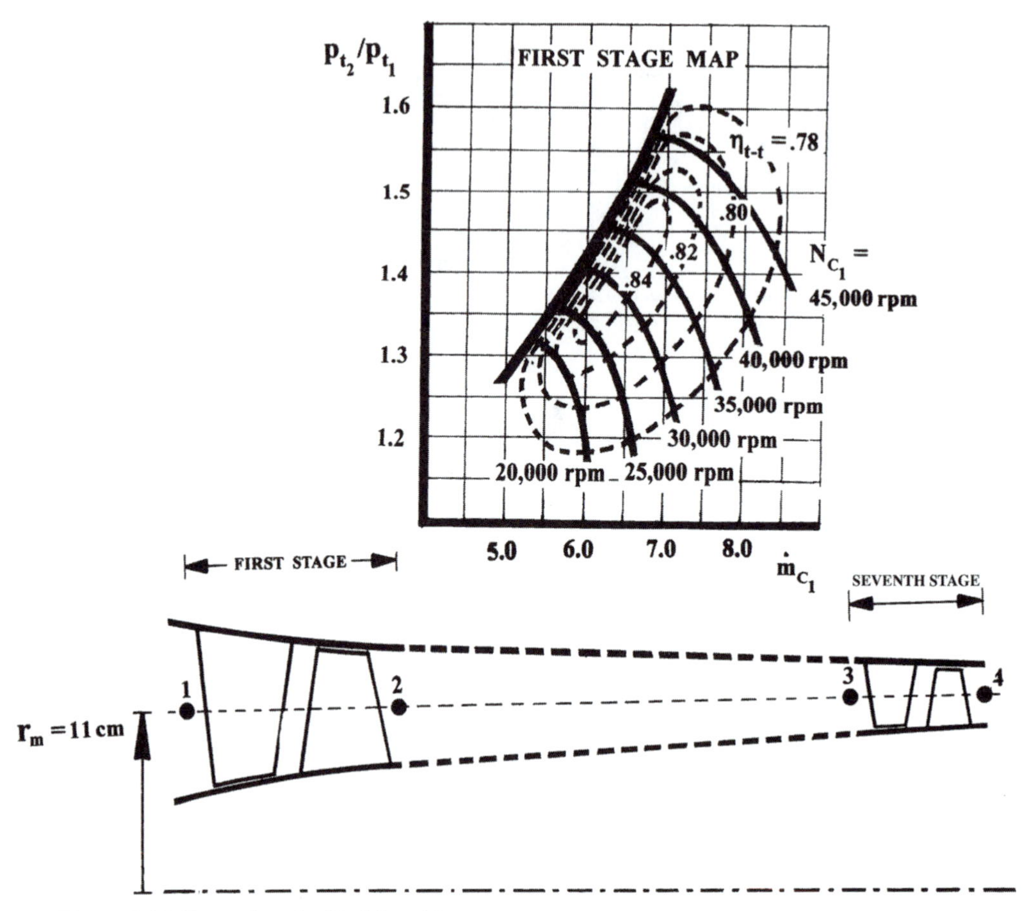

Figure 9.37. Input data for Problem 13.

13) Figure 9.37 shows an adiabatic seven-stage axial-flow compressor along with the first-stage map. All stages share the same total-to-total magnitudes of pressure ratio and efficiency. The operating conditions are as follows:

- Inlet total temperature (T_{t1}) = 302.6 K
- Inlet total pressure (p_{t1}) = 1.15 bars
- First-stage exit total temperature (T_{t2}) = 344.2 K
- Shaft speed (N) = 41,000 rpm
- Flow coefficient (ϕ) = 0.4 (constant)

Calculate the seventh-stage corrected mass-flow rate ($\dot{m}_{C,7}$).

14) In the preceding problem, consider a *four-stage* configuration instead. All four stages share the same total-to-total values of pressure ratio and efficiency. Combined together, these stages will have to give rise to the total-to-total magnitudes of pressure and temperature ratios as in Problem 13.

Calculate the stage efficiency in the new compressor configuration.

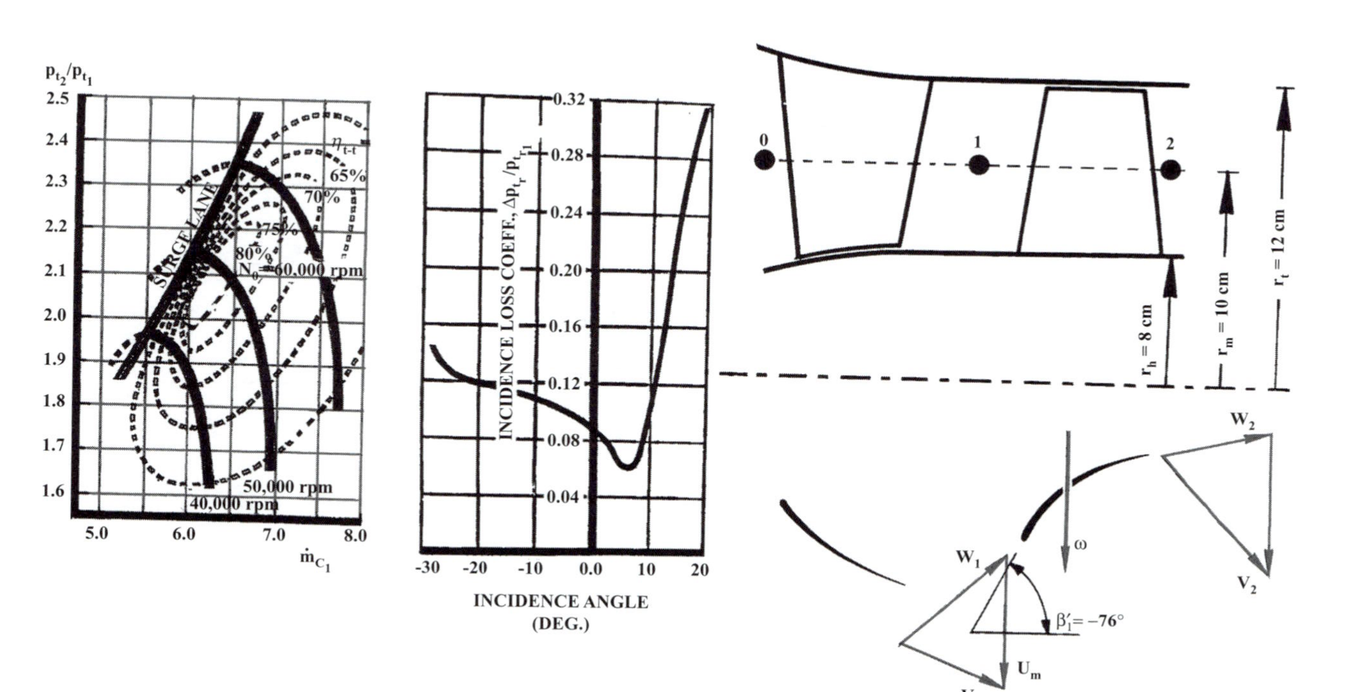

Figure 9.38. Input data for Problem 15.

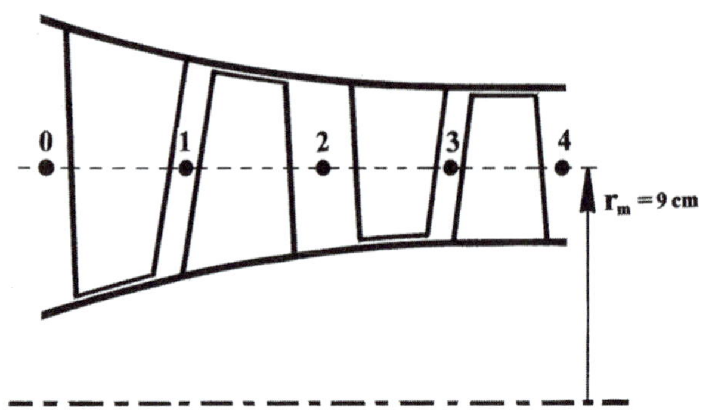

Figure 9.39. Input data for Problem 16.

15) Figure 9.38 shows a single compressor stage, the rotor incidence-loss characteristics, and the stage map. The stage is operating under the following conditions:

- Rotor-inlet total pressure $(p_{t1}) = 3.05$ bars
- Rotor-inlet total temperature $(T_{t1}) = 423.4$ K
- Flow coefficient $(\phi) = 0.41$
- Rotor-exit static density $(\rho_2) = 2.89$ kg/m^3

a) Calculate the stage-exit critical Mach number (M_{cr2}).

b) Calculate the rotor entropy-based efficiency $e^{\frac{-\Delta s}{c_p}}$ arising from the incidence losses.

16) Figure 9.39 shows a two-stage axial-flow compressor that is operating under the following conditions:

- Shaft-work split is 55:45
- Inlet total pressure $(p_{t0}) = 5.7$ bars
- Inlet total temperature $(T_{t0}) = 482$ K
- Shaft speed $(N) = 51{,}000$ rpm
- Flow coefficient $(\phi) = 0.42$ (constant)
- Compressor-exit total temperature $(T_{t4}) = 840$ K

First stage

- Stator total pressure loss = 8%
- Rotor total relative pressure loss = 14%
- Stage reaction $(R) = 50\%$
- Exit relative-flow angle $(\beta_2) = -41^\circ$

Second stage

- Stator-exit static temperature $(T_3) = 648$ K
- Stator-exit swirl angle $(\alpha_3) = 22^\circ$
- Stator flow process is isentropic
- Total-to-total efficiency $(\eta_{t-t}) = 82\%$

a) Calculate the entropy production across the second stage.

b) Calculate the second-stage reaction.

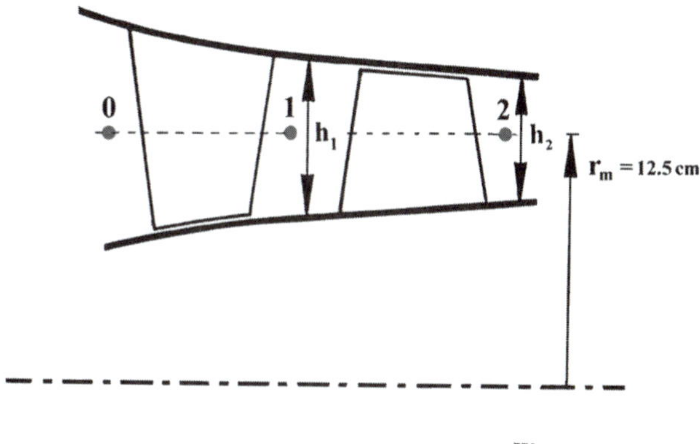

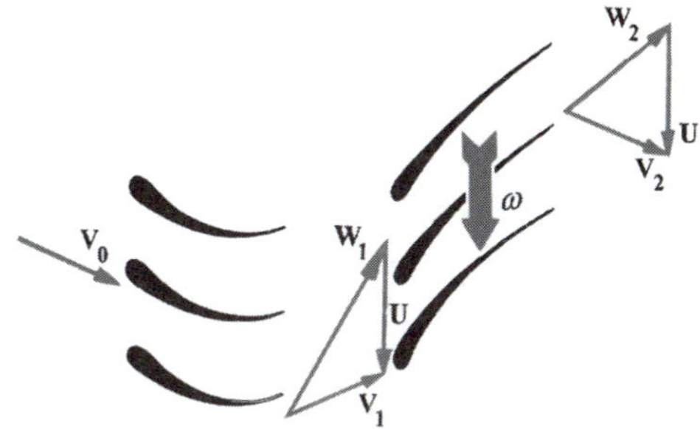

Figure 9.40. Input data for Problem 17.

17) Figure 9.40 shows a high-pressure compressor stage and a sketch of the rotor inlet and exit velocity triangles. The stage operating conditions are as follows:

- Stage reaction (R) = 100%
- Mass-flow rate ($\dot{m}$) = 8.6 kg/s
- Shaft speed (N) = 33,600 rpm
- Stage-inlet total pressure (p_{t0}) = 9.2 bars
- Stage-inlet total temperature (T_{t0}) = 612.0 K
- Stator total-pressure loss coefficient ($\bar{\omega}_{stator}$) = 0.06
- Stator-exit swirl angle (α_1) = −24.0°
- Stator-exit critical Mach number (M_{cr_1}) = 0.42
- Stage total-to-total efficiency (η_C) = 78.0%

Assuming an adiabatic flow throughout the stage, a stagewise fixed magnitude of axial-velocity component, and a γ magnitude of 1.4, calculate:

a) The stage flow and work coefficients (ϕ and ψ);
b) The percentage of total relative pressure loss $\{(\Delta p_{tr})_{rotor}/p_{tr1}\}$;
c) The specific speed (N_s), and evaluate the axial-stage choice;
d) The rotor inlet and exit annulus heights (h_1 and h_2).

Hint: See whether the rotor is choked by computing W_1/W_{cr1}. Should this be the case, make the necessary changes. The changes in this case will not stop at previously computed results but will include some of the input variables as well.

CHAPTER TEN

Radial-Inflow Turbines

Introduction

For more than three decades now, radial-inflow turbines have been established as a viable alternative to its axial-flow counterpart, specifically in power-system applications. Despite its relatively-primitive means of fabrication, radial turbines are capable of extracting a large per-stage shaft work in situations with low mass-flow rates. This turbine category also offers little sensitivity to tip clearances, in contrast to axial-flow turbines. Nevertheless, the turbine large envelope, bulkiness and heavy weight (Fig. 10.1), virtually prohibits its use in propulsion devices.

Components of Energy Transfer

Figure 10.2 shows the velocity diagrams at the rotor inlet and exit stations within a typical radial-turbine stage. As derived in Chapter 4, the combined Euler/energy-transfer equation can be expressed, for the specific shaft work (w_s), as

$$w_s = h_{t1} - h_{t2} = \frac{\left(U_1{}^2 - U_{2m}{}^2\right)}{2} + \frac{\left(W_{2m}{}^2 - W_1{}^2\right)}{2} + \frac{\left(V_1{}^2 - V_{2m}{}^2\right)}{2} \tag{10.1}$$

where the subscripts 1 and $2m$ refer to the rotor inlet and mean-radius exit stations, respectively. The velocity components in expression (10.1) are all shown in Figure 10.2.

For all three terms in expression 10.1 to contribute positively to the shaft-work production, the following velocity-component relationships must be satisfied:

$$U_1 > U_{2m} \tag{10.2}$$

This implies a streamwise decline in radius across the rotor. It also underscores the use of the phrase "inflow" in referring to this turbine category.

$$W_{2m} > W_1 \tag{10.3}$$

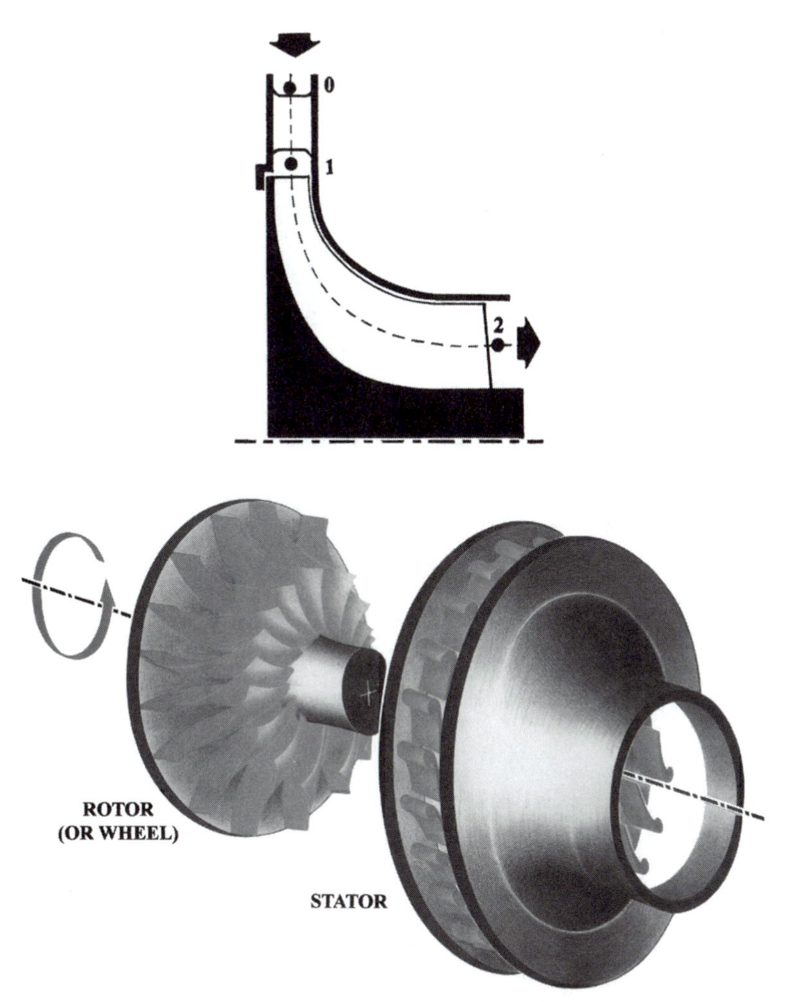

Figure 10.1. Isometric and meridional views of a typical radial-inflow turbine stage.

This implies an accelerating (nozzle-like) blade-to-blade passage, as shown in Fig. 10.2.

$$V_1 > V_{2m} \tag{10.4}$$

This condition calls for a large stator-exit velocity (Fig. 10.2) or, equivalently, a large stator-exit swirl angle (α_1).

Flow Angles

With the rotor inlet segment being radial, the absolute and relative flow angles (α_1 and β_1) are referenced to the local radial direction. The rotor exit segment, by reference to Fig. 10.2, indeed looks like that of an axial-flow turbine rotor. It follows that the datum to which the flow exit angles (α_2 and β_2) are referenced is the axial direction.

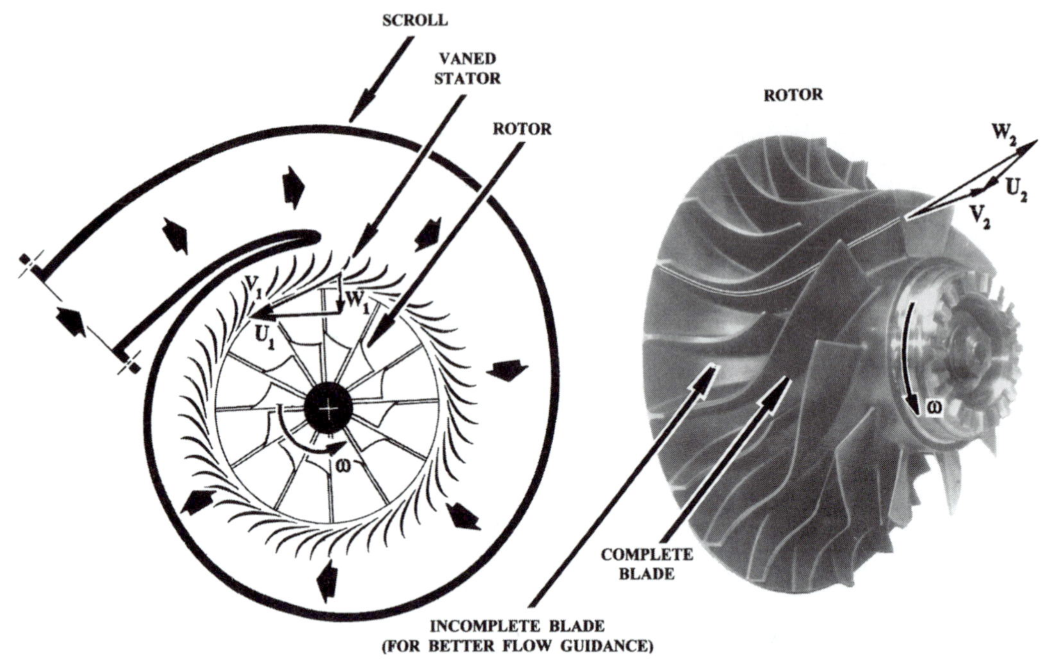

Figure 10.2. Component identification and velocity triangles.

The sign convention of these angles is the same as stated in Chapter 3: positive in the direction of rotation and negative otherwise.

Stage Reaction

As defined in Chapter 4, the stage reaction (R) is the ratio between the static and total enthalpy changes across the rotor. Referring to the energy-transfer components in (10.2) through (10.4) and dropping, for simplicity, the subscript m, we obtain the following reaction expression:

$$R = \frac{\left(W_2^2 - W_1^2\right) + \left(U_1^2 - U_2^2\right)}{\left(W_2^2 - W_1^2\right) + \left(U_1^2 - U_2^2\right) + \left(V_1^2 - V_2^2\right)} \qquad (10.5)$$

Note that for the terms in equation (10.5) to represent an enthalpy change (static or dynamic), each parenthesized term in this expression should be divided by 2. With this in mind, one may understandably have some doubt as far as categorizing the term $(U_1^2 - U_2^2/2)$ to represent a *static* enthalpy change. First, one should recall (by reference to the total-enthalpy definition in Chapter 3) that the only term representing the dynamic enthalpy change is $(V_1^2 - V_2^2/2)$. It therefore follows that both of the two other energy-transfer components (10.2) and (10.3) will have to represent static enthalpy.

Other Performance-Related Dimensionless Variables

In addition to the stage reaction, there exists a set of performance-influencing variables. Compared with their axial-turbine counterparts, these variables are hardly new, but they are now defined in a manner that is consistent with the stage geometrical complexity.

Reynolds Number

As indicated in Chapter 3, the Reynold number is a measure of the domination of the viscosity caused shear forces relative to the inertia forces. Magnitudes of this variable in excess of roughly 5×10^5 imply a lower loss level.

$$Re = \frac{\rho U_t D}{\mu}$$

where

ρ is the rotor-tip static density,
U_t is the rotor-tip rotational velocity,
D is the rotor-tip diameter, and
μ is the rotor-tip dynamic viscosity coefficient.

The subscript t refers to the rotor-tip radius.

Flow Coefficient

This is a velocity ratio that is indicative of the mass-flow rate, which, in radial turbines, has its own design implications. With the specific speed definition in mind (Chapter 5), the mass-flow rate is one of two key input variables (the other being the shaft work) that govern the selection of a turbine type. Given the symbol ϕ, the flow coefficient is defined as

$$\phi = \frac{V_{r1}}{U_t}$$

where the subscript 1 denotes the rotor inlet station (Fig. 10.2). Because V_{r1} is the inlet through-flow velocity, it is naturally indicative of the stage mass-flow rate. A radial-turbine stage would typically possess a flow coefficient that is notably less than 0.5.

Work Coefficient

Recognized as ψ, this variable implies the stage total-to-total pressure ratio, which is considerably higher than that of a typical axial-flow stage. The variable is defined as

$$\psi = \frac{w_s}{U_t^2} = \frac{\Delta h_t}{U_t^2}$$

The magnitude of ψ in a radial turbine stage is typically high (normally in excess of 1.0).

Total Relative Properties and Critical Mach Number

At any point in the rotor subdomain, the total relative properties are dependent on the local velocity diagram. Of these, the total relative temperature, T_{tr}, is defined as follows:

$$T_{tr} = T + \frac{W^2}{2c_p}$$

In a more applicable form, this expression can be rewritten as

$$T_{tr} = \left(T_t - \frac{V^2}{2c_p}\right) + \frac{W^2}{2c_p} \tag{10.6}$$

Applying this definition at the rotor inlet and exit stations, and employing equation (10.1), we obtain the following result:

$$T_{tr2} - T_{tr1} = \frac{(U^2{}_2 - U^2{}_1)}{2c_p} = \frac{\omega^2}{2c_p}(r^2{}_2 - r^2{}_1) \tag{10.7}$$

which simply reaffirms the fact that T_{tr} declines across the radial-turbine rotors as a result of the streamwise radius decline. Note, by reference to Fig. 10.2, that a radial-inflow turbine rotor is where the radius continually declines in the streamwise direction and so does the total relative temperature, as is implied by equation (10.7).

The relative critical velocity, W_{cr}, can be consistently defined as

$$W_{cr} = \sqrt{\left(\frac{2\gamma}{\gamma+1}\right) RT_{tr}} \tag{10.8}$$

As defined in Chapter 3, the relative-critical velocity ratio (or the relative critical Mach number) can be consistently defined as follows:

$$M_{crr} = \frac{W}{W_{cr}} = \frac{W}{\sqrt{\left(\frac{2\gamma}{\gamma+1}\right) RT_{tr}}} \tag{10.9}$$

The total relative pressure, p_{tr}, may now be defined as follows:

$$p_{tr} = \frac{p}{\left[1 - \frac{\gamma-1}{\gamma+1}\left(\frac{W}{W_{cr}}\right)^2\right]^{\frac{\gamma}{\gamma-1}}} \tag{10.10}$$

where the static pressure (p) can itself be calculated as

$$p = p_t\left[1 - \frac{\gamma-1}{\gamma+1}\left(\frac{V}{V_{cr}}\right)^2\right]^{\frac{\gamma}{\gamma-1}} \tag{10.11}$$

Referring to Figure 9.8 in particular, a given point in the rotor subdomain can be represented by its static, total, or total relative states, all vertically above one another on the T-s diagram. This simply means that any two of these properties at any thermophysical state can be related to one another via isentropic relationships.

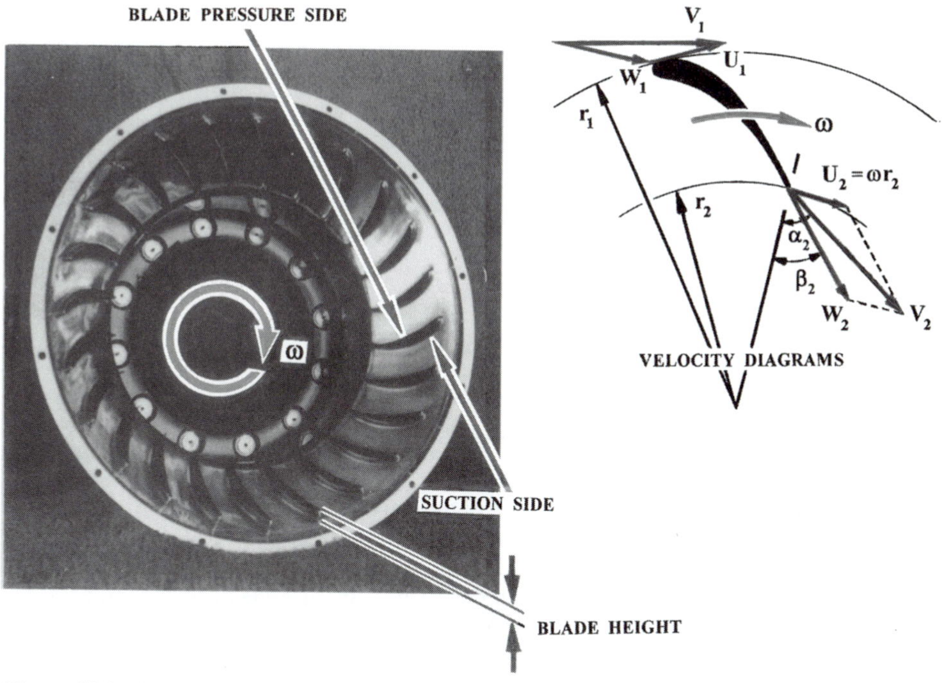

Figure 10.3. A purely radial turbine rotor and velocity diagrams.
Notes
- Blade and flow angles are measured from the "local" radial direction.
- Excessive inlet swirl velocity suggests the presence of upstream guide vanes.
- Large exit swirl component is a waste of much-desired angular momentum.

It is therefore possible to express the total relative pressure (p_{t_r}) in terms of the often-known total pressure as follows:

$$p_{t_r} = p_t \left(\frac{T_{t_r}}{T_t} \right)^{\frac{\gamma}{\gamma - 1}} \tag{10.12}$$

Another reminder at this point has to do with changes in the total relative pressure (p_{t_r}) within the rotor flow region. In Chapter 3, the point was made that the decline in this particular variable is indicative of real-life irreversibilities (e.g., friction) only in the case of *axial*-flow rotors. In the current radial rotor, irreversibilities will present themselves as part of a larger drop in this variable. The difference, as implied earlier, is due to the mere change of radius.

Conventional-Stage Geometrical Configurations

Figures 10.1 and 10.2 show different components of what traditionally defines a conventional radial-inflow turbine stage (as contrasted with that in Figure 10.3). The *T-s* representation of the conventional-stage flow process is shown in Figure 10.4. The stator flow process representation in this figure makes it clear that the process

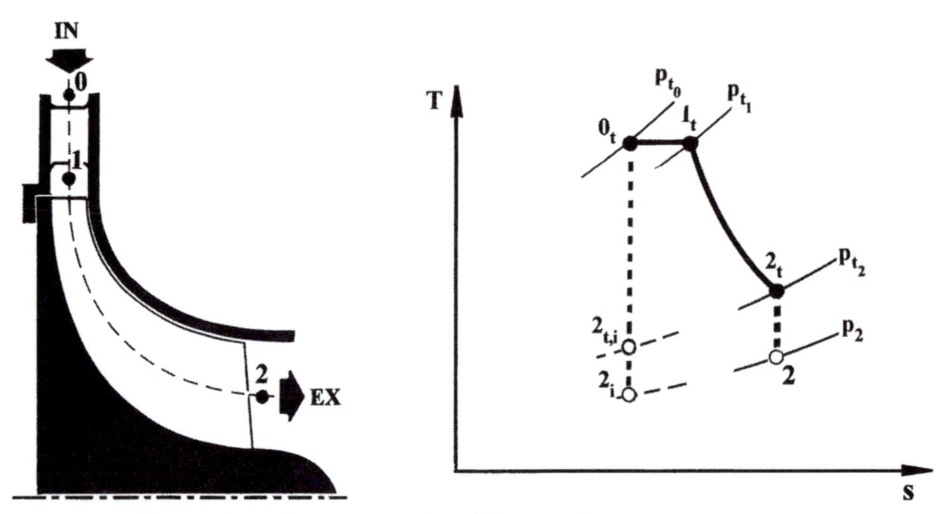

Figure 10.4. Thermodynamic representation of the stage flow process.

is assumed adiabatic, as evidenced by the statorwise invariance of total temperature. Of course, turbine stators can be subject to cooling, in which case heat energy would be exchanged, with the result being a usually minor drop in the primary-flow total temperature. Should a high computational precision be sought, the primary/cooling flow interaction can be separately investigated in a heat-exchanger type of analysis and the corresponding drop in the stator-flow total temperature invoked. This magnitude is subsequently substituted in the energy-conservation relationship (3.2). For practical purposes, this heat-transfer subproblem will be brief in this text, for it adds a normally negligible heat-transfer dimension to what is overwhelmingly an aerothermodynamics flow problem.

The nonrotating component of a radial turbine may, or may not, contain a vaned stator, which is shown in Figure 10.5. In either case, however, the first segment of such a component will typically be the stage "scroll." This is basically a conical flow passage that is "wrapped" around the stage, with an inner flow-permeable station. It is the scroll responsibility to essentially produce, in the "loose" sense of the word, a circumferentially uniform exit flow in terms of mass flux and direction. As shown in Figure 10.6, the tangential uniformity of mass flux is, to a certain extent, attained by the gradual reduction of the scroll cross section in the streamwise direction. This accounts for the continuous mass discharge around the circumference. In practice, unfortunately, a much-simplified approach to do so, is through a linearly declining circumferential distribution of the scroll cross-flow area (Fig. 10.7). As proven by many researchers, such an overly simplified approach is hardly the means to achieve the above-stated task. This is particularly true in the region around the scroll "tongue," as is clearly shown in Figure 10.8. It is perhaps safe to say that it is the normally terrible surface roughness of this passage (Fig. 10.9) that makes it tempting to excessively oversimplify the problem from a design viewpoint. Nevertheless, an uncarefully designed scroll, particularly in the absence of a vaned stator, will have

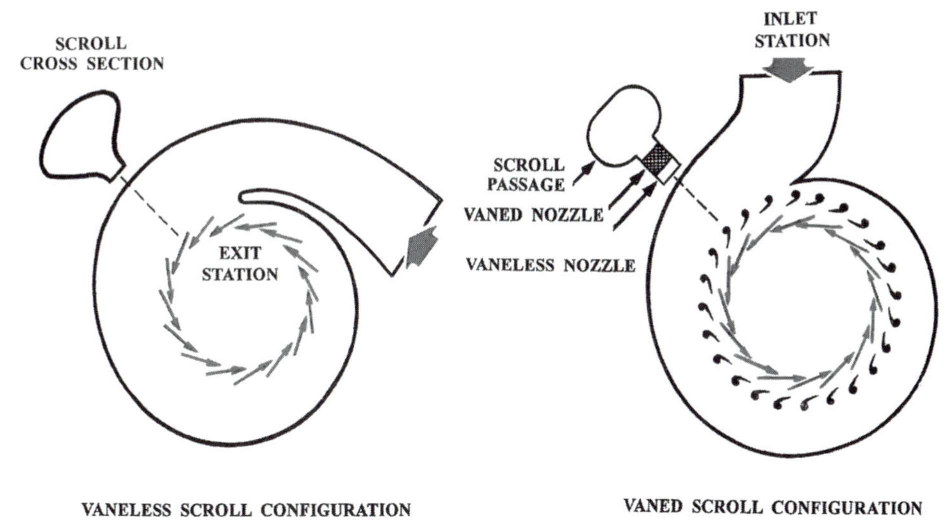

Figure 10.5. Vaneless versus vaned stator and typical scroll cross-section configurations.

the negative effects of producing nonuniform flow streams in the rotor passages and creating a cyclic rotordynamic force on the rotor blades.

The cyclic stresses (above) result from the lack of static pressure uniformity around the circumference. The situation just described is potentially capable of

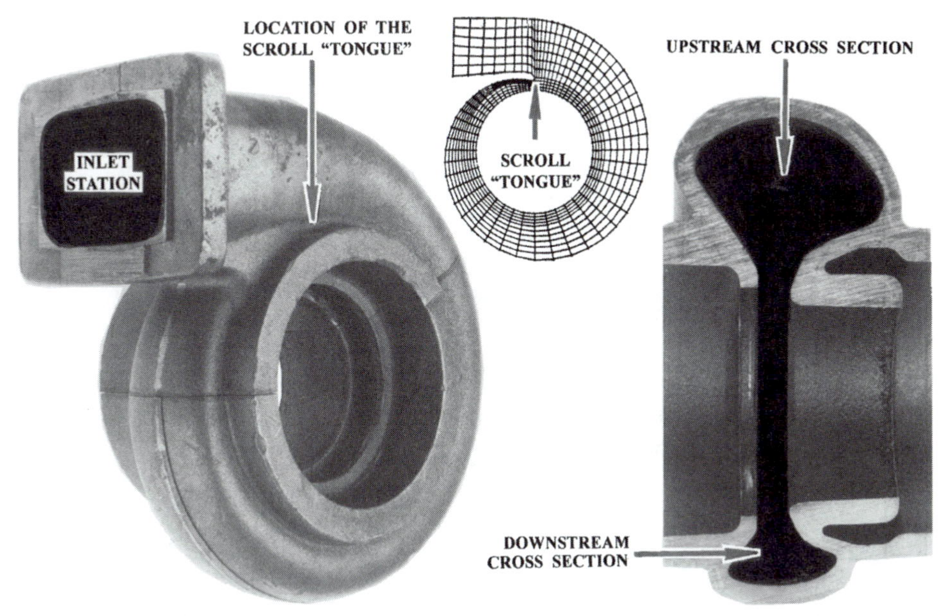

Figure 10.6. Decline of the scroll cross-sectional area in the circumferential direction. Typical scroll configuration in power-system applications.

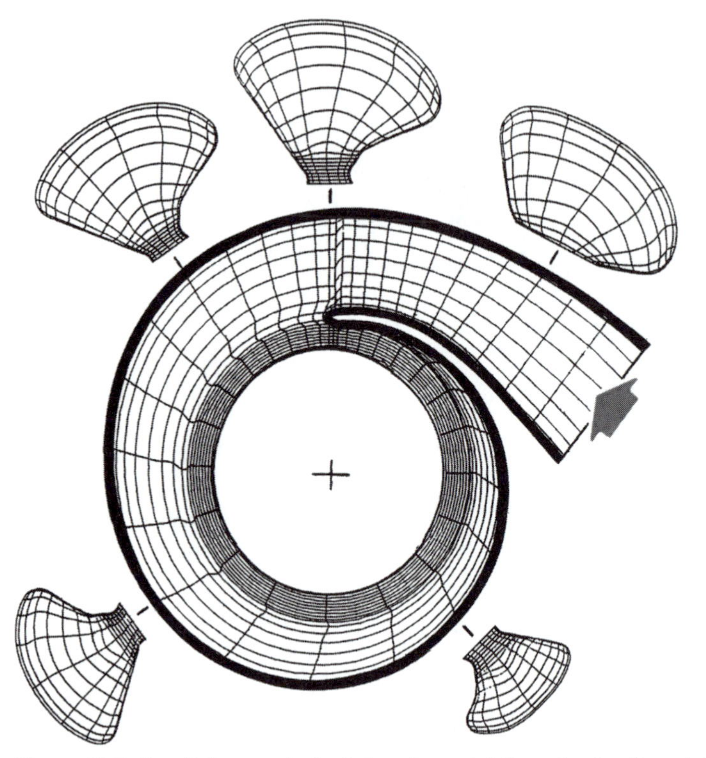

Figure 10.7. Scroll flow analysis: finite-element "discretization" model.

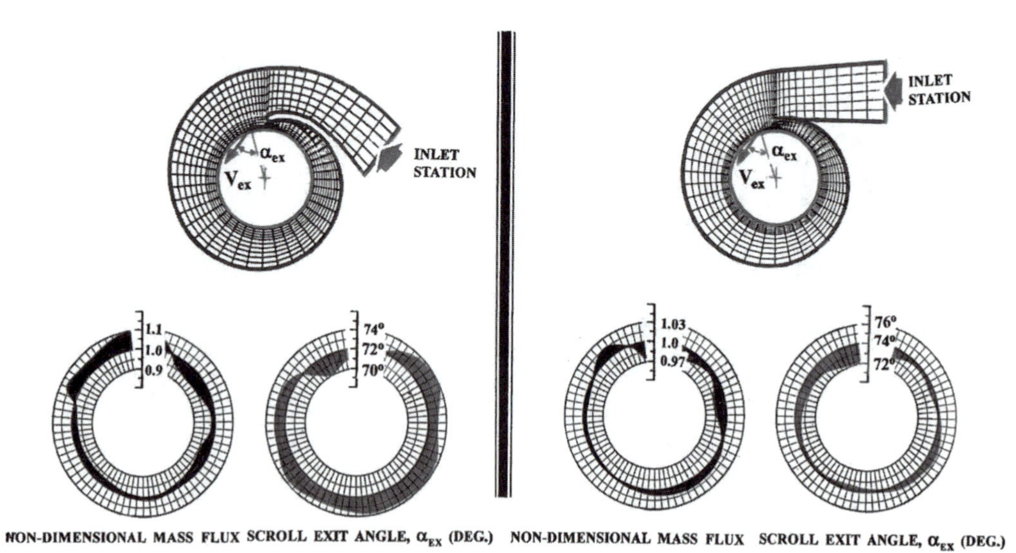

Figure 10.8. Scroll flow analysis: circumferential distributions of mass flux and exit angle.

Figure 10.9. Excessive surface roughness of the scroll interior surface leads to aerodynamic degradation.

causing a premature fatigue failure of the rotor blades, depending on the closeness of the blades to the scroll tongue (Fig. 10.6).

If so equipped, the next component of the nonrotating assembly will be a radial stator of the type shown in Figure 10.10. This cascade of airfoils can be bound by either straight or contoured endwalls, as shown in Figure 10.11. In addition to the nozzle-like shape of the stator passage (in the sense of the streamwise radius decline and vane-to-vane passage shape), contouring the endwalls in the manner displayed in this figure provides an added contributor to the flow-acceleration process. Examination of the two stator aerodynamic loading graphs representing the straight versus contoured sidewall configurations reveals that the latter gives rise to less diffusion (i.e., flow deceleration) near the stator exit station.

Vaned or not, an influential part of the stationary segment in this turbine type is the radial gap upstream from the rotor (Fig. 10.11). This simple component helps accelerate the flow even more, creating, at virtually no expense, a larger swirl-velocity component (V_θ) at the rotor inlet station. This, in view of Euler's equation, will increase the shaft work produced within the rotor subdomain. As indicated in Chapter 3, the flow structure over this gap is close to that of the free-vortex pattern, where both V_r and V_θ are inversely proportional to the local radius, at least midway between the sidewalls. As noted in Chapter 3, the flow swirl angle, in this case, remains constant across the entire gap.

Figure 10.12 shows the traditional configuration of a radial-turbine rotor. As seen in the figure, the rotor inlet segment is perfectly radial, and so is the relative velocity vector, $\boldsymbol{W}_1$. Note that the incidence angle in this figure is zero. This, at the design point, requires a stator-exit swirl-velocity component ($V_{\theta 1}$) that is identical to the tip speed (U_t). As for the rotor-exit segment, Figure 10.12 shows it to be axial-like by reference to the meridional projection in the figure. The gradual hub-to-casing tangential deflection of this exit segment (shown in the rotor isometric

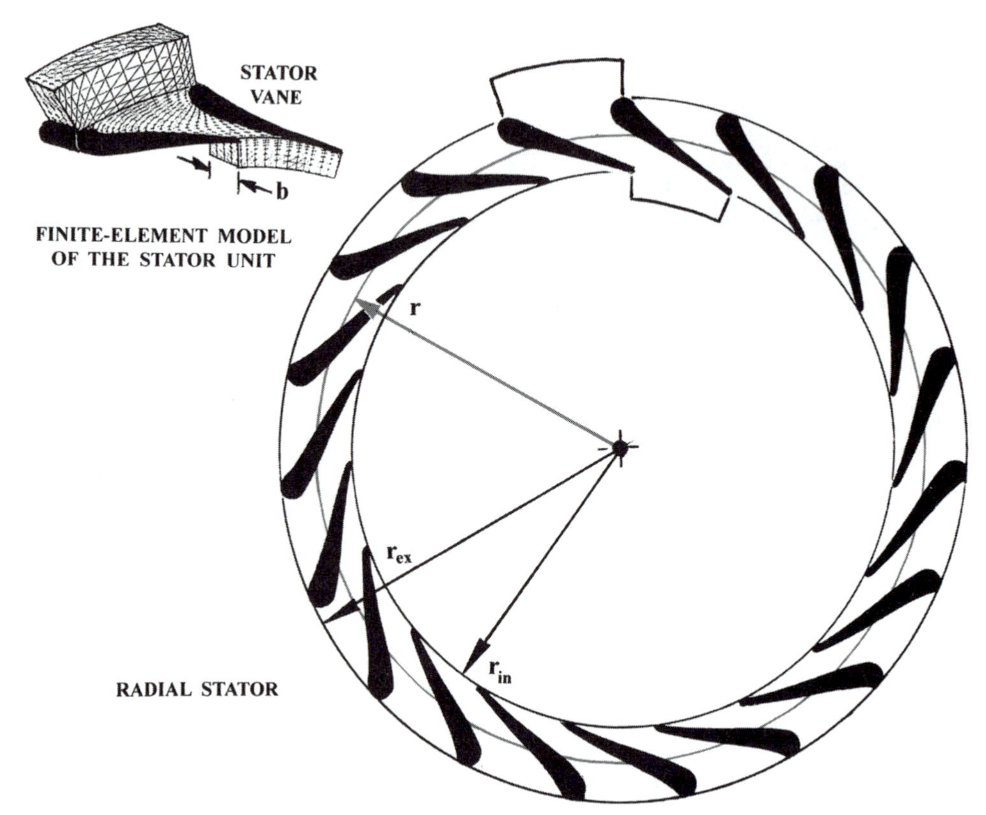

Figure 10.10. Vaned stator as part of the turbine nonrotating assembly.

sketch) is clearly in response to the change in U across the exit blade height. The exit velocity triangle in the figure corresponds to the exit mean-radius location and will be considered representative of the entire exit station from a computational standpoint.

Compressibility Effects

The point was emphasized in Chapter 3 that the flow compressibility is gauged by the critical Mach number. In viewing the stage as a whole, however, it is perhaps appropriate to define a general Mach-number-like variable (M^*), as follows:

$$M^* = \frac{U_t}{V_{cr0}} = \frac{U_t}{\sqrt{\left(\frac{2\gamma}{\gamma+1}\right) RT_{t0}}} \tag{10.13}$$

where the subscript 0 denotes the stage inlet station. Termed the *stage* critical Mach number, the M^* effect on the overall stage efficiency is shown in Figure 10.13. Superimposed on the same figure is the Reynolds number influence as a way of including the flow-viscosity effects as well. The common (perhaps anticipated) trend in this figure is a rather notable increase in efficiency with the increase in Reynolds number. While

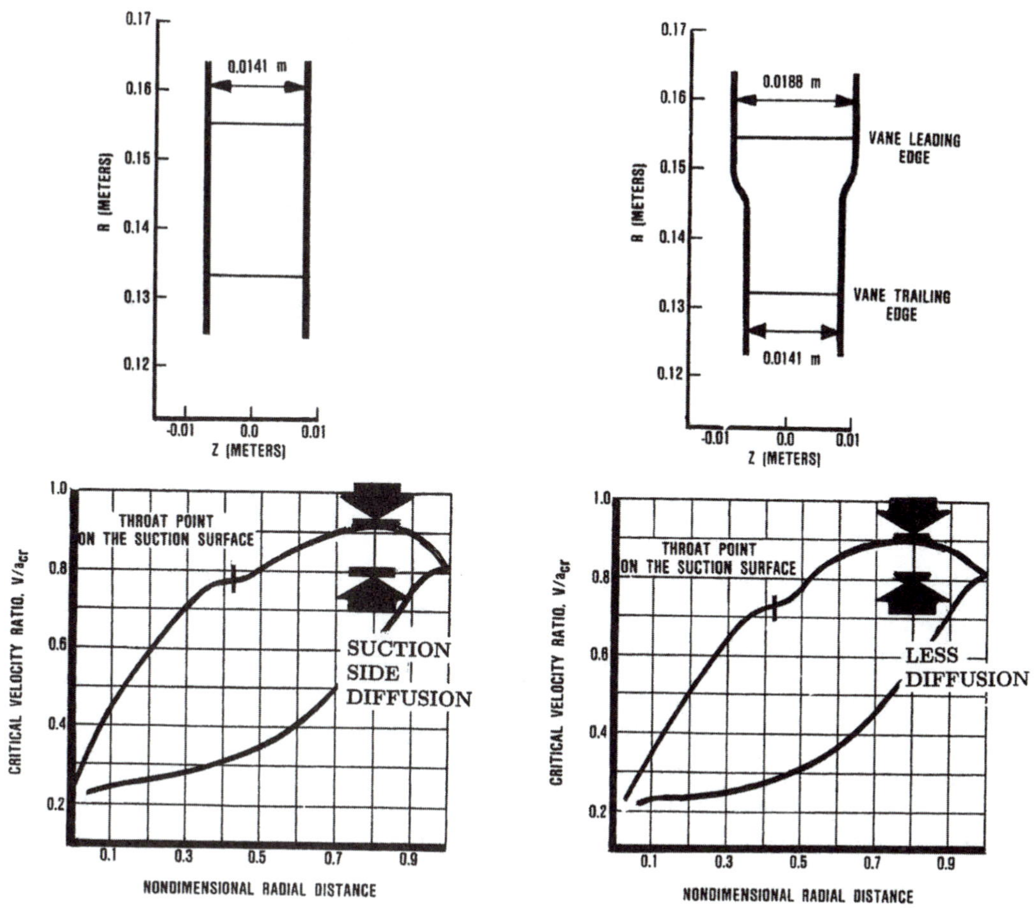

Figure 10.11. Straight versus contoured sidewalls across the vaned nozzle. (Numerical results from Baskharone 1984 for the GT601 turbocharger, a product of Garrett Turbine Engine Co.)

the proportionality here is almost linear at high Reynolds number magnitudes, the region where

$$10^5 \leq Re \leq 5 \times 10^5$$

is characterized by a comparatively much more sensitivity to the Reynolds number magnitude. The reason is that this range is where the viscosity-related forces dominate the flow behavior.

Specific physical implications of the condition $M^* = 1.0$ have to do with, and stop at, the stator flow field. Under an adiabatic-flow situation, and referring to the rotor-inlet velocity triangle in Figure 10.12, this condition means that the stator-exit tangential-velocity component ($V_{\theta 1}$) has itself reached a sonic magnitude in the absence of an incidence angle. With the stator being incapable of producing a supersonic flow stream, the condition simply means that the stator in this case is choked. A worthy question here is what happens in the downstream *flow-accelerating* stator/rotor radial gap in this case. In practical terms, the flow stream immediately

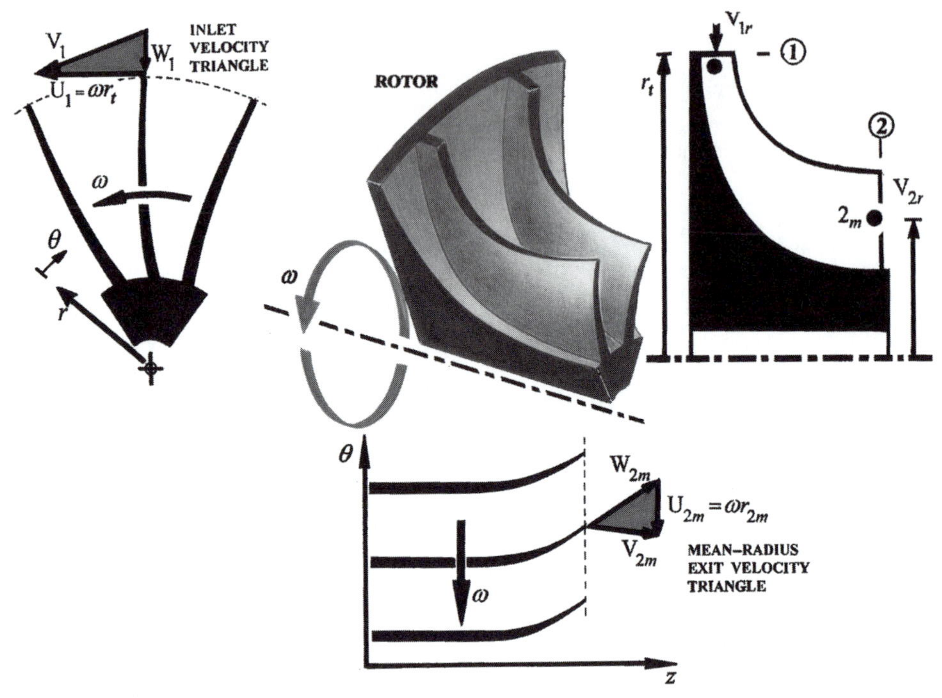

Figure 10.12. Rotor inlet and exit velocity diagrams.

downstream from the stator trailing edges is subject to a sudden expansion as the blockage caused by, at least, the airfoil trailing-edge thicknesses is left behind. This so-called "dump" effect will cause an abrupt local drop in velocity and therefore a return to a subsonic flow stream.

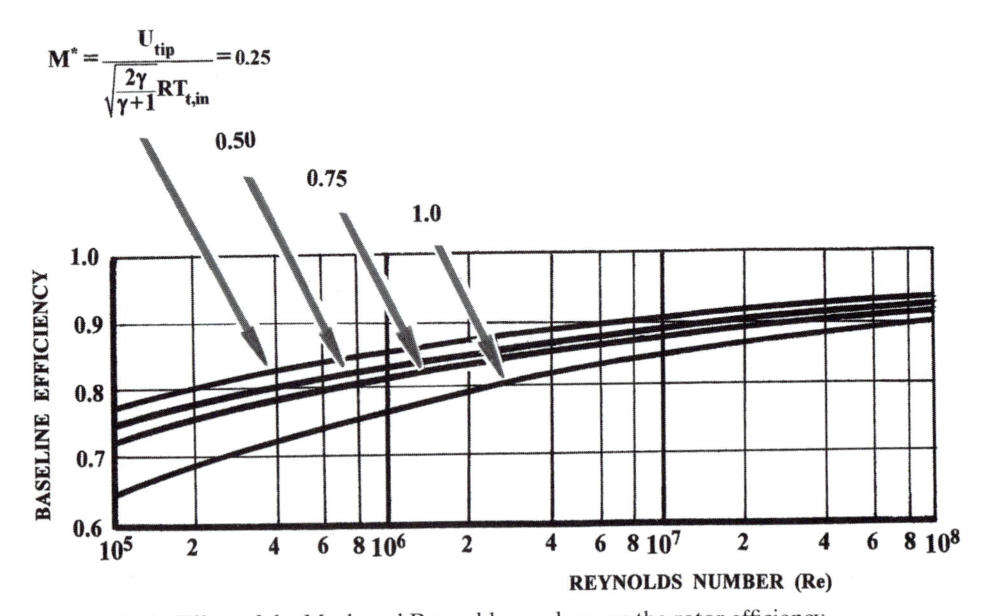

Figure 10.13. Effect of the Mach and Reynolds numbers on the rotor efficiency.

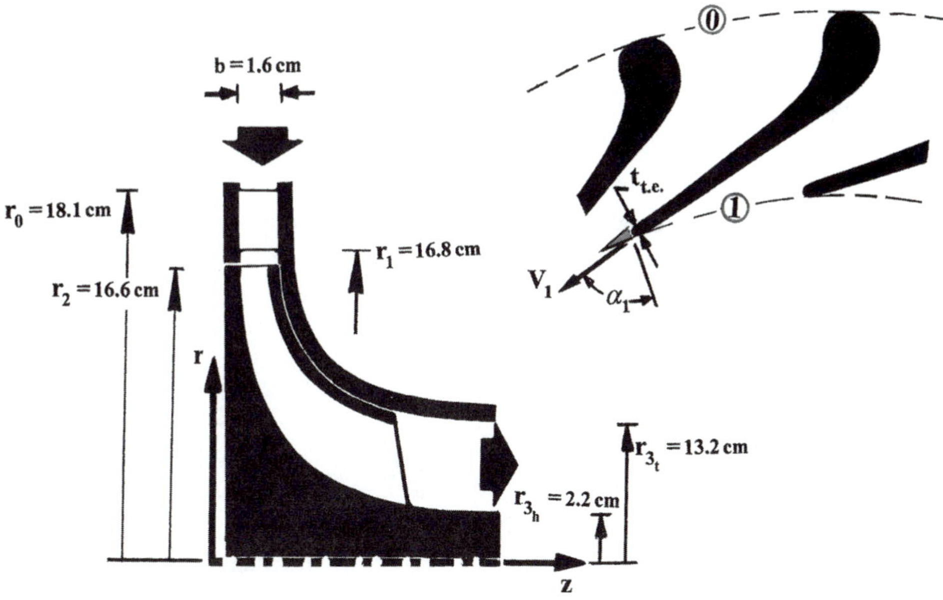

Figure 10.14. Input variables for Example 1.

Referring to the M^* effect in Figure 10.13, one line stands out – namely that labeled $M^* = 1.0$ – as it relates to the rotor subdomain. This condition is hardly indicative of the onset of a sonic state anywhere within the rotor component, for it is the relative critical Mach number (W/W_{cr}) which is the controlling variable in this case. However, it is the rotor-flow environment under such a condition that is most likely to negatively impact the rotor efficiency. The reason is that the transonic flow, which is likely to prevail in this case, is hardly "immune" to local supersonic "pockets" at locations along solid walls, where the relative velocity locally peaks. Within such pockets, the threat is there that oblique shocks may very well exist. In terms of entropy production, these shocks are fortunately weak but are, nevertheless, degrading.

EXAMPLE 1

Figure 10.14 shows an adiabatic radial-inflow turbine stage together with its major dimensions. The number of stator vanes (N_v) is 37, and each vane has a trailing-edge thickness ($t_{t.e.}$) of 1.5 mm. The stator flow field is assumed isentropic, and the stage operating conditions are as follows:

- Shaft speed (N) = 48,000 rpm
- Inlet total temperature (T_{t0}) = 1280 K
- Inlet total pressure (p_{t0}) = 8.6 bars
- Stator is choked
- Stator-exit (swirl) flow angle (α_1) = 70°
- Rotor-exit critical Mach number = 0.47
- Rotor-exit average total temperature (T_{t3}) = 800 K

- Rotor-exit (absolute) flow angle (α_{3m}) is negative
- Stage total-to-total efficiency (η_T) = 88%

Assuming an average γ magnitude of 1.33, calculate:

a) The mass-flow rate ($\dot{m}$);
b) The change in M_{cr} in crossing the trailing-edge station;
c) The rotor-exit absolute and relative flow angles (α_{3m} and β_{3m});
d) The variable indicating whether the rotor is choked;
e) The change in total relative temperature ($T_{t,r2} - T_{t,r3m}$);
f) The stage specific speed (N_s);
g) The stage reaction (R);
h) The work coefficient (ψ).

SOLUTION

Part a: At the stator exit station, let the symbol $(L)_{act.}$ be the actual trailing-edge-blocked segment of the circumference and $(L)_{id.}$ the ideal zero-trailing-edge-thickness circumference. Let us now compute the ratio Γ between the two:

$$\Gamma = \frac{(L)_{act.}}{(L)_{id.}} = \frac{2\pi r_1 - N_v\left(\frac{t_{t.e.}}{\cos\alpha_1}\right)}{2\pi r_1} = 0.846$$

where N_v refers to the stator number of vanes. Let us now apply the continuity equation at a station just inside the vane trailing edge:

$$\frac{\dot{m}\sqrt{T_{t1}}}{p_{t1}A_{1\,act.}} = \sqrt{\frac{2\gamma}{(\gamma+1)R}}M_{cr1}\left[1 - \left(\frac{\gamma-1}{\gamma+1}\right)M_{cr1}{}^2\right]^{\frac{1}{\gamma-1}}$$

where the actual stator-exit cross-flow (projected) area $[(A_1)_{act.}]$ can be expressed as

$$(A_1)_{act.} = \Gamma(2\pi r_1 b)\cos\alpha_1 = 0.00489\,\text{m}^2$$

Substituting this and other known variables into the continuity equation, we obtain:

$$\dot{m} = 4.66\,\text{kg/s}$$

Note that the magnitude of M_{cr1} is 1.0, because the stator, according to the problem statement, is choked.

Part b: Because M_{cr1} is equal to 1.0,

$$V_1 = V_{cr1} = 647.6\,\text{m/s}$$

This is the velocity magnitude immediately upstream from the trailing edge and can be resolved as follows:

$$V_{\theta 1} = V_1 \sin\alpha_1 = 608.5\,\text{m/s}$$

$$V_{r1} = V_1 \cos\alpha_1 = 221.5\,\text{m/s}$$

As soon as the flow stream departs this trailing-edge-blocked circumference (actually a cylindrical surface), a sudden geometrical expansion takes place now that the stator trailing edges are no longer in the picture. The just-before/just-after cross-flow area ratio has already been computed as the variable Γ.

Of the two velocity components (above), only the radial component (V_{r1}), which contributes to the mass-flow rate, will be affected by this "sudden" passage expansion. As a result, the radial-velocity component will suffer a decline and by a factor of Γ. Referring to this new component by $V_{r1}{}'$, we have

$$V_{r1}{}' = \Gamma V_{r1} = 187.4\,\text{m/s}$$

Now we can calculate the corresponding critical Mach number ($M_{cr1}{}'$) just downstream from the trailing-edge station:

$$M_{cr1}{}' = \frac{\sqrt{V_{\theta 1}{}^2 + (V_{r1}{}')^2}}{V_{cr1}} = 0.98\,(\text{a } 2\%\text{ decline})$$

Part c: With the total-to-total stage efficiency being 88%, we can calculate the stage-exit total pressure (p_{t3}) as

$$\eta_T = 0.88 = \frac{1 - \left(\frac{T_{t3}}{T_{t0}}\right)}{1 - \left(\frac{p_{t3}}{p_{t0}}\right)^{\frac{\gamma-1}{\gamma}}}$$

which, upon substitution, yields

$$p_{t3} = 1.06\,\text{bars}$$

In preparation for applying the continuity equation at the rotor-exit station (which is the next step), we are required to compute a few more variables, as follows:

$$\rho_{t3} = \frac{p_{t3}}{RT_{t3}} = 0.461\,\text{kg/m}^3$$

$$\rho_3 = \rho_{t3}\left[1 - \left(\frac{\gamma-1}{\gamma+1}\right) M_{cr3}{}^2\right]^{\frac{1}{\gamma-1}} = 0.419\,\text{kg/m}^3$$

$$V_3 = M_{cr3} V_{cr3} = 240.6\,\text{m/s}$$

Now, applying the continuity equation at the rotor-exit station, we get

$$V_{z3} = \frac{\frac{\dot{m}}{\rho_3}}{[\pi(r_{t3}{}^2 - r_{h3}{}^2)]} = 209.0\,\text{m/s}$$

The rest of the rotor-exit variables can be computed as follows:

$$\alpha_3 = \cos^{-1}\left(\frac{V_{z3}}{V_3}\right) = -29.7^\circ\ (\text{negative sign set in problem statement})$$

$$W_{\theta 3} = V_{\theta 3} - U_3 = (V_3 \sin\alpha_3) - (\omega r_3) = -506.3\,\text{m/s}$$

$$W_3 = \sqrt{W_{\theta 3}{}^2 + V_{z3}{}^2} = 547.7\,\text{m/s}$$

$$\beta_3 = \tan^{-1}\left(\frac{W_{\theta 3}}{V_{z3}}\right) = -67.6^\circ$$

Part d: In order to verify the rotor choking status, we need to compute the relative critical Mach number $[(W/W_{cr})_3]$ at the rotor exit station. To this end, we proceed as follows:

$$T_{tr3} = T_{t3} - \frac{V_3{}^2 - W_3{}^2}{2c_p} = 904.6\,\text{K}$$

$$W_{cr3} = \sqrt{\left(\frac{2\gamma}{\gamma+1}\right) R T_{tr3}} = 544.4\,\text{m/s}$$

$$(M_{crr})_3 = \frac{W_3}{W_{cr3}} = 1.006$$

This relative critical Mach number magnitude is impossible, because the rotor blade-to-blade passage is essentially a subsonic nozzle in the rotating frame of reference.

We therefore conclude that the rotor passage is choked and that

$$W_3 = W_{cr3} = 544.4\,\text{m/s}$$

Generally speaking, the majority of the rotor-exit flow properties must also be corrected. However, noting that the computed relative-critical Mach number is very slightly above unity, such changes are certain to be minor and will therefore be ignored. Note that this type of corrective action was previously made in the solution of Example 8 in Chapter 4 for an axial-flow turbine rotor.

Part e: Across a radial-turbine rotor, an adiabatic flow will still be associated with a total relative temperature decline as a result of the streamwise radius change. In order to compute the change in this total relative property, we proceed as follows:

$$V_2 = M_{cr2} V_{cr2} = 634.7\,\text{m/s}$$

$$W_{r2} = V_{r2} = V_2 \cos\alpha_2 = 217.1\,\text{m/s}$$

$$W_{\theta 2} = V_{\theta 2} - U_2 = -238.0\,\text{m/s}$$

$$W_2 = \sqrt{W_{r2}{}^2 + W_{\theta 2}{}^2} = 322.1\,\text{m/s}$$

$$T_{tr2} = T_{t2} - \frac{V_2{}^2 - W_2{}^2}{2c_p} = 1150.7\,\text{K}$$

$$(\Delta T_{tr})_{rotor} = T_{tr3} - T_{tr2} = -246.1\,\text{K}$$

Part f: Substituting in expression (5.28), for the specific speed, we get

$$N_s = 2.50\,\text{radians}$$

Referring to Figure 5.9, we see that this N_s magnitude places the stage well within the *axial* stage range. Therefore, had this been a real-life design problem, the radial-stage choice would definitely be unwise.

Part g: Substituting in expression (10.5), for the stage reaction, we get

$$R = 67.0\%$$

Part h: The stage work coefficient (ψ) can easily be computed as follows:

$$\psi = \frac{c_p(T_{t0} - T_{t3})}{U_2{}^2} = 0.797$$

Stage Design Approach

In the following, a sequence of simple computational steps toward the creation of a first-order radial turbine stage is presented. The process begins with the justification of the radial-stage-type selection.

Justification of a Radial-Turbine Stage Choice

To a trained designer, two specific conditions trigger, almost instantly, the radial-stage category as the appropriate stage choice. These are a low mass-flow rate magnitude and a high per-stage pressure ratio, both of which are usually provided in the given set of data. In justifying a radial-stage choice, one may adopt (for instance) a combined flow-coefficient/work-coefficient approach, making sure that the former, say, is much less than 0.5, and the latter is significantly higher than unity. However, the single most reliable variable in making the decision is the specific speed (N_s), as it embraces (by definition) both design parameters. Placing the N_s magnitude on Figure 5.9 will, almost conclusively, indicate whether the radial-stage choice is justified.

The Scroll Passage

Also termed the *distributor*, the scroll flow passage is basically a stationary flow passage that has a streamwise-decreasing cross-flow area, with the objective of providing nearly uniform flow properties at the exit radius (Fig. 10.8). The passage inner radius, can be calculated by first calculating the rotor tip radius (r_t) as follows:

$$r_t = \frac{(U_t)_a}{N\left(\frac{2\pi}{60}\right)}$$

where $(U_t)_a$ is the allowable magnitude of rotor-tip rotational velocity and N is the given shaft speed. Depending on the radial extent of the radial gap and vaned stator (if any), the scroll exit radius can then be computed (Fig. 10.2).

Within the scroll passage, and over the plane of symmetry in Figure 10.15, the flow behavior is sufficiently close to that of the free-vortex flow type. As shown in this figure, the flow trajectory in this case is a logarithmic spiral, where the flow angle α, measured from the local radial direction, remains constant over the entire trajectory.

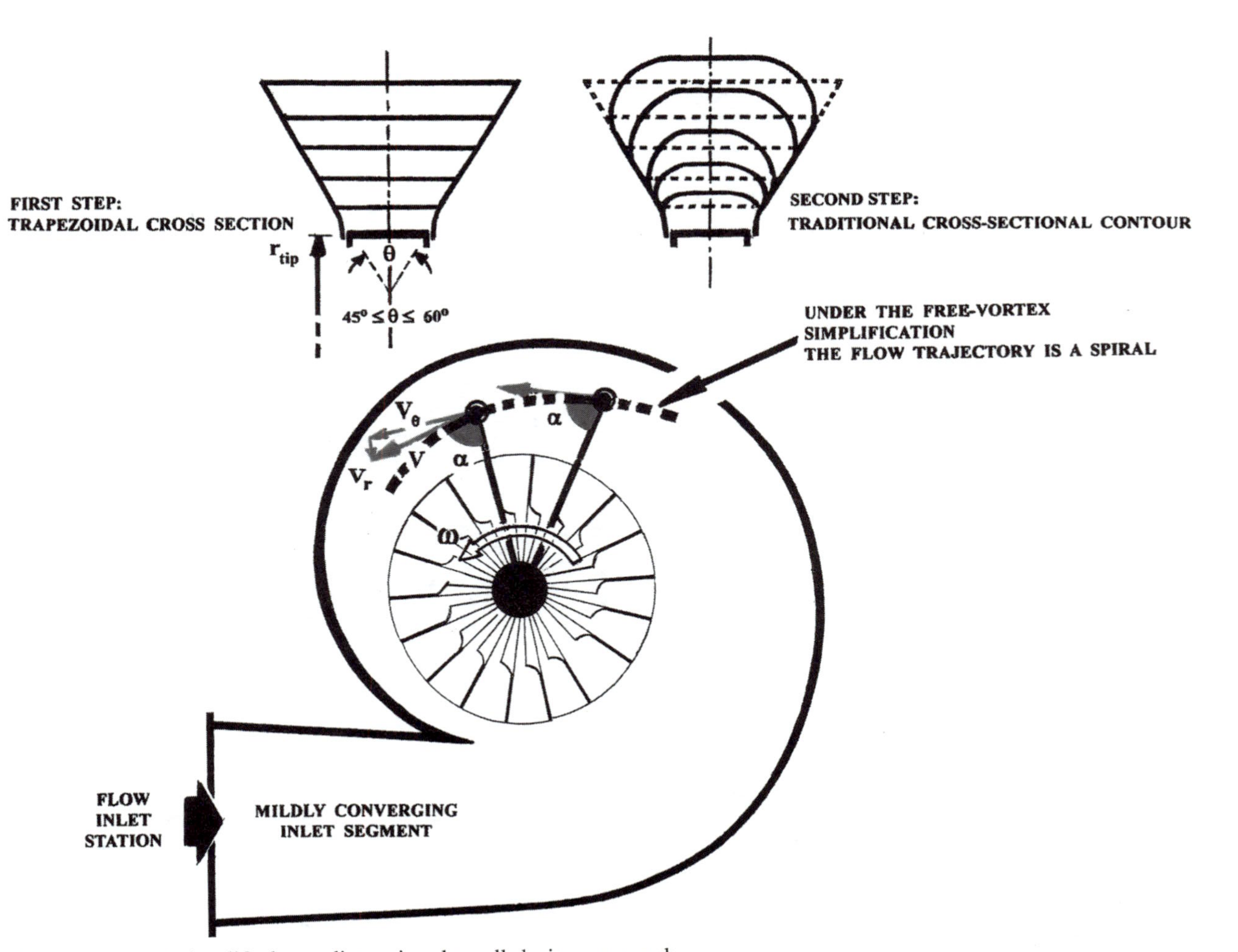

Figure 10.15. A simplified one-dimensional scroll-design approach.

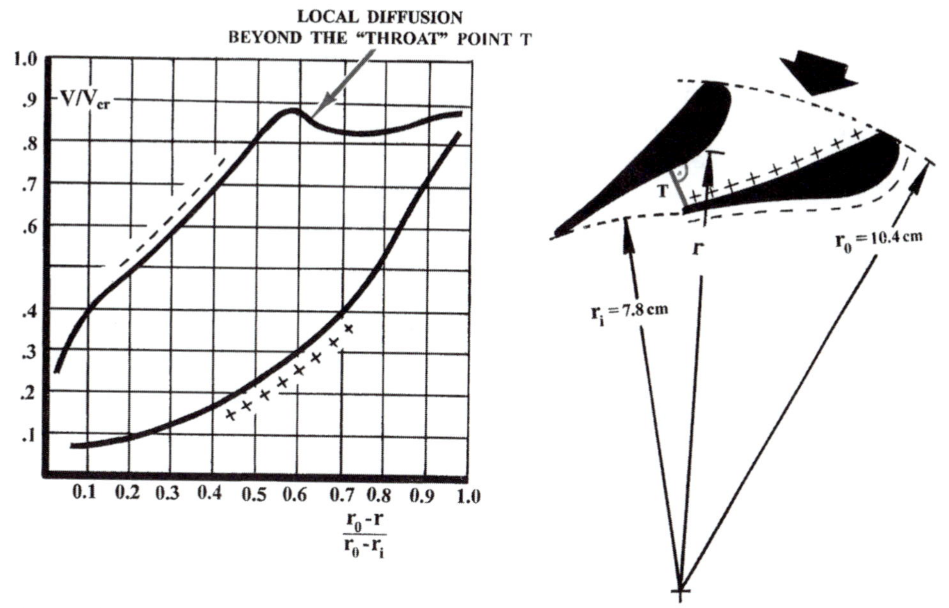

Figure 10.16. Typical aerodynamic loading of a radial stator vane.

A uniform scroll inlet velocity, coupled with the converging inlet segment in the figure, will help maintain the flow angle almost fixed in the circumferential direction.

As for the scroll cross-flow area, the tangential distribution of this area, usually linear, is then established, and a suitable included angle θ defined (Fig. 10.15). Next, rough approximations of the cross sections, in the form of characteristically similar trapezoids, are created to cover the 360° span. As shown in the figure, these sections are then modified to look like the characteristic shape in the figure while preserving the same cross-sectional area. The final product is a sequence of sections, each with what is agreeably the optimum shape (shown in Figure 10.15) for this flow-passage type.

The Vaned Stator

Notwithstanding the vane shape arbitrariness, Figure 10.16 depicts one of the typical cascades of radial stator vanes. Examination of this figure reveals a large exit/inlet velocity ratio as a result of the streamwise radius decline as well as the nozzle-like shape of the vane-to-vane passage. In view of the highly favorable pressure gradient under such circumstances, the boundary-layer buildup over the vanes and sidewalls would be next to nonexistent. Perhaps one of very few flaws in this case is what often appears as a reverse curvature close to the "throat" point on the suction side *T*. As shown in Figure 10.16, this will fortunately have a limited local impact on the vane aerodynamic loading. To summarize, the stator vane-to-vane subdomain is a good example of one where an ideally potential flow analysis is fitting. The flow model in this case can be two- or three-dimensional (e.g., Baskharone 1979 & 1984), with the former type of analysis being sufficiently accurate in most cases. The solution accuracy in this case is enhanced as the shape of sidewalls is simulated within a stream-filament type of approach (e.g., Katsanis and McNally 1973). The design process therefore involves repetitive execution of a potential-flow numerical code in what is essentially a trial-and-error iterative procedure. The ultimate objective of this procedure is to reach the vaned-stator configuration (in terms of the vaned-stator cascade and sidewall geometry) that is closest to produce the best possible aerodynamic loading, one that is, ideally, diffusion-free.

The Rotor

The flow field in the blade-to-blade passage of a radial turbine rotor is highly complicated because of the complexity of the flow domain (Fig. 10.12). Potentially damaging, from both aerodynamics and heat-transfer standpoints, is the 90° flow turning within the rotor subdomain. Results of this direction shift include a secondary flow stream, normal to the primary flow in the blade-to-blade rotor passage, as well as thermal-stress concentration.

Building a flow solver that lends itself to the complex geometry of the blade-to-blade hub-to-casing flow passage in the case at hand is hardly an easy task. Understandably so, full-blown three-dimensional viscous and compressible-flow models are prohibitively costly. Slightly simpler is an analytical approach, where the rotational, but inviscid, flow field is computed over two mutually orthogonal families of "stream" surfaces: a hub-to-casing set of surfaces between two successive blades and another blade-to-blade set beginning with the hub surface and ending with that of the casing. The two sets of flow-governing equations associated with both families are in this case coupled, and the solution procedure is iterative. In the area of turbomachinery flow analysis, this method is referred to as the "quasi-three-dimensional" flow-analysis approach and was originally devised by Wu (1952).

Aimed at optimizing the blade shape at different hub-to-casing locations, the simpler flow code by Katsanis and McNally (1973) was applied to the rotor in Figure 10.17. The corresponding aerodynamic-loading diagrams are shown in the same figure. The highlighted rotor segments, in the figure is that where flow reversal over the blade pressure side occurs. This early pressure-side flow reversal in Figure 10.17 is obviously the result of a large positive incidence angle at the rotor-tip radius. Despite its noncommonality and unfavorable stress consequences, the blade inlet segment of this particular rotor was, in the end, chosen to be nonradial, in such a way to minimize the tip incidence angle.

Closed-Form Loss Correlations

This section offers a set of simple loss-estimation relationships spanning over most components of a traditional radial-turbine stage. These relationships are selected in such a way to suit the preliminary design phase. Aiding us in the process, Figure 10.18 shows the station-designation pattern. In reality, however, these may very well define the ultimate point to which a typical designer would go in invoking real-life flow degradations in the final stage design. Part of the reason here is the virtual absence of this turbine type in propulsion applications, which rules it out of the range where "heavyweight" computational models have historically been developed.

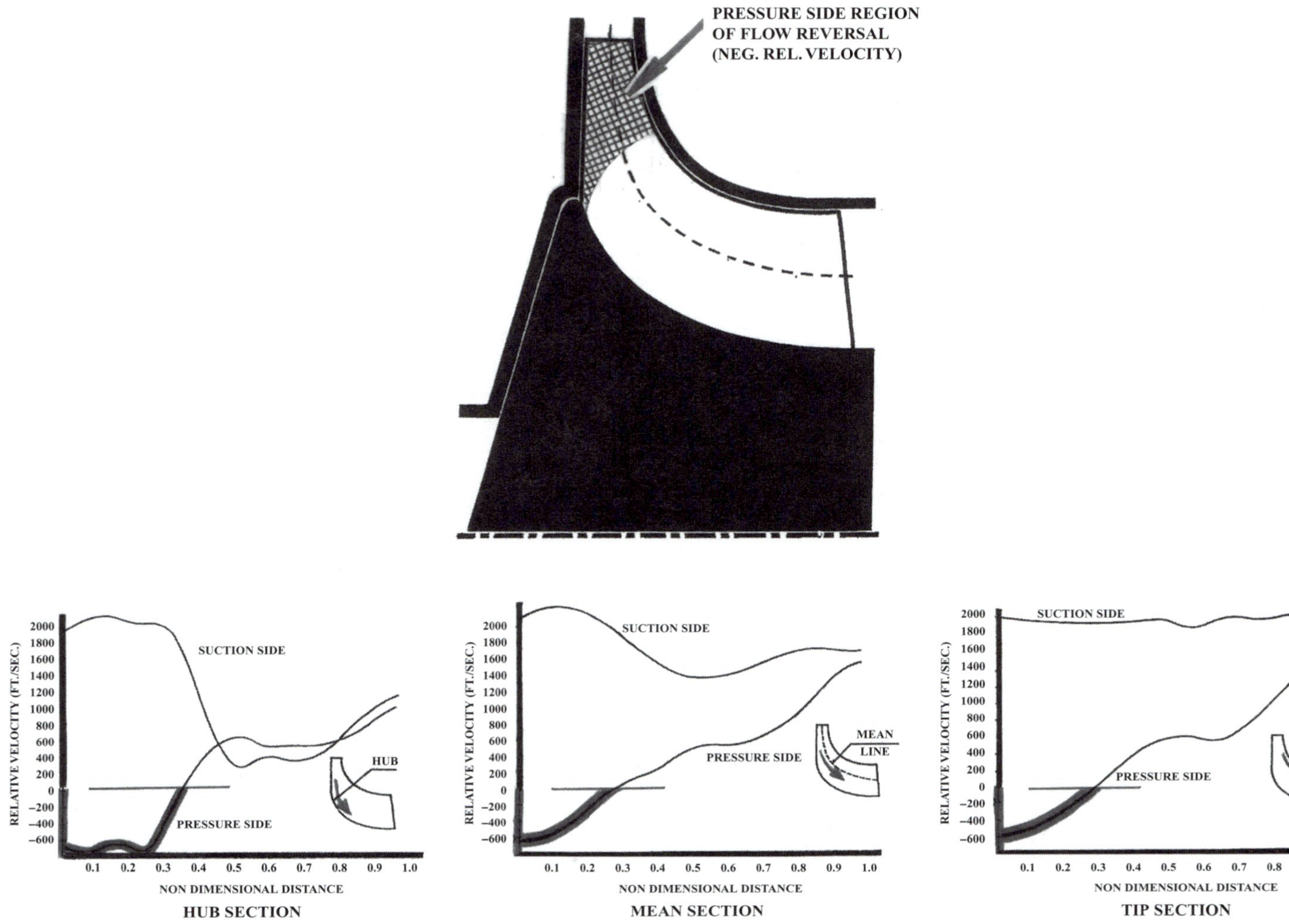

Figure 10.17. Aerodynamic loading of a rotor blade, including inlet segments of flow reversal over the pressure side.

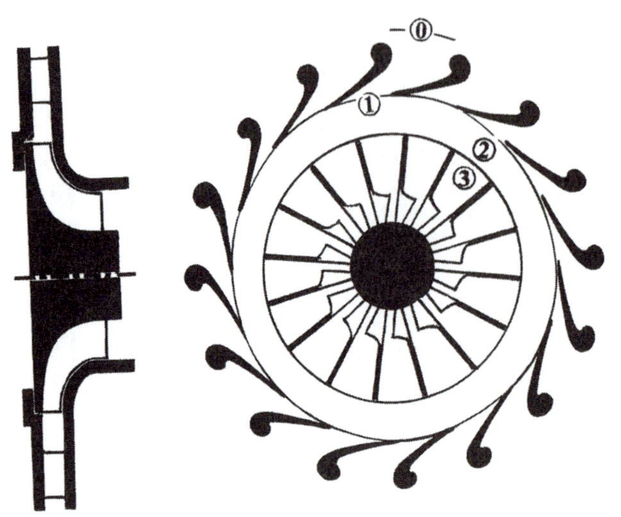

Figure 10.18. Station designation in the aerodynamic-loss calculations.

Of the stage nonrotating components, the scroll passage, in particular, is omitted. To date, there has not been a reliably accurate relationship that would account for such aspects as the scroll cross-section geometry, surface roughness, and tongue thickness (Fig. 10.2). Attempts to optimize the cross-sectional area distribution around the circumference have been made using large-size finite-element "discretization" models, as shown in Figure 10.7 (Baskharone 1984), on a potential-flow basis. Viscous-flow models, on the other hand, have been lagging in areas such as the near-wall flow resolution, in a way that would not overwhelm the computational resources.

The Vaned Stator

Perhaps the most comprehensive means of estimating the kinetic-energy loss coefficient ($\bar{e}_S$) across the stator (Fig. 10.19) is the following relationship developed by NASA-Lewis Research Center:

$$\bar{e}_S = E\left(\frac{\theta_{tot}}{S\cos\alpha_{ex} - t_{t.e.} - \delta^*_{tot}}\right)\left(1 + \frac{S\cos\alpha_{av}}{b}\right)\left(\frac{Re}{Re_{ref}}\right) \tag{10.14}$$

where

E is the so-called energy factor, and is equal to 1.8,
S is the stator-exit vane-to-vane spacing (Fig. 10.19),
α_{ex} is the stator-exit flow angle,
b is the sidewall axial spacing (Fig. 10.19),
α_{av} is the average of inlet and exit flow angles,
$t_{t.e.}$ is the trailing-edge thickness,
Re is the Reynolds number based on C and V_{ex} in Figure 10.19,
Re_{ref} is a reference Reynolds' number = 2.74×10^6,
δ^*_{tot} is the combined (suction plus pressure) displacement thickness, and
θ_{tot} is the combined (suction plus pressure) momentum thickness.

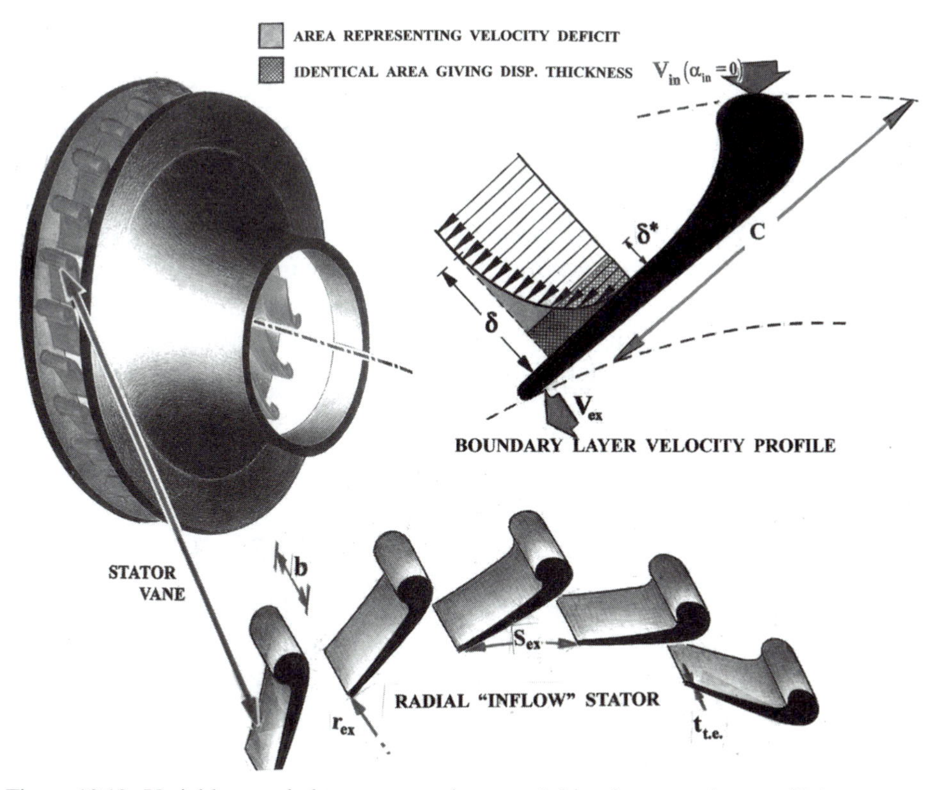

Figure 10.19. Variables needed to compute the stator's kinetic-energy loss coefficient.

The last two variables are primary characteristics of the vane-exit boundary layers on the airfoil suction and pressure sides. These were defined in Chapter 3, where simple closed-form approximations were provided on a flat-plate simplification basis. Note that the simplification here is uncharacteristically on the conservative side of what is actually a highly favorable pressure gradient across the stator cascade. A flat plate, by comparison, provides a surface that is theoretically free of any pressure gradient, which is what its aerodynamic definition entails.

The Vaneless Nozzle

Highlighted in Figure 10.18, this is the radial gap extending from the stator-exit to the rotor-inlet flow stations. As part of the nonrotating stage assembly, this gap contributes to the flow-acceleration process in a nearly loss-free fashion.

By reference to the station-designation pattern in Figure 10.18, let us define the following variables:

$$\bar{M} = \frac{V_1}{\sqrt{\gamma R T_{t1}}} \tag{10.15}$$

$$B = \frac{r_1}{b} \tag{10.16}$$

$$\bar{R} = \frac{r_2}{r_1} \tag{10.17}$$

$$\bar{Y} = \frac{1 - (p_{t,2}/p_{t,1})}{1 - (p_2/p_{t,1})} \tag{10.18}$$

In addition to the vaneless-gap inlet flow angle (α_1), the total-pressure loss parameter ($\bar{Y}$) is a function of the dimensionless variables in (10.15) through (10.17). Using experimental data in this radial gap (Khalil et al. 1976), the following relationship was obtained by choosing key points on the experimental property profiles and then using simple quadratic interpolation:

$$\begin{aligned} \bar{Y}_{gap} &= (0.193 - 0.193\bar{R}) \\ &[1 + 0.0641(\alpha_1 - \alpha_{1,ref}) + 0.0023(\alpha_1 - \alpha_{1,ref})^2] \\ &[1 + 0.6932(\bar{M}_1 - \bar{M}_{1,ref}) + 0.4427(\bar{M}_1 - \bar{M}_{1,ref})^2] \\ &[1 + 0.0923(B - B_{ref}) + 0.0008(B - B_{ref})^2] \end{aligned} \tag{10.19}$$

where

$\alpha_{1,ref} = 70°$,
$\bar{M}_{1,ref} = 0.8$, and
$B_{ref} = 10.0$.

As an example, consider the stator/rotor radial gap with the following set of contributing variables:

$\bar{R} = 0.85$,
$\alpha_1 = 76°$,
$B = 15$, and
$\bar{M}_1 = 0.88$.

Substitution of these magnitudes into (10.19) yields a total-pressure loss coefficient ($\bar{Y}_{gap}$) of 11.6%. For an unvaned gap with a naturally favorable pressure gradient, this $\bar{Y}$ value may seem on the high side. However, close examination of the input variables reveals a rather small sidewall spacing and a large flow-inlet swirl angle. The latter would elongate the flow trajectory over the sidewalls, providing a long path for boundary-layer growth.

The Rotor Admission Segment

Referring to Figure 10.20, this is the small subregion inside the rotor leading edge where the impact of the incidence angle presents itself. Perhaps the most widely accepted strategy here is Futral's "tangential-kinetic-energy-destruction" model. As the title may imply, this model is supposed to fulfill the idea that the part of kinetic

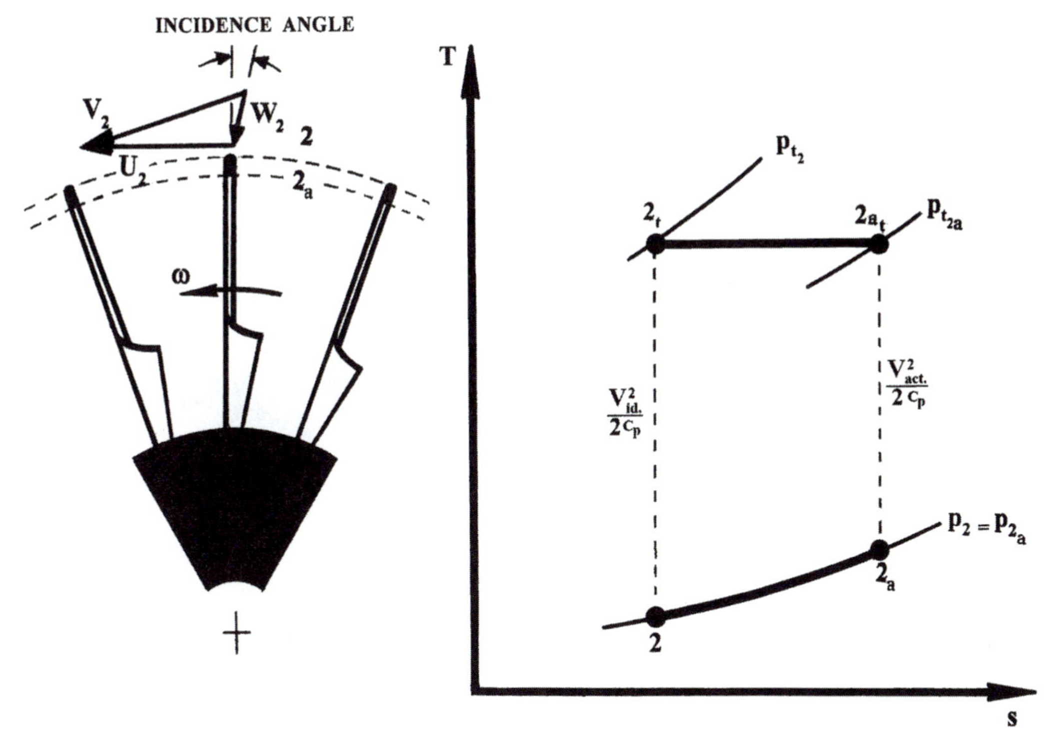

Figure 10.20. Rotor incidence-loss model.

energy that is based on the tangential velocity component is lost once the flow stream impacts the rotor blades. However, the model is "smart" enough to recognize the fact that this energy destruction process is dependent on the number of rotor blades. Considering the station-designation pattern in Figure 10.20, and referring to the incidence-caused loss of kinetic energy by $L_{incid.}$, Futral model provides the following expression:

$$L_{incid.} = \frac{1}{2}(V_{\theta 2} - V_{\theta 2,id})^2 \tag{10.20}$$

where

$$V_{\theta 2,id} = \left(1 - \frac{2}{N_b}\right)(\omega r_{tip}) \tag{10.21}$$

with N_b being the rotor blade count.

The Rotor Blades

The rotor profile losses are calculated in a manner that is essentially similar to that of the vaned stator presented earlier, and is NASA devised as well. The exception, however, is that the combined (suction side plus pressure side) momentum thickness

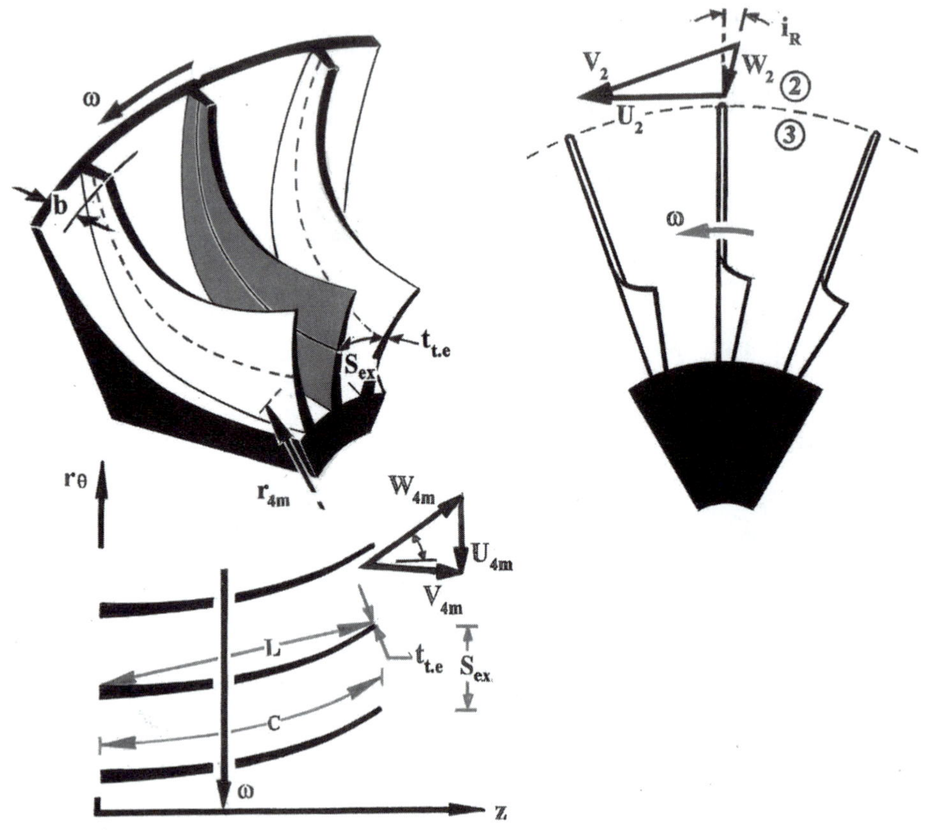

Figure 10.21. Variables needed to compute the rotor kinetic-energy loss coefficient.

(θ_{tot}) is now expressed in an empirical form that is particularly tailored for radial rotors,

$$\theta_{tot} = 0.11595 L Re^{-0.2} \tag{10.22}$$

where L is the mean camberline length (Fig. 10.21). The rotor kinetic-energy-loss coefficient, $\bar{e}_R$, is expressed as

$$\bar{e}_R = \left(\frac{\theta_{tot}}{S_m \cos\beta_{ex} - t_{t.e.} - \delta^*{}_{tot}} \right) \left(\frac{A_{3D}}{A_{2D}} \right) \left(\frac{Re}{Re_{ref}} \right)^{-0.2} \tag{10.23}$$

where

β_{ex} is the exit mean-radius relative flow angle,
S_m is the exit mean-radius blade-to-blade spacing,
A_{2D} is the summed-up blade surface areas,
A_{3D} is A_{2D} plus the hub and casing areas,
Re is the Reynolds no., based on W_{ex} and C (Fig. 10.21), and
Re_{ref} is a reference Reynolds number $= 7.57 \times 10^6$.

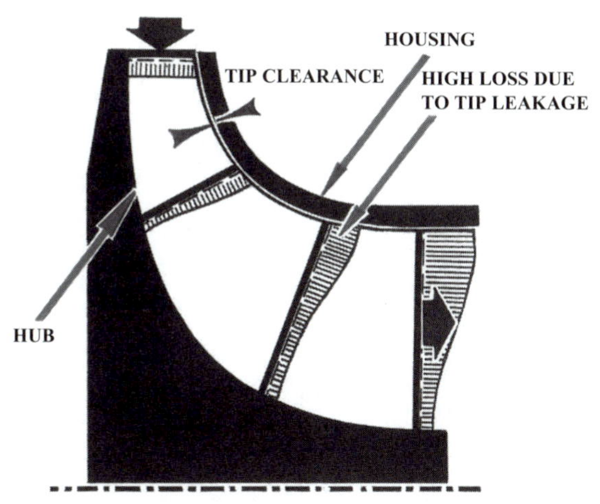

Figure 10.22. Typical kinetic-energy loss profiles at several streamwise locations.

Figure 10.22 offers a closer look at the accumulative loss in kinetic energy, based on flow measurements, in the streamwise direction. The rotor in this figure is unshrouded, a feature that tolerates the pressure-to-suction-side flow migration within the tip clearance gap. As shown in the figure, this magnifies the rotor-tip losses when compared with those along the hub.

The Downstream Duct

Normally an annular diffuser, the duct downstream from the radial-turbine rotor can be a source of a significant total pressure loss. This would naturally give rise to a proportionally large total-to-static efficiency decrement of the combined stage/ diffuser unit. As was previously indicated (Fig. 3.11), the performance degradation of what was then an exhaust diffuser was a strong function of the rotor-exit (diffuser-inlet) swirl angle. This fact remains valid, regardless of the shape and function of the downstream duct. As shown in Figure 10.23, the total-pressure loss coefficient ($\bar{\omega}$) is seen to rise almost exponentially as the rotor-exit swirl angle gets to exceed a magnitude of roughly 20°. The results in this figure are the outcome of flow measurements in the interstage duct, which is shown in the same figure, with L being the duct length and D its average diameter.

Effect of the "Scallop" Radius and Backface Clearance

In presenting radial-turbine rotors, so far, the simplification has been made that the rotor backplate extends radially outward up to the leading-edge radial location (e.g., Figures 10.1, 10.2, and 10.4). In all cases, of course, there will have to be a finite backplate/housing clearance, where a backface flow leakage is virtually unavoidable. The problem, however, is further magnified in the case of the "scalloped"-rotor category, which is conceptually represented in Figure 10.24.

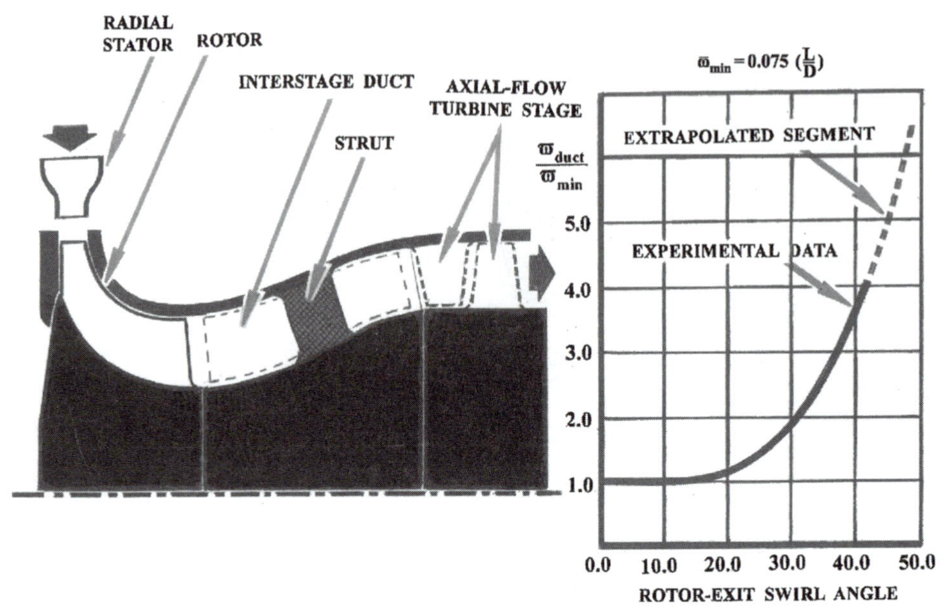

Figure 10.23. Total pressure loss across the stage downstream duct. Effect of the rotor exit swirl on the interstage duct losses.

Mostly for mechanical considerations, the blade inducer (or inlet segment) is frequently designed to be detached from the rotor backplate (Fig. 10.24). One may think of such a rotor-configuration choice as a means of alleviating some of the stress-concentration problems. This, in particular, is what the hub surface would otherwise experience as the meridional flow stream changes direction from purely radial to purely axial. Regardless of the validity of this or any other interpretation, the fact remains that this rotor-design option will worsen the through-flow and pressure-to-suction flow migration through the backface clearance gap. The fact nevertheless, remains that such a rotor configuration is more common than one would expect or indeed wish (e.g., Galligan 1979).

In an experimental study, Galligan investigated the leakage-related efficiency decrement as a function of the clearance width and scallop radius. The results of this study are chart-form adapted, and the results are shown in Figure 10.24. Examination of these results confirms the fact that the lower the scallop/tip radius ratio, the worse the rotor performance will be. (Note the descending manner in which the horizontal axis is labeled in this figure.)

EXAMPLE 2

Shown in Figure 10.25 is a radial-inflow turbine stage. The stage operating conditions are as follows:

- Flow process is adiabatic throughout the entire stage
- Average specific-heat ratio (γ) is 1.35

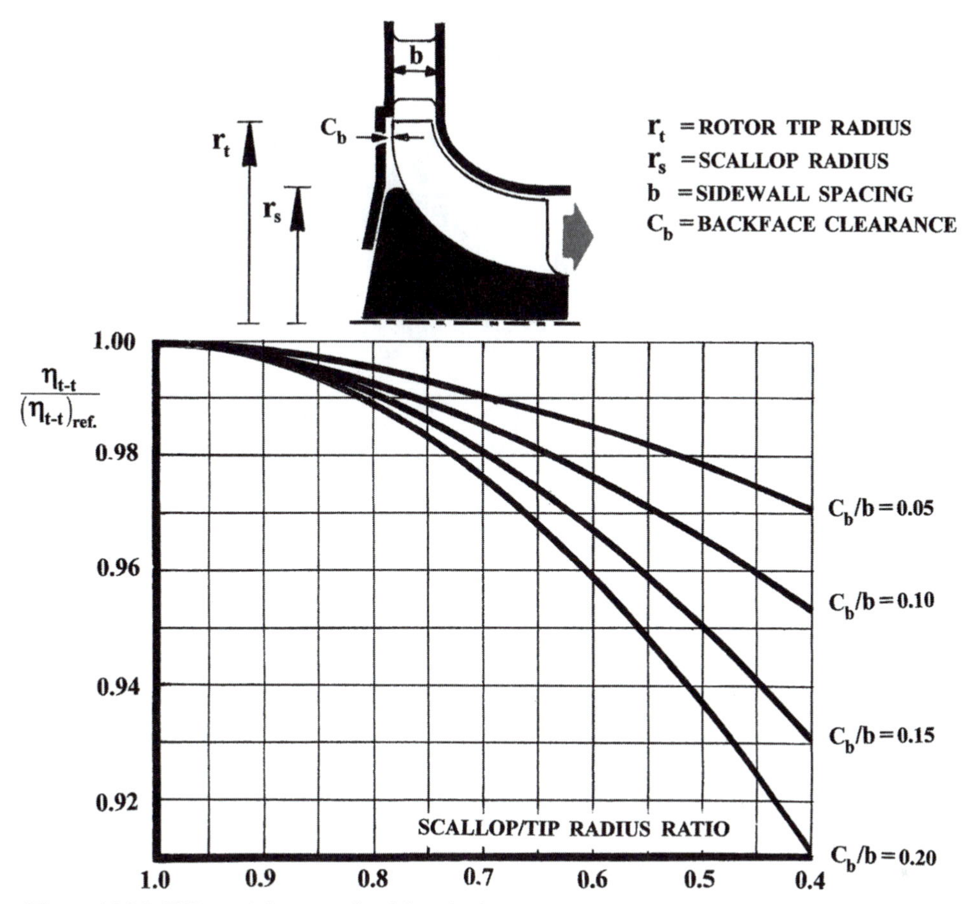

Figure 10.24. Effect of the rotor backface leakage.

- Stator-inlet flow angle $(\alpha_0) = 0$
- Stator-inlet total pressure $(p_{t0}) = 10$ bars
- Stator-inlet total temperature $(T_{t0}) = 1250$ K
- Stator-inlet critical Mach number $(V_0/V_{cr0}) = 0.22$
- Stator-exit swirl angle $(\alpha_1) = 72°$
- Stator-exit static pressure $(p_1) = 6.52$ bars
- Mass-flow rate $(\dot{m}) = 2.4$ kg/s
- Shaft speed $(N) = 38{,}000$ rpm
- Average kinematic viscosity coefficient $(\nu) = 7.2 \times 10^{-6}\text{m}^2/\text{s}$
- Rotor-tip static pressure $(p_2) = 4.85$ bars

The following geometrical variables have the assigned magnitudes:

- Number of stator vanes $(N_v) = 32$
- Number of rotor blades $(N_b) = 29$
- Stator-vane mean camberline length $(L_v) = 0.032$ m

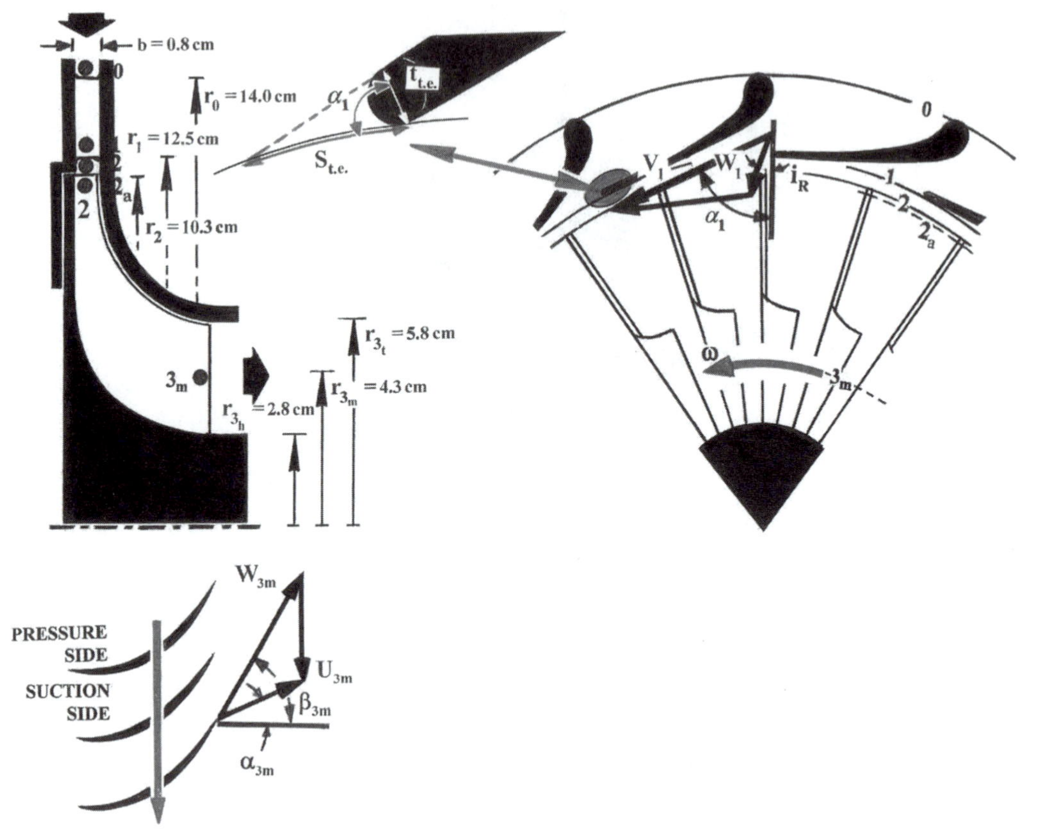

Figure 10.25. Input variables for Example 2.

- Stator-vane true chord (C_v) = 0.025 m
- Stator-vane trailing-edge thickness ($t_{t.e.}$) = 0.0002 m

Calculate the loss in total pressure for all of the stage components ending at the rotor inlet station.

SOLUTION

This problem is constructed to use the majority of loss correlations stated so far. The problem targets (in particular) the stage nonrotating components, with the exception of the scroll passage. The reason, in part, is the virtually total arbitrariness of both the scroll cross-section choice and the cross-flow area distribution around the circumference.

VANED STATOR

In the following, we calculate the magnitudes of all variables that contribute to the stator kinetic-energy loss coefficient ($\bar{e}_S$) in expression (10.14).

First, we compute the stator-exit cascade pitch (S_1), meaning the arc length between two successive vanes, as follows:

$$S_1 = \frac{2\pi r_1}{N_v} = 0.025\,\text{m}$$

where N_v is the number of stator vanes (32 in the problem statement).

Using the displacement-thickness approximation in Chapter 3, we have

$$\delta^*_{tot.} = 0.008 S_1 = 0.002\,\text{m}$$

where the subscript "*tot.*" refers to the combined (suction plus pressure) displacement thicknesses at the vane exit station.

In order to calculate the combined (suction plus pressure) momentum thickness, we need to compute the stator inlet and exit velocities. Noting that the specific-heat ratio here is 1.35, we begin with the expression for the stator-exit static pressure:

$$p_1 = p_{t1}\left[1 - \left(\frac{\gamma - 1}{\gamma + 1}\right)\left(\frac{V_1}{V_{cr1}}\right)^2\right]^{\frac{\gamma}{\gamma-1}}$$

Substitution of p_{t0} for p_{t1} in this equation implies an isentropic process across the stator. This is clearly at odds with our very task of computing an entropy-production-related loss in kinetic energy. However, short of any loss indicator, at this computational stage, an assumption will have to be made. The stator exit velocity can now be computed as

$$V_1 \approx (V_1)_{id.} = 539.3\,\text{m/s}$$

Let us proceed with this ideal magnitude to compute a rough estimate of the vane Reynolds number. With C_v being the vane true chord, this Reynolds number magnitude is

$$(Re)_{stator} = \frac{V_1 C_v}{\nu} = 1.87 \times 10^6$$

It is important to realize that the task at hand requires a sequence of assumptions and approximations. One of these simplifications has already been made in identifying the exit velocity as we computed the stator Reynolds number.

The important question, in this context, is whether there are means of going back and properly adjusting this and other approximations. The answer to this is yes more often than not. Take, for instance, the approximation we have just made. Realizing that the objective of the current computational segment is to find (as we will) the stator kinetic-energy loss coefficient, we can always go back, using this very variable, to determine a more realistic stator-exit velocity toward a more accurate magnitude of the stator Reynolds number. Theoretically speaking, one could keep executing such a corrective action in an iterative procedure toward convergence. However, noting that the loss-correlating expressions are themselves rough estimates of what prevails in real life, a wiser approach (perhaps) is to proceed with simplifications, particularly those which are perceived to cause acceptable inaccuracies in the final results.

The stator inlet velocity (V_0), however, is much easier to calculate:

$$V_0 = M_{cr0} V_{cr0} = 141.2\,\text{m/s}$$

At this point, we are prepared to calculate the trailing-edge magnitude of total momentum thickness ($\theta_{tot.}$) and the stator kinetic-energy loss coefficient ($\bar{e}_S$) thereafter.

As implemented in earlier chapters, we will first calculate the pressure-side magnitude ($\theta_{press.}$) at the vane trailing edge. To this end, we will again use the already computed ideal exit velocity for the actual magnitude. Assuming, in view of the large Reynolds number magnitude, a turbulent boundary-layer flow structure, we will employ expression (3.81) to achieve our objective. As for the suction side trailing-edge momentum thickness ($\theta_{suc.}$), we will utilize a previously stated rationale whereby this variable is set to 3.5 times its pressure-side counterpart:

$$\theta_{press.} = 0.022 C_v Re^{-\frac{1}{6}} = 6.34 \times 10^{-5}\,\text{m}$$

Note that a more accurate value of the airfoil characteristic length would be its mean camberline length, which is not defined in the problem statement. The next best length, in this case, is the airfoil true chord (C_v), which we have already utilized. It follows that

$$\theta_{suc.} = 3.5\theta_{press.} = 2.22 \times 10^{-4}\,\text{m}$$

The total momentum thickness may now be calculated:

$$\theta_{tot.} = 0.285\,\text{mm}$$

Proceeding with the same rationale in computing the total displacement thickness, on the basis of expression (3.80), we get

$$\delta^*{}_{tot.} = 0.739\,\text{mm}$$

We are now in a position to calculate the stator kinetic-energy loss coefficient ($\bar{e}_s$) by direct substitution in expression (10.14):

$$\bar{e}_s = 0.183$$

which is on the high side. We can convert this $\bar{e}_s$ magnitude to determine the stator total pressure loss as follows:

$$V_{1act.} = \sqrt{(1-\bar{e}_s) V_{1id.}{}^2} = 487.4\,\text{m/s}$$

$$M_{cr1} = \frac{V_{1act.}}{V_{cr1}} = 0.759$$

$$p_{t1} = \frac{p_1}{\left[1 - \left(\frac{\gamma-1}{\gamma+1}\right) M_{cr1}{}^2\right]^{\frac{\gamma}{\gamma-1}}} = 9.22\,\text{bars}$$

$$(\Delta p_t)_{stator} = 0.785\ \text{bars (a total pressure loss of } 7.85\%)$$

STATOR/ROTOR VANELESS NOZZLE

The vehicle to quantify the losses in this subdomain is expression (10.19). First, we calculate all variables appearing in this expression:

$$T_1 = T_{t1} - \left(\frac{V_{1\,act.}{}^2}{2c_p}\right) = 1129.8\,\text{K}$$

$$\bar{M}_1 = \frac{V_{1\,act.}}{\sqrt{\gamma R T_{t1}}} = 0.74$$

The relatively small value of $\bar{M}_1$ (above) simply indicates that the flow compressibility is not a real loss-contributing factor (the cut-off magnitude is 0.8). We will therefore ignore its effect altogether.

We now pursue the computational procedure by considering the geometry-related variables:

$$B = \frac{r_1}{b} = 15.63$$

$$\bar{R} = \frac{r_2}{r_1} = 0.827$$

As for the swirl-angle change, across the radial gap, there is none. This is because the assumed free-vortex flow trajectories maintain the same angle across the gap.

At this point, we are in a position to calculate the loss variable $\bar{Y}$ by direct substitution in expression (10.19):

$$\bar{Y}_{vaneless} = 0.0587 = \frac{1 - \left(\frac{p_{t2}}{p_{t1}}\right)}{1 - \left(\frac{p_1}{p_{t1}}\right)}$$

which yields

$$p_{t2} = 9.34\,\text{bars}$$

Thus, the total pressure loss across the vaneless nozzle can now be calculated:

$$(\Delta p_t)_{vaneless} = p_{t2} - p_{t1} = -0.28\,\text{bars}$$

Finally, the combined total pressure loss across the stage nonrotating unit is

$$\text{Combined total pressure loss} = p_{t0} - p_{t2} = 1.065\,\text{bars}$$

Comparing this to the stator-inlet total pressure, we see that the stationary-unit total pressure loss is approximately 11%, which is notably high.

EXAMPLE 3

Figure 10.26 shows two views of a radial-inflow turbine stage, where the rotor is of the purely radial type. The stator and rotor exit radii are 8.1 and 6.8 cm, respectively, and the stator/rotor radial gap is negligibly small. The common stator and rotor

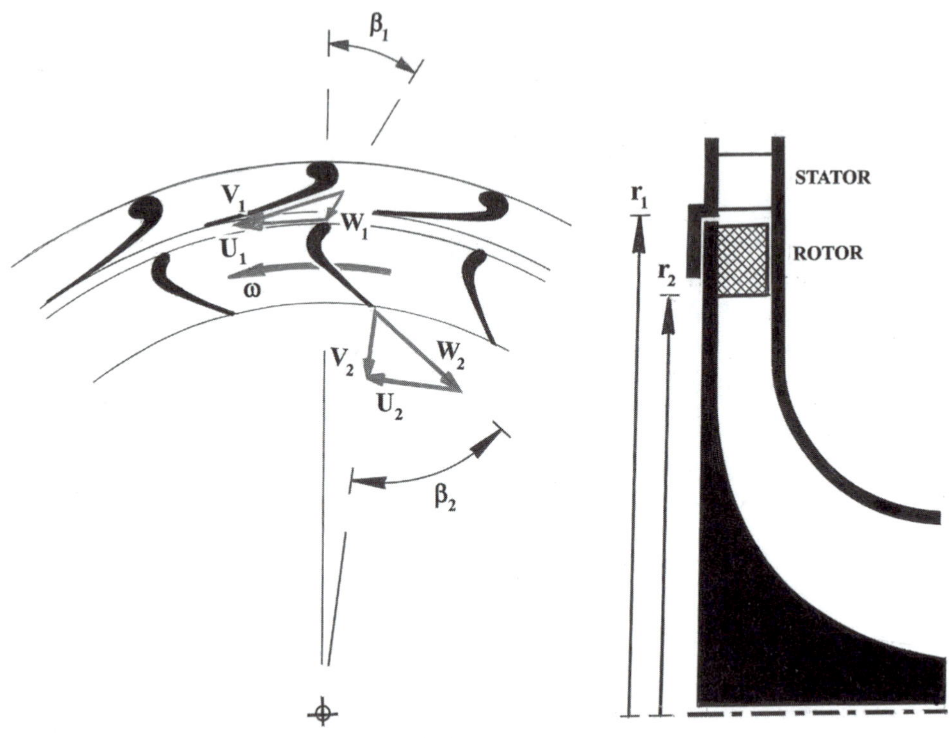

Figure 10.26. Input variables for Example 3.

sidewall width (b) is constant and equal to 0.6 cm. The stage operating conditions are as follows:

- Mass-flow rate ($\dot{m}$) = 3.8 kg/s
- Rotor speed (N) = 64,000 rpm
- Stator-exit total pressure (p_{t1}) = 5.2 bars
- Stator-exit total temperature (T_{t1}) = 1206 K
- Choked stator
- Rotor total-to-total efficiency (η_{t-t}) = 87.0%
- Rotor-exit (absolute) kinetic energy is minimum
- Entire flow process is adiabatic
- Average specific-heat ratio (γ) = 1.33

Assuming zero incidence and deviation angles, sketch the rotor blade shape.

SOLUTION

Let us first apply the continuity equation at the stator exit station:

$$\frac{\dot{m}\sqrt{T_{t1}}}{p_{t1}(2\pi r_1 b)\cos\alpha_1} = \sqrt{\frac{2\gamma}{(\gamma+1)R}} M_{cr1}\left[1 - \left(\frac{\gamma-1}{\gamma+1}\right) M_{cr1}{}^2\right]^{\frac{1}{\gamma-1}}$$

where

$$M_{cr1} = 1.0\,(\text{choked stator})$$

which, upon substitution, yields

$$\alpha_1 = 77.9^\circ$$

Now we proceed to calculate the rotor-inlet relative flow angle (β_1):

$$V_1 = V_{cr1} = \sqrt{\left(\frac{2\gamma}{\gamma+1}\right) R T_{t1}} = 628.6\,\text{m/s}$$

$$V_{\theta 1} = V_1 \sin \alpha_1 = 614.6\,\text{m/s}$$

$$V_{r1} = V_1 \cos \alpha_1 = 131.8\,\text{m/s}$$

$$U_1 = \omega r_1 = 542.9\,\text{m/s}$$

$$W_{\theta 1} = V_{\theta 1} - U_1 = 71.7\,\text{m/s}$$

$$\beta_1 = \tan^{-1}\left(\frac{W_{\theta 1}}{V_{r1}}\right) = 28.5^\circ$$

Proceeding to the rotor exit station, we recall that the minimum kinetic-energy condition simply means the following:

$$\alpha_2 = 0$$

$$V_{\theta 2} = 0$$

$$V_2 = V_{r2}$$

Preparing to apply the continuity equation at the rotor exit station, we first calculate the local total properties:

$$T_{t2} = T_{t1} - \frac{U_1 V_{\theta 1}}{c_p} = 917.5\,\text{K}$$

$$p_{t2} = p_{t1}\left\{1 - \frac{1}{\eta_{t-t}}\left[1 - \left(\frac{T_{t2}}{T_{t1}}\right)\right]\right\}^{\frac{\gamma}{\gamma-1}} = 1.42\,\text{bars}$$

Applying the continuity equation at the rotor exit station in the stationary frame of reference (i.e., using absolute flow properties), with the unknown being M_{cr2}, we get

$$0.501 = M_{cr2}\left(1 - 0.1416 M_{cr2}{}^2\right)^{3.0303}$$

Highly nonlinear, this equation can only be solved iteratively. The final result is:

$$M_{cr2} \approx 0.58$$

Now we can calculate the rotor-exit relative flow angle (β_2) as follows:

$$V_2 = M_{cr2} V_{cr2} = 318.0\,\text{m/s}$$

$$U_2 = \omega r_2 = 455.7\,\text{m/s}$$

$$\beta_2 = \tan^{-1}\left(\frac{-U_2}{V_{r2}}\right) = -55.1^\circ$$

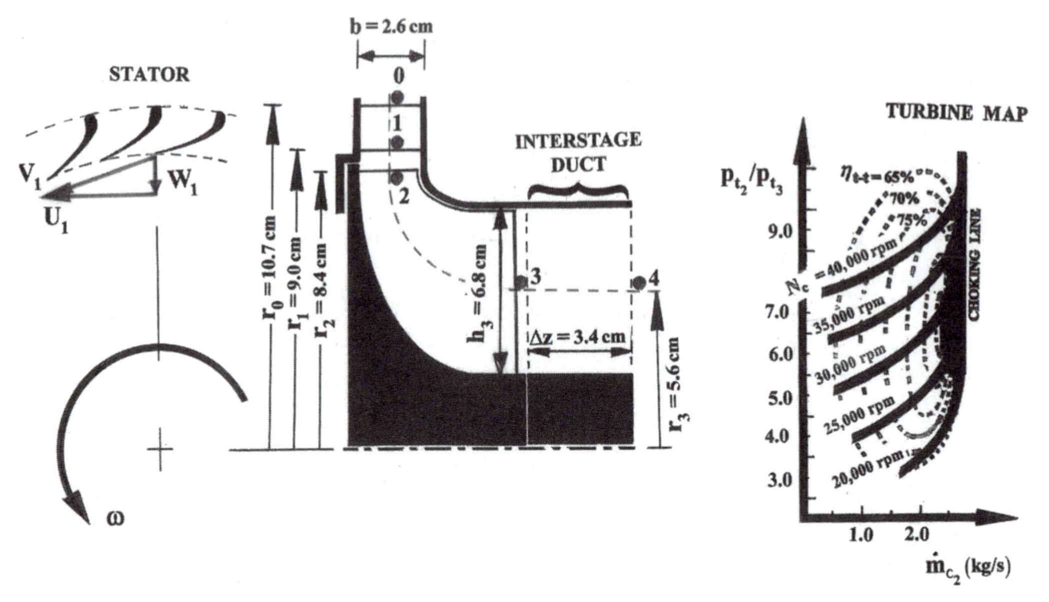

Figure 10.27. Input variables for Example 4.

The negative sign (above) simply means that β_2 is in the direction that is opposite to the direction of rotation.

With the rotor-blade incidence and deviation angles being zero, both pairs of flow and blade angles will be identical; that is,

$$\beta_1{}' = \beta_1 = 28.5^\circ$$
$$\beta_2{}' = \beta_2 = -55.1^\circ$$

With these angles, the rotor blade should look very much like that in Figure 10.26.

EXAMPLE 4

Figure 10.27 shows a radial-inflow turbine stage, a constant-area interstage duct, and the stage map. The stage operating conditions are as follows:

- Mass-flow rate ($\dot{m}$) = 4.16 kg/s
- Rotor speed (N) = 71,600 rpm
- Stator-inlet total temperature (T_{t0}) = 1205.0 K
- Stator-inlet total pressure (p_{t0}) = 8.84 bars
- Stator total-pressure loss coefficient ($\bar{\omega}_{stator}$) = 0.038
- Stator-exit critical Mach number (M_{cr1}) = 0.94

The following approximations also apply:

- Adiabatic flow throughout the stage and interstage duct
- Free-vortex flow structure in the stator/rotor radial gap
- Interstage-duct friction coefficient (f) = 0.0298
- Average specific-heat ratio (γ) = 1.33

Considering these data items and using the stage map:

a) Calculate the stage specific speed (N_s).
b) Comment on the result in item "a" by referring to Figure 5.9.
c) Calculate the stage reaction (R).
d) Calculate the interstage duct total pressure loss.

SOLUTION

Part a: First, we calculate the stator-exit total pressure using the definition of the total pressure loss coefficient $\bar{\omega}$ (3.57) as follows:

$$p_{t1} = p_{t0} - \bar{\omega} p_{t0} = 8.50\,\text{bars}$$

Now we calculate the stator-exit swirl angle by applying the continuity equation as follows:

$$\frac{\dot{m}\sqrt{T_{t1}}}{p_{t1}(2\pi r_1 b)\cos\alpha_1} = \sqrt{\frac{2\gamma}{\gamma + 1R}} M_{cr1}\left[1 - \left(\frac{\gamma - 1}{\gamma + 1}\right) M_{cr1}{}^2\right]^{\frac{1}{\gamma-1}}$$

which, upon substitution, yields

$$\alpha_1 = 73.0^\circ$$

It follows that

$$V_1 = M_{cr1} V_{cr1} = 590.6\,\text{m/s}$$

$$V_{r1} = V_1 \cos\alpha_1 = 172.5\,\text{m/s}$$

$$V_{\theta 1} = V_1 \sin\alpha_1 = 564.8\,\text{m/s}$$

Applying the free-vortex flow conditions over the stator/rotor radial gap, we get

$$V_{r2} = \left(\frac{r_1}{r_2}\right) V_{r1} = 184.8\,\text{m/s}$$

$$V_{\theta 2} = \left(\frac{r_1}{r_2}\right) V_{\theta 1} = 605.1\,\text{m/s}$$

$$V_{cr2} = V_{cr1} = V_{cr0} = \sqrt{\left(\frac{2\gamma}{\gamma + 1}\right) R T_{t0}} = 628.3\,\text{m/s (adiabatic flow)}$$

$$M_{cr2} = \frac{V_2}{V_{cr2}} = 1.007$$

Just like the vaned stator (or nozzle), the stator/rotor radial gap is conceptually a subsonic nozzle (Chapter 3). In this sense, the above-computed critical Mach number, being higher than unity, is refused. In reality, this passage is indeed choked, with the critical Mach number being unity just upstream from the rotor inlet station (i.e, $M_{cr2} = 1.0$).

In view of the above, we need to go back and correct the velocity components at the rotor inlet station:

$$V_{r2} = V_{cr2} \cos \alpha_2 = 183.7 \text{ m/s}$$

$$V_{\theta 2} = V_{cr2} \sin \alpha_2 = 600.8 \text{ m/s}$$

Let us now calculate the corrected magnitudes of mass-flow rate and speed:

$$\dot{m}_{C_2} = \frac{\dot{m}\sqrt{\theta_2}}{\delta_2} \approx 1.0 \text{ kg/s}$$

$$N_{C_2} = \frac{N}{\sqrt{\theta_2}} \approx 35,000 \text{ rpm}$$

With these two variables, the stage map yields

$$p_{t3} = p_{t2}/6.7 = 1.27 \text{ bars}$$

$$\eta_{t-t} = 75.0\%$$

Applying the Euler/energy-transfer equation, we get

$$V_{\theta 3} = V_{\theta 2} - \eta_{t-t} c_p T_{t2} \left[1 - \left(\frac{p_{t3}}{p_{t2}} \right)^{\frac{\gamma-1}{\gamma}} \right]$$

Now we can calculate the rotor-exit tangential-velocity component:

$$V_{\theta 3} = -35.4 \text{ m/s}$$

We can also calculate the rotor-exit total properties as follows:

$$T_{t3} = T_{t2} \left\{ 1 - \eta_{t-t} \left[1 - \left(\frac{p_{t3}}{p_{t2}} \right)^{\frac{\gamma-1}{\gamma}} \right] \right\} = 865.0 \text{ K}$$

$$\rho_{t3} = \frac{p_{t3}}{R T_{t3}} = 0.512 \text{ kg/m}^3$$

Approximation: Referring to the small magnitude of $(V_\theta)_3$ (above), and the relatively large rotor-exit annulus height (promising a small magnitude of V_{z3}), we could practically ignore the critical Mach number effect there. In other words, we may very well assume that

$$\rho_3 \approx \rho_{t3} = 0.512 \text{ kg/m}^3$$

Application of the continuity equation at the rotor exit station will now provide V_{z3} as follows:

$$V_{z3} = \frac{\dot{m}}{\rho_3 (2\pi r_3 h_3)} = 339.6 \text{ m/s}$$

We are now in a position to verify the above-stated approximation, meaning to actually compute M_{cr3}, as follows:

$$M_{cr3} = \frac{V_3}{V_{cr3}} = 0.64$$

Referring to our approximation, this magnitude is unfortunately too high to ignore. We are now forced to do the inevitable, which is to iteratively improve the magnitude of M_{cr3} by continually changing V_{z3}, calculating the corresponding M_{cr3}, recomputing ρ_3, and then calculating the corresponding mass-flow rate. We will then have to repeat the entire procedure if the latter is different from the given $\dot{m}$ magnitude.

The final set of results are as follows:

$$V_{z3} = 493.9\,\text{m/s}$$
$$M_{cr3} = 0.93$$
$$\rho_3 = 0.452\,\text{kg/m}^3$$

The magnitude of M_{cr3} not only invalidates the approximation we made earlier but also means that a real-life design that is based on the given and computed data would be flawed. The excessively large rotor-exit velocity is nearly (and unnecessarily) causing the rotor passage to choke. One of the ways to reduce what will certainly be a huge dynamic pressure at exit is to increase the rotor-exit annulus height. In fact, the magnitude of specific speed (to be computed next) will indicate that an axial-flow stage type would, in this case, be a wiser choice.

We are now in a position to calculate the stage specific speed (N_s) by applying equation (5.28):

$$N_s = 1.32\,\text{radians}$$

Part b: Referring to Figure 5.9, it is clear that the stage stated tasks can be achieved more efficiently through one or more axial-flow stages instead. Worth noting is the fact that it is the combination of high speed and relatively high mass-flow rate that led to this result.

Part c: In order to calculate the stage reaction, we first calculate the magnitudes of all absolute and relative velocities at the rotor inlet and exit stations:

$$W_2 = \sqrt{(V_{\theta 2} - U_2)^2 + V_{r2}^2} = 186.0\,\text{m/s}$$

$$W_3 = \sqrt{(V_{\theta 3} - U_3)^2 + V_{z3}^2} = 671.7\,\text{m/s}$$

$$U_2 = \omega r_2 = 629.8\,\text{m/s}$$

$$U_3 = \omega r_3 = 419.9\,\text{m/s}$$

$$V_2 = V_{cr2} = 628.3\,\text{m/s (computed earlier)}$$

$$V_3 = \sqrt{V_{\theta 3}^2 + V_{z3}^2} = 495.2\,\text{m/s}$$

Applying equation (4.21), for a radial-turbine stage, we get

$$\text{stage reaction } (R) = 80.9\%$$

Part d: The adiabatic, frictional, and constant-area interstage duct is one that is governed by the Fanno-process relationships (Chapter 3). Because these are based on the traditional (as opposed to the critical) Mach number, we should convert the duct-inlet critical Mach number M_{cr3} into M_3. Using equation (3.73), we get

$$M_3 = 0.92$$

In order to calculate the total pressure loss across the duct, we first calculate the duct-exit Mach number (M_4) using equation (3.71), in which

$$f = 0.0298 \text{ (given)}$$

$$D_h = 2h_3 = 13.6 \text{ cm}$$

$$\alpha_3 = \tan^{-1}\frac{V_{\theta 3}}{V_{z3}} = -4.1^\circ$$

$$L = \frac{\Delta z}{\cos\alpha_3} = 3.41 \text{ cm}$$

As we substitute these variables in the Fanno-flow relationship (3.71), we end up with a highly nonlinear relationship in the duct-exit Mach no. (M_4). Executing a trial-and-error procedure, we achieve the following result:

$$M_4 = 0.995$$

At this point, we use equation (3.70) to calculate the total-to-total pressure ratio across the interstage duct:

$$\frac{p_{t4}}{p_{t3}} = 0.995$$

It follows that

$$\frac{(\Delta p_t)_{duct}}{p_{t3}} = 0.43\%$$

EXAMPLE 5

Figure 10.28 shows the turbine section in a power system, which is composed of two, radial and axial, stages, with a constant-area interstage duct. Also shown in the figure is the map of the radial stage and the major dimensions. Contrary to established design guidelines, the second stage has a purely axial flow path in the sense of a constant annulus area throughout the stage. The turbine section is operating under the following conditions:

- Shaft speed (N) = 37,390 rpm
- Mass-flow rate ($\dot{m}$) = 2.9 kg/s
- Radial-stage inlet total pressure (p_{t0}) = 10.1 bars
- Radial-stage inlet total temperature (T_{t0}) = 1260 K
- Radial-stator exit critical Mach no. (M_{cr1}) = 0.98
- Radial-stage total-to-total efficiency ($\eta_{rad.}$) = 83%

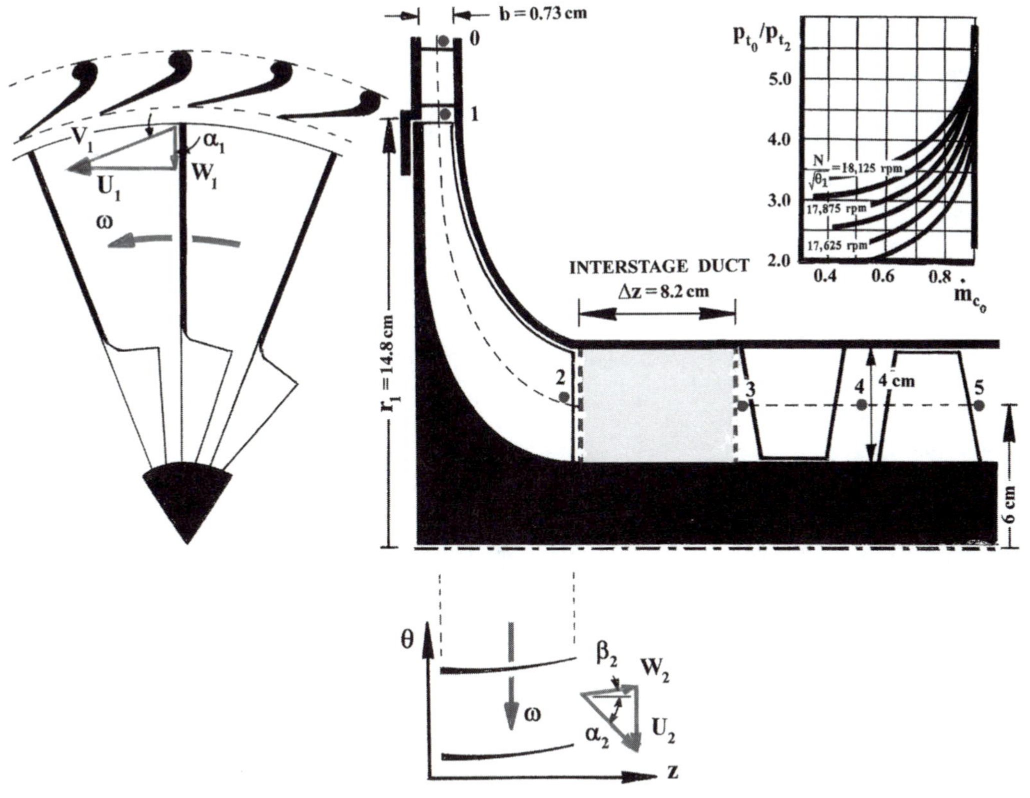

Figure 10.28. Input variables for Example 5.

- Radial-stage exit swirl angle (α_2) = 39.5°
- Axial-stage total-to-total efficiency ($\eta_{ax.}$) = 85%
- Interstage-duct friction coefficient (f) = 0.11
- Axial-flow stator is choked
- Axial-flow stage is symmetric (i.e., $R_{ax.\ stg.}$ = 50%)

The following approximations also apply:

- Isentropic flow across both radial and axial stators
- Adiabatic flow throughout the entire turbine section
- V_z is constant across the axial-flow stage
- Zero incidence and deviation angles over the axial stage
- Average specific-heat ratio (γ) = 1.33

Considering the input data, the major dimensions in the figure, and the radial stage map:

a) Sketch the radial-rotor inlet and exit velocity diagrams.
b) Calculate the total pressure loss over the interstage duct.
c) Calculate the axial-stage exit critical Mach number (M_{cr5}).
d) Calculate the total-to-static efficiency of the entire turbine.
e) Calculate the specific speed for each turbine stage.

SOLUTION

Part a: In order to make use of the radial stage map, we first compute the corrected magnitudes of mass-flow rate and speed as follows:

$$\dot{m}_{C_0} = \frac{\dot{m}\sqrt{\theta_0}}{\delta_0} = 0.6\,\text{kg/s}$$

$$N_{C_0} = \frac{N}{\sqrt{\theta_0}} = 17,875\,\text{rpm}$$

Using the given map, we get

$$\frac{p_{t0}}{p_{t2}} = 3.5\,(\text{obtained by extrapolation})$$

meaning that

$$p_{t2} = 2.886\,\text{bars}$$

Applying the continuity equation at the radial-stator exit station, we get

$$\alpha_1 = 68.0^\circ$$

Also

$$U_1 = \omega r_1 = 579.5\,\text{m/s}$$

$$U_2 = \omega r_2 = 234.9\,\text{m/s}$$

Let us now calculate the rotor inlet and exit thermophysical properties:

$$T_{t2} = T_{t0}\left\{1 - \eta_{rad.\ stg.}\left[1 - \left(\frac{p_{t2}}{p_{t0}}\right)^{\frac{\gamma-1}{\gamma}}\right]\right\} = 1010.5\,\text{K}$$

$$V_{cr2} = \sqrt{\left(\frac{2\gamma}{\gamma+1}\right)RT_{t2}} = 575.4\,\text{m/s}$$

$$V_{\theta 2} = \frac{1}{U_2}\left[U_1 V_{\theta 1} - c_p(T_{t1} - T_{t2})\right] = 211.6\,\text{m/s}$$

$$V_1 = V_{cr1} M_{cr1} = 583.8\,\text{m/s}$$

$$V_2 = \frac{V_{\theta 2}}{\sin\alpha_2} = 332.8\,\text{m/s}$$

$$V_{\theta 1} = V_1 \sin\alpha_1 = 583.8\,\text{m/s}$$

$$W_{\theta 1} = V_{\theta 1} - U_1 = 4.3\,\text{m/s}$$

$$W_{\theta 2} = V_{\theta 2} - U_2 = -23.3\,\text{m/s}$$

$$V_{r1} = V_1 \cos\alpha_1 = 218.7\text{m/s}$$

$$M_{cr2} = V_2 / V_c r_2 = 0.57$$

$$p_2 = p_{t2}\left[1 - \left(\frac{\gamma-1}{\gamma+1}\right) M_{cr2}{}^2\right]^{\frac{\gamma}{\gamma-1}} = 2.387\,\text{bars}$$

$$T_2 = T_{t2} - \frac{V_2{}^2}{2c_p} = 962.6\,\text{K}$$

$$\rho_2 = \frac{p_2}{RT_2} = 0.864\,\text{kg/m}^3$$

$$V_{z2} = \frac{\dot{m}}{\rho_2(2\pi r_2 h_2)} = 222.6\,\text{m/s}$$

$$\beta_1 = \tan^{-1}\left(\frac{W_{\theta 1}}{V_{r1}}\right) = 1.1^\circ$$

$$\beta_2 = \tan^{-1}\left(\frac{W_{\theta 2}}{V_{z2}}\right) = -6.0^\circ$$

$$\alpha_2 = \tan^{-1}\left(\frac{V_{\theta 2}}{V_{z2}}\right) = 43.5^\circ$$

With these kinematical variables, the radial-rotor inlet and exit velocity diagrams will look like those in Fig. 10.28.

Part b: To calculate the interstage-duct total pressure loss, we follow the same computational procedure in the preceding example, including, unfortunately, the iterative process to achieve the duct-exit Mach number. The final results of this procedure are as follows:

$$M_2 = 0.54$$

$$M_3 \approx 0.57$$

$$\frac{p_{t3}}{p_{t2}} = 0.965$$

$$\frac{\Delta p_t}{p_{t2}} = 3.5\%$$

Part c: Let us first apply equation (3.73) to convert the traditional into the critical Mach no. at the duct exit station, with the result being

$$M_{cr3} = 0.60$$

With the axial-stator flow process being isentropic, we have

$$p_{t4} = p_{t3} = 2.79\,\text{bars}$$

$$T_{t4} = T_{t3} = 1010.5\,\text{K}$$

Furthermore, with the axial stator being choked (i.e., $M_{cr4} = 1.0$), application of the continuity equation at the stator exit station yields the swirl angle at this location:

$$\alpha_4 = 56.4^\circ$$

Let us now proceed to calculate all of the relevant stator-exit kinematical variables:

$$V_4 = V_{cr4} = 575.4\,\text{m/s}$$

$$U_4 = U_2 = 234.9\,\text{m/s}$$

$$V_{\theta 4} = V_4 \sin\alpha_4 = 479.3\,\text{m/s}$$

$$W_{\theta 4} = V_{\theta 4} - U_4 = 244.4\,\text{m/s}$$

$$W_{z4} = V_{z4} = V_4 \cos\alpha_4 = 318.4\,\text{m/s}$$

$$\beta_4 = \tan^{-1}\left(\frac{W_{\theta 4}}{W_{z4}}\right) = 37.5^\circ$$

Because the axial-flow stage reaction is 50%, the following relationships automatically apply:

$$\alpha_5 = -\beta_4 = -37.5^\circ$$

$$\beta_5 = -\alpha_4 = -56.4^\circ$$

$$V_{\theta 5} = -W_{\theta 4} = -244.4\,\text{m/s}$$

$$W_{\theta 5} = -V_{\theta 4} = -479.3\,\text{m/s}$$

$$W_5 = V_4 = 575.4\,\text{m/s}$$

$$V_5 = W_4 = 401.4\,\text{m/s}$$

Furthermore,

$$T_{t5} = T_{t4} - \frac{U_4}{c_p}(V_{\theta 4} - V_{\theta 5}) = 962.8\,\text{K}$$

$$T_{tr5} = T_{t5} - \left(\frac{V_5^2 - W_5^2}{2c_p}\right) = 1036.3\,\text{K}$$

$$V_{cr5} = \sqrt{\left(\frac{2\gamma}{\gamma+1}\right) R T_{t5}} = 531.9\text{m/s}$$

$$W_{cr5} = \sqrt{\left(\frac{2\gamma}{\gamma+1}\right) R T_{tr5}} = 582.7\text{m/s}$$

$$M_{cr5} = \frac{V_5}{V_{cr5}} = 0.754$$

$$M_{cr,r5} = \frac{W_5}{W_{cr5}} = 0.987\,\text{(acceptably subsonic)}$$

Part e: The procedure to calculate each stage specific speed, by reference to expression (5.28), is as follows:

$$\rho_2 = 0.864\,\text{kg/m}^3\ \text{(computed earlier)}$$

$$T_5 = T_{t5} - \frac{V_5^2}{2c_p} = 893.2\,\text{K}$$

$$p_5 = 1.58\,\text{bars (computed earlier)}$$

$$\rho_5 = \frac{p_5}{RT_5} = 0.616\,\text{kg/m}^3$$

$$(N_s)_{rad.\ stg.} = 0.66\,\text{radians}$$

$$(N_s)_{ax.\ stg.} = 2.67\,\text{radians}$$

Referring to Figure 5.9, we see that the first and second stages indeed belong to the radial and axial stage types, respectively.

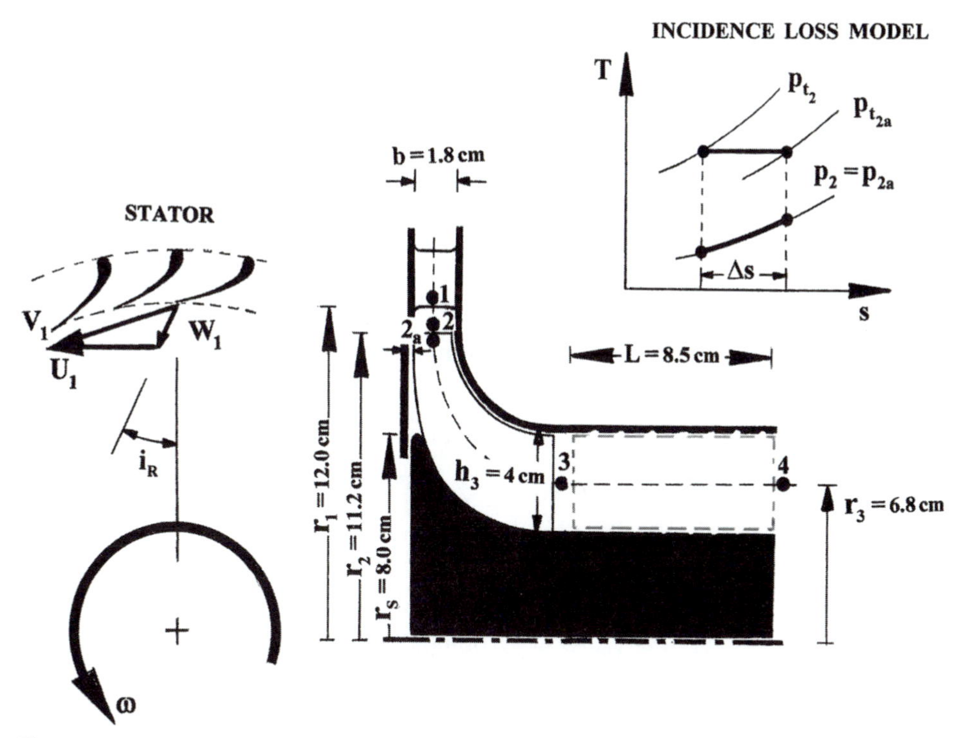

Figure 10.29. Input variables for Example 6.

EXAMPLE 6

Figure 10.29 shows a radial-turbine stage followed by an interstage constant-area annular duct. The operating conditions of the combined turbine/duct system are as follows:

- Stator-exit total pressure $(p_{t1}) = 10.6$ bars
- Stator-exit total temperature $(T_{t1}) = 1165$ K
- Stator-exit critical Mach number $(V_1/V_{cr1}) = 0.97$
- Mass-flow rate $(\dot{m}) = 5.3$ kg/s
- Rotor speed $(N) = 44{,}500$ rpm
- Stage work coefficient $(\psi) = 0.9$
- Duct-wall friction coefficient $(f) = 0.099$
- Initial total-to-total efficiency estimate $(\eta_{ref.}) = 92\%$

The total number of rotor blades (N_b) is 14, each with a perfectly radial admission segment, and the backplate clearance (C_b) is 1.8 mm. The following simplifications are also deemed reasonable:

- Futral model of rotor-incidence loss is applicable [equations (10.20) and (10.21)]
- Fanno-flow duct relationships govern the annular duct
- Entire flow process is adiabatic
- Average specific-heat ratio $(\gamma) = 1.33$

Calculate the following variables:

a) The total pressure (p_{t2a}) just inside the rotor leading edge;
b) The rotor-exit critical Mach number (M_{cr3});
c) The duct total pressure loss;
d) The duct static-pressure decline.

SOLUTION

Part a: Application of the continuity equation at the stator exit station yields

$$\alpha_1 = 71.5^\circ$$

Proceeding to the vaneless nozzle, we have

$$V_{cr1} = \sqrt{\left(\frac{2\gamma}{\gamma+1}\right)RT_{t1}} = 617.8\,\text{m/s}$$

$$V_{cr2} = V_{cr1} = 617.8\,\text{m/s (adiabatic flow)}$$

$$V_1 = M_{cr1}V_{cr1} = 599.3\,\text{m/s}$$

$$V_{\theta 1} = V_1 \sin\alpha_1 = 568.4\,\text{m/s}$$

$$V_{r1} = V_1 \cos\alpha_1 = 189.9\,\text{m/s}$$

$$V_{\theta 2} = \left(\frac{r_1}{r_2}\right)V_{\theta 1} = 609.0\,\text{m/s (free-vortex flow pattern)}$$

$$V_{r2} = \left(\frac{r_1}{r_2}\right)V_{r1} = 203.5\,\text{m/s (free-vortex flow pattern)}$$

$$\alpha_2 = \alpha_1 = 71.5^\circ\ \text{(free-vortex flow pattern)}$$

$$M_{cr2} = V_2/V_{cr2} = 1.04$$

Being categorically a subsonic nozzle, the stator/rotor radial gap is incapable of producing an exit critical Mach number that is in excess of unity. Instead, the flow passage is termed "choked," and the gap-exit critical Mach number will be 1.0. As a result,

$$V_2 = V_{cr2} = 617.8\,\text{m/s}$$

$$V_{\theta 2} = V_2 \sin\alpha_2 = 585.9\,\text{m/s}$$

$$V_{r2} = V_2 \cos\alpha_2 = 196.0\,\text{m/s}$$

Incidence loss calculations: With the rotor blade being perfectly radial at inlet (i.e., $\beta_2{}' = 0$), the blade incidence angle can be computed as follows:

$$i_R = \tan^{-1}\left(\frac{V_{\theta 2} - U_2}{V_{r2}}\right) - \beta_2{}' = 18.1^\circ$$

Let us now apply the Futral incidence-loss model to the rotor-inlet segment:

$$(V_{\theta 2})_{id.} = \left(1 - \frac{2}{N_b}\right) U_2 = 447.3\,\text{m/s}$$

$$L_{incid.} = \frac{1}{2}(V_{\theta 2} - V_{\theta 2id.}) = 9605\,\text{J/kg}$$

$$(K.E.)_2 = \frac{1}{2} V_2^{\,2} = 190{,}838\,\text{J/kg}$$

$$(K.E.)_{2a} = (K.E.)_2 - L_{incid.} = 181{,}233\,\text{J/kg}$$

In terms of static temperatures, we have

$$T_2 = T_{t2} - \left[\frac{(K.E.)_2}{c_p}\right] = 1000.0\,\text{K}$$

$$T_{2a} = T_{t2a} - \left[\frac{(K.E.)_{2a}}{c_p}\right] = 1008.3\,\text{K}$$

Referring to the process *T-s* representation in Fig. 10.29, and noting that this loss mechanism occurs at a constant static pressure (i.e., $p_{2a} = p_2$), we can calculate the entropy production as a result of this flow/blade interaction mechanism as follows:

$$\Delta s = c_p \ln\left(\frac{T_{2a}}{T_2}\right) = 9.56\,\text{J/(kgK)}$$

Furthermore, as the flow process occurs adiabatically, the total temperature will remain constant (i.e., $T_{t2a} = T_{t2}$). As a result, Δs can be reexpressed as

$$9.56 = \Delta s = -R \ln\left(\frac{p_{t2a}}{p_{t2}}\right)$$

which gives p_{t2a}, the total pressure just inside the rotor-blade leading edge:

$$p_{t2a} = 10.25\,\text{bars}$$

This means that the incidence-related total-pressure-loss percentage is

$$\frac{\Delta p_{t\,incid.}}{p_{t2}} = 3.30\%$$

Part b: As we proceed to the rotor exit station, and ignoring the profile losses, there is one more loss mechanism to consider, namely that caused by the backplate clearance. First, we calculate the tangential velocity component just inside the blade leading edge as follows:

$$V_{2a}^{\,2} = 2(K.E.)_{2a} = 362{,}467\,\text{J/kg} = V_{r2a}^{\,2} + V_{\theta 2a}^{\,2}$$

Recalling that the radial velocity component under the Futral incidence-loss model remains constant (i.e., $V_{r2a} = V_{r2} = 203.5$ m/s), we can now find $V_{\theta 2a}$:

$$V_{\theta 2a} = 569.3\,\text{m/s}$$

which, by comparison, indicates an incidence-related loss of 2.8% in the tangential-velocity component.

Backface-clearance loss calculations: As represented in Figure 10.29 and in the problem statement, the relevant variables are provided to compute the following two variables:

$$C_b/b = 0.10$$
$$r_s/r_2 = 0.71$$

where r_s is the scallop radius.

Referring to Figure 10.24, and noting that the baseline efficiency ($\eta_{ref.}$) is 92%, we get the clearance-penalized magnitude of total-to-total efficiency:

$$(\eta_{t-t})_{rotor} = 89.7\%$$

We now proceed to calculate the rotor-exit flow properties as follows:

$$U_3 = \omega r_3 = 316.9\,\text{m/s}$$

$$T_{t3} = T_{t2} - \left(\frac{\psi U_2^{\,2}}{c_p}\right) = 953.1\,\text{K}$$

$$V_{cr3} = \sqrt{\left(\frac{2\gamma}{\gamma+1}\right) R T_{t3}} = 558.8\,\text{m/s}$$

$$V_{\theta 3} = \frac{1}{U_3}\left(U_2 V_{\theta 2} - \psi U_2^{\,2}\right) = 164.0\,\text{m/s}$$

$$p_{t3} = p_{t2}\left\{1 - \frac{1}{\eta_{t-t}}\left[1 - \left(\frac{T_{t3}}{T_{t2}}\right)\right]\right\}^{\frac{\gamma}{\gamma-1}} = 4.25\,\text{bars}$$

$$\rho_{t3} = \frac{p_{t3}}{R T_{t3}} = 1.55\,\text{kg/m}^3$$

At this point, we have a rather difficult subproblem to solve. When applying the continuity equation at the rotor exit station, we end up with two unknowns. These are the axial-velocity component (V_{z3}) and the static density (ρ_3).

To overcome this difficulty, we proceed as follows:

1) Assume the static density magnitude (ρ_3).
2) Compute V_{z3} by substituting in the continuity equation.
3) With $V_{\theta 3}$ already known, compute M_{cr3}.
4) With M_{cr3} and ρ_{t3} known, calculate ρ_3.
5) Compare the newly computed ρ_3 with that in step 1.
6) Repeat steps 1 through 5 toward convergence.

Executing this computational procedure, the final results are as follows:

$$\rho_3 = 1.39\,\text{kg/m}^3$$
$$V_{z3} = 223.0\,\text{m/s}$$
$$V_3 = 276.8\,\text{m/s}$$

$$M_{cr3} = 0.495$$

$$\alpha_3 = \cos^{-1}\left(\frac{V_{z3}}{V_3}\right) = 39.3°$$

$$p_3 = 3.69 \text{ bars}$$

$$M_3 = 0.467$$

The next step is to substitute these results in the Fanno-flow relationship (3.71), looking for the duct-exit Mach number. The equation is highly nonlinear, and an iterative solution is in this case necessary. The final result of this step is

$$M_4 = 0.480$$

The duct total-to-total pressure ratio can then be obtained by direct substitution in equation (3.70), and the result is

$$\frac{p_{t4}}{p_{t3}} = 0.98$$

$$\frac{(\Delta p_t)_{duct}}{p_{t3}} = 2.0\%$$

Part d: With p_3 already computed (3.69 bars), we now compute the duct-exit static pressure:

$$p_4 = \frac{p_{t4}}{\left[1 + \left(\frac{\gamma-1}{2}\right) M_4{}^2\right]^{\frac{\gamma}{\gamma-1}}} = 3.58 \text{ bars}$$

Finally, we can compute the percentage of static-pressure decline:

$$\frac{(\Delta p)_{duct}}{p_3} = 2.98\%$$

Note that both the total and static pressures experience declines of varied magnitudes across the constant-area interstage duct. Of these, the loss in total pressure is a direct reflection of the friction along the endwall. This very loss contributes to the decline in static pressure, as the two pressure components are interrelated. The static pressure decline is further magnified by the rise in Mach number along the duct as a result of the boundary layer displacement (or blockage) effect.

EXAMPLE 7

The turbine section of an auxiliary power unit is shown in Fig. 10.30. This is composed of two stages radial- and axial-flow, with an interstage constant-area duct in between. The turbine operating conditions and approximations are as follows:

- Mass-flow rate ($\dot{m}$) = 7.7 kg/s
- Shaft speed (N) = 42,800 rpm
- Radial-stator exit total pressure (p_{t1}) = 14.6 bars

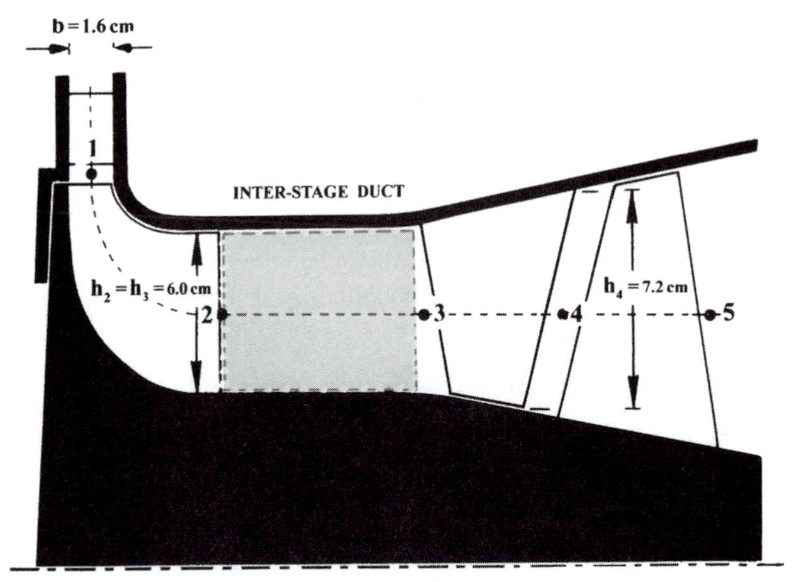

Figure 10.30. Input variables for Example 7.

- Radial-stator exit total temperature $(T_{t1}) = 1392$ K
- Radial-stage stator is choked
- Radial-stage work coefficient $(\psi_{rad.}) = 1.02$
- Total-to-total radial-stage efficiency $(\eta_{rad.}) = 83\%$
- Isentropic flow processes across both stators
- Axial-stator exit critical Mach no. $(M_{cr4}) = 0.94$
- M_{cr} jumps by 18% over the interstage gap
- Axial-stage exit total pressure $(p_{t5}) = 1.08$ bars
- Axial-stage total-to-total efficiency $(\eta_{ax.}) = 93\%$
- V_z is constant across the axial-flow stage
- Average specific-heat ratio $(\gamma) = 1.33$
- Adiabatic flow process throughout the entire turbine

The following geometrical variables are applicable:

- Rotor-blade admission segment is perfectly radial
- Axial-rotor-blade inlet (metal) angle $(\beta_4{}') = 22.3°$

Considering the preceding input data and the turbine major dimensions in Figure 10.30, calculate the following variables:

a) Radial-rotor incidence angle, $(i_R)_{rad.}$;
b) Axial-rotor incidence angle, $(i_R)_{ax.}$;
c) Axial-stage work coefficient $(\psi_{ax.})$;
d) Axial-rotor exit swirl angle (α_5).

SOLUTION

Part a: Applying the continuity equation at the radial-stator exit station, the exit swirl angle (α_1) can be determined:

$$\alpha_1 = 69.1^\circ$$

The procedure to compute the rotor-blade incidence angle, $(i_R)_{rad.}$, is as follows:

$$V_1 = V_{cr1} = \sqrt{\left(\frac{2\gamma}{\gamma+1}\right) R T_{t1}} = 675.3\,\text{m/s (choked stator)}$$

$$V_{\theta 1} = V_1 \sin\alpha_1 = 630.9\,\text{m/s}$$

$$V_{r1} = V_1 \cos\alpha_1 = 240.9\,\text{m/s}$$

$$W_{\theta 1} = V_{\theta 1} - U_1 = 12.4\,\text{m/s}$$

$$\beta_1 = \tan^{-1}(W_{\theta 1}/V_{r1}) = 2.95^\circ$$

$$(i_R)_{rad.} = \beta_1 - \beta_1{}' = 2.95^\circ$$

Part b:

$$V_{\theta 2} = \left(\frac{1}{U_2}\right)\left[U_1 V_{\theta 1} - \psi_{rad.} U_1{}^2\right] = 0.07\,\text{m/s} \approx 0$$

$$\alpha_2 \approx 0$$

$$T_{t2} = T_{t1} - \left(\frac{\psi_{rad.} U_1{}^2}{c_p}\right) = 1054.7\,\text{K}$$

$$p_{t2} = p_{t1}\left\{1 - \left(\frac{1}{\eta_{rad.}}\right)\left[1 - \left(\frac{T_{t2}}{T_{t1}}\right)\right]^{\frac{\gamma}{\gamma-1}}\right\} = 3.63\,\text{bars}$$

$$\rho_{t2} = \frac{p_{t2}}{R T_{t2}} = 1.20\,\text{kg/m}^3$$

$$V_{cr2} = \sqrt{\left(\frac{2\gamma}{\gamma+1}\right) R T_{t2}} = 587.9\,\text{m/s}$$

Let us now apply the continuity equation at the radial-rotor exit station:

$$\dot{m} = \rho_2 V_{z2}(2\pi r_2 h_2)$$

which, upon substitution, gives rise to the following compact expression:

$$\rho_2 V_{z2} = 227.0$$

The two variables appearing on the left-hand side are unknown but, yet, interrelated. The following is a trial-and-error procedure to solve this equation:

1) Assume the value of V_{z2}.
2) With $V_{\theta_2} \approx 0$, $V_2 \approx V_{z_2}$.
3) Calculate M_{cr2}.

4) Calculate ρ_2.
5) Calculate the product $\rho_2 V_{z2}$.
6) Compare the result of item 5 with the targeted value of 227.0.
7) Repeat steps 1 through 6 toward convergence.

The above-outlined computational procedure was executed with the following final results:

$$V_2 \approx V_{z2} = 200\,\text{m/s}$$
$$M_{cr2} = 0.34$$

Now the axial-rotor inlet relative flow angle (β_4) can be calculated:

$$\beta_4 = \tan^{-1}\left(\frac{W_{\theta 4}}{V_{z4}}\right) = 15.8^\circ$$

The rotor-blade incidence angle can now be obtained:

$$(i_R)_{ax.\ stg.} = \beta_4 - \beta_4{}' = -6.5^\circ$$

Part c: The procedure to calculate the axial-stage work coefficient is as follows:

$$T_{t4} = T_{t3} = 1054.7\,\text{K (adiabatic flow)}$$
$$p_{t4} = p_{t3} = 3.15\,\text{bars (isentropic stator flow)}$$
$$p_{t5} = 1.08\,\text{bars (given)}$$
$$T_{t5} = T_{t4}\left\{1 - \eta_{t-t}\left[1 - \left(\frac{p_{t5}}{p_{t4}}\right)^{\frac{\gamma-1}{\gamma}}\right]\right\} = 825.9\,\text{K}$$
$$(\psi)_{ax.\ stg.} = \frac{c_p(T_{t4} - T_{t5})}{U_4{}^2} = 1.63$$

Part d:

$$V_{\theta 5} = V_{\theta 4} - \frac{c_p}{U_4}(T_{t4} - T_{t5}) = -175.6\,\text{m/s}$$
$$\alpha_5 = \tan^{-1}\left(\frac{V_{\theta 5}}{V_{z5}}\right) = -32.8^\circ$$

EXAMPLE 8

Figure 10.31 shows different views of a radial turbine stage, the stage inlet and exit velocity diagrams, and the stage major dimensions, with the stator/rotor radial-gap length being negligible. The stage operating conditions are as follows:

- Rotational speed $(N) = 46{,}500$ rpm
- Mass-flow rate $(\dot{m}) = 5.04$ kg/s
- Inlet total pressure $(p_{t1}) = 11.8$ bars
- Inlet total temperature $(T_{t1}) = 1265.0$ K
- Both the stator and rotor passages are choked

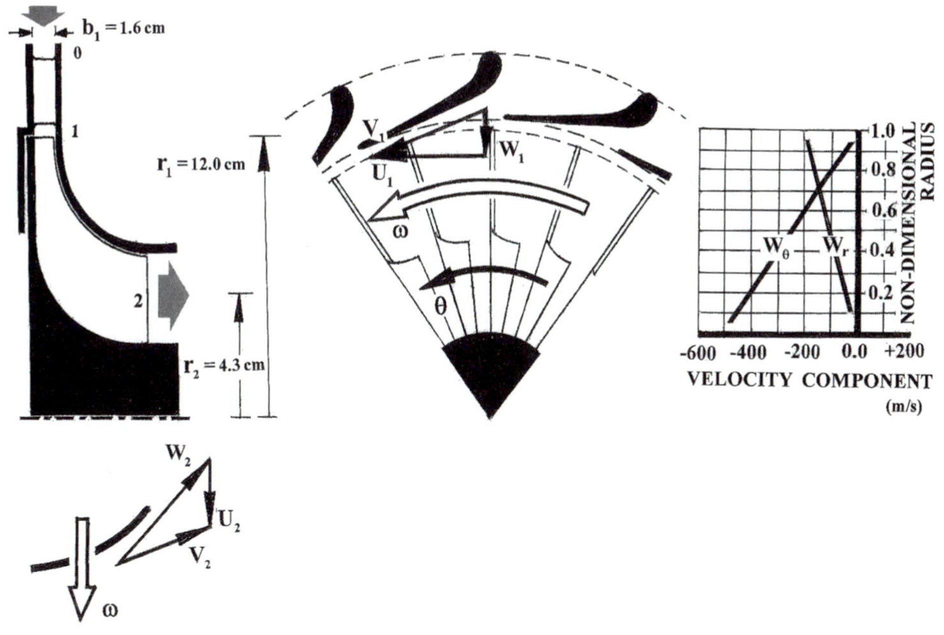

Figure 10.31. Input data for Example 8.

- Rotor-inlet incidence angle (i_R) = 0
- Specific shaft work produced (w_s) = 404.56 kJ/kg
- Total-to-total pressure ratio ($Pr = p_{t1}/p_{t2}$) = 4.85

By assuming a linear variation of W_r and W_θ with radius across the rotor, calculate the combined magnitudes of the centripetal and Coriolis acceleration components at several radial locations. Using the resulting tangential component of these two accelerations, sketch the relative trajectory of a fluid particle as it traverses the rotor subdomain. Also comment on the radial variation of the radial-acceleration component.

SOLUTION

The inlet and exit solid-body rotational velocities are calculated first:

$$\omega = 4869.5\,\text{m/s}$$

$$U_1 = \omega r_1 = 584.3\,\text{m/s}$$

$$U_2 = \omega r_2 = 209.4\,\text{m/s}$$

Applying the continuity equation at the stator-exit radius, we get

$$\alpha_1 = 71.5^\circ$$

The rotor-inlet absolute-velocity components can now be computed:

$$V_{\theta 1} = U_1 = 584.3\,\text{m/s}\,(\beta_1 = 0)$$

$$V_{r1} = \frac{V_{\theta 1}}{\tan\alpha_1} = -195.5\,\text{m/s (negative sign inserted)}$$

The exit total properties can be computed as follows:

$$p_{t2} = \frac{p_{t1}}{Pr} = 2.43\,\text{bars}$$
$$T_{t2} = T_{t1} - \frac{w_s}{c_p} = 915.2\,\text{K}$$

The rest of the stage-exit thermophysical properties can be calculated as follows:

$$V_{\theta 2} = \left(\frac{U_1}{U_2}\right) V_{\theta 1} - \frac{c_p}{U_2}(T_{t1} - T_{t2}) = 301.8\,\text{m/s}$$
$$W_{\theta 2} = V_{\theta 2} - U_2 = -511.2\,\text{m/s}$$
$$T_{tr2} = T_{t2} + \frac{{W_{\theta 2}}^2 - {V_{\theta 2}}^2}{2c_p} = 988.8\,\text{K}$$
$$W_{cr2} = \sqrt{\left(\frac{2\gamma}{\gamma+1}\right) R T_{tr2}} = 569.2\,\text{m/s}$$
$$W_2 = W_{cr2} = 569.2\,\text{m/s (rotor is choked)}$$
$$\beta_2 = \sin^{-1}\left(\frac{W_{\theta 2}}{W_2}\right) = -63.9^\circ$$
$$V_{z2} = W_{z2} = 250.3\,\text{m/s}$$
$$\alpha_2 = \tan^{-1}\left(\frac{V_{\theta 2}}{V_{z2}}\right) = -50.3^\circ$$

Noting that $W_r = V_r$, and using the inlet and exit magnitudes as boundary conditions, we can express W_r as a linear function of radius as follows:

$$W_r = 109.2 - 2539.0\,r \text{ (where } r \text{ is in meters)}$$

Now, noting that $W_\theta = V_\theta - \omega r$, and using the inlet and exit magnitudes of W_θ as boundary conditions, we can similarly establish the following linear relationship:

$$W_\theta = -796.6 + 6638.5\,r \text{ (where } r \text{ is in meters)}$$

These are graphically represented in Figure 10.31 along with the inlet and exit velocity diagrams.

With the preceding two expressions, let us calculate both relative-velocity components at six equidistant radial locations between the rotor inlet and exit radii (r_1 and r_2), beginning at the rotor-inlet radius, as follows:

$$\text{at } r = 0.1200\,\text{m}, \quad W_\theta = 0.0000\,\text{m/s} \quad \text{and} \quad W_r = -195.5\,\text{m/s}$$
$$\text{at } r = 0.1046\,\text{m}, \quad W_\theta = -102.2\,\text{m/s} \quad \text{and} \quad W_r = -156.4\,\text{m/s}$$
$$\text{at } r = 0.0892\,\text{m}, \quad W_\theta = -204.4\,\text{m/s} \quad \text{and} \quad W_r = -117.3\,\text{m/s}$$
$$\text{at } r = 0.0738\,\text{m}, \quad W_\theta = -306.7\,\text{m/s} \quad \text{and} \quad W_r = -78.20\,\text{m/s}$$
$$\text{at } r = 0.0584\,\text{m}, \quad W_\theta = -408.9\,\text{m/s} \quad \text{and} \quad W_r = -39.10\,\text{m/s}$$
$$\text{at } r = 0.0430\,\text{m}, \quad W_\theta = -511.1\,\text{m/s} \quad \text{and} \quad W_r = 0.0000\,\text{m/s}$$

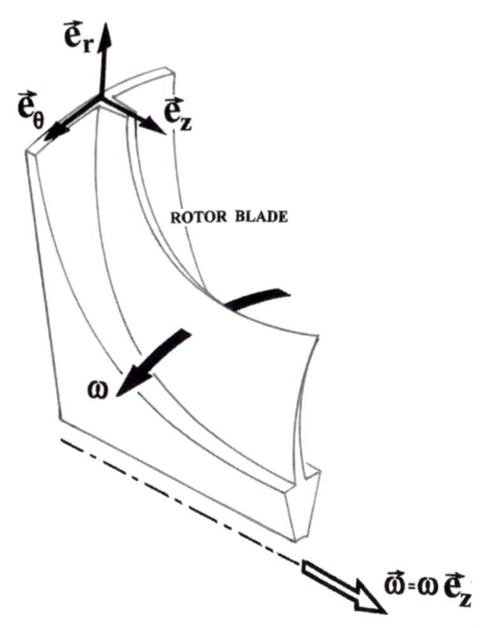

Figure 10.32. Relative cylindrical frame of reference.

At this point, we have to define our relative (spinning) frame of reference, which we will use in the acceleration calculations (to follow). The reference frame here is defined in Figure 10.32 in the form of unit vectors in the r, θ, and z directions. Using this frame of reference, the Coriolis-acceleration vector (the nature of which is discussed in Chapter 11) can be written in the form of the following vector product:

$$\mathbf{a}_C = 2\omega \times \mathbf{W} = [(0.0)\mathbf{e}_r + (0.0)\mathbf{e}_\theta + (2\omega)\mathbf{e}_z] \times [(W_r)\mathbf{e}_r + (W_\theta)\mathbf{e}_\theta + (W_z)\mathbf{e}_z]$$

The above "vector" product is usually expressed in the form of a determinant that, upon expansion, yields the following *nonzero* components:

$$\mathbf{a}_C = (-2\omega W_\theta)\mathbf{e}_r + (2\omega W_r)\mathbf{e}_\theta$$

The centripetal acceleration, however, can consistently be written as

$$\mathbf{a}_{cent.} = (\omega^2 r)\mathbf{e}_r$$

Utilizing the foregoing dependencies of W_r and W_θ on the local radius, as well as the different acceleration components, we are now in a position to compute the net acceleration components acting on a typical fluid particle. To this end, let us choose the same six radial locations that we defined earlier to compute the relative-velocity components. The net-acceleration components (in m/s^2) at these radii are as follows:

$$\text{at}\, r = 0.1200\,\text{m}, \quad a_r = +2.85 \times 10^6 \quad \text{and} \quad a_\theta = -1.90 \times 10^6$$

$$\text{at}\, r = 0.1046\,\text{m}, \quad a_r = +3.48 \times 10^6 \quad \text{and} \quad a_\theta = -1.52 \times 10^6$$

$$\text{at}\, r = 0.0892\,\text{m}, \quad a_r = +4.11 \times 10^6 \quad \text{and} \quad a_\theta = -1.14 \times 10^6$$

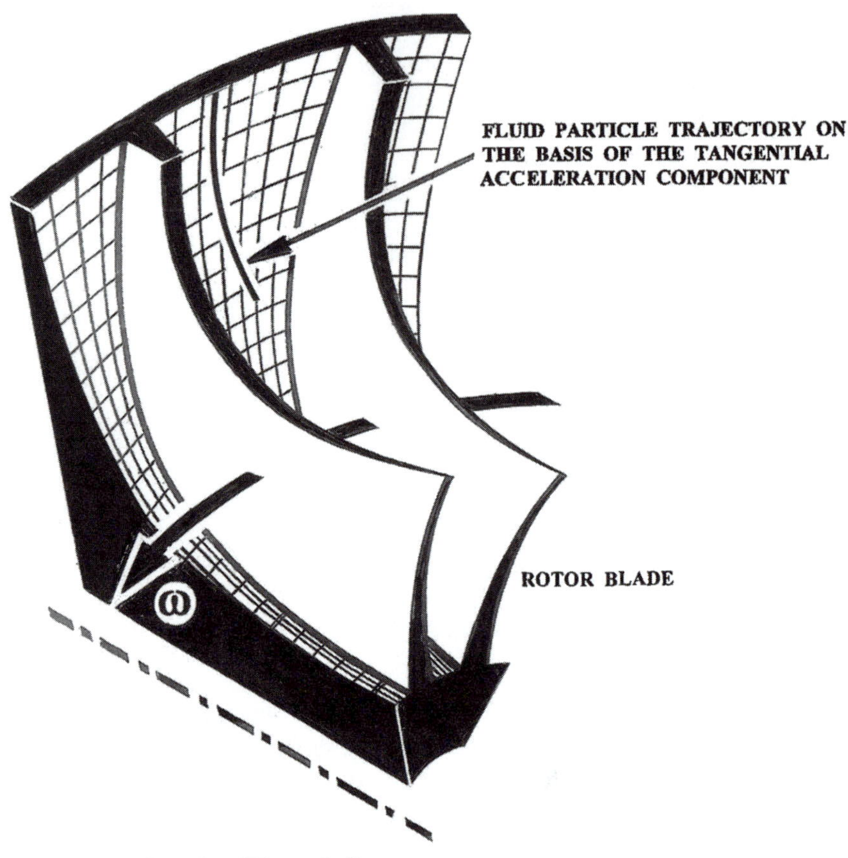

Figure 10.33. Results of Example 8.

$$\text{at } r = 0.0738\,\text{m}, \quad a_r = +4.74 \times 10^6 \quad \text{and} \quad a_\theta = -0.76 \times 10^6$$
$$\text{at } r = 0.0584\,\text{m}, \quad a_r = +5.36 \times 10^6 \quad \text{and} \quad a_\theta = -0.38 \times 10^6$$
$$\text{at } r = 0.0430\,\text{m}, \quad a_r = +6.00 \times 10^6 \quad \text{and} \quad a_\theta = 0.0$$

Using the tangential-acceleration component magnitudes (above), a trajectory sketch of a typical fluid particle, within the rotor, should look as that in Figure 10.33. Also shown in the figure is the radius $\bar{r}$ (approximately 8.9 cm) at which the acceleration radial component switches sign. Above this radius, the radially outward centripetal acceleration is predominant, whereas the Coriolis radial-acceleration component dictates the sign within the $r_2 > r > \bar{r}$ subdomain. This Coriolis component is clearly negative as W_θ continues to be negative all the way down to the exit station.

Stage Placement in a Multistage Turbine

In a multistage turbine sequence, a radial stage (if present) would normally be the first. This is only logical because any other arrangement would add a loss-generating upstream duct to lead the flow stream to the stage inlet station. Aside from aerodynamic differences, such a duct is "functionally" similar to the return duct between the two centrifugal compressor stages in Figure 2.12. The major difference, however, is that the latter involves a negative radius shift, giving it a favorable pressure gradient. On the contrary, a duct leading to the inlet station of a radial-turbine stage from an upstream axial-flow stage is one where the radius is rising and so is the cross-flow area. Such an environment creates a highly unfavorable pressure gradient along the return duct and is certain to cause a significant loss in total pressure. Although this may provide a sound aerodynamic argument, we should also realize that using a radial stage as the first recipient of the combustor flow stream will expose such a thermally sensitive stage to the highest temperature in the entire engine. This may be mechanically challenging, particularly over the hub segment, where a radial-to-axial flow-path deflection occurs. Considered an example of how severe stress concentration can get, this particular surface segment, under excessively high temperatures and temperature gradients, could very well pose a mechanical threat. With the added aspect of high bending stress at the root, a typical designer would normally entertain the idea of cooling the stage rotor in particular.

Cooling Techniques

In the following, two (external and internal) cooling methods are discussed in light of a sequence of numerical results reported by Baskharone (1975 & 1977). The object then was the rotor of a radial-turbine stage in a turbocharger.

Cooling Effectiveness

In assessing the different cooling configurations in Figures 10.34 and 10.35, a meaningful measure of cooling effectiveness has to be established. With the focus being on the circled hub segment in Fig. 10.34, it is perhaps appropriate to adopt the temperature-reduction percentage, in this subregion, as a nontraditional, yet helpful, cooling-effectiveness gauge.

External Cooling

With no mechanical price to pay, external (or veil) cooling (depicted in Fig. 10.34) is a viable cooling strategy. Figure 10.34 begins with the uncooled-rotor temperature contours serving as a datum. The cooling air, in Figure 10.34, enters the stage at such a high radius that it cools the stator backplate as well. The coolant is tangentially injected into the rotor primary-flow stream over the hub surface. The two cooling models in Figure 10.34 differ only in the cooling-to-primary mass-flow percentage.

With coolant percentages of 1.5 and 3.0, the post-processing calculations yield cooling-effectiveness magnitudes of 9.1 and 12.3%, respectively. The closeness of these numbers proves the preexisting belief that the effect of increasing the coolant-flow rate has its own cooling-effectiveness limit at which further elevation in the coolant-flow rate is unwise, a feature that exclusively applies to this particular cooling technique.

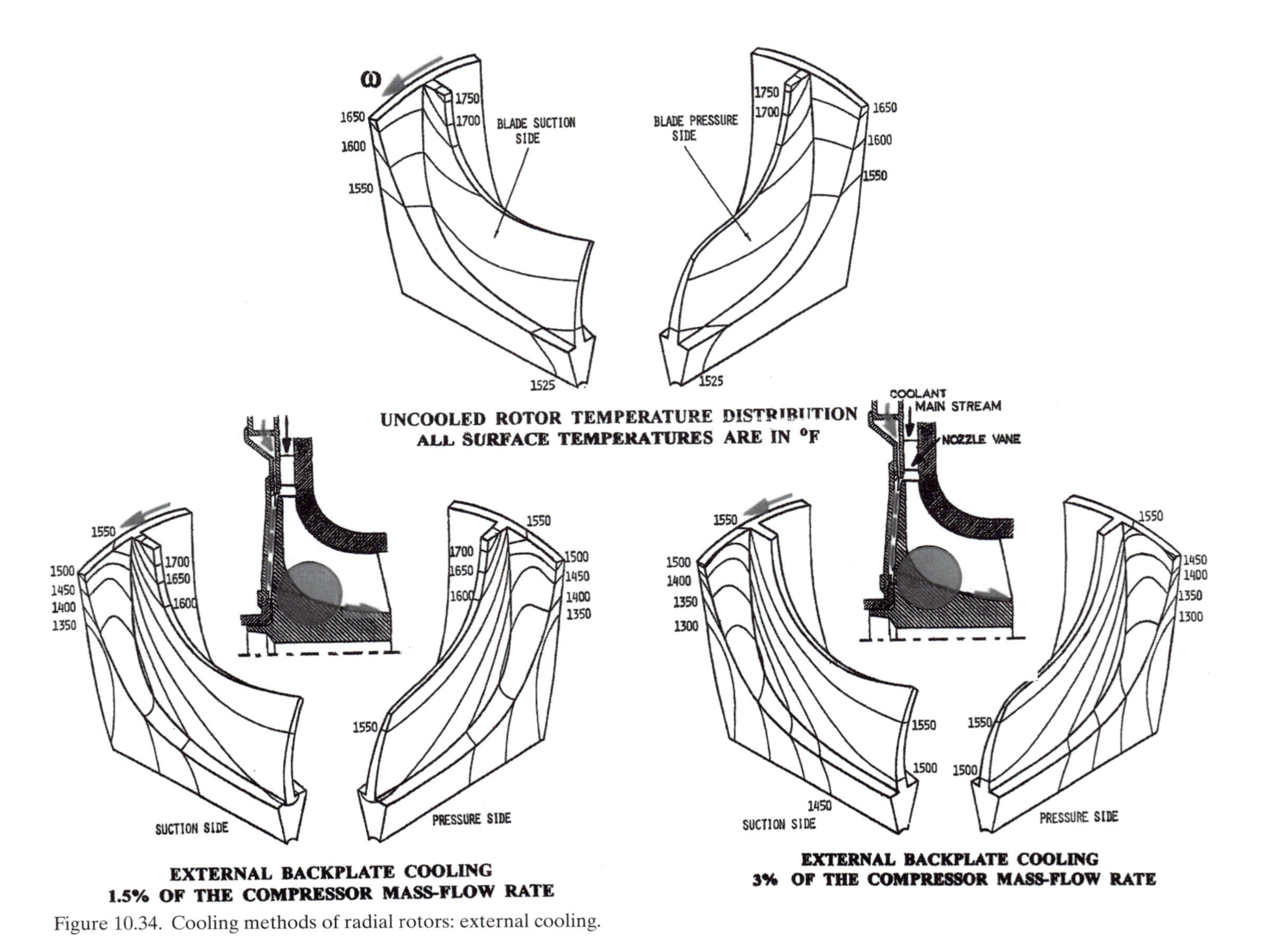

Figure 10.34. Cooling methods of radial rotors: external cooling.

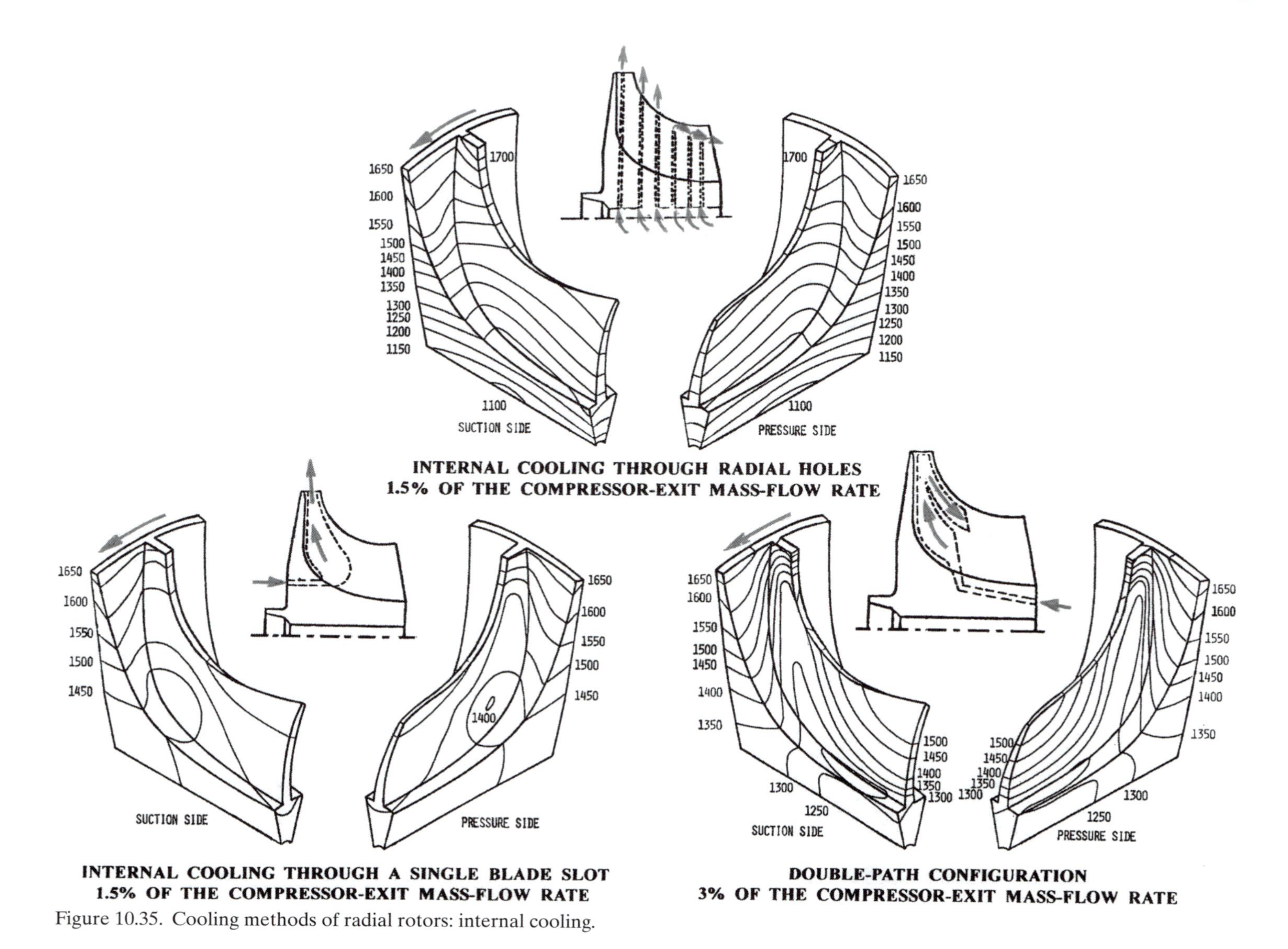

Figure 10.35. Cooling methods of radial rotors: internal cooling.

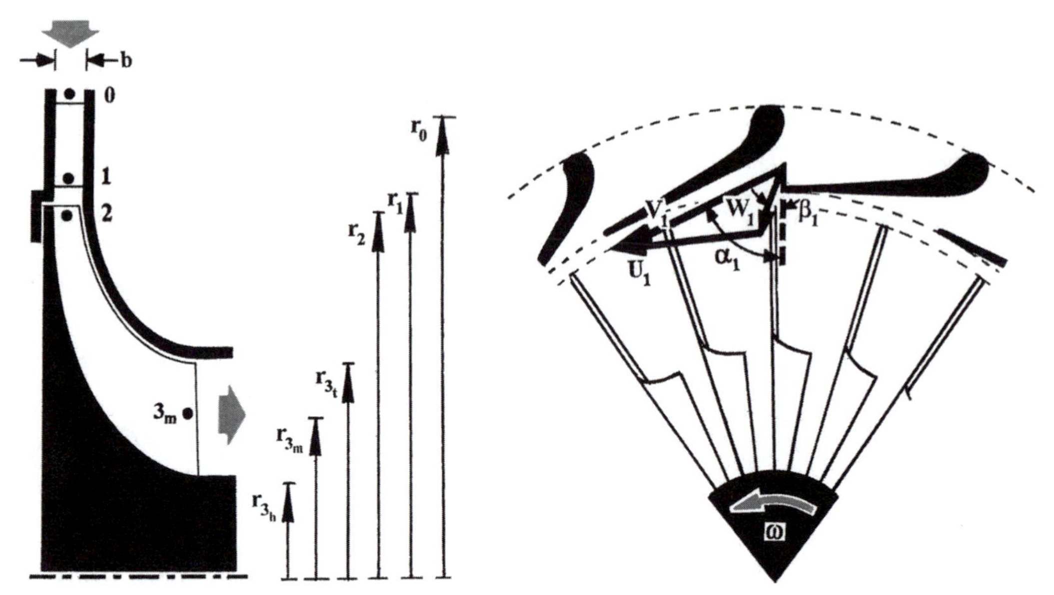

Figure 10.36. Input variables for Problem 1.

Internal Cooling

Shown in Figure 10.35 are different internal cooling configurations, all of which have been historically employable. First, in this figure, is a cooling technique where the exchange of heat energy is effected through a sequence of radial holes, which extend radially from the blade root. Although the cooling-air percentage here is low, the cooling effectiveness in this case is as high as 18.8%. Next is a cooling-flow passage in the form of a single slot inside the rotor blade with the same 1.5% cooling-flow rate. A modest cooling-effectiveness magnitude of 7.5% is produced in this case. The last of these cooling configurations features a double-slot passage with a coolant-flow rate of 3%. The combined effects of the coolant-passage geometry and relatively high coolant percentage give rise to an impressive cooling effectiveness of 15.6% in this case.

PROBLEMS

1) Figure 10.36 shows a radial turbine stage, of which the major dimensions are as follows:

   - Stator-inlet radius $(r_0) = 14.8$ cm
   - Stator-exit radius $(r_1) = 12.5$ cm
   - Stator sidewall spacing $(b) = 1.4$ cm
   - Rotor-inlet radius $(r_2) = 11.5$ cm
   - Rotor-exit mean radius $(r_{3m}) = 6.5$ cm
   - Rotor-blade inlet segment is perfectly radial

The stage operating conditions are as follows:

- Mass-flow rate ($\dot{m}$) $= 1.8$ kg/s
- Rotor speed (N) $= 44{,}000$ rpm
- Stator-exit total pressure (p_{t1}) $= 15$ bars
- Stator-exit total temperature (T_{t1}) $= 1385$ K
- Stator-exit critical Mach no. (V_1/V_{cr1}) $= 0.82$
- Stator-exit swirl angle (α_1) $= 72°$
- Free-vortex flow across the vaneless nozzle
- Rotor-exit total temperature (T_{t3m}) $= 1105$ K
- Rotor-exit axial-velocity component (V_{z3m}) $= 150$ m/s
- Stage total-to-total efficiency (η_T) $= 86\%$

Assuming an adiabatic flow and a γ value of 1.33, calculate:

a) The rotor-exit absolute critical Mach number $(V/V_{cr})_{3m}$;
b) The rotor-blade incidence angle (i_R);
c) The rotor-exit absolute and relative flow angles (α_{3m}, β_{3m});
d) The stage reaction (R);
e) The stage specific speed (N_s).

2) Redraw the rotor blade shape in Example 3 considering a 15% rotor-speed elevation.

3) Figure 10.37 shows a well-insulated radial turbine stage, where

- Stator-inlet radius (r_0) $= 15$ cm
- Stator-exit radius (r_1) $= 13.5$ cm
- Rotor-inlet radius (r_2) $= 12.5$ cm
- Rotor-exit mean radius (r_{3m}) $= 4.4$ cm
- Stator sidewall spacing (b) $= 0.8$ cm

The stage operating conditions are as follows:

- Scroll-inlet total pressure ($p_{t\,in}$) $= 9.5$ bars
- Scroll-inlet total temperature ($T_{t\,in}$) $= 1200$ K
- Scroll total pressure loss ($\Delta p_{t\,scr.}$) $= 2.5\%$
- Shaft speed (N) $= 43{,}000$ rpm
- Mass-flow rate ($\dot{m}$) $= 1.74$ kg/s
- Stator-exit critical Mach no. (V_1/V_{cr1}) $= 0.85$
- Rotor-exit total pressure (p_{t3m}) $= 2.9$ bars
- Rotor-exit critical Mach no. (V_{3m}/V_{cr3m}) $= 0.36$
- Rotor-exit swirl angle (α_{3m}) $=$ zero
- Isentropic flow across the vaned and vaneless nozzles
- Free-vortex flow behavior across the vaneless nozzle

Assuming an average specific heat ratio (γ) of 1.33, calculate:

a) The stator-exit swirl angle (α_1);
b) The rotor-exit annulus height (h_3);

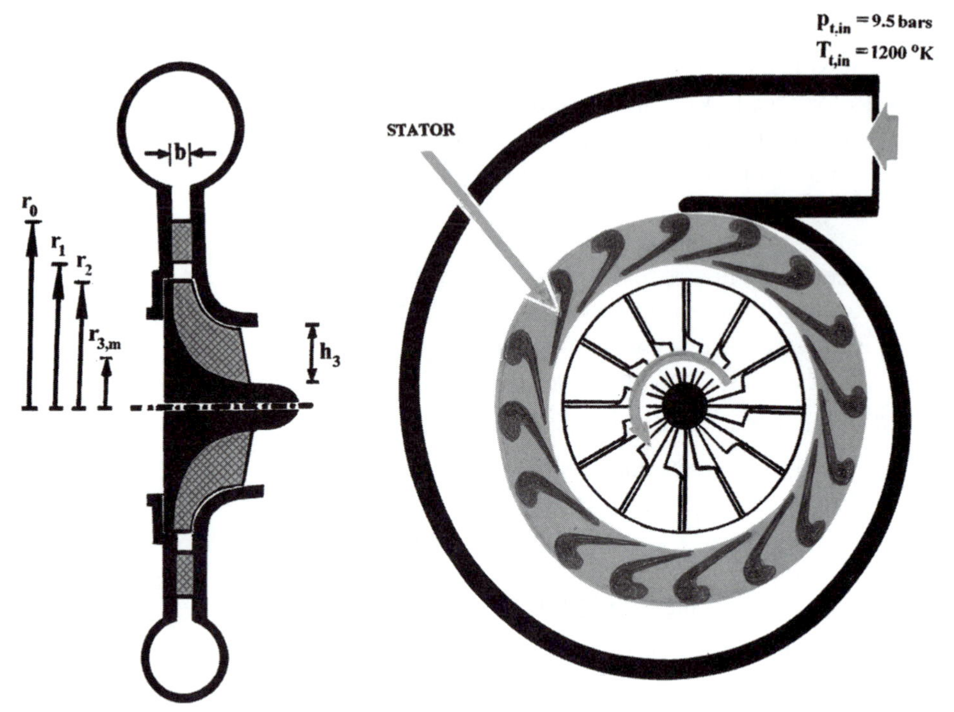

Figure 10.37. Input variables for Problem 2.

c) The rotor-inlet critical Mach no. (V_2/V_{cr2});
d) The actual magnitude of specific shaft work (w_s);
e) The scroll-inlet to rotor-exit total-to-total efficiency (η_T);
f) The stage specific speed (N_s).

Hint: With zero exit swirl angle, the rotor-exit total temperature (T_{t3m}) depends only on the rotor-inlet velocity diagram.

4) In the preceding problem, consider the case where the stator trailing-edge segment is rotatable in the manner illustrated in Fig. 3.18. Through this variable-geometry means, and starting from a stator-exit critical Mach number (V_1/V_{cr1}) of 0.85, the vanes are rotated-closed toward the stator-choking status. Ignoring the radial extension (Δr) of the vaneless nozzle, calculate:

   a) The rotation angle $(\Delta\alpha_1)$ of the vane trailing-edge segment (Fig. 3.18);
   b) The magnitude and angle of the rotor-inlet relative velocity;
   c) The rotor-inlet total relative temperature.

5) An uncooled radial-turbine stage has the following major dimensions:

   - Stator-exit radius $(r_1) = 12.2$ cm
   - Rotor-inlet radius $(r_2) = 11.5$ cm
   - Sidewall spacing $(b) = 0.64$ cm
   - Rotor-exit mean radius $(r_{3m}) = 6.6$ cm
   - The rotor inducer is totally radial

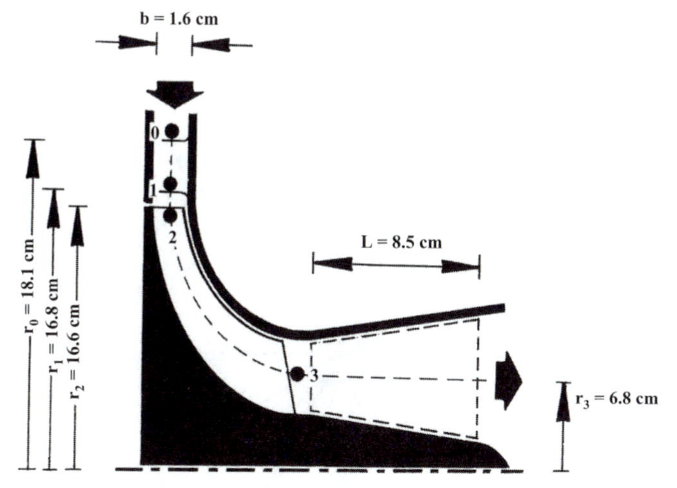

Figure 10.38. Input variables for Problem 7.

The stage operating conditions are as follows:

- Shaft speed (N) = 38,000 rpm
- Stator-exit swirl angle (α_1) = 75°
- Stage-inlet total pressure (p_{t0}) = 15 bars
- Stage-inlet total temperature (T_{t0}) = 1390 K
- Stator-exit radial-velocity component (V_{r1}) = 137.2 m/s
- Rotor-exit absolute flow angle (α_{3m}) = zero
- Rotor-exit absolute velocity (V_{3m}) = 91.4 m/s
- Isentropic flow across the vaned and vaneless nozzles
- Stage total-to-total efficiency (η_T) = 91%
- Free-vortex flow across the vaneless nozzle
- Rotor flow process is adiabatic
- Average specific-heat ratio (γ) = 1.33

Calculate the following variables:

a) The rotor-blade incidence angle (i_R);
b) The stage reaction (R);
c) The rotor-wise change in total relative temperature (ΔT_{t_r});
d) The rotor-wise change in total relative pressure (Δp_{t_r});

6) Referring to Example 5, consider the following geometrical change:

- Axial-stator-exit annulus height (h_4) = 4.5 cm = h_5.

With this change, calculate:

a) The axial-stage specific speed $(N_s)_{ax}$;
b) The decline in total relative pressure ($\Delta p_{t,r}$) across the axial-flow rotor.

7) Figure 10.38 shows the major dimensions of a radial-inflow turbine stage in an auxiliary power unit (APU). The stage is followed by an exhaust diffuser that

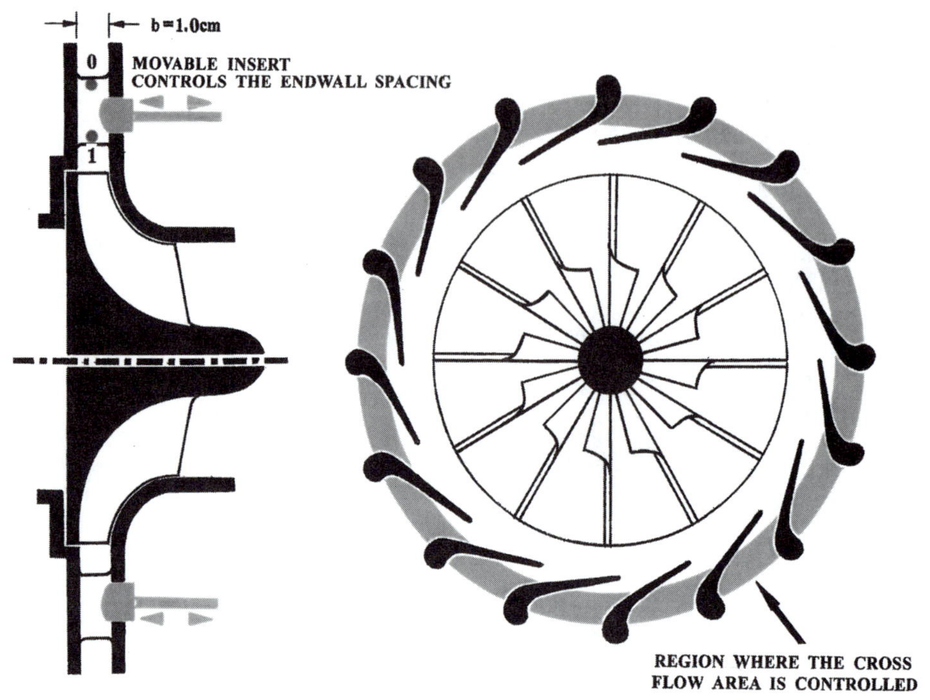

Figure 10.39. Input variables for Problem 8.

is designed in light of Sovran and Klomp chart (Fig. 3.37). The stage operating conditions are as follows:

- Rotational speed $(N) = 27{,}000$ rpm
- Mass-flow rate $(\dot{m}) = 5.6$ kg/s
- Stator-exit total temperature $(T_{t1}) = 1280$ K
- Stator-exit total pressure $(p_{t1}) = 11.5$ bars
- Stator-exit critical Mach no. $(M_{cr1}) = 0.83$
- Stage work coefficient $(\psi) = 1.4$
- Rotor-exit static temperature $(T_3) = 1003$ K
- Rotor-exit total pressure $(p_{t3}) = 3.8$ bars

In the interest of computational simplicity, the following approximations are also applicable:

- Isentropic flow through the vaned and vaneless nozzles
- Free-vortex flow structure in the vaneless nozzle

Assuming an adiabatic flow process throughout the rotor and diffuser, and taking the average γ magnitude to be 1.33, calculate:

a) The stage total-to-total efficiency (η_{t-t});
b) The ratio of static density decline across the rotor;
c) The overall total-to-static efficiency (η_{t-s});

8) Figure 10.39 highlights the stator of a radial-inflow turbine. The stator is equipped with an axially translatable insert, with the objective of varying the exit cross-flow

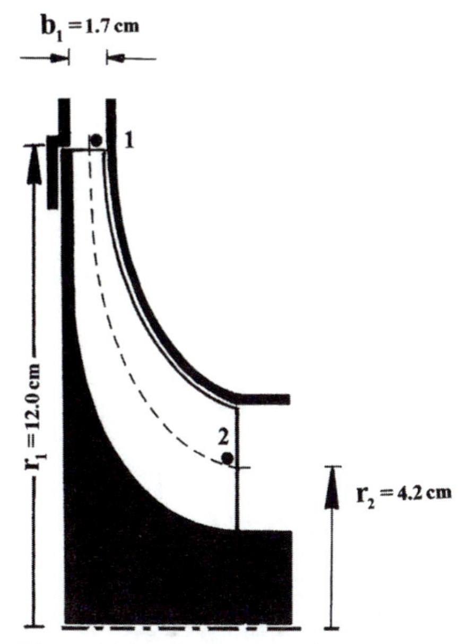

Figure 10.40. Geometry variables for Problem 9.

area. The stator operating conditions are as follows:

- Mass-flow rate ($\dot{m}$) = 5.2 kg/s
- Inlet total temperature (T_{t0}) = 1150 K
- Inlet total pressure (p_{t0}) = 10.45 bars
- Exit swirl angle (α_1) = 62°

Assuming an isentropic stator flow process and a γ magnitude of 1.33, calculate:

a) The stator-exit critical Mach number (M_{cr1});
b) The translating-insert displacement that will choke the stator.

9) Figure 10.40 shows the meridional view of a radial-turbine stage and its major dimensions. The stage operating conditions are as follows:

- Rotor-inlet total pressure (p_{t1}) = 12 bars
- Rotor-inlet total temperature (T_{t1}) = 1350 K
- Mass-flow rate ($\dot{m}$) = 4.8 kg/s
- Rotor speed (N) = 32,500 rpm
- Rotor-inlet critical Mach no. (M_{cr1}) = 0.86
- Stage-exit total temperature (T_{t2}) = 1144 K
- Stage total-to-total efficiency ($\eta_{stg.}$) = 81%
- Rotor-exit axial velocity component (V_z) = 215 m/s
- Perfectly radial blade-inlet segment (i.e., $\beta_1{}' = 0$)

The stage flow process is assumed adiabatic, and the specific-heat ratio (γ) is fixed at 1.33. Calculate the following variables:

a) The rotor-inlet swirl angle (α_1);
b) The stage flow coefficient (ϕ);

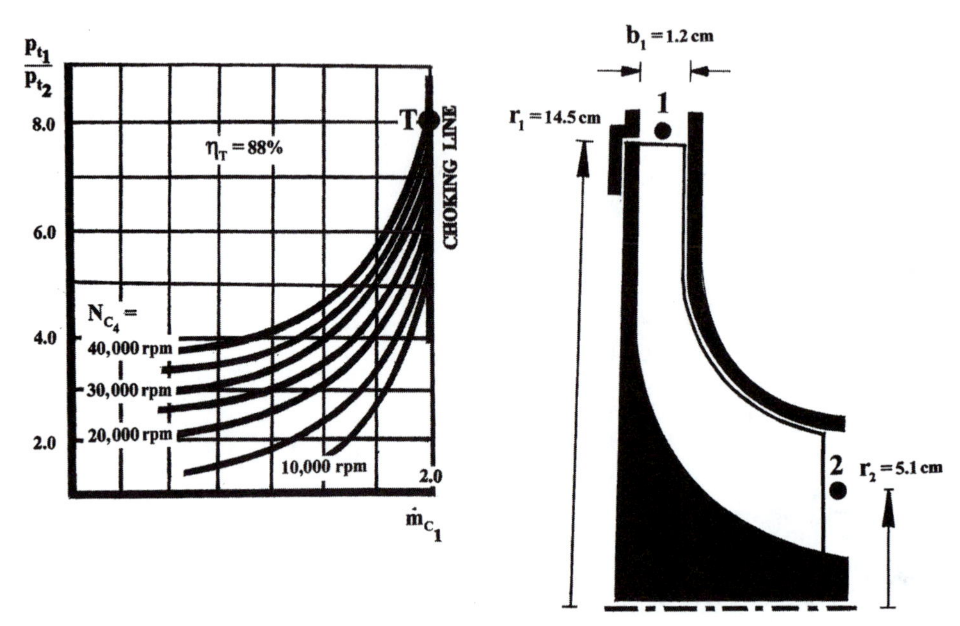

Figure 10.41. Input variables for Problem 10.

c) The rotor-blade incidence angle (i_R);
d) The stage-exit swirl angle (α_2);
e) The static temperature decline (ΔT) across the rotor;
f) The stage specific speed (N_s).

10) Figure 10.41 shows the major dimensions of a radial-turbine stage along with its map. On the map, the stage operating mode is identified by the point "T." In addition, the following conditions apply:

- Stator is choked
- Rotor-inlet total pressure (p_{t1}) = 10.8 bars
- Rotor-inlet total temperature (T_{t1}) = 1150 K
- Rotor speed (N) = 31,000 rpm
- Rotor total-to-total efficiency (η_{t-t}) = 88%
- Rotor-exit critical Mach no. (M_{cr2}) = 0.41
- Rotor-exit swirl angle (α_2) = −24°

Assuming an adiabatic flow process and a specific-heat ratio (γ) of 1.33, calculate the change in total relative temperature across the rotor.

11) Figure 10.42 shows the radial stage in a two-stage turbine section of a power system. This particular stage is operating under the following conditions:

- Inlet total pressure (p_{t0}) = 13.6 bars
- Inlet total temperature (T_{t0}) = 1295 K
- Shaft speed (N) = 33,000 rpm
- Stator-exit critical Mach number (M_{cr1}) = 0.93
- Stator-exit swirl angle (α_1) = 72.5°

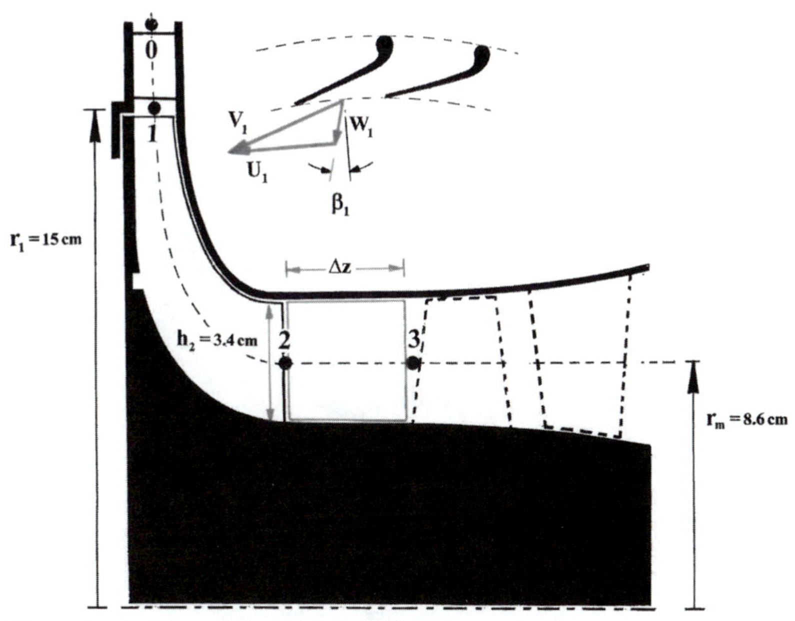

Figure 10.42. Input variables for Problem 11.

- Mass-flow rate ($\dot{m}$) = 6.77 kg/s
- Rotor-exit total temperature (T_{t2}) = 1018 K
- Rotor-exit axial-velocity component (V_{z2}) = 274 m/s
- Interstage-duct friction coefficient (f) = 0.15
- Duct-exit critical Mach number (M_{cr3}) = 0.93

The radial rotor blade has a perfectly radial admission segment (i.e., $\beta_1{}' = 0$), and the stator flow process is assumed isentropic. Calculate the following:

a) The radial stator sidewall spacing (b_1);
b) The radial rotor incidence angle (i_R);
c) The variable(s) that would indicate whether the rotor is choked;
d) The interstage-duct axial length (Δz).

12) Figure 10.43 shows a radial-turbine stage together with its map. The turbine operating conditions are as follows:

- Stator-exit total temperature (T_{t1}) = 1152 K
- Stator-exit total pressure (p_{t1}) = 12.8 bars
- Mass-flow rate ($\dot{m}$) = 4.48 kg/s
- Stator-exit critical Mach number (M_{cr1}) = 0.92
- Rotor-inlet relative flow angle (β_1) = 0
- Rotor-exit critical Mach no. (M_{cr2}) = 0.62

In addition, the entire flow process is adiabatic, and the specific-heat ratio (γ) is 1.33.

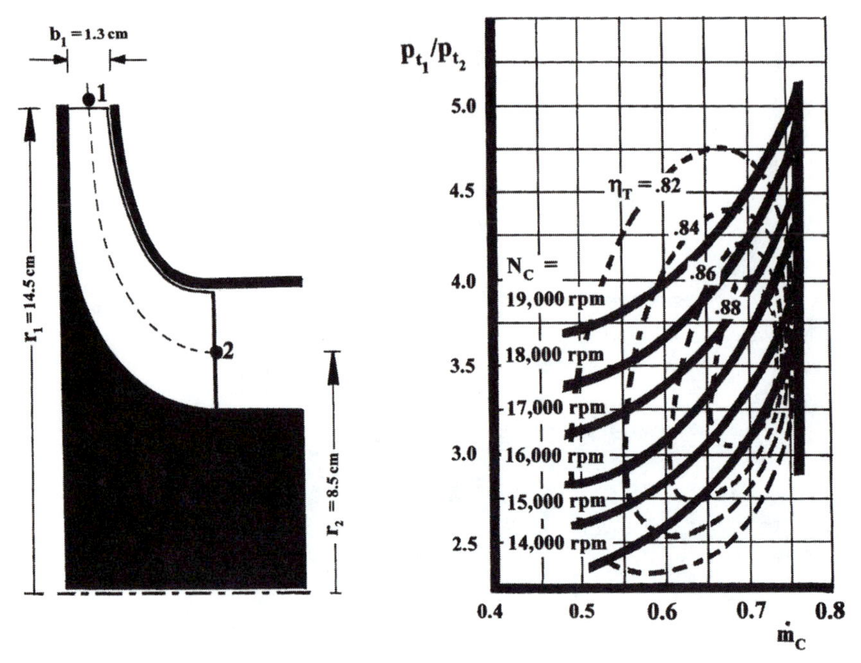

Figure 10.43. Input variables for Problem 12.

Part I:

a) Calculate the decline in total relative pressure across the rotor.
b) Calculate the stage specific speed.

Part II: Due to stricter envelope constraints, the radial stage (above) is to be replaced by four axial-flow stages, which share the same total-to-total magnitudes of pressure ratio and efficiency. Combined together, these axial stages will have to produce the same overall total-to-total values of pressure and temperature ratios as the radial turbine stage. Calculate the common total-to-total efficiency of the four axial-flow stages.

13) Figure 10.44 shows the meridional projection of a radial-inflow turbine stage. The stage operating conditions are as follows:

- Mass-flow rate ($\dot{m}$) = 5.1 kg/s
- Stator-exit total temperature (T_{t1}) = 1380 K
- Stator-exit total pressure (p_{t1}) = 14.2 bars
- Stator-exit critical Mach number (M_{cr1}) = 0.93
- Rotor speed (N) = 48,500 rpm
- Free-vortex flow behavior in the vaneless nozzle

The entire flow process is adiabatic, and the specific-heat ratio (γ) is 1.33. Calculate the change in total relative pressure (Δp_{t_r}) across the rotor caused by:

a) The streamwise decline in radius;
b) The rotor irreversibility sources (e.g., friction).

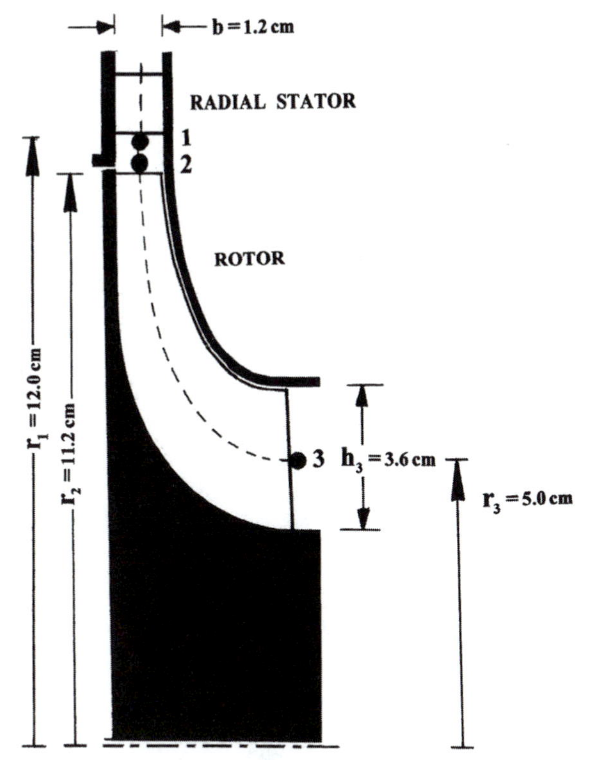

Figure 10.44. Radial-turbine stage geometry for Problem 13.

Hint: Calculate Δp_{t_r} using the given stage efficiency. The result will be the combined contributions of both radius change and irreversibilities. Next, repeat the computational steps that were (directly or indirectly) influenced by the given efficiency as you compute (only) the radius-change contribution, but with a stage efficiency of 100% this time. The contribution of irreversibilities will be the difference between the computed magnitudes of Δp_{t_r}. In computing the two magnitudes of Δp_{t_r}, you will have to conduct an iterative procedure to compute the rotor-exit critical Mach number (M_{cr3}) so that the continuity equation is satisfied at this particular station. This iterative procedure may consist of the following steps:

- Assume the magnitude of the rotor-exit critical Mach number (M_{cr3}).
- Calculate the exit static density (ρ_3).
- Apply the continuity equation to find V_{z3}.
- Knowing that $V_3 = V_{z3}$, recalculate M_{cr3}.
- Repeat the preceding steps toward convergence.

14) Figure 10.45 shows the meridional view of a radial-turbine stage that is operating under the following conditions:

- Mass-flow rate ($\dot{m}$) = 1.6 kg/s
- Rotor-inlet total pressure (p_{t1}) = 15.2 bars

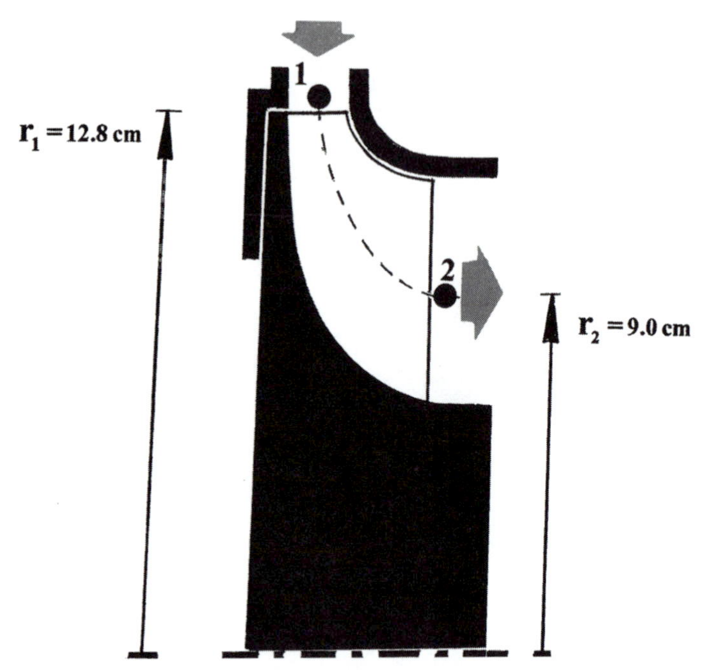

Figure 10.45. Input variables for Problem 14.

- Rotor-inlet total temperature $(T_{t1}) = 1420$ K
- Rotor-inlet "absolute" critical Mach number $(M_{cr1}) = 0.98$
- Rotor-inlet relative flow angle $(\beta_1) = 0$
- Shaft speed $(N) = 47{,}500$ rpm
- Rotor-exit total pressure $(p_{t2}) = 3.4$ bars
- Rotor-exit swirl angle $(\alpha_2) = 0$
- Rotor-exit critical Mach number $(M_{cr2}) = 0.22$
- Rotor total-to-total efficiency $(\eta_{t-t}) = 82.5\%$

The flow process is adiabatic in its entirety, and the average specific-heat ratio (γ) is 1.33.

a) Calculate the specific speed (N_s) of the stage.
b) Consider the replacement of the radial stage by three axial-flow stages sharing the same total-to-total values of pressure ratio and efficiency. These are intended to give rise to the same exit conditions as the radial stage. Now calculate the shared total-to-total efficiency magnitude of the axial-flow stages.

15) Figure 10.46 shows an industrial-turbine section that is composed of two geometrically identical radial-inflow stages. The figure also shows the first-stage map and the stage operation point on the map. The turbine operating conditions are as follows:

- Inlet total temperature $(T_{t1}) = 1380.0$ K
- Inlet total pressure $(p_{t1}) = 14.8$ bars
- First rotor-inlet relative flow angle $(\beta_2) = 0$
- First rotor-exit swirl angle $(\alpha_3) = 0$

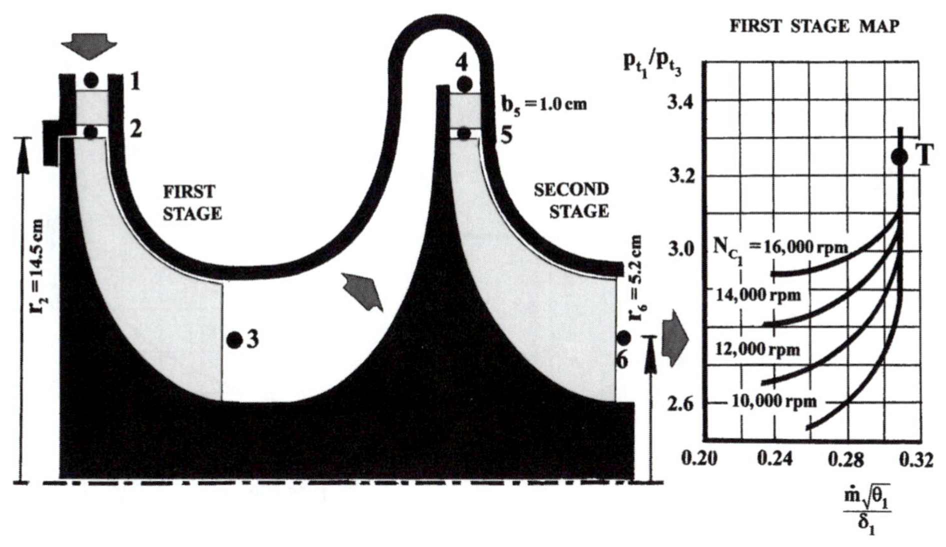

Figure 10.46. Input data for Problem 15.

- First-stage total-to-total efficiency $(\eta_{stg.1}) = 91.0\%$
- Turbine-section work split $[(w_s)_{stg.1}{:}(w_s)_{stg.2}] = 54{:}46$
- Return-duct total-pressure loss coefficient $(\bar{\omega}_{ret.duct}) = 0.08$
- Second rotor-inlet critical Mach number $(M_{cr5}) = 0.96$
- Second rotor-exit critical Mach number $(M_{cr6}) = 0.38$
- Second-stage total-to-total efficiency $(\eta_{stg.2}) = 89.0\%$

Assuming an isentropic flow process through each of the two stators and an adiabatic flow elsewhere, calculate:

a) The physical mass-flow rate $(\dot{m})$ and "physical" speed (N);
b) The second rotor-exit annulus height (h_6);
c) The second-stage specific speed $[(N_s)_{stg.2}]$;
d) The second-stage total-to-static efficiency $[(\eta_{t-s})_{stg.2}]$.

16) Figure 10.47 shows a radial-inflow turbine stage together with its map. The stage operating conditions are as follows:

- Stator-exit total temperature $(T_{t1}) = 1410$ K
- Stator-exit total pressure $(p_{t1}) = 13.8$ bars
- Mass-flow rate $(\dot{m}) = 9.05$ kg/s
- Power output $(P) = 2993$ kW
- Free-vortex flow in the stator/rotor gap
- Stator-exit critical Mach number $(M_{cr1}) = 0.88$

Assuming a specific-heat ratio of 1.33, an isentropic flow over the stator and the stator/rotor gap, and an adiabatic flow field elsewhere, calculate:

a) The rotor-blade incidence angle (i_R).
b) The stage specific speed (N_s).
c) The irreversibility-caused magnitude of ΔT_{tr}.

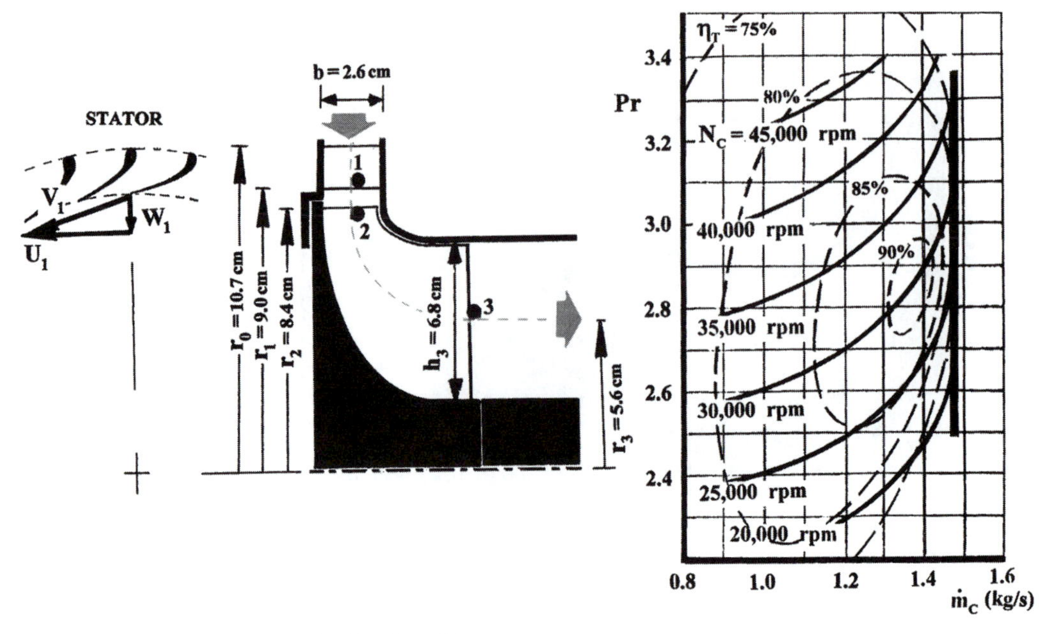

Figure 10.47. Input data for Problem 16.

Hint

- With the total-to-total efficiency, calculate the overall total relative temperature decline. This magnitude corresponds to the combined effects of radius shift and irreversibility sources.
- Keeping the rotor-inlet properties fixed (including T_{tr1}), calculate the drop in total relative temperature caused by the radius shift only. This requires a new magnitude of T_{tr2} that is associated with a stage efficiency of 100%.
- The effect of the flow-process irreversibility sources can be obtained by subtracting it from the former change in total relative temperature.
- In carrying out the 100% efficiency step (above), you do not have to resolve the continuity equation at the rotor exit station to calculate V_{z3}. The exit total-relative temperature in this case can be calculated as follows:

$$T_{tr2} = T_{t2} + \frac{(W_{\theta 2}{}^2 - V_{\theta 2}{}^2)}{2c_p}$$

CHAPTER ELEVEN

Centrifugal Compressors

From a historical viewpoint, the centrifugal compressor configuration was developed and used well before axial-flow compressors, even in the propulsion field. The common belief that such a "bulky" compressor type, because of its large envelope and weight (Fig. 11.1), has no place except in ground applications is not exactly accurate. For example, with a typical total-to-total pressure ratio of, say, 5:1, it would take up to three axial-compressor stages to absorb similar amounts of shaft work that a single centrifugal compressor stage would. In fact, the added engine length, with so many axial stages, would increase the skin-friction drag on the engine exterior almost as much as the profile drag, which is a function of the frontal area.

Despite the preceding argument, the tradition remains that the centrifugal-compressor propulsion applications are unpopular. Exceptions to this rule include turboprop engines and short-mission turbofan engines, as shown in Figure 11.2.

An attractive feature of centrifugal compressors has to do with their off-design performance. Carefully designed, a centrifugal compressor will operate efficiently over a comparatively wider shaft speed range. This exclusive advantage helps alleviate some of the problems associated with the turbine-compressor matching within the gas generator.

One of the inherent drawbacks of centrifugal compressors has to do with multiple staging. As illustrated in Figure 2.12, the excessive 180° flow-turning angle of the annular return duct, in this case, will increase the flow rotationality (in terms of vorticity) and encourage the cross-stream secondary flow migration. This simply sets the stage for high magnitudes of total pressure loss and boundary-layer separation. These and other irreversibility sources concern, perhaps exclusively, apply to the return duct early segment, where the flow direction is radially outward in an environment of an adverse pressure gradient. The elbow-type flow deflection that follows magnifies the total pressure loss even more, due to the secondary cross-flow stream which is associated with this type of abrupt flow deflection. The remainder of the duct, however, is one of decreasing radius and, therefore, a favorable pressure gradient.

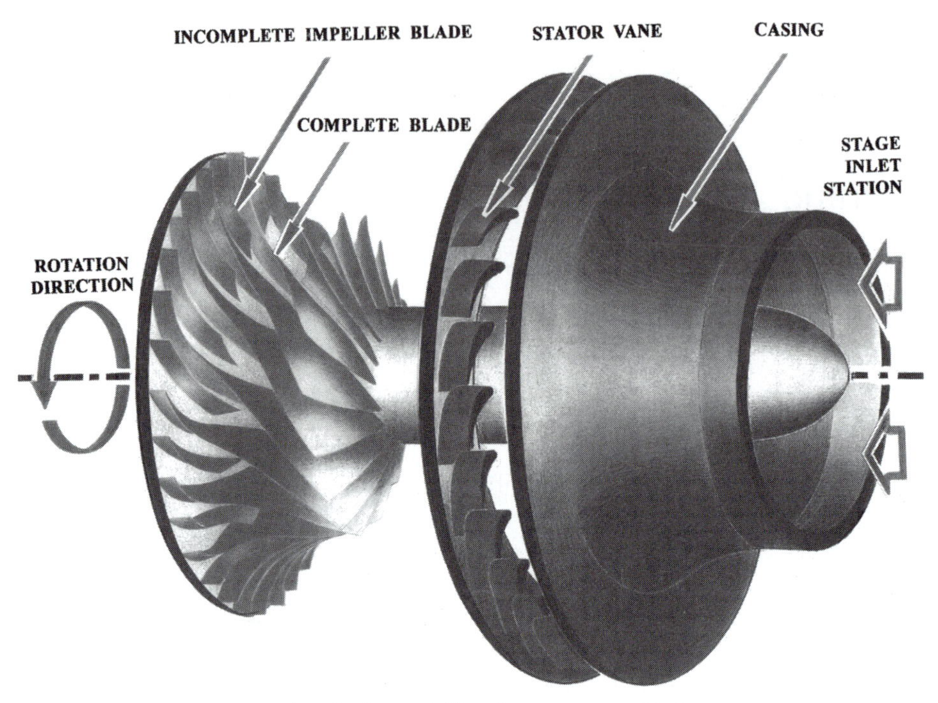

Figure 11.1. An isometric sketch of a centrifugal-compressor stage.

Component Identification

Depicted in Figure 11.3 is a traditional centrifugal compressor stage, together with the station-designation pattern. Not shown in the figure, however, is an optional "preswirl" cascade of stationary vanes. Discussed later in this chapter, this stationary axial-flow cascade would offset the effect of radius change along the impeller inlet station.

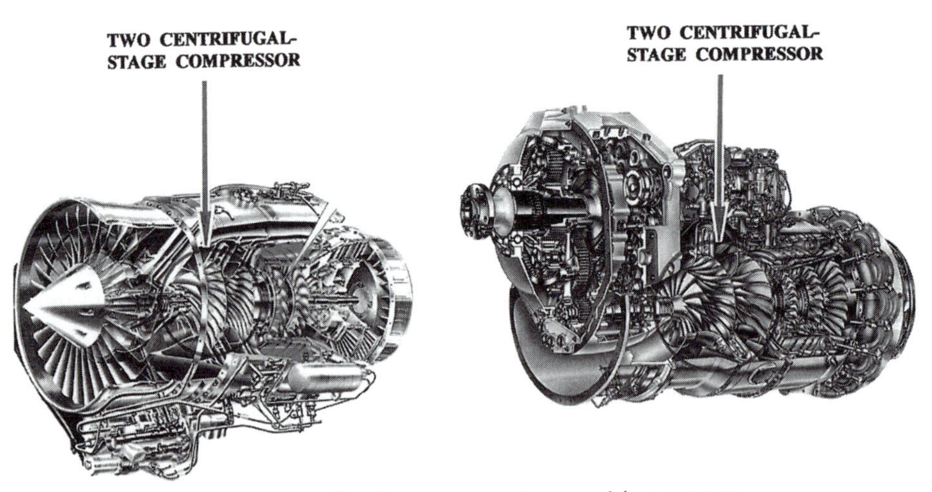

Figure 11.2. Utilization of centrifugal compressors in propulsion systems.

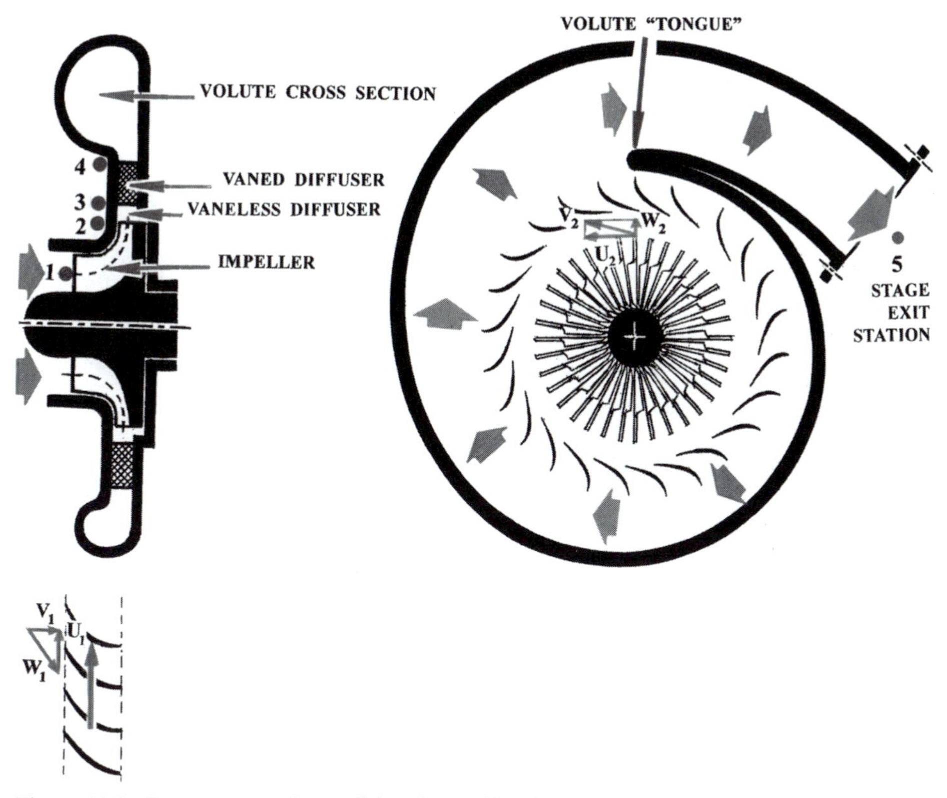

Figure 11.3. Components of a traditional centrifugal-compressor stage and velocity triangles.

Extending between stations 1 and 2 in Figure 11.3 is the stage impeller. In the absence of preswirl guide vanes, the absolute velocity (V_1) will, in most cases, be axial. Based on the average radius at this station, the inlet velocity triangle will characteristically look like that in Fig. 11.3. For a perfectly radial exit segment, the impeller-exit velocity diagram will be consistent with that in Figure 11.3.

The next component in Figure 11.3 is the vaneless diffuser. As indicated in Chapter 3, the mere radius rise across this component makes it a flow-decelerating vessel. From a computational standpoint, the flow structure in this subdomain is often assumed to be that of the free-vortex type. However, the distinction should be clear between this and its radial-turbine nozzle-counterpart. In the latter, the streamwise pressure gradient is favorable, a factor that will suppress the sidewall boundary-layer buildup. This makes the free-vortex flow assumption more or less justifiable. On the contrary, the compressor vaneless diffuser in Fig. 11.3 is a source of total pressure loss, for it is associated with an unfavorable pressure gradient. In reality, the performance of this component is a function of the circumferential velocity distribution at the impeller outlet and will be the worst should boundary-layer flow separation occur within the impeller. Contrary to the vaneless-nozzle function in a radial turbine, the role of this radial gap is to slow down the flow stream. As a result, a fraction of the dynamic pressure is converted into a static pressure rise.

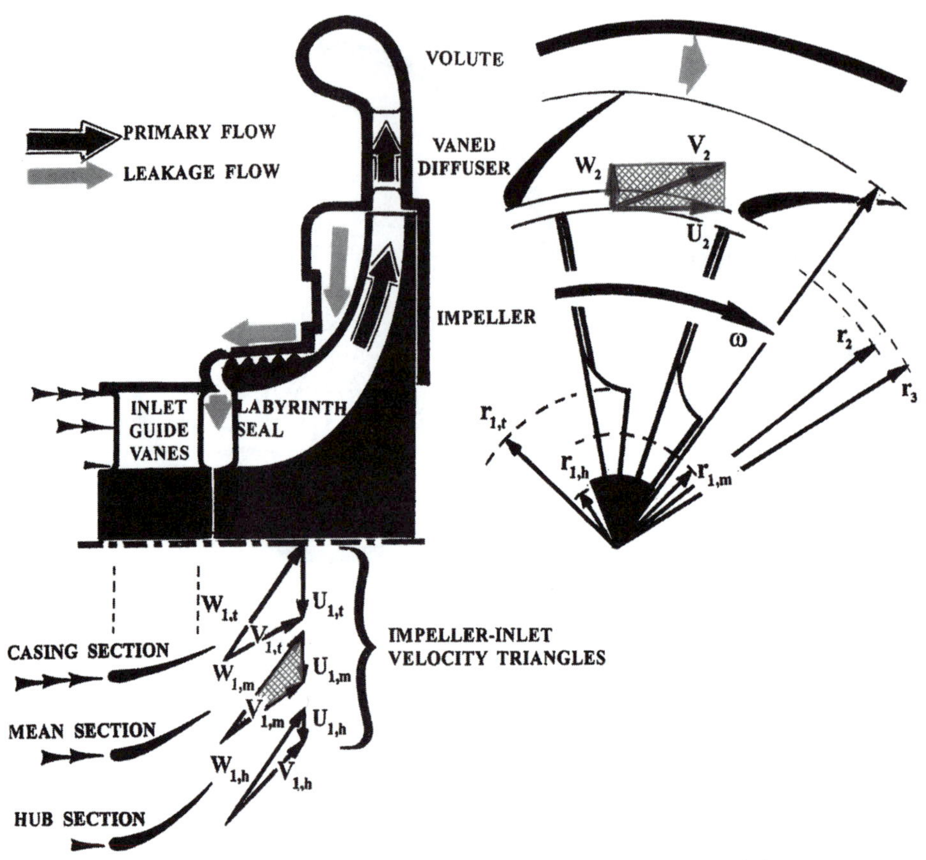

Figure 11.4. Preswirl guide vanes eliminate the need to twist the impeller inlet segment.

The vaned diffuser in Figure 11.4 is also aimed at total-to-static pressure conversion (or recovery). As represented in Chapter 3, both the diffuser-like vane-to-vane passage and the radius increase in this region contribute to the static-pressure recovery across this component. Note that the mere presence of vanes in this component provides the boundary layer with additional surface areas on which to grow. As a result, it would be logical to go with the minimum possible number of vanes without severely affecting the component-wise flow guidedness.

The last of the compressor nonrotating components in Fig. 11.3 is termed the volute (or collector). This is geometrically similar, but functionally opposite, to the scroll (or distributor) component in radial-inflow turbines. In the presence of an annular combustor (Fig. 1.2), as is normally the case, the compressor volute is essentially non-existing. However, it is the norm in stand-alone compressor units to include such a component as the flow-delivery passage. In the event it exists, the volute will be responsible for circumferentially collecting the diffuser-exit flow increments while ideally maintaining a constant through-flow velocity, on an average basis. As a result, and opposite to its scroll counterpart, the volute cross-flow area keeps increasing in the streamwise direction (Fig. 11.4).

Impeller Inlet System

This is normally a purely axial duct that may be equipped with a cascade of preswirl guide vanes (Fig. 11.4). The function of this stationary cascade, if present, is to give rise to a spanwise swirl-velocity variation which offsets the spanwise variation of the solid-body rotational velocity U. In effect, this will ensure the optimum impeller-inlet incidence angle. From a mechanical viewpoint, the inlet guide vanes will facilitate an impeller-inlet segment (or inducer) that is untwisted. This, in turn, will have the effect of reducing the blade-root bending stresses.

Let us now consider the configuration of a vaneless inlet system which, in this case, becomes a simple annular duct. It is important to realize that there could still be a nonzero distribution of swirl velocity in this duct. This would be the outcome of an upstream compressor stage or a ducted fan.

One of the inlet-duct flow-governing equations is that of mass conservation:

$$\dot{m} = \rho_1 V_{z1} A_1 \tag{11.1}$$

Depending on which is more convenient, the same equation can be rewritten as follows:

$$\frac{\dot{m}\sqrt{T_{t1}}}{p_{t1}(A_1 \cos\alpha_1)} = \sqrt{\frac{2\gamma}{(\gamma+1)R}} M_{cr1}\left(1 - \frac{\gamma-1}{\gamma+1} {M_{cr1}}^2\right)^{\frac{1}{\gamma-1}} \tag{11.2}$$

where

ρ_1 is the duct-exit static density,
V_{z1} is the duct-exit axial-velocity component,
A_1 is the duct-exit annular area $= \pi({r_{o1}}^2 - {r_{i1}}^2)$,
T_{t1} is the duct-exit total temperature,
p_{t1} is the duct-exit total pressure,
α_1 is the duct-exit average swirl angle,
R is the gas constant, and
M_{cr1} is the duct-exit critical Mach number.

Inlet-Duct Total Pressure Loss

A vaneless inlet duct will typically give rise to a small total pressure loss, unless a significant magnitude of swirl angle exists at the duct inlet station. Figure 11.5 provides a simple means of determining the total-pressure loss coefficient $\bar{\omega}$, where

$$\bar{\omega} = \frac{\Delta p_t}{p_{t\,in}} \tag{11.3}$$

with $p_{t\,in}$ referring to the duct-inlet total pressure. As indicated in Figure 11.5, the loss coefficient embraces the effects of both the duct length and the inlet swirl angle as independent variables, with L and D being the duct length and average diameter, respectively. For example, an inlet duct that is 7 cm long, has an average diameter

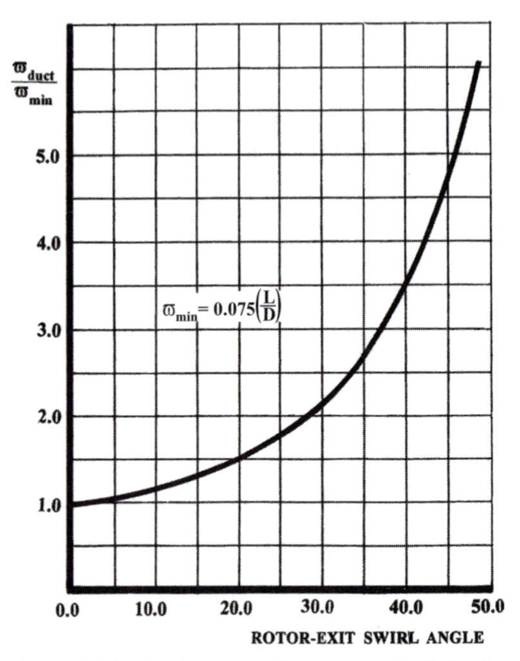

Figure 11.5. Total pressure loss across the inlet duct.

of 20.4 cm, and is operating under an inlet swirl angle of 20° would produce a total-pressure loss coefficient $\bar{\omega}$ that is calculable, with the aid of Figure 11.5, as follows:

$$\bar{\omega}_{min} = 0.075 \times \left(\frac{0.07}{0.204}\right) = 0.026 \tag{11.4}$$

Direct utilization of the chart in Figure 11.5 yields

$$\frac{\bar{\omega}}{\bar{\omega}_{min}} = 1.50 \tag{11.5}$$

with which the total-pressure loss coefficient $\bar{\omega}$ is calculated as

$$\bar{\omega} = 0.039 \text{ (or } 3.9\%).$$

Compressor Thermodynamics

Figure 11.6 shows the stage flow process from a thermodynamic standpoint. Let us first consider the case where the exit-diffuser flow process is isentropic. As seen in Figure 11.6, this will lead to the conservation of both total temperature and pressure, resulting in a single point on the *T*-*s* diagram representing the entire diffuser flow process. In reality, however, the diffuser flow process can be nearly adiabatic, in which case only the total temperature remains constant (Fig. 11.3). The last *T*-*s* representation in Figure 11.6 offers a different way of viewing the rotor (or impeller) flow process. In this case, the process is viewed on a total relative-to-total relative

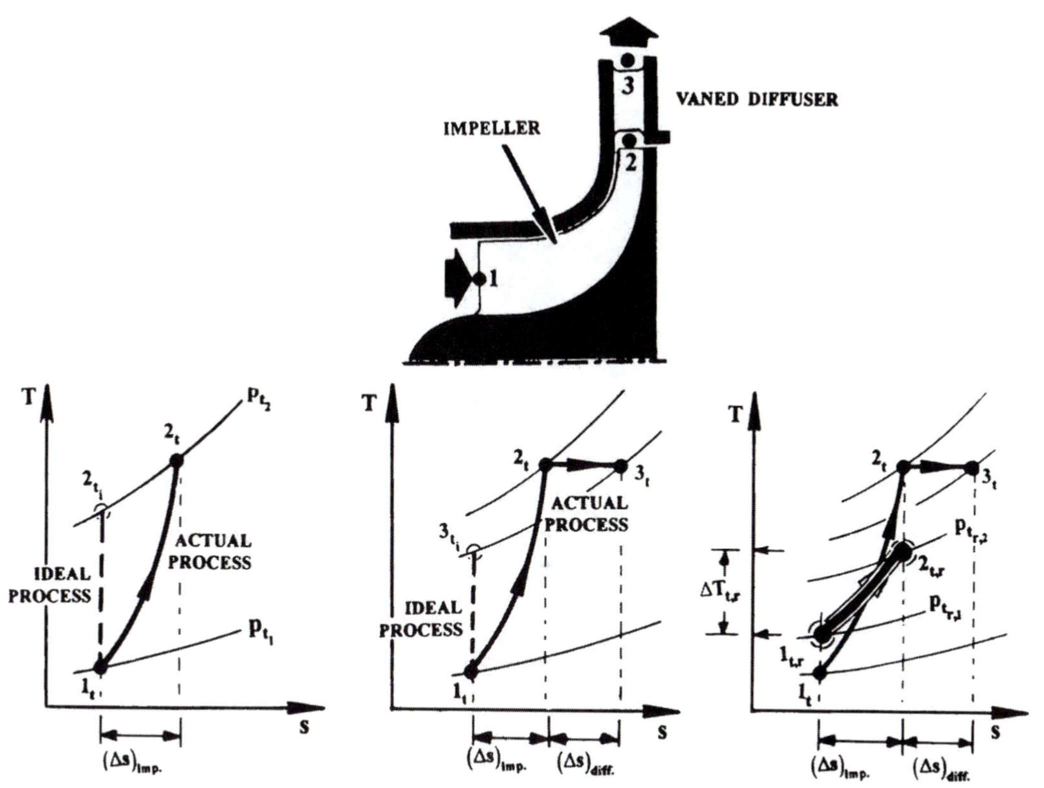

Figure 11.6. The impeller flow process on a total-to-total and total-relative-to-total-relative bases.

basis. As is always the case with radial or centrifugal turbomachines, the total relative temperature and pressure will both change across the rotor. As explained in Chapter 10, for its turbine-rotor counterpart, this particular feature is a consequence of the streamwise radius change across the rotor. The difference, in centrifugal compressors, is that both of these total properties will increase across the impeller.

Impeller Blading Options

Figure 11.7 shows three impeller design configurations, each with its own exducer sign of the blade exit angle. The assumption in this figure is full guidedness of the relative flow by the impeller blades. Of the different impeller configurations in Figure 11.7, the first features blades which are curved opposite to the direction of rotation. The middle configuration in the figure is one where the exducer is composed of purely radial blade segments. The last of these configurations features impeller blades which are curved in the direction of rotation. Termed *backward-curved*, *radial*, and *forward-curved* blading types, respectively, the advantages and drawbacks of each blading option are instrumental. Included here are both the impeller performance and the dynamic integrity.

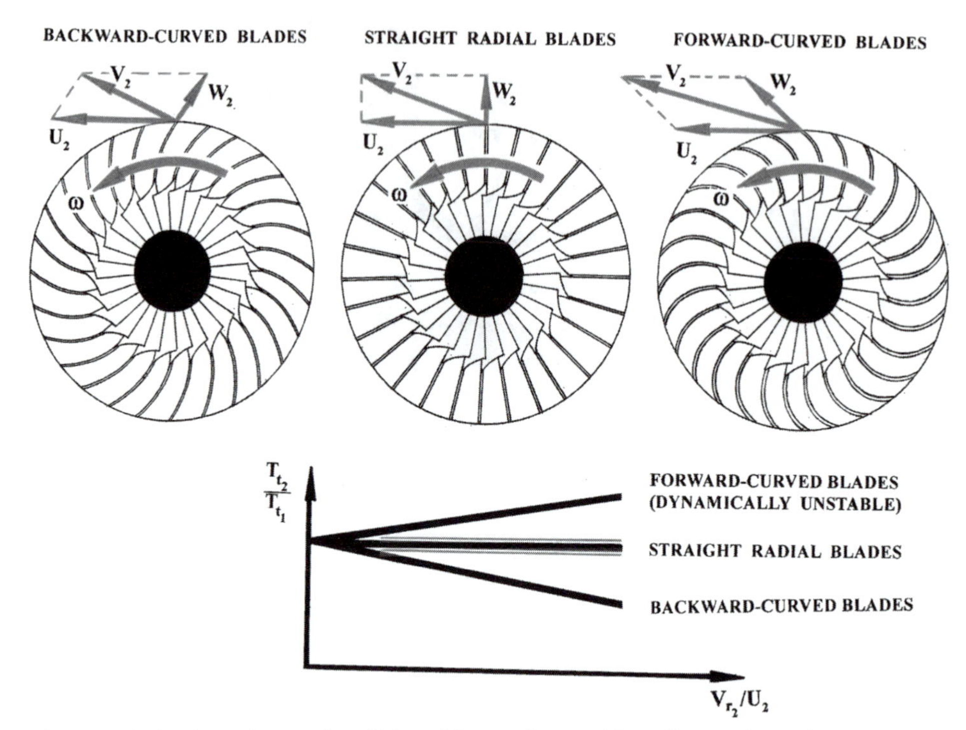

Figure 11.7. Backward-curved, radial, and forward-curved impeller configurations.

The graph in Figure 11.7, together with Figure 11.8, addresses the all-important question of which of the impeller blading options is operationally stable. The stability concern here is, by all means, justified, considering what may otherwise be a catastrophic mechanical failure. As explained in Chapter 5, this would be the potential consequence of entering the surge mode of compressor operation, as shown in Fig. 11.8. The key issue here, by reference to this figure, is the slope of the dotted (constant N_C) lines to the left of the surge line, which is clearly positive. Worded differently, it is a matter of how, in terms of pressure-ratio change, the impeller will respond to an increase in the mass-flow rate. Instability in this sense would prevail once a positive increase of pressure ratio becomes the impeller response. In the following, a comparison between the three impeller blading configurations will be conducted on the basis of this very statement as a stability criterion.

For simplicity, let us make the following assumptions:

1) The impeller flow process is adiabatic.
2) The inlet total properties, p_{t1} and T_{t1}, are fixed.
3) The inlet flow possesses no swirl-velocity component ($V_{\theta 1} = 0$).
4) Full guidedness of the relative flow stream within the impeller.

Now, let us begin with Euler's energy-transfer equation:

$$\Delta h_t = c_p(T_{t2} - T_{t1}) = U_2 V_{\theta 2} \tag{11.6}$$

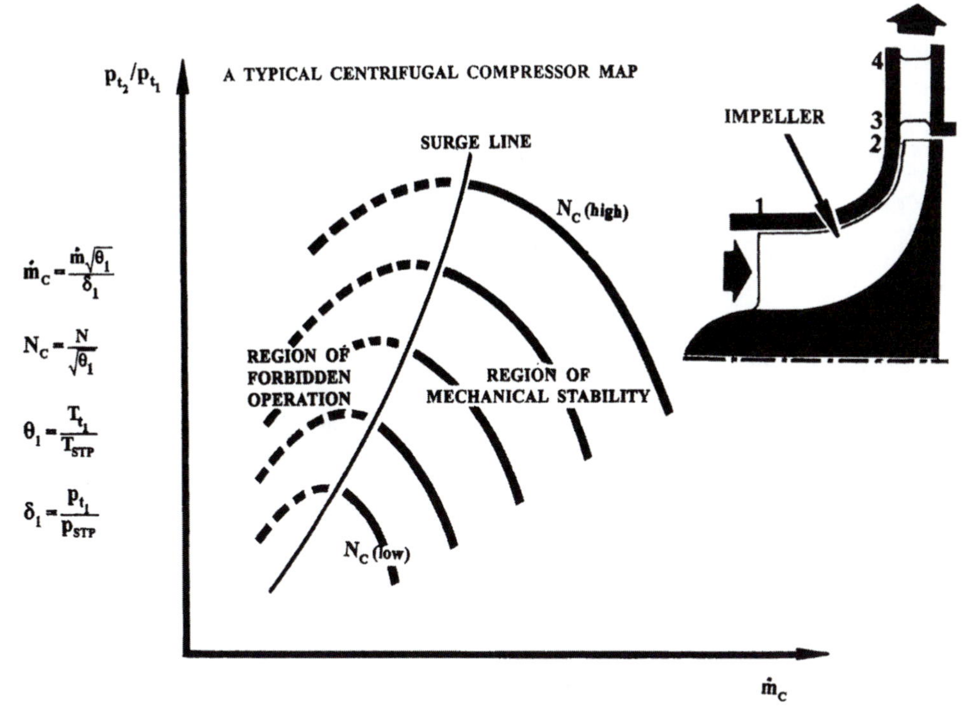

Figure 11.8. Impeller stability prevails in the domain where the pressure ratio and flow rate are inversely proportional to each other.

which can be rewritten as

$$\frac{T_{t2}}{T_{t1}} = 1 + \frac{U_2 V_{\theta 2}}{c_p T_{t1}} \tag{11.7}$$

where

$$c_p = \left(\frac{\gamma}{\gamma - 1}\right) R \tag{11.8}$$

which gives rise to the expression

$$\frac{T_{t2}}{T_{t1}} = 1 + (\gamma - 1)\left(\frac{U_2}{a_{t1}}\right)^2 \left(\frac{V_{\theta 2}}{U_2}\right) \tag{11.9}$$

where

$$a_{t1} = \sqrt{\gamma R T_{t1}} \tag{11.10}$$

and

$$\frac{V_{\theta 2}}{U_2} = 1 + \frac{W_{\theta 2}}{U_2} \tag{11.11}$$

Substituting (11.10) and (11.11) into (11.9), we get

$$\frac{T_{t2}}{T_{t1}} = 1 + (\gamma - 1)\left(\frac{U_2}{a_{t1}}\right)^2 \left[1 + \left(\frac{W_{r2}}{U_2}\right)\tan\beta'_2\right] \tag{11.12}$$

Using the assumptions stated earlier, equation (11.10) can be compacted as

$$\frac{T_{t2}}{T_{t1}} - 1 = K\left[1 + \left(\frac{W_{r2}}{U_2}\right)\right]\tan\beta'_2 \tag{11.13}$$

where

$$K = (\gamma - 1)\left(\frac{U_2}{a_{t1}}\right)^2 \tan\beta'_2 \tag{11.14}$$

Referring to equation (11.13), the two terms T_{t2}/T_{t1} and W_{r2}/U_2 are interpreted as follows:

The term (T_{t2}/T_{t1}): This term is indicative of the stage-supplied shaft work. Subject to the stage efficiency, the same term is proportional to the total-to-total pressure ratio and will be viewed as such.

The term (W_{r2}/U_2): With a constant denominator, and in reference to Figure 11.4, this term is representative of the impeller-exit volumetric-flow rate, for it is implied that the impeller is one of fixed geometry. Subject to the magnitude of static density at the same location, the same term is indicative of the mass-flow rate.

In viewing the preceding two terms as dependent and independent variables, respectively, equation (11.13) is seen to represent a straight line (Fig. 11.7). With these interpretations, and in light of the typical compressor map in Figure 11.8, equation (11.13) can now be used as a dynamic-stability litmus test. The rationale here is that the dotted (positive slope) line segments on the compressor map in Figure 11.8 result from a purely theoretical aerodynamic analysis. In reality, the compressor operation represented by these segments cannot be dynamically sustained. The fact is that an increase in the high frequency of mechanically dangerous events, as a result of elevating the pressure ratio, would cause a catastrophic mechanical failure. Essentially stalled blade-to-blade channels in this case would cause a dangerously close sequence of flow reversal and redirection with even the slightest change in pressure ratio. This would give rise to a strong sequence of fluid-induced vibrations, which the impeller cannot mechanically endure. To summarize, the compressor operating mode where the mass-flow rate is directly proportional to the pressure ratio can be dangerously unstable.

Let us begin with the backward-curved impeller configuration in Figure 11.7, where β'_2 is negative (opposite to the direction of rotation). As a result, the corresponding straight line in Figure 11.7 will have a negative slope. This classifies the backward-curved impeller configuration as dynamically stable and therefore desirable. Applying the same criterion to the forward-curved impeller category, it is clear,

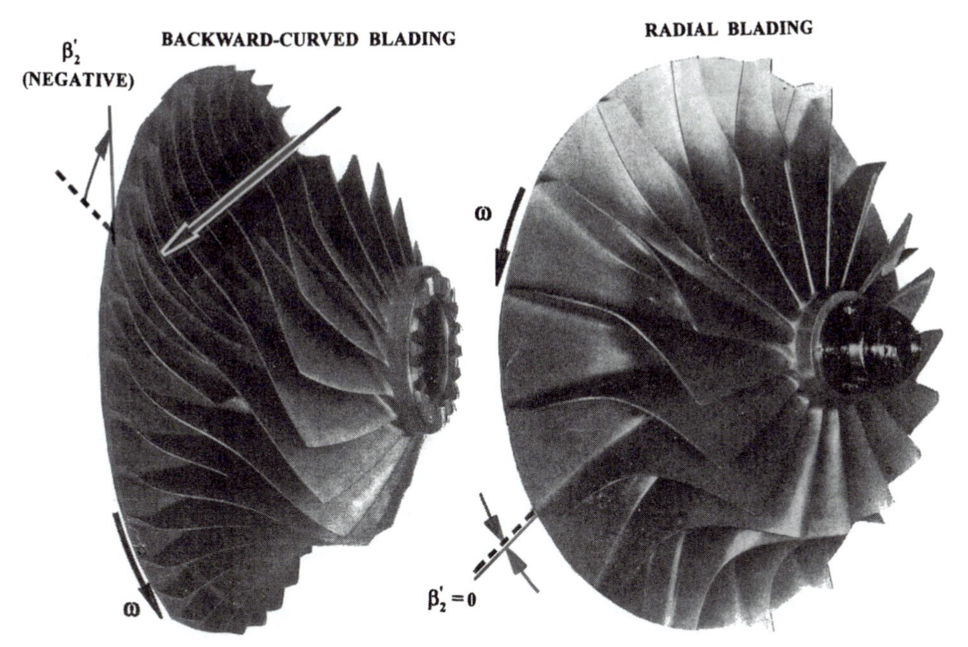

Figure 11.9. Examples of backward-swept and radial blading configurations. Common blading configurations of centrifugal-impeller exducers.

through the same reasoning process, that such an impeller geometry is inherently unstable from a dynamic standpoint. With these two extremes, the radial impeller blading seems to be, and indeed is, a compromise between the two. This and the backward-curved blading configurations are both shown in Figure 11.9. An isometric view of the latter, together with the inlet and exit velocity diagrams, is shown in Fig. 11.10.

Components of Energy Transfer and Stage Reaction

Following a logic that is similar to that in Chapter 10 (for a radial-turbine stage), the shaft work that is consumed by a centrifugal compressor stage can be expressed as

$$w_s = \frac{(W_1{}^2 - W_2{}^2)}{2} + \frac{(U_2{}^2 - U_1{}^2)}{2} + \frac{(V_2{}^2 - V_1{}^2)}{2} \tag{11.15}$$

In expression (11.15), the first two terms constitute an impeller static head and the remaining term is, by definition, the dynamic head. With this in mind, the stage reaction (R) can be expressed as follows:

$$R = \frac{(W_1{}^2 - W_2{}^2) + (U_2{}^2 - U_1{}^2)}{(W_1{}^2 - W_2{}^2) + (U_2{}^2 - U_1{}^2) + (V_2{}^2 - V_1{}^2)} \tag{11.16}$$

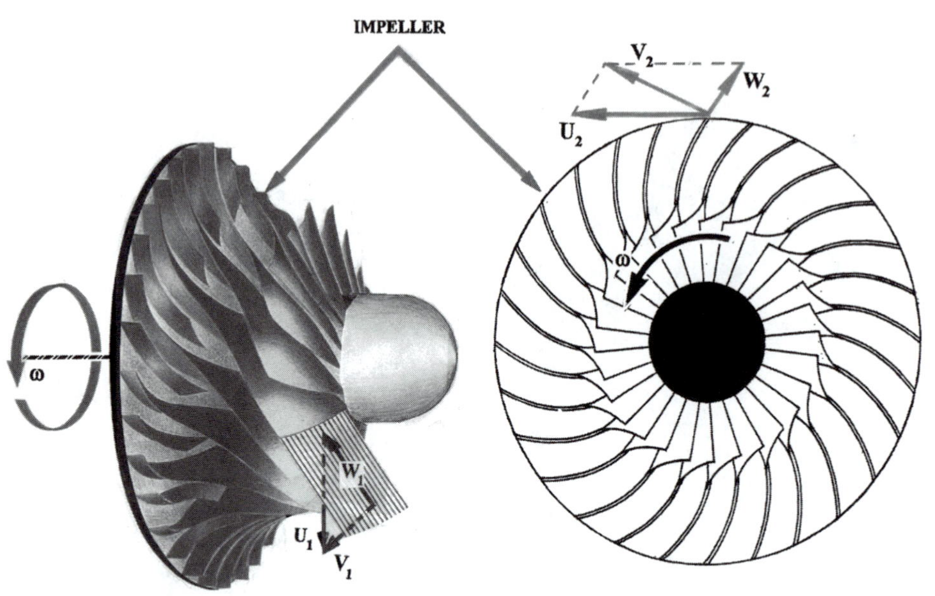

Figure 11.10. Velocity diagrams associated with an impeller with backward-curved blades.

Performance Consequences of the Static Head

Corresponding to the impeller static-enthalpy rise, there will always be a proportional increase in the static pressure that, in part, is caused by the streamwise increase in radius. Although this is an important component of the compressor shaft-work absorption capacity, the static-pressure gradient here is, by definition, unfavorable. In the following, two aerodynamically harmful effects of the positive pressure differential across the impeller are discussed.

1. *Boundary-layer buildup*: The impeller-wise unfavorable static-pressure gradient will dictate substantial boundary-layer growth both over the blade surfaces and the hub and casing endwalls. The worst possible consequence here is a premature boundary-layer separation, which would give rise to a wasteful recirculatory motion (Fig. 3.1). As a result, the impeller-exit relative-flow stream in this case will be significantly underturned. Extending downstream, the impeller-exit flow deviation angle will induce an excessively negative incidence angle as the vaned diffuser is approached. While it would be desirable under such conditions to increase the blade count for better flow guidedness, the added blades themselves would increase the flow-exposed surfaces and therefore the skin friction.

2. *Direct tip leakage*: In discussing the topic of axial-turbine flow leakage (Chapter 8), two distinct tip-leakage mechanisms were identified. One of these was the "indirect" leakage, in the form of a pressure-to-suction side flow migration, which will always plague unshrouded rotors. Under focus here is the other, namely the direct, tip-leakage mechanism. This is a result of the streamwise static-pressure differential in the secondary-stream passage. Consequences of this leakage category are

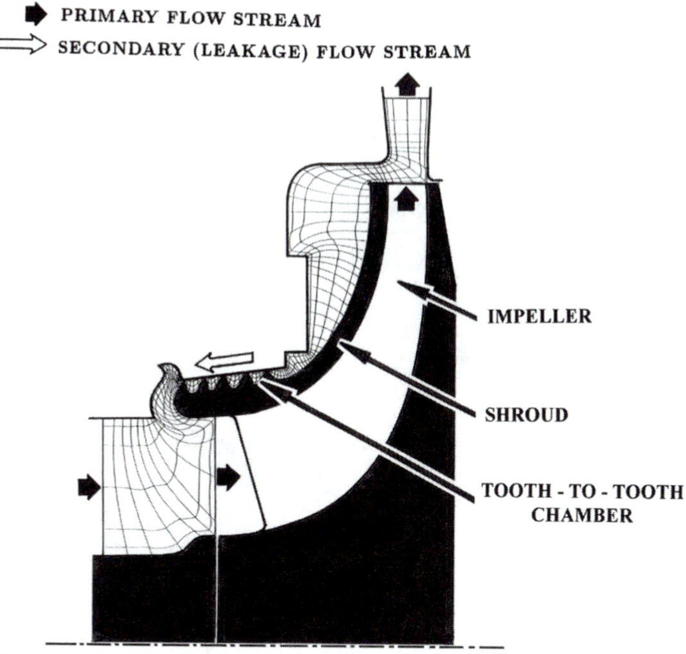

Figure 11.11. Labyrinth seal as a leakage-control device.

comparatively less pronounced in turbine rotors, as the sign of the static-pressure differential gives rise to a leakage-flow direction that is the same as that of the primary flow. In compressors, however, the leakage flow proceeds in the direction which is opposite to that of the primary flow, meaning a flow reversal within the tip-clearance gap. In terms of the leakage/primary ratio of mass-flow rates, the problem is even worse in centrifugal compressors. The basic reason here is that the high total-to-total pressure ratio across the impeller is in part composed of a high static pressure rise across the same. To the secondary (or leakage) flow, the same static-pressure differential acts as a strong driving force to the leakage flow stream. Moreover, the streamwise radius variation within the leakage passage provides a highly favorable pressure gradient across this secondary-flow passage.

In an attempt to control the centrifugal-compressor direct leakage, for shrouded impellers, shroud-mounted seals of different configurations are typically inserted. Of these, the so-called labyrinth seals are the most popular (Fig. 11.11). In the following, however, the effectiveness of simpler seals is discussed in light of numerical results reported by Baskharone and Hensel (1989, 1993) and Baskharone and Wyman (1999), where the rotordynamic consequences of the leakage-passage geometry were also investigated.

At its inception, the above-referenced research program was focused on improving the rotordynamic characteristics of the booster impeller being one of the Space Shuttle Main Engine turbopumps. This task entailed a thorough examination of the shroud-to-casing leakage-flow structure over this impeller. Given the unstable

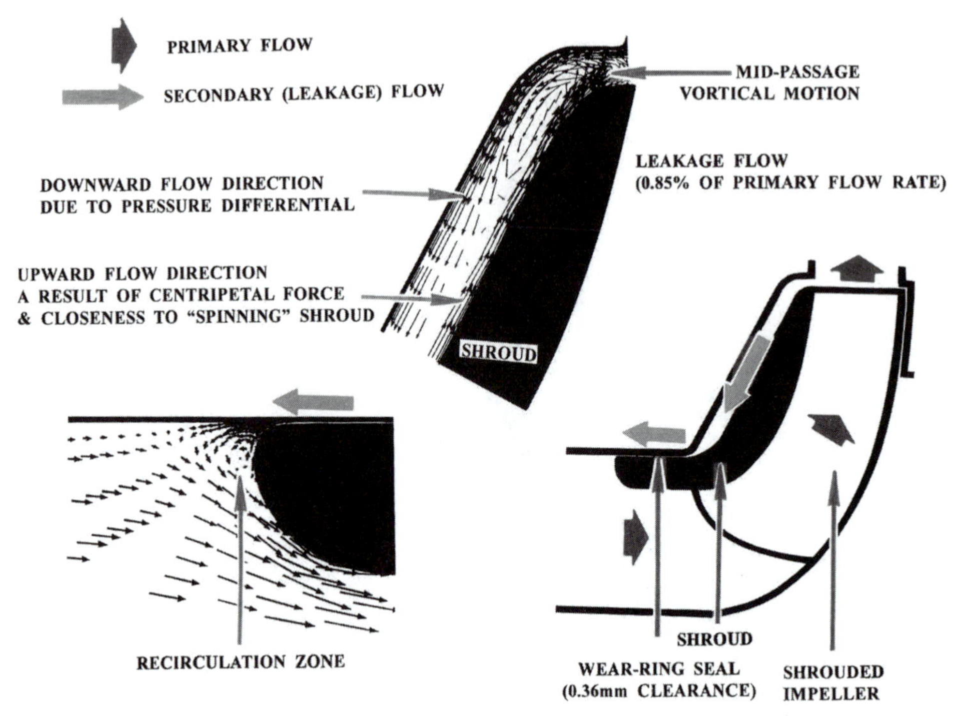

Figure 11.12. Leakage-flow consequences with a wear-ring seal in the secondary passage.

operation of the impeller, early on, the problem was actually centered around the fluid-induced vibrations caused by specific features of the secondary flow in the leakage passage. The fluid forces in this passage caused the impeller to spin off-center in what is referred to as a "whirling" motion. In executing the major task of this study, several leakage-control devices were explored and the leakage-flow behavior examined.

Figure 11.12 shows the velocity-vector plot that is associated with the so-called wear-ring seal in a similarly configured centrifugal pump. Because the leakage flow direction here is opposite to that of the primary flow, the location where these flow streams mix is characterized by a subregion of flow vortical motion, as shown in the Figure. Such energy-wasting motion has the added effect of distorting the primary-flow stream as the impeller subdomain is approached.

Common in Figures 11.12 and 11.13 is the flow recirculation in the leakage-flow passage segment leading to the seal subregion. The flow recirculation here is produced by two opposing flow tendencies near the shroud and housing surfaces. First, the near-shroud flow direction will be in the outward direction as a result of the centrifugal force that is a consequence of the shroud spinning motion. On the other hand, the near-housing flow motion is inward, for it is the result of the leakage-passage static-pressure differential. As would be expected, the outcome of these two opposite secondary-flow motions is the vortical flow field in this segment of the leakage passage. Better resolution of this same flow field revealed the fact that it is a

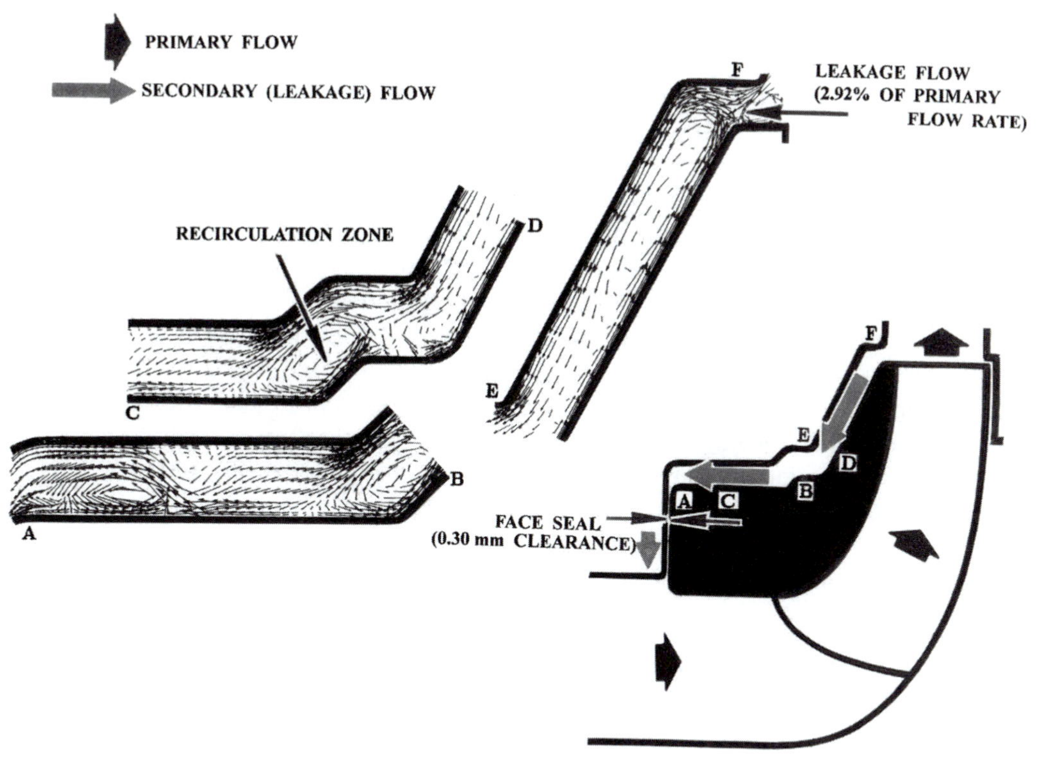

Figure 11.13. Leakage-flow behavior under a face-seal configuration of leakage discourager.

sequence of vortices that exists along this particular leakage passage segment. This observation was found to be equally valid under the two different seal configurations in Figures 11.12 and 11.13. As indicated in these two figures, the first (wear-ring) seal type gives rise to a small tip leakage of 0.85% of the primary mass-flow rate. This ratio is as high as 2.92% in the case of an even tighter face-seal configuration, as shown in Fig. 11.13.

Performance Consequences of the Dynamic Head

Among the impeller-blading configurations discussed earlier, the forward-curved blade option, in Fig. 11.7, is an example of an excessively high dynamic head. Referring to the continuous flow deceleration across the vaneless and vaned-diffuser segments (Fig. 11.3), it may appear that the choking status, if achieved, would take place in the immediate vicinity of the vaneless-diffuser inlet station. However, this impression is not necessarily true, for it is indeed possible to attain this status at the vaned-diffuser inlet station. This is due to the blockage posed by the vane leading edge thickness, which gives rise to a sudden cross-flow area contraction, that leads to an abrupt increase in the local Mach number and makes it, at least theoretically, possible to reach the choking status there.

In reference to the vaned-diffuser flow nature, the design of this stationary component can be a challenging task. On one hand, and in the interest of leading-edge tolerance to different incidence-angle magnitudes, the vane leading edge will have to be particularly thick. Depending on other leading-edge geometrical features, as well as the mean camber-line shape, a substantially thick leading edge can itself cause an early flow separation, which would call for rigorous limitations in the vane design process. For the dynamic-to-static pressure recovery process to be efficient, the streamwise cross-flow area increase must be gradual. This may not impose any significant restrictions on the vane-to-vane passage geometry should the sidewalls be used as a supplemental means of ensuring a smooth distribution of the cross-flow area.

Complexity of the vaned-diffuser flow field and design process are functions of the impeller-exit dynamic head. This is due to the fact that the dynamic-to-static pressure conversion will largely take place across this diffuser. In the interest of fairness between the impeller and diffuser aerodynamic loadings, the impeller and diffuser static-pressure differentials should be close to one another. Ideally, these two static-pressure contributions should be identical.

Acceleration Components within the Impeller

Across the impeller blade-to-blade passage, the flow behavior is heavily influenced by many acceleration components. These give rise to a set of corresponding force components, which stem from:

1) The streamwise static pressure gradient
2) The inertia effect along the relative streamlines
3) The near-wall viscosity-related effects
4) The centrifugal effect as a function of the shaft speed and local radius

In the above it is important to realize that the term "velocity" refers to the relative velocity vector W. It is equally important to comprehend the reason for singling out this velocity vector, as we just did. In matters related to simple one-dimensional bulk-flow kinematics, the adoption of either the absolute or relative velocity vector is usually convenience-related. However, the topic at hand involves detailed fluid/structure interaction mechanisms (e.g., boundary layer growth), making it dependent on the fluid/structure relative motion.

Within the impeller subdomain, another fluid-imposed effect is due to the Coriolis acceleration component. In general terms, this acceleration will present itself once the passage of flowing medium is physically part of a larger rotating domain. Perhaps the most obvious example of such a situation is the flow in a blood vessel as the human body moves in a nonstraight path. With such a path, the body is viewed as instantaneously rotating around the local center of curvature at a finite rotational speed. Under such circumstances, the blood-vessel flow field is part of a larger rotating domain, namely the rotating human body. In this case, the physically meaningful blood velocity is that relative to the body itself.

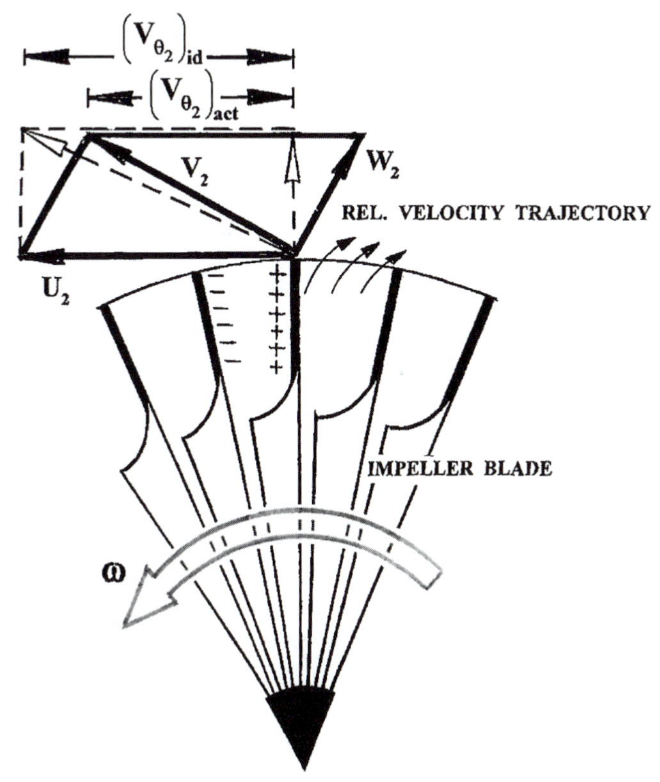

Figure 11.14. The slip phenomenon: Deterioration of Coriolis acceleration as the impeller tip is approached.

At any point within the impeller blade-to-blade passage (Fig. 11.14), the Coriolis-acceleration vector (a_C) is defined as follow:

$$a_C = 2(\boldsymbol{\omega} \times \boldsymbol{W}) \tag{11.17}$$

Expression (11.17) defines a vector that is perpendicular to ω (i.e., parallel to the axial direction) while also being perpendicular to the relative velocity vector. The right-hand side of expression (11.17) can be resolved in the r and θ directions as follows:

$$a_C = -2\omega(W_\theta \mathbf{e}_r - W_r \mathbf{e}_\theta) \tag{11.18}$$

Being perpendicular to the relative-velocity vector ($\boldsymbol{W}$), the Coriolis acceleration will produce a cross-flow blade-to-blade static-pressure gradient. Moreover, the negative sign in expression (11.18) indicates an increasing pressure opposite to the direction of rotation (i.e., toward the blade pressure side; Figure 11.14).

Slip Phenomenon

This phenomenon refers to the flow migration over the blade tip from the blade pressure to the suction side. It also refers to the flow gradual deflection, in the same direction, elsewhere in the impeller subdomain as the blade tip is approached. The blade-to-blade secondary flow that is produced by this phenomenon has the degrading effect of unloading the impeller blades.

To understand the slip phenomenon, we go back to the expression of Coriolis acceleration, focusing on the high-radius region close to the blade tip. For simplicity, we will be confined to the case of straight radial blades, as shown in Fig. 11.14.

Near the impeller exit station, the influence of Coriolis acceleration will begin to diminish. This is because of the decline of the relative-velocity magnitude in the characteristically flow-decelerating blade-to-blade passage. In fact, the Coriolis-acceleration effect will almost vanish as the impeller exit station is approached, depending on the shaft speed. As a result, a typical midpassage fluid particle will be unable to proceed in a purely radial direction, as the blade exit segment would imply. The fluid particle will therefore "slip" back in a direction that is opposite to the direction of rotation. In the case of a small blade count, the impeller-exit relative streamlines will have the pattern shown in Fig. 11.14. As a result, the fluid will have the tendency to migrate (or slip) over the blade tip. As indicated earlier, this aerodynamically damaging motion will not only affect the tip location but will also help unload the near-exit blade segment.

Worth noting here is the effect of flow guidedness, in terms of the impeller blade count, in the impeller exducer (or exit) region. Assuming a controllable friction over the blade surfaces, an increased number of blades will help alleviate a large-scale slip mechanism. In fact, most of the carefully designed centrifugal impellers will have incomplete blades near the impeller exit station, as shown in Fig. 11.9. These are aimed at providing better guidance in the near-tip region, where the complete blades gradually become farther from one another.

Slip Factor

Symbolized as σ_s, the slip factor is defined, by reference to Fig. 11.14, as

$$\sigma_s = \frac{(V_{\theta 2})_{act.}}{(V_{\theta 2})_{id.}} \tag{11.19}$$

where $(V_{\theta 2})_{id}$ is the ideal impeller-exit swirl-velocity magnitude that is associated with perfect flow guidedness. For straight radial blading (Fig. 11.14), and for an impeller blade count that is in excess of 10, the following empirical expression provides an acceptable approximation of the slip factor:

$$\sigma_s = 1 - \frac{2}{N_b} \tag{11.20}$$

where N_b is the number of complete and incomplete impeller blades. As for the backward-curved blade configuration (Fig. 11.6), the slip-factor expression is

$$\sigma_s = \left[1 - \frac{2}{N_b}\sqrt{\cos\beta_2{}'}\right]\left[1 + \frac{W_{r2}}{U_2}\tan\beta_2{}'\right] \tag{11.21}$$

where $\beta_2{}'$ is the blade angle, measured from the local radial direction, at the impeller exit station. Note that $\beta_2{}'$ is negative for backward-curved impeller blades.

Stage Total-to-Total Efficiency

Being a power-absorbing turbomachine, the total-to-total efficiency (η_{t-t}) is defined as the ideal-to-actual shaft-work ratio. In this statement, the former (ideal) shaft work is the minimum magnitude that would lead to the same total-to-total pressure ratio the actual process does. Assuming an adiabatic flow and a fixed specific-heat ratio (γ), the total-to-total efficiency can be expressed, in light of Figure 11.6, as

$$\eta_{t-t} = \frac{\left(\frac{p_{t3}}{p_{t1}}\right)^{\frac{\gamma-1}{\gamma}} - 1}{\left(\frac{T_{t3}}{T_{t1}}\right) - 1} \tag{11.22}$$

where stations 1 and 3 are the impeller inlet and diffuser exit stations, respectively (Fig. 11.4). By focusing on these two stations, we are implicitly assuming that the volute is nonexistent or its effect negligible, which is not necessarily the case.

Volute Flow Field

An alternate and more suiting name for this component is "collector." This is in reference to the component major function, as it "collects" the impeller-outlet flow segments around the circumference. Figure 11.15 illustrates this function, showing one of the typical shapes of the volute cross section. Referring to the same figure, note that the volute exit segment is that of the flow-decelerating type. This is effected in the hope of converting more of the remaining kinetic energy into a static pressure rise.

Figure 11.16 highlights the difference between the core flow and the near-wall viscosity-dominated flow. As a starting point, we realize that any vortical motion of the fluid particles is a waste of kinetic energy and, therefore, the total pressure. However, this is as good a point as any to introduce another loss-related variable, namely the vorticity. Unlike the other loss-contributing variables, the vorticity presents itself in the form of a specific flow motion that is unmistakable.

Despite being a vector, the focus on vorticity in this section is hardly a pure exercise in vector algebra. Although the flow field in virtually any turbomachinery

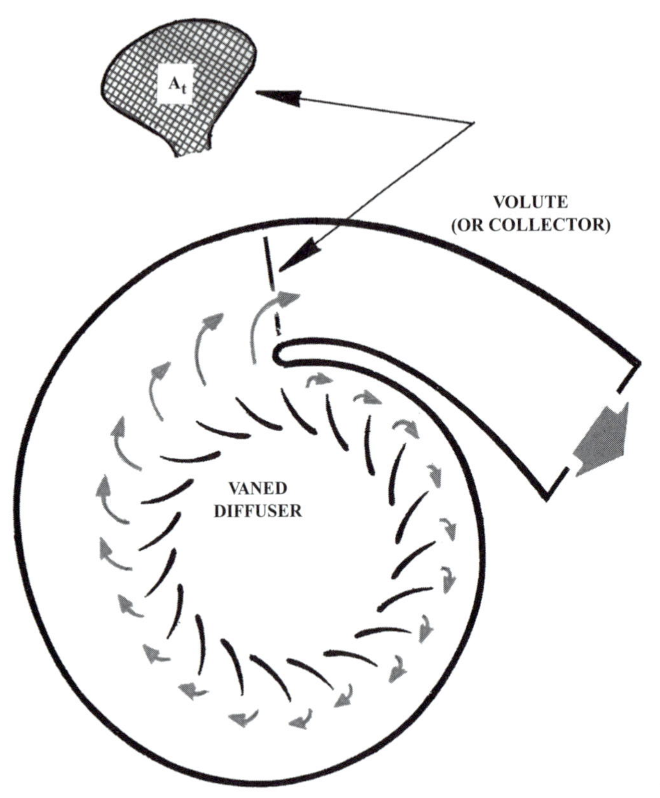

Figure 11.15. Role of the volute as a mass-flux collector around the circumference.

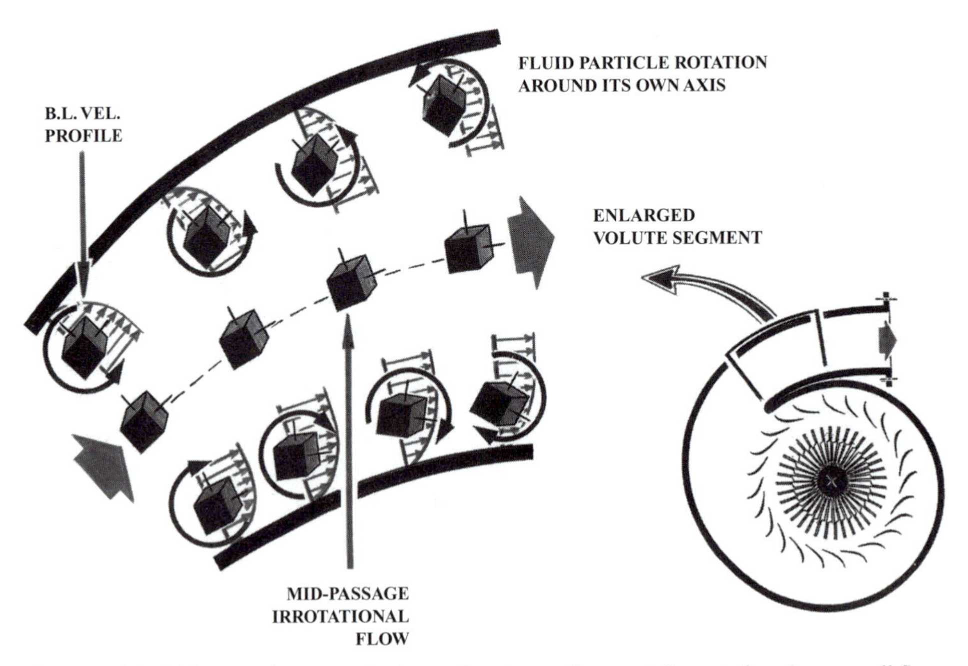

Figure 11.16. Difference between the irrotational core flow and the rotational near-wall flow.

component is three-dimensional, the approach in this book has been to break it up into one or more in-plane flow-field subproblems. In this case, the vorticity-vector direction will always be perpendicular to the stream surface and is therefore immaterial from a computational standpoint.

The vorticity magnitude (Ω) is defined as

$$\Omega = |\nabla \times \mathbf{V}| \tag{11.23}$$

In order to understand Ω components note that the operator "∇" will produce *all* of the r, θ and z derivatives of *all* of the r, θ and z components of velocity. As a result, we can safely say that *any* gradient of *any* velocity component will contribute to the vorticity magnitude.

In physical terms, the vorticity (Ω) is a measure of the fluid particle tendency to spin around its *own* axis, as in the near-wall subregion shown in Figure 11.16. In the following, a volute segment (highlighted in Figure 11.16) will be taken as an example, where the flow field is separated into a middle-of-the-duct "core" flow subdomain and a near-wall viscosity-dominated zone. As is traditional in gas dynamics, we will proceed with the assumption that the core flow is "vorticity-clean," but with a reservation. The fact of the matter is that the core flow could enter the flow domain possessing a great deal of vorticity caused by an upstream component or a complex-flow curved passage, as is the case here. The flow in this case is normally referred to as *rotational*, a term that is used to identify an inertia-dominated flow but with a significant vortical motion that is produced upstream from the subdomain of interest. An example of how an upstream component may export vorticity to the flow domain at hand is shown in Figure 11.17. The figure shows a vaned diffuser which, at inlet, is receiving the wakes of the upstream-impeller blades. These wakes are free-shear layers that, just like boundary layers, carry a great deal of vorticity into the diffuser core flow, and possibly the volute flow stream as well. Once admitted to the downstream component, vorticity, just like turbulence, is practically impossible to get rid of.

Within the flow domain of concern, being the volute segment in Fig. 11.16, the overwhelming vorticity generators are the bounding solid walls. As is seen in the figure, the viscosity-dominated boundary-layer zone grows in the streamwise direction. Over the solid wall itself, the so-called "no-slip" condition applies, which means that the fluid particle will adhere to the wall, possessing the same velocity (zero in this case) as the solid wall itself. The transition in this case from a zero magnitude to the large core-flow velocity occurs over an extremely thin boundary-layer thickness. As a result, extremely high velocity gradients will exist, in the lateral direction, and over such a thin zone, with the maximum gradients being along the solid wall itself.

Superimposed on Figure 11.16 is a sequence of fluid particles, all existing within the boundary-layer zone. As they proceed forward, these particles are illustrated to be spinning around their own axes. A simple way to explain this loss-causing motion is to examine the local velocity distribution where the particle physically exists.

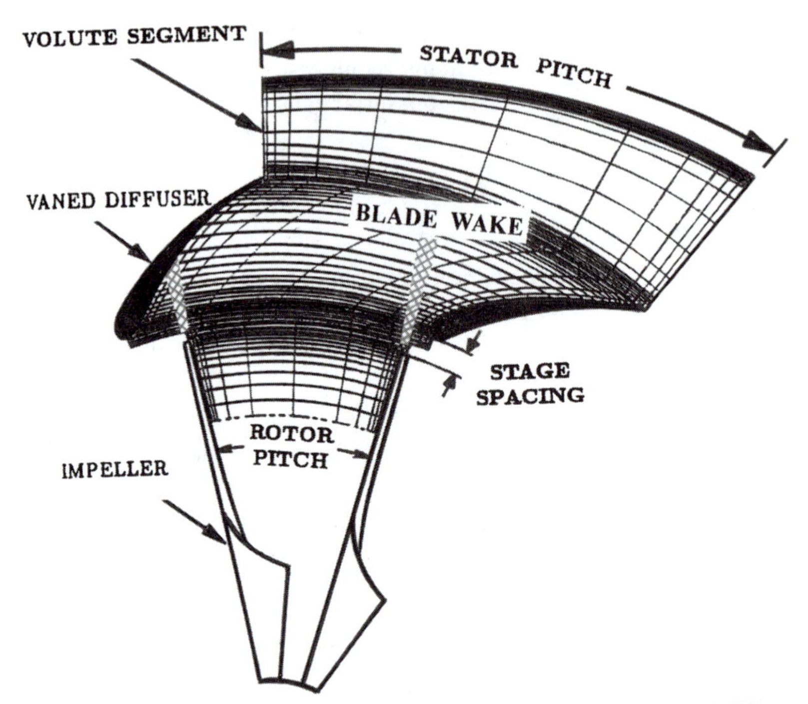

Figure 11.17. Wakes of the impeller blades create an impeller/vaned-diffuser unsteady flow interaction between the two subdomains.

Obviously, the particle face closer to the wall is associated with a through-velocity magnitude that is less than that of the opposite face. Depending on the local velocity gradient, the fluid particle will respond to this situation by spinning around its own axis. This energy-wasting motion is therefore viscosity-generated, and the fluid-particle rotation is the result of shear forces within the boundary layer.

A Common Mistake

Referring to the midpassage "master" streamline in Fig. 11.16, the fluid-particle motion can be viewed as one of spinning. This is obvious by the nonstraight flow trajectory there. Some may look at a picture such as this and judge the flow behavior in this region to be rotational. Such a conclusion is simply *wrong*.

The fluid particle traversing the mid-passage line, in this case, is *not* rotating around its own axis. The particle in fact is spinning around an external axis. Perpendicular to the paper plane, this external axis is located at the flow-trajectory center of curvature. Neither the existence of this axis, nor the flow-stream spinning motion around it, have anything to do with any vortical motion.

One-Dimensional Approach to Volute Design

The simplest design procedure of the volute passage is that based on a linear cross-flow area distribution around the circumference. Note that the volute inlet station is downstream from at least one flow-decelerating component. Therefore, it is in most cases safe to assume a volute-wise average Mach number that is sufficiently small to facilitate a simple design procedure on an incompressible-flow basis.

Referring to Figure 11.15, the volute tongue segment, where the cross-flow area is A_t, is also where the volute passage ends, by definition. The area A_t can be calculated as follows:

$$A_t = \frac{\dot{m}}{(p_{t2}/RT_{t2})V_t} \tag{11.24}$$

where

$\dot{m}$ is the mass-flow rate,
p_t is the diffuser-exit total pressure minus an estimate of the volute total pressure loss,
T_t is the diffuser-exit total temperature, and
V_t is an average velocity over the tongue plane.

The chosen tongue-plane velocity magnitude V_t can very well be the object of a simple trial-and-error procedure aimed at securing an A_t magnitude that is consistent with a given compressor-envelope constraint, if any. Next, the minimum area (A_{min}) of the volute cross section (immediately underneath the tongue) is selected. This should be consistent with the tongue thickness, which is usually dependent on the manufacturing (or casting) precision and is known a priori. At any angle (θ), the volute cross-sectional area (A_θ) can be calculated using the following linear relationship:

$$A_\theta = A_{min} + \left(\frac{\theta}{2\pi}\right) A_t \tag{11.25}$$

where θ is the angle (in radians) of the plane at hand.

Total-to-Static Efficiency

Maximizing the static pressure recovery, across the compressor diffuser and volute, is one of the most important tasks. This holds true regardless of whether the outlet stream feeds into the combustor or a discharge vessel. In this sense, it is the total-to-static efficiency that is naturally the focus of attention, more so than the total-to-total efficiency. Following a logic that is similar to the derivation of the same in the turbine/exhaust diffuser case, the total-to-static efficiency here can be expressed, by reference to Fig. 11.6, as follows:

$$\eta_{t-s} = \frac{\left(\frac{p_3}{p_{t1}}\right)^{\frac{\gamma-1}{\gamma}} - 1}{\left(\frac{T_{t3}}{T_{t1}}\right) - 1} \tag{11.26}$$

Comparing the contents of (11.26) to those of the total-to-total efficiency, it is clear that

$$\eta_{t-s} < \eta_{t-t} \tag{11.27}$$

This is identical to the conclusion reached in the turbine/exhaust-diffuser counterpart. The efficieny difference in expression 11.27 is indicative of the designer's skill and carefulness.

Tip-Clearance Effect

In unshrouded impellers, both the direct and indirect tip leakage will simultaneously exist in the clearance gap. These two simultaneous motions are conceptually illustrated in Figure 8.28 for an axial-flow turbine rotor. Of these, the direct tip leakage has an obvious efficiency-degradation effect, for it involves a fraction of the mass-flow rate that is unavailable to absorb shaft work. The bigger problem, however, is that this leakage flow stream is in the direction opposite the primary flow stream. Having the potential of further disturbing the near-tip inlet flow structure, this type of leakage will typically produce a larger efficiency decrement in compressors by comparison with turbines. As for the indirect (pressure-to-suction) flow migration, it is always an existing source of shaft-work decline for any unshrouded rotor.

One of the most reliable means of assessing the leakage-related centrifugal-impeller efficiency decrement is due to Mashimo et al. (1979) and is represented in Figure 11.18. Invoking the Reynolds number as a viscosity representative, Mashimo's expression is

$$\eta_L/\eta_{cr} = 1 - (1 - \eta_{cr})\left(\frac{R_{e,cr}}{R_e}\right)^n \tag{11.28}$$

where

η_L is the stage efficiency including tip clearance, effect
η_{cr} is the efficiency corresponding to the critical Reynolds number,
R_e is the Reynolds number $(= 2r_t{}^2\omega/\nu)$,
r_t is the impeller tip radius, and
$R_{e,cr}$ is the critical Mach number $(= 1.15 \times 10^6)$.

Employing polynomial fitting, the charts representing η_{cr} and n, being part of Mashimo's work, were lumped together, and the following two expressions were obtained:

$$\eta_{cr} = 0.7113 - 0.1415\frac{c}{h} - 0.6189\left(\frac{c}{h}\right)^2$$

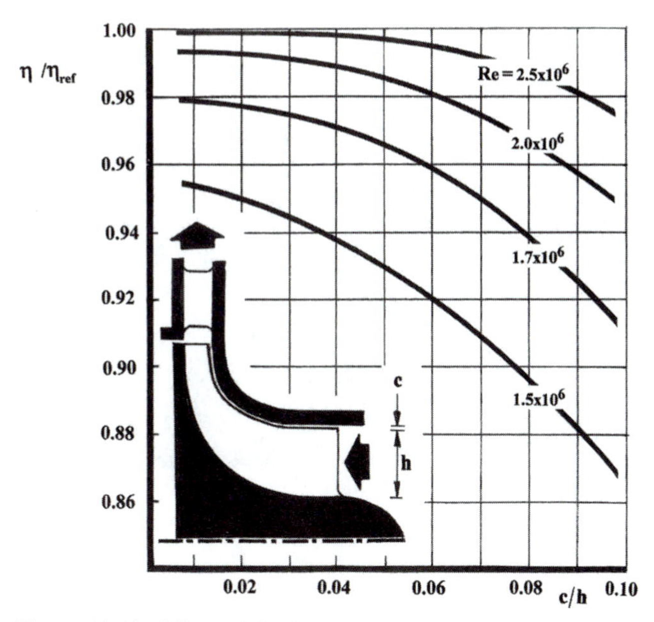

Figure 11.18. Effect of tip clearance on the overall efficiency.

and

$$n = 7.189 - 30.389\frac{c}{h} - 101.583\left(\frac{c}{h}\right)^2$$

where

c is the impeller tip clearance, and
h is the impeller-inlet radial height.

Multiple Staging

A general problem in radial turbomachines has to do with carrying the flow stream from one stage to the next. Figure 11.19 shows a gas turbine engine where the compressor section is composed of two centrifugal stages. Highlighted in the figure is the flow path in the interstage return duct, an annular passage which directs the flow stream to the second stage. Although the flow-redirecting process here is as smooth as the interstage axial gap allows, it is still one where the flow-direction change is basically 180° close to the first-stage outlet. On top of this, the swirl velocity component out of the first-stage impeller is typically high, giving rise to overly elongated streamlines, feared exclusively over the bounding surfaces. When added to the secondary cross-flow stream produced at each of the two 90° bends, the result is an aerodynamically-damaging environment. In other words, one should expect, and account for, a practically unavoidable total pressure loss across this annular duct. In

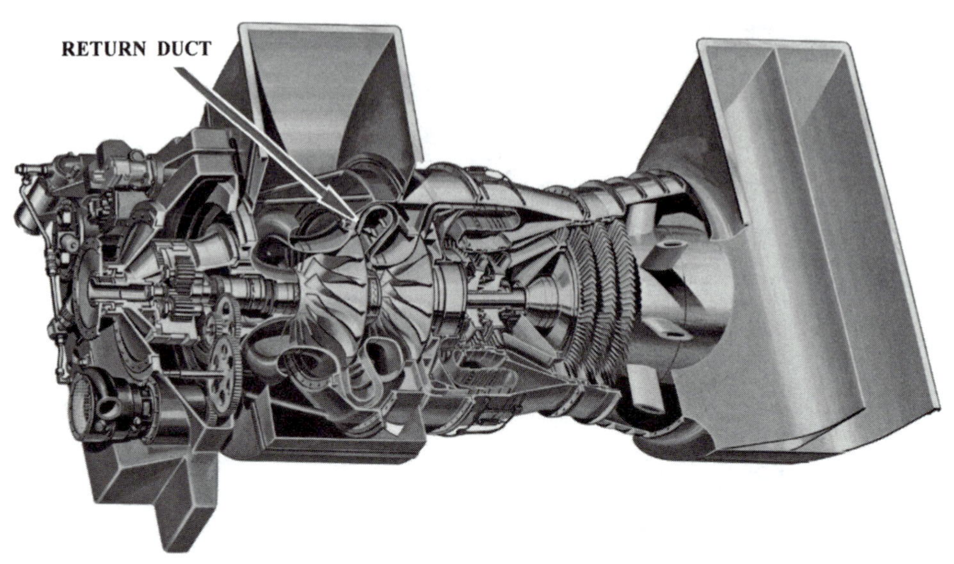

Figure 11.19. Interstage return duct between two centrifugal stages. ME9901-3 Marine gas turbine engine (Garrett Turbine Engine Co.).

most cases, the large power-absorption capacity that is attainable through centrifugal compressor stages, outweighs such an interstage total pressure loss.

Impeller/Stator Unsteady Flow Interaction

As indicated in Chapter 8, the close proximity of two blade cascades, one of which is rotating, creates a cyclic stress pattern (on both sets of blades), with potentially catastrophic consequences. A conceptually similar situation can and does prevail within a centrifugal compressor stage should the latter be equipped with a vaned stator (Fig. 11.17). In this case, the "spinning" wakes of the impeller blades will periodically impinge upon the stator vanes, as shown in the figure. As a result, the stationary vanes could reach a level of fatigue stress which may very well chip off material at and beyond the vane leading edge. Assuming a subsonic flow field, the presence of the downstream stator vanes would very much be felt near and at the blades' trailing edges. This will cause a similar cyclic-stress pattern that is imposed on the impeller blades within the region where the blades are customarily the thinnest.

Ways to reduce, but not alleviate, the impact of the impeller/stator flow interaction include the increase of the impeller/stator radial gap. In addition, it has long been held, in the turbomachinery industry community, that certain combinations of the impeller and stator blade counts would improve the fatigue-stress characteristics in both cascades. Quantifying such a topic would naturally be difficult, for it concerns individual experiences among designers and with specific classes of turbomachines and design settings.

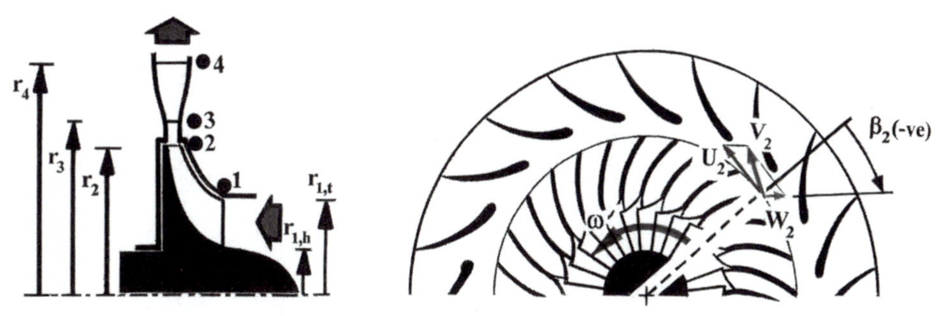

Figure 11.20. Input variables for Example 1.

EXAMPLE 1

The design point of a centrifugal-compressor stage (shown in Fig. 11.20) is defined as follows:

- Inlet total pressure $(p_{t1}) = 1.08$ bars
- Inlet total temperature $(T_{t1}) = 302$ K
- Mass-flow rate $(\dot{m}) = 5.2$ kg/s
- Shaft speed $(N) = 12,000$ rpm
- Inlet absolute velocity is totally axial
- Inlet critical Mach number $= 0.30$
- Impeller static head $(h_2 - h_1) = 84.1$ kJ/kg
- Impeller total-to-total efficiency $(\eta_{t-t}) = 78\%$
- Vaned-diffuser exit flow angle $(\alpha_4) = 0°$
- Vaneless-diffuser flow is assumed isentropic
- Vaned-diffuser total pressure loss $[(p_{t3} - p_{t4})/p_{t3}] = 6.3\%$
- Free-vortex flow structure across the vaneless diffuser
- Average specific-heat ratio $(\gamma) = 1.4$

The following geometrical data are also applicable:

- Impeller-inlet inner radius $(r_{1h}) = 4.3$ cm
- Impeller exit radius $(r_2) = 32.5$ cm
- Vaned-diffuser inlet radius $(r_3) = 37.2$ cm
- Vaned-diffuser exit radius $(r_4) = 41.6$ cm
- Diverging sidewalls across the vaned diffuser
- Vaned-diffuser-exit sidewall spacing $(b_4) = 0.7$ cm
- Backward-curved impeller blades, $\beta'_2 = -21°$
- Fully guided flow across bladed components

Assuming an adiabatic flow throughout the stage, calculate:

a) The impeller-inlet outer radius (r_{1h});
b) The impeller-inlet hub-to-tip twist angle $(\beta'_{1t} - \beta'_{1h})$;

c) The impeller dynamic head ($\Delta h_{dyn.}$);
d) The stage reaction (R);
e) The impeller-exit sidewall spacing (b_2);
f) The impeller-wise rise in total relative pressure ($p_{tr2} - p_{tr1}$);
g) The vaned-diffuser static-pressure rise ($p_4 - p_3$);
h) The stage total-to-static efficiency (η_{t-s})

SOLUTION

Part a: Let us begin by computing the impeller-inlet static density:

$$\rho_1 = \rho_{t1}\left[1 - \left(\frac{\gamma - 1}{\gamma + 1}\right) M_{cr1}{}^2\right]^{\frac{1}{\gamma-1}}$$

$$= \left(\frac{p_{t1}}{RT_{t1}}\right)\left[1 - \left(\frac{\gamma - 1}{\gamma + 1}\right) M_{cr1}{}^2\right]^{\frac{1}{\gamma-1}} = 1.23\,\text{kg/m}^3$$

Next, we apply the continuity equation at the impeller inlet station, knowing that V_1 is totally axial:

$$V_{z1} = V_1 = M_{cr1} V_{cr1} = M_{cr1}\sqrt{\left(\frac{2\gamma}{\gamma + 1}\right) RT_{t1}} = 95.4\,\text{m/s}$$

The impeller-inlet tip radius can now be calculated:

$$r_{1t} = \sqrt{r_{1h} + \left(\frac{\dot{m}}{\pi \rho_1 V_{z1}}\right)} = 0.126\,\text{m}$$

Part b: In order to calculate the difference in the hub-to-tip blade inlet (metal) angle ($\Delta\beta_1{}'$), we recall that the flow is fully guided (at inlet) by the impeller blades. The procedure to calculate this variable is as follows:

$$U_{1m} = \omega r_{1m} = 106.2\,\text{m/s}$$

$$W_{1m} = \sqrt{U_{1m}{}^2 + V_{z1}{}^2} = 142.7\,\text{m/s}$$

$$U_{1h} = \omega r_{1h} = 54.0\,\text{m/s}$$

$$U_{1t} = \omega r_{1t} = 158.3\,\text{m/s}$$

$$\beta_{1h} = \tan^{-1}\left(\frac{U_{1h}}{V_{z1}}\right) = 29.5^\circ$$

$$\beta_{1t} = \tan^{-1}\left(\frac{U_{1t}}{V_{z1}}\right) = 58.9^\circ$$

$$\Delta\beta_1{}' = \beta_{1t} - \beta_{1h} = 29.4^\circ$$

Part c: Let us calculate the impeller-exit relative velocity (W_2):

$$U_2 = \omega r_2 = 408.4\,\text{m/s}$$

$$h_2 - h_1 = 84,100\,\text{J/kg} = \frac{\left({W_{1m}}^2 - {W_2}^2\right)}{2} + \frac{\left({U_2}^2 - {U_{1m}}^2\right)}{2}$$

Upon substitution, we get

$$W_2 = 87.6\,\text{m/s}$$

As for the impeller-exit absolute velocity (V_2), we proceed as follows:

$$V_{r2} = W_{r2} = W_2 \cos\beta_2 = 81.8\,\text{m/s}$$

$$V_{\theta 2} = W_{\theta 2} + U_2 = W_2 \sin[-21^\circ] + U_2 = 377.0\,\text{m/s}$$

$$V_2 = \sqrt{{V_{\theta 2}}^2 + {V_{r2}}^2} = 385.8\,\text{m/s}$$

Now, the impeller dynamic head can be calculated as follows:

$$(\Delta h)_{dyn.} = \frac{\left({V_2}^2 - {V_{1m}}^2\right)}{2} = 69{,}860\,\text{J/kg}$$

Part d: The stage reaction can easily be calculated as follows:

$$\text{Stage Reaction}\,(R) = \frac{(\Delta h)_{static}}{(\Delta h)_{static} + (\Delta h)_{dynamic}} = 54.6\%$$

Part e: In order to calculate the impeller-exit sidewall spacing (b_2), we first calculate the static-density magnitude as follows:

$$T_{t2} = T_{t1} + \frac{\Delta h_t}{c_p} = T_{t1} + \frac{\left[(\Delta h)_{static} + (\Delta h)_{dynamic}\right]}{c_p} = 455.3\,\text{K}$$

$$p_{t2} = p_{t1}\left\{1 + \eta_{imp.}\left[\left(\frac{T_{t2}}{T_{t\,1}}\right) - 1\right]\right\}^{\frac{\gamma}{\gamma-1}} = 3.47\,\text{bars}$$

$$\rho_2 = \left(\frac{p_{t2}}{RT_{t2}}\right)\left\{1 - \left(\frac{\gamma-1}{\gamma+1}\right)\left[\frac{{V_2}^2}{\left(\frac{2\gamma}{\gamma+1}\right)RT_{t2}}\right]\right\}^{\frac{1}{\gamma-1}} = 1.69\,\text{kg/m}^3$$

Now we calculate the impeller-exit endwall spacing as follows:

$$b_2 = \frac{\dot{m}}{(\rho_2 V_{r2} 2\pi r_2)} = 0.42\,\text{cm}$$

Part f: In order to calculate the impeller-wise change in total relative pressure $[(\Delta p_{tr})_{imp}]$, we proceed as follows:

$$(T_{tr})_{1m} = T_{t1} + \left(\frac{{W_{1m}}^2 - {V_{1m}}^2}{2c_p}\right) = 307.6\,\text{K}$$

Similarly,

$$T_{tr2} = 385.0\,\text{K}$$

$$(p_{tr})_{1m} = p_{t1}\left(\frac{T_{tr1m}}{T_{t1}}\right)^{\frac{\gamma}{\gamma-1}} = 1.15\,\text{bars}$$

$$(p_{tr})_2 = p_{t2}\left(\frac{T_{tr2}}{T_{t2}}\right)^{\frac{\gamma}{\gamma-1}} = 1.93\,\text{bars}$$

$$(\Delta p_{tr})_{imp.} = p_{tr2} - p_{tr1m} = 0.78\,\text{bars}$$

In reference to the preceding results, note the following:

1) Although $(\Delta p_{tr})_{imp.}$ encompasses the profile (or skin friction) losses, it also reflects the fact that the total relative pressure will rise as a result of the radius change along the impeller (master) streamline.

2) In referring to the inlet station, it was necessary to use the subscript "1m," signifying the mean radius, since the total relative temperature is a function of velocity-diagram variables, where the solid-body rotational velocity U is radius-dependent. The total relative pressure, by reference to the last three computational steps (above), is (in turn) a function of the total-relative temperature.

Part g: Let us now perform the critical computational step of verifying that the impeller-exit (absolute) critical Mach number is not greater than unity. Choking, if present, will now take place immediately outside the impeller exit station. Note that the impeller-exit magnitude of relative velocity (W_2) will always be too small to warrant verification of the exit relative critical Mach number, a compressor-rotor feature that is quite the opposite when it comes to turbine aerodynamics.

$$M_{cr2} = \frac{V_2}{\sqrt{\left(\frac{2\gamma}{\gamma+1}\right)RT_{t2}}} = 0.988\ \text{(acceptably subsonic)}$$

The impeller-exit static pressure can now be calculated:

$$p_2 = p_{t2}\left[1 - \left(\frac{\gamma-1}{\gamma+1}\right)M_{cr2}{}^2\right]^{\frac{\gamma}{\gamma-1}} = 1.86\,\text{bars}$$

Under the free-vortex flow-structure assumption across the vaneless diffuser, we can calculate the flow variables at the vaned-diffuser inlet station (station 3) as follows:

$$V_{r3} = \left(\frac{r_2}{r_3}\right)V_{r2} = 66.7\,\text{m/s}$$

$$V_{\theta 3} = \left(\frac{r_2}{r_3}\right)V_{\theta 2} = 307.5\,\text{m/s}$$

$$V_3 = \sqrt{V_{r3}{}^2 + V_{\theta 3}{}^2} = 314.6\,\text{m/s}$$

$$V_{cr3} = \sqrt{\left(\frac{2\gamma}{\gamma+1}\right) R T_{t3}} = 390.4 \text{ m/s}$$

$$p_3 = p_{t3}\left[1 - \left(\frac{\gamma-1}{\gamma+1}\right) M_{cr,3}^{\ 2}\right]^{\frac{\gamma}{\gamma-1}} = 2.32 \text{ bars}$$

Proceeding to the vaned-stator exit station, we have

$$p_{t4} = (1 - 0.063) p_{t3} = 3.25 \text{ bars}$$
$$T_{t4} = T_{t3} = T_{t2} = 455.3 \text{ K} \text{ (adiabatic flow)}$$

In this part of the problem, we are required to calculate the exit value of the vaned-diffuser static pressure (p_4). Unfortunately, there is no direct way of achieving this.

With the vaned-stator exit value of total pressure now known, we first have to compute the exit magnitude of the critical Mach number (M_{cr4}) or, equivalently, the exit velocity (V_4), which we know to be totally radial (from the problem statement).

Naturally, in this case, we would think of applying the continuity equation at the vaned-stator exit station. However, such a step requires knowledge of the exit static density magnitude (ρ_4), which itself is a function of the critical Mach number.

The procedure, under these circumstances, has to be iterative, whereby the following computational procedure is repeated towards convergence:

- Assume the M_{cr4} magnitude.
- Calculate the corresponding magnitude of static density (ρ_4).
- Apply the continuity equation at station 4, and find V_{r4} ($=V_4$).
- Now calculate the exit critical Mach no. (M_{cr4}).
- Compare the computed M_{cr4} to the assumed value (above).

The foregoing computational procedure should be repeated until the point is reached where the assumed and computed critical Mach numbers are sufficiently close to one another.

The iterative procedure (above) was executed and the final results obtained:

$$M_{cr4} = 0.303$$
$$\rho_4 = 2.399 \text{ kg/m}^3$$
$$V_4 = V_{r4} = 118.5 \text{ m/s}$$

The exit magnitude of static pressure (p_4) can now be calculated as

$$p_4 = p_{t4}\left[1 - \left(\frac{\gamma-1}{\gamma+1}\right) M_{cr4}^{\ 2}\right]^{\frac{\gamma}{\gamma-1}} = 3.08 \text{ bars}$$

The static-pressure rise across the vaned stator can now be determined:

$$\Delta p_{stator} = p_4 - p_3 = 0.76 \text{ bars}$$

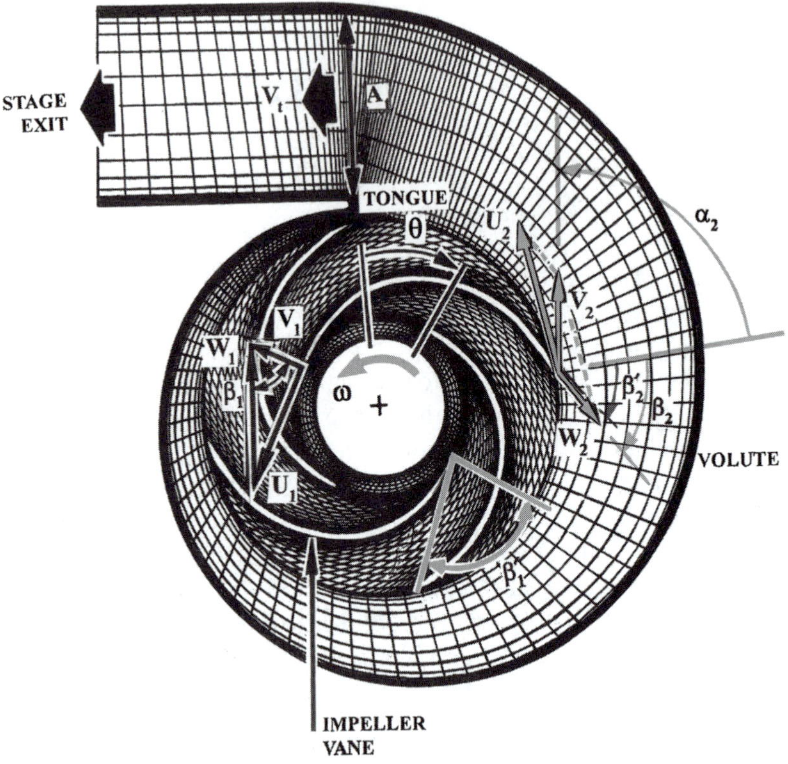

Figure 11.21. Input variables for Example 2.

Part h: Having computed the stage-exit static pressure (p_4), determination of the stage total-to-static efficiency is straightforward:

$$\eta_{t-s} = \frac{\left(\frac{p_4}{p_{t1}}\right)^{\frac{\gamma-1}{\gamma}} - 1}{\left(\frac{T_{t4}}{T_{t1}}\right) - 1} = 68.6\%$$

EXAMPLE 2

Figure 11.21 shows a centrifugal compressor stage, which is equipped with a totally radial impeller (i.e., one with no appreciable axial-velocity component anywhere). The stage geometrical data and operating conditions are as follows:

- Mass-flow rate ($\dot{m}$) = 2.5 kg/s
- Shaft speed (N) = 22,000 rpm
- Impeller-inlet total pressure (p_{t1}) = 1.09 bars
- Impeller-inlet total temperature (T_{t1}) = 300 K
- Sidewall spacing (b) = 2.2 cm
- Impeller-inlet blade angle (β'_1) = -52°
- Impeller-exit blade angle (β'_2) = -55°
- Impeller-inlet flow is assumed incompressible
- Impeller-inlet velocity is totally radial (i.e., $\alpha_1 = 0$)

- Impeller-inlet radius (r_1) = 8.0 cm
- Impeller-exit radius (r_2) = 20 cm
- Impeller total-to-total pressure ratio (p_{t2}/p_{t1}) = 3.6
- Impeller total-to-total efficiency (η_{t-t}) = 78%

Assuming a specific-heat ratio (γ) of 1.4, calculate:

a) The incidence-caused total-pressure loss using Fig. 9.15;
b) The blade-exit deviation angle (ϵ_2);
c) The specific speed (N_s);
d) The torque (τ) transmitted to the impeller;
e) The exit/inlet total relative temperature ratio (T_{tr2}/T_{tr1}).

SOLUTION

Part a: Let us begin by computing the impeller-inlet thermophysical properties, noting that the inlet flow is stated to be incompressible:

$$\rho_1 \approx \rho_{t1} = \frac{p_{t1}}{RT_{t1}} = 1.27\,\text{kg/m}^3$$

$$V_1 = V_{r1} = \frac{\dot{m}}{\rho_1(2\pi r_1 b)} = 178.0\,\text{m/s}$$

$$U_1 = \omega r_1 = 184.3\,\text{m/s}$$

$$W_1 = \sqrt{{V_{r1}}^2 + {U_1}^2} = 256.2\,\text{m/s}$$

$$\beta_1 = \tan^{-1}\left(\frac{U_1}{V_{r1}}\right) = 46.0^\circ$$

$$i_{imp.} = \beta_1 - \beta_1' = -6.0^\circ$$

Using the graph in Figure 9.15, we get

$$(\Delta p_t)_{incid.} = 0.022 p_{t1} = 0.024\,\text{bars}$$

Part b:

$$U_2 = \omega r_2 = 460.0\,\text{m/s}$$

$$T_{t2} = T_{t1}\left\{1 + \frac{1}{\eta_{t-t}}\left[\left(\frac{p_{t2}}{p_{t1}}\right)^{\frac{\gamma-1}{\gamma}} - 1\right]\right\} = 470.0\,\text{K}$$

$$V_{\theta 2} = \frac{c_p}{U_2}(T_{t2} - T_{t1}) = 370.5\,\text{m/s}$$

$$\rho_{t2} = \frac{p_{t2}}{RT_{t2}} = 2.9\,\text{kg/m}^3$$

$$V_{cr2} = \sqrt{\left(\frac{2\gamma}{\gamma+1}\right)RT_{t2}} = 396.7\,\text{m/s}$$

In order to compute V_{r2}, we need to apply the continuity equation at the impeller exit station. However, the application of this equation requires knowledge of the static density ρ_2. This, in turn, requires knowledge of the impeller-exit critical Mach number (M_{cr2}), a step we cannot conduct until we determine V_{r2}, which (at this point) is unknown.

The computational difficulty just outlined is precisely the same as that encountered in Example 1. We will therefore follow the same iterative procedure we executed then. The final results come out to be as follows:

$$V_{r2} = 46.3 \text{ m/s}$$

$$V_2 = \sqrt{{V_{r2}}^2 + {V_{\theta 2}}^2} = 373.4 \text{ m/s}$$

$$M_{cr2} = 0.941$$

$$\rho_2 = 1.95 \text{ kg/m}^3$$

It follows that

$$\beta_2 = \tan^{-1}\left(\frac{W_{\theta 2}}{V_{r2}}\right) = -62.5^\circ$$

$$\text{Deviation angle } \epsilon_{imp.} = \beta_2{}' - \beta_2 = -55^\circ - (-62.5^\circ) = 7.8^\circ$$

Part c: The specific speed can now be easily computed as follows:

$$N_s = \frac{N\left(\frac{2\pi}{60}\right)\sqrt{\frac{\dot{m}}{\rho_2}}}{[\eta_C c_p(T_{t2} - T_{t1})]^{\frac{3}{4}}} = 0.37 \text{ radians}$$

Referring to Figure 5.8, we see that this magnitude places the stage well within the centrifugal-stage range.

Part d: Supplied torque $(\tau) = r_2 V_{\theta 2} = 74.1$ N/m (zero inlet swirl)

Part e: The exit/inlet total relative temperature ratio can be computed as follows:

$$T_{tr1} = T_{t1} + \left(\frac{{W_1}^2 - {V_1}^2}{2c_p}\right) = 316.9 \text{ K}$$

$$T_{tr2} = T_{t2} + \left(\frac{{W_2}^2 - {V_2}^2}{2c_p}\right) = 405.7 \text{ K}$$

$$\frac{T_{tr2}}{T_{tr1}} = 1.28$$

EXAMPLE 3

Figure 11.22 shows the last stage of a high-pressure compressor section where the impeller is of the radial type. The stage design point is defined as follows:

- Impeller speed $(N) = 64,290$ rpm
- Impeller-inlet total pressure $(p_{t1}) = 5.2$ bars
- Impeller-inlet total temperature $(T_{t1}) = 587.0$ K

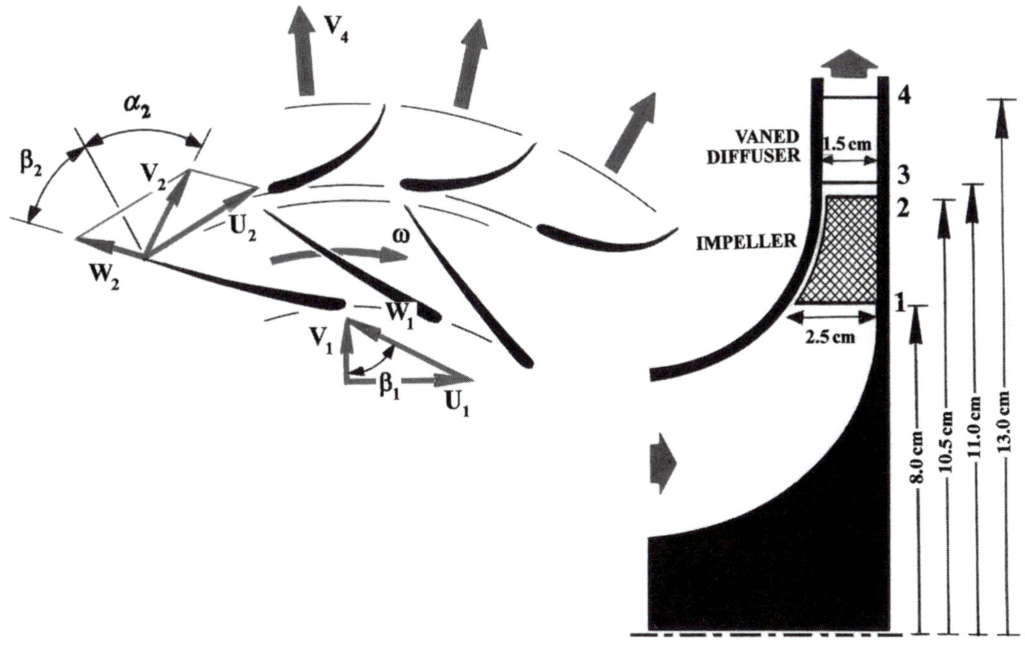

Figure 11.22. Input variables for Example 3.

- Impeller-inlet swirl angle (α_1) $= 0$
- Stage efficiency (η_C) $= 62\%$
- Impeller-exit total temperature (T_{t2}) $= 884.6\,\text{K}$
- Impeller-exit Mach number (M_2) $= 0.91$
- Vaned-diffuser exit flow stream is nonswirling (i.e., $\alpha_4 = 0$)
- Adiabatic flow throughout the stage
- Free-vortex flow in the impeller/diffuser gap

Assuming a specific-heat ratio (γ) of 1.4:

a) Calculate the stage specific speed (N_s).
b) Sketch both the impeller blade and the diffuser vane.

SOLUTION

Part a:

$$U_1 = \omega r_1 = 538.6\,\text{m/s}$$

$$U_2 = \omega r_2 = 706.9\,\text{m/s}$$

$$V_{\theta 2} = \frac{c_p}{U_2}(T_{t2} - T_{t1}) = 422.9\,\text{m/s (note that } V_{\theta 1} = 0)$$

$$T_2 = \frac{T_{t2}}{\left[1 + \left(\frac{\gamma-1}{\gamma}\right) M_2{}^2\right]} = 758.9\,\text{K}$$

$$\text{sonic speed}\,(a_2) = \sqrt{\gamma R T_2} = 552.2\,\text{m/s}$$

$$V_2 = M_2 a_2 = 502.5\,\text{m/s}$$

$$V_{r2} = \sqrt{{V_2}^2 - {V_{\theta 2}}^2} = 271.4\,\text{m/s}$$

$$W_{\theta 2} = V_{\theta 2} - U_2 = -284.0\,\text{m/s}$$

$$\beta_2 = \tan^{-1}\left(\frac{W_{\theta 2}}{V_{r2}}\right) = -46.3^\circ$$

$$\alpha_2 = \tan^{-1}\left(\frac{V_{\theta 2}}{V_{r2}}\right) = 57.3^\circ$$

$$p_{t2} = p_{t1}\left\{1 + \eta_C\left[\left(\frac{T_{t2}}{T_{t1}}\right) - 1\right]\right\}^{\frac{\gamma}{\gamma-1}} = 13.5\,\text{bars}$$

$$p_2 = \frac{p_{t2}}{\left[1 + \left(\frac{\gamma-1}{2}\right){M_2}^2\right]^{\frac{\gamma}{\gamma-1}}} = 7.90\,\text{bars}$$

$$\rho_2 = \frac{p_2}{RT_2} = 3.63\,\text{kg/m}^3$$

$$\dot{m} = \rho_2 V_{r2}(2\pi r_2 b_2) = 9.74\,\text{kg/s}$$

In order to calculate the impeller-inlet relative flow angle (β_1), we need to calculate V_1 or, equivalently, M_1. To this end, we apply the continuity equation at the impeller inlet station, using expression (3.36), as follows:

$$\frac{\dot{m}\sqrt{T_{t1}}}{p_{t1}A_1} = \sqrt{\left(\frac{\gamma}{R}\right)}M_1\left[1 + \left(\frac{\gamma-1}{2}\right){M_1}^2\right]^{\frac{(1+\gamma)}{2(1-\gamma)}}$$

Upon substitution, we get

$$M_1\left(1 + 0.2{M_1}^2\right)^{-3} = 0.517$$

This equation is clearly nonlinear. One way of solving it is the trial-and-error method, whereby M_1 is repeatedly assumed, and the equation satisfaction verified, toward convergence. Such an approach was implemented, and the final result came out as

$$M_1 = 0.67$$

It follows that

$$T_1 = \frac{T_{t1}}{\left[1 + \left(\frac{\gamma-1}{2}\right){M_1}^2\right]} = 538.6\,\text{K}$$

$$a_1 = \sqrt{\gamma R T_1} = 465.2\,\text{m/s}$$

$$V_{r1} = V_1 = a_1 M_1 = 311.7\,\text{m/s}$$

$$\beta_1 = \tan^{-1}\left(\frac{-U_1}{V_{r1}}\right) = -59.9^\circ$$

Let us now calculate the stage specific speed:

$$N_s = \frac{\omega\sqrt{\frac{\dot{m}}{\rho_2}}}{[\eta_C c_p(T_{t2} - T_{t1})]^{\frac{3}{4}}} = 1.23\,\text{radians}$$

Referring to Figure 5.8, this N_s magnitude places the compressor stage within the centrifugal-stage "dome" but close to the centrifugal/axial stage interface.

Part b: With no incidence or deviation angles anywhere, we have the following impeller-blade and diffuser-vane (metal) angles:

$$\beta_1{}' = \beta_1 = -59.9^\circ$$

$$\beta_2{}' = \beta_2 = -46.3^\circ$$

$$\alpha_3{}' = \alpha_3 = \alpha_2 = 57.3^\circ \text{ (free-vortex flow structure)}$$

$$\alpha_4{}' = \alpha_4 = 0 \text{ (cited in the problem statement)}$$

These angles are those reflected (within reasonable accuracy) in Figure 11.22.

EXAMPLE 4

Figure 11.23 shows a centrifugal compressor stage that is geometrically identical to that in Example 3. The impeller is radial, meaning there is no appreciable axial velocity component anywhere within this subdomain. The stage operating conditions are as follows:

- Inlet total pressure $(p_{t0}) = 11.4$ bars
- Inlet total temperature $(T_{t0}) = 649.5$ K
- Inlet critical Mach number $(M_{cr0}) = 0.46$
- Zero swirl velocity in the inlet duct $(\alpha_0 = \alpha_1 = 0)$
- Impeller-inlet relative flow angle $(\beta_1) = -54^\circ$
- Impeller-exit relative flow angle $(\beta_2) = -46^\circ$
- Impeller-exit total pressure $(p_{t2}) = 18.2$ bars
- Impeller total-to-total efficiency $(\eta_{t-t}) = 78\%$

The following simplifications are also valid:

- Isentropic flow in the unbladed inlet duct
- Adiabatic flow throughout the stage
- A constant specific-heat ratio (γ) of 1.4

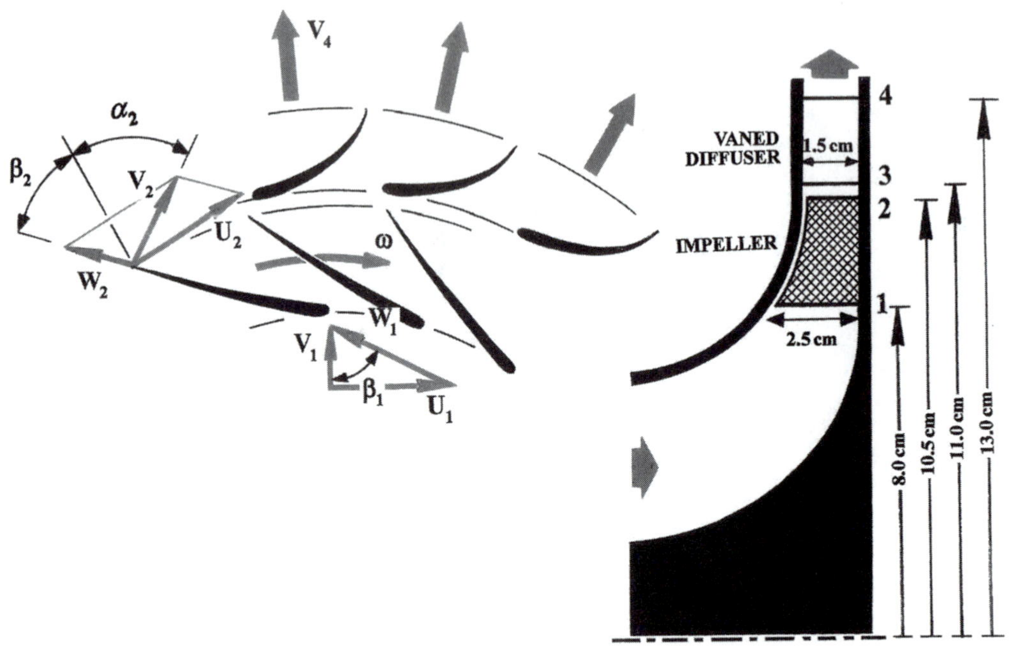

Figure 11.23. Input variables for Example 4.

With the preceding data, calculate:

a) The mass-flow rate ($\dot{m}$);
b) The impeller-inlet critical Mach number (M_{cr1});
c) The impeller "physical" speed (N);
d) The impeller-exit absolute flow angle (α_2);
e) The stage specific speed (N_s).

SOLUTION

Part a: Knowing that the stage-inlet swirl angle (α_0) is zero, we can apply the continuity equation there, with the result being

$$\dot{m} = 12.08\,\text{kg/s}$$

This mass-flow rate is on the high side for a centrifugal stage. However, the specific-speed magnitude also depends on the ideally supplied shaft work. The latter will have to be sufficiently large to justify a centrifugal-stage choice.

Part b: Let us reapply the continuity equation, but at the impeller-inlet station this time:

$$\frac{\dot{m}\sqrt{T_{t1}}}{p_{t1}(2\pi r_1 b_1)} = \sqrt{\frac{2\gamma}{(\gamma+1)R}}M_{cr1}\left[1-\left(\frac{\gamma-1}{\gamma+1}\right)M_{cr1}{}^2\right]^{\frac{1}{\gamma-1}}$$

where

$$T_{t1} = T_{t0} \text{ (adiabatic flow through the inlet passage)}$$

$$p_{t1} = p_{t0} \text{ (isentropic flow through the same passage)}$$

Upon substitution, the preceding continuity equation can be rewritten in the following compact form:

$$M_{cr1}(1 - 0.1667 M_{cr1}{}^2) = 0.561$$

This nonlinear equation is very similar to that in Example 3. A similar trial-and-error procedure was performed in this case, and the final result is

$$M_{cr1} = 0.69$$

Part c:

$$V_{r1} = V_1 = M_{cr1} V_{cr1} = 321.7 \text{ m/s}$$

$$U_1 = \omega r_1 = V_{r1} \tan \beta_1 = 442.8 \text{ m/s}$$

$$\omega = \tfrac{U_1}{r_1} = 5534.8 \text{ radians/s}$$

$$N = \left(\frac{60}{2\pi}\right)\omega = 52{,}853 \text{ rpm}$$

Part d:

$$T_{t2} = T_{t1}\left\{1 + \frac{1}{\eta_{t-t}}\left[\left(\frac{p_{t2}}{p_{t1}}\right)^{\frac{\gamma-1}{\gamma}} - 1\right]\right\} = 768.6 \text{ K}$$

$$V_{\theta 2} = \frac{c_p}{U_2}(T_{t2} - T_{t1}) = 200.1 \text{ m/s}$$

$$W_{\theta 2} = V_{\theta 2} - U_2 = -397.7 \text{ m/s}$$

$$V_{r2} = W_{r2} = \frac{W_{\theta 2}}{\tan \beta_2} = 384.1 \text{ m/s}$$

$$\alpha_2 = \tan^{-1}\left(\frac{V_{\theta 2}}{V_{r2}}\right) = 27.5^\circ$$

Part e: To calculate the stage specific speed, we proceed as follows:

$$V_2 = \sqrt{V_{\theta 2}{}^2 + V_{r2}{}^2} = 433.1 \text{ m/s}$$

$$M_{cr2} = \frac{V_2}{V_{cr2}} = \frac{V_2}{\sqrt{\left(\frac{2\gamma}{\gamma+1}\right) R T_{t2}}} = 0.854$$

$$\rho_2 = \left(\frac{p_{t2}}{R T_{t2}}\right)\left[1 - \left(\frac{\gamma - 1}{\gamma + 1}\right) M_{cr2}{}^2\right]^{\frac{1}{\gamma-1}} = 5.97 \text{ kg/m}^3$$

$$N_s = \frac{\omega\sqrt{\frac{\dot{m}}{\rho_2}}}{[\eta_C c_p (T_{t2} - T_{t1})]^{\frac{3}{4}}} = 1.47 \text{ radians}$$

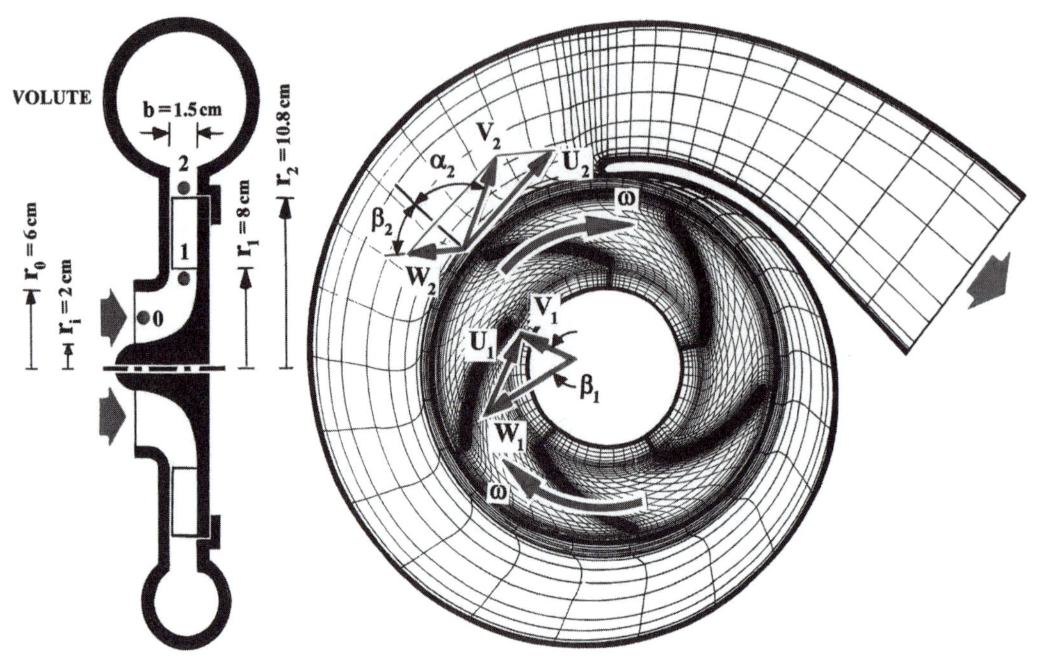

Figure 11.24. Input variables for Example 5.

EXAMPLE 5

The geometrical details of a centrifugal-compressor stage are shown in Figure 11.24. The stage operating conditions are as follows:

- Mass-flow rate ($\dot{m}$) = 2.36 kg/s
- Impeller speed (N) = 35, 600 rpm
- Inlet total pressure (p_{t1}) = 1.8 bars
- Inlet total temperature (T_{t1}) = 366 K
- Impeller dynamic head [$(\Delta h)_{dyn.}$] = 57,921 J/kg
- Impeller-exit static pressure (p_2) = 2.56 bars
- Impeller-exit swirl angle (α_2) = 68°

The following approximations are also made:

- No slip or deviation angles at the impeller outlet
- Adiabatic flow throughout the stage
- A constant specific-heat ratio (γ) of 1.4

With the preceding data, Calculate:

a) The impeller-inlet critical Mach number (V_1/V_{cr1});
b) The impeller total-to-total efficiency $(\eta_{t-t})_{imp.}$;
c) The impeller-exit sidewall spacing (b);
d) The percentage of impeller-wise increase in total relative pressure.

SOLUTION

Part a: Let us calculate the impeller-inlet static density, assuming M_{cr1} to be sufficiently small to justify an incompressible flow at this flow station:

$$\rho_1 \approx \rho_{t1} = \frac{P_{t1}}{RT_{t1}} = 1.71\,\text{kg/m}^3$$

$$V_1 = V_{r1} = \frac{\dot{m}}{\rho_1 \pi ({r_0}^2 - {r_i}^2)} = 190.7\,\text{m/s}$$

$$M_{cr1} = 0.54$$

This critical Mach number magnitude is disappointing, for it is too high to be consistent with our impeller-inlet incompressible-flow assumption (above). To overcome this difficulty, we take the newly computed critical Mach number (above) as an initial guess, re-calculate the static density, re-apply the continuity equation to find a new critical Mach number, repeating the entire procedure until convergence is attained. In the end, the critical Mach number came out to be:

$$M_{cr1} = 0.65$$

It follows that

$$V_1 = V_{r1} = 228.9\,\text{m/s}$$

$$\rho_1 = 1.425\,\text{kg/m}^3$$

Part b:

$$U_2 = \omega r_2 = 380.3\,\text{m/s}$$

Knowing that the impeller dynamic head is 57,921 J/kg (given), we can proceed to calculate the impeller-exit critical Mach number and the impeller total-to-total efficiency as follows:

$$V_2 = \sqrt{{V_1}^2 + 2(\Delta h)_{dyn.}} = 410.2\,\text{m/s}$$

$$V_{\theta 2} = V_2 \sin\alpha_2 = 380.3\,\text{m/s}$$

$$W_{\theta 2} = V_{\theta 2} - U_2 = 0$$

This means that the exit relative velocity (W_2) is totally in the radial direction:

$$W_2 = V_{r2} = V_2 \cos\alpha_2 = 153.7\,\text{m/s}$$

With $V_{\theta 1}$ being zero, the impeller-exit total temperature can now be calculated:

$$T_{t2} = T_{t1} + \frac{U_2 V_{\theta 2}}{c_p} = 509.9\,\text{K}$$

$$M_{cr2} = \frac{V_2}{V_{cr2}} = 0.99\ \text{(subsonic, as it should be)}$$

$$p_{t2} = \frac{p_2}{\left[1 - \left(\frac{\gamma-1}{\gamma+1}\right) M_{cr2}{}^2\right]^{\frac{\gamma}{\gamma-1}}} = 4.78\,\text{bars}$$

$$\eta_C = \frac{\left(\frac{p_{t2}}{p_{t1}}\right)^{\frac{\gamma-1}{\gamma}} - 1}{\left(\frac{T_{t2}}{T_{t1}}\right) - 1} = 81.8\%$$

Part c:

$$\rho_2 = \rho_{t2}\left[1 - \left(\frac{\gamma - 1}{\gamma + 1}\right) M_{cr2}{}^2\right]^{\frac{1}{\gamma-1}} = 2.09\,\text{kg/m}^3$$

$$b_2 = \frac{\dot{m}}{\rho_2 V_{r2}(2\pi r_2)} = 1.15\,\text{cm}$$

Part d: Let us first calculate the impeller-exit total relative pressure (p_{tr2}):

$$T_{tr2} = T_{t2} - \left(\frac{V_2{}^2 - W_2{}^2}{2c_p}\right) = 437.9\,\text{K}$$

$$p_{tr2} = p_{t2}\left(\frac{T_{tr2}}{T_{t2}}\right)^{\frac{\gamma}{\gamma-1}} = 2.81\,\text{bars}$$

As for the impeller inlet station, we have

$$W_{\theta 1} = V_{\theta 1} - U_1 = 0 - (\omega r_{1m}) = -134.2\,\text{m/s}$$

$$W_{z1} = V_{z1} = V_1 = 228.9\,\text{m/s}$$

$$T_{tr1} = T_{t1} - \left(\frac{V_1{}^2 - W_1{}^2}{2c_p}\right) = 375.0\,\text{m/s}$$

$$p_{tr1} = p_{t1}\left(\frac{T_{tr1}}{T_{t1}}\right)^{\frac{\gamma}{\gamma-1}} = 1.96\,\text{bars}$$

Now, we can calculate the percentage of the impeller-wise increase in total relative pressure as follows:

$$\frac{\Delta p_{tr}}{p_{tr1}} = \frac{(p_{tr2} - p_{tr1})}{p_{tr1}} = 43.4\%$$

EXAMPLE 6

The objective of this example is to numerically verify the theoretical rationale with which the "slip-phenomenon" section was introduced. The computational procedure here is consistent with this rationale. In other words, we will calculate each of the centripetal and Coriolis acceleration components, these being the two predominant components, weighing each magnitude, as we follow a fluid particle in the blade-to-blade passage. In doing so, it is important to define a right-handed cylindrical

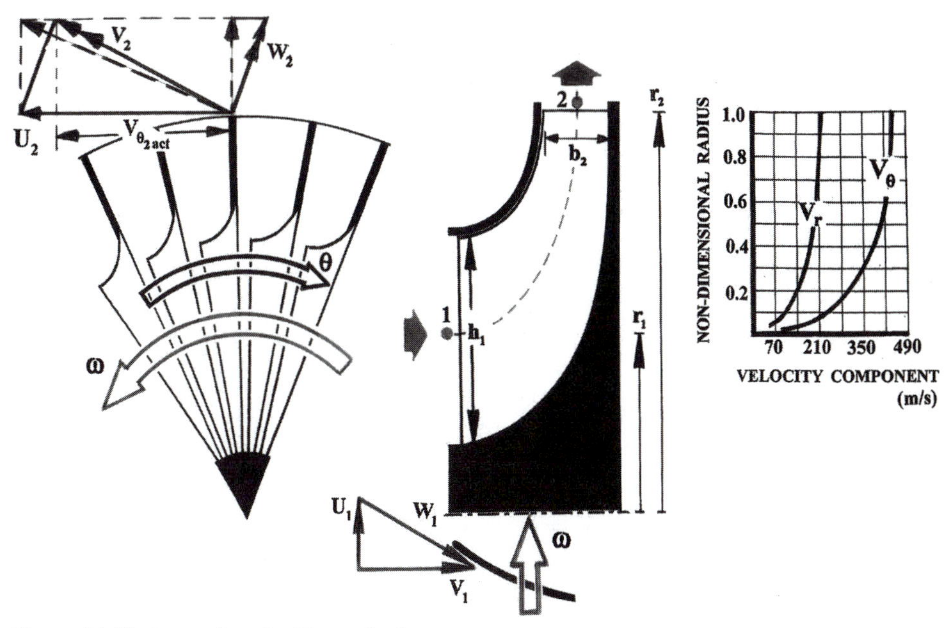

Figure 11.25. Input data for Example 6.

frame of reference in order to ensure the correct signs of the different acceleration components (Fig. 11.26).

Figure 11.25 shows a centrifugal compressor stage and the major geometrical variables. These variables have the following magnitudes:

- Mean inlet radius $(r_1) = 4.6\,\text{cm}$
- Exit radius $(r_2) = 10.5\,\text{cm}$
- Inlet blade height $(h_1) = 5.2\,\text{cm}$
- Exit blade width $(b_2) = 1.8\,\text{cm}$

The stage design point gives rise to the following set of data:

- Rotational speed $(N) = 41{,}600\,\text{rpm}$
- Inlet total pressure $(p_{t1}) = 1.88\,\text{bars}$
- Inlet total temperature $(T_{t1}) = 490\,\text{K}$
- Inlet swirl angle $(\alpha_1) = 0$
- Exit total pressure $(p_{t2}) = 6.02\,\text{bars}$
- Exit total temperature $(T_{t2}) = 683\,\text{K}$
- Radial blading, with no deviation angle (i.e., $\beta_2 = 0$)
- Slip factor $(\sigma_s) = 92.6\%$
- Choked flow at the impeller inlet station (i.e., $W_1/W_{cr1} = 1.0$)
- Sonic flow at the impeller outlet (i.e., $V_2/V_{cr2} = 1.0$)

Assuming a parabolic dependency of the velocity components (V_r and V_θ) on the radius (r) (Fig. 11.25), calculate the components of both the Coriolis and centripetal accelerations at key radial locations of your choice. Use the results to explain the tangential shifting of a fluid particle which is traversing the impeller subdomain.

SOLUTION

Referring to the inlet velocity diagram in Figure 11.25, we get

$$W_{\theta 1} = U_1 = 209.1 \text{ m/s}$$

$$T_{tr1} = T_{t1} + \frac{(W_{\theta 1}{}^2 - V_{\theta 1}{}^2)}{2c_p} = T_{t1} + \frac{W_{\theta 1}{}^2}{2c_p} = 496.8 \text{ K}$$

$$W_{cr1} = \sqrt{\left(\frac{2\gamma}{\gamma+1}\right) RT_{tr1}} = 407.8 \text{ m/s}$$

$$W_1 = W_{cr1} = 407.8 \text{ m/s}$$

$$V_1 = V_{z1} = \sqrt{W_1{}^2 - U_1{}^2} = 350.2 \text{ m/s}$$

$$V_{cr1} = \sqrt{\left(\frac{2\gamma}{\gamma+1}\right) RT_{t1}} = 405.1 \text{ m/s}$$

$$M_{cr1} = \frac{V_1}{V_{cr1}} = 0.88$$

$$\rho_1 = \frac{p_{t1}}{RT_{t1}} \left[1 - \left(\frac{\gamma-1}{\gamma+1}\right) M_{cr1}{}^2\right]^{\frac{1}{\gamma-1}} = 0.976 \text{ kg/m}^3$$

$$\dot{m} = \rho_1 V_{z1} (2\pi r_1 h_1) = 5.14 \text{ kg/s}$$

Referring to the exit velocity diagram in Fig. 11.25, we can calculate the following exit variables:

$$V_{cr2} = \sqrt{\left(\frac{2\gamma}{\gamma+1}\right) RT_{t2}} = 478.2 \text{ m/s}$$

$$V_2 = V_{cr2} = 478.2 \text{ m/s}$$

$$(V_{\theta 2})_{id.} = U_2 = 457.4 \text{ m/s}$$

$$(V_{\theta 2})_{act.} = \sigma_s (V_{\theta 2})_{id.} = 423.6 \text{ m/s}$$

$$V_{r2} = \sqrt{V_2^2 - V_{\theta 2}{}^2} = 222.0 \text{ m/s}$$

PARABOLIC INTERPOLATION OF V_r AND V_θ:

Let us define the nondimensional radial coordinate $\bar{r}$ as follows:

$$\bar{r} = \frac{r - r_1}{r_2 - r_1}$$

Using the impeller-exit magnitudes of V_r and V_θ as boundary conditions, we get the following two parabolic relationships:

$$V_r = 222.0\,\bar{r}^2$$
$$V_\theta = 423.6\,\bar{r}^2$$

Referring back to Example 7 in Chapter 9, and noting that the rotational speed ω is opposite to the θ direction, we can express the combined centripetal and Coriolis acceleration components as

$$(a)_{net} = [(\omega^2 r) - 2(\omega W_\theta)]\mathbf{e}_r + [2\omega W_r]\mathbf{e}_\theta$$

Noting that $W_\theta = V_\theta - \omega r$, we can now calculate the net acceleration components at the following five radial locations:

$$(a_r)_{\bar{r}=0.2} = 3.14 \times 10^6\ \text{m/s}^2 \quad \text{and} \quad (a_\theta)_{\bar{r}=0.2} = 0.08 \times 10^6\ \text{m/s}^2$$
$$(a_r)_{\bar{r}=0.4} = 3.38 \times 10^6\ \text{m/s}^2 \quad \text{and} \quad (a_\theta)_{\bar{r}=0.4} = 0.31 \times 10^6\ \text{m/s}^2$$
$$(a_r)_{\bar{r}=0.6} = 3.30 \times 10^6\ \text{m/s}^2 \quad \text{and} \quad (a_\theta)_{\bar{r}=0.6} = 0.70 \times 10^6\ \text{m/s}^2$$
$$(a_r)_{\bar{r}=0.8} = 2.95 \times 10^6\ \text{m/s}^2 \quad \text{and} \quad (a_\theta)_{\bar{r}=0.8} = 1.24 \times 10^6\ \text{m/s}^2$$
$$(a_r)_{\bar{r}=1.0} = 2.29 \times 10^6\ \text{m/s}^2 \quad \text{and} \quad (a_\theta)_{\bar{r}=1.0} = 1.94 \times 10^6\ \text{m/s}^2$$

A plot of these results is shown in Figure 11.26, where the following conclusions can be drawn:

- The radial-acceleration component continues to decline as the impeller exit station is approached. This is due to the decline in the Coriolis-acceleration radial component. As a result, a fluid particle in this region will continually lose radial momentum.
- Over the same exit subregion, the tangential component of Coriolis acceleration continuously grows, causing the tangential shift of a fluid particle that is sketched and labeled "particle trajectory" in Figure 11.26.
- Combination of the two flow behavioral characteristics (above) gives rise to the slip phenomenon, which was qualitatively discussed earlier in this chapter.

PROBLEMS

1) The last stage in a turboprop compressor section is centrifugal. The stage, which lacks a vaned diffuser, handles an entirely adiabatic flow. The following is a list of its major dimensions and operating conditions:
   - Impeller-inlet inner radius (r_{1h}) = 5.2 cm
   - Impeller-inlet outer radius (r_{1t}) = 9.4 cm
   - Impeller exit radius (r_2) = 12.8 cm
   - Vaneless-diffuser exit radius (r_3) = 16.8 cm
   - Constant vaneless-diffuser sidewall spacing
   - Shaft speed (N) = 48,000 rpm
   - Impeller-inlet absolute velocity (V_1) = 174 m/s

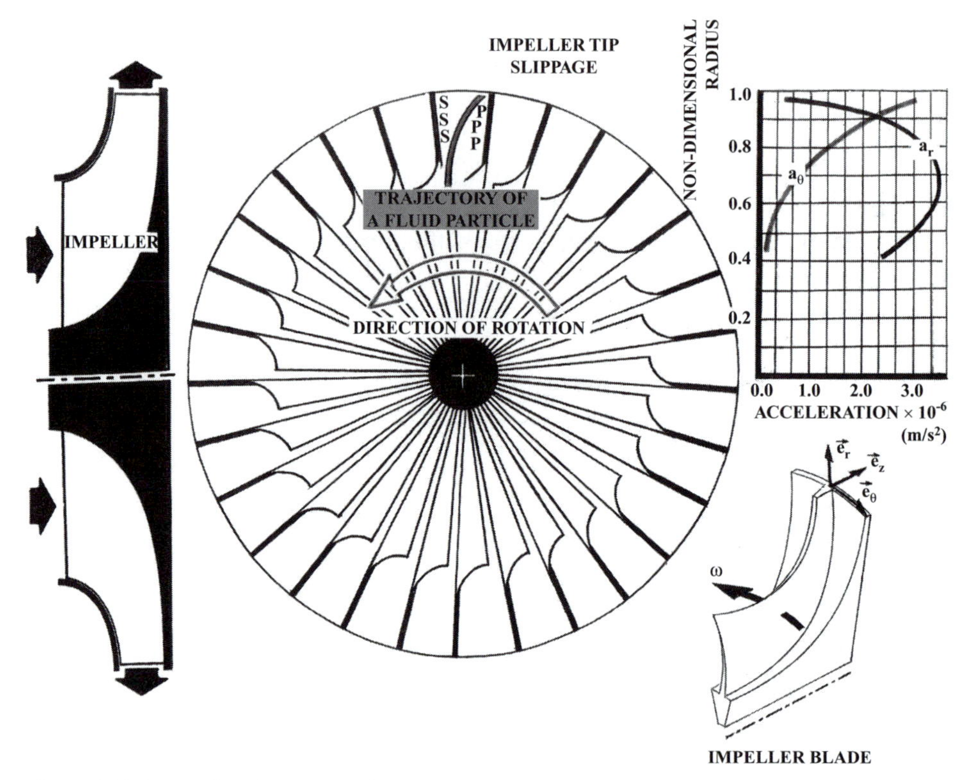

Figure 11.26. Results of Example 6.

- Impeller-inlet absolute flow angle $(\alpha_1) = +26^\circ$
- Impeller-inlet total pressure $(p_{t1}) = 5.2\,\text{bars}$
- Impeller-inlet total temperature $(T_{t1}) = 510\,\text{K}$
- Impeller-exit total temperature $(T_{t2}) = 730\,\text{K}$
- Impeller total-to-total efficiency $(\eta_{t-t}) = 85\%$
- Impeller-exit absolute velocity $(V_2) = 462\,\text{m/s}$
- Vaneless-diffuser free-vortex path
- Vaneless-diffuser total pressure loss $= 5.4\%$

Assuming a specific-heat ratio (γ) of 1.4, calculate:

a) The mass-flow rate $(\dot{m})$;
b) The impeller-exit sidewall spacing (b_2);
c) Impeller-inlet relative critical Mach no. (W_1/W_{cr1});
d) Impeller-exit relative critical Mach no. (W_2/W_{cr2});
e) The stage total-to-static efficiency (η_{t-s})

2) Referring to Example 1, consider the following operational change:

- Impeller efficiency is 86.5% (instead of 78%)

By building on the previously-attained results, calculate:

a) The impeller-exit sidewall spacing (b_2);
b) The rise in total relative pressure over the impeller.

3) The compressor section in an auxiliary power unit (APU) is composed of a single centrifugal stage. The impeller geometrical and operational variables are as follows:

   - Impeller-inlet inner radius $(r_{1h}) = 4.5\,\text{cm}$
   - Impeller-inlet outer radius $(r_{1t}) = 7.5\,\text{cm}$
   - Impeller exit radius $(r_2) = 16\,\text{cm}$
   - Impeller-exit sidewall spacing $(b_2) = 0.8\,\text{cm}$
   - Backward-curved impeller blades
   - Shaft speed $(N) = 22,700\,\text{rpm}$
   - Impeller-inlet total pressure $(p_{t1}) = 1.02\,\text{bars}$
   - Impeller-inlet total temperature $(T_{t1}) = 291\,\text{K}$
   - Impeller-inlet absolute velocity is totally axial
   - Impeller-inlet absolute velocity $(V_1) = 78\,\text{m/s}$
   - Impeller-exit total pressure $(p_{t2}) = 3.24\,\text{bars}$
   - Impeller-exit total temperature $(T_{t2}) = 428\,\text{K}$
   - Impeller-exit absolute velocity $(V_2) = 367\,\text{m/s}$
   - Full flow guidedness across the impeller

   Assuming a specific-heat ratio (γ) of 1.4, calculate:

   a) The mass-flow rate $(\dot{m})$;
   b) The impeller-exit blade angle $(\beta_2{}')$;
   c) The impeller static head $(h_2 - h_1)$;
   d) The stage specific speed (N_s).

4) Preceded by a ducted fan, the first stage of a small turbofan compressor is centrifugal. The impeller major dimensions and operating conditions are as follows:

   - Impeller-inlet inner radius $(r_{1h}) = 5.0\,\text{cm}$
   - Impeller-inlet outer radius $(r_{1t}) = 11.0\,\text{cm}$
   - Impeller exit radius $(r_2) = 18.0\,\text{cm}$
   - Impeller-exit sidewall spacing $(b_2) = 1.5\,\text{cm}$
   - Backward-curved impeller blades, where $\beta'_2 = -25^\circ$
   - Number of impeller blades $(N_b) = 12$
   - Shaft speed $(N) = 28{,}000\,\text{rpm}$
   - Inlet total pressure $(p_{t1}) = 1.18\,\text{bars}$
   - Inlet total temperature $(T_{t1}) = 310\,\text{K}$
   - Absolute inlet velocity (V_1) is totally axial
   - Absolute inlet velocity $(V_1) = 64\,\text{m/s}$
   - Impeller-exit total pressure $(p_{t2}) = 5.24\,\text{bars}$
   - Impeller-exit static density $(\rho_2) = 3.1\,\text{kg/m}^3$
   - Average specific-heat ratio $(\gamma) = 1.4$

   Taking the slip factor (σ_s) into account, calculate:

   a) The stage total-to-total efficiency (η_{t-t});
   b) The impeller-exit critical Mach number (V_2/V_{cr2});
   c) The stage reaction (R);
   d) The stage specific speed (N_s).

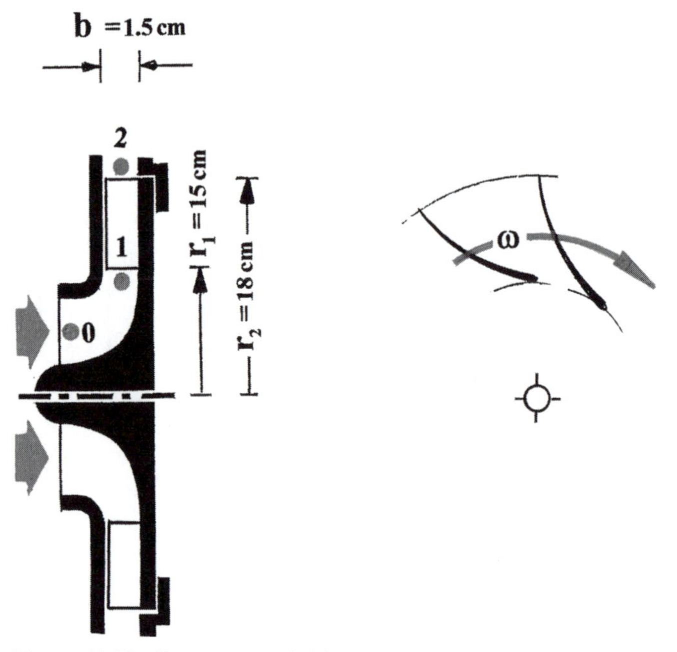

Figure 11.27. Geometry variables for Problem 5.

5) Figure 11.27 shows the inlet duct and impeller of a centrifugal compressor stage along with its major dimensions. The operating conditions are summarized as follows:

- Shaft speed $(N) = 26,500$ rpm
- Inlet total temperature $(T_{t0}) = 322$ K
- Inlet total pressure $(p_{t0}) = 1.12$ bars
- Inlet-duct total pressure loss $(\Delta p_t) = 4.5\%$
- Constant impeller-blade angle (i.e., $\beta_2' = \beta_1'$)
- Impeller-blade incidence angle $(i_R) = -7.0°$
- Impeller-exit deviation angle $(\epsilon) = 0$
- Impeller-inlet absolute velocity $(V_1) = 162$ m/s (totally radial)
- Impeller-exit total temperature $(T_{t2}) = 455$ K
- Impeller total-to-total efficiency $(\eta_{imp.}) = 78\%$

Assuming an incompressible flow field at the impeller exit station and a γ magnitude of 1.4, calculate:

a) The impeller-blade inlet (metal) angle (β_1');
b) The stage's specific speed (N_s).

6) Figure 11.28 shows the major dimensions of a centrifugal compressor stage along with its map. The stage operating conditions are as follows:

- Inlet critical Mach no. $(M_{cr1}) = 0.65$
- Inlet swirl velocity $(V_{\theta 1}) = 0$
- Inlet total temperature $(T_{t1}) = 330$ K

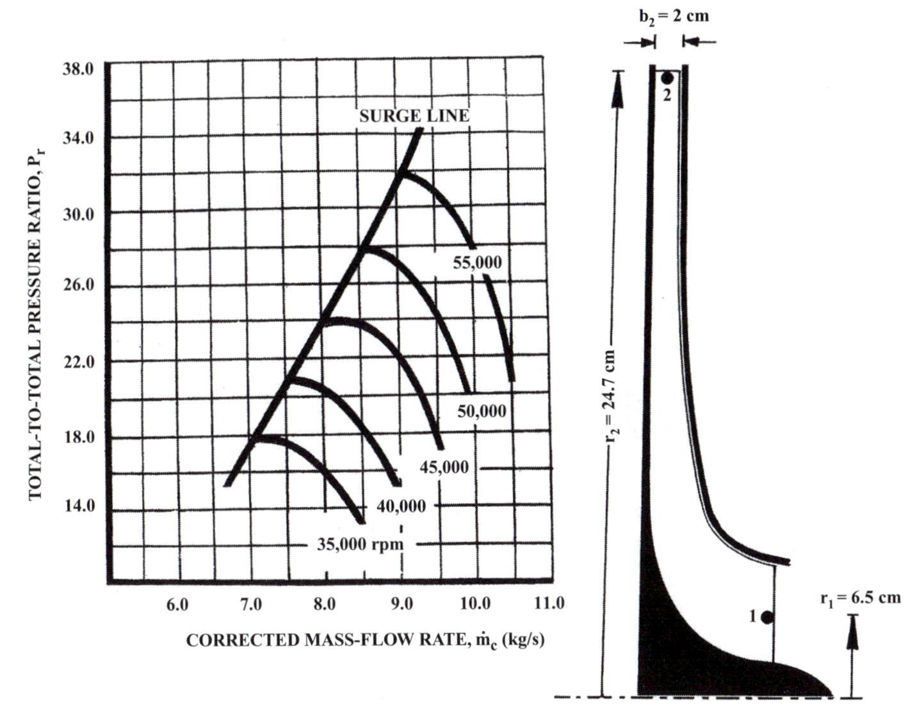

Figure 11.28. Geometry variables for Problem 6.

- Inlet total pressure (p_{t1}) = 1.6 bars
- Mass-flow rate ($\dot{m}$) = 12.0 kg/s
- Shaft speed (N) = 37,450 rpm
- Stage total-to-total efficiency (η_C) = 86%
- Impeller-exit swirl angle (α_2) = 78.5°

Using the stage map, and assuming a constant specific heat ratio of 1.4, calculate:

a) The impeller-inlet annulus height (h_1);
b) The change in total relative pressure across the impeller.

7) Figure 11.29 shows a centrifugal-compressor stage. The stage operating conditions are as follows:

- Inlet total temperature (T_{t1}) = 492 K
- Inlet total pressure (p_{t1}) = 4.4 bars
- Exit total temperature (T_{t2}) = 690 K
- Impeller total-to-total efficiency ($\eta_{imp.}$) = 81%
- Mass-flow rate ($\dot{m}$) = 5.4 kg/s
- Shaft speed (N) = 54,120 rpm
- Inlet swirl-velocity component ($V_{\theta 1}$) = 0
- Stator leading-edge blockage = 8.5%

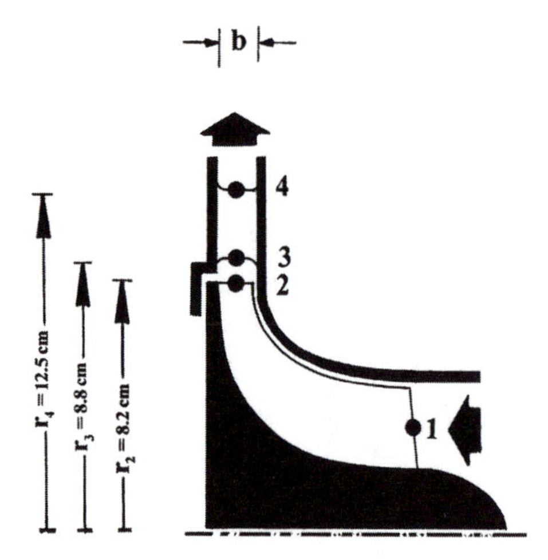

Figure 11.29. Geometry variables for Problem 7.

- Stator-exit swirl angle (α_4) = 0
- Impeller-exit swirl angle (α_3) = 69°
- Loss in total pressure across the stator [$(\Delta p_t)_{st.}$] = 4.8%
- Impeller-exit critical Mach number (M_{cr2}) = 0.97
- Free-vortex flow structure in the impeller/stator radial gap

Assuming an adiabatic flow throughout the stage and a γ magnitude of 1.4, calculate:

a) The impeller-exducer slip factor (σ_s);
b) The impeller-exit sidewall spacing (b_2);
c) The stator-vane incidence angle (i_S);
d) The static-pressure recovery coefficient across the stator.

8) Figure 11.30 shows a centrifugal compressor stage with a varying sidewall spacing. The stage operating conditions are as follows:

- Impeller-inlet total pressure (p_{t1}) = 5.0 bars
- Impeller-inlet total temperature (T_{t1}) = 468 K
- Mass-flow rate ($\dot{m}$) = 5.6 kg/s
- Rotational speed (N) = 31,000 rpm
- Inlet swirl angle (α_1) is positive
- Inlet critical Mach number (M_{cr1}) = 0.46
- Impeller-exit total temperature (T_{t2}) = 735 K
- Impeller-exit static pressure (p_2) = 11.6 bars
- Impeller total-to-total efficiency ($\eta_{imp.}$) = 83%

Considering an adiabatic flow field and a specific-heat ratio (γ) of 1.4, calculate:

a) The impeller-inlet swirl angle (α_1);
b) The impeller-exit critical Mach number (M_{cr2});
c) The impeller-wise change in total relative pressure (Δp_{t_r});
d) The stage specific speed (N_s).

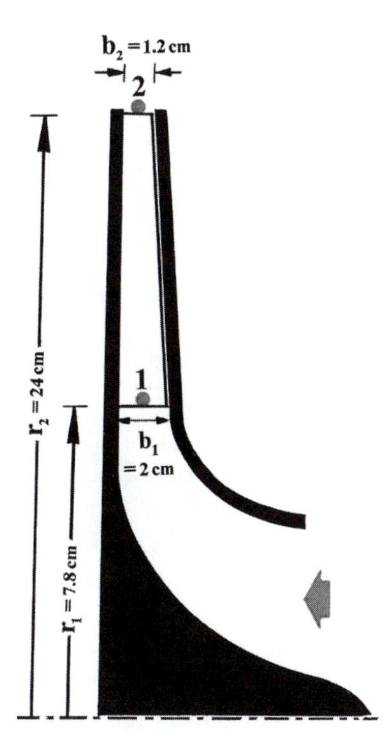

Figure 11.30. Geometry variables for Problem 8.

9) Figure 11.31 shows a compressor section that is composed of two centrifugal stages that are connected by a return duct. The two stages are geometrically identical. The compressor operating conditions are as follows:

- Shaft speed (N) = 27,000 rpm
- Inlet total pressure (p_{t1}) = 1.2 bars
- Inlet total temperature (T_{t1}) = 320 K
- Inlet swirl angle (α_1) = 0
- First-stage total-to-total pressure ratio $[(Pr)_{Stg.1}] = 2.4$
- First-stage total-to-total efficiency $[(\eta)_{Stg.1}] = 82\%$
- Second-stage total-to-total pressure ratio $[(Pr)_{Stg.2}] = 1.23$
- Second-stage total-to-total efficiency $[(\eta)_{Stg.2}] = 77\%$
- $\beta_2 = \beta_4$
- $M_{cr2} = 0.92$
- $M_{cr3} << 0.3$
- $M_{cr4} = 0.58$
- V_θ remains constant across the return duct
- $(\Delta p_t)_{Ret.Duct} = 8.6\%$

Assuming an adiabatic flow stream, and a fixed γ of 1.4, calculate:

a) The mass-flow rate ($\dot{m}$);
b) The overall total-to-total efficiency (η_{t-t});
c) The compressor-exit swirl angle (α_4);
d) The overall total-to-static efficiency (η_{t-s}).

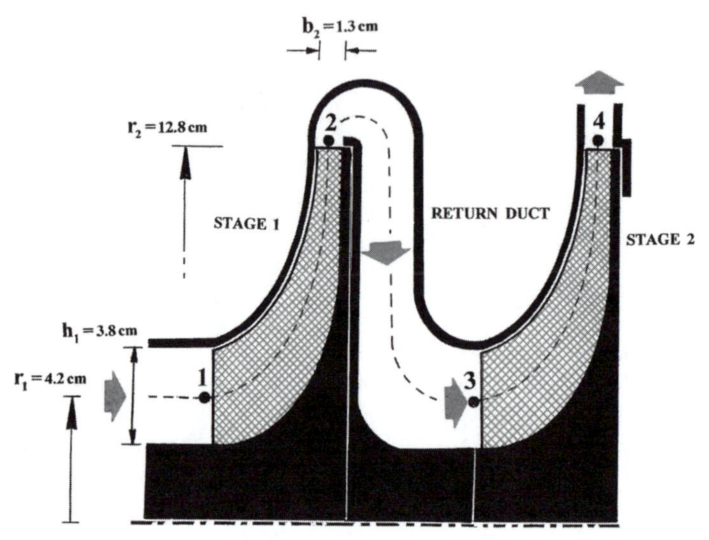

Figure 11.31. Input variables for Problem 9.

10) Figure 11.32 shows an axial-flow compressor stage followed by a centrifugal one, with a moderately-long axial gap in between. The two-stage compressor is operating under the following conditions:
   - Mass-flow rate ($\dot{m}$) = 2.1 kg/s
   - Rotational speed (N) = 31,517 rpm
   - Inlet total pressure (p_{t0}) = 1.13 bars
   - Inlet total temperature (T_{t0}) = 318 K

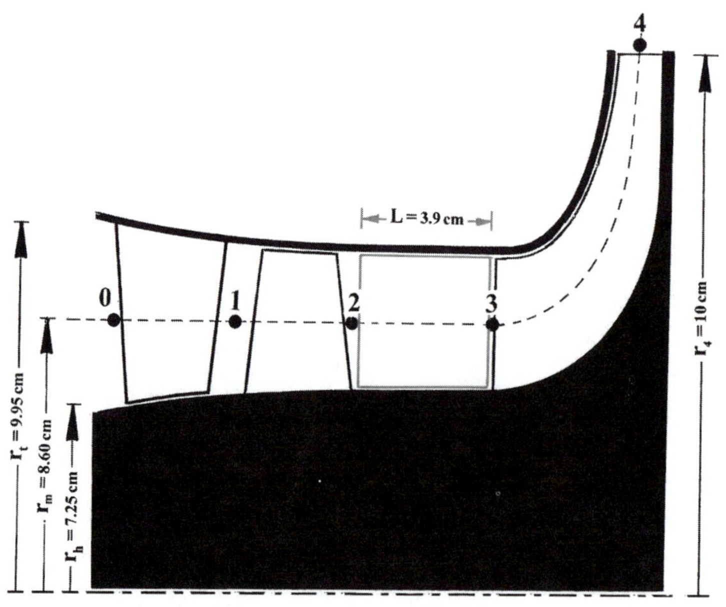

Figure 11.32. Geometry variables for Problem 10.

- Inlet swirl angle $(\alpha_0) = 0$
- Inlet critical Mach number $(M_{cr0}) = 0.46$
- Axial-stator total pressure loss $[(\Delta p_t)_{st.}] = 0.08$ bars
- Stator-exit relative flow angle $(\beta_1) = -48°$
- Axial-rotor inlet relative Mach number $(W_1/W_{cr1}) = 0.68$
- Constant axial-velocity component (V_z) across the axial stage
- Axial-stage total-to-total efficiency = 76%
- Axial-stage-exit relative critical Mach number $(W_2/W_{cr2}) = 0.64$
- Axial-stage-exit relative flow angle (β_2) is negative
- Constant swirl angle across the gap (i.e., $\alpha_3 = \alpha_2$)
- Gap-exit critical Mach number $(M_{cr3}) = 0.88$
- Impeller-exit relative flow angle $(\beta_4) = 0$
- Impeller-exit total temperature $(T_{t4}) = 785$ K
- Impeller-exit critical Mach number $(M_{cr4}) = 0.97$
- Impeller total-to-total efficiency $[(\eta_{t-t})_{imp.}] = 81\%$

Assuming an adiabatic flow throughout the compressor, and a fixed specific heat ratio (γ) of 1.4, calculate:

a) The mass-flow rate $(\dot{m})$;
b) The axial-rotor-exit critical Mach number (M_{cr2});
c) The gap-wise total-to-total pressure ratio (i.e., p_{t3}/p_{t2});
d) The friction coefficient (f) over the interstage gap;
e) The centrifugal stage reaction $[(R)_{Cent.Stg.}]$.

11) Figure 11.33 shows a centrifugal compressor stage where the impeller blades are of the backward-curved type. The geometrical variables in the figure have the following magnitudes:

- $r_1 = 8.2$ cm
- $r_2 = 17.5$ cm
- $h_1 = 10.1$ cm
- $b = 1.5$ cm
- $\beta_2 = -38°$

The stage's operating conditions are as follows:

- Inlet total temperature $(T_{t1}) = 312$ K
- Inlet total pressure $(p_{t1}) = 1.25$ bars
- Mass-flow rate $(\dot{m}) = 5.7$ kg/s
- Consumed specific shaft work $(w_s) = 107.4$ kJ/kg
- Inlet swirl angle $(\alpha_1) = 0$

Assuming an adiabatic flow through the impeller and a specific heat ratio (γ) of 1.4, calculate:

a) The shaft speed (N);
b) The stage reaction $(R_{stg.})$;
c) The stage specific speed (N_s);
d) The impeller total-to-static efficiency (η_{t-s}).

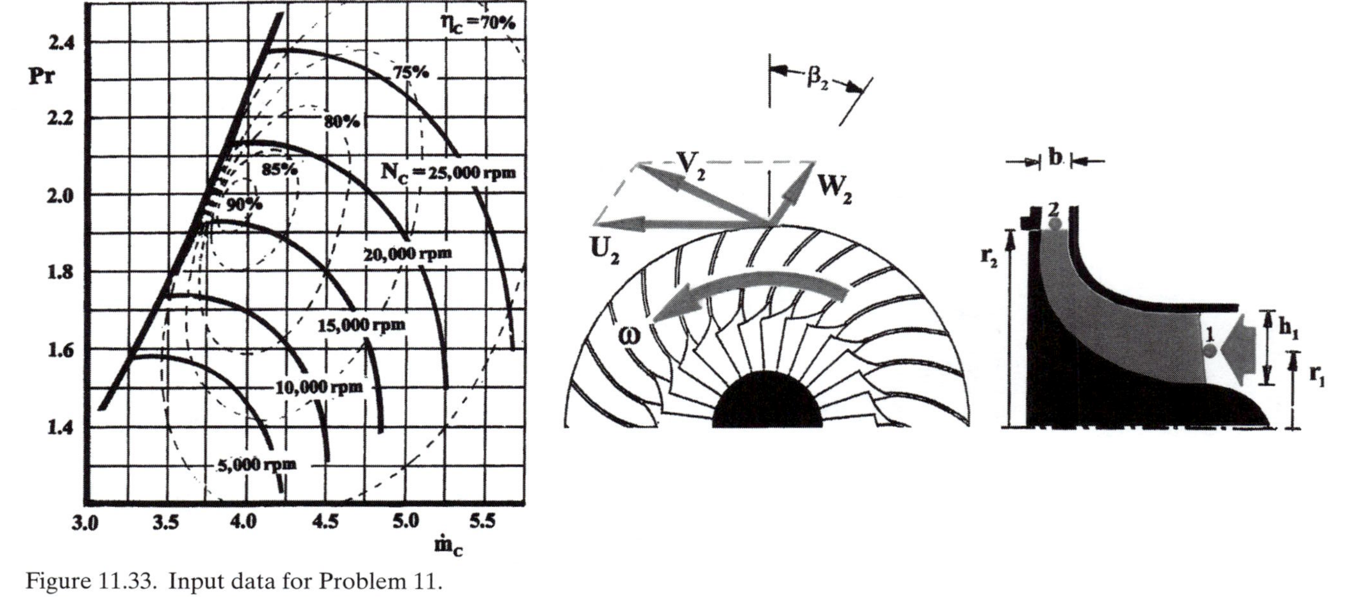

Figure 11.33. Input data for Problem 11.

12) Figure 11.34 shows a centrifugal compressor stage where the blades are of the forward-curved type. Also shown in the figure is the stage map. The stage operating conditions are as follows:

- Inlet total pressure $(p_{t1}) = 4.6\,\text{bars}$
- Inlet total temperature $(T_{t1}) = 485.0\,\text{K}$
- Physical speed $(N) = 23{,}400\,\text{rpm}$
- Stage total-to-total pressure ratio $(Pr) = 2.8$
- Impeller-inlet swirl angle $(\alpha_1) = -32^\circ$

Assuming an adiabatic flow through the impeller and a specific-heat ratio (γ) of 1.4, calculate:

a) The mass-flow rate $(\dot{m})$;
b) The stage specific speed (N_s);
c) The impeller-exit relative flow angle (β_2);
d) The loss in p_{t_r} caused by the irreversibility sources $[(\Delta p_{t_r})_{irrev.}]$.

13) Figure 11.35 shows two views of a radial-blading centrifugal-compressor stage, which is composed of 28 blades. The stage design point is defined as follows:

- Inlet total temperature $(T_{t1}) = 412.0\,\text{K}$
- Inlet total pressure $(p_{t1}) = 2.8\,\text{bars}$
- Impeller-inlet swirl angle $(\alpha_1) = -28.0^\circ$
- Impeller-inlet critical Mach number $(M_{cr1}) = 0.42$
- Mass-flow rate $(\dot{m}) = 2.1\,\text{kg/s}$
- Shaft speed $(N) = 43{,}800\,\text{rpm}$
- Stage total-to-total efficiency $(\eta_C) = 82.5\%$

Assuming an adiabatic flow through the impeller, a constant specific heat ratio (γ) of 1.4, and knowing that each axial-flow stage (to be discussed later in the problem statement) has total-to-total pressure ratio and efficiency magnitudes of 1.343 and 71%, respectively, calculate:

a) The change in total relative temperature $[(\Delta T_{t_r})_{imp.}]$;
b) The specific speed (N_s), based on the exit total density (ρ_{t2});
c) The no. of axial-flow stages, to replace the centrifugal stage, giving rise to:

   i) The same total-to-total pressure ratio (p_{t2}/p_{t1});
   ii) The same total-to-total temperature ratio (T_{t2}/T_{t1}).

14) Figure 11.36 shows a centrifugal-compressor stage with backward-curved impeller blades. The geometrical variables cited in this figure have the following magnitudes:

- $(r_1)_h = 5.2\,\text{cm}$
- $(r_1)_t = 8.4\,\text{cm}$
- $r_2 = 13.7\,\text{cm}$
- $\beta_2{}' = \beta_2 = -32.0^\circ$

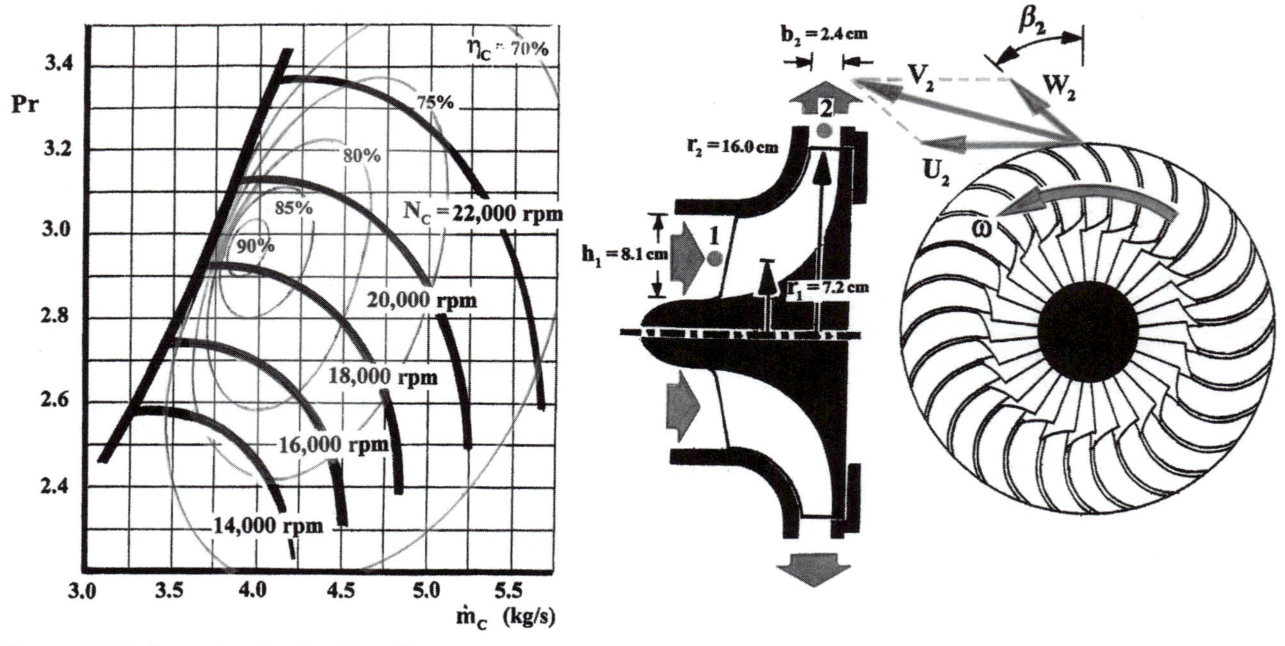

Figure 11.34. Input data for Program 12.

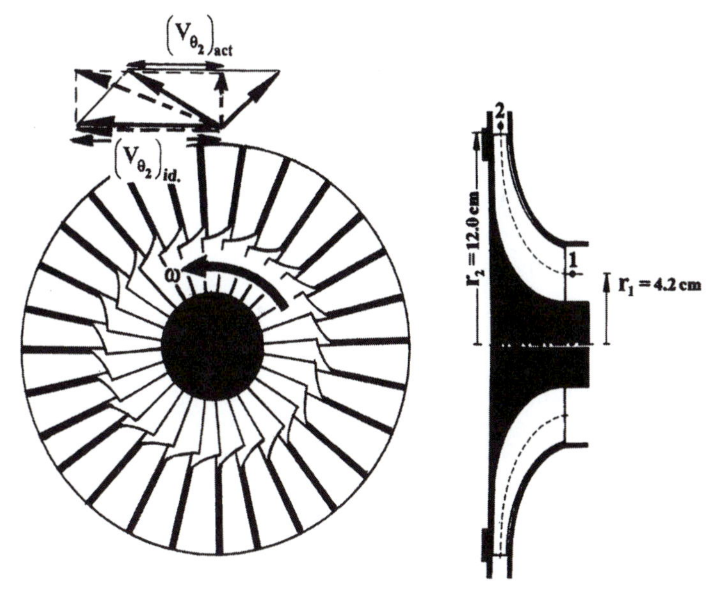

Figure 11.35. Input data for Problem 13.

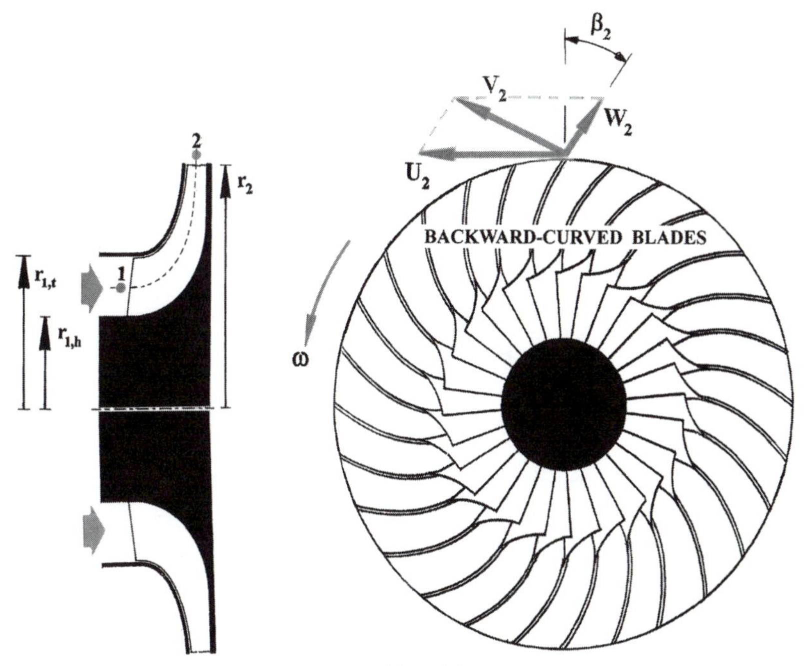

Figure 11.36. Input variables for Problem 14.

The stage operating conditions are as follows:

- Inlet total pressure $(p_{t1}) = 2.3$ bars
- Inlet total temperature $(T_{t1}) = 386.0$ K
- Exit total temperature $(T_{t2}) = 577.0$ K
- Total-to-total efficiency $(\eta_C) = 81.5\%$
- Inlet critical Mach number $(M_{cr1}) = 0.47$
- Inlet swirl angle $(\alpha_1) = -42.0°$
- Rotational speed $(N) = 36{,}500$ rpm

Assuming an adiabatic flow field and a γ magnitude of 1.4, calculate:

a) The stage reaction $(R_{stg.})$;
b) The stage specific speed (N_s);
c) The stage total-to-static efficiency (η_{t-s}).

CHAPTER TWELVE

Turbine-Compressor Matching

Consider the simple nonafterburning, single-spool turbojet engine, which is shown schematically in Figure 12.1. Assuming a viable (i.e., stable compressor) operation mode, there are obvious constraints relating the gas-generator components to one another. These generally enforce the uniformity of shaft speed as well as ensure the mass- and energy-conservation principles (Fig. 12.2). In terms of physical variables, these can be expressed as follows:

$$N_T = N_C \tag{12.1}$$

$$\dot{m}_T = (1 + f)\dot{m}_C \tag{12.2}$$

$$(w_s)_C = \eta_m[1 + f](w_s)_T \tag{12.3}$$

where f is the fuel-to-air ratio and η_m is the torque-transfer mechanical efficiency, with the subscripts T and C referring to the turbine and compressor sections, respectively. The mechanical efficiency η_m in equation (12.3) accounts for such contributors as the shaft length and the bearings effect on the shaft.

The basic problem at hand, by reference to Figure 12.3, is to be able to find the thermophysical state T, on the turbine map, once the corresponding state C is placed on the compressor map, or vice versa. To this end, it is assumed that the so-called "pumping" characteristics, primarily represented by the compressor and turbine maps, of the gas generator are given. In addition to the maps, these normally include a burner (or combustor) chart. The chart will provide, among other variables, the total pressure loss across this component. The loss in this case is a function of such variables as the combustor inlet magnitude of the Mach number, the inlet magnitude of the swirl-velocity component, the Reynolds number, and the inlet level of turbulence. Recognizing that the combustor-exit temperature is much higher than that at inlet, and that the exit flow stream is that of combustion products, the energy-transfer expression (12.3) can be rewritten as follows:

$$(1 + f)\eta_m\eta_T(c_p)_T T_{t4}\left(1 - \pi_T^{\frac{\gamma_T-1}{\gamma_T}}\right) = \left[\frac{(C_p)_C T_{t2}}{\eta_C}\right]\left(\pi_C^{\frac{\gamma_C-1}{\gamma_C}} - 1\right) \tag{12.4}$$

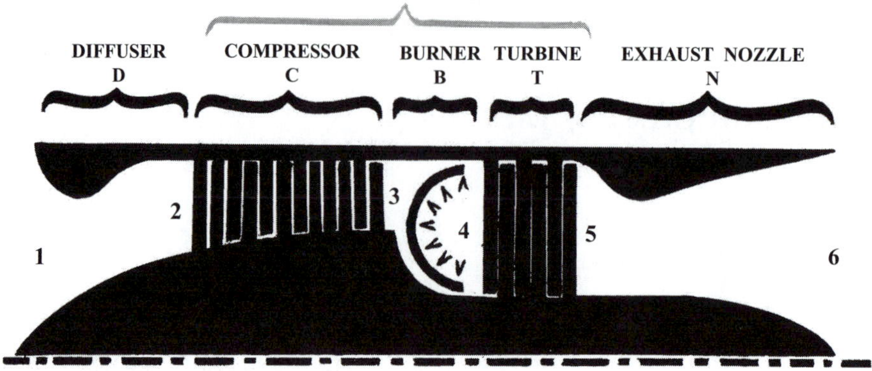

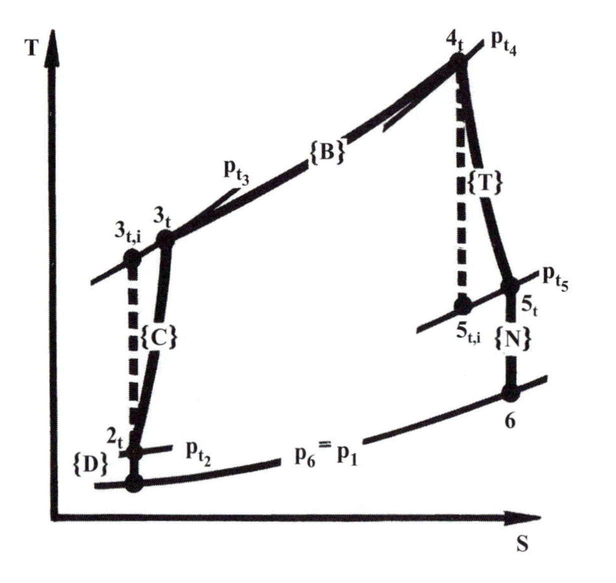

DIFFUSER, NOZZLE FLOW PROCESSES ARE ASSUMED ISENTROPIC.

$\mathbf{P_1}$&$\mathbf{P_6}$ ARE BOTH EQUAL TO AMBIENT PRESSURE.

$\tau_r = \frac{T_{t_2}}{T_1}$

$\tau_D = \frac{T_{t_2}}{T_1} = \left[1 + \left(\frac{\gamma - 1}{2}\right) M_1^2\right]$

$\tau_C = \frac{T_{t_3}}{T_{t_2}}$

$\tau_B = \frac{T_{t_4}}{T_{t_3}}$

$\tau_T = \frac{T_{t_5}}{T_{t_4}} < 1.0$

$\pi_D = \frac{p_{t_2}}{p_1}$

$\pi_C = p_t$

$\pi_B = \frac{p_{t_4}}{p_{t_3}}$

$\pi_T = \frac{p_{t_5}}{p_{t_4}}$

Figure 12.1. Special nomenclature and station designation.

where

$$c_{pC} = \left(\frac{\gamma_C}{\gamma_C - 1}\right) R_{air} \tag{12.5}$$

and

$$c_{pT} = \left(\frac{\gamma_T}{\gamma_T - 1}\right) R_{mix.} \tag{12.6}$$

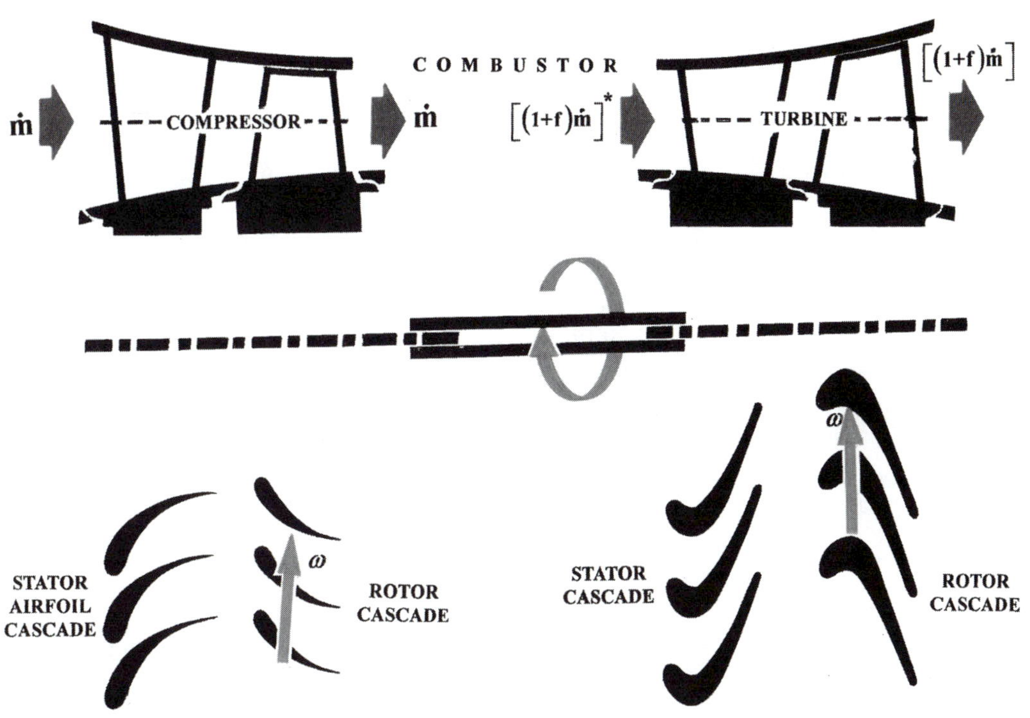

Figure 12.2. Rules of turbine/compressor mass-flow matching. *The fuel-to-air ratio "f" accounts for the presence of combustion products in the turbine flow stream.

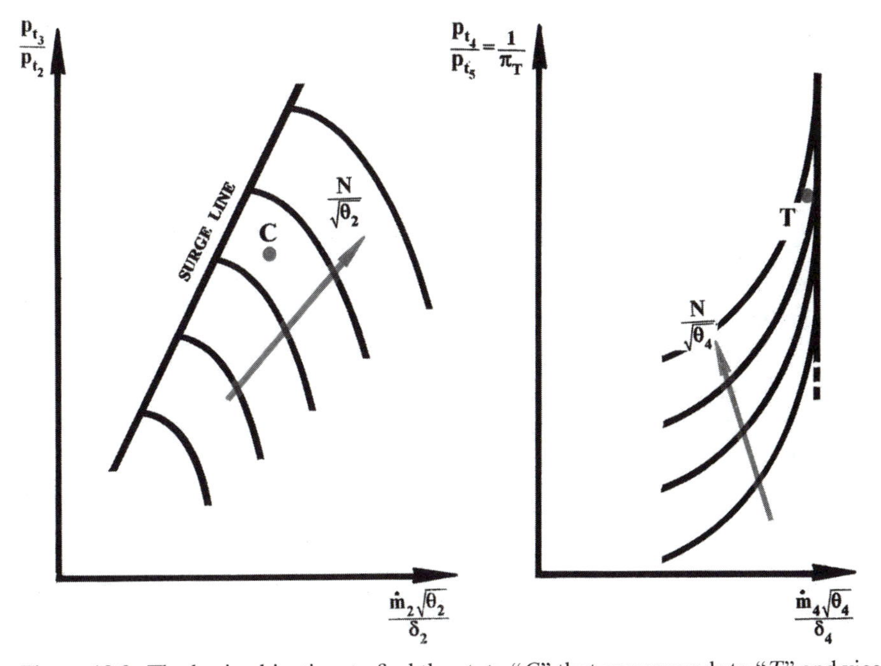

Figure 12.3. The basic objective: to find the state "*C*" that corresponds to "*T*" and vice versa.

with R_{air} and $R_{mix.}$ referring to the gas constants of air and combustion products, respectively. In the absence of afterburning, the latter is assumed equal to the air gas constant. Introducing the nondimensional variables defined in Fig. 12.1, equation (12.4) can be rewritten as follows:

$$\tau_T = 1 - \left[\frac{1}{\eta_m(1+f)}\right]\left(\frac{c_{p_C}}{c_{p_T}}\right)\left(\frac{T_{t2}}{T_{t4}}\right)(\tau_C - 1) \tag{12.7}$$

Equations (12.1), (12.2), and (12.7) establish the basic rules by which corresponding points on the compressor and turbine maps are related to one another. First, it is perhaps appropriate to reintroduce the expressions for the corrected speeds and flow rates, in view of the station-designation pattern in Figure 12.1, as follows:

$$N_{c,2} = \frac{N}{\sqrt{\theta_2}} \tag{12.8}$$

$$\dot{m}_{c,2} = \dot{m}_2 \frac{\sqrt{\theta_2}}{\delta_2} \tag{12.9}$$

$$N_{c,4} = \frac{N}{\sqrt{\theta_4}} \tag{12.10}$$

$$\dot{m}_{c,4} = \dot{m}_4 \frac{\sqrt{\theta_4}}{\delta_4} \tag{12.11}$$

where the nondimensional variables θ and δ are defined as follows:

$$\theta = \frac{T_t}{T_{STP}} \tag{12.12}$$

$$\delta = \frac{p_t}{p_{STP}} \tag{12.13}$$

with the subscript "STP" referring to the standard sea-level temperature and pressure (approximately 288 K and 1 bar, respectively). In addition, the compressor-inlet total properties can be cast in terms of the ambient conditions and flight Mach number M_1 as follows:

$$T_{t2} = T_1\left[1 + \left(\frac{\gamma_C - 1}{2}\right) M_1{}^2\right] \tag{12.14}$$

$$p_{t2} = p_1\left[1 + \left(\frac{\gamma_C - 1}{2}\right) M_1{}^2\right]^{\frac{\gamma_C}{\gamma_C - 1}} \tag{12.15}$$

where the flow field in the engine inlet segment is assumed isentropic for simplicity. Now, using each map-suited corrected variable in expressions (12.8) through (12.15), the turbine-compressor matching process becomes as simple as illustrated by the example to follow.

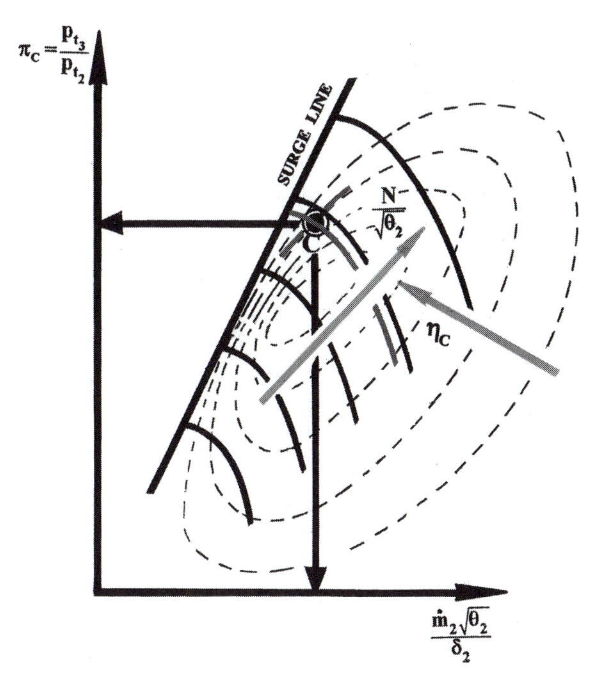

Figure 12.4. Definition of state "C," including the compressor efficiency.

Problem Category 1

Referring to Figure 12.4, consider the situation where the state C, on the compressor map, is defined. You are then required to find the corresponding state T on the turbine map. In doing so, the following variables are also provided:

- The ambient conditions (T_1 and p_1)
- The flight Mach number (M_1)
- The "physical" values of the speed (N) and flow rate ($\dot{m}_C$)
- The fuel-to-air ratio (f)
- The system mechanical efficiency (η_m)
- The turbine-inlet total temperature (T_{t4})

Note that the corresponding turbine operation point T could very well exist on the choking line, as shown in Fig. 12.5. Knowing the turbine corrected speed in this case is hardly helpful. The reason, as illustrated in Chapter 5, is that the choking line is where all corrected-speed lines collapse.

Exploiting the above-stated data, the following procedure is suggested:

Step 1:

If not explicitly provided, read off the corrected mass-flow rate ($\dot{m}_c$), corrected speed (N_c), total-to-total pressure ratio (π_C), and total-to-total efficiency (η_C) at the state C on the compressor map (Fig. 12.4).

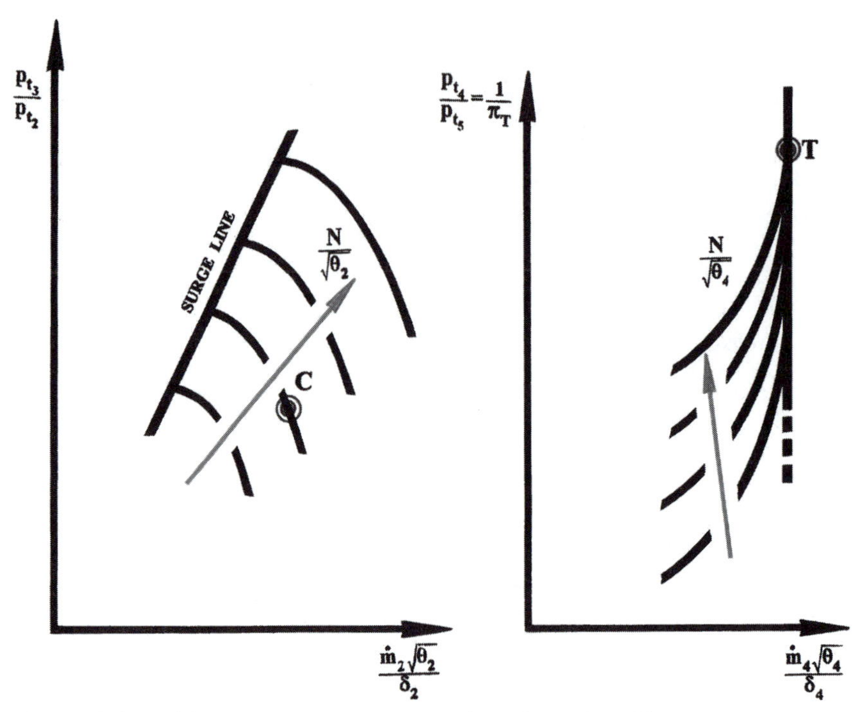

Figure 12.5. Problem category no. 1: move point "C" to the tubine map.

Step 2:

Knowing the ambient conditions (p_1 and T_1) and the flight Mach number (M_1), calculate the compressor-inlet total properties as follows:

$$T_{t2} = T_1\left[1 + \left(\frac{\gamma_C - 1}{2}\right) M_1{}^2\right] \tag{12.16}$$

$$p_{t2} = p_1\left[1 + \left(\frac{\gamma_C - 1}{2}\right) M_1{}^2\right]^{\frac{\gamma}{\gamma-1}} \tag{12.17}$$

Step 3:

Calculate the compressor total-to-total temperature ratio $[\tau_C \equiv T_{t3}/T_{t2}]$ as follows:

$$\tau_C = 1 + \frac{1}{\eta_C}\left[\pi_C{}^{\frac{\gamma_C-1}{\gamma_C}} - 1\right]$$

Step 4:

Calculate the ratio c_{p_C}/c_{p_T} as follows:

$$\frac{c_{p_C}}{c_{p_T}} = \left[\frac{\left(\frac{\gamma_C}{\gamma_C-1}\right)}{\left(\frac{\gamma_T}{\gamma_T-1}\right)}\right]\left(\frac{R_{air}}{R_{mix.}}\right) \tag{12.18}$$

Because the mixture of combustion products is predominantly air, the latter ratio in expression (12.18) can simply be ignored unless directed otherwise. Moreover, with the "cold" and "hot" specific-heat ratios being roughly 1.4 and 1.33, respectively, the magnitude of the fraction in (12.18) is approximately 0.868.

Step 5:
Calculate the total-to-total temperature ratio $[\tau_T \equiv T_{t5}/T_{t4}]$ by direct substitution in equation (12.7).

Step 6:
Calculate the turbine corrected speed (N_{c_4}) as follows:

$$N_{c_4} = N_{c_2}\sqrt{\frac{T_{t2}}{T_{t4}}} \tag{12.19}$$

At this point, there exist the following two possibilities:

a) *The combustor chart is available:* In this case, the combustor characteristic chart can be used to find the total pressure loss, $\Delta p_t = p_{t3} - p_{t4}$, and, consequently, the turbine-inlet total pressure (p_{t4}). In doing so, note that the compressor-exit total pressure (p_{t3}) is essentially known, at this point, since both p_{t2} and the total-to-total pressure ratio (π_C) are known. With p_{t4} being known, the turbine corrected mass-flow rate ($\dot{m}_{c,4}$) can be computed as follows:

$$\dot{m}_{c,4} = \dot{m}_{c,2}\sqrt{\frac{T_{t4}}{T_{t2}}}\left(\frac{p_{t2}}{p_{t4}}\right) \tag{12.20}$$

Finally, the state "T," on the turbine map (Fig. 12.6) can be located in terms of the turbine magnitudes of corrected speed and mass-flow rate in equations (12.19) and (12.20), respectively.

b) *The combustor chart is unavailable:* This is the more likely and more practical scenario. The reason is that such compressor-exit variables as the circumferential swirl-velocity distribution and turbulence level are generally neither given nor easily calculable. Under such circumstances, the procedure to define the thermophysical state "T," on the turbine map (Fig. 12.3) will have to be iterative.

The following trial-and-error procedure is recommended:

1) With the turbine corrected speed known from equation (12.19), use the turbine map, identify the constant corrected-speed line, and pick an arbitrary point "T_1" on it (Fig. 12.6).
2) Now, read off the corresponding magnitudes of $(\pi_T)_{T_1}$, $(\dot{m}_{c4})_{T_1}$, and $(\eta_T)_{T_1}$, then calculate the corresponding total-to-total temperature ratio (τ_{T_1}) as follows:

$$\tau_{T1} = 1 - \eta_{T1}\left(1 - \pi_{T1}^{\frac{\gamma_T - 1}{\gamma_T}}\right) \tag{12.21}$$

3) Because the state "C", on the compressor map is fully defined (Fig. 12.4), you are now in a position to compute the turbine total-to-total temperature ratio (τ_T) using the energy-balance equation (12.7).

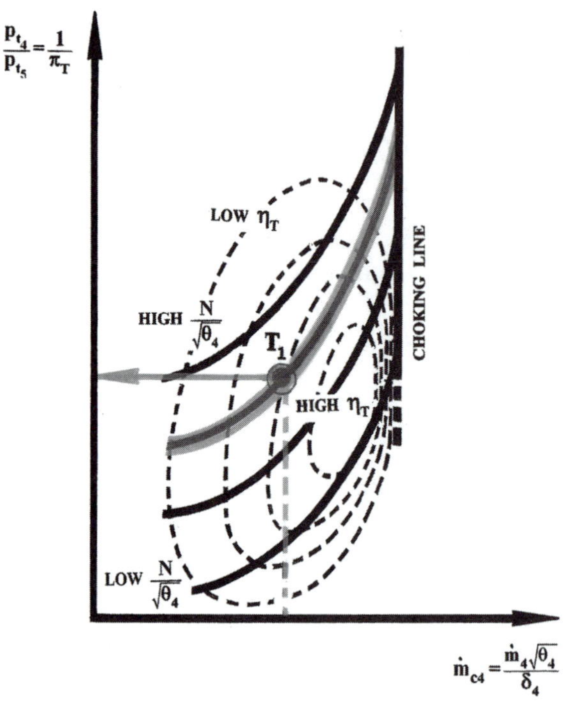

Figure 12.6. An intermediate step in the iterative procedure.

4) Comparing the two magnitudes of τ_T, above, you are likely to find them different.
5) Pick another point on the turbine map and on the same corrected speed line as in step 1, then repeat steps 2 and 3.
6) Repeat the entire procedure until you get to a point where the two magnitudes of τ_T are sufficiently close to one another. At this point, you have determined the thermophysical state "T" on the turbine map (Fig. 12.7), which corresponds to the state C on the compressor map (Fig. 12.5).

Problem Category 2

Another category of turbine-compressor matching is one where the turbine-operation point "T" (Fig. 12.8) is on the choking line. The difficulty here is that the turbine corrected speed cannot be found since corrected-speed lines in this case are undistinguishable. The requirement here is to locate point "C" on the compressor map, which corresponds to point "T" in Fig. 12.8. The proposed computational procedure under such conditions is as follows:

1) Calculate the compressor-inlet total temperature (T_{t2}) in the same manner as in step 2 of the previous example. The compressor corrected speed (N_{c2}) can then be calculated as follows:

$$N_{c2} = \frac{N}{\sqrt{T_{t2}/T_{STP}}} \tag{12.22}$$

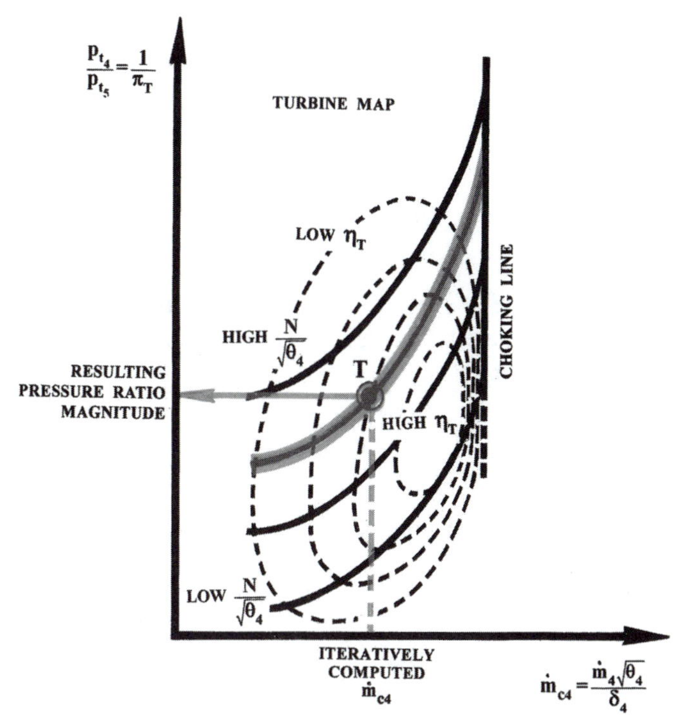

Figure 12.7. Final determination of the state "T" on the turbine map.

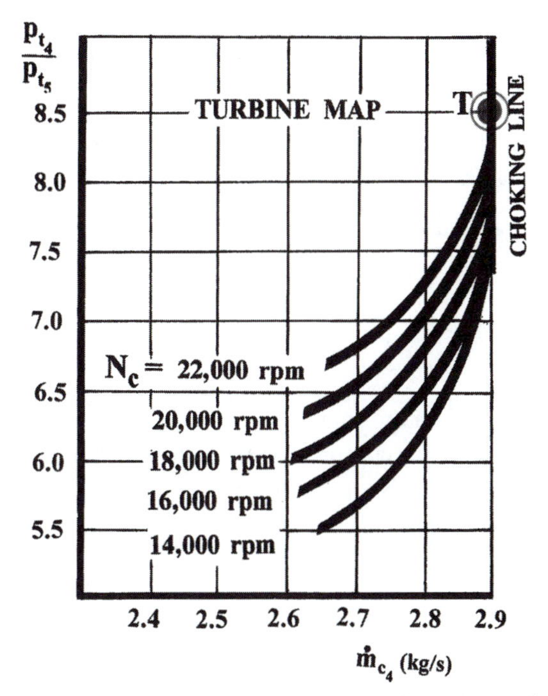

Figure 12.8. Problem category no. 2: the state "T" is on the turbine choking line.

where "N" is the shaft physical speed. This creates one "locus" of the compressor point of operation (referred to as state "C"), which is simply the corrected-speed line on the compressor map. The second locus of point "C" is developed next.

2) Consider the equality

$$\dot{m}_{c,2} = \dot{m}_{c,4} \frac{1}{(1+f)} \left(\frac{p_{t4}}{p_{t2}} \right) \sqrt{\frac{T_{t2}}{T_{t4}}} \tag{12.23}$$

where the turbine-inlet total temperature (T_{t4}) is given. If unavailable, the fuel-to-air ratio can always be assumed as somewhere between approximately 0.02 and 0.025, and expression (12.23) can be rewritten as follows:

$$\dot{m}_{c,2} = \left[\dot{m}_{c,4} \frac{1}{(1+f)} \left(\frac{p_{t4}}{p_{t3}} \right) \sqrt{\frac{T_{t2}}{T_{t4}}} \right] \frac{p_{t3}}{p_{t2}} \tag{12.24}$$

The contents of the bracketed quantity in equation (12.24) are individually discussed next:

- The term $\dot{m}_{c,4}$ is a known variable, for it is the choking magnitude of the corrected mass-flow rate on the turbine map (Fig. 12.8).
- The term $1/(1+f)$ depends on the fuel-to-air ratio, which is either given or assumed.
- The term (p_{t4}/p_{t3}) is the combustor total-to-total pressure ratio. Again, this term is either supplied (through the combustor chart) or assumed.
- The ratio (T_{t2}/T_{t4}) is known at this point. With T_{t4} always given, the compressor-inlet temperature T_{t2} is the result of applying expression 12.16.

Reviewing the foregoing remarks, it is clear that the bracketed group in expression (12.24) is simply a constant, say "K", that is,

$$\dot{m}_{c,2} = K \left(\frac{p_{t3}}{p_{t2}} \right)$$

or

$$\dot{m}_{c,2} = K \pi_C \tag{12.25}$$

Equation (12.25) represents a straight line on the compressor map in Figure 12.9. This line is the second locus for the state "C," on the compressor map. The intersection of this line and the corrected-speed line (previously identified) gives the required thermophysical state "C" on the compressor map (Fig. 12.9).

Performance-Related Variables in Propulsion Systems

In propulsion applications, once the turbine-compressor matching problem is resolved, many performance variables can then be calculated. In the following, expressions for the specific thrust force and the specific fuel consumption are provided.

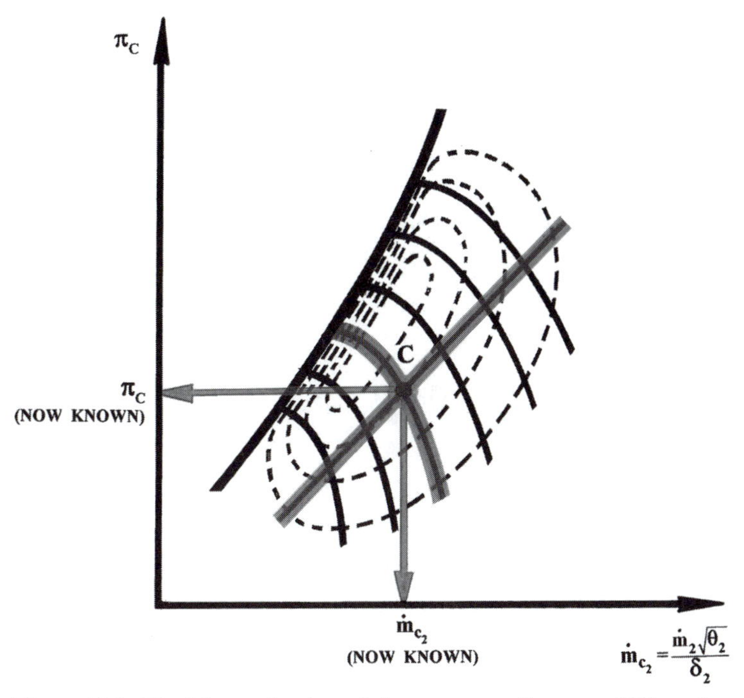

Figure 12.9. Final determination of the corresponding state "C" on the compressor map.

The Specific Thrust

Not very far from reality, the assumption of an isentropic exhaust nozzle flow process is normally made. This is particularly applicable in the event where the turbine-exit swirl velocity is considerably small. In this case, both of the total magnitudes of temperature and pressure remain constant across the nozzle; that is,

$$T_{t6} = T_{t5} \tag{12.26}$$

and

$$p_{t6} = p_{t5} \tag{12.27}$$

For a complete expansion (or acceleration) process across the nozzle, the nozzle-exit static pressure, by definition, will be equal to the back pressure, with the latter being the ambient pressure, namely

$$p_6 = p_1 \tag{12.28}$$

In this case, the nozzle-exit Mach number (M_6) can be calculated as a result of the following relationship:

$$p_{t6} = p_6\left(1 + \frac{\gamma_T - 1}{2} {M_6}^2\right)^{\frac{\gamma_T}{\gamma_T - 1}} \tag{12.29}$$

in which the only unknown is the nozzle-exit Mach number (M_6). Once computed, M_6 will aid in computing the nozzle-exit static temperature (T_6) as follows:

$$T_6 = \frac{T_{t6}}{\left(1 + \frac{\gamma_T - 1}{2} {M_6}^2\right)} \tag{12.30}$$

The magnitude of T_6 makes it possible to compute the engine-exit sonic speed (a_6) as

$$a_6 = \sqrt{\gamma_T R T_6} \tag{12.31}$$

This, together with the exit Mach number (M_6), gives rise to the engine exit velocity (V_6) as follows:

$$V_6 = M_6 a_6 \tag{12.32}$$

The same procedure can be utilized to calculate the flight speed, V_1. In the end, the specific thrust (i.e., the thrust force per unit mass of the flowing air stream) can be computed as

$$\frac{F}{\dot{m}_2} = (1 + f)V_6 - V_1 \tag{12.33}$$

Specific Fuel Consumption

In order to compute this variable, we first compute the fuel-to-air ratio. To this end, we apply the energy-balance equation across the combustor as follows:

$$\dot{m}_f H = [(\dot{m}_2 + \dot{m}_f) c_{p_T} T_{t4}] - \dot{m}_2 c_{p_C} T_{t3} \tag{12.34}$$

where H is the fuel heating value and $\dot{m}_f$ is the fuel mass-flow rate. Dividing through by $\dot{m}_2$, equation (12.34) will lead to the fuel-to-air ratio as follows:

$$f \equiv \frac{\dot{m}_f}{\dot{m}_2} = \frac{(c_{p_T} T_{t4} - c_{p_C} T_{t3})}{(H - c_{p_T} T_{t4})} \tag{12.35}$$

Defined as the mass of fuel consumed per unit thrust force, the specific fuel consumption may now be calculated as

$$SFC = \frac{f}{F/\dot{m}_2} \tag{12.36}$$

where the denominator is the specific thrust from equation (12.33). With the design point of a pure propulsion device being the cruise-altitude operation mode, the specific fuel consumption is most relevant where the minimum possible magnitude is at the forefront of the optimization tasks.

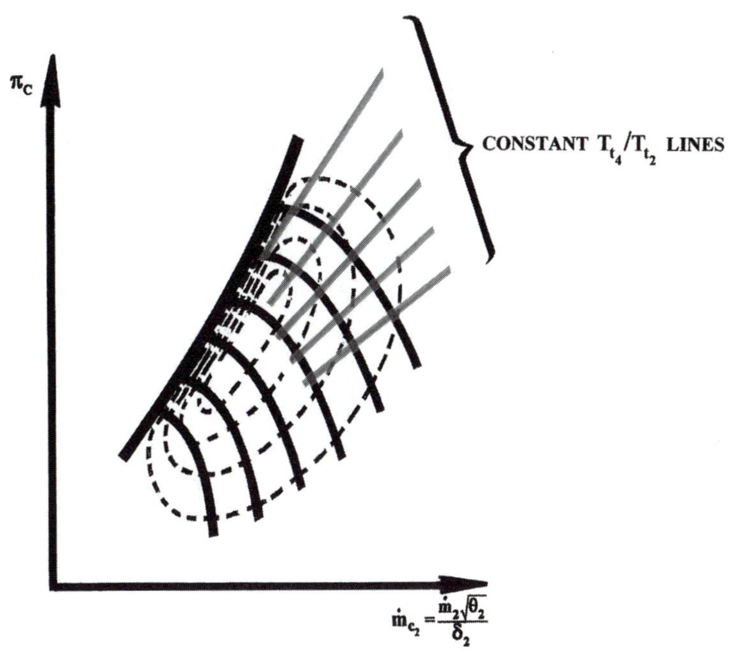

Figure 12.10. Constant T_{t_4}/T_{t_2} lines are crucial in facilitating the turbine/compressor matching.

Gas Generator Operating Lines on Compressor Maps
Constant T_{t4}/T_{t2} Lines

This is perhaps the most important exercise in matching the turbine and compressor operational modes. The importance of these lines stems from the fact that the temperature ratio they represent is the single most persistent linkage in the turbine-compressor matching process. Figure 12.10 shows a family of these lines on a compressor map. As a starting point, it is assumed that the ambient conditions (p_1 and T_1), the flight Mach number (M_1), the system mechanical efficiency (η_m), and the fuel-to-air ratio (f) are all known. The procedure to create such a family of lines is indeed lengthy and tedious. Summarized below is an iterative two-phase computational procedure that is composed of an outer loop, within which an inner loop is executed toward convergence before the outer loop is reentered.

Outer Loop

Step 1:
Select a point "C" on the compressor map (Fig. 12.11). It is this very point for which we will eventually compute the ratio T_{t4}/T_{t2}. Associated with point "C," read off such variables as π_C, N_{c_2}, $\dot{m}_{c,2}$, and η_C.

Step 2:
Assume a reasonable magnitude of T_{t4}/T_{t2}, an assumption that the upcoming inner loop is meant to verify and modify. While at it, however, you may also compute

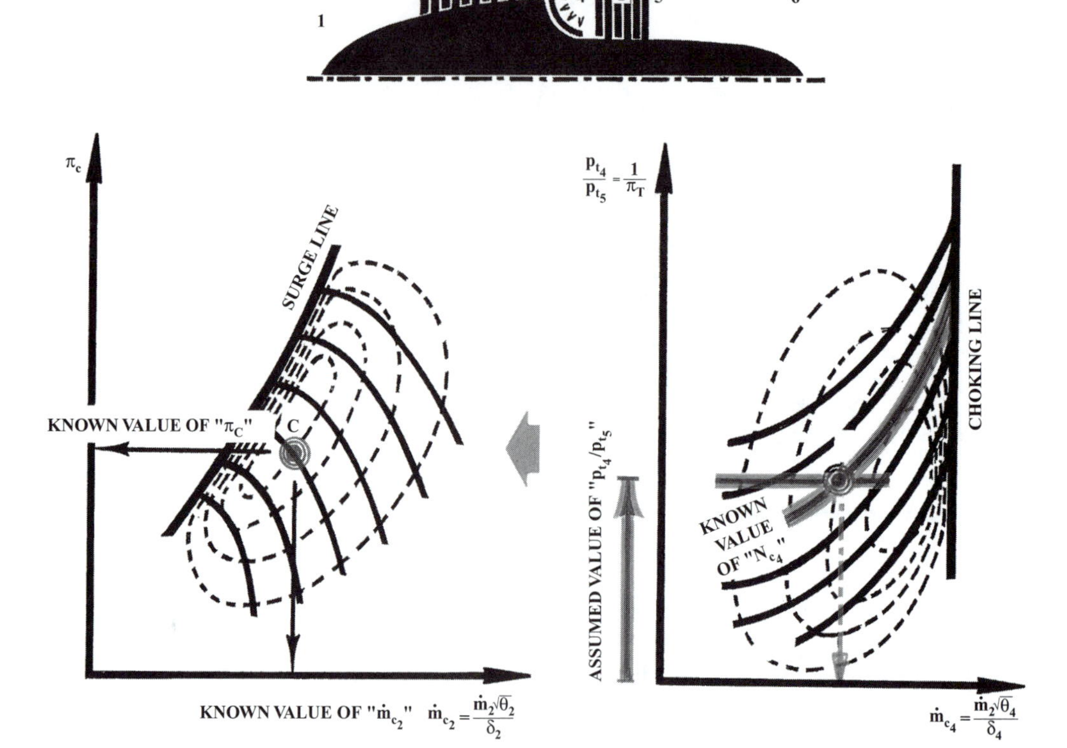

Figure 12.11. An intermediate step in the double-loop iterative process to construct a constant T_{t4}/T_{t2} line.

the turbine corrected speed, N_{c4}, as follows:

$$N_{c4} \equiv \frac{N}{\sqrt{T_{t4}}} = \frac{N}{\sqrt{T_{t2}}}\sqrt{\frac{T_{t2}}{T_{t4}}} = N_{c2}\sqrt{\frac{T_{t2}}{T_{t4}}} \tag{12.37}$$

with N_{c2} already known in step 1.

Inner Loop

We enter this loop with all of the known and assumed variables, particularly the total-to-total temperature ratio T_{t4}/T_{t2}.

Step 1:
Assume the turbine exit/inlet total-to-total pressure ratio $\pi_T \equiv p_{t5}/p_{t4}$.

Step 2:
With this magnitude and the above-computed corrected speed N_{c4} of the turbine (step 2 of the outer loop), you can locate the turbine operation point "T" on the turbine map in Fig. 12.11.

Step 3:

Read off the turbine corrected flow rate $\dot{m}_{c,4}$ and the turbine total-to-total efficiency η_T.

Step 4:

Compute the corresponding magnitude of the compressor corrected flow rate $\dot{m}_{c,2}$ as follows:

$$\dot{m}_{c,2} = \frac{\dot{m}_{c,4}}{(1+f)}\sqrt{\frac{T_{t2}}{T_{t4}}}\left(\frac{p_{t4}}{p_{t2}}\right) \tag{12.38}$$

or

$$\dot{m}_{c,2} = \frac{\dot{m}_{c,4}}{(1+f)}\sqrt{\frac{T_{t2}}{T_{t4}}}\left(\frac{p_{t4}}{p_{t3}}\right)\left(\frac{p_{t3}}{p_{t2}}\right) \tag{12.39}$$

The term p_{t4}/p_{t3} is particularly known. As explained earlier, the term is indicative of the total pressure loss across the combustor and is either given (through a combustor chart) or simply assumed. All other right-hand-side terms in equation (12.39) are also known, as a result of the preceding assumptions and calculations.

Step 5:

Now we have a *new* magnitude of the compressor corrected flow rate (12.39). Compare it with the magnitude in the first step of the outer loop. If these magnitudes are identical (which is unlikely), then terminate the inner loop and proceed with the remainder of the outer loop (to be summarized next). If not, then assume another value of the turbine total-to-total pressure ratio and repeat the inner loop in its entirety until the point comes where conducting another iteration produces practically the same magnitude of the corrected flow rate. At this point, get out of the inner loop.

Back to the Outer Loop

Step 3:

We now recalculate the magnitude of the temperature ratio (T_{t4}/T_{t2}) for comparison with the assumed value in the second step of the outer loop (above).

To this end, we utilize the following turbine-to-compressor shaft-work transfer expression:

$$\eta_m\eta_T(1+f)c_{p_T}T_{t4}\left[1-\left(\frac{p_{t5}}{p_{t4}}\right)^{\frac{\gamma_T-1}{\gamma_T}}\right] = \frac{c_{p_C}T_{t2}}{\eta_C}\left[\left(\frac{p_{t3}}{p_{t2}}\right)^{\frac{\gamma_C-1}{\gamma_C}}-1\right] \tag{12.40}$$

Through algebraic manipulation, equation 12.40 can be rewritten in the following convenient form:

$$\frac{T_{t4}}{T_{t2}} = \left[\frac{1}{(1+f)\eta_m\eta_T\eta_C}\right]\left(\frac{c_{p_C}}{c_{p_T}}\right)\frac{\left[\left(\frac{p_{t3}}{p_{t2}}\right)^{\frac{\gamma_C-1}{\gamma_C}}-1\right]}{\left[1-\left(\frac{p_{t5}}{p_{t4}}\right)^{\frac{\gamma_T-1}{\gamma_T}}\right]} \tag{12.41}$$

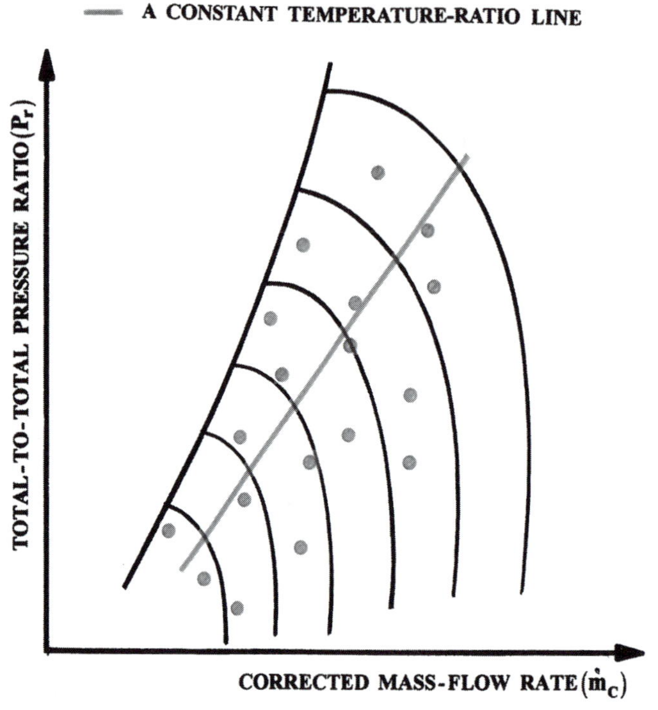

Figure 12.12. Determination of a constant-temperature line by interpolation.

With all other terms known, equation (12.41) yields a *new* magnitude of the temperature ratio, T_{t4}/T_{t2}.

Step 4:

Compare the new magnitude of T_{t4}/T_{t2} with the value assumed in step 2 of the outer loop. Should these two magnitudes be sufficiently close to one another, then this is the correct magnitude to assign to the point (on the compressor map) that was selected in step 1 of the outer loop. In this case, select another point on the compressor map, and repeat the entire procedure.

Should the two magnitudes of temperature ratio be far from one another, then go to step 2 of the outer loop. The only difference this time is the newly computed (or any other) magnitude of T_{t4}/T_{t2}.

Required Postprocessing Work

The preceding two-phase computational procedure will typically produce scattered points (Fig. 12.12) on the compressor map, with each point carrying its own magnitude of the temperature ratio (T_{t4}/T_{t2}). Separating each constant-temperature-ratio line from the others requires a great deal of interpolation between one of these points and all surrounding points. To simplify this postprocessing task (in the absence of suitable graphics software), one may proceed with the simplest (meaning linear)

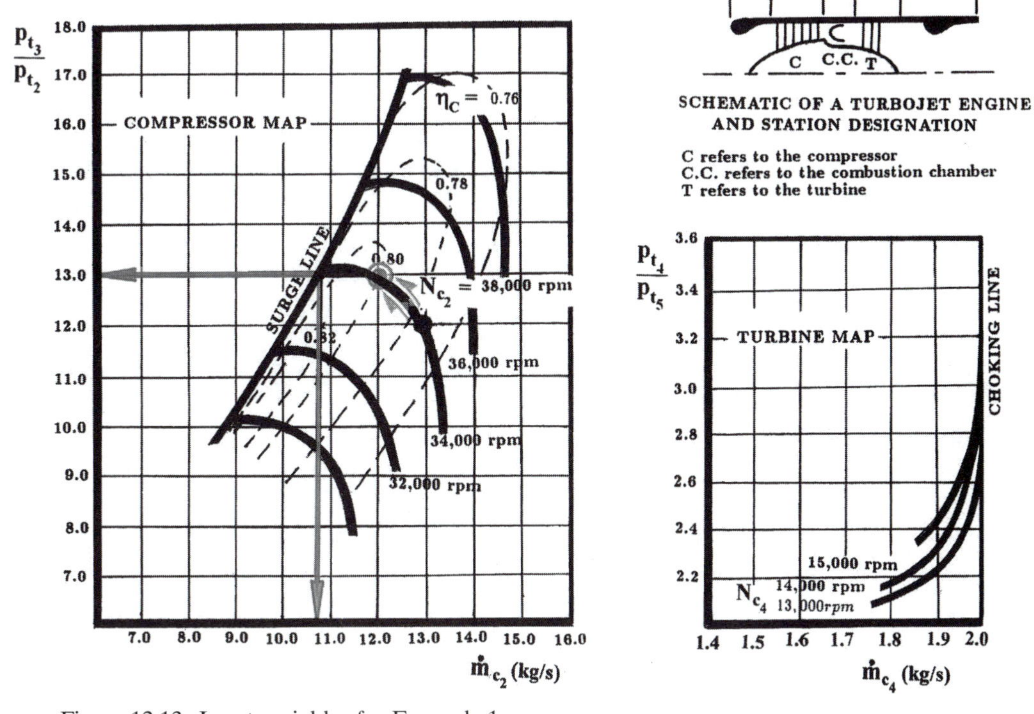

Figure 12.13. Input variables for Example 1.

interpolation means. To the trained eye, however, the required family of lines can be obtained through nothing but simple inspection.

EXAMPLE 1

The cruise operation of a single-spool turbojet engine (Fig. 12.13) is defined as follows:

- Flight Mach number (M_1) = 0.85
- Ambient temperature (T_1) = 251.6 K
- Ambient pressure (p_1) = 0.45 bars
- Shaft speed (N) = 34,000 rpm
- Turbine-inlet total temperature (T_{t4}) = 1152 K
- Combustor total-to-total pressure ratio (p_{t4}/p_{t3}) = 0.93
- Fuel heating value (H) = 4.42×10^7 J/kg
- Mechanical efficiency (η_m) = 98%
- Turbine total-to-total efficiency (η_T) = 84%
- Turbine stator is choked

Assuming a negligible fuel-to-air ratio only in items Ia through Id, calculate:

Ia) The turbine-exit total pressure (p_{t5});
Ib) The turbine-exit total temperature (T_{t5});

Ic) The physical mass-flow rate ($\dot{m}$) through the compressor;
Id) The nozzle-exit Mach number and the specific thrust;
Ie) The actual fuel-to-air ratio (f);
If) The specific fuel consumption (SFC).

II) Consider the case where the mass-flow rate ($\dot{m}$) and shaft speed (N) are gradually decreased, while maintaining the same compressor total-to-total pressure ratio, compressor inlet conditions, and turbine-inlet total temperature (T_{t4}). Calculate the minimum magnitude of physical shaft speed (N) below which the compressor operation becomes unstable. Also calculate the corresponding magnitude of the compressor physical mass-flow rate ($\dot{m}_2$).

SOLUTION

Part Ia:

$$T_{t2} = T_1\left[1 + \left(\frac{\gamma - 1}{2}\right)M_1{}^2\right] = 288.0\ \text{K}$$

$$p_{t2} = p_1\left[1 + \left(\frac{\gamma - 1}{2}\right)M_1{}^2\right]^{\frac{\gamma}{\gamma-1}} = 0.722\ \text{bars}$$

$$N_{C2} = \frac{N}{\sqrt{\theta_2}} = 34{,}000\ \text{rpm}$$

$$N_{C4} = \frac{N}{\sqrt{\theta_4}} = 17{,}000\ \text{rpm}$$

Because the turbine is choked,

$$\dot{m}_{C4} = 2.0\ \text{kg/s}$$

Let us now consider the following equality:

$$\dot{m}_{C4} = \dot{m}_{C2}\sqrt{\frac{T_{t4}}{T_{t2}}}\left(\frac{p_{t2}}{p_{t4}}\right)$$

In this equality, both $\dot{m}_{C2}$ and p_{t4} are unknown. Knowing that N_{C2} will always be 34,000 rpm ($\theta_2 = 1.0$), we can proceed to solve this equality through a trial-and-error procedure, as follows:

First attempt – Set $\dot{m}_{C2}$ to 13.0 kg/s: Now

$$\dot{m}_{C4} = 2.0\ \text{kg/s} = \dot{m}_{C2}\sqrt{\frac{T_{t4}}{T_{t2}}}\left(\frac{p_{t2}}{p_{t4}}\right)$$

This will provide us with a magnitude of 9.39 bars for p_{t4}.

On the other hand, we can compute the same variable (say $p_{t4}{}'$), where

$$p_{t4}{}' = \frac{p_{t4}}{p_{t3}}\frac{p_{t3}}{p_{t2}}p_{t2} = 8.06\ \text{bars}$$

Comparing the two magnitudes of p_{t4}, this attempt has produced an error of 15.2%, which is *unacceptable*.

Second attempt – Set "$\dot{m}_{C2}$" to 12.0 kg/s: Following the same approach as on the first attempt, we get

$$p_{t4} = 8.66 \text{ bars}$$
$$p_{t4}' = 8.72 \text{ bars}$$

This yields an error of 0.7%, which is acceptable. Let us now proceed with an average p_{t4} magnitude of 8.69 bars.

At this point, the compressor operating conditions can be read-off the compressor map as follows:

$$\dot{m}_{C2} = 12.0 \text{ kg/s}$$
$$\pi_C = \frac{p_{t3}}{p_{t2}} = 13.0$$
$$N_{C2} = 34{,}000 \text{ rpm}$$
$$\eta_C = 80\%$$

The compressor total-to-total temperature ratio (τ_C) can be computed as

$$\tau_C = 1 + \frac{1}{\eta_C}\left[(\pi_C)^{\frac{\gamma_C - 1}{\gamma_C}} - 1\right] = 2.351$$

With the fuel-to-air ratio $f \approx 0$, we can compute the turbine total-to-total temperature ratio (τ_T) as

$$\tau_T = 1 - \left[\frac{1}{\eta_m}\frac{c_{p_C}}{c_{p_T}}\frac{T_{t2}}{T_{t4}}(\tau_C - 1)\right] = 0.701$$

Just as easily, we can calculate the turbine total-to-total pressure ratio (π_T) as

$$\pi_T = \left[1 - \frac{1}{\eta_T}(1 - \tau_T)\right]^{\frac{\gamma_T}{\gamma_T - 1}} = 0.169$$

It follows that

$$p_{t5} = \pi_T p_{t4} = 1.472 \text{ bars}$$

Part Ib:

$$T_{t5} = \tau_T T_{t4} = 807.6 \text{ K}$$

Part Ic:

$$\dot{m}_2 = \dot{m}_{C2}\frac{\delta_2}{\sqrt{\theta_2}} = 8.66 \text{ kg/s}$$

Part Id:

$$p_{t5} = p_{t6} = p_6\left[1 + \left(\frac{\gamma_T - 1}{2}\right)M_6{}^2\right]^{\frac{\gamma_T}{\gamma_T - 1}}$$

which, upon substitution, yields

$$M_6 = 1.307$$

Noting that T_{t5} is equal to T_{t6}, we can calculate the exit static temperature as follows:

$$T_6 = \frac{T_{t6}}{\left[1 + (\frac{\gamma_T - 1}{2}) M_6^2\right]} = 630.0 \text{ K}$$

We can now proceed to calculate the engine specific thrust as follows:

$$V_6 = M_6 a_6 = M_6 \sqrt{\gamma_T R T_6} = 641.0 \text{ m/s}$$

$$V_1 = M_1 a_1 = M_1 \sqrt{\gamma_C R T_1} = 270.3 \text{ m/s}$$

$$F/\dot{m} = V_6 - V_1 = 370.7 \text{N/kg}$$

Part Ie: We now calculate the accurate value of the fuel-to-air ratio (by applying the energy equation to the combustor) as follows:

$$f = \frac{(c_{p_T} T_{t4} - c_{p_C} T_{t3})}{(H - c_{p_T} T_{t4})} = 0.0152$$

Part If: The specific fuel consumption is now easy to calculate:

$$SFC = \frac{f}{F/\dot{m}_2} = 4.11 \times 10^{-5} \text{ kg/N}$$

Part II: Following a constant-π_C (i.e., horizontal) line on the compressor map until we get to the surge line, the compressor instability prevails at the point where

$$(N_C)_{min.} = 33{,}700 \text{ rpm}$$

$$(\dot{m}_C)_{min.} = 10.7 \text{ kg/s}$$

which, in terms of physical variables, translate into

$$(N)_{min.} = \sqrt{\theta_2}(N_C)_{min.} = 33,700 \text{ rpm}$$

$$(\dot{m}_2)_{min.} = \frac{\delta_2}{\sqrt{\theta_2}}(\dot{m}_C)_{min.} = 7.73 \text{ kg/s}$$

EXAMPLE 2

Figure 12.14 shows a single-stage axial-flow compressor together with its map and major dimensions. The compressor operating conditions are as follows:

- Physical mass-flow rate $(\dot{m}) = 1.5$ kg/s
- Stage-inlet total pressure $(p_{t1}) = 0.25$ bars
- Stage-inlet total temperature $(T_{t1}) = 200$ K
- Stator flow process is assumed isentropic
- Constant axial-velocity component (i.e., $V_{z3}=V_{z2}$)
- Stage-exit total pressure $(p_{t3}) = 0.35$ bars
- Stator-exit critical Mach number $(V_2/V_{cr2}) = 0.62$

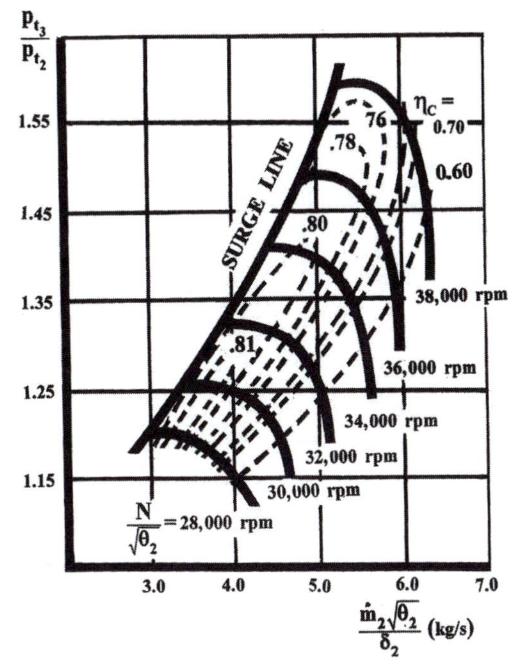

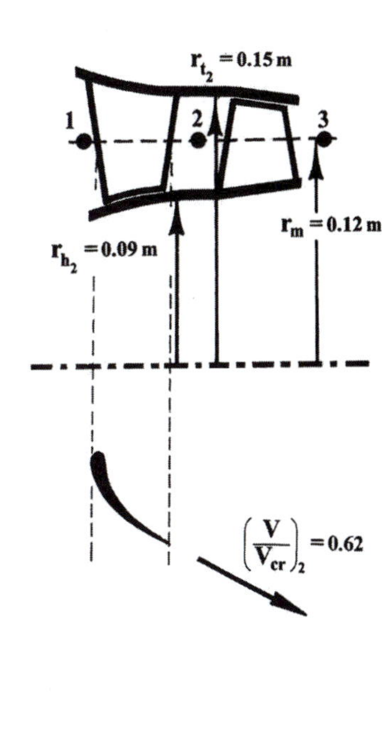

Figure 12.14. Input variables for Example 2.

Calculate the following variables:

a) The rotor-inlet relative critical Mach number (W_2/W_{cr2});
b) The stage-exit static pressure (p_3).

SOLUTION

Part a: Because the stator flow is given as isentropic, it follows that

$$p_{t2} = p_{t1} = 0.25 \text{ bars}$$
$$T_{t2} = T_{t1} = 200 \text{ K}$$

Let us now locate the compressor point of operation on the map by computing the following variables:

$$\dot{m}_{C2} = \frac{\dot{m}\sqrt{\theta_2}}{\delta_2} = 5.0 \text{ kg/s}$$
$$\pi_C = \frac{p_{t3}}{p_{t2}} = 1.4$$

Referring to the compressor map in Figure 12.14, we get

$$N_{C2} = 34{,}000 \text{ rpm}$$
$$\eta_C = 80.0\%$$

We can also calculate the "physical" variables (flow rate and speed) as follows:

$$\dot{m} = \frac{\delta_2}{\sqrt{\theta_2}}\dot{m}_{C2} = 1.5 \text{ kg/s}$$

$$N = \sqrt{\theta_2}N_C = 28{,}333 \text{ rpm}$$

Applying the continuity equation at the stator-exit station, we have

$$\frac{\dot{m}\sqrt{T_{t2}}}{p_{t2}[\pi\,(r_{t2}{}^2 - r_{h2}{}^2)\cos\alpha_2]} = \sqrt{\frac{2\gamma}{(\gamma+1)R}}M_{cr2}\left[1 - \left(\frac{\gamma-1}{\gamma+1}\right)M_{cr2}{}^2\right]^{\frac{1}{\gamma-1}}$$

which, upon substitution, yields

$$\alpha_2 = 56.0^\circ$$

Making use of the preceding results and the given data, we get

$$V_2 = M_{cr2}V_{cr2} = 0.62\sqrt{\left(\tfrac{2\gamma}{\gamma+1}\right)RT_{t2}} = 160.4 \text{ m/s}$$

$$U_m = \omega r_m = 356.0 \text{ m/s}$$

$$V_{z2} = V_{z3} = V_2\cos\alpha_2 = 89.7 \text{ m/s}$$

$$V_{\theta 2} = V_2\sin\alpha_2 = 133.0 \text{ m/s}$$

$$W_{\theta 2} = V_{\theta 2} - U_m = -223.0 \text{ m/s}$$

$$W_2 = \sqrt{W_{\theta 2}{}^2 - V_z^2} = 240.4 \text{ m/s}$$

$$T_{tr2} = T_{t2} - \left(\frac{V_2{}^2 - W_2{}^2}{2c_p}\right) = 216.0 \text{ m/s}$$

$$W_{cr2} = \sqrt{\left(\frac{2\gamma}{\gamma+1}\right)RT_{tr2}} = 269.0 \text{ m/s}$$

$$\frac{W_2}{W_{cr2}} = 0.894$$

Part b: To calculate the stage-exit static pressure, we proceed as follows:

$$T_{t3} = T_{t2}\left\{1 + \tfrac{1}{\eta_C}\left[\left(\frac{p_{t3}}{p_{t2}}\right)^{\frac{\gamma-1}{\gamma}} - 1\right]\right\} = 225.2 \text{ K}$$

$$V_{\theta 3} = V_{\theta 2} - \frac{c_p}{U_m}(T_{t3} - T_{t2}) = 204.1 \text{ m/s}$$

$$V_3 = \sqrt{V_{\theta 3}{}^2 + V_{z3}{}^2} = 222.9 \text{ m/s}$$

$$M_{cr3} = \frac{V_3}{V_{cr3}} = \frac{V_3}{\sqrt{\left(\tfrac{2\gamma}{\gamma+1}\right)RT_{t3}}} = 0.812$$

$$p_3 = p_{t3}\left[1 - \left(\frac{\gamma-1}{\gamma+1}\right)M_{cr}{}^2\right]^{\frac{\gamma}{\gamma-1}} = 0.233 \text{ bars}$$

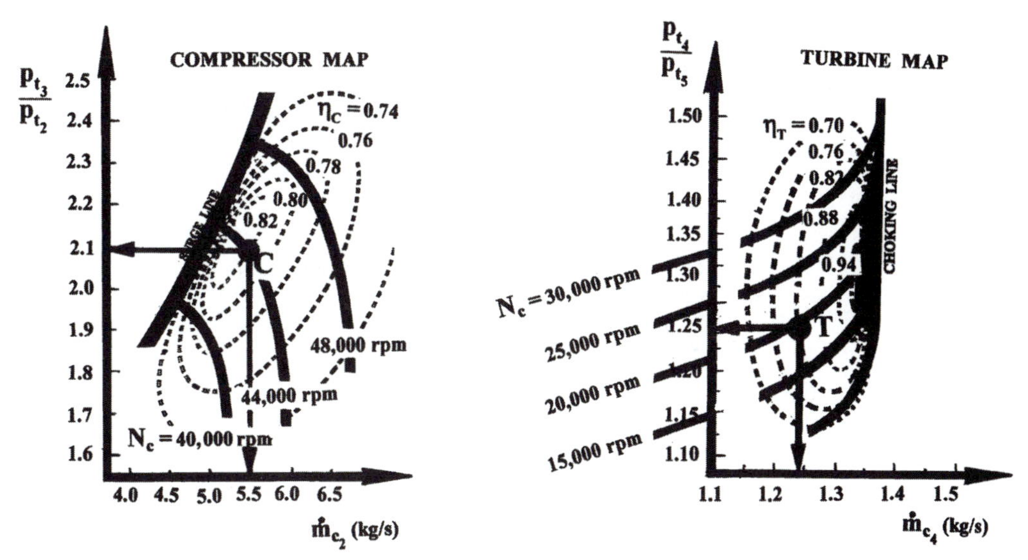

Figure 12.15. Input variables for Example 3.

EXAMPLE 3

Figure 12.15 shows the compressor and turbine maps in a simple turbojet engine. The engine design point is its 11.3 km (37,000 ft) cruise operation, where the flight Mach number is 0.71. The ambient pressure and temperature at this altitude, are 0.21 bars and 218 K, respectively. The compressor operating state is identified on the map as point "*C*," and the station-designation scheme is the same as that in Fig. 12.1. The engine cruise operation is also subject to the following:

- Fuel-to-air ratio $(f) = 0.024$
- Mechanical efficiency $(\eta_m) = 94\%$
- Combustor total pressure loss $(\Delta p_t/p_{t3}) = 12.5\%$
- Turbine-inlet total temperature $(T_{t4}) = 1244$ K
- The cold and hot γ magnitudes are 1.4 and 1.33, respectively

Calculate the turbine total-to-total pressure ratio.

SOLUTION

Using the compressor map, we get

$$N_{C2} = 44{,}000 \text{ rpm}$$

$$\pi_C = 2.1$$

$$\eta_C = 82.0\%$$

$$\dot{m}_{C2} = 5.5 \text{ kg/s}$$

In order to calculate the turbine total-to-total pressure ratio, we need to locate its point of operation on the turbine map (note that point T on the turbine map is actually the final result of this example). To this end, we proceed as follows:

$$T_{t2} = T_{t1} = T_1\left[1 + \left(\frac{\gamma - 1}{2}\right)M_1{}^2\right] = 240.0 \text{ K}$$

$$p_{t2} = p_{t1} = p_1\left[1 + \left(\frac{\gamma - 1}{2}\right)M_1{}^2\right]^{\frac{\gamma}{\gamma-1}} = 0.294 \text{ bars}$$

$$\tau_C \equiv \frac{T_{t3}}{T_{t2}} = 1 + \frac{1}{\eta_C}\left[(\pi_C)^{\frac{\gamma-1}{\gamma}} - 1\right] = 1.29$$

$$\tau_T \equiv \frac{T_{t5}}{T_{t4}} = 1 - \frac{1}{\eta_m(1 + f)}\frac{c_{p_C}}{c_{p_T}}\left(\frac{T_{t2}}{T_{t4}}\right)(\tau_C - 1) = 0.949$$

$$N_{C4} = N_{C2}\sqrt{\frac{T_{t2}}{T_{t4}}} = 19{,}326 \text{ rpm}$$

$$p_{t3} = \pi_C p_{t2} = 0.617 \text{ bars}$$

$$(\Delta p_t)_{combustor} = 0.125 p_{t3} = 0.077 \text{ bars}$$

$$p_{t4} = p_{t3} - (\Delta p_t)_{combustor} = 0.54 \text{ bars}$$

$$\dot{m}_{C4} = \sqrt{\frac{T_{t4}}{T_{t2}}}\frac{p_{t2}}{p_{t4}}\dot{m}_{C2} = 1.24 \text{ kg/s}$$

With the corrected magnitudes of speed and mass-flow rate, we are now in a position to locate the turbine operation point on its map (point "T" on the map). Associated with this point is the following total-to-total pressure ratio:

$$\frac{p_{t4}}{p_{t5}} = 1.25$$

EXAMPLE 4

The compressor and turbine maps of a turbojet engine are shown in Figure 12.16, with the latter carrying the turbine operation point (circled on the map). The engine cruise operation is defined as follows:

- Ambient pressure $(p_1) = 0.21$ bars
- Ambient temperature $(T_1) = 218$ K
- Flight Mach number $(M_1) = 0.81$
- Physical speed $(N) = 64{,}850$ rpm
- Turbine-inlet total temperature $(T_{t4}) = 1182$ K
- Fuel-to-air ratio $(f) = 0.022$
- Combustor total-to-total pressure ratio $(p_{t4}/p_{t3}) = 0.91$

Calculate the torque supplied to the compressor.

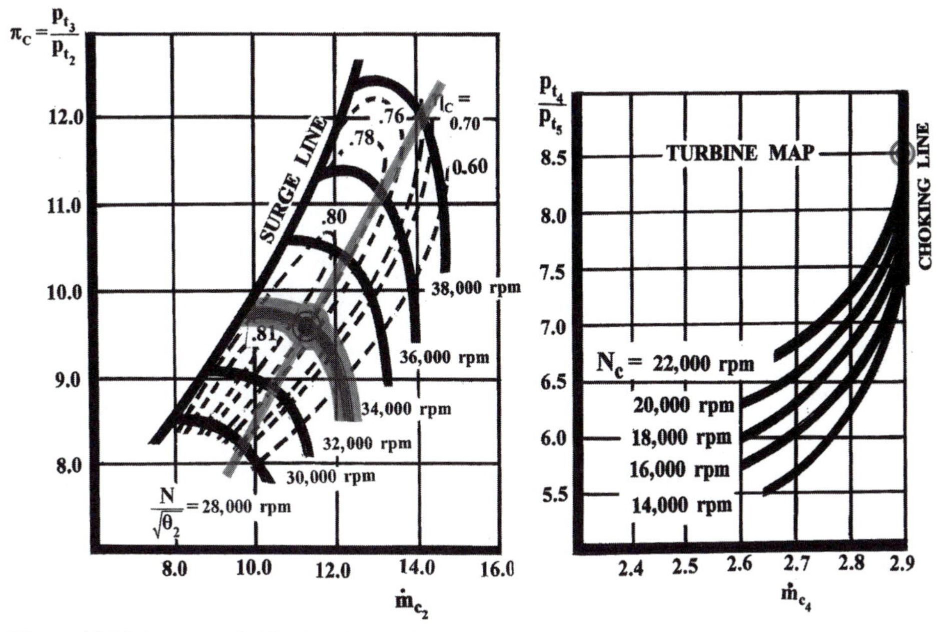

Figure 12.16. Input variables for Example 4.

SOLUTION

Let us first calculate the compressor-inlet total properties:

$$T_{t2} = T_{t1} = T_1\left[1 + \left(\frac{\gamma - 1}{2}\right)M_1{}^2\right] = 246.6 \text{ K}$$

$$p_{t2} = p_{t1} = p_1\left[1 + \left(\frac{\gamma - 1}{2}\right)M_1{}^2\right]^{\frac{\gamma}{\gamma-1}} = 0.323 \text{ bars}$$

We can also calculate the compressor corrected speed as follows:

$$N_{C2} \equiv \frac{N}{\sqrt{\theta_2}} = 32{,}000 \text{ rpm}$$

The turbine map, on the other hand, produces the following:

$$(\Delta\dot{m})_{C4} = 2.9 \text{ kg/s}$$

$$\frac{p_{t4}}{p_{t5}} = 8.5$$

Let us now establish a functional relationship between $(\dot{m})_{C2}$ and π_C as follows:

$$(\dot{m})_{C2} = \left[(\dot{m})_{C4}\left(\frac{1}{1+f}\right)\sqrt{\frac{T_{t2}}{T_{t4}}\left(\frac{p_{t4}}{p_{t3}}\right)}\right]\frac{p_{t3}}{p_{t2}}$$

which, upon substitution, reduces to

$$(\dot{m})_{C2} = 1.18\left(\frac{p_{t3}}{p_{t2}}\right)$$

This relationship represents a straight line, shown on the compressor map in Fig. 12.16, intersecting the corrected-speed line (where $N_{C2} = 32{,}000$ rpm) at a point where

$$(\dot{m})_{C2} = 11.2 \text{ kg/s}$$
$$\frac{p_{t3}}{p_{t2}} = 9.49$$
$$\eta_C = 78\%$$

Let us now calculate the "physical" magnitudes of speed and mass-flow rate:

$$N = \sqrt{\theta_2} N_{C2} = 29{,}611 \text{ rpm}$$
$$\dot{m}_2 = \frac{\delta_2}{\sqrt{\theta_2}} (\dot{m})_{C2} = 3.33 \text{ kg/s}$$

Finally, we can calculate the torque transmitted to the compressor as follows:

$$\tau_{comp.} = \frac{\dot{m}_2 c_p T_{t2} \left[\left(\frac{p_{t3}}{p_{t2}} \right)^{\frac{\gamma-1}{\gamma}} - 1 \right]}{\eta_C \omega} = 307.6 \text{ N m}$$

EXAMPLE 5

Figure 12.17 shows the turbine and compressor maps for a single-spool no-load gas turbine engine. The engine is operating under the following conditions:

- Fuel-to-air ratio $(f) = 0.024$
- Mechanical efficiency $(\eta_m) = 0.95\%$

Select any reasonable magnitude for the T_{t4}/T_{t2} ratio. Now select a point on the compressor map and determine its "conjugate" point on the turbine map.

NOTE

In solving this problem, do *not* use equation (12.41) to verify the actual magnitude of T_{t4}/T_{t2}, as this is the major requirement in the next example. This is where we will verify and subsequently modify this temperature ratio in a rather tedious procedure, as will be obviously clear.

SOLUTION

Let us arbitrarily set the ratio (T_{t4}/T_{t2}) to 5.4. As part of the outer loop (in the last segment of the text), let us (also arbitrarily) consider the point C_1 (shown on the compressor map in Fig. 12.17), which is associated with the following variables:

$$\pi_C = 9.5$$
$$(\dot{m})_{C2} = 12.0 \text{ kg/s}$$
$$N_{C2} = 32{,}400 \text{ rpm}$$

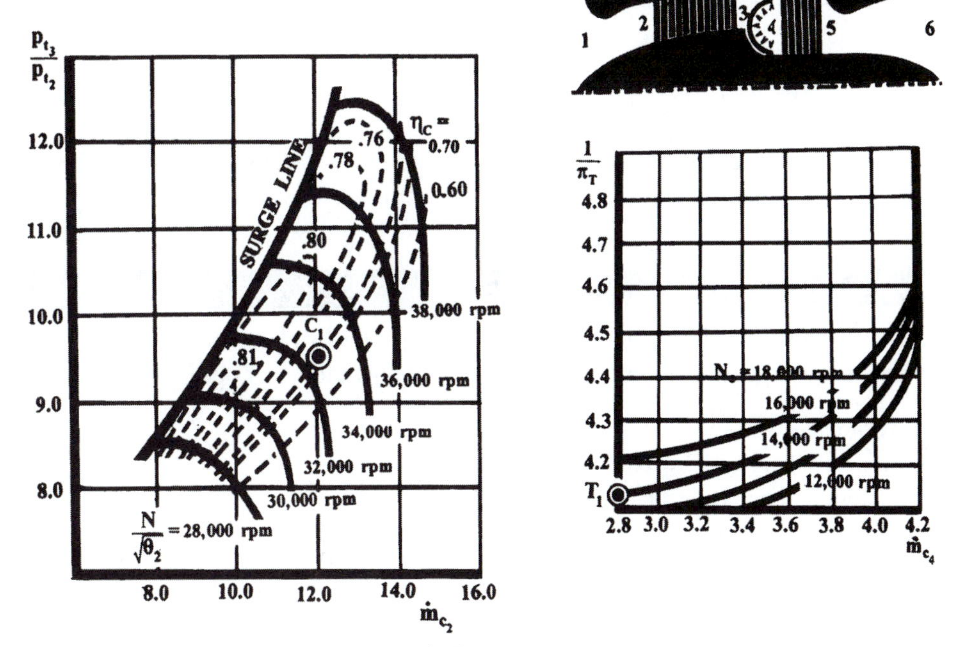

Figure 12.17. Input variables for Example 5.

With this choice, we can now proceed to calculate the turbine corrected speed as follows:

$$N_{C4} = N_{C2}\sqrt{\frac{T_{t2}}{T_{t4}}} = 14{,}000 \text{ rpm}$$

Referring again to the same double-loop procedure at the end of the text, we are now looking for the point "T_1," on the turbine map, which corresponds to the previously chosen point "C_1" (on the compressor map). In doing so, note that the only turbine variable of which we are aware is the corrected speed computed earlier. The process is indeed iterative. In the following, only the first step of this process is presented, followed by the final result, for the purpose of brevity.

Step 1:

Let us pick the point on the turbine map where $p_{t4}/p_{t5} = 4.2$ and $(\dot{m})_{C4} = 3.4$ kg/s. With this choice, we are now able to calculate the compressor corrected mass-flow rate as follows:

$$(\dot{m})_{C2} = (\dot{m})_{C4}\left(\frac{1}{1+f}\right)\pi_C\sqrt{\frac{T_{t2}}{T_{t4}}} = 13.6 \text{ kg/s}$$

which is far from that associated with point "C_1," with the magnitude sought after being 12.0 kg/s.

Repeating this computational step, the final results are

$$\frac{p_{t4}}{p_{t5}} = 4.12$$

$$(\dot{m})_{C4} = 2.8 \text{ kg/s}$$

These two variables are sufficient to place the point "T_1" on the turbine map to correspond to the previously-selected point "C_1" on the compressor map.

EXAMPLE 6

Figure 12.18 shows the compressor and turbine maps for a turbojet engine. The following variables are assumed fixed under all viable modes of operation:

- Combustor total-to-total pressure ratio $(p_{t4}/p_{t3}) = 0.94$
- Fuel-to-air ratio $(f) = 0.024$
- System mechanical efficiency $(\eta_m) = 97\%$

Beginning with an assumed magnitude of T_{t4}/T_{t2}, carry out the entire outer/inner-loop iterative procedure (outlined earlier in this chapter). Your objective is to find two points, C and T (on the compressor and turbine maps), with both points corresponding to one unique value of T_{t4}/T_{t2}.

SOLUTION

OUTER LOOP

Step 1: Let us arbitrarily pick a point (not shown on the compressor map) where

$$\frac{p_{t3}}{p_{t2}} = 2.2$$

$$(\dot{m})_{C2} = 7.0 \text{ kg/s}$$

$$N_{C2} = 58{,}000 \text{ rpm}$$

$$\eta_C = 75.0\%$$

Step 2: Let us pick a T_{t4}/T_{t2} magnitude of 3.5. Now the turbine corrected speed can be computed:

$$N_{C4} = N_{C2}\sqrt{\frac{T_{t2}}{T_{t4}}} = 31,000 \text{ rpm}$$

INNER LOOP

1) Let (p_{t4}/p_{t5}) be 2.4:

$$(\dot{m})_{C4} = 4.3 \text{ kg/s (from the turbine map)}$$

$$(\dot{m})_{C2} = \frac{(\dot{m})_{C4}}{(1+f)}\sqrt{\frac{T_{t2}}{T_{t4}}}\left(\frac{p_{t4}}{p_{t3}}\right)\left(\frac{p_{t3}}{p_{t2}}\right) = 4.64 \text{ kg/s} < 7.0 \text{ kg/s (the target value)}$$

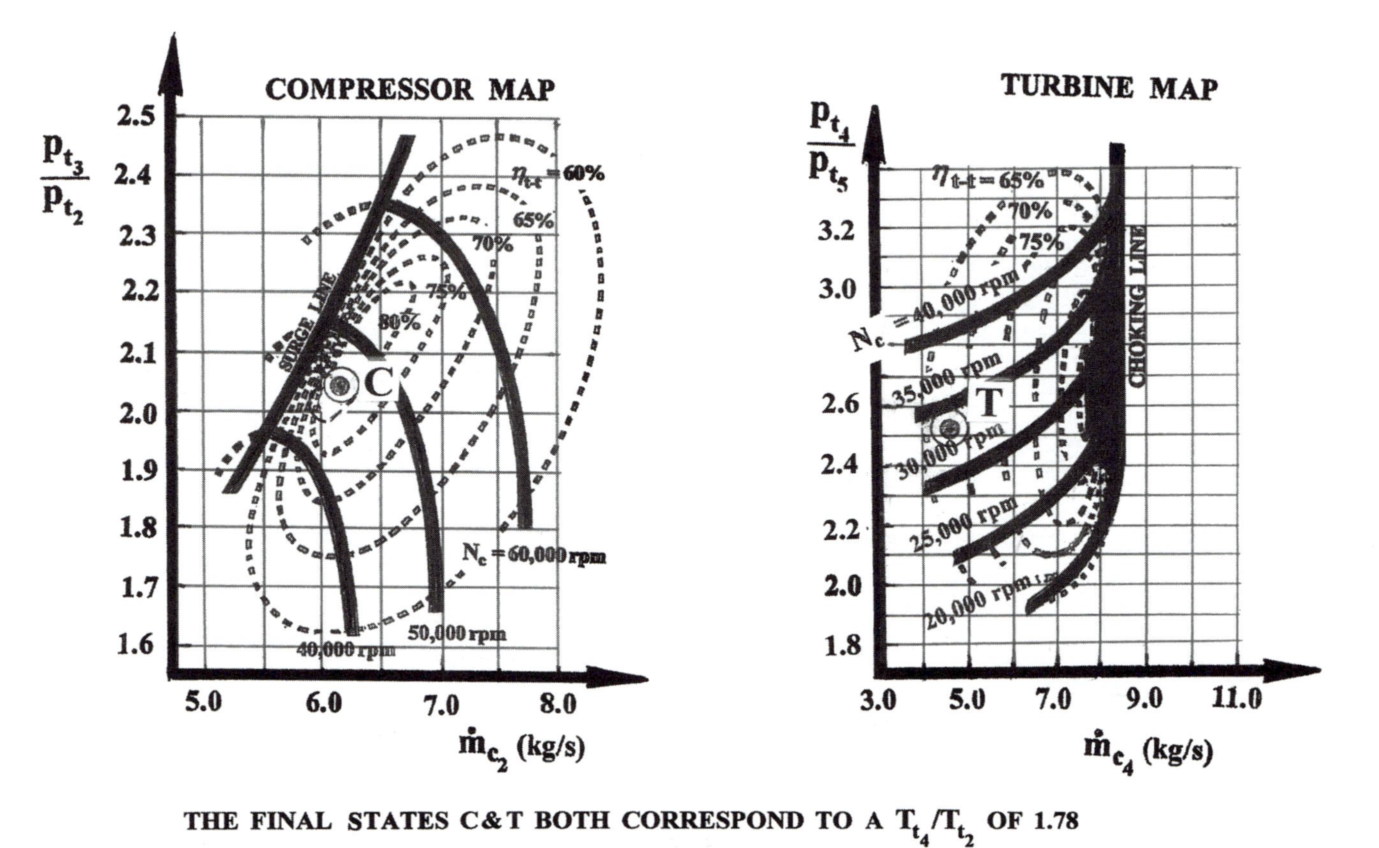

Figure 12.18. Input variables for Example 6.

2) Let (p_{t4}/p_{t5}) be 2.6:

$$(\dot{m})_{C4} = 6.5 \text{ kg/s}$$
$$(\dot{m})_{C2} = 7.02 \text{ kg/s} \approx 7.0 \text{ (the target value)}$$

BACK TO THE OUTER LOOP

With the preceding results, let us now recalculate the previously assumed temperature ratio (T_{t4}/T_{t2}) by substituting in equation (12.41), which yields

$$\frac{T_{t4}}{T_{t2}} = 1.74 \neq 3.5 \text{ (the targeted temperature-ratio magnitude)}$$

Now, we have no choice but to complete the entire (double-loop) procedure with a new magnitude of the temperature ratio.

Let us now pick a point on the compressor map where $p_{t3}/p_{t2} = 2.05$, $(\dot{m})_{C2} = 6.5$ kg/s, and $N_{C2} = 50{,}000$ rpm.

OUTER LOOP

Let us select the temperature ratio (T_{t4}/T_{t2}) to be <u>1.74</u> this time. In doing so, note that:

1) The problem statement does not confine us to a specific value of T_{t4}/T_{t2}.
2) The vast difference in the assumed and then computed values of T_{t4}/T_{t2} may make it harder to match the turbine to the compressor under such a magnitude.
3) The newly selected temperature ratio is excessively low. A comment at the end of the solution will shed some light on this particular choice.

$$N_{C4} = 37{,}900 \text{ rpm}$$

INNER LOOP

1) Let (p_{t4}/p_{t5}) be 2.8:

$$(\dot{m})_{C4} = 6.0 \text{ kg/s}$$
$$(\dot{m})_{C2} = 8.56 \text{ kg/s} \neq 6.5 \text{ kg/s (our current target value)}$$

2) Let (p_{t4}/p_{t5}) be 2.7:

$$(\dot{m})_{C4} = 5.0 \text{ kg/s}$$
$$(\dot{m})_{C2} = 7.13 \text{ kg/s} \neq 6.5 \text{ kg/s}$$

3) Let (p_{t4}/p_{t5}) be 2.63:

$$(\dot{m})_{C4} = 4.55 \text{ kg/s}$$
$$(\dot{m})_{C2} = 6.5 \text{ kg/s (identical to the target value)}$$

BACK TO THE OUTER LOOP

Using equation (12.41), we update the assumed temperature-ratio magnitude:

$$\frac{T_{t4}}{T_{t2}} = 1.91 \neq 1.74 \text{ (the target temperature-ratio magnitude)}$$

NOTE

As is clear by now, we have to reexecute the double-loop procedure, taking the temperature ratio (T_{t4}/T_{t2}) to be the newly computed magnitude or, generally speaking, any other magnitude we wish, knowing that our choice may not be exactly consistent with our wish to shorten this iterative procedure.

It seems appropriate, however, to skip all of these intermediate steps indicating, only the final results. The "finality" here simply means that the magnitudes of temperature ratio (T_{t4}/T_{t2}) at the beginning of the outer loop and at the end of the inner loop are sufficiently close to one another.

FINAL RESULTS

With a temperature ratio of 1.78 (at both the outer-loop beginning and inner-loop end), we get the two points "C" and "T" on the compressor and turbine maps, respectively. The operational points "C" and "T" shown on the corresponding maps in Fig. 12.18 both correspond to a (T_{t_4}/T_{t_2}) of 1.78, and are defined as follows:

Point C:

$$\frac{p_{t3}}{p_{t2}} = 2.05$$

$$(\dot{m})_{C2} = 6.2 \text{ kg/s}$$

$$N_{C2} = 45{,}500 \text{ rpm}$$

Point T:

$$\frac{p_{t4}}{p_{t5}} = 2.55$$

$$(\dot{m})_{C4} = 4.40 \text{ kg/s}$$

$$N_{C4} = 34{,}100 \text{ rpm}$$

COMMENTS

This example is meant to provide a closer look at the tedious procedure to produce the engine operating lines on (particularly) the compressor map (Fig. 12.10). It is clear that our chosen T_{t4}/T_{t2} magnitude is excessively small. To make some practical "sense" out of the results, let us compute the combustor temperature ratio (T_{t4}/T_{t2}) to at least prove it to be above unity:

$$\frac{T_{t4}}{T_{t2}} = 1 + \frac{1}{\eta_C}\left(\pi_C^{\frac{\gamma-1}{\gamma}} - 1\right) = 1.27$$

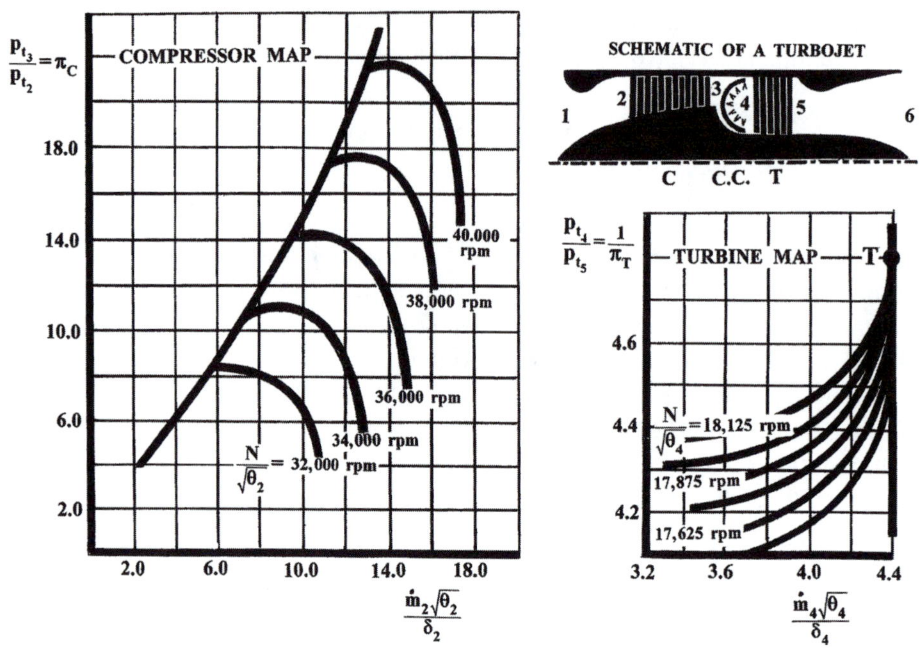

Figure 12.19. Input variables for Problem 1.

The combustor total-to-total temperature ratio can now be computed:

$$\frac{T_{t4}}{T_{t3}} = \frac{T_{t4}}{T_{t2}}\frac{T_{t2}}{T_{t3}} = \frac{1.78}{1.27} = 1.40 > 1.0$$

A suitable description therefore is that the chosen magnitude of T_{t4}/T_{t2} represents a far off-design engine-operation mode.

PROBLEMS

1) Figure 12.19 shows the compressor and turbine maps for a single-shaft turbojet engine at sea-level takeoff. The turbine operation mode is shown on the map as point "T." Details of the engine operating conditions are as follows:

- Ambient pressure (p_1) = 1.0 bar
- Ambient temperature (T_1) = 288 K
- Flight Mach number (M_1) = 0.85
- Fuel-to-air ratio (f) = 0.028
- Turbine total-to-total efficiency (η_T) = 81%
- Mechanical efficiency (η_m) = 91%
- Turbine-inlet total temperature (T_{t4}) = 1300 K
- Constant total pressure across the combustor (i.e., $p_{t4} = p_{t3}$)
- Complete nozzle-flow expansion (i.e., $p_6 = p_1$)
- Isentropic diffuser and nozzle flow processes

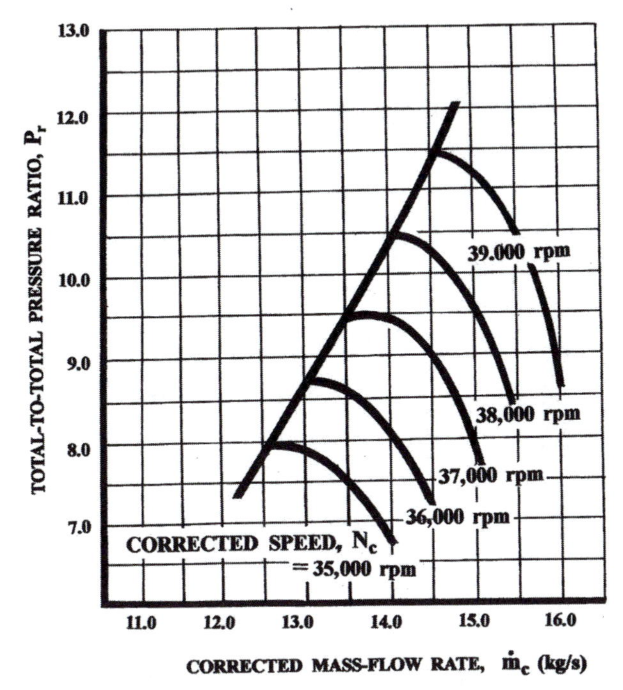

Figure 12.20. Input variables for Problem 2.

- Cold-section specific heat ratio $(\gamma_C) = 1.4$
- Hot-section specific heat ratio $(\gamma_T) = 1.33$

Calculate the following variables:

a) The compressor total-to-total efficiency (η_C);
b) The thrust force (f) produced by the engine.

2) Figure 12.20 shows the compressor map in a turbojet engine. The compressor cruise operation point is defined as follows:

- Supplied specific shaft work $(w_s) = 220$ kJ/kg
- Shaft speed $(N) = 31{,}177$ rpm
- Inlet total temperature $(T_{t\,in}) = 216$ K
- Inlet total pressure $(p_{t\,in}) = 0.235$ bars
- Compressor total-to-total efficiency $(\eta_C) = 80\%$

The preceding operating conditions are simulated in a test rig utilizing air at 288 K and 1.0 bar. The compressor average specific heat ratio (γ) is 1.4 under both sets of operating conditions.

Ia) Calculate the cruise-operation mass-flow rate $(\dot{m})$.
Ib) Calculate the supplied power (P) in the test rig.

II) Consider the situation where a gradual increase in the inlet total pressure takes place in the test rig. During the process, the physical mass-flow

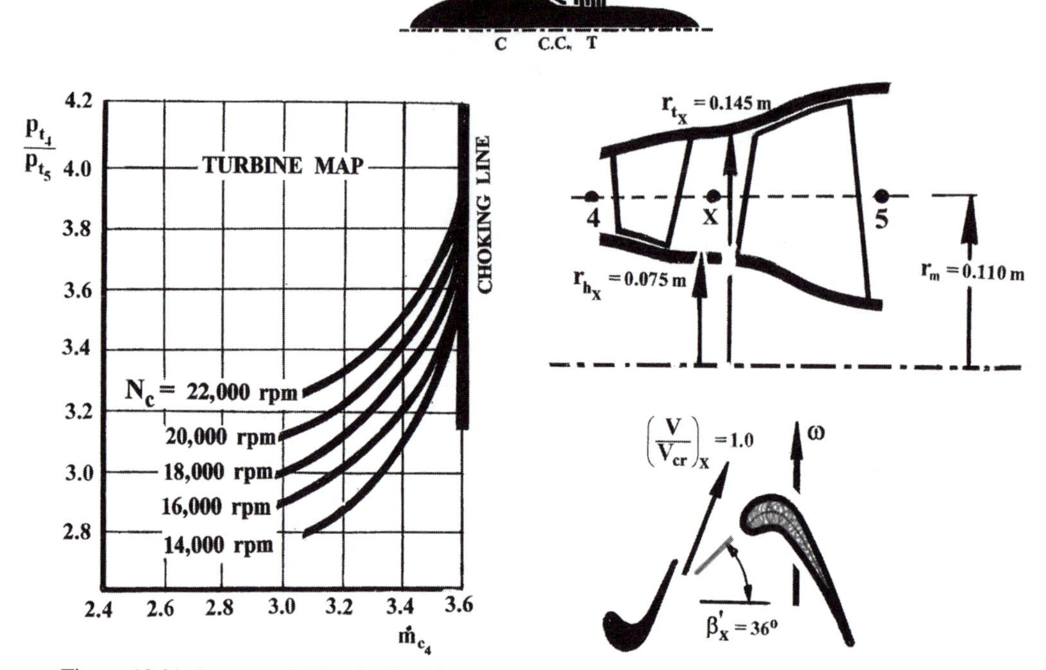

Figure 12.21. Input variables for Problem 3.

rate, physical speed, and inlet total temperature all remain constant at the above-stated magnitudes. Calculate the critical value of inlet total pressure above which the compressor operation becomes unstable.

3) A small turbojet engine is flying at a cruise altitude of 10.7 km (35,000 ft), where the ambient temperature and pressure are 219 K and 0.236 bars, respectively, and the flight Mach number is 0.78. The engine turbine section is composed of a single axial-flow stage as shown in Fig. 12.21. The engine operating conditions are as follows:

- Ambient pressure (p_1) = 0.304
- Ambient temperature (T_1) = 205 K
- Turbine total-to-total efficiency (η_T) = 86.3%
- The turbine-stator exit critical Mach number ($V_X/V_{cr\,X}$) = 1.0
- Compressor corrected mass-flow rate ($\dot{m}_{c,2}$) = 17.0 kg/s
- Compressor corrected speed (N_{c2}) = 46,000 rpm
- Compressor total-to-total pressure ratio (p_{t3}/p_{t2}) = 11.5
- Compressor total-to-total efficiency (η_C) = 81%
- System mechanical efficiency (η_m) = 88%
- Turbine-inlet total temperature (T_{t4}) = 1205 K
- Full expansion across the nozzle (i.e., p_6=p_1)
- Nozzle-exit Mach number (M_6) = 1.49

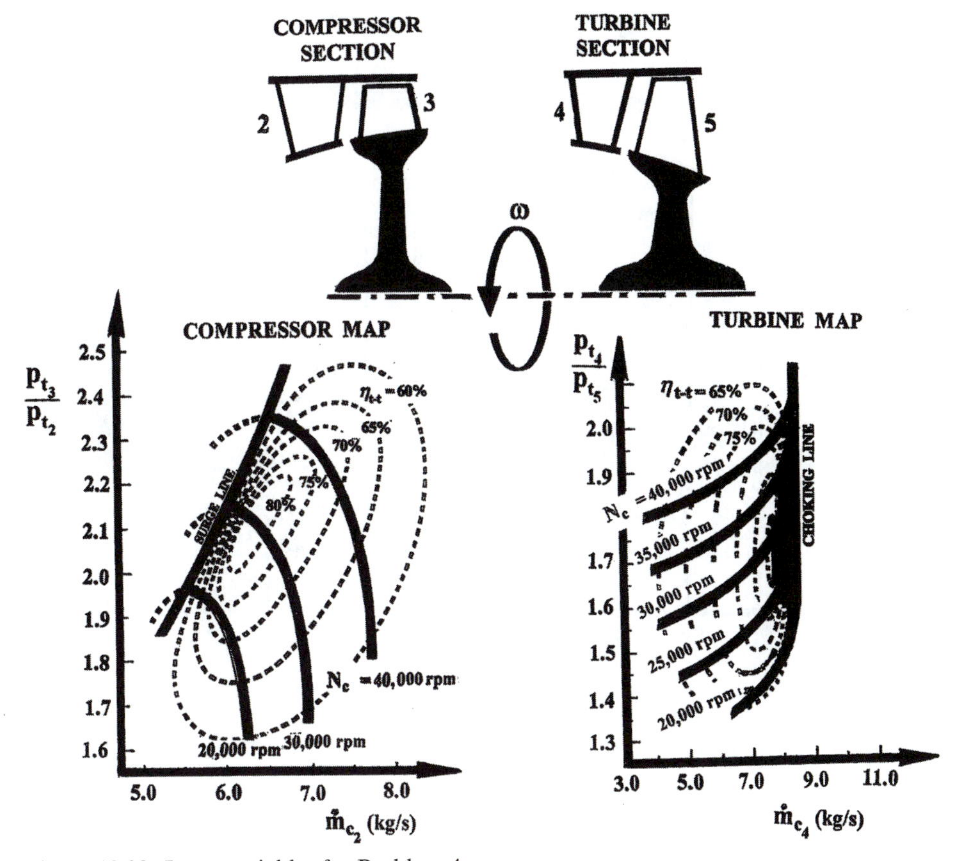

Figure 12.22. Input variables for Problem 4.

- Rotor-blade inlet (metal) angle $(\beta')_X = 36°$
- Combustor total-to-total pressure ratio $(p_{t4}/p_{t3}) = 0.93$

The following simplifications also apply:

- Isentropic turbine-stator flow process
- Isentropic diffuser and nozzle flow fields
- Constant cold and hot values of "γ" (1.4 and 1.33, respectively)

Calculate the following variables:

a) The fuel-to-air ratio (f);
b) The turbine corrected mass-flow rate;
c) The specific fuel consumption (SFC);
d) The turbine-rotor incidence angle (i_R).

4) Figure 12.22 shows the compressor and turbine maps for a no-load gas turbine engine. The engine operating conditions are as follows:

- Compressor corrected speed (N_{c2}) = 50,000 rpm
- Compressor corrected mass-flow rate ($\dot{m}_{c2}$) = 6.5 kg/s

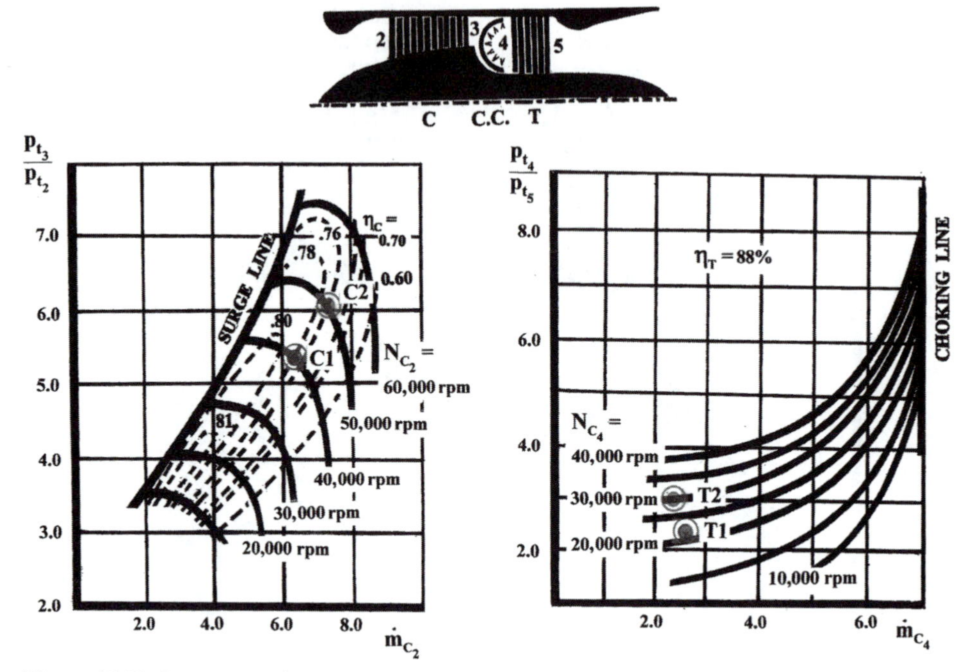

Figure 12.23. Input maps for Problem 5.

- Compressor-inlet total pressure (p_{t2}) = 1.41 bars
- Compressor-inlet total temperature (T_{t2}) = 320 K
- Turbine-inlet total temperature (T_{t4}) = 1100 K
- Fuel-to-air ratio (f) = 0.022
- Combustor total pressure loss $[(\Delta p_t)_{combustor}] = 0.47$ bars
- System mechanical efficiency (η_m) = 92%
- Cold and hot magnitudes of "γ" are 1.4 and 1.33, respectively

Calculate the following variables:

a) The turbine-delivered specific shaft work (w_s);
b) The turbine specific speed (N_s).

In item "b," you may assume a single turbine-stage configuration and a negligibly small turbine-exit critical Mach number.

5) Figure 12.23 shows the compressor and turbine maps for a single-spool turbojet engine. The engine operating conditions are established in such a way that the following variables remain fixed at all times:

- Mechanical efficiency (η_m) = 96%
- Combustor total-to-total pressure ratio (p_{t4}/p_{t3}) = 94%
- Fuel-to-air ratio (f) = 0.023

Also, the turbine efficiency (η_T) is 88% and is assumed to remain constant regardless of the engine operation mode.

Create, on both maps, two pairs of conjugate points, with each pair corresponding to a specific T_{t4}/T_{t2} ratio.

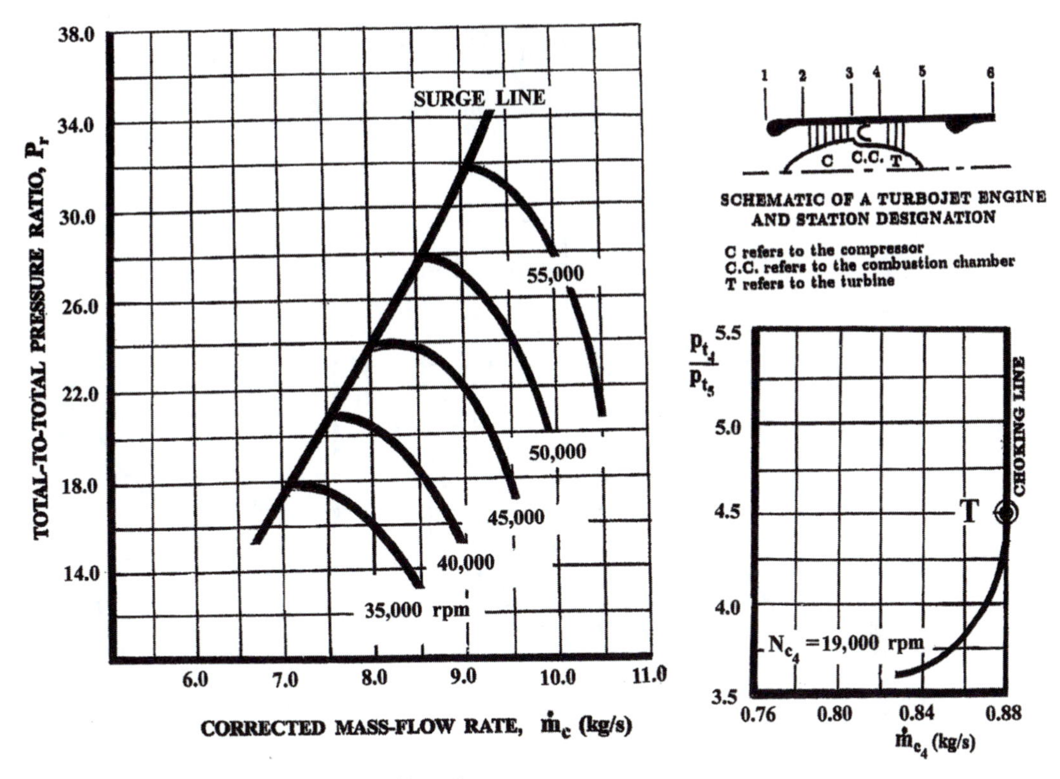

Figure 12.24. Input variables for Problem 6.

6) Figure 12.24 shows the compressor and turbine maps belonging to a simple single-shaft turbojet engine. Shown on the turbine map is the turbine operation point (point "T") that causes the compressor to begin to enter the state of total surge. Referring to the station-designation pattern in the figure, the engine operating conditions at this particular point are as follows:

- $T_0 = 200$ K
- $p_0 = 0.26$ bars
- $M_0 = 0.715$
- $T_{t4} = 1400$ K
- $T_{t5} = 1010$ K
- $p_{t5} = 2.21$ bars
- turbine *corrected* speed $(N/\sqrt{\theta_4}) = 19{,}000$ rpm

The following simplifying assumptions are also applicable:

- $\eta_m \approx 100\%$
- $p_{t5} \approx p_{t4}$
- Fuel-to-air ratio $(f) \approx 0$
- Diffuser and nozzle flow processes are assumed isentropic
- $\gamma_D \approx \gamma_C \approx 1.4$

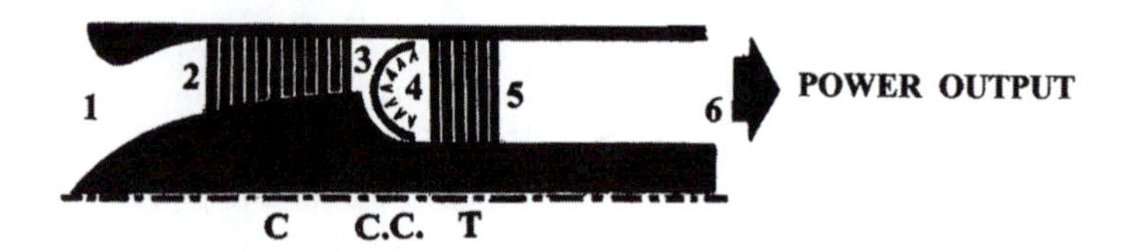

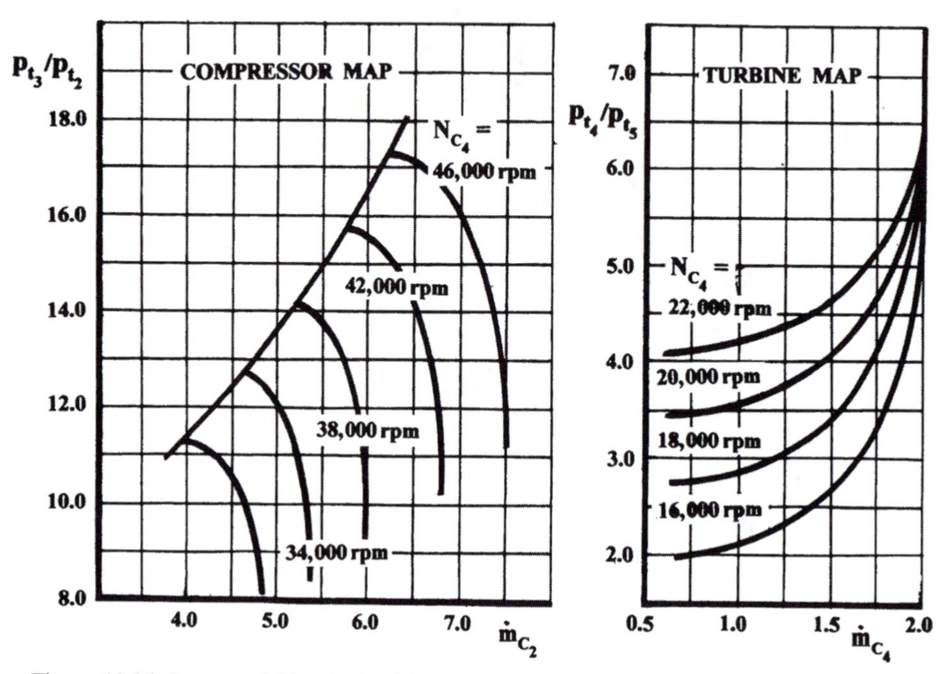

Figure 12.25. Input variables for Problem 7.

- $\gamma_T \approx \gamma_N \approx 1.33$
- $p_6 \approx p_0$

Under these operating conditions and assumptions, calculate:

a) The compressor total-to-total efficiency (η_C);
b) The magnitude of torque (τ) transmitted to the compressor;
c) The thrust force (F) produced by the engine.

7) Figure 12.25 shows a schematic of a turbojet engine along with its compressor and turbine maps.

Compressor data

- Number of stages $(N_s) = 14$
- Stage total-to-total pressure ratio $[(Pr)_s] = 1.2$
- Stage total-to-total efficiency $(\eta_s) = 84\%$

Turbine data

- Stage inlet/exit pressure ratio $[(Pr)_s] = 1.36$
- Stage total-to-total efficiency $(\eta_s) = 93.2\%$

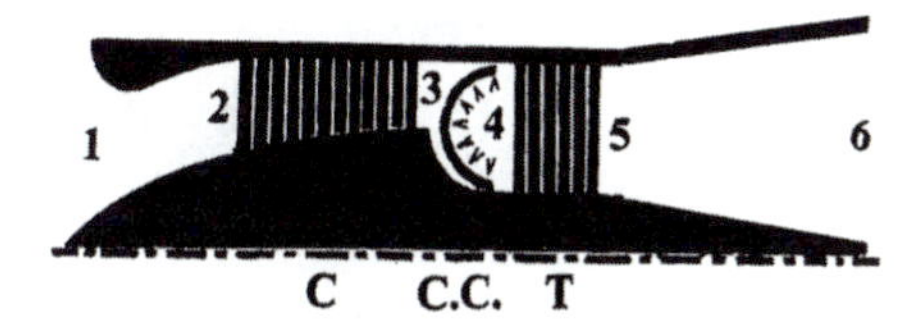

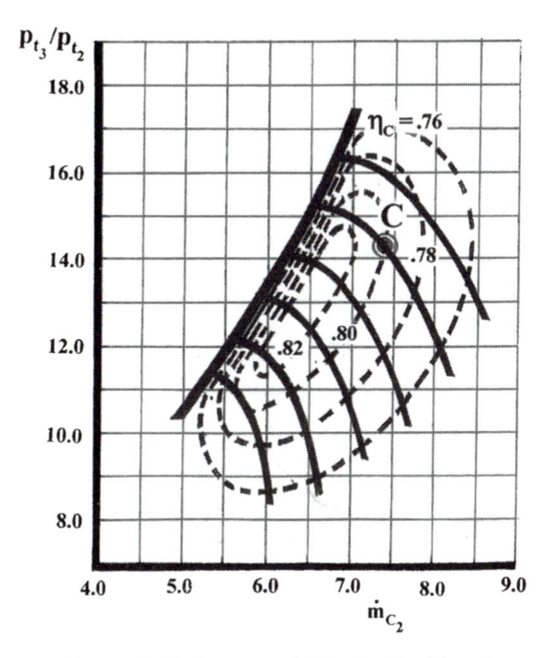

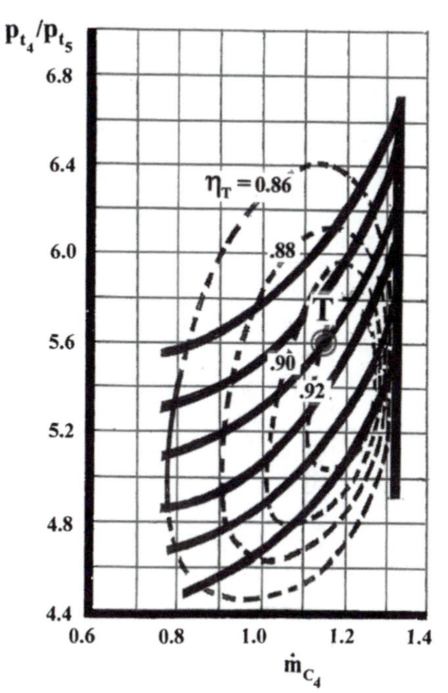

Figure 12.26. Input variables for Problem 8.

Other conditions and simplifications

- $T_{t2} \approx T_1 = 288$ K
- $p_{t2} \approx p_1 = 1.0$ bar
- $T_{t4} = 1350$ K
- Mechanical efficiency $(\eta_m) = 91\%$
- Rotational speed $(N) = 42{,}000$ rpm
- $p_{t4} \approx p_{t3}$
- Fuel-to-air ratio (f) is negligible
- $p_{t6} \approx p_6 = p_1$

Taking γ_C and γ_T as 1.4 and 1.33, respectively, calculate:

a) The mass-flow rate $(\dot{m})$;
b) The turbine overall total-to-total efficiency (η_T);
c) The power output (P).

8) Referring to Figure 12.26, the points "C" and "T" represent the compressor and turbine operation modes in a single-shaft turboprop engine at sea-level takeoff.

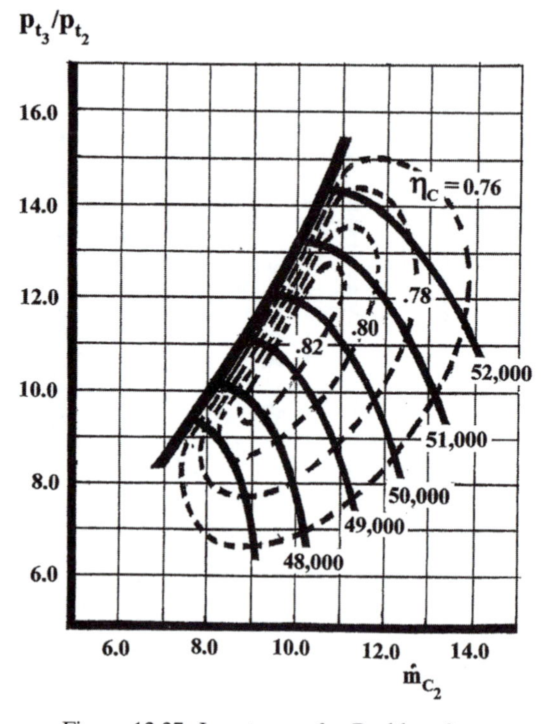

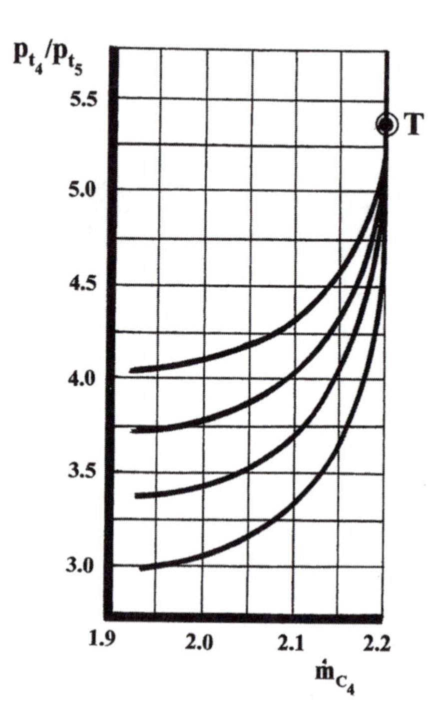

Figure 12.27. Input maps for Problem 9.

In addition, the following conditions apply:

- $p_1 = 1.0$ bar
- $T_1 = 288$ K
- $M_1 = 0.51$
- Inlet diffuser total-pressure loss coefficient $(\bar{\omega}) = 0.06$
- Fuel heating value $(H) = 30{,}000$ kJ/kg
- Fuel-to-air ratio $(f) = 0.026$
- Actual power consumed by the propeller $(P_{prop.}) = 325$ kW

Calculate the following variables:

a) Turbine-inlet total temperature (T_{t4}) and pressure (p_{t4});
b) The combustor percentage of total pressure loss;
c) The gas generator mechanical efficiency (η_m);

9) Figure 12.27 shows a schematic of a single-spool turbojet engine along with the compressor and turbine maps. The sea-level takeoff engine operation mode is

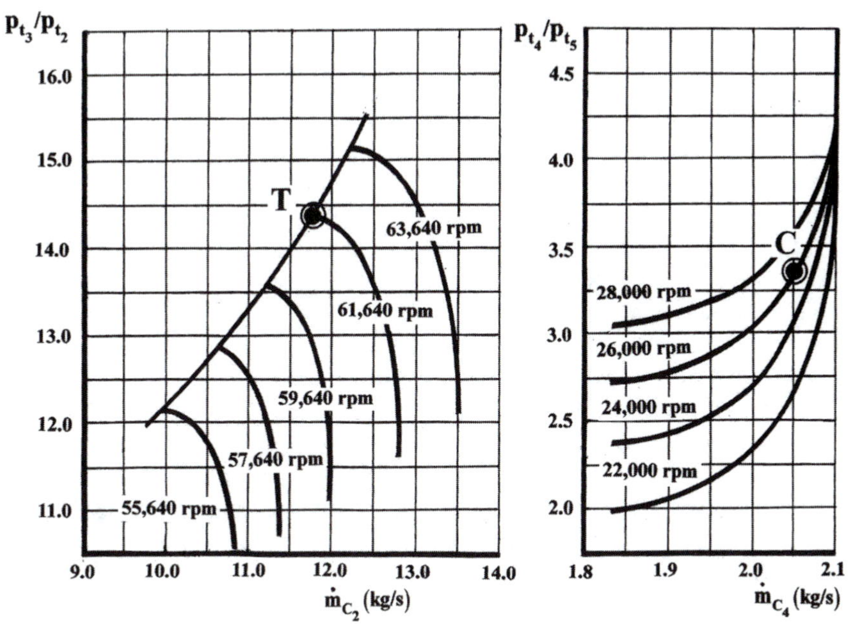

Figure 12.28. Compressor and turbine maps for Problem 10.

defined as follows:

- Mass-flow rate ($\dot{m}$) = 7.3 kg/s
- Shaft speed (N) = 52,860 rpm
- Airframe Mach number (M_1) = 0.61
- Ambient pressure (p_1) = 1.0 bar
- Ambient temperature (T_1) = 288 K
- Fuel-to-air ratio (f) = 0.024
- Combustor total-to-total pressure ratio (p_{t4}/p_{t3}) = 0.945
- Turbine-inlet total temperature (T_{t4}) = 1275 K
- Complete nozzle expansion (i.e., $p_6 = p_1$)
- Mechanical efficiency (η_m) = 88%

Assuming γ_C and γ_T to be 1.4 and 1.33, respectively, calculate the following variables:

a) The turbine total-to-total efficiency (η_T);
b) The fuel heating value (H);
c) The specific fuel consumption (SFC).

10) Figure 12.28 shows the compressor and turbine maps which belong to a single-shaft turbojet engine. The engine cruise-altitude operation mode was

complicated by adverse transient conditions, which placed the compressor at the verge of full-scale surge (as seen in the figure). The engine operating conditions, under such conditions, are as follows:

- Ambient pressure $(p_1) = 0.42$ bars
- Ambient temperature $(T_1) = 230$ K
- Flight Mach number $(M_1) = 0.78$
- Shaft speed $(N) = 58{,}340$ rpm
- Mechanical efficiency $(\eta_m) = 87\%$
- Turbine total-to-total efficiency $(\eta_T) = 93\%$
- Fuel-to-air ratio $(f) = 0.026$
- $p_{t4} \approx p_{t3}$

Calculate the following variables:

a) The percentage of total pressure loss across the combustor;
b) The compressor total-to-total efficiency (η_C);
c) The specific thrust $(F/\dot{m})$.

11) Referring back to Problem 5, adopting the maps on which it is based, and accepting all the stated simplifications, create on the compressor map two operating lines. These will correspond to two T_{t4}/T_{t2} ratios of your choice.

COMMENT

The large number of iterative subprocesses in solving this problem, as well as the lengthy postprocessing interpolation effort (Fig. 12.12), will both contribute to what will truly be a large-scale numerical procedure. It is wise, under such circumstances, to start with a flowchart as a basis for developing a program in an appropriate computer language to undertake the entire numerical solution task.

12) Figure 12.29 shows the turbine and compressor maps, which belong to a single-spool turbojet engine. The different subscripts in the figure are consistent with the station designation in Figure 12.1. At sea-level takeoff, the engine operating conditions are as follows:

- Compressor-inlet total pressure $(p_{t2}) = 1.516$ bars
- Compressor-inlet total temperature $(T_{t2}) = 324.3$ K
- Mass-flow rate $(\dot{m}) = 6.22$ kg/s
- Shaft speed $(N) = 55{,}700$ kg/s
- Turbine-inlet total temperature $(T_{t4}) = 1430.0$ K
- Combustor total-to-total pressure ratio $(p_{t4}/p_{t3}) = 0.93$
- Mechanical efficiency $(\eta_m) = 96\%$
- Complete expansion across the nozzle (i.e., $p_6 = p_1 = 1$ bar)

Part I: Calculate the fuel-to-air ratio (f).

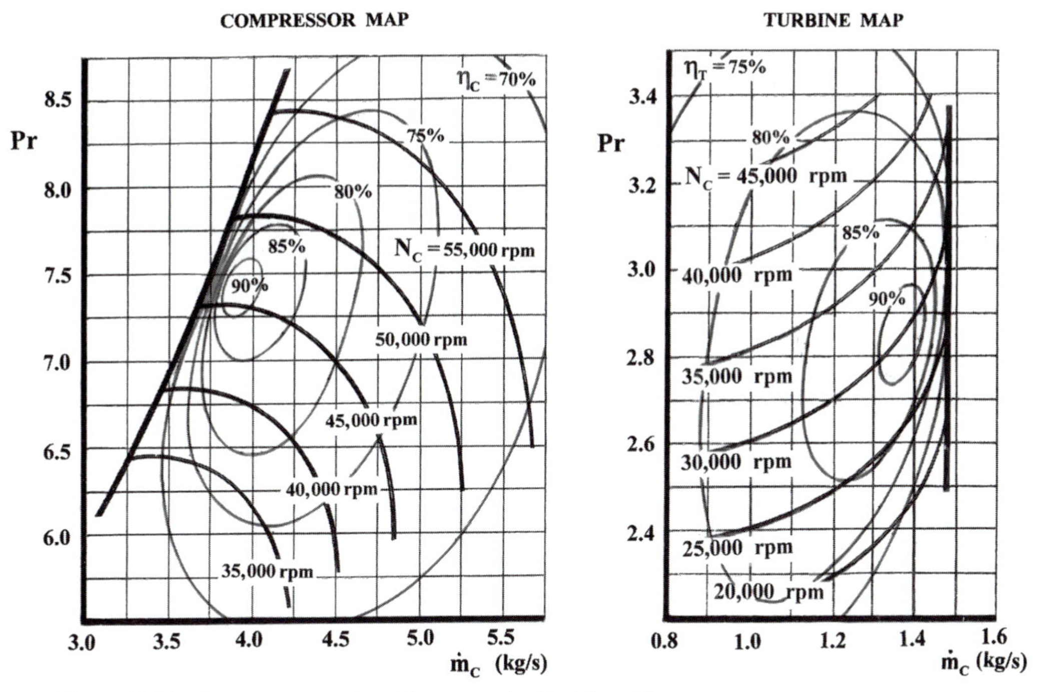

Figure 12.29. Compressor and turbine maps for Problem 12.

Part II: During takeoff, a cooling mechanism was effected whereby the cooling air was gradually extracted at the compressor *inlet* station and separately pumped before being delivered at the turbine inlet station. Meanwhile, the shaft speed was gradually decreased in such a way to maintain fixed both the compressor and turbine total-to-total pressure ratios. Determine which component (compressor or turbine) will terminate this process.

Hint: As the turbine-cooling process continues, the changes in both the compressor and turbine mass-flow rates will direct the engine-operation point toward the surge line (on the compressor map) and, simultaneously, the choking line (on the turbine map). Reaching either one of these two lines will terminate the process.

References

1. Baskharone, E. A., and Hamed, A., "A New Approach in Cascade Flow Analysis Using the Finite Element Method," AIAA Journal, Vol. 19, No. 1, January 1981, pp. 65–71.
2. Baskharone, E. A., and Hamed, A., "Flow in Non-Rotating Passages of Radial-Inflow Turbines," NASA-CR-159679, September 1979.
3. Baskharone, E. A., and Hensel, S. J., "A New Model for Leakage Prediction in Shrouded-Impeller Turbopumps," Journal of Fluids Engineering (ASME Transactions), Vol. 111, No. 2, June 1989, pp. 118–123.
4. Baskharone, E. A., and Hensel, S. J., "Flow Field in the Secondary, Seal-Containing Passages of Centrifugal Pumps," Journal of Fluids Engineering (ASME Transactions), Vol. 115, No. 4, September 1993, pp. 702–709.
5. Baskharone, E. A., and McArthur, D. R., "A Comprehensive Analysis of the Viscous Incompressible Flow in Quasi-Three-Dimensional Aerofoil Cascades," Int. J. for Numerical Methods in Fluids, Vol. II, No. 2, July 1990.
6. Baskharone, E. A., and Wyman, N. J., "Primary/Leakage Flow Interaction in a Pump Stage," Journal of Fluids Engineering (ASME Transactions), Vol. 121, No. 1, March 1999, pp. 133–138.
7. Baskharone, E. A., "F109 High Pressure Turbine First-Stage Rotor Redesign," Report No. 22-2138, Garrett Turbine Engine Co., Phoenix, AZ. March 1984.
8. Baskharone, E. A., "Finite-Element Analysis of Turbulent Flow in Annular Exhaust Diffusers of Gas Turbine Engines," Journal of Fluids Engineering (ASME Transaction), Vol. 113, No. 1, March 1991, pp. 104–110.
9. Baskharone, E. A., "Investigation of Different Cooling Categories in a Radial Inflow turbine Rotor Using the Finite Element Technique," M.S. thesis, University of Cincinnati, August 1975. Also published in the AIAA Journal of Aircraft, Vol. 14, No. 2, February 1977.
10. Baskharone, E. A., "Optimization of the Three-Dimensional Flow Path in the Scroll-Nozzle Assembly of a Radial Inflow Turbine," Journal of Engineering for Gas Turbine and Power (ASME Transactions), Vol. 106, No. 2, April 1984, pp. 511–515.
11. Bathie, W. W., *Fundamentals of Gas Turbines*, John Wiley & Sons, New York, 1984.
12. Booth, T. C., "Low Aspect Ratio Turbine (LART) – Phase V Final Report," Design Report No. 75-211701(5), Airesearch Manufacturing Co., Phoenix, AZ, May 1980.
13. Church, A. H., *Centrifugal Pumps and Blowers*, John Wiley & Sons, New York, 1944.
14. Cohen, H., Rogers, G. F. C., and Saravanamuttoo, H. I. H., *Gas Turbine Theory*, Longman, Essex, U.K., 1972.
15. Galligan, J. E., "Advanced Technology Components for Model GTCP 305-2 Aircraft Auxiliary Power System," Design Report No. 31-2874, May 1979. AiResearch Manufacturing Co. of Arizona, Phoenix, AZ.

16. Glassman, A. J., "Turbine Design and Application," NASA SP No. 290, National Aeronautics and Space Administration, Washington, DC, 1973.
17. Hamed, A., and Baskharone, E., "Analysis of the Three-Dimensional Flow in a Turbine Scroll," Journal of Fluids Engineering (ASME Transactions), Vol. 102, No. 3, September 1980, pp. 297–301.
18. Hamed, A., Baskharone, E., and Tabakoff, W., "Temperature Distribution Study in a Cooled Radial Turbine Rotor," AIAA Journal of Aircraft, Vol. 14, No. 2, February 1977, pp. 173–176.
19. Hill, P. G., and Peterson, C. R., *Mechanics and Thermodynamics of Propulsion*, Addison-Wesley, Reading, MA, 1992.
20. Hinch, D. V., "Axial-Turbine Tip Clearance Study," Report No. 22-1648, Garrett Turbine Engine Co., April 1982.
21. Hlavaty, S. T., "A Finite-Element Model of the Turbulent Flow Field in a Centrifugal Impeller," M. S. Thesis, Texas A & M University, Aug. 1993.
22. Horlock, J. H., *Axial Flow Turbines: Fluid Mechanics and Thermodynamics*, Krieger, Malabar, FL, 1985.
23. Katsanis, T., and McNally, W., "Fortran Program for Calculating Velocities and Streamlines on the Hub, Shroud and Mid-Channel Flow Surface of Axial or Mixed-Flow Turbomachines," NASA TN D-7343, July 1973.
24. Kavanough, P., and Ye, Z. Q., "Axial-Flow Turbine Design Procedure and Sample Design Cases," Technical Report No. TCRL-28, Iowa State University, Ames, IA, January 1984.
25. Khalil, I. M., Tabakoff, W., and Hamed, A., "Losses in Radial Inflow Turbines," Journal of Fluids Engineering (ASME Transactions), September 1976, pp. 364–373.
26. Kovats, A., *Design and Performance of Centrifugal and Axial Flow Pumps and Compressors*, MacMillan, New York, 1964.
27. Lakshminarayana, B., "Loss Evaluation Methods in Axial-Flow Compressors," Proceedings of the Workshop on Flow in Turbomachines, Naval Postgraduate School, Report No. NPS-57VA71111A, U.S. Naval Academy, Annapolis, MD, November 1971.
28. MacGregor, J. D., and Baskharone, E. A., "A Finite-Element Model of the Cyclic Aerodynamic Loading of a Turbine Rotor Due to an Upstream Stator," International Journal of Computational Fluid Dynamics, Vol. 6, 1996, pp. 291–306.
29. Mashimo, T., Watanabe, I., and Ariga, I., "Effects of Fluid Leakage on the Performance of a Centrifugal Compressor," Journal of Engineering for Power, Vol. 101, No. 3, July 1979.
30. Mattingly, J. D., *Elements of Gas Turbine Propulsion*, McGraw-Hill, New York, 1996.
31. McFarland, E. R., "A Rapid Blade-to-Blade Solution for Use in Turbomachinery Design," Journal of Engineering for Gas Turbine and Power (ASME Transactions), Vol. 106, No. 2, April 1984, pp. 376–382.
32. Schlichting, H., *Boundary Layer Theory,* McGraw-Hill, New York, 1979.
33. Shepherd, D. G., *Principles of Turbomachinery*, MacMillan, New York, 1956.
34. Sovran, G., and Klomp, E. D., "Experimentally Determined Optimum Geometries for Diffusers with Rectangular, Conical or Annular Cross Sections," General Motors Research Publication GMR-511, General Motors Corporation, November 1965.
35. Szanca, E., Behning, F., and Schum, H., "Research Turbine for High Temperature Core Engine Application: Part II – Effect of Rotor Tip Clearance on Overall Performance," NASA TN D-7639, April 1974.
36. Thwaites, B., *Incompressible Aerodynamics*, Clarendon Press, Oxford, 1960.
37. Vavra, M. H., *Aerothermodynamics and Flow in Turbomachines*, John Wiley & Sons, New York, 1960.
38. Wu, C. H., "A General Theory of Three-Dimensional Flow in Subsonic and Supersonic Turbomachines of Axial-, Radial-, and Mixed-Flow Types," NACA TN 2604, January 1952.
39. Zucker, R. D., *Fundamentals of Gas Dynamics*, Matrix Publishers, Beaverton, OR, 1977.

Index

RECEIVED
OCT - 5 2006